SELECTED PAPERS OF
YU. I. MANIN

World Scientific Series in 20th Century Mathematics

Published

Vol. 1 The Neumann Compendium
edited by F. Bródy and T. Vámos

Vol. 2 40 Years in Mathematical Physics
by L. D. Faddeev

Vol. 3 Selected Papers of Yu I Manin
by Y. Manin

Forthcoming

A Mathematician and His Mathematical Work
— Selected Papers of S. S. Chern
edited by S. Y. Cheng, P. Li and G. Tian

World Scientific Series in 20th Century Mathematics Vol. 3

SELECTED PAPERS OF YU. I. MANIN

Yu. I. Manin
Max-Planck-Institut für Mathematik
Germany

World Scientific
Singapore • New Jersey • London • Hong Kong

Published by

World Scientific Publishing Co. Pte. Ltd.
5 Toh Tuck Link, Singapore 596224
USA office: 27 Warren Street, Suite 401-402, Hackensack, NJ 07601
UK office: 57 Shelton Street, Covent Garden, London WC2H 9HE

Library of Congress Cataloging-in-Publication Data
Manin, Yu. I.
[Selections, English]
Selected papers of Yu. I. Manin / Yu. I. Manin.
p. cm. -- (World Scientific series in 20th century mathematics ; vol. 3)
Includes bibliographical references.
ISBN-13 9789810224981 -- ISBN-10 9810224982
1. Mathematics. I. Title. II. Series.
QA3.M316213 1996
510--dc20 96-18705
CIP

British Library Cataloguing-in-Publication Data
A catalogue record for this book is available from the British Library.

PREFACE

I am grateful to the World Scientific Publishers for their suggestion to publish my Selected Papers.

This volume contains about one fourth of my articles written before 1994. The choice was difficult and it was not governed by any systematic criteria. Probably, I simply picked those which I enjoyed most.

It was my joy and privilege to collaborate with many friends and colleagues, and I am thankful to all the co-authors of the papers reprinted here: V. V. Batyrev, A. A. Beilinson, V. G. Drinfeld, J. Franke, V. A. Iskovskikh, M. M. Kapranov, D. R. Lebedev, A. A. Panchishkin, A. O. Radul, V. V. Schechtman, Yu. Tschinkel, M. M. Vishik, Ju. G. Zarkhin.

A considerable part of these texts was initially published in Russian. The English translations are not always of the best quality, but I did not try to improve them, restricting myself to correcting a few mistakes.

The Publishers suggested that I supply some comments. For the most part, I have chosen to give references for further reading and to sketch the subsequent development of the ideas presented in my works rather than reminisce about their conception. The reason is that most mathematical texts have a transient value, if any. Their survival is secured only if their content becomes a part of oral tradition, even a commonplace. The author may lament that his ideas were misrepresented, misattributed, and misnamed, but who could care less.

Yu. I. Manin.

Bonn, October 1995.

CONTENTS

Preface vii

PART I. Algebraic Geometry

The Hasse–Witt matrix of an algebraic curve 3
Russian: Izv. AN SSSR, ser. mat. **25**:*1 (1961), 153-172.*
English: Amer. Math. Soc. Transl. Ser. 2, **45** *(1965), 245-264.*

Rational points of algebraic curves over functional fields 23
Russian: Izv. AN SSSR, ser. mat. **27**:*6 (1963), 1395-1440.*
English: Amer. Math. Soc. Transl. Ser. 2, **50** *(1966), 189-234.*

Letter to the editor 69
Russian: Izv. AN SSSR, ser. mat. **53**:*2 (1989).*
English: Math. USSR Izvestija **34**:*2 (1990), 465-466.*

Correspondences, motives, and monoidal transformations 71
Russian: Mat. Sbornik **77**:*4 (1968), 475-507.*
English: Math. USSR-Sb. **6** *(1968), 439-470.*

Three-dimensional quartics and counterexamples to the Lüroth problem (with V. A. Iskovskih) 103
Russian: Mat. Sbornik **86**:*1 (1971), 140-166.*
English: Math. USSR-Sb. **15** *(1971), 141-166.*

New directions in geometry 129
Russian: Uspekhi Mat. Nauk **39**:*6 (1984), 47-73.*
English: Russian Math. Surveys **39**:*6 (1984), 51-83.*

Arrangements of hyperplanes, higher braid groups and higher Bruhat orders (with V. V. Schechtman) 162
Algebraic Number Theory, Advanced Studies in Pure Math.
17 (1989), 289–308, *Academic Press*

PART II. Modular Forms and Diophantine Equations

The p-torsion of elliptic curves is uniformly bounded 185
Russian: Izv. AN SSSR, ser. mat. **33**:*3 (1969), 459-465.*
English: Math. USSR-Izv. **3** *(1969), 433-438.*

Le groupe de Brauer–Grothendieck en géométrie diophantienne 191
Actes Congr. Int. Math. Nice, Vol. 1 (1971), 401-411,
Gauthier-Villars.

Parabolic points and zeta-functions of modular curves 202
Russian: Izv. AN SSSR, ser. mat. **36**:*1 (1972), 19–66.*
English: Math. USSR-Izv. **6**:*1 (1972), 19–64.*

Height on families of abelian varieties (with Ju. G. Zarhin) 248
Russian: Mat. Sbornik **89**:*2 (1972), 171–181.*
English: Math. USSR-Sb. **18**:*2 (1972), 169–179.*

Periods of p-adic Schottky groups (with V. G. Drinfeld) 259
J. Reine Angew. Math. **262/263** *(1973), 239–247.*

Periods of parabolic forms and p-adic Hecke series 268
Russian: Mat. Sbornik **92**:*3 (1973), 378–401.*
English: Math. USSR-Sb. **21**:*3 (1973), 371–393.*

The values of p-adic Hecke series at integer points of the critical strip 291
Russian: Mat. Sbornik **93**:*4 (1974), 621–626.*
English: Math. USSR-Sb. **22**:*4 (1974), 631–637.*

p-adic Hecke series of imaginary quadratic fields (with M. M. Vishik) 298
Russian: Mat. Sbornik **95**:*3 (1974), 357–383.*
English: Math. USSR-Sb. **24**:*3 (1974), 345–371.*

Convolutions of Hecke series and their values at lattice points (with A. A. Panĉhiŝhkin) 325
Russian: Mat. Sbornik **104**:*4 (1977), 617–651.*
English: Math. USSR-Sb. **33**:*4 (1977), 539–571.*

Rational points of bounded height on Fano varieties (with J. Franke, Yu. Tschinkel) 358
Inv. Math. **95** *(1989), 421–435.*

Sur le nombre des points rationnels de hauteur bornée des variétés algébriques (with V. V. Batyrev) 373
Math. Ann. **286** *(1990), 27–43.*

Points of bounded height on del Pezzo surfaces (with Yu. Tschinkel) 390
Compositio Math. **85** *(1993), 315–332.*

Notes on the arithmetic of Fano threefolds 408
Compositio Math. **85** *(1993), 37–55.*

PART III. Differential Equations and Mathematical Physics

Conservation laws and Lax representation of Benney's long wave equations (with D. R. Lebedev) 429
Phys. Lett. A **74** *(1979), 154–156.*

Gelfand–Dikii Hamiltonian operator and the coadjoint representation of the Volterra group (with D. R. Lebedev) 432
Russian: Funkts. Anal. i ego Prilozhen **13***:4 (1979), 40–46.*
English: Functional Anal. Appl. **13***:4 (1979), 268–273.*

A supersymmetric extension of the Kadomtsev–Petviashvili hierarchy (with A. O. Radul) 438
Comm. Math. Phys. **98** *(1985), 65–77.*

The twistor transformation and algebraic-geometric constructions of solutions of the equations of field theory (with M. M. Kapranov) 451
Russian: Uspekhi Mat. Nauk **41***:5 (1986), 85–107.*
English: Russian Math. Surveys **41***:5 (1986), 33–61.*

Critical dimensions of string theories and the dualizing sheaf on the moduli space of (super)curves 480
Russian: Funkts. Anal. i ego Prilozhen **20***:3 (1986), 88–89.*
English: Functional Anal. Appl. **20***:3 (1986), 244–246.*

The partition function of the Polyakov string can be expressed in terms of theta-functions 483
Phys. Lett. B **172***:2 (1986), 184–185.*

The Mumford form and the Polyakov measure in string theory (with A. A. Beilinson) 485
Comm. Math. Phys. **107***:3 (1986), 359–376.*

Some remarks on Koszul algebras and quantum groups 503
Ann. Inst. Fourier **37***:4 (1987), 191–205.*

Reflections on arithmetical physics 518
Conformal Invariance and String Theory (Poiana Brasov, 1987), 293–303, Academic Press, 1989.

Notes on quantum groups and quantum de Rham complexes 529
Teoreticheskaya i Matematicheskaya Fizika **92***:3 (1992), 425–450.*

PART IV. Miscellany

Mathematics as Metaphor 557
Proc. of ICM, Kyoto 1990, vol. 2, 1665–1671, The AMS of Japan and Springer-Verlag.

Arakelov's Drawing 565

Related References and Comments 567

List of Publications by Yu. I. Manin 589

PART I. ALGEBRAIC GEOMETRY

Reprinted from *Amer. Math. Soc. Transl. Ser.* 2, **45** (1965).

THE HASSE-WITT MATRIX OF AN ALGEBRAIC CURVE

Ju. I. MANIN

The characteristic polynomials of the Hasse-Witt matrix and of some related matrices are computed. It is shown that the rows of the Hasse-Witt matrix are solutions of the classical systems of differential equations for the periods.

§1. Characteristic polynomial

1. **The Hasse-Witt Matrix.** Let k be a field of finite characteristic and Γ a complete nonsingular curve defined over k on genus $g > 0$. The field of functions on Γ defined over k is designated by $R_k(\Gamma)$. We suppose that there is on the curve Γ a set $P_1, \cdots, P_g$ of distinct k-rational points such that the divisor $D = \Sigma_{i=1}^g P_i$ is nonspecial. For each P_i we choose a local uniformizer t_i and define a repartition r_i in the sense of Chevalley of the field $R_k(\Gamma)$ by putting $(r_i)_P = 1/t_i$ if $P = P_i$ and $(r_i)_P = 0$ otherwise. Let $\mathfrak{R}(D)$ be the space of all repartitions which are multiples of D. Then the $r_1, \cdots, r_g$ are a basis of

$$\mathfrak{R}/\mathfrak{R}(0) + R_k(\Gamma)),$$

since the divisor $\Sigma_{i=1}^g P_i$ was chosen so as to be nonspecial. In particular, there is a congruence

$$r_i^p \equiv \sum_{j=1}^g a_{ij} r_j \bmod (\mathfrak{R}(0) + R_k(\Gamma)),$$

where $A = (a_{ij})$ is some (g, g) matrix with elements in k. It was first introduced by Hasse and Witt in [7] in the way just described, and there they established two of its basic properties:

1°. If the nonspecial system P_i of points is replaced by another such system and the local uniformizers t_i are replaced by others, then A is transformed according to the law

$$A \longrightarrow S^{-1} A S^{(p)},$$

where S is a nonsingular (g, g) matrix with elements in k and $S^{(p)}$ is obtained from S by raising each of its elements to the pth power.

2°. If k is algebraically closed, the cyclic unramified extensions of $R_k(\Gamma)$ of degree p are in one-to-one correspondence with equivalence classes of vectors $\bar{c}$ satisfying

$$\bar{c}^{(p)} A = \bar{c}$$

modulo multiplication by elements of the prime field. The $\bar{c}$ form a linear space whose dimension s is the rank of the matrix

$$AA^{(p)} \dots A^{(p^{g-1})}.$$

The number of classes of divisors of $R_k(\Gamma)$ of order p is precisely p^s (for a proof of this last fact see also [12]).

2. **Example. Formulation of the theorem.** Let $g = 1$, $p > 3$. For Γ we take the projective closure of an affine elliptic curve in Weierstrass normal form:

$$y^2 = 4x^3 + ax + b.$$

We denote by P_1 the point at infinity on the curve and put $t_1 = x/y$. In this case the Hasse-Witt matrix is simply an element A of k, which is called the (relative) Hasse invariant of Γ. Deuring ([5], p. 255) has computed A for the basis corresponding to our choice of P_1 and t_1:

$$A = \sum_{2h+3i=\frac{p-1}{2}} \frac{\left(\frac{p-1}{2}\right)!}{i!\,h!\left(\frac{p-1}{2}-i-h\right)!} a^h b^i 4^{\frac{p-1}{2}-i-h} \tag{1}$$

By property 1° above the absolute invariant is a class of elements of k of the type $c^{1-p}A$, $c \in k$. when k is algebraically closed there are thus only the two cases $A \neq 0$ and $A = 0$. The second case corresponds to the absence on Γ of points of order p and the nonexistence of unramified cyclic extensions of $R_k(\Gamma)$ of degree p.

At the other extreme is the case when k is the prime field, so that $c^{1-p} = 1$ for every $c \in k$, and hence A itself is the absolute invariant of the k-curve Γ. The author noticed that in this case A has a very simple meaning. Let π be the Frobenius endomorphism of $R_k(\Gamma)$. Its characteristic polynomial $\pi(\lambda)$ is of the form

$$\pi(\lambda) = \lambda^2 + S\lambda + p,$$

where S is an integer satisfying Hasse's inequality $S^2 < 4p$ (cf. [6]). The number of points on Γ defined over the prime field is

$$N = p + 1 + S.$$

On the other hand, it is easy to see that the number of points on the affine Weierstrass curve defined over k is given by

$$p + \sum_x \left(\frac{4x^3 + ax + b}{p}\right),$$

where the brackets denote the Legendre symbol, a and b are interpreted as classes of residues mod p and x runs through a complete system of residues mod p. Hence on recalling that the unique point at infinity on Γ is rational over k, we have

$$\bar{S} = \sum_{x \in k} (4x^3 + ax + b)^{\frac{p-1}{2}},$$

where $\bar{S}$ is the element of k corresponding to the residue class of $S \bmod p$. Since it is easily seen that

$$\sum_{x \in k} x^i = 0 \text{ for } i > 0,\ i \not\equiv 0 \bmod (p-1),$$

and

$$\sum_{x \in k} x^{m(p-1)} = -1,\ m \geq 1,$$

the sum in question is equal to the negative of the coefficient of x^{p-1} in the expansion of

$$(4x^3 + ax + b)^{\frac{p-1}{2}}$$

in powers of x, i.e., to the negative of the coefficient of $x^{-\frac{1}{2}(p-1)}$ in the expansion of

$$(4 + ax^{-2} + bx^{-3})^{\frac{p-1}{2}}.$$

But by (1), this coefficient is just A. Hence

$$\bar{S} = -A.$$

Thus if we denote by $\bar{\pi}(\lambda)$ the polynomial obtained from $\pi(\lambda)$ by reducing its coefficients modulo p we have

$$\bar{\pi}(\lambda) = -\lambda(A - \lambda).$$

The main purpose of this part of the paper is to prove the following generalization of the equation just obtained.

Theorem 1. *Let Γ be a curve of genus $g > 0$ defined over the field k_0 of characteristic $p \neq 0$ with $q = p^a$ elements and suppose that there is on it a non-special system of g points defined over k_0. By property 1° the Hasse-Witt matrix A of Γ is determined up to a transformation of the type $A \to S^{-1}AS^{(p)}$, where S is a nonsingular matrix with elements in k_0, and thus the characteristic polynomial $|A_\pi - \lambda E|$ of the matrix $A_\pi = AA^{(p)} \ldots A^{(p^{a-1})}$ is an absolute invariant of Γ. Then*

$$\bar{\pi}(\lambda) = (-1)^g \lambda^g |A_\pi - \lambda E|,$$

where $\bar{\pi}(\lambda)$ is the reduction mod p of the characteristic polynomial of the endomorphism π: $(x) \to (x^q)$ of the Jacobian j of Γ.

3. **Witt Vectors.** To prove the theorem we need to use results of Serre about

the cohomology of algebraic varieties with coefficients in sheaves of Witt vectors. In this section and the next we give a brief résumé of the concepts and results required.

Let p be a fixed prime and O a commutative ring with unit of characteristic p. We denote by $W_n(O)$ the ring of Witt vectors of length n with components in O (cf. [15]). The composition laws of the commutative ring $W_n(O)$ are given by certain polynomials with coefficients in the prime field. The rings $W_n(O)$ are mapped onto one another by the following equations:

a) **The Frobenius endomorphism** F: $W_n(O) \to W_n(O)$. By definition

$$F(a_0, \ldots, a_{n-1}) = (a_0^p, \ldots, a_{n-1}^p).$$

This is a ring homorphism.

b) **The shift** V: $W_n(O) \to W_{n-1}(O)$. By definition

$$V(a_0, \ldots, a_{n-1}) = (0, a_0, \ldots, a_{n-1}).$$

This is an additive operator. If O is a k-algebra, where k is a perfect field of characteristic p, then V is an F^{-1}-linear transformation of the structure of $W_n(O)$ as $W_n(k)$-module.

c) **The restriction** R: $W_{n+1}(O) \to W_n(O)$. By definition

$$R(a_0, \cdots, a_n) = (a_0, \cdots, a_{n-1}).$$

This is a ring homomorphism and commutes with the Frobenius endomorphism. Further, we have

$$RVF = FRV = RFV = p$$

(multiplication by p).

The projective limit of the system $W_n(O)$ of rings with respect to restriction is denoted by $W(O)$. It is a ring of characteristic zero on which the operators F and V are defined and satisfy the relation $FV = VF = p$.

If $O = k$ is a perfect field of characteristic p, then $W(k)$ is a complete discretely normed ring with the unique maximal prime ideal $pW(k)$. In this case

$$W_n(k) \cong W(k)/p^n W(k).$$

If k is the prime field, then $W(k)$ is isomorphic to the ring Z_p of p-adic integers, F is the identical isomorphism and $Vw = pW$ for every $w \in W(k)$.

4) **Cohomology with coefficients in a sheaf of Witt vectors.** Let X be an algebraic k-variety, where k is any algebraically closed field of characteristic p, and let O be the sheaf of local rings on X. Each fibre O_y is a ring of characteristic p. The union of the rings $W_n(O_x)$ for each $x \in X$ has a natural structure as

a sheaf of rings over X, which we shall denote by $\mathfrak{B}_n$. The operations F, V, R extend to $\mathfrak{B}_n$. The sheaves $\mathfrak{B}_n$ are sheaves of Λ-modules, where $\Lambda = W(k)$, which are annihilated by the ideals $p^n\Lambda$. Following Serre we define cohomology groups $H^q(X, \mathfrak{B}_n)$ which can be operated on by F, V, R. For any $q \geq 0$ the Λ-modules $H^q(X, \mathfrak{B}_n)$ and the homomorphisms R: $H^q(X, \mathfrak{B}_{n+1}) \to H^q(X, \mathfrak{B}_n)$ form a projective system. The projective limit, which we denote by $H^q(\mathfrak{B})$, is also a q-module and admits the operators V and R. The exact sequence

$$0 \to \mathfrak{B}_N \xrightarrow{V^n} \mathfrak{B}_{N+n} \xrightarrow{R^N} \mathfrak{B}_n \to 0$$

of sheaves gives rise to the exact cohomology sequence

$$\ldots \to H^q(X, \mathfrak{B}_N) \xrightarrow{V^n} H^q(X, \mathfrak{B}_{N+n}) \xrightarrow{R^N} H^q(X, \mathfrak{B}_n) \xrightarrow{\delta^q_n} H^{b+1}(X, \mathfrak{B}_N) \to \ldots$$

and so, on going to the projective limit as $N \to \infty$ (cf. [12]), we obtain the exact sequence

$$\to H^q(X, \mathfrak{B}) \xrightarrow{V^n} H^q(X, \mathfrak{B}) \xrightarrow{\rho^n} H^q(X, \mathfrak{B}_n) \xrightarrow{\delta^q_n} H^{q+1}(X, \mathfrak{B}) \to \ldots$$

We shall need the following fundamental proposition proved in [13].

Proposition 1. *Let X be an abelian variety. Then:*

1) *The coboundary operators δ^q_n are identically zero. In other words R^N and ρ_n are epimorphisms.*

2) *The group $H^1(X, \mathfrak{B})$ is a free Λ-module of finite rank $r = 2\dim X - s$, where p^s is the number of points on x of order p.*

3) *Let $T(X^*)$ be the group of Tate vectors on the dual abelian variety X^* with respect to p, i.e., the Z_p-module of vectors $(x_1, x_2, \cdots, x_n, \cdots)$, where $x_i \in X^*$, $px_1 = 0$, $px_{i+1} = x_i$, and let $T'(X^*)$ be the Λ-module $T(X^*) \otimes \Lambda$. Then $T'(X^*)$ is a free Λ-module with the finite rank s* (cf. (9), Chapter VII).

4. *The ring of endomorphisms of X has an inverse representation $\sigma \to \sigma^*$ on the direct sum*

$$L(X) = H^1(X, \mathfrak{B}) + T'(X^*).$$

The characteristic equation of any endomorphism in this representation is the same as its characteristic equation determined à la Weil for any prime $l \neq p$.

5. **Proof of Theorem 1.** Let X be an abelian variety of dimension g defined over the field k_0 of $q = p^a$ elements and let k be the algebraic closure of k_0. By 1) of Proposition 1 we have an exact sequence

$$H^1(X, \mathfrak{W}) \xrightarrow{V} H^1(X, \mathfrak{W}) \xrightarrow{\rho} H^1(X, O) \to 0. \tag{1}$$

The dimension of the k-space $H^1(X, O)$ is g. We choose a basis $\overline{h}_1, \cdots, \overline{h}_g$ of $H^1(X, O)$ and then a basis $h_1, \cdots, h_r$ of the Λ-module $H^1(X, \mathfrak{W})$ for which $\rho(h_i) = \overline{h}_i$ $(1 \le i \le g)$ and $\rho(h_i) = 0$ $(g < i \le r)$. The endomorphism π of X determines Δ-homomorphisms π_1^*: $H^1(X, O) \to H^1(X, O)$ and π^*: $H^1(X, \mathfrak{W}) \to H^1(X, \mathfrak{W})$ such that

$$\pi_1^* \rho = \rho \pi^*.$$

Let

$$\pi^*(h_1, \cdots, h_r) = (h_1, \cdots, h_r) B,$$

where B is an (r, r) matrix with elements in Λ, and let

$$\pi_1^*(\overline{h}_1, \cdots, \overline{h}_g) = (\overline{h}_1, \cdots, \overline{h}_g) B',$$

where B' is a (g, g) matrix with elements in $k \cong \Lambda/p\Lambda$. Then

$$\overline{B} = \left(\begin{array}{c|c} B' & * \\ \hline 0 & 0 \end{array}\right),$$

where $\overline{B}$ is the mod p reduction of B, i.e., the image under $\Lambda \to \Lambda/p\Lambda \cong k$. Indeed, it is enough to show that $\pi^* h_i \in pH^1(X, \mathfrak{W})$ for $g < i \le r$. Since $\rho h_i = 0$ it follows from the exactness of (1) that $h_i \in V(H^1(X, \mathfrak{W}))$ and hence

$$\pi^* h_i \in V \pi^*(H^1(X, \mathfrak{W})).$$

Hence, and because k is perfect, we have

$$\pi(R_k(X)) = R_k(X))^q \subset (R_k(X))^p,$$

and so

$$\pi^*(H^1(X, \mathfrak{W})) \subset F(H^1(X, \mathfrak{W})).$$

This implies

$$\pi^* h_i \in VF(H^1(X, \mathfrak{W})) = pH^1(X, \mathfrak{W}).$$

From what has just been proved we have the relation

$$(-1)^{r-g} \lambda^{r-g} |B' - \lambda E| = \overline{|B - \lambda E|}$$

between characteristic polynomials. By 4) of Proposition 1 this gives

$$\overline{\pi}(\lambda) = (-1)^{r-g} \lambda^{r-g} |B' - \lambda E| \cdot \pi'(\lambda), \tag{2}$$

where $\pi'(\lambda)$ is the characteristic polynomial of the endomorphism ${}^t\pi$ on the group X_p^* of points of order p of X^* regarded as a module over the prime field of characteristic p. We shall show that

$$\pi'(\lambda) = (-1)^s \lambda^s.$$

This clearly follows from

Lemma 1. $X_p^* \subset \operatorname{Ker} {}^t\pi$.

Proof. Let $\sigma: X \to Y$ denote the isogeny of X induced by raising all the elements of the universal domain to the pth power. It is clearly enough to show that $Y_p \subset \operatorname{Ker} {}^t\sigma$. We use the Barsotti-Serre theory of purely inseparable isogenies (cf. [13]). According to this theory every purely inseparable isogeny $\phi: X \to X'$ is the product of elementary isogenies of height 1 (i.e., such that $(R_k(X))^p \subset R_k(X'))$ of one of two types i_1, i_2 defined as follows. By Lemma 6 of [13] it is enough to describe the subfield of $R_k(X)$ corresponding to isogenies of the two types. Let $\mathfrak{n}$ be the p-Lie-algebra of invariant differentiations of $R_k(X)$. Subfields corresponding to isogenies of type i_1 consist of all elements annihilated by some $\partial \in \mathfrak{n}$ for which $\partial^p = 0$, but subfields corresponding to isogenies of type i_2 consist of all elements annihilated by some $\partial \in \mathfrak{n}$ for which $\partial^p = \partial$.

By Lemma 11 of [13], if ϕ is of type i_2, then ${}^t\phi$ is separable and has a kernel of order p, but if ϕ is of type i_1, then ${}^t\phi$ is purely inseparable. By Cartier's duality theorem [3] we have

$$\nu(\phi) = \nu({}^t\phi),$$

so that the isogeny ${}^t\phi$ is also of type i_1 or i_2 but it cannot be of type i_2, since ${}^t({}^t\phi) = \phi$ and thus, by 1), the isogeny ϕ would be separable. Finally, if ϕ is a separable isogeny with kernel of order p, a similar argument that ${}^t\phi$ must be of type i_2.

Let $y \in Y_p^*$ and let ϕ_y be the isogeny of degree p of Y^* with kernel generated by Y. Then, by what has just been said ${}^t\phi_y$ is an inseparable isogeny of type i_2.

For any isogeny $X \to Y$ we agree to identify $R_k(Y)$ with a subfield of $R_k(X)$. Now consider the isogeny ϕ_y: $Y^* \to Z$ and its conjugate ${}^t\phi_y$: $Z^* \to Y$. Since

$$(R_k(Z^*))^p \subset R_k(Y) \subset R_k(Z^*)$$

and

$$R_k(Y) = (R_k(X))^p,$$

we have

$$(R_k(X))^p \subset R_k(Z^*) \subset R_k(X).$$

Let τ: $X \to Z^*$ be the corresponding isogeny. Then

$${}^t\phi_y \circ \tau = \sigma.$$

The dual sequence gives

$${}^t\sigma = {}^t\tau \circ \phi_y,$$

and so

$$Y_p^* \subset \operatorname{Ker} {}^t\sigma,$$

as required.

Remark. One can show that $Y_p^* = \operatorname{Ker} {}^t\sigma$. It is enough to show that the order of $\operatorname{Ker} {}^t\sigma$ is p^s. It cannot be greater than p^s because the number of elementary isogenies of type i_2 occurring in any decomposition of σ is equal to the dimension of the "semisimple component" of the space $\mathfrak{n}$ or of its dual $H^0(X, \Omega^1)$; and this has dimension s by Proposition 10 of [12].

Returning to formula (2) we now have

$$\bar{\pi}(\lambda) = (-1)^g \lambda^g \,|\, B' - \lambda E \,| \tag{3}$$

(since $s + r = 2g$), where B' is the matrix representing the endomorphism π of the abelian variety X on some basis of the k-space

$$H^1(X, \mathfrak{M}_1) = H^1(X, O).$$

Now let Γ be a complete nonsingular curve defined over the field k_0 of $q = p^a$ elements and let $P_1, \cdots, P_g$ be a nonspecial system of rational points on it. As Chow has shown [4] we may suppose that the jacobian J of Γ and the canonical embedding ϕ: $\Gamma \to J$ are defined over k_0.

Denote by O the sheaf of local rings on J and by O_Γ the sheaf of local rings on Γ. Rosenlicht [11] has shown that the induced cohomology map ϕ^*: $H^1(J, O) \to H^1(\Gamma, O_\Gamma)$ is an isomorphism (cf. also [14], Chapter VII, §4). We recall that the exact sequence

$$0 \to O_\Gamma \to R \to R/O_\Gamma \to 0$$

of sheaves on Γ (where R is the constant sheaf with fibre $R_k(\Gamma)$) and the corresponding exact sequence

$$R_k(\Gamma) \to H^0(\Gamma, R/O_\Gamma) \to H^1(\Gamma, O_\Gamma) \to 0$$

of cohomology allows us to identify the k-spaces $H^1(\Gamma, O_\Gamma)$ and $\mathfrak{R}/\mathfrak{R}(0) + R_k(\Gamma))$. We choose a basis $h_1, \cdots, h_g$ of $H^1(J, O)$ which under this identification is mapped by ϕ^* onto the basis of $\mathfrak{R}/\mathfrak{R}(0) + R_k(\Gamma)$ constructed in subsection 1. As Cartier [1] was the first to remark (cf. also section 9 of [12]) the Hasse-Witt matrix is the matrix of the endomorphism F of $H^1(\Gamma, O_\Gamma)$ induced by the Frobenius endomorphism on the sheaf O_Γ of local rings with respect to the basis $\phi^*(h_1), \cdots, \phi^*(h_g)$. Let π denote the endomorphism $(x) \to (x^q)$ of J. Then

$$\varphi^{*-1} \circ F^a \circ \varphi^* (h_i) = \pi^* (h_i),$$

as follows immediately from the fact that the map ϕ and the cohomology classes h_i are rational over k_0 (cf. [14], Chapter VI, §1, Proposition 1). Since ϕ^* is linear but F is p-linear the matrix

$$A_\pi = AA^{(p)} \dots A^{(p^{a-1})}$$

represents π with respect to the basis $(h_1, \cdots, h_g)$. Consequently (3) implies

$$\overline{\pi}(\lambda) = (-1)^g \lambda^g |A_\pi - \lambda E|,$$

which concludes the proof of Theorem 1.

6. **Remark.** The Hasse-Witt matrix is an object which (at least in principle) can be computed as a function of the modules of the curve (in any reasonable sense of the word). On the other hand, the polynomial $\pi(\lambda)$ is essentially just the ζ-function of the field $R_k(\Gamma)$ and so is the most important arithmetic invariant of Γ. Hence Theorem 1 gives some information about the dependence of the ζ-function of Γ on its modules. The nature of this information is, for example, like the classical theorem which states that

$$\left(\frac{a}{p}\right) \equiv a^{\frac{p-1}{2}} \bmod p,$$

and this analogy actually goes to the heart of the matter (cf. subsection 2).

On the other hand, there is no answer to the question of the dependence of the Hasse-Witt matrix on the modules of Γ which shows that in some sense A is like the matrix of periods of Γ. More precisely A is a matrix of solutions of the system of differential equations which in the classical case like the periods of differentials of the first kind with modules. The next part of this paper is devoted to a proof of this result.

§2. Differential equations of the Hasse-Witt matrix

1. **Example.** Let us again turn to the case $g = 1$. Consider an affine elliptic curve whose point at infinity is rational and which is given by an equation of the type

$$y^2 = x(x-1)(x-t),$$

where t is a parameter. Put

$$r = \frac{p-1}{2},$$

where p is the characteristic of the field of definition of Γ. A simple calculation like that of subsection 2 of 1 shows that the Hasse invariant is given as a function of t by the equation

$$A(t) = (-1)^r \sum_{i=0}^{r} \binom{r}{i}^2 t^i.$$

In order to show that $A(t)$ has no multiple roots, Igusa drew attention in [8] to the fact that it satisfies the classical differential equation

$$t(1-t)\frac{d^2A}{dt^2} + (1-2t)\frac{dA}{dt} - \frac{1}{4}A = 0.$$

The same differential equation is satisfied by the period of the abelian integral

$$\int \frac{dx}{\sqrt{x(x-1)(x-t)}}$$

taken over any 1-dimensional cycle of the Riemann surface of Γ regarded as a function of t (in the case when the field of constants is the field of complex numbers). We shall generalize Igusa's result to any genus $g \geq 1$.

First, some general remarks. First of all, for a general curve Γ of genus g the author does not know any canonical form, any canonical choice of basis for the differentials of the first kind or of the variety of modules. Hence we shall consider an arbitrary basis of the space of differentials of the first kind and consider a curve Γ defined over any field which has nontrivial abstract differentiations. The algebraic construction of the differential equations for the periods of Γ was given by the author [17] for the case of zero characteristic. We give them below in a somewhat more general form than is required for our immediate purpose, and for arbitrary characteristic. Then we show that for $p \neq 0$ the Hasse-Witt matrix is the matrix of solutions of the system of differential equations for some choice of basis for the space of differentials of the first kind. Finally, we shall make some remarks on the universality of this system of differential equations (in the sense of its independence of p).

2. Differentiations. Let A be a commutative ring. A differentiation ∂ of A is an endomorphism of the additive group of A which satisfies the further condition

$$\partial(ab) = \partial a \cdot b + a\partial b$$

for all $a, b \in A$. The endomorphisms $a\partial$ and $[\partial, \partial'] = \partial \circ \partial' - \partial' \circ \partial$ are differentiations if ∂, ∂' are. Hence the set $\mathfrak{G}(A)$ of all differentiations of A is an A-Lie-algebra. The usual rules of differentiation continue to hold; in particular, Leibnitz's formula

$$\partial^m(ab) = \sum_{i=0}^{m} \binom{m}{i} \partial^i a \partial^{m-i} b.$$

If the characteristic of A is p, it follows that ∂^p is a differentiation whenever ∂ is and so $\mathfrak{G}(A)$ is a p-Lie-algebra (a restricted algebra in Jacobson's sense). For any Lie subalgebra $\mathfrak{H} \subset \mathfrak{G}(A)$ the set $A^{\mathfrak{H}}$ of elements of A annihilated by all elements of $\mathfrak{H}$ is a subring of A.

Let M be a left A-module and let $\partial \in \mathfrak{G}(A)$. An endomorphism $r(\partial)$: $M \longrightarrow M$ of the additive group is said to be a differentiation of M extending ∂ if it satisfies the further condition

$$r(\partial)(am) = \partial a \cdot m + ar(\partial)m, \ a \in A.$$

In particular, every A-endomorphism of M is a differentiation of M extending $\partial = 0$. The set $\mathfrak{G}(M)$ of differentiations of M has the structure of an A-module. Besides, a map $M \longrightarrow M$ may be iterated and so the set has an auxiliary structure as an A-Lie-algebra (a restricted Lie algebra if the characteristic of A is a prime p). In an analogous manner, one may also define a differentiation of a ring A with values in an A-module M.

3. **The module of differential relations.** Let A be a ring, M a left A-module, and $\mathfrak{H}$ a subalgebra of the Lie algebra $\mathfrak{G}(M)$. Denote by $U = U(\mathfrak{H})$ the universal enveloping algebra of the Lie algebra $\mathfrak{H}$ (which in applications always exists). The A-module M has a natural structure as left U-module.

Let $m_1, \cdots, m_g$ be a finite set of elements of the U-module M. An equation of the type

$$u_1 m_1 + \cdots + u_g m_g = 0$$

(which in the future we shall sometimes write as a symbolic scalar product $(\bar{u}, \bar{m}) = 0$) is called a linear differential (or just a differential) relation between $m_1, \cdots, m_g$ (with respect to the algebra $\mathfrak{H}$ of differentiations). The set of all differential relations between elements $m_1, \cdots, m_g$ has a natural structure

$$R(m_1, \ldots, m_g) \subset U^g = U \oplus \ldots \oplus U$$

as left U-module.

For convenience of reference we enunciate sufficient conditions under which the U-module $R(m_1, \cdots, m_g)$ is finitely generated:

Proposition 2. *Let $A = K$ be a field and suppose that the algebra $\mathfrak{H}$ is a finite-dimensional vector space over K. Then the U-module $R(m_1, \cdots, m_g)$ is finitely generated for any set $m_1, \cdots, m_g$ of elements of M.*

Proof. For characteristic zero an algorithm for the construction of a generating system was given in the author's paper [17]. It is easy to see that the hypothesis about the characteristic was not actually used there. The construction itself depends only on the fact that under the conditions stated the algebra U is finitely generated over K. The concrete interpretation of U as an algebra of differential operators is essential for the investigation of the invariant properties of the U-module M, when in the passage from one system of elements to another a new system of relations is obtained from an old one by means of the commutative relation

$$\partial(am) = \partial a \cdot m + a\partial m.$$

4. **Differential forms.** Following Cartier [2] we give the definition and some facts about abstract differential forms.

Let $A = K$ be a field which is a regular extension of its subfield k and let $(x_1, \cdots, x_k)$ be a separably generating transcendence basis for K/k. The maximal Lie-algebra $\mathfrak{H}$ of differentiations of K for which $k \subset K^{\mathfrak{H}}$ is an n-dimensional space over K with a basis consisting of the differentiations ∂_i which are uniquely defined by the equations

$$\partial_i x_j = \delta_{ij}, \quad i \leq 1, \quad j \leq n.$$

As is well known, $K^{\mathfrak{G}} = K^p k$ if K has characteristic $p \neq 0$ and $K^{\mathfrak{G}} = k$ otherwise.

For an arbitrary integer $r \geq 0$ denote by Ω^r the K-linear space of K-multilinear skewsymmetric forms in r variables on the algebra $\mathfrak{G}$. The space Ω^r is naturally isomorphic to the rth exterior power of the space Ω^1 of K-linear forms on $\mathfrak{G}$: the element

$$\omega_1 \wedge \cdots \wedge \omega_r, \quad \omega_i \in \Omega^1,$$

being identified with the form

$$(\partial_1', \ldots, \partial_r') \rightarrow |\omega_i(\partial_j')|.$$

The elements of Ω^r are called (homogeneous) differential forms of degree r of the field K over the field k of constants.

For any element $x \in K$ we denote by $dx \in \Omega^1$ the map $\partial \rightarrow \partial x$. This map is $K^{\mathfrak{G}}$-linear in x and satisfies

$$d(xy) = dx \cdot y + x\,dy;$$

so that dx belongs to the space of differentiations of the field K into the K-module Ω^1. It has a unique extension as a $K^{\mathfrak{G}}$-linear operator $d\colon\ \Omega^r \rightarrow \Omega^{r+1}$, the fundamental properties of which are given by

$$\begin{gathered} d(\omega \wedge \omega') = d\omega \wedge \omega' + (-1)^r \omega \wedge d\omega', \\ \omega \in \Omega^r, \quad d(d\omega) = 0. \end{gathered} \tag{4}$$

Put

$$Z^r = \operatorname{Ker}(d\colon\ \Omega^r \rightarrow \Omega^{r+1}), \quad B^r = \operatorname{Im}(d\colon\ \Omega^{r-1} \rightarrow \Omega^r),$$

where $\Omega^0 = K$, by definition. Clearly $B^r \subset Z^r$ and the factor space $H^r(K) =$

Z^r/B^r is the space of "de Rham cohomology" of K/k of dimension r.

5. **The Cartier operator.** The direct sum

$$\sum_r Z^r = Z$$

has a structure as $K^{\mathfrak{G}}$-subalgebra of the exterior algebra of Ω^1. It follows at once from (4) that $\Sigma B^r = B$ is an ideal in Z, so that

$$H(K) = \sum Z^r / B^r$$

is a $K^{\mathfrak{G}}$-algebra of cohomology. When the characteristic p of K is different from zero its structure is described by the following theorem of Cartier [2]. Let $dx_1, \cdots, dx_n$ be a K-basis of Ω^1. Then the $K^{\mathfrak{G}}$-algebra Z is the direct sum of the ideal B and the $K^{\mathfrak{G}}$-algebra generated by the $f_i = x_i^{p-1}\,dx_i$.

Hence any closed form $\omega \in Z^r$ is uniquely representable in the form

$$\omega = d\varphi + \sum_{i_1 < \cdots < i_r} a_{i_1 \ldots i_r} f_{i_1} \wedge \ldots \wedge f_{i_r},$$

where $a_{i_1, \cdots, i_r} \in K^{\mathfrak{G}} = K^p k$. Cartier's operator $C\colon Z^r \to k^{1/p}\Omega^r$ is defined by

$$C\omega = \sum_{i_1 < \cdots < i_r} a_{i_1 \ldots i_r}^{\frac{1}{p}}\, dx_{i_1} \wedge \ldots \wedge dx_{i_r}$$

and is independent of the choice of basis $dx_1, \cdots, dx_n$, so that C induces a homomorphism

$$H(K) \to k^{\frac{1}{p}}\Omega.$$

By definition $C(d\phi) = 0$ for any form ϕ.

6. **Differentiations of $H^r(K)$.** (In this subsection the characteristic is arbitrary.) We suppose that the field k of constants possesses nontrivial differentiations. Let $\partial \in \mathfrak{G}(k)$. For an arbitrary separating transcendence base $x = (x_1, \cdots, x_n)$ of K/k there is a uniquely defined differentiation ∂_x of K such that: 1) $\partial_x x_i = 0$ $(1 \leq i \leq n)$ and 2) ∂_x coincides with ∂ on $k \subset K$. In turn, the differentiation ∂_x extends to a differentiation $r(\partial_x)$ of the K-module Ω^r by the formula

$$r(\partial_x)\Big(\sum_{i_1 < \cdots < i_r} a_{i_1 \ldots i_r} dx_{i_1} \wedge \ldots \wedge dx_{i_r}\Big) = \sum_{i_1 < \cdots < i_r} \partial_x a_{i_1 \ldots i_r} dx_{i_1} \wedge \ldots \wedge dx_{i_r}.$$

Lemma 2. $r(\partial_x) B \subset B$; $r(\partial_x) Z \subset Z$.

Proof. We shall show that

$$r(\partial_x)(d\omega) = d(r(\partial_x)\omega).$$

It is clearly enough to show that the map

$$r(\partial_x)\circ d - d\circ\partial_x : K \to \Omega^1$$

is zero. But it belongs to $\mathfrak{G}(K, \Omega^1)$ and vanishes on the subfield $k \subset K$ and on the x_i. Since $K/k(x_1, \cdots, x_n)$ is separable a standard argument shows that

$$r(\partial_x)\circ d - d\circ\partial_x = 0.$$

The differentiations $r(\partial_x)$ depend not only on the differentiation ∂ but also on the choice of basis x. Generalizing a lemma of Chevalley to the n-dimensional case we shall show now that for $\omega \in Z^r$ the class $r(\partial_x)\omega + B$ modulo B depends only on ∂.

Proposition 3. *Let $\partial \in \mathfrak{G}(k)$ and let $(x_1, \cdots, x_n)$, (y_1, y_n) be separably generating transcendence bases of K/k.*

Let

$$\omega = \sum_{i_1<\ldots<i_r} a_{i_1\ldots i_r} dx_{i_1}\wedge\ldots\wedge dx_{i_r}, \quad d\omega = 0.$$

Then

$$(r(\partial_y) - r(\partial_x))\,\omega = d\varphi,$$

where

$$\varphi = \sum_{i_1<\ldots<i_r}\sum_{\alpha=1}^{r}(-1)^{\alpha-1} a_{i_1\ldots i_k} dx_{i_1}\wedge\ldots\wedge\partial_y x_{i_\alpha}\wedge\ldots\wedge dx_{i_r}.$$

Proof. First of all $\partial_y - \partial_x \in \mathfrak{G}(K/k)$ and so there exist $z_i \in K$ such that

$$\partial_y - \partial_x = \sum_{i=1}^{n} z_i\partial_i,$$

where the ∂_i are the basis for $\mathfrak{G}(K/k)$ dual to the basis dx_i, i.e., $\partial_i x_j = \delta_{ij}$. On applying $\partial_y - \partial_x$ to x_i we at once get

$$z_i = \partial_y x_i.$$

Further, it follows easily from the definitions that

$$\partial_y(\omega\wedge\omega') = \partial_y\omega\wedge\omega' + \omega\wedge\partial_y(\omega'),$$

and so

$$(r(\partial_y) - r(\partial_x))\,\omega = \sum_{i_0=1}^{n}\sum_{i_1<\ldots<i_r}\partial_y x_{i_0}\partial_{i_0} a_{i_1\ldots i_r} dx_{i_1}\wedge\ldots\wedge dx_{i_r}$$

$$+ \sum_{i_1<\ldots<i_r}\sum_{\alpha=1}^{r} a_{i_1\ldots i_r} dx_{i_1}\wedge\ldots\wedge d(\partial_y x_{i_\alpha})\wedge\ldots\wedge dx_{i_r}.$$

On the other hand

$$d\varphi = \sum_{i_1<\dots<i_r} \sum_{\alpha=1}^{r} (-1)^{\alpha-1} da_{i_1\dots i_r} \wedge dx_{i_1} \wedge \dots \wedge \partial_y x_{i_\alpha} \wedge \dots \wedge dx_{i_r}$$

$$+ \sum_{i_1<\dots<i_r} \sum_{\alpha=1}^{r} a_{i_1\dots i_r} dx_{i_1} \wedge \dots \wedge d(\partial_y x_{i_\alpha}) \wedge \dots \wedge dx_{i_r}.$$

Consequently, we need to show that

$$\sum_{i_0=1}^{n} \sum_{i_1<\dots<i_r} \partial_y x_{i_0} \cdot \partial_{i_0} a_{i_1\dots i_r} dx_{i_1} \wedge \dots \wedge dx_{i_r}$$

$$= \sum_{i_1<\dots<i_r} \sum_{\alpha=1}^{r} (-1)^{\alpha-1} da_{i_1\dots i_r} \wedge dx_{i_1} \wedge \dots \wedge \partial_y x_{i_\alpha} \wedge \dots \wedge dx_{i_r}.$$

For this it is enough to check that the coefficients of $\partial_y x_m$ on both sides are the same, i.e.,

$$\sum_{i_1<\dots<i_r} \partial_m a_{i_1\dots i_r} dx_{i_1} \wedge \dots \wedge dx_{i_r}$$

$$= \sum_{(i_1\dots i_r)\ni m=i_\alpha} (-1)^{\alpha-1} da_{i_1\dots i_r} \wedge dx_{i_1} \wedge \dots \wedge \widehat{dx_{i_\alpha}} \wedge \dots \wedge dx_{i_r}, \qquad (5)$$

where the sum on the right is over all ordered combinations of indexes containing m and where the roof, as usual, indicates that the corresponding term is to be omitted. But the sum on the right is

$$\sum_{(i_1\dots i_r)\ni m} \partial_m a_{i_1\dots i_r} dx_{i_1} \wedge \dots \wedge dx_{i_r}$$

$$+ \sum_{i_0\neq m} \sum_{(i_1\dots i_r)\ni m=i_\alpha} (-1)^{\alpha-1} \partial_{i_0} a_{i_1\dots i_r} dx_{i_0} \wedge dx_{i_1} \wedge \dots \wedge \widehat{dx_{i_\alpha}} \wedge \dots \wedge dx_{i_r}. \qquad (6)$$

On comparing (5) and (6) we see that we have to check the equation

$$\sum_{(i_1\dots i_r)\ni m} \partial_m a_{i_1\dots i_r} dx_{i_1} \wedge \dots \wedge dx_{i_r}$$

$$= \sum_{i_0\neq m} \sum_{(i_1\dots i_r)\ni m=i_\alpha} (-1)^{\alpha-1} \partial_{i_0} a_{i_1\dots i_r} dx_{i_0} \wedge dx_{i_1} \wedge \dots \wedge \widehat{dx_{i_\alpha}} \wedge \dots \wedge dx_{i_r}.$$

Let us fix the combination $(j_1, \cdots, j_r)$ and check that the coefficients of $dx_{j_1} \wedge \cdots \wedge dx_{j_r}$ on both sides are the same. If $m \in (j_1, \cdots, j_r)$, these coefficients both vanish. If

$$m \notin (j_1 \dots j_r), \quad j_\alpha < m < j_{\alpha+1},$$

then the coefficient on the left is $\partial_m a_{j_1 \dots j_r}$. A simple combinatorial computation shows that the coefficient on the right is

$$\sum_{1 \leqslant \beta \leqslant \alpha} (-1)^{\alpha+\beta} \partial_{j_\beta} a_{j_1 \dots \hat{j}_\beta \dots m \dots j_r} + \sum_{\alpha+1 \leqslant \beta \leqslant r} (-1)^{\alpha+\beta+1} \partial_{j_\beta} a_{j_1 \dots m \dots \hat{j}_\beta \dots j_r}.$$

That this is just $\partial_m a_{j_1 \dots j_r}$ follows at once from the fact that $d\omega = 0$ (it is necessary to decompose this equation into its components and use the fact that the coefficient of a term of the shape $dx_{j_1} \wedge \dots \wedge dx_m \wedge \dots \wedge r_{j_r}$ is zero). This completes the proof.

This proposition together with the lemma shows that $H^r(K)$ has a natural structure as $\mathfrak{G}(k)$-module.

When the characteristic is zero, this structure can be interpreted as follows. Let K be the field of functions of an algebraic variety X which is defined over a finitely generated extension k of the field of complex numbers, i.e., is the generic member of an algebraic system of varieties. As is well known, in this case a closed differential form $\omega \in \Omega^r(K)$ defines an r-dimensional cocycle on X whose value on any r-dimensional cycle Z is just integral of ω over Z. This integral (a period of ω) depends on the parameters of the algebraic system and thus we may ask about its "derivative with respect to a parameter" $\partial(\int_Z \omega)$ for $\partial \in \mathfrak{G}(k)$. Proposition 3 shows that the differences arising from different ways of carrying ∂ past the sign of integration are compensated for by the fact that the difference is the integral of a complete differential over the cycle, i.e., zero. Thus the differentiation of an abstract class of de Rham cohomology corresponds to taking the derivative with respect to a parameter of a vector whose components are the periods of the differential form with respect to a basis of the topological cycles of the corresponding dimension.

7. **Differentiations and Cartier's operator.** In this section we suppose that the the characteristic $p \neq 0$. Let $\partial \in \mathfrak{G}(k)$. The isomorphism $F^{-1}: k \to k^{1/p}$ gives rise to an isomorphism $F: \mathfrak{G}(k) \to \mathfrak{G}(k^{1/p})$ defined by

$$(F^* \partial) a = (\partial a^p)^{\frac{1}{p}}.$$

Any separably generating transcendence base $(x_1, \dots, x_n)$ of K/k is also a separably generating transcendence base of $Kk^{1/p}/k^{1/p}$ and so we may define the differentiation $F^* \partial_x$.

Now let Γ be a complete nonsingular curve of genus $g \geq 1$ defined over a field k of characteristic $p \neq 0$, let $\mathfrak{G}$ be a finite-dimensional Lie algebra of differentiations of k and let $(r_1, \cdots, r_g)$ be a canonical basis of the space

$$\mathfrak{R}/\mathfrak{R}(0) + R_k(\Gamma)) = H^1(\Gamma, O_\Gamma).$$

It is well known that the spaces $H^1(\Gamma, O_\Gamma)$ and $H^0(\Gamma, \Omega^1)$, where $\Omega^1 = \Omega^1(R_k(\Gamma)/k)$, are dual. Choose a basis $\omega_1, \cdots, \omega_g$ for the space $H^0(\Gamma, \Omega^1)$ of differentials of the first kind on Γ and denote by $K = \|(\omega_i, r_j)\|$ the corresponding matrix of scalar products. From the definition of the Hasse-Witt matrix A it follows readily that

$$KA = \|(\omega_i, Fr_j)\|.$$

In particular, if $(\omega_1, \cdots, \omega_g)$ is a dual basis to $(r_1, \cdots, r_g)$, as we shall suppose henceforth, we have

$$A = \|(\omega_i, Fr_j)\|.$$

Let M be the factor space of differentials of the first kind on Γ modulo complete differentials, considered as a k-module. Rosenlicht [10] showed that M has dimension g. Let m_i denote the image of ω_i in M. Proposition 3 allows us to regard M as a $\mathfrak{G}$-module. We shall consider the module

$$R(m_1, \ldots, m_g) \subset U^g$$

of differential relations. Further, $\mathfrak{G}$ acts on k^g, and we also consider the module

$$R(n_1, \ldots, n_g) \subset U^g$$

of differential relations, where n_i is the ith row of A.

Theorem 2. *$R(m_1, \cdots, m_g) \subset R(n_1, \cdots, n_g)$. In other words the rows n_i of the Hasse-Witt matrix satisfy every differential relation which is satisfied by the elements of a basis of the space of differentials of the first kind (dual to a canonical basis of $H^1(\Gamma, O_\Gamma)$) moaulo the space of complete differentials.*

Proof. Let

$$\sum_{i=1}^{g} u_i m_i = 0, \quad u_i \in U.$$

This means that for any separably generating $x \in R_k(\Gamma)/k$ we have

$$\sum_{i=1}^{g} r(u_i)_x \omega_i = d\varphi,$$

where $r(u_i)_x$ is composed of the $r(\partial_x)$ in precisely the same way as u_i is composed of the ∂, and where the differential $d\phi$ depends on x. Hence for any canonical basis r_j, $j = 1, \cdots, g$, we have

$$\sum_{i=1}^{g}(r(u_i)_x\omega_i,\ Fr_j)=(d\varphi,\ Fr_j). \tag{7}$$

By Cartier's Lemma (cf. [13], §9) we have

$$(\omega,\ Fr_j)=(C\omega,\ r_j)^p$$

for every $\omega \in \Omega^1$. Hence from the fact that $C(d\phi)=0$ we have

$$\sum_{i=1}^{g}(Cr(u_i)_x(\omega_i),\ r_j)^p=0, \quad j=1,\ldots,g, \tag{8}$$

by (7). By Proposition 3 the operator $C\circ r(u_i)_x$ is independent of the choice of x. For each point P_i $(i=1,\cdots,g)$ we thus can choose the local uniformizer t_i for x, when it is easily seen that

$$\begin{aligned}(Cr(u_i)_x(\omega_i),\ r_j)^p &= (F^*u_i(C\omega_i,\ r_j))^p, \\ &= u_i((C\omega_i,\ r_j)^p)=u_i(\omega_i,\ Fr_j).\end{aligned}$$

Thus (8) implies

$$\sum_{i=1}^{g}u_i(\omega_i,\ Fr_j)=0, \quad j=1,\ldots,g,$$

which is just the assertion of the theorem.

8. **Remark on universality.** The assertion that the classical system of differential equations for the periods of a curve Γ are satisfied by the rows of the Hasse-Witt matrix *independently of the characteristic* p can be correctly interpreted in the following fashion. We take a curve of genus $g \geq 1$ defined over some field of algebraic functions in characteristic zero (e.g., "the generic curve of genus g"). On it we take a basis for the differentials of the first kind dual to a canonical basis and set up the system of differential relations between them. This will be a set of formulae of the form

$$\sum a_{i_1\cdots i_k}\partial_1^{i_k}\ldots\partial_k^{i_k}u_j\,dv=dv_j,$$

where $a_{i_1\cdots i_k}\in R$, u_j, $v_j\in R_k(\Gamma)$. Let X be an algebraic variety defined over an absolutely algebraic field k_0 of constants of finite degree over the rationals whose field of functions is abstractly isomorphic to the field of functions on Γ. The curve Γ lies in X. For almost all prime divisors $\mathfrak{p}$ of k_0 the reductions $X\to\overline{X}$ and $\Gamma\to\overline{\Gamma}$ are nondegenerate, do not decrease the genus of g. Transform differentials of the first kind into differentials of the first kind and in general have every "nice" property. The fact that we may also reduce the differentiations ∂ follows from the fact that the formulae for differentiations are universal, but the existence of a result follows from the regularity of the reduction (these arguments do not pretend to be a rigorous demonstration).

Returning to the example considered at the beginning of this section, we may deduce Igusa's result from Theorem 2 on using the fact that the differential $\omega = dx/y$ on

$$y^2 = x(x-1)(x-t)$$

satisfies

$$t(1-t)\partial_x^2\omega + (1-2t)\partial_x\omega - \frac{1}{4}\omega = d\frac{y}{(x-t)^2}, \tag{9}$$

where ∂ is differentiation with respect to t: $\partial t = 1$.

Igusa remarks that although the equation for the periods obtained from (9) has two linearly independent solutions in characteristic zero the corresponding equation in the finite characteristic case has only one independent solution A. This fact is intimately connected with the Rosenlicht theorem quoted above about the dimension of the space of differentials of the first kind modulo complete differentials. Indeed, if in characteristic zero the differentials ω and $\partial_x\omega$ are linearly independent modulo the complete differentials (cf. [17], section 11), then in the case of finite characteristic there must be an equation of the form

$$\partial_x\omega + a(t)\omega = d\varphi,$$

i.e., an equation such as

$$\frac{dA}{dt} + a(t)A = 0.$$

But the coefficient $a(t)$ here clearly depends on the characteristic p, so that the reduction in the order of the equation has been obtained at the price of a loss of universality.

The author is deeply grateful for advice and assistance to I. R. Šafarevič, under whose guidance the work was done.

BIBLIOGRAPHY

[1] P. Cartier, *Une nouvelle operation sur les formes différentielles*, C. R. Acad. Sci. Paris 244 (1957), 426–428. MR 18, 870.

[2] ———, *Questions de rationalité des diviseurs en géomètrie algébrique*, Bull. Soc. Math. France 86 (1958), 177–251. MR 21 #4957.

[3] ———, *Dualité des variétés abéliennes*, Séminaire Bourbaki; (1957/58), Textes des Conferences: Exposés 152–168; 2nd corrected ed., Exposé 164,

pp. 1–13, Sécrétariat Mathématique, Paris, 1958 (mimeographed). MR 21 #4161.

[4] W. L. Chow, *The Jacobian variety of an algebraic curve*, Amer. J. Math. 76 (1954), 453–476. MR 15, 823.

[5] M. Deuring, *Die Typen der Multiplikatorenringe elliptischer Functionenkörper*, Abh. Math. Sem. Univ. Hamburg 14 (1941), 197–272. MR 3, 104.

[6] H. Hasse, *Zur Theorie der abstrakten elliptischer Functionenkörper*, J. Reine Angew. Math. 175 (1936), 55–62, 69–88, 193–208.

[7] H. Hasse and E. Witt, *Zyklische unverzweigte Erweiterungskörper vom Primzahlgrade p über einem algebraischen Functionenkörper der Charakteristik p*, Monatsh. Math. Phys. 43 (1936), 477–492.

[8] J. I. Igusa, *Class number of a definite quaternion with prime discriminant*, Proc. Nat. Acad. Sci. U. S. A. 44 (1958), 312–314. MR 20 #5183.

[9] S. Lang, *Abelian varieties*, Interscience, New York, 1959. MR 21 #4959.

[10] M. Rosenlicht, *Differentials of the second kind for algebraic function fields of one variable*, Ann. of Math (2) 57 (1953), 517–523. MR 14, 1012.

[11] ———, *Extensions of vector groups by abelian varieties*, Amer. J. Math. 80 (1958), 685–714. MR 20 #5780.

[12] J. P. Serre, *Sur la topologie des variétés abéliennes en caracteristique p*, Sympos. Internac. Topologia Algebraica, Mexico City (1956), pp. 24–53, Univ. Nac. Aut. Mexico and UNESCO, Mexico, 1958. MR 20 #4559.

[13] ———, *Quelques propriétés des variétés abéliennes en caracteristique p*, Amer. J. Math. 80 (1958), 715–739. MR 20 #4562.

[14] ———, *Groupes algébriques et corps de classes*, Hermann, Paris. MR 21 #1973.

[15] E. Witt, *Zyklische Körper und Algebren der Charakteristik p vom Grade p^m*, J. Reine Angew. Math. 176 (1936), 126–140.

[16] C. Chevalley, *Introduction to theory of algebraic functions of one variable*, Mathematical Surveys, No. 6, Amer. Math. Soc., Providence, R. I., 1951; Russian transl., Fizmatgiz, Moscow, 1959. MR 13, 64; MR 22 #2611.

[17] Ju. I. Manin, *Algebraic curves over fields with differentations*, Izv. Akad. Nauk. SSSR Ser. Mat. 22 (1958), 737–756; English transl., Amer. Math. Soc. Transl. (2) 37 (1964, 59–78. MR 21 #2652.

Translated by:
J. W. S. Cassels

Reprinted from *Amer. Math. Soc. Transl. Ser.* 2, **50** (1966).

RATIONAL POINTS OF ALGEBRAIC CURVES OVER FUNCTION FIELDS

Ju. I. MANIN

Summary. Mordell's conjecture is proved for algebraic curves with function fields as ground fields.

Introduction

1. The principal goal of this paper is the proof of the following result.

Theorem. *Let K be a regular extension of the field k of characteristic* 0 *and let C be a curve of genus ≥ 2 defined over K. If there are infinitely many points on C rational over K, then it is birationally equivalent to a curve C_0 defined over k and all except possibly finitely many of the points correspond to points of C_0 defined over k.*

In particular, if C is a "nonconstant" curve, i.e., if its least field of definition is not absolutely algebraic, then it has only a finite number of points defined over any given field of finite type.

This theorem gives a proof of a part of Mordell's conjecture (cf. Lang [5], p. 29). In fact, Mordell [11] enunciated the conjecture only for curves over the field of rational numbers: the subsequent extension of the Mordell-Weil theorem to algebraic fields and then to function fields makes it natural to pose the question of the truth of the conjecture for all possible ground fields of finite type.

As Lang remarked, Mordell's conjecture is equivalent to the assertion that a curve lying in an abelian variety has only a finite number of points in common with any subgroup of finite type of the variety. This formulation contains no explicit reference to the ground field. Nevertheless, it seems probable that the proof of Mordell's conjecture cannot be achieved without using the arithmetic (in the wide sense of the word) of the ground field. If that is so, then our result can be regarded as a reduction of the general problem to the case when the ground field is an algebraic number field: a case which has so far remained inviolate to known methods. We leave it to the reader to judge how far the ideas of this paper can be useful when applied to that case.

2. This paper naturally falls into three parts, corresponding to the three chapters. The first chapter is devoted to the working out of a technique which is essential to the rest of the exposition; the second chapter proves the theorem enunciated above; and the third contains the proof of some strong finiteness theorems for elliptic curves.

We shall illustrate the fundamental idea of this paper by the example of the "general" elliptic curve $C: Y^2 = X(X-1)(X-t)$ and will take as our ground field a finite extension of $\mathbf{C}(t)$, where $\mathbf{C}$ is the field of complex numbers. Let γ be a 1-dimensional cycle on the Riemann surface of C and let $\omega = Y^{-1}\,dX$ be a differential of the first kind on C. The period $\eta(t) = \int_\gamma \omega$, considered as a function of t, is an infinitely many-valued analytic function on the t-plane punctured at $t = 0, 1$. (For the precise meaning of this and the following assertions see §6.) The nature of these functions has been well known for a long time: they are solutions of the Gauss linear differential equation

$$4t(1-t)\frac{d^2\eta}{dt^2} + 4(1-2t)\frac{d\eta}{dt} - \eta = 0. \tag{1}$$

The space of solutions of this equation is 2-dimensional (over the complex field), having as basis an arbitrary pair of periods

$$\eta_1(t) = \int_{\gamma_1}\omega, \quad \eta_2(t) = \int_{\gamma_2}\omega$$

over linearly independent cycles γ_1, γ_2 generating the integral 1-dimensional homology of the Riemann surface of C.

On the other hand, functions of the type $\int_O^P \omega$ where O is the point at infinity, play a large role in the investigation of points on C with coordinates in K. This function is defined by P only up to a integral linear combination $m_1\eta_1(t) + m_2\eta_2(t)$ and satisfies the equation

$$\int_O^P \omega + \int_O^Q \omega = \int_O^{P+Q} \omega,$$

modulo such a linear combination, where $P + Q$ is the sum according to the group law on C which has O as its zero. It is natural to attempt to remove the indeterminacy of this last functional equation by acting on both sides with the Gauss differential operator which annihilates the periods $\eta_1(t)$ and $\eta_2(t)$. Then the function

$$\mu(P) = \left(4t(1-t)\frac{d^2}{dt^2} + 4(1-2t)\frac{d}{dt} - 1\right)\int_O^P \omega$$

becomes a finitely many-valued meromorphic function on the Riemann t-sphere. It becomes single-valued on the Riemann surface of K and thus may be identified with an element of K. A remarkable property of this function, which follows from the linearity of the Gauss operator, is that it defines a homomorphism of the group of points of C defined over K into the additive group of K:

$$\mu(P+Q)=\mu(P)+\mu(Q).$$

It is not difficult to express $\mu(P)$ explicitly in terms of the coordinates $(X(t), Y(t))$ of P:

$$\mu(P)=\frac{1}{2}\frac{Y(t)}{(X(t)-t)^2}-\frac{d}{dt}\left[2t(t-1)\frac{\frac{d}{dt}X(t)}{Y(t)}\right]-2t(t-1)\frac{d}{dt}X(t)\cdot\frac{d}{dt}\frac{1}{Y(t)}. \quad (2)$$

This formula is instructive in that it shows the nature of the function $\mu(P)$ obtained from the coordinates of P by applying a linear differential operator (in this case differentiation with respect to t). The proof of (2) depends on the following argument.

First of all the Gauss equation (1) is a consequence of the purely algebraic equation

$$\left[4t(1-t)\frac{\partial^2}{\partial t^2}+4(1-2t)\frac{\partial}{\partial t}-1\right](Y^{-1})\,dX=\frac{1}{2}\,d\frac{Y}{(X-t)^2}. \quad (3)$$

Here $\partial/\partial t$ denotes the operator "partial differentiation with respect to t keeping X fixed" and d is the differential in the field $K(X, Y)/K$, i.e., "keeping t fixed". The application of the Gauss operator to the period $\int_\gamma\omega$ is carried out by taking the operator under the integral sign; and then it is necessary only to note that the integral round a closed contour of the exact differential $d(Y/(X-t)^2)$ is zero.

The formula (2) is obtained analogously, but here it is necessary to consider the dependence on the parameter t of the limits of integration in

$$\int_0^P\omega=\int_\infty^{X(t)}Y^{-1}\,dX.$$

The first term in the right hand of (2) is the integral of $-d(Y/(X-t)^2)$ between the given limits, the remaining terms are the contribution of the change of the limits of integration. We note that if the value of X at P does not depend on t, then all the terms except the first vanish.

The construction described here can be carried out in a completely general situation: for an arbitrary system of algebraic varieties depending on parameters with respect to which one may differentiate in some sense. The generalization of the Gauss differential equation to periods of abelian integrals on curves of arbitrary genus was obtained by Picard and Fuchs (cf. [12]). We give a similar generalization to the case of arbitrary algebraic varieties. The key idea is that relations similar to (3) can be found for closed 1-dimensional differential forms of the first kind on arbitrary algebraic varieties on whose field of definition there is a derivation ∂ (corresponding to the differentiation d/dt). The first section of Chapter I is devoted to the construction and theory of such relations, which we propose to call Picard-Fuchs equations. We should remark that Picard and Fuchs considered only equations of the type

$$\mathfrak{L} \int_{\gamma} \omega = 0,$$

where $\mathfrak{L}$ is a linear differential operator and ω is an abelian differential. There are good algebraic reasons, however, for considering the more general equations

$$\mu : \sum_{i=1}^{n} \mathfrak{L}_i \int_{\gamma} \omega_i = 0$$

since this allows us to endow the totality of equations with the structure of a module over the ring of differential operators and to find a finite number of generators.

§§2, 3 and 4 contain the definition and investigation of functions analogous to $\mu(P)$. Heuristically they are given by

$$\mu(P, Q) = \sum_{i=1}^{n} \mathfrak{L}_i \int_{P}^{Q} \omega_i,$$

but up to §6 we employ only an algebraic definition. Besides having essential methodological advantages our definition works in situations in which it would be unnatural to introduce analytical limitations (e.g., when the ground field is a field of formal power series).

The content of the fifth and sixth sections runs parallel to the theory described above for elliptic curves and consists essentially in carrying this theory over to abelian varieties of arbitrary dimension depending on parameters. We illustrated earlier only the construction of the homomorphism $P \longrightarrow \mu(P)$: a less trivial part of the theory is to investigate how far the homomorphism is nontrivial, i.e., the investigation of the kernel. Let us return to the curve $Y^2 = X(X-1)(X-t)$. Clearly all points of finite order are in the kernel. The following argument shows that the kernel contains no further points. If $\mu(P) = 0$, the integral $\int_O^P \omega$ must be a linear combination with constant coefficients of $\eta_1(t)$ and $\eta_2(t)$:

$$\int_{0}^{P} \omega = c_1\eta_1(t) + c_2\eta_2(t), \quad c_1, c_2 \in \mathrm{C}.$$

To establish that P is of finite order we must show that c_1, c_2 are *rational*. For this we make essential use of the fact that the coordinates of P are *algebraic* functions of t. The idea of the proof is as follows. The cycles γ_1, γ_2 can be chosen in such a way that circuits σ_1, σ_2 round the points $t = 0$, $t = 1$ respectively produce the following changes in the many-valued functions η_1, η_2:

$$\sigma_1: \begin{cases} \eta_1(t) \to \eta_1(t), \\ \eta_2(t) \to \eta_2(t) + \eta_1(t); \end{cases} \qquad \sigma_2: \begin{cases} \eta_1(t) \to \eta_1(t) + \eta_2(t), \\ \eta_2(t) \to \eta_2(t). \end{cases}$$

Hence one circuit adds $c_2\eta_1(t)$ or $c_1\eta_2(t)$ respectively to the integral $\int_O^P \omega$. But since the coordinates of P are algebraic functions, they must return to their original values after a finite number, say d, of circuits, and so $\int_O^P \omega$ must take a value differing from the original one by a linear combination of the periods η_1, η_2. Hence dc_1 and dc_2 are integral; that is, c_1 and c_2 are rational. This proves our assertion about the kernel of μ. The analogous result for abelian varieties is Theorem 2, which is formulated in §5 and proved in §6.

It is important to emphasize that Theorem 2, which describes the intersection of the kernels of all possible homomorphisms μ on an abelian variety, is then used to give information about the functions μ on algebraic curves. This is possible because the construction of these functions has functorial character and, in particular, for the points of a curve immersed in its jacobian the functions μ give the same result whether we start with the Picard-Fuchs equations for the whole variety, or only with their restrictions to the curve.

3. The second and third chapters are of a significantly more technical nature. The possibility of using functions of the type μ to prove finiteness theorems about points on curves is by no means obvious, and the argument is different in the cases genus ≥ 2 and genus 1.

In very general terms the contents of the second chapter are as follows. We suppose that there are infinitely many points on the curve C defined over K. We embed C in its jacobian and consider the values of the functions μ at points P of C. By the Mordell-Weil theorem (cf. [6]) and our Theorem 2 every such function takes its values in a subgroup of finite rank in the additive group of K.

On the other hand, as for an elliptic curve, each function $\mu(P)$ results from the operation of some differential operator on the coordinates of P. A rather obvious investigation of these operators leads to linear relations of the type

$$\Sigma z_i(P)\mu_i(P) = 0,$$

where the z_i are rational functions on C. If there are sufficiently many of these relations, we can deduce relations of the type

$$\mu_i(P) = c_i \in K$$

and apply the theorem about the intersection of the kernels of the μ (on the jacobian variety). To obtain sufficiently many relations we have to use not only C itself but certain unramified coverings, which also contain infinitely many points. The form of the function μ depends on the structure of the module of Picard-Fuchs equations on the curve and consequently one has to consider separately two cases.

It is difficult to sketch further particulars without going into the treatment in full.

In the third chapter, which treats the elliptic case, we have only the one function $\mu(P)$ and everything depends on it. This is connected with the fact that the module of Picard-Fuchs equations has only one generator. The idea of Theorem 4 (cf. its formulation in §11) consists in establishing a uniform continuity of the map $P \longrightarrow \mu(P)$ in the local topologies of the field K. Since the values of $\mu(P)$ lie in a subgroup of finite type of the additive group of K, which is discrete in every local topology, it follows that the group of rational points is discrete. In particular, in every local topology the point P cannot approach too closely to O. For the curve $Y^2 = X(X-1)(X-t)$ this means that the denominator of $X(t)$ cannot have zeros of arbitrary large multiplicity. This gives a variant for the function-field case of Mahler's theorem about the number of "integral" points on an elliptic curve. Actually Theorem 4, as is shown in §11, is a pretty strong statement and its analogue for algebraic number fields is false.

It is possible that the methods of the last two chapters can be applied to Lang's conjectures (cf. [5]) about integral points on affine subvarieties and rational points on closed subvarieties respectively of abelian varieties (of course, with a function field as ground field). Partly for these reasons the constructions of the first chapter have been carried out in somewhat greater generality than was immediately necessary for the investigation of curves. We also remark that it would be of interest to connect our arguments with the technique of principal homogeneous spaces (cf. [1]).

4. The fundamental ideas of the first chapter are already contained in [8] and [9]. The second of these explains the meaning of the Picard-Fuchs equations for curves over fields of finite characteristic, thus giving a general answer to the question posed by Igusa [13] and solved by him for elliptic curves [14]. The last chapter of this paper is a reworking of [10]. The essentially new part of this paper is the second chapter containing the proof of the function-field case of Mordell's conjecture. Nevertheless the paper has been written so that it can be read independently of the previous publications, which can be regarded as of a preliminary nature.

We shall use constantly and without special reference algebro-geometric concepts and results (in particular from the theory of abelian varieties) which are expounded in the two books of Lang [2], [3]. Besides, we shall need the elements of the theory of heights described in the papers of Lang [5] and Lang and Néron [6]. Further, we shall need the fundamental facts of the theory of abstract derivations (the theorems about extension and existence) and of differential forms. Finally, in §6 we shall use some simple topologico-analytic tools.

The numeration of sections, theorems, propositions, lemmas and formulas is continuous throughout the paper.

CHAPTER I

Picard-Fuchs equations and additive functions on abelian varities

§1. Picard-Fuchs equations

1. Let K be a field of characteristic 0 and let L be a regular extension (in the sense of Weil) of transcendence degree n. For any derivation $\partial : K \longrightarrow K$ and any transcendence base $\mathbf{x} = (x_1, \cdots, x_n)$ of L/K we shall denote by $\partial_{\mathbf{x}}$ the unique derivation of L which vanishes on all the x_i, $1 \leq i \leq n$, and coincides with ∂ on K. In its turn, we extend $\partial_{\mathbf{x}}$ to a derivation of the L-module Ω of differential forms of L/K by putting

$$\partial_{\mathbf{x}} \left(\sum u_{i_1 \ldots i_r} dx_{i_1} \wedge \ldots \wedge dx_{i_r} \right) = \sum \partial_{\mathbf{x}} u_{i_1 \ldots i_r} dx_{i_1} \wedge \ldots \wedge dx_{i_r}.$$

Let B, $Z \subset \Omega$ be the K-subspaces of exact and closed forms respectively.

Lemma 1. $\partial_{\mathbf{x}}(B) \subset B$, $\partial_{\mathbf{x}}(Z) \subset Z$.

Proof. It is enough to check that $\partial_{\mathbf{x}}(d\omega) = d(\partial_{\mathbf{x}}\omega)$ for any form $\omega \subset \Omega$. In turn, for this it is enough to check that the map $\partial_{\mathbf{x}} \circ d - d \circ \partial_{\mathbf{x}} : L \longrightarrow \Omega$ is zero. But this is a derivation of L with values in the L-module Ω vanishing on $K \subset L$ and on the x_i, $1 \leq i \leq n$. Since L is a separable extension of $K(x_1, \cdots, x_n)$ a standard argument shows that $\partial_{\mathbf{x}} \circ d - d \circ \partial_{\mathbf{x}}$ is trivial on L. This proves the lemma.

The result just proved shows that $\partial_{\mathbf{x}}$ induces a derivation on the K-space Z/B of "de Rham cohomology" of L. We show now that it depends only on ∂ and not on the choice of transcendence basis $\mathbf{x}$ of L/K.

Lemma 2. *Let* $\mathbf{x} = (x_1, \cdots, x_n)$, $\mathbf{y} = (y_1, \cdots, y_n)$ *be two transcendence bases of* L/K *and let*

$$\omega = \sum_{i_1 < \ldots < i_r} u_{i_1 \ldots i_r} dx_{i_1} \wedge \ldots \wedge dx_{i_r} \subset Z.$$

Then

$$(\partial_{\mathbf{y}} - \partial_{\mathbf{x}}) \, \omega = d\phi$$

where

$$\varphi = \sum_{i_1 < \ldots < i_r} \sum_{\alpha=1}^{r} (-1)^{\alpha-1} u_{i_1 \ldots i_r} dx_{i_1} \wedge \ldots \wedge \partial_{\mathbf{y}} x_{i_\alpha} \wedge \ldots \wedge \partial x_{i_r}.$$

Proof. This lemma was proved in [9]. For completeness we reproduce the proof for the only case, namely $r = 1$, which is needed below.

On putting

$$\omega=\sum_{i=1}^{n} u_i\,dx_i;$$

and since $d\omega=0$ we have $\partial^{(i)}u_j=\partial^{(j)}u_i$ for all $1\le i,\ j\le n$, where $\partial^{(i)}$ is the derivation of L which is trivial on K and satisfies $\partial^{(i)}x_j=\delta_{ij}$ (in other words, partial differentiation with respect to x_i).

On the other hand, $\partial_{\mathbf{y}}-\partial_{\mathbf{x}}$ is a derivation on L which is trivial on K and so is a linear combination of the $\partial^{(j)}$. It is easy to see that

$$\partial_{\mathbf{y}}-\partial_{\mathbf{x}}=\sum_{j=1}^{n}(\partial_{\mathbf{y}}x_j)\,\partial^{(j)}.$$

In fact both sides coincide on K and the x_j, $1\le j\le n$. Hence

$$(\partial_{\mathbf{y}}-\partial_{\mathbf{x}})\,\omega=\sum_{i=1}^{n}\sum_{j=1}^{n}\partial_{\mathbf{y}}x_j\cdot\partial^{(j)}u_i\,dx_i+\sum_{i=1}^{n}u_i\,d(\partial_{\mathbf{y}}x_i)$$

$$=\sum_{j=1}^{n}\partial_{\mathbf{y}}x_j\sum_{i=1}^{n}\partial^{(i)}u_j\,dx_i+\sum_{i=1}^{n}u_i\,d(\partial_{\mathbf{y}}x_i)$$

$$=\sum_{j=1}^{n}\partial_{\mathbf{y}}x_j\,du_j+\sum_{i=1}^{n}u_i\,d(\partial_{\mathbf{y}}x_i)=d\Big(\sum_{i=1}^{n}u_i\,\partial_{\mathbf{y}}x_i\Big).$$

This concludes the proof of the lemma.

2. Let V be a model of L/K and V_K the set of points of V defined over K. For any point $P\in V_K$ the symbol o_P will denote the local ring, which we identify with a subring of L. For every $x\in o_P$ the symbol x_P will denote the value of x at P and $m_P\subset o_P$ will be the maximal ideal of o_P.

A set $\mathbf{x}=(x_1,\cdots,x_n)$ of elements of $L=K(V)$ will be called a *system of quasiparameters* at $P\in V_K$ if P is simple, if $x_i\in o_P$, $1\le i\le n$, and if the functions $(x_1-x_{1P},\cdots,x_n-x_{nP})$ form a system of parameters of the local ring o_P, i.e.,

$$m_P=\sum_{i=1}^{n}o_P\,(x_i-x_{iP}).$$

Clearly every system of quasiparameters is a transcendence basis of L/K.

Let $\mathbf{x}$ be a fixed transcendence basis, ∂ a derivation of K, and define $\partial^{(i)}$, $\partial_{\mathbf{x}}$ as above.

Lemma 3.[1] (a) *Let* $\mathbf{x}$ *be a system of quasiparameters at the simple point*

---

1) I am obliged to S. Lang for having found an error in the first version of the proof of this lemma.

$P \in V_K$. Then $\partial_{\mathbf{x}} o_P \subset o_P$ and $\partial^{(i)} o_P \subset o_P$, $1 \le i \le n$.

(b) *Suppose, further, that $\partial(x_{iP}) = 0$, $1 \le i \le n$. Then $(\partial_{\mathbf{x}} u)_P = \partial(u_P)$ for every $u \in o_P$.*

Proof. Put $o = o_P \cap K(x_1, \cdots, x_n)$ so that o consists of functions of the form FG^{-1}, where $F, G \in K[x_1, \cdots, x_n]$, $G(x_{1P}, \cdots, x_{nP}) \ne 0$, and an immediate verification shows that $\partial_{\mathbf{x}} o \subset o$, $\partial^{(i)} o \subset o$.

The ring o_P is a localization of the integral closure $\bar{o}$ of o in L. Hence it is enough to check that $\partial_{\mathbf{x}} \bar{o} \subset o_P$, $\partial^{(i)} \bar{o} \subset o_P$. But $\bar{o}$ is generated as an o-module by a finite number of elements $y_1, \cdots, y_r$. We shall denote by $F_i(X_1, \cdots, X_n, Y)$ irreducible polynomials with coefficients in K for which

$$F_i(x_1, \cdots, x_n, y_i) = 0, \quad 1 \le i \le r.$$

The symbol $F_i^{\tilde{\partial}}$ will denote the polynomial which is obtained from F_i by applying to all the coefficients a derivation $\tilde{\partial}$ of L which maps K into itself. Then we have identically $(1 \le i \le r)$:

$$F_i^{\tilde{\partial}}(x_1, \ldots, x_n, y_i) + \sum_{j=1}^{n} \frac{\partial F_i}{\partial X_j}(x_1, \ldots, x_n, y_i)\,\tilde{\partial} x_j + \frac{\partial F_i}{\partial Y}(x_1, \ldots, x_n, y_i)\,\tilde{\partial}_i y_i = 0.$$

If $\tilde{\partial} = \partial_{\mathbf{x}}$, we have $\tilde{\partial} x_j = 0$; and if $\tilde{\partial} = \partial^{(i)}$, we have $\tilde{\partial} x_j = \delta_{ij}$: and in both cases we have $\tilde{\partial}(K) \subset K$. Hence there is a $z \in o_P$ such that $z\tilde{\partial} y_i \in o_P$ for all $1 \le i \le r$, if when $\tilde{\partial}$ is one of the derivations $\partial_{\mathbf{x}}$, $\partial^{(i)}$ (for example, we may take z to be $\prod_{i=1}^{r} \partial F_i / \partial Y\, (x_1, \cdots, x_n, y_i)$).

Since $\bar{o} = \sum_{i=1}^{r} o y_i$, the derivation $z\tilde{\partial}$ takes o_P into itself. From the elementary properties of derivations it follows that $(z\tilde{\partial})\,(m_P^2) \subset m_P$ and so by induction on N that $(z\tilde{\partial})\,(m_P^N) \subset m_P^{N-1}$ for all $N \ge 1$.

Let $u \in o_P$ be arbitrary and put

$$u \equiv \sum_{i_1 + \ldots + i_n \le N} u_{i_1 \ldots i_r} (x_1 - x_{1P})^{i_1} \ldots (x_n - x_{nP})^{i_n} \bmod m_P^{N+1}, \qquad u_{i_1 \ldots i_n} \in K.$$

Then

$$z\tilde{\partial} u \equiv z \sum_{i_1 + \ldots + i_n \le N} \tilde{\partial}\left(u_{i_1 \ldots i_n} (x_1 - x_{1P})^{i_1} \ldots (x_n - x_{nP})^{i_n}\right) \bmod m_P^N.$$

On going to the limit as $N \longrightarrow \infty$ and using the absence of divisors of zero in the completion of o_P we find that z can be cancelled in the last relation. Hence to compute $\tilde{\partial} u$ we may derive term by term the development of u in a formal power series in the parameters $x_i - x_{iP}$. Assertion (a) of the lemma follows almost immediately. Assertion (b) follows from the remark that $\partial(x_{iP}) = 0$ implies

$$\partial_{\mathbf{x}}(x_i - x_{iP}) = 0$$

and so $\partial_{\mathbf{x}} m_P \subset m_P$. This completes the proof of the lemma.

Corollary 1. *Let ω be a 1-dimensional closed differential form regular at the simple point $P \in V_K$ and let the transcendence basis $\mathbf{x}$ be a system of quasiparameters at P. Then the differential form $\partial_{\mathbf{x}}\omega$ is regular at P.*

Proof. Let $\omega = \sum_{i=1}^{n} u_i dx_i$. Then $u_i \in o_P$ and the assertion follows from Lemma 3.

Corollary 2. *Let ω be a 1-dimensional differential form of the second kind on V. Then $\partial_{\mathbf{y}}\omega$ is also of the second kind for any transcendence basis $\mathbf{y}$.*

Proof. We have to check that at every simple point P of the model of L/K there is a decomposition $\partial_{\mathbf{y}}\omega = \omega' + dw$, where the form ω' is regular at P and $\omega \in L$. Let $\omega = \omega'' + du$ be the corresponding decomposition when $\mathbf{x}$ is a system of quasiparameters at P. Then

$$\omega = \partial_{\mathbf{x}}\omega + \{(\partial_{\mathbf{y}} - \partial_{\mathbf{x}})\,\omega + du\}$$

does what is required by Corollary 1 and Lemma 2.

3. Now let $\{\partial_i\}$ be a family of derivations of K. The elements of the free multiplicative monoid with unit generated by the symbols $\{\partial_i\}$ can be regarded as operators acting on K. These elements have the form $D = \partial_{i_1}^{m_1} \cdots \partial_{i_s}^{m_s}$ and will be called *differential monomials*. The number $m_1 + \cdots + m_s$ is called the *order* of D. The K-linear space U generated by the differential monomials can also be regarded as a space of operators on K. The space U can be given a structure of associative ring. For this it is enough to define the product $D_1(aD_2)$, where D_1, D_2 are differential monomials and $a \in K$. If $D_1 = \partial$ is a derivation we put

$$\partial(aD_2) = (\partial a)\,D_2 + a(\partial D_2);$$

and induction on the order of D_1 shows how the definition is to be extended. The verification of the axioms and of the compatibility with the structure of a space of operators is carried out as usual. The elements of the algebra U will be called differential operators (with respect to the set of derivations $\{\partial_i\}$). We define the order of a differential operator $\mathfrak{L} \in U$ to be the maximal order of a differential monomial occurring in $\mathfrak{L}$ with a nonzero coefficient.

For any $\mathfrak{L} \in U$ and any transcendence basis $\mathbf{x} = (x_1, \ldots, x_n)$ of L/K we denote by $\mathfrak{L}_{\mathbf{x}}$ the extension of $\mathfrak{L}$ to the field L and to the L-module Ω obtained by replacing all the ∂_i in $\mathfrak{L}$ by the derivations $\partial_{i\mathbf{x}}$ of L and Ω defined in subsection 1 above. The map $\mathfrak{L} \to \mathfrak{L}_{\mathbf{x}}$ gives a representation of the ring U in the ring of endomorphisms of the additive groups L and Ω. Further, $\mathfrak{L}_{\mathbf{x}}$ induces by Lemmas 1 and 2 an operator on the factor space Z/B and this operator is independent of the choice of $\mathbf{x}$. Hence we have defined a representation of U in the additive group Z/B and we shall denote by $\mathfrak{L}\bar{\omega}$ the class $\mathfrak{L}_{\mathbf{x}}\omega \bmod B$, where

$\bar{\omega} = \omega \mod B$, $\omega \in Z$, $\mathfrak{L} \in U$.

Definition. *We define a Picard-Fuchs equation* μ *in the field* L *(relative to a family* $\{\partial_i\}$ *of derivations of* K*) to be any relation of the form*

$$\mu: \sum_{i=1}^{k} \mathfrak{L}_i \bar{\omega}_i = 0,$$

where $\bar{\omega}_i$ *are the classes of the* 1*-dimensional differential forms of the first kind of* L/K, *and* $\mathfrak{L}_i \in U$.

4. To every Picard-Fuchs equation μ there corresponds a class of relations in the group Ω of the type

$$\sum_{i=1}^{k} \mathfrak{L}_{i\mathbf{x}} \omega_i = dz_{\mathbf{x}},$$

where $\mathbf{x}$ runs through all transcendence bases of L/K. We shall call each such relation a *representation* of the Picard-Fuchs equation.

Two Picard-Fuchs equations

$$\sum_{i=1}^{k} \mathfrak{L}_i \bar{\omega}_i = 0, \quad \sum_{i=1}^{l} \mathfrak{L}'_i \bar{\omega}'_i = 0$$

will be called *equivalent* if

$$\sum_{i=1}^{k} \mathfrak{L}_{i\mathbf{x}} \omega_i = \sum_{i=1}^{l} \mathfrak{L}'_{i\mathbf{x}} \omega'_i$$

for all bases $\mathbf{x} = (x_1, \cdots, x_n)$.

Let $\omega = (\omega_1, \cdots, \omega_g)$ be any basis of the 1-dimensional differential forms of the first kind of L/K. We shall say that a Picard-Fuchs equation is *reduced to the basis* ω if it is a relation between the classes of the forms of the basis, i.e., has the form $\sum_{i=1}^{g} \mathfrak{L}_i \bar{\omega}_i = 0$. On replacing in any Picard-Fuchs equation the classes of the forms involved by their expressions as linear combinations of $\bar{\omega}_1, \cdots, \bar{\omega}_g$ and combining similar terms we obviously obtain an equivalent equation. For convenience of reference we formulate this result as:

Lemma 4. *Every Picard-Fuchs equation is equivalent to a reduced equation.*

A reduced Picard-Fuchs equation $\sum_{i=1}^{g} \mathfrak{L}_i \bar{\omega}_i = 0$ can be identified with its set of coefficients $\mu = (\mathfrak{L}_1, \cdots, \mathfrak{L}_g) \in U^g$. Thus on the space M of all such equations there is a structure of left U-module and so of K-space. We shall show that if the K-space generated by the derivations $\{\partial_i\}$ is finite-dimensional, then the U-module M has a finite number of generators. (We put this result here only for aesthetic reasons: in what follows we shall need it only in the case of one derivation ∂ and then we shall establish a much more precise property of the module M from which the finite number of generators follows immediately.)

For we may suppose that there is a finite number of derivations $\{\partial_i\}$. Let g be the dimension of the space of 1-dimensional differential forms of the first kind on L. Then the K-space $\Sigma_{i=1}^{g} U\bar{\omega}_i$ has dimension at most $2g$ since by Lemma 3 Corollary 2 it is spanned by the classes mod B of forms of the second kind. Let $U_i \subset U$ be the space of differential operators of order $\leq i$. There is a j such that

$$\sum_{i=1}^{g} U\,\bar{\omega}_i = \sum_{i=1}^{g} U_j\,\bar{\omega}_i.$$

The Picard-Fuchs equations of the type

$$\sum_{i=1}^{g'} \mathfrak{L}_i\,\bar{\omega}_i = 0,$$

with $\mathfrak{L}_i \in U_{j+1}$ clearly form a finite-dimensional space over K. We assert that any base b of this space spans the U-module M. Suppose that it has already been shown that every equation

$$\sum_{m=1}^{g} \mathfrak{L}_i\bar{\omega}_i = 0, \qquad \mathfrak{L}_i \in U_k,$$

is a consequence of the equations of b. We shall show that this is true for $\mathfrak{L}_i \in U_{k+1}$ (the assertion being trivial for $k = j+1$). First of all, the equation $\Sigma_{i=1}^{g} \mathfrak{L}_i\bar{\omega}_i = 0$ is a linear combination over K of equations of the type

$$D\bar{\omega}_i + \sum_{m=1}^{g} \mathfrak{L}_{im}\,\bar{\omega}_m = 0, \qquad \mathfrak{L}_{im} \in U_j, \tag{4}$$

and elements of b, where D is a monomial of order $k+1$. Let $D = \partial D'$ where D' is of order k. Then (4) can be put in the form

$$\partial\left(D'\bar{\omega}_i + \sum_{m=1}^{g} \mathfrak{L}'_{im}\,\bar{\omega}_m\right) + \sum_{m=1}^{g} (\mathfrak{L}_m - \partial\mathfrak{L}'_m)\,\bar{\omega}_m = 0, \qquad \mathfrak{L}'_m \in U_j,$$

where $D'\bar{\omega}_i + \Sigma_{m=1}^{g} \mathfrak{L}'_{im}\bar{\omega}_m = 0$ is an equation of order k. This allows us to use induction.

5. The following technical result will be necessary in what follows.

Let K' be some extension of K linearly disjoint from L and put $L' = K'L$. We suppose that the derivations $\{\partial_i\}$ of K are somehow extensible to derivations of K'. Let U' be the corresponding algebra of differential operators, and M' the module of Picard-Fuchs equations of L'/K' reduced to the basis $\omega = (\omega_1, \cdots, \omega_g)$ of 1-dimensional differential forms of the first kind, which is also a base for L/K. As above, M is the U-module of equations for L/K and we may clearly consider it as naturally embedded in M'.

Lemma 5. *As a linear space over K, M' is generated by its K-subspace M.*

Proof. The result becomes obvious if we consider that the K'-space

$\sum_{i=1}^{g} U'\bar{\omega}_i$ is generated by its K-subspace $\sum_{i=1}^{g} U\bar{\omega}_i$ and interpret every element $\mu \in M$ (or $\mu' \in M'$) as a *linear* relation with coefficients in K (or in K') between elements of the type $D\bar{\omega}_i \in \sum_{i=1}^{g} U\bar{\omega}_i$, where D runs through all differential monomials. This proves the lemma.

Finally, a last remark.

Let L', L'' be independent regular extensions of K, let $\{\partial_i\}$ be a family of derivations of K and let U be the corresponding algebra. We shall consider the Picard-Fuchs equations for $L'L''/K$ reduced relative to a basis ω which is the union of bases ω', ω'' of 1-dimensional differential forms of the first kind for L'/K, L''/K respectively. We shall denote by M', M'', M the corresponding U-modules of Picard-Fuchs equations reduced relative to the respective bases ω', ω'', ω in the respective fields L', L'', $L'L''$. Then the following statement holds: M', M'' can be identified with U-submodules of M and $M = M' + M''$ (direct sum). Besides, the same statement is true for the representatives of the Picard-Fuchs equations taken relative to a transcendence base $\mathbf{x}$ of $L'L''/K$ which is a union of transcendence bases $\mathbf{x}'$, $\mathbf{x}''$ of L'/K, L''/K respectively. More precisely, let

$$\sum_{\omega_i \in \omega} \mathfrak{L}_{i\mathbf{x}}\, \omega_i = dz$$

be a representative of some $\mu \in M$; Then we have

$$\sum_{\omega_i \in \omega'} \mathfrak{L}_{i\mathbf{x}'}\, \omega_i = dz', \qquad \sum_{\omega_i \in \omega''} \mathfrak{L}_{i\mathbf{x}''}\, \omega_i = dz'', \qquad z' \in L', \qquad z'' \in L'',$$

and $dz = dz' + dz''$. (Indeed if $dz = \omega_1 + \omega_2$ where ω_1 (or ω_2) is a 1-dimensional closed differential form of L' (or L''), then ω_1, ω_2 are themselves exact differentials.)

§2. The functions $\mu(P, Q)$

1. Let the transcendence basis $\mathbf{x}$ be a system of quasiparameters at $P \in V_K$. We shall say that this system is *admissible* at P if $\partial(x_{iP}) = 0$ for any derivation ∂ of K and $1 \leq i \leq n$.

Lemma 6. *Let P, Q, $R \in V_K$ be three simple points. If there is a common system of quasiparameters at P, Q, R, then there is a common system admissible at those points.*

Proof. Let $(x'_1, \cdots, x'_n)$ be a common system of quasiparameters. For each i, $1 \leq i \leq n$, we can find an x_i which is a fractional linear function of x'_i with coefficients in K, and which is such that the three values x_{iP}, x_{iQ}, x_{iR} lie in the prime subfield of K, and so are annihilated by any derivation. It is easily verified that $(x_1, \cdots, x_n)$ again form a system of quasiparameters at each of P, Q, R. This completes the proof of the lemma.

We now fix attention on some Picard-Fuchs equation

$$\mu : \sum_{i=1}^{r} \mathfrak{L}_i \overline{\omega}_i = 0, \quad \mathfrak{L}_i \in U$$

in L (we sometimes say that μ is a Picard-Fuchs equation on V).

We shall suppose that the set of simple points in V_K is nonempty. Let P, $Q \in V_K$ be two simple points having a common admissible system of quasiparameters $\mathbf{x}$. By Lemma 3, Corollary 1, the differential $dz_{\mathbf{x}}$ defined by the representative

$$\sum_{i=1}^{r} \mathfrak{L}_{i\mathbf{x}} \omega_i = dz_{\mathbf{x}}$$

of μ is regular at P and Q. Hence the value

$$\mu_{\mathbf{x}}(P, Q) = z_{\mathbf{x}Q} - z_{\mathbf{x}P} \in K \tag{5}$$

is well-defined. For any third point R at which $\mathbf{x}$ is an admissible system of quasiparameters we have the identity

$$\mu_{\mathbf{x}}(P, Q) = \mu_{\mathbf{x}}(P, R) + \mu_{\mathbf{x}}(R, Q). \tag{6}$$

The following important result holds.

Theorem 1. (a) *The value of $\mu_{\mathbf{x}}(P, Q)$ does not depend on the choice of the admissible system $\mathbf{x}$ of quasiparameters but is defined uniquely by μ, P, Q.*

(b) *On the set of all pairs of simple points of V_K there is a unique function $\mu(P, Q)$ with values in K which coincides with $\mu_{\mathbf{x}}(P, Q)$ for points having a common admissible system of quasiparameters $\mathbf{x}$ and satisfying the identity*

$$\mu(P, Q) = \mu(P, R) + \mu(R, Q) \tag{7}$$

throughout its field of definition.

2. The proof of this result is the aim of the following section. The plan of the proof is as follows. First we verify (a) directly. Then we use (7) to *define* a function of pairs of points without a common system of quasiparameters (I did not manage to find out whether such pairs actually exist. If not, the proof is somewhat simplified). For this, we have to choose a point R such that each of the pairs (P, R), (R, Q) has a common system of quasiparameters. To make such a choice possible we may have to extend the field of constants temporarily and so one must study the behavior of $\mu_{\mathbf{x}}(P, Q)$ under such an extension. When the existence and uniqueness of $\mu(P, Q)$ have been proved the verification of (7) presents no difficulty.

§3. Proof of Theorem 1

1. **Lemma 7.** *Let P, Q be simple points of V_K, and let $\mathbf{x}$, $\mathbf{y}$ be two admissible systems of quasiparameters common to them. Suppose that $\omega \in Z$ is a 1-dimensional differential form of the first kind regular at P and Q and let $\mathfrak{L}$ be any element of U. Put $(\mathfrak{L}_{\mathbf{y}} - \mathfrak{L}_{\mathbf{x}})\omega = dz$, $z \in L$. Then $z \in o_P \cap o_Q$ and $z_P = z_Q$.*

Proof. The inclusion $z \in o_P \cap o_Q$ follows from Lemma 3, Corollary 1. If Lemma 7 is true for $\mathfrak{L}'$, $\mathfrak{L} \in U$, it is true for any linear combination of them over K. Hence we may suppose without loss of generality that $\mathfrak{L}$ is a differential monomial. The proof is by induction on the order of the monomial.

Suppose that the order is unity, so that we have $\mathfrak{L} = \partial$, $\omega = \Sigma_{i=1}^{n} u_i\, dx_i$. Then

$$(\partial_{\mathbf{y}} - \partial_{\mathbf{x}})\,\omega = d\left(\sum_{i=1}^{n} u_i\, \partial_{\mathbf{y}}\, x_i\right) = dz.$$

The conditions of the lemma imply that $u_i \in o_P \cap o_Q$ and

$$(\partial_{\mathbf{y}} x_i)_P = (\partial_{\mathbf{y}} x_i)_Q = \partial(x_{iP}) = \partial(x_{iQ}) = 0$$

and so $z_P = z_Q$.

Suppose now that the lemma has already been proved for monomials D of order m and consider the monomial ∂D of order $m+1$. We have

$$(\partial_{\mathbf{y}} D_{\mathbf{y}} - \partial_{\mathbf{x}} D_{\mathbf{x}})\,\omega = \partial_{\mathbf{y}}\,(D_{\mathbf{y}} - D_{\mathbf{x}})\,\omega + (\partial_{\mathbf{y}} - \partial_{\mathbf{x}})\, D_{\mathbf{x}}\,\omega = dz.$$

Put

$$(D_{\mathbf{y}} - D_{\mathbf{x}})\,\omega = dz_1, \qquad (\partial_{\mathbf{y}} - \partial_{\mathbf{x}})\,\omega = dz_2.$$

By the inductive hypothesis we have $z_{1P} = z_{1Q}$ and $z_{2P} = z_{2Q}$, since $D_{\mathbf{x}}\omega$ is regular at P, Q together with ω. Further

$$(\partial_{\mathbf{y}} z_1)_P = \partial z_{1P} = \partial z_{1Q} = (\partial_{\mathbf{y}} z_1)_Q$$

because $\mathbf{y}$ is an admissible system of quasiparameters. Hence $z_P = z_Q$ because $z = \partial_{\mathbf{y}} z_1 + z_2 + c$ with $c \in K$.

This concludes the proof of the lemma.

Corollary. *Assertion* (a) *of Theorem* 1 *is true.*

Proof. For on putting

$$\sum_{i=1}^{r} \mathfrak{L}_{i\,\mathbf{x}}\,\omega_i = dz_{\mathbf{x}}, \quad \sum_{i=1}^{r} \mathfrak{L}_{i\,\mathbf{y}}\,\omega_i = dz_{\mathbf{y}};$$

we have

$$d\,(z_{\mathbf{x}} - z_{\mathbf{y}}) = \sum_{i=1}^{r} (\mathfrak{L}_{i\,\mathbf{x}} - \mathfrak{L}_{i\,\mathbf{y}})\,\omega_i$$

and so, by the lemma,

$$(z_{\mathbf{x}} - z_{\mathbf{y}})_P = (z_{\mathbf{x}} - z_{\mathbf{y}})_Q.$$

This equation is equivalent to the identity

$$\mu_{\mathbf{x}}(P, Q) = \mu_{\mathbf{y}}(P, Q).$$

2. Let $\mathbf{x}$ be any transcendence basis of L/K. Then there is a set $V_{\mathbf{x}} \subset V$ open in the Zariski K-topology such that $\mathbf{x}$ is a system of quasiparameters at any point of $V_K \cap V_{\mathbf{x}}$. For we may take for $V_{\mathbf{x}}$ the complement of the carrier of the divisor of the differential form $dx_1 \wedge \cdots \wedge dx_n$.

Let K' be a finite algebraic extension of K and suppose that $P, Q \in V_K \cap V_{\mathbf{x}}$. Let $\mathbf{y}$ be a common system of admissible quasiparameters at P, Q with $y_i \in K'L$. (Such a system exists by Lemma 6.) Let σ be an automorphism of K'/K. We use the same symbol to denote its extension to $K'L/L$. Further, σ acts on the points of $V_{K'}$ in such a way that $z^{\sigma}_{P\sigma} = (z_P)^{\sigma}$ for all $z \in K'L$, $P \in V_{K'}$ for which the right-hand side has a meaning. Finally, σ acts in the obvious way on the differential forms of $K'L/K'$.

Every derivation ∂ of K can be uniquely extended to a derivation of K' and then

$$\partial_{\mathbf{y}^{\sigma}} z^{\sigma} = (\partial_{\mathbf{y}} z)^{\sigma}$$

for any transcendence basis $\mathbf{y}$ of $K'L/K'$ and any $z \in K'L$. To prove this, it is enough to note that the map $z \longrightarrow (\partial_{\mathbf{y}^{\sigma}} z^{\sigma})^{\sigma-1}$ is a derivation and that it coincides with $\partial_{\mathbf{y}}$ on $K(y_1, \cdots, y_n)$.

We now have

Lemma 8. $\mu_{\mathbf{y}}(P, Q)^{\sigma} = \mu_{\mathbf{y}^{\sigma}}(P^{\sigma}, Q^{\sigma})$.

Proof. For if $\mathbf{y}$ is an admissible system of quasiparameters at P, Q then $\mathbf{y}^{\sigma}$ is a similar system at P^{σ}, Q^{σ}. Hence

$$\mu_{\mathbf{y}^{\sigma}}(P^{\sigma}, Q^{\sigma}) = (z_{\mathbf{y}^{\sigma}})_{Q^{\sigma}} - (z_{\mathbf{y}^{\sigma}})_{P^{\sigma}} = (z^{\sigma}_{\mathbf{y}})_{Q^{\sigma}} - (z^{\sigma}_{\mathbf{y}})_{P^{\sigma}} = \mu_{\mathbf{y}}(P, Q)^{\sigma}.$$

This completes the proof of the lemma.

3: We now go over to the proof of (b) of Theorem 1. Lemma 6, formula (6) and (a) of Theorem 1 together imply the existence and uniqueness of $\mu(P, Q)$ on the set of pairs P, Q of simple points in $V_K \cap V_{\mathbf{x}}$, where $\mathbf{x}$ is any transcendence basis of L/K. Suppose now that $P \in V_{\mathbf{x}}$ and $Q \in V_{\mathbf{y}} \setminus V_{\mathbf{x}}$. As stated above, we define $\mu(P, Q)$ by

$$\mu(P, Q) = \mu_{\mathbf{x}'}(P, R') + \mu_{\mathbf{y}'}(R', Q), \tag{8}$$

where R' is any point of $V_{\mathbf{x}} \cap V_{\mathbf{y}}$ defined over some finite normal extension K'/K and $\mathbf{x}'$, $\mathbf{y}'$ are systems of quasiparameters admissible respectively for the pairs (P, R') and (R', Q). (They exist by Lemma 6.)

We must show that if the systems $\mathbf{x}''$, $\mathbf{y}''$ and the point R'' satisfy similar conditions, then they give the same value of $\mu(P, Q)$.

First of all we may replace R' by any other point $R \in V_{x'} \cap V_{y'}$ since

$$\mu_{x'}(P, R') + \mu_{x'}(R', R) = \mu_{x'}(P, R),$$
$$\mu_{y'}(R, R') + \mu_{y'}(R', Q) = \mu_{y'}(R, Q)$$

and

$$\mu_{x'}(R', R) = -\mu_{y'}(R, R')$$

by (a) of Theorem 1. Hence we may suppose that

$$R' = R'' = R \in V_{x'} \cap V_{y'} \cap V_{x} \cap V_{y},$$

when the equation

$$\mu_{x'}(P, R) + \mu_{y'}(R, Q) = \mu_{x''}(P, R) + \mu_{y''}(R, Q)$$

follows immediately from (a). Hence our function is well defined and we may omit the suffix indicating the system of quasiparameters used in computing $\mu(P, Q)$.

We shall show now that $\mu(P, Q) \in K$. For suppose that R is defined over the normal extension K' of K and let σ be a relative automorphism. Then Lemma 8, part (a) of Theorem 1 and the independence of the choice of R together imply that

$$\mu(P, Q)^{\sigma} = \mu(P, R^{\sigma}) + \mu(R^{\sigma}, Q) = \mu(P, Q).$$

Finally, we may verify the identity (7) by using an auxiliary point

$$S \in V_{\mathbf{x}} \cap V_{\mathbf{y}} \cap V_{\mathbf{z}}$$

where x, y, z are admissible systems of quasiparameters at P, Q, R respectively. The argument is similar to the one above and is left to the reader.

This completes the proof of Theorem 1.

§4. Some properties of $\mu(P, Q)$

Let V' be an algebraic variety over K of the same dimension as V with field of functions L' and let $\phi: V' \to V$ be an epimorphism. We identify L with a subfield of L'. Then every Picard-Fuchs equation μ in L gives a Picard-Fuchs equation $\phi^*\mu$ in L'.

Let $P, Q \in V'_K$ be simple points and let their images $\phi(P)$, $\phi(Q)$ be simple in V_K. Then we have

Lemma 9. $\phi^*\mu(P, Q) = \mu(\phi(P), \phi(Q))$.

Proof. We may suppose that the points $\phi(P)$, $\phi(Q)$ have a common system $\mathbf{x}$ of admissible quasiparameters and similarly that there is a system y for P, Q, since the general case is obtained from this one by the use of intermediate points.

As in the proof of (a) of Theorem 1 it is enough to check the following analogue of Lemma 7. Suppose that $\omega \in Z$ is a form on V regular at P, Q, that $\mathfrak{L}$ is any element of U and that $dz = (\mathfrak{L}_y - \mathfrak{L}_x)\omega$. Then $z \in o_P \cap o_Q$ and $z_P = z_Q$. The verification of this is a word-for-word repetition of the proof of Lemma 7.

This concludes the proof of the lemma.

Lemma 10. *Let V_1, V_2 be two algebraic varieties over K, let $V = V_1 \times V_2$ and let μ be a Picard-Fuchs equation on V. Then there are Picard-Fuchs equations μ_1, μ_2 on V_1, V_2 respectively such that*

$$\mu(P, Q) = \mu_1(P_1, Q_1) + \mu_2(P_2, Q_2)$$

for every pair of points $P = (P_1, P_2) \in V_K$, $Q = (Q_1, Q_2) \in V_K$ for which μ is defined.

(This is a translation into geometric language of the remark made at the end of subsection 3 of the previous section.)

Finally, our last result is a slightly disguised version of the obvious fact that the application of our construction to the zero derivation gives the zero function μ.

Lemma 11. *In the notation of subsection 1 of §2 we suppose that V is defined over some subfield $k \subset K$ on which the differential algebra U acts trivially. Then for every Picard-Fuchs equation μ on V and for any simple points P, $Q \in V$ defined over k we have $\mu(P, Q) = 0$.*

Proof. If the Picard-Fuchs equations μ_1, μ_2 are equivalent in the sense of subsection 4 of §1, then

$$\mu_1(P, Q) \equiv \mu_2(P, Q).$$

Hence it is enough by Lemma 4 to prove the lemma for equations reduced with respect to some basis of the forms of the first kind. We choose a basis which is defined over k. Then the representative of any Picard-Fuchs equation with respect to a transcendence basis $\mathbf{x}$ of the function field of V defined over k has the shape

$$\sum_{i=1}^{g} \mathfrak{L}_{i\mathbf{x}}\omega_i = 0.$$

The assertion of the lemma now follows almost immediately.

§5. Additive functions on abelian varieties. The kernel theorem

1. From now until the end of this chapter L will denote the field of rational functions on an abelian variety A defined over K. We fix attention on some Picard-Fuchs equation μ on A and consider the corresponding function $\mu(P, Q)$ on $A_K \times A_K$.

Lemma 12. *$\mu(P+R, Q+R) = \mu(P, Q)$ for any $P, Q, R \in A_K$ (the addition of points being understood in the sense of the group law on A).*

Proof. We may suppose that P, Q have a common system $\mathbf{x}$ of admissible parameters. Let $\sigma = \sigma_R : L \longrightarrow L$ be the automorphism of L/K defined by $(u^\sigma)_P = u_{P-R}$ for any points P, Q and function u regular at $P-R$. As in subsection 2 of §4 we have

$$\partial_{\mathbf{x}^\sigma} u^\sigma = (\partial_{\mathbf{x}} u)^\sigma$$

and

$$\partial_{\mathbf{x}^\sigma} \omega^\sigma = (\partial_{\mathbf{x}} \omega)^\sigma$$

for any differential form ω on A. Since differential forms of the first kind are invariant with respect to translations on A, on applying σ to the representative

$$\sum_{i=1}^{k} \mathfrak{L}_{i\mathbf{x}}\, \omega_i = dz$$

of μ we obtain

$$\sum_{i=1}^{k} \mathfrak{L}_{i\mathbf{x}^\sigma}\, \omega_i = dz^\sigma.$$

The functions $(x_1^\sigma, \ldots, x_n^\sigma)$ are a system of admissible quasiparameters at the points $P+R$, $Q+R$. Hence

$$\mu(P+R, Q+R) = z^\sigma_{Q+R} - z^\sigma_{P+R} = z_Q - z_P = \mu(P, Q).$$

This proves the lemma.

In what follows we shall adopt the following notation for brevity: we shall put

$$\mu(P) = \mu(O, P)$$

for a Picard-Fuchs equation μ and $P \in A_K$, where $O \in A_K$ is the zero of the group law.

Proposition 1. $\mu(P+Q) = \mu(P) + \mu(Q)$.

Proof. By (7) and Lemma 12 we have

$$\mu(O, P+Q) = \mu(O, P) + \mu(P, P+Q) = \mu(O, P) + \mu(O, Q),$$

which is what was required.

Our fundamental aim to the end of this chapter is the description of the intersection A_K^0 of the kernels of all the homomorphisms corresponding to all the Picard-Fuchs equations (for a fixed set of derivations $\{\partial_i\}$ generating U).

The following results are easily obtained by combining the known properties of $\mu(P, Q)$.

Lemma 13. (a) *Let $K' \supset K$ be any extension of K which is linearly free with* *respect to L. Let the derivations* $\{\partial_i\}$ *of K which generate U be extended to K'. Then* $A^0_K = A^0_{K'} \cap A_K$.

(b) *Let* $\phi : A \to B$ *be an isogeny defined over K. Then*

$$A^0_K = \varphi^{-1}(B^0_K), \qquad B^0_K = \varphi(A^0_{\bar K}) \cap B_K,$$

where $\bar K$ *is the algebraic closure of K.*

(c) *Let* $A = B \times C$. *Then* $A^0_K = B^0_K \times C^0_K$.

Proof. Assertion (a) follows from Lemma 5 and the fact that the map $\mu \to \mu(P)$ of the U' module of Picard-Fuchs equations into the U' module of functions $A_{K'} \to K'$ is a module homomorphism, and so, in particular, is K'-linear.

Assertion (b) follows from Lemma 9, from (a) and from the fact that an isogeny of abelian varieties induces an isomorphism of the spaces of differential forms of the first kind compatible with the action of U.

The assertion (c) follows directly from Lemma 10. This proves the lemma. It implies

Proposition 2. *Suppose that U acts trivially on some subfield* $k \subset K$. *Let* (B, τ) *denote the K/k-trace of A. Suppose that for a* $P \in A_K$ *there is an integer* $d \neq 0$ *such that* $dP \in \tau(B_k)$. *Then* $P \in A^0_K$.

Proof. Let $(\tau, \phi) : B \times C \to A$ be an isogeny defined over the algebraic closure $\bar K$ of K. By Lemma 13 we have

$$A^0_K = (\tau, \varphi)(B^0_{\bar K} \times C^0_{\bar K}) \cap A_K$$

and so $B^0_k \subset B^0_{\bar K}$ by Lemma 11, since B is defined over k. Hence $\tau(B_k) \subset A^0_K$; and if $dP \in A^0_K$, then $P \in A^0_K$, since the additive group of a field of zero characteristic is torsion-free. This proves the proposition.

Considerably more difficult to prove is the converse of this assertion, which we shall refer to below as the "kernel theorem".

Theorem 2. *Let K be a regular extension of its subfield k and let the algebra U contain all the derivations of the field K that are trivial on k. Then the group* $A^0_K = A^0_{K/k}$ *consists of those points* $P \in A_K$ *for which there exists an integer* $d \neq 0$ *such that* $dP \in \tau(B_k)$.

2. We shall show now that it is enough to prove the kernel theorem in the case when K is the field of rational functions on an algebraic curve with the complex field as constant field. This case presents fundamental difficulties requiring topologico-algebraic tools and will be treated in the next section.

In the proof of the following reduction theorem we shall, without special reference to them, make use of properties of the K/k-trace established in Lang's book [3].

Proposition 3. *If Theorem 2 is true when K is the field of rational functions on an algebraic curve with the complex field as constant field, then it is universally true.*

Proof. Let the hypothesis of the proposition be satisfied.

(a) Theorem 2 is true if k is finitely generated over its prime subfield and the transcendence degree $\dim_k K$ is unity.

For then we may regard k as a subfield of the complex field C. We can extend the action of U to the compositum $K' = KC$ by making it act trivially on $C \subset K'$. Let (B', τ') be the K'/C-trace of A. If $P \in A^0_{K/k}$, then $P \in A^0_{K'/C}$ and so by hypothesis $dP \in \tau'(B'_C)$. But $\tau'(B'_C) \cap A_K = \tau(B_k)$ and the required result follows.

(b) Theorem 2 is true if k is finitely generated over its prime subfield and K/k is any regular extension.

We proceed by induction on the transcendence degree m of K/k, the case $m = 1$ being already proved. Suppose that the assertion is true for dimension m and let K/k have dimension $m + 1$. Consider a subfield K', $k \subset K' \subset K$, which is algebraically closed in K and such that $\dim_{K'} K = 1$. Let (B', τ') be the K/K'-trace of A and (B'', τ'') the K'/k-trace of B'. If $P \in A^0_{K/k}$, then P is annihilated by all homomorphisms μ obtained from Picard-Fuchs equations relating to derivations which are trivial on K' and so $P \in A^0_{K/K'}$. By part (a) of the proof we have $dP \in \tau'(B'_{K'})$. Let $dP = \tau'(P')$, so that by Lemma 13 we have $P' \in (B')^0_{K'/k}$, and thus, by the hypothesis of the induction, $d'P' \in \tau''(B''_k)$. Hence $d'dP \in \tau(\bar{B}_k)$, which is what we had to show.

(c) Theorem 2 is true in general. For there exists a field $K_0 \subset K$ finitely generated over its prime subfield which is a field of definition of A and satisfies $K = kK_0$. From Lemma 13 it follows without difficulty that $A^0_{K/k}$ is the union of all $A^0_{K'/K' \cap k}$ when K' runs through all subfields of finite type of K containing K_0. The required result now follows from (b) above and the transitivity of the trace.

§6. Proof of the kernel theorem

1. In this section k is the field of complex functions and K is the field of meromorphic functions on the Riemann surface Γ of a complete nonsingular algebraic curve. Further, A is an abelian variety defined over K and L the field of functions on A defined over K. We shall prove Theorem 2 for A.

It is convenient to regard A as an algebraic family of abelian varieties parametrized by Γ. More precisely, we take a complete model $\mathfrak{A}$ of the field L/k such that its canonical projection $p:\mathfrak{A} \longrightarrow \Gamma$ induced by the inclusion $K \subset L$ has the two following properties:

(1) There is a finite number of points $a_1, \cdots, a_s \in \Gamma$ such that for any point $a \in \Gamma \setminus \bigcup_{i=1}^s a_i$ the fiber $A_a = p^{-1}(a)$ is an abelian variety with a group law induced by that on A.

(2) On the set $\mathfrak{A} \setminus \bigcup_{i=1}^s p^{-1}(a_i)$ the projection p onto $\Gamma \setminus \bigcup_{i=1}^s a_i$ induces a structure of topological fiber space.

We shall suppose the model $\mathfrak{A}$ chosen once and for all. Let $\Gamma_0 = \Gamma \setminus \bigcup_{i=1}^s a_i$, and denote by $\widetilde{\Gamma}$ the universal covering of Γ_0 with the induced complex-analytic structure. We denote the field of meromorphic functions on $\widetilde{\Gamma}$ by $\widetilde{K}$. The natural projection $\widetilde{\Gamma} \longrightarrow \Gamma$ induces an embedding $K \subset \widetilde{K}$. Every derivation ∂ of K/k extends canonically to a derivation of $\widetilde{K}$ which we shall denote by the same symbol ∂.

We shall fix once and for all a point $a \in \Gamma_0$ and denote by $\pi_1(\Gamma_0)$ the fundamental group of Γ_0 relative to a. Instead of $\mathfrak{A}$ we shall consider its restriction $\mathfrak{A}_0$ to Γ_0; that is, we take out the singular fibers. Since the projection $p:\mathfrak{A}_0 \longrightarrow \Gamma_0$ is topologically a fiber bundle, we have the following:

(a) Every point $\widetilde{b} \in \widetilde{\Gamma}$ can be regarded as a homotopy class β of paths on Γ_0 with beginning at a and end $b = p(\widetilde{b})$. Hence a point $\widetilde{b}$ determines a homomorphism by "carrying a homology class along β":

$$\widetilde{b}: H_1(A_a, Z) \longrightarrow H_1(A_b, Z).$$

We shall denote the image of a $\gamma \in H_1(A_a, Z)$ under $\widetilde{b}$ by $\gamma\widetilde{b}$.

(b) Let $\sigma \in \pi_1(\Gamma_0)$. The map $\beta \longrightarrow \sigma^{-1}\beta$, which makes correspond to every class of paths β on Γ_0 with beginning at a the class $\sigma^{-1}\beta$ of paths with the same beginning and end, defines an action of $\pi_1(\Gamma_0)$ on $\widetilde{\Gamma}$. If β corresponds to the point $\widetilde{b}$, then by definition the class $\sigma^{-1}\beta$ corresponds to the point $\sigma^{-1}\widetilde{b}$.

On the other hand, let $\widetilde{s} \in \widetilde{\Gamma}$ be a point corresponding to the class $\sigma^{-1} \in \pi_1(\Gamma_0)$ of paths. By (a) above to every $\gamma \in H_1(A_a, Z)$ there corresponds a $\gamma\widetilde{s} \in H_1(A_a, Z)$. It easily follows from the definitions that

$$(\gamma\widetilde{s})\widetilde{b} = \gamma(\sigma^{-1}\widetilde{b}).$$

(c) Let $f \in \widetilde{K}$ and $\sigma \in \pi_1(\Gamma_0)$. We define an automorphism $f \longrightarrow f^\sigma$ of $\widetilde{K}$ by putting $f^\sigma(\widetilde{b}) = f(\sigma^{-1}\widetilde{b})$.

(d) Let $\gamma \in H_1(A_a, Z)$ and let ω be a closed 1-dimensional differential form of L/K. For almost all fibers A_b, $b \in \Gamma_0$ the form ω induces a closed 1-dimensional form ω_b which is invariant if ω is invariant; of the second kind if ω is of the second kind; and exact if ω is exact.

We shall denote by $\int_\gamma \omega$ the element of $\widetilde{K}$ defined by

$$\left(\int_{\gamma}\omega\right)(\widetilde{b}) = \int_{\gamma\widetilde{b}} \omega_b,$$

where the right-hand side is the period of the form ω_b on the homology class $\gamma\widetilde{b}$ in the fiber A_b. The element σ of $\pi_1(\Gamma_0)$ acts on $\int_\gamma \omega$ as follows:

$$\left(\int_{\gamma}\omega\right)^{\sigma} = \int_{\gamma\widetilde{s}} \omega,$$

where $\widetilde{s}$, as above, is the point of $\widetilde{\Gamma}$ corresponding to σ^{-1}. In fact

$$\left(\int_{\gamma}\omega\right)^{\sigma}(\widetilde{b}) = \left(\int_{\gamma}\omega\right)(\sigma^{-1}\widetilde{b}) = \int_{\gamma(\sigma^{-1}\widetilde{b})} \omega_b = \int_{(\widetilde{\gamma s})\widetilde{b}} \omega_b = \left(\int_{\gamma\widetilde{s}}\omega\right)(\widetilde{b}).$$

(e) Let K' be a finite normal extension of K unramified outside $a_1, \dots, a_s$ and let Γ' be its Riemann surface with the inverse images of those points removed. We shall regard Γ' as an intermediate covering $\widetilde{\Gamma} \to \Gamma' \to \Gamma_0$ corresponding to a normal divisor of finite index $\pi_1(\Gamma') \subset \pi_1(\Gamma_0)$. The field K' is contained in $\widetilde{K}$ and consists precisely of those elements which are σ-invariant for all $\sigma \in \pi_1(\Gamma')$. The points $b' \in \Gamma'$ may be identified with classes $\pi_1(\Gamma')\beta$, where β runs through the homotopy classes of paths on Γ_0 starting at a.

The points $P \in A_{K'}$ are in 1-1 correspondence with the analytic Γ_0-morphisms $P: \Gamma' \to \mathfrak{A}_0$. If $b' \in \Gamma'$ we shall denote by $P_{b'} \in A_b$, where b is the projection of b' on Γ_0, the image of b' under P. In particular, the zero O of the group $A_{K'}$, being already determined over K, satisfies the equations $O_{b'} = O_b$ (in the earlier notation).

Let $P \in A_{K'}$ and let $a' \in \Gamma'$ be a point corresponding to a class of paths in $\pi_1(\Gamma')$ (so it lies over $a \in \Gamma_0$). Denote by $\delta_P \subset A_a$ the homology class in $A_a \setminus \{O_a \cup P_{a'}\}$ of a path in A_a with origin O_a and endpoint $P_{a'}$. For any $\widetilde{b} \in \widetilde{\Gamma}$ we shall denote by $\delta_P\widetilde{b}$ the homology class in $A_b \setminus \{O_b \cup P_{b'}\}$ with origin O_b and endpoint $P_{b'}$, where b' corresponds to $\pi_1(\Gamma')\beta$, of the path obtained from a path of δ_P by transport along β in such a way that the origin and end in each fiber A_c are correspondingly O_c and $P_{c'}$.

As above, we denote by $\int_{\delta_P}\omega$ the element of $\widetilde{K}$ defined by

$$\left(\int_{\delta_P}\omega\right)(\widetilde{b}) = \int_{\delta_P b} \omega_b$$

(in the same notation as in (d) above). Similarly, for every $\sigma \in \pi_1(\Gamma_0)$ we have

$$\left(\int_{\delta_P}\omega\right)^{\sigma} = \int_{\delta_P\widetilde{s}} \omega.$$

The path $\delta_P\widetilde{s}$ lies in A_a, has its origin at O_a and endpoint at $P_{a''}$, where a'' corresponds to $\pi_1(\Gamma')\sigma^{-1}$. It is easy to see that the path $\delta_P\widetilde{s}$ may be interpreted as δ_{P^σ}, where P^σ is the point conjugate to P under the action of σ in the field K'. Hence, if we allow for the indeterminacy in the choice of the path δ_Q given Q and do not intend to give a firm prescription for the choice, we at last obtain

$$\Big(\int_{\delta_P}\omega\Big)^{\sigma}=\int_{\delta_{P^\sigma}}\omega+\int_{\gamma}\omega,\quad \gamma=\gamma(\sigma)\in H_1(A_a,\mathbf{Z}).$$

(f) Finally we note the following facts. Let $P,\ Q\in A_{K'}$, and suppose there is a $\gamma\in H_1(A_a,\ \mathbf{Z})$ such that

$$\int_{\delta_P}\omega=\int_{\delta_Q}\omega+\int_{\gamma}\omega$$

for any 1-dimensional form ω of the first kind for L/K. Then $P=Q$ (Abel's theorem).

Let $\int_\gamma\omega=0$ for a given form ω and all $\gamma\in H_1(A_a,\ \mathbf{Z})$. Then ω is exact (the converse is also true).

Let ∂ be a derivation of $\widetilde{K}$ which induces a derivation of K into itself. Then

$$\partial\int_{\gamma}\omega=\int_{\gamma}\partial_{\mathbf{x}}\omega$$

for any choice of transcendence basis $\mathbf{x}=(x_1,\cdots,x_n)$ of L/K provided that ω is of the second kind.

If, further, $\mathbf{x}$ is a system of quasiparameters at O and P and $\partial(x_{iO})=\partial(x_{iP})=0$, $1\le i\le n$ (the limits of integration are independent of the parameters) then also

$$\partial\int_{\delta_P}\omega=\int_{\delta_P}\partial_{\mathbf{x}}\omega$$

provided that ω is a form of the second kind regular at O and P. In particular for every equation $\mu:\sum_{i=1}^{n}\mathfrak{L}_i\bar{\omega}_i=0$ we have

$$\mu(P)=\sum_{i=1}^{n}\mathfrak{L}_i\int_{\delta_P}\omega_i.$$

Finally, we have the usual linearity properties

$$\int_{\gamma_1+\gamma_2}\omega=\int_{\gamma_1}\omega+\int_{\gamma_2}\omega;\quad \int_{\gamma}(\alpha\omega_1+\beta\omega_2)=\alpha\int_{\gamma}\omega_1+\beta\int_{\gamma}\omega_2,\quad \alpha,\beta\in\widetilde{K},$$

and similarly for $\int_{\delta_P}\omega$.

2. After these preparatory remarks we proceed to the proof of the kernel theorem.

We shall fix some nontrivial derivation ∂ of K/k, so that U consists of differential operators of the type $\sum_{i=0}^{m}a_i\partial^i$, $a_i\in K$. Let $(\omega_1,\cdots,\omega_n)$ be a fixed basis of the space of invariant differential forms of L/K, where $n = \dim A$. As in subsection 3 of §2 we denote by M the left U-module of Picard-Fuchs equations on A reduced to the basis $(\omega_1, \cdots, \omega_n)$. Let $\mu \in M$ have the form

$$\mu : \sum_{i=1}^{n} \mathfrak{L}_i \bar{\omega}_i = 0,$$

and let $P \in A_k$. The condition $\mu(P) = 0$ is equivalent to

$$\sum_{i=1}^{n} \mathfrak{L}_i \int_{\delta_P} \omega_i = 0.$$

Let us denote by X the linear space over the complex field of vectors $(\xi_1, \cdots, \xi_n)$, $\xi_i \in \widetilde{K}$ such that

$$\sum_{i=1}^{n} \mathfrak{L}_i \xi_i = 0$$

for any Picard-Fuchs equation $\mu \in M$. The space X contains the subspace X_0 spanned by vectors $(\int_\gamma \omega_1, \cdots, \int_\gamma \omega_n)$, $\gamma \in H_1(A_a, \mathbb{Z})$, since

$$\sum_{i=1}^{n} \mathfrak{L}_i \int_\gamma \omega_i = \sum_{i=1}^{n} \int_\gamma \mathfrak{L}_{ix} \omega_i = \int_\gamma dz = 0.$$

A fundamental role in the proof of the kernel theorem is played by

Proposition 4. $X = X_0$.

Before proving this we show how to deduce the kernel theorem from it. Let $\gamma = (\gamma_1, \cdots, \gamma_{2n})$ be a fixed basis for $H_1(A, \mathbb{Z})$. Proposition 4 implies that for every $P \in A^0_{K/k}$ and path δ_P there exists a complex column-vector

$$\mathbf{c} = \begin{pmatrix} c_1 \\ \vdots \\ c_{2n} \end{pmatrix}$$

such that

$$\int_{\delta_P} \omega = \sum_{i=1}^{2n} c_i \int_{\gamma_i} \omega = \left(\int_\gamma \omega\right) \mathbf{c}, \quad \left(\int_\gamma \omega\right) = \left(\int_{\gamma_1} \omega, \ldots, \int_{\gamma_{2n}} \omega\right). \tag{9}$$

The representation (9) is, in general, not unique. Hence we replace it by one of the form

$$\int_{\delta_P} \omega = \sum_{i=1}^{2n} r_i \int_{\gamma_i} \omega = \left(\int_\gamma \omega\right) \mathbf{r}, \quad \mathbf{r} = \begin{pmatrix} r_1 \\ \vdots \\ r_{2n} \end{pmatrix}, \tag{10}$$

where the r_i are the *real* functions on $\widetilde{\Gamma}$ uniquely determined by

$$\sum_{i=1}^{2n} r_i \int_{\gamma_i} \omega_j = \sum_{i=1}^{2n} c_i \int_{\gamma_i} \omega_j$$

$$(j = 1, \ldots, n).$$

$$\sum_{i=1}^{2n} r_i \overline{\int_{\gamma_i} \omega_j} = \sum_{i=1}^{2n} \bar{c}_i \overline{\int_{\gamma_i} \omega_j}$$

(the bar denoting the complex conjugate). Hence the map

$$(c_1, \cdots, c_{2n}) \longrightarrow (r_1(b), \cdots, r_{2n}(b)),$$

for any point $b \in \widetilde{\Gamma}$, is a linear map of linear spaces over the reals.

We shall apply $\sigma \in \pi_1(\Gamma_0)$ to the equation (9). Let

$$(\gamma_1\sigma^{-1}, \ldots, \gamma_{2n}\sigma^{-1}) = (\gamma_1, \ldots, \gamma_{2n}) B_\sigma,$$

where B_σ is a real matrix of order $2n$. Then, since $P^\sigma = P$, we have

$$\Big(\int_{\delta p} \omega\Big)^\sigma = \int_{\delta p} \omega + \int_{\gamma(\sigma)} \omega, \quad \gamma(\sigma) \in H_1(A_a, \mathbf{Z}).$$

On the other hand, the right-hand side transforms to

$$\Big(\int_{\gamma} \omega\Big) B_\sigma \mathbf{c} = \Big(\int_{\gamma} \omega\Big) B_\sigma \mathbf{r},$$

if we take into account the linearity of the map $\mathbf{c} \longrightarrow \mathbf{r}$ and the uniqueness of the representation (10). Consequently, by computing $(\int_{\delta p} \omega)^\sigma - \int_{\delta p} \omega$ two ways we obtain

$$\int_{\gamma(\sigma)} \omega = \Big(\int_{\gamma} \omega\Big) (B_\sigma - E)\, \mathbf{r},$$

where E is the unit matrix. If

$$\gamma(\sigma) = (\gamma_1, \ldots, \gamma_{2n}) \begin{pmatrix} k_1(\sigma) \\ \cdot \\ \cdot \\ \cdot \\ k_{2n}(\sigma) \end{pmatrix}$$

and $\mathbf{k}(\sigma)$ denotes the last column, we obtain the following system of equations for the vector-function $\mathbf{r}$:

$$(B_\sigma - E)\, \mathbf{r} = \mathbf{k}(\sigma), \qquad \sigma \in \pi_1(\Gamma_0).$$

Let

$$\mathbf{r}^0 = \begin{pmatrix} r_1^0 \\ \cdot \\ \cdot \\ \cdot \\ r_n^0 \end{pmatrix}$$

be one rational solution of this system (i.e., the r_i^0 are rational numbers), which certainly exists since the system is solvable and has integral coefficients. On putting $\mathbf{r} = \mathbf{r}^0 + \mathbf{r}^1$ we obtain the real vector-function $\mathbf{r}^1$ which satisfies the homogeneous system

$$(B_\sigma - E)\mathbf{r}^1 = 0, \quad \sigma \in \pi_1(\Gamma_0).$$

Let $d \neq 0$ be a common divisor of the coordinates of $\mathbf{r}^0$ and consider the point $dP \in A^0_{K/k}$ and some path $\delta_{dP} \subset A_a$. Then

$$\int_{\delta_{dP}} \omega = d\int_{\delta_P} \omega + \int_{\gamma'} \omega = \int_{\gamma''} \omega + \left(\int_\gamma \omega\right) d\mathbf{r}^1,$$

where $\gamma'' = \gamma' + \gamma d\mathbf{r}^0 \in H_1(A_a, \mathbf{Z})$. We shall show that the point dP is infinitely divisible in A_K. For certainly for any integer $m \neq 0$ there is a finite normal extension K' of K unramified outside $a_1, \cdots, a_s \in \Gamma$ such that dP can be divided by m in $A_{K'}$:

$$dP = mQ, \qquad Q \in A_{K'}.$$

We may even suppose that all solutions of this equation are defined over K'. We shall show that one of them is defined over K. Let us take that one of the solutions for which there is a path $\delta_Q \subset A_a$ such that

$$\int_{\delta_Q} \omega = m^{-1}\left(\int_\gamma \omega\right) d\mathbf{r}^1.$$

On applying $\sigma \in \pi_1(\Gamma_0)$ to both sides and using the fact that $B_\sigma \mathbf{r}^1 = \mathbf{r}^1$ we obtain

$$\int_{\delta_{Q^\sigma}} \omega = \int_{\delta_Q} \omega + \int_{\gamma'(\sigma)} \omega, \quad \gamma'(\sigma) \in H_1(A_a, \mathbf{Z}).$$

By Abel's theorem this means that $Q^\sigma = Q$ for all σ. Hence $Q \in A_K$.

The above proof also shows that for any $P \in A^0_{K/k}$ there is an integer $d \neq 0$ such that dP is infinitely divisible in A_K. By the functional variant of the Mordell-Weil theorem it follows that $dP \in \tau(B_k)$ where (B, τ) is the K/k-trace of A. (The idea of using the infinite divisibility was suggested to me by I. R. Šafarevič. I could not directly prove the inclusion $dP \in \tau(B_k)$ without recourse to a strong theorem like the Mordell-Weil.)

This completes the proof of finiteness.

3. It remains to prove Proposition 4. To this end we shall describe a special system of generators of the U-module W which will be convenient for computing the complex dimension of X.

We shall put

$$\Omega_j = \sum_{i=1}^{n} U_j \bar{\omega}_i,$$

where $U_j \subset U$ is the space of operators of order $\leq j$, and

$$\Omega = \bigcup_{j=0}^{\infty} \Omega_j = \Omega_t,$$

where $t < \infty$ since the dimension of Ω does not exceed $2n$ and $\Omega_j \subset \Omega_{j+1}$. We shall suppose that t is taken as small as possible.

By the "Leibnitz formula" the map $\partial^i : \Omega_0 \longrightarrow \Omega$ induces a K-linear map $\partial(i) : \Omega_0 \longrightarrow \Omega/\Omega_{i-1}$. Put

$$\Omega^{(i)} = \operatorname{Ker} \partial (i) \subset \Omega_0.$$

Setting $\Omega_{-1} = \{0\}$, we have the increasing sequence of subspaces

$$\{0\} = \Omega^{(0)} \subset \Omega^{(1)} \subset \Omega^{(2)} \subset \ldots \subset \Omega^{(t)} \subset \Omega^{(t+1)} = \Omega_0.$$

On putting $r_i = \dim \Omega_0/\Omega^{(i)} = \dim \Omega_i/\Omega_{i-1}$ we clearly have

$$n = r_0 \geqslant r_1 \geqslant \ldots \geqslant r_t > r_{t+1} = 0.$$

We shall prove Proposition 4 for some special basis of Ω_0 which is enough for Theorem 2, although it is not difficult to see that it implies the corresponding result for any basis.

We choose the numbering of the forms $\omega_1, \cdots, \omega_n$ of the first kind, so that the classes $\bar{\omega}_n, \cdots, \bar{\omega}_{r_i+1}$ generate $\Omega^{(i)}$ for $i = 1, \cdots, t+1$. By the definition of the $\Omega^{(i)}$ we have equations of the form:

$$\mu_{ij} : \partial^i \bar{\omega}_j + \sum_{k=1}^{n} \mathfrak{L}_{ijk} \bar{\omega}_k = 0, \quad \bar{\omega}_j \in \Omega^{(i)} \setminus \Omega^{(i-1)}, \quad \mathfrak{L}_{ijk} \in U_{i-1}, \tag{11}$$

where $i = 1, \cdots, t+1$ and for each value of i there are $r_i - r_{i-1}$ equations of order i corresponding to the j with $r_{i-1} \geq j \geq r_i + 1$.

Lemma 14. *The equations μ_{ij} span the U-module of Picard-Fuchs equations reduced to the basis $\omega_1, \cdots, \omega_n$.*

Proof. We first suppose that there is an equation of the type

$$\sum_{k=1}^{n} \mathfrak{L}_k \bar{\omega}_k = 0, \quad \mathfrak{L}_k \in U_{i_k - 1},$$

where the integers i_k, $1 \leq k \leq n$ are defined by the conditions $\bar{\omega}_k \in \Omega^{(i_k)} \setminus \Omega^{(i_k - 1)}$. We shall show that then $\mathfrak{L}_k = 0$ for all $k = 1, \cdots, n$. For this it is enough to show that the elements

$$\bar{\omega}_1, \ldots, \bar{\omega}_n; \partial \bar{\omega}_1, \ldots, \partial \bar{\omega}_{r_1}; \ldots; \partial^k \bar{\omega}_1, \ldots, \partial^k \bar{\omega}_{r_k}; \ldots; \partial^t \bar{\omega}_1, \ldots, \partial^t \bar{\omega}_{r_t} \tag{12}$$

are linearly independent, because the left-hand side of our equation is a linear combination of these elements over K in which "similar terms" corresponding to each $\bar{\omega}_k$ are collected together.

Thus we assume that the elements (12) are linearly dependent over K. Then this linear dependence has the form:

$$\sum_{i=1}^{r_k} \alpha_i \partial^k \bar{\omega}_i \in \Omega_{k-1}, \quad k \leqslant t, \text{ not all } \alpha_i = 0,$$

(where on the left-hand side we have put the terms of maximal order), from which we have

$$\partial^k \Big(\sum_{i=1}^{r_k} \alpha_i \bar{\omega}_i \Big) \in \Omega_{k-1},$$

and so

$$\sum_{i=1}^{r_k} \alpha_i \bar{\omega}_i \in \Omega^{(k)},$$

which is impossible because $\Omega^{(k)}$ is generated by $\bar{\omega}_{r_k+1}, \cdots, \bar{\omega}_n$. Hence there are no nontrivial equations of the type (11). Let us now consider any equation $\mu \in M$. It may be put in the form

$$\sum_{i=1}^{n} (\mathfrak{L}_k + M_k \partial^{i_k}) \bar{\omega}_k = 0, \quad \mathfrak{L}_k \in U_{i_k-1}, \quad M_k \in U. \tag{13}$$

On subtracting from this a linear combination of equations (11) of the type $\Sigma_{k=1}^{n} M_k \mu_{i_k,k}$ we come to an equation of the same type as (13):

$$\sum_{i=1}^{n} (\mathfrak{L}'_k + M'_k \partial^{i_k}) \bar{\omega}_k = 0, \quad \mathfrak{L}'_k \in U_{i_k-1}, \quad M'_k \in U,$$

in which, however, the order of M'_k is less by at least one than the order of M_k ($M'_k = 0$ if the order of M_k was zero, i.e., it was a constant). After a finite number of steps, on subtracting an element of $\Sigma_{i,j} U\mu_{ij}$ we thus ultimately come to the trivial equation. This proves the lemma.

Let us now consider the matrix W with $2n$ columns and infinitely many rows in which at the intersection of the ith column, $1 \leq i \leq 2n$ and the $(kn+j)$th row, $0 \leq k < \infty$, $1 \leq j \leq n$ is the element $\partial^k \int_{\gamma_i} \omega_j$ of $\widetilde{K}$. Let W_0 denote the matrix consisting of the first n rows of W.

Let p denote the number of columns of W_0 linearly independent over the complex number field k and let p_1 be the rank of W. Then we have the following inqualities:

(a) $p_1 \leq p$. For any linear dependence over k between the columns of W_0 implies the same linear dependence between the columns with the same indices in W.

(b) $p \leq \Sigma_{i=0}^t r_i$. For every column of W_0 belongs to the space X of solutions of the equations (11). The complex dimension of X does not exceed $\Sigma_{i=0}^t r_i$, since it is easy to see that the system (11) is equivalent to a system of the type

$$\partial\,(\overline{\omega}_1, \ldots, \partial^t\overline{\omega}_{r_t}) = (\overline{\omega}_1, \ldots, \partial^t\overline{\omega}_{r_t})\, N$$

in matrix notation, where N is a matrix of order $\Sigma_{i=0}^t r_i$ with elements in K and ∂ acts on the row (12). The assertion about the dimension of the space of such solutions is well known (and essentially belongs to differential algebra).

(c) $p_1 \geq \Sigma_{i=0}^t r_i$. To prove this inequality we shall show that the rows of W with the indices $kn + j$; $1 \leq j \leq r_k$; $k = 0, \cdots, t$ are linearly independent over $\widetilde{K}$. In fact, a linear dependence of the type

$$\sum_{k=0}^{t} \sum_{j=1}^{r_k} a_{kj}\, \partial^k \int_\gamma \omega_j = 0, \quad a_{kj} \in \widetilde{K},$$

valid for all $\gamma \in H_1(A_\alpha, \mathbb{Z})$ is equivalent to a Picard-Fuchs equation in the field $L\widetilde{K}/\widetilde{K}$ of the type

$$\sum_{k=0}^{t} \sum_{j=1}^{r_k} a_{kj}\, \partial^k \overline{\omega}_j = 0.$$

But the elements of (12) being linearly independent over K remain so over $\widetilde{K}$. Hence $a_{kj} = 0$, and (c) is proved.

On comparing (a), (b) and (c) we see that the complex dimension of X is precisely $p = \Sigma_{i=0}^t r_i$ and that X is generated over k by columns of W_0, i.e., by the space X_0. This completes the proof of Proposition 4 and thus of the kernel theorem.

4. We note that our result is essentially of a global character. It would be interesting to investigate the group $A^0_{K/k}$ by regarding K as a locally discrete normed field with field k of residues and considering only Picard-Fuchs equations relative to continuous derivations of K.

The structure of the U-module M in this case, as is shown by direct computation for elliptic curves, must reflect essential features of the "infinitesimal" behavior of A. It may be that one can also interpret the numbers r_i, $1 \leq i \leq t$ in other terms. It is easy to see that they are independent of the choice of ∂ and so depend only on A. For elliptic curves the only possible cases $r_1 = 0$ and $r_1 = 1$ are determined by whether ∂ annihilates the absolute invariant of the curve A or not.

CHAPTER II

Rational points on curves of genus ≥ 2

§7. Formulation of the problem

The aim of this chapter is the proof of the following result (the analogue of Mordell's conjecture for function fields).

Theorem 3. *Let K be a regular extension of the field k of characteristic zero and let C be a curve of genus ≥ 2 defined over K. If the set of points of C defined over K is infinite, then there is a curve C_0 which is birationally equivalent to C over K and defined over k. All the points C_K except a finite number are images of points of C_{0k}.*

For technical reasons it is convenient to suppose that the transcendence degree of K/k is one. The following simple result shows that we may confine consideration to this case.

Proposition 5. *If Theorem 3 is true for* $\dim_k K = 1$, *then it is true universally.*

Proof. Let $K' \subset K$ be a relatively algebraically closed subfield such that $\dim_{K'} K = 1$. If the set C_K is infinite, then by the hypothesis that Theorem 3 is true the curve C is birationally equivalent to a curve C' defined over K'. Since almost all the points C_K are images of points of $C'_{K'}$, this last set must also be infinite. Further, $\dim_k K' = \dim_k K - 1$ and so we may prove the proposition by induction on the transcendence degree.

Hence in future we confine attention to the case $\dim_k K = 1$. It is enough to establish the existence of the curve C_0 defined over k and birationally equivalent to C in order to get the whole of Theorem 3; for Theorem 3 then follows immediately from the theorem of de Franchis reproduced, for example, in Lang's book [4] (historical remarks to Chapter VII).

Since $\dim_k K = 1$ we can canonically define the height in the set of points defined over K of any projective algebraic variety V defined over K (cf. Lang [4]). We shall need only the most elementary facts about heights. More precisely, all that we need are the following properties of the concept "a set of points of bounded height".

If the set $F \subset V_K$ of points on the projective variety V has bounded height, then this property is preserved for any other projective embedding of V. Further, F remains a set of bounded height if we measure the height with respect to any finite extension K' of K or with respect to a field of the type Kk', where $k' \supset k$ is an extension of the constant field linearly disjoint to K.

Let $f: C' \longrightarrow C$ be an epimorphism of algebraic curves (everything defined over K). If the set $f(C'_K) \subset C_K$ has bounded height, then C'_K also has bounded height.

Let A be an abelian variety defined over K, and let (B, τ) be its K/k-trace. Then the set $a + \tau(B_k) \subset A_k$ (where a is any point of A_K) has bounded height.

All these results are easily obtained from the properties of height described in Lang's book [4] and also in [5] and [6].

The following assertion is also proved without difficulty but because of its importance and for ease of reference we formulate it separately.

Proposition 6. *Let C be a curve over the field K and put $L = K(C)$. Let E be a finite-dimensional subspace of K over k. Then the set F of points $P \in C_K$ for which there is a $f^{(P)} \in E$ with $f^{(P)} \neq 0$ such that $f_P^{(P)} = 0$, has bounded height.*

Proof. Let $(y_1, \cdots, y_n)$ be a k-basis of E. We may suppose without loss of generality that $L = k(y_1, \cdots, y_n)$, on extending E if necessary, which only strengthens the assertion. Then the locus of the point $(1, y_1, \cdots, y_n)$ over K is birationally equivalent to C. We shall measure heights with respect to this embedding.

Let V be the locus of $(1, y_1, \cdots, y_n)$ over k in n-dimensional projective space and let κ be its degree in this embedding. Further, for any $P \in F$ we denote by $\tilde{P}$ the locus of $(1, y_{1P}, \cdots, y_{nP})$ over k in the same projective space. Then $\tilde{P}$ is either a point (and then the logarithmic height of P is zero) or a curve, and then the logarithmic height $h(P)$ is equal to the product of the degree d of the curve $\tilde{P}$ by $e_P = [K : k(y_{1P}, \cdots, y_{nP})]$ (cf. Lang and Néron [6], formula 2 on p. 104).

By the definition of F, the curve $\tilde{P}$ is a component of a hyperplane section of V and so $d \leq \kappa$. Hence if Proposition 6 is false, there must be a sequence of $P \in F$ for which e_P increases indefinitely. But it is possible to avoid this by adding an element of $K \backslash k$ to $(y_1, \cdots, y_n)$ (if E contains no such element, then we must again extend E). This proves the proposition.

Finally, we remark that if C satisfies the conditions of Theorem 3 and if C_K contains infinitely many points of bounded height, then Theorem 3 is true. This is shown, for example, in Lang's book [4] (Chapter VII, §2, Corollary to Proposition 2).

In view of these results it is enough for the proof of Theorem 3 to deduce a contradiction from the following statement.

(H) *There is a curve C of genus ≥ 2 defined over K and having points defined over K of arbitrarily large height.*

This will occupy us in what follows. In preparation we investigate more carefully the function $\mu(P, Q)$ on curves.

§8. The function $\mu(P, Q)$ on curves

1. Let C be a complete nonsingular curve of genus $g \geq 2$ over the field K and let J be its jacobian. We shall suppose that C_K is nonempty. We shall consider the canonical map $\phi : C \longrightarrow J$ defined over K and normalized by the condition that the image of some fixed $O \in C_K$ is the zero of the group law on J. We shall identify C with its image under ϕ.

Lemma 15. *Let $\mu : \Sigma_{i=1}^{n} \mathfrak{L}_i \bar{\omega}_i = 0$ be a Picard-Fuchs equation on J, where ω_i are invariant differentials. Then on C there is a Picard-Fuchs equation $\mu' : \Sigma_{i=1}^{n} \mathfrak{L}_i \bar{\eta}_i = 0$, where the η_i are the differentials of the first kind on C induced by the forms ω_i. Conversely to every equation $\mu' : \Sigma_{i=1}^{n} \mathfrak{L}_i \bar{\eta}_i = 0$ on C there corresponds an equation $\mu : \Sigma_{i=1}^{n} \mathfrak{L}_i \bar{\omega}_i = 0$ on J, where the ω_i are the unique invariant forms inducing the η_i on C. Further, $\mu(P, Q) = \mu'(P, Q)$ for all $P, Q \in C_K$.*

Proof. The map $\phi : C \longrightarrow J$ can be identified with the composition

$$C \xrightarrow{i} C^g \xrightarrow{j} J,$$

where C^g is the g-fold direct product of C with itself, $i(P) = (P, 0, \cdots, 0)$ for any $P \in C$; and $j(P_1, \cdots, P_g) = \phi(P_1) + \cdots + \phi(P_g)$. For any differential η on C we denote by $\eta^{(i)}$ the differential form on C^g, which is the inverse image of η under the projection $p_i : C^g \longrightarrow C$ onto the ith component. It is well known that if the differential form of the first kind ω on J induces the differential η on C, then its inverse image on C^g is $\Sigma_{k=1}^{g} \eta^{(k)}$.

The equation $\mu : \Sigma_{k=1}^{n} \mathfrak{L}_k \bar{\omega}_k = 0$ on J corresponds to some equation $j^*\mu$ on C^g and by Lemma 9 we have the identity

$$j^*\mu(P, Q) = \mu(j(P), j(Q))$$

for all $P, Q \in C^g$.

By Lemma 10 there are Picard-Fuchs equations $\mu_1, \cdots, \mu_g$ on C such that

$$j^*\mu(P, Q) = \sum_{i=1}^{g} \mu_i(P_i, Q_i), \quad P = (P_1, \ldots, P_g), \quad Q = (Q_1, \ldots, Q_g).$$

It is easy to see that the μ_i are the inverse images under the projection $p_i : C^g \longrightarrow C$ of one and the same equation $\mu' : \Sigma_{k=1}^{n} \mathfrak{L}_k \bar{\eta}_k = 0$ as defined in the enunciation of the lemma. More precisely, let v be a nonconstant function on C and $v^{(i)}$ the inverse image on C^g of v with respect to p_i, so that $\mathrm{v} = (v^{(1)}, \cdots, v^{(g)})$ is a transcendence basis for the function field on C^g. Then as a representative of the equation $j^*\mu$ on C^g and μ' on C we may take respectively

$$\sum_{i=1}^{n} \mathfrak{L}_{i\mathrm{v}}\Bigl(\sum_{k=1}^{g} \eta_i^{(k)}\Bigr) = \sum_{k=1}^{n} dz^{(k)}, \qquad \sum_{i=1}^{n} \mathfrak{L}_{iv}\eta_i = dz.$$

It is easy to check that the subfield $K(J) \subset K(C^g)$, which may be identified with the subfield of the symmetric functions on C^g, is mapped into itself by any derivation of the type ∂_v of $K(C^g)$. Hence the argument may be reversed, which allows us to recover the Picard-Fuchs equation μ on J from the equation μ' on C. The last assertion of the lemma follows from

$$\mu(P, Q) = \mu(j(P, 0, \ldots, 0), j(Q, 0, \ldots, 0)) = \mu_1(P, Q) = \mu'(P, Q).$$

This completes the proof of the lemma.

Corollary 1. *Let μ be a fixed Picard-Fuchs equation on C. Then the $\mu(P)$, $P \in C_K$ all lie in a finite-dimensional k-subspace of K (which, of course, depends on μ).*

(For, in fact by the Mordell-Weil theorem for function fields it follows from the kernel theorem that the values of $\mu(P)$ belong to an abelian subgroup of K of finite type.)

Corollary 2. *Let $c_\mu \in K$ be some choice of elements as μ runs through some system of generators of the U-module of Picard-Fuchs equations on C reduced to some basis or other. Then the set of $P \in C_K$ for which $\mu(P) = c_\mu$ has bounded height.*

(For by the Mordell-Weil theorem and the kernel theorem those points belong to a finite number of cosets modulo $\tau(B_k)$, where (B, τ) is the K/k-trace of J, and so have bounded height.)

2. In future, in conformity with the notation for abelian varieties we shall write $\mu(P)$ instead of $\mu(0, P)$ $(P \in C \subset J)$ for any Picard-Fuchs equation μ on C.

We shall show now that the function $\mu(P)$ may be computed for almost all $P \in C_K$ using only *one* representative of μ. ("Almost all" here and henceforward means "all except a finite number".)

Lemma 16. *Let ∂ be a derivation of K/k, let $\omega_i = u_i dv$ and let*

$$\sum_{i=1}^{n} \mathfrak{L}_{iv}\,\omega_i = dz \quad \left(\mathfrak{L}_i = \sum_{j=0}^{m} a_{ij}\partial^j, \quad a_{ij} \in K\right)$$

be a representative of μ, where v is a local parameter at O and $z_0 = 0$. Then

$$\mu(P) = z_P + \sum_{i=1}^{n}\sum_{j=1}^{m} a_{ij} M_i^j(P), \tag{14}$$

$$M_i^j(P) = \sum_{k=0}^{j-1} \partial^{j-1-k}\,[(\partial_v^k u_i)_P\, \partial v_P]. \tag{15}$$

We may use these formulas to compute $\mu(P)$ for all $P \in C_K$ except finitely many.

Proof. Indeed if P does not belong to the poles of v, the zeros of v or the zeros of dv, then the function $w = vv_P^{-1}$ serves as a permissible local parameter at O and P. Hence

$$\mu(P) = z'_P - z'_O, \quad dz' = \sum_{i=1}^{n} \mathfrak{L}_{iw}\,\omega_i.$$

We shall compute first $\partial^i_w\,(udv)$. We have

$$(\partial_w - \partial_v)(u\,dv) = d\,(uw\,\partial v_P)$$

(using Lemma 2 and the fact that $v = wv_P$). Let us suppose that we have already proved the identity

$$(\partial^j_w - \partial^j_v)(u\,dv) = d\left(w \sum_{k=0}^{j-1} \partial_w^{j-1-k}\,[\partial^k_v u \cdot \partial v_P]\right) = N^j\,(u\,dv) \tag{16}$$

for some $j \geq 1$. It is true for $j = 1$; let us check it for $j + 1$. We have

$$\begin{aligned}(\partial^{j+1}_w - \partial^{j+1}_v)(udv) &= (\partial^j_w - \partial^j_v)\,\partial_v u \cdot dv + \partial^j_w\,(\partial_v - \partial_w)(udv)\\ &= d\left(w \sum_{k=0}^{j-1} \partial_w^{j-1-k}\,[\partial^{k+1}_P u \cdot \partial v_v]\right) + d\,(w\partial^j_w\,(u\,\partial v_P))\\ &= d\left(w \sum_{k=0}^{j} \partial_w^{j-k}\,[\partial^k_v u \cdot \partial v_P]\right) = N^{j+1}\,(u\,dv),\end{aligned}$$

which completes the induction. Hence

$$\sum_{i=1}^{n} \mathfrak{L}_{iw}\,\omega_i = \sum_{i=1}^{n} \mathfrak{L}_{iv}\,\omega_i + \sum_{i=1}^{n}\sum_{j=1}^{m} a_{ij} N^j\,(u_i\,dv).$$

On comparing this with the formulas (14), (15) and (16) and noting that $w_O = 0$, $w_P = 1$ we see that it is enough to check the formula

$$(\partial_w^{j-1-k}[\partial^k_v\,u \cdot \partial v_P])_P = \partial^{j-1-k}[(\partial^k_v\,u)_P\,\partial v_P].$$

But this follows from Lemma 3 (b) since w is an admissible quasiparameter at P. This proves the lemma.

§9. Proof of Theorem 2. First case

From now until the end of the chapter we shall suppose that C satisfies the statement (H). The notation ∂, O are the same as in §8 and v denotes a function satisfying the conditions of Lemma 16.

In this section we reduce the statement (H) to an absurdity under the auxiliary supposition that *in the space of differentials of the first kind on C there are two linearly independent ones $\omega_1 = u_1 dv$, $\omega_2 = u_2 dv$, which satisfy equations of the type*

$$\begin{cases}\partial_v \omega_1 + \omega'_1 = dw_1,\\ \partial_v \omega_2 + \omega'_2 = dw_2,\end{cases} \tag{17}$$

where ω'_1, ω'_2 *are differentials of the first kind* (this supposition characterises our "first case").

Let μ_1, μ_2 be the Picard-Fuchs equation on C of which the first and second equation in (17) respectively are representatives. By Lemma 16 we have

$$\begin{aligned}\mu_1(P) &= u_{1P}\partial v_P + w_{1P},\\ \mu_2(P) &= u_{2P}\partial v_P + w_{2P}\end{aligned} \tag{18}$$

for almost all P. On multiplying the first of these equations by u_{2P} and the second by $-u_{1P}$ and adding, we have

$$\mu_1(P)\,u_{2P} - \mu_2(P)\,u_{1P} - (w_1u_2 - w_2u_1)_P = 0 \tag{19}$$

for almost all $P \in C_K$.

We shall show now that $w_1u_2 - w_2u_1$ can be expressed K-linearly in terms of u_1 and u_2. For otherwise the function

$$f^{(P)} = \mu_1(P)\,u_2 - \mu_2(P)\,u_1 - (w_1u_2 - w_2u_1),$$

would be distinct from zero for all $P \in C_K$ and would lie by Lemma 15, Corollary 1 in a finite-dimensional k-subspace of L. From Proposition 6 and (19) it would then follow that the set C_k is a set of points of bounded height in contradiction to (H).

Hence

$$w_1u_2 - w_2u_1 = c_1u_2 - c_2u_1, \qquad c_1, c_2 \in K.$$

and so

$$f^{(P)} = (\mu_1(P) - c_1)\,u_2 - (\mu_2(P) - c_2)\,u_1.$$

Since u_1, u_2 are linearly independent over K, a similar argument using Proposition 6 shows that *for almost all* $P \in C_K$ we have

$$\mu_1(P) = c_1, \quad \mu_2(P) = c_2. \tag{20}$$

Now it is not difficult to establish that for *any* equation μ on C there is a $c = c_\mu \in K$ such that $\mu(P) = c_\mu$ for almost all $P \in C_K$. We shall do that in a moment.

From (18) and (20) follows the existence of a rational function w on C such that

$$\partial v_P = w_P \tag{21}$$

for almost all $P \in C_K$.

Lemma 17. *For any function* x *on* C *and any integer* $j \geq 2$ *there is a function* y *depending only on* x *and* j *such that* $\partial^j x_P = y_P$ *for all those* $P \in C_K$ *for which* (21) *holds.*

Proof. It is enough to clear up the case $j = 1$, since induction on j will then provide the necessary proof.

Let $F(X, V)$ be an irreducible polynomial over K such that $F(x, v) = 0$. The identity

$$(F^{\partial}(x, v))_P + \left(\frac{\partial F}{\partial X}(x, v)\right)_P \partial x_P + \left(\frac{\partial F}{\partial V}(x, v)\right)_P \partial v_P = 0,$$

obtained by differentiating the equation $F(x_P, v_P) = 0$ together with (21) proves the lemma.

On applying this to (14) and (15) we easily show that for any Picard-Fuchs equation μ on the curve there is a rational function x^{μ} depending only on μ such that

$$\mu(P) = x^{\mu}_P$$

for almost all $P \in C_K$. The hypothesis that $x^{\mu} \notin K$ leads immediately to a contradiction with assertion (H) in view of Proposition 6, since the function $f^{(P)} = \mu(P) - x^{\mu}$, which would be different from zero, belongs (by Lemma 15, Corollary 1) to a finite-dimensional k-linear subspace of L. The alternative, that $x^{\mu} \in K$ for all μ leads to a contradiction to assertion (H) by Lemma 15, Corollary 2.

This concludes the proof ot the first case of Theorem 3.

§10. Proof of Theorem 2. Second case

Let C again be a curve for which the assertion (H) is true. Then it is true also for field $K\bar{k}$, where $\bar{k}$ is the algebraic closure of k. Hence from now on we shall suppose that k is algebraically closed.

Amongst the curves C satisfying (H) we shall take one such that the dimension of the space of differentials ω of the first kind satisfying

$$\partial\bar{\omega} + \bar{\omega}' = 0 \tag{22}$$

(where ω' is also a differential of the first kind) is maximal. As was shown in the preceding section, this dimension must be either zero or 1, on the assumption that (H) can be satisfied at all.

In Lang's paper [5] (Proposition 1 on p. 40) it is shown that there is an unramified covering $p: C' \longrightarrow C$ of arbitrarily high degree defined over K and such that the truth of (H) for C implies its truth for C'. Let J, J' be the jacobians of C, C' respectively. Then there exists an abelian variety A defined over K and an isogeny $j: J' \longrightarrow J \times A$ also defined over K, the "first coordinate" of which coincides with the map induced by p. (Here and above we regard C' and C as embedded in their jacobians J' and J and passing through the zero points which are chosen so as to agree with p.) The isogeny induces an isomorphism of the spaces of differential forms of the first kind on J' and $J \times A$ respectively. The last space splits as the direct sum of the spaces of forms which induce zero on J and A respectively. Interpreting this in terms of forms on C' we see that the subspace H of those differential forms of the first kind on C' which are induced by the differential forms on C has a uniquely defined complement H_1. The differentials of H_1 are induced on the image of C' by the sequence

$$C' \xrightarrow{\phi'} J' \xrightarrow{j} J \times A$$

of maps from the differential forms which vanish on J. There is a 1-1 correspondence between the Picard-Fuchs equations on A and the Picard-Fuchs equations on C' which connect differentials belonging to H_1. This correspondence preserves the values of the μ-function, as follows from Lemma 15 and Lemmas 10 and 11.

We shall show that the Picard-Fuchs equations on H_1 can be described very simply.

Lemma 18. *For any differential* $\omega \in H_1$ *there are uniquely determined* ω', $\omega'' \in H_1$ *such that there is an equation of the type*

$$\mu_\omega : \partial^2\overline{\omega} + \partial\overline{\omega}' + \overline{\omega}'' = 0. \tag{23}$$

The equations (23) *reduced to any basis of* H_1 *generate the* U*-module of Picard-Fuchs equations in* H_1 *reduced to that basis. For any two differentials* $\omega_1, \omega_2 \in H_1$ *and any two points* $P, Q \in C'_K$ *we have the identity*

$$\mu_{\omega_1-\omega_2}(P, Q) = \mu_{\omega_1}(P, Q) - \mu_{\omega_2}(P, Q). \tag{24}$$

Proof. The classes $(\overline{\omega}_1, \cdots, \overline{\omega}_n, \partial\overline{\omega}_1, \cdots, \partial\overline{\omega}_n)$, where $(\omega_1, \cdots, \omega_n)$ is any basis of H_1, are linearly independent over K. For otherwise we should have an equation of the type

$$\overline{\omega} + \sum_{i=1}^{n} \alpha_i \partial\overline{\omega}_i = 0, \quad \omega \in H_1,$$

where not all the α_i are zero, i.e., an equation

$$\overline{\omega}' + \partial\left(\sum_{i=1}^{n} \alpha_i \overline{\omega}_i\right) = 0, \quad \omega' = \omega - \sum_{i=1}^{n} \partial\alpha_i \omega_i.$$

But this would mean that on C' the dimension of the space of differentials satisfying (22) would be greater than the number for C, contrary to our choice of C.

Since H_1 is induced by forms on the variety A, it follows that the classes $\partial^j\overline{\omega}$ for any $j \geq 2$ and $\omega \in H_1$ must depend linearly on the classes $\overline{\omega}_i, \partial\overline{\omega}_i$, $i = 1, \cdots, n$. Any relation of the type

$$\mu_\omega : \partial^2\overline{\omega} + \sum_{i=1}^{n} (\alpha_i \partial\overline{\omega}_i + \beta_i \overline{\omega}_i) = 0$$

is of the shape (23) on putting $\omega' = \sum_{i=1}^n a_i\omega_i$, $\omega'' = \sum_{i=1}^n (\beta_i - \partial a_i)\omega_i$. The uniqueness of ω' and ω'' is clear.

The equations μ_{ω_i}, $i = 1, \cdots, n$ generate the whole module of equations in H_1 reduced to the basis $(\omega_1, \cdots, \omega_n)$ by Lemma 14 and since in this case $n = r_0 = r_1$, $r_2 = 0$. Finally (24) follows from the uniqueness of (23). This concludes the proof of the lemma.

Let us consider (23). As above, let v be an element of $K(C)$ which is a local parameter at $O \in C'_K$ (it exists since the covering $C' \to C$ is unramified). By Lemma 16 we have

$$\mu_\omega(P) = \omega_P + u'_P\,\partial v_P + (\partial_v u)_P\,\partial v_P + \partial\,(u_P\,\partial v_P) \tag{25}$$

for almost all $P \in C'_K$, where

$$\partial_v^2\,\omega + \partial_v\omega' + \omega'' = dw, \quad w_O = 0, \quad \omega = u\,dv, \quad \omega' = u'\,dv.$$

In the proof of Lemma 17 we established the existence of $x^u, y^u \in K(C')$ such that

$$\partial u_P = x_P^u + y_P^u\,\partial v_P \tag{26}$$

for all $P \in C'_{\bar{K}}$, where $\bar{K}$ is the algebraic closure of K (except at the poles of u and v, where the equation becomes meaningless). It is easy to see that these functions are uniquely determined, for otherwise we would have an equation of the type $\partial v_P = v'_P$, $v' \in K(C')$ for all $P \in C'_{\bar{K}}$ and v' would have to take each of its values v'_P infinitely often, namely at all the Q where $v_Q = v_P + c$, $c \in k$. This is impossible because ∂v_P by the nontriviality of ∂ takes more than one value on $C'_{\bar{K}}$.

By (26) and (25) we have

$$\mu_\omega(P) = u_P\,\partial^2 v_P + y_P^u\,(\partial v_P)^2 + (x^u + \partial_v u + u')_P\,\partial v_P + w_P \tag{27}$$

for almost all $P \in C'_K$.

Take a basis $\omega_i = u_i dv$, $i = 1, \cdots, n$ of H_1 and consider the matrix T with three columns and n rows, the ith row having the form

$$T_i = (u_i,\, y^{u_i},\, x^{u_i} + \partial_v u_i + u'_i). \tag{28}$$

The rank of this matrix is at most three. Hence *there is a covering $C' \to C$ defined over K for which the rank of T constructed with respect to some basis of H_1 takes its maximal value $R \le 3$* (the maximum being taken, of course, only over those unramified coverings for which C' satisfied (H)).

From now on we shall suppose that C' has been so chosen that the rank R of T is maximal.

Let us now consider a further unramified covering $C'' \to C'$ of degree $m > 1$

such that C'' also satisfies (H). As above, the space $H + H_1$ of differentials of the first kind on C' has a uniquely determined complement H_2. The elements of H_2 correspond to forms on an abelian variety B which is determined by an isogeny

$$j' : J'' \longrightarrow J' \times B.$$

The "first coordinate" $J'' \longrightarrow J'$ of this isogeny is generated by the covering $C'' \longrightarrow C'$.

Let $\omega = udv$ be any element of H_2. We shall apply (27) to the relation μ_ω of the type (23) *in the space* H_2 and to the point $P \in C''_K$. By the choice of C' the row

$$T_u = (u, y^u, x^u + \partial_v u + u')$$

is a linear combination of the rows T_i, $i = 1, \cdots, R$ (cf. (28)):

$$T_u = \sum_{i=1}^{R} z_i T_i, \quad z_i \in K(C'').$$

Hence a relation of the type

$$\mu_\omega(P) = \sum_{i=0}^{R} z_{iP} \mu_i(P), \quad (\mu_0(P) = 1, \quad \mu_i(P) = \mu_{\omega_i}(P), \quad i = 1, \ldots, R), \tag{29}$$

is true for almost all $P \in C''_K$. Let $1, t_1, \cdots, t_S$ be a maximal subset of $1, z_1, \cdots, z_R$ linearly independent over K. On expressing the z_i in terms of the t_j and substituting in (29) we obtain

$$\mu_\omega(P) + \sum_{i=1}^{R} c_i \mu_i(P) + \sum_{j=1}^{S} t_{jP} \sum_{i=1}^{R} c_{ij} \mu_i(P) = 0, \quad c_i, c_{ij} \in K, \tag{30}$$

for almost all $P \in C''_K$. By Lemma 15, Corollary 1, the function

$$f^{(P)} = \mu_\omega(P) + \sum_{i=1}^{R} c_i \mu_i(P) + \sum_{j=1}^{S} t_j \sum_{i=1}^{R} c_{ij} \mu_i(P)$$

lies in a finite k-dimensional subspace of $K(C'')$ for fixed ω as P runs through the points of C''_K. Hence if C'' satisfies (H) we must have $f^{(P)} = 0$ for almost all $P \in C''_K$, i.e.,

$$\mu_\omega(P) + \sum_{i=1}^{R} c_i \mu_i(P) = 0, \quad c_i \in K, \tag{31}$$

for such P. We emphasize that c_i depends on ω but not on P.

We shall now suppose that the covering $C'' \longrightarrow C'$ is normal and has a cyclic Galois group generated by an automorphism $\sigma : C'' \longrightarrow C''$. Such a covering always exists over some finite extension of K and since the truth of (H) and all the previous constructions are compatible with such an extension we may suppose that the covering is actually defined over K.

Since the $\mu_i(P)$, $i = 1, \cdots, R$ depend only on the projection of $P \in C''$ onto C' it follows from (31) that

$$\mu_\omega (P^{\sigma^{-1}}) = \mu_\omega (P) \tag{32}$$

for almost all $P \in C''_K$. The automorphism σ acts on $K(C'')$ and since $v^\sigma = v$ it is easy to see that $(x^u)^\sigma = x^{u^\sigma}$, $(y^u)^\sigma = y^{u^\sigma}$. Further, for any $z \in K(C'')$ we have

$$z_P^{\sigma^{-1}} = z_{P^\sigma}.$$

Hence (27) implies

$$\mu_\omega (P^{\sigma^{-1}}) = \mu_{\omega^\sigma} (P).$$

On taking this with (32) and using (24) we have

$$\mu_{\omega-\omega^\sigma}(P) = 0 \tag{33}$$

for all $\omega \in H_2$ and almost all $P \in C''_K$.

The linear mapping $\omega \longrightarrow \omega^\sigma - \omega$ of H_2 into itself is clearly epimorphic. Since the μ_ω generate the module of Picard-Fuchs equations in H_2 it follows that $\mu(P) = 0$ for all such equations μ and for almost all $P \in C''_K$. But this means that the set $j'(C''_K)$ of rational points on the curve $j(C'')$ in the abelian variety B determined by the diagram $J'' \longrightarrow J' \times B$ has bounded height. This set does not reduce to a point because C'' generates J''. Hence the curve C'' cannot satisfy (H), which has led us to our required contradiction.

This completes the proof of Theorem 3.

CHAPTER III

Rational points on elliptic curves

§11. Formulation of the result

On an elliptic curve C over a field K of finite type (and also over a field of finite relative type K/k if C is not isomorphic to a curve defined over k) the points C_K form a finitely generated group, which may sometimes actually be finite. The functional variant of the Siegel-Mahler theorem shows that in any affine model of C there are only finitely many points with "integral coordinates" (i.e., coordinates lying in a given subring of finite type of K) (cf. [5]).

In this chapter we shall prove by means of the function $\mu(P)$ a finiteness theorem from which the Siegel-Mahler theorem follows and which is so strong that the analogue over number fields is false.

From the very beginning we restrict ourselves to a ground field K which is a field of algebraic functions of a single variable over an algebraically closed field $k \subset K$. The standard reduction, which we shall not repeat, permits one to make deductions for the case of fields of finite type or finite relative type.

Besides, we shall suppose that the curve C of genus one is not isomorphic to a curve defined over k even in any finite extension of K; the contrary case from our present point of view being trivial.

Let $L = K(C)$. We shall use the letters P, Q for prime divisors of L/K and will identify prime divisors of the first degree with points of C_K, which we shall consider infinite. We denote prime divisors (points) of K/k by $\mathfrak{p}$.

For any $v \in L$ we denote by $(v), (v)_0, (v)_\infty$ respectively the divisor, divisor of zeros and divisor of poles of v, and by $\nu_P(v)$ the order of v relative to the divisor P. As usual, v_P is the value of v at P. We shall use completely analogous notation for the elements of K. There can only be confusion in relation to symbols of the type (a) where $a \in K$ if we have not agreed whether a divisor in L or in K is meant, and usually the exact sense will be clear from the context.

It is convenient to introduce a function $\epsilon(x)$ of a real argument:

$$\epsilon(x) = \begin{cases} 0 & \text{if } x \leq 0, \\ x & \text{if } x > 0. \end{cases}$$

Theorem 4. *Let $y \in L \setminus K$ and let $r \geq 1$ be the maximum order of a zero of y. Then there is a constant h, depending only on y and K, such that*

$$\sum_{\mathfrak{p}} \epsilon(\nu_{\mathfrak{p}}((y_Q)) - 2r) < h \tag{34}$$

for all $Q \in C_K$.

(If $y_Q = 0, \infty$ we put $\nu_{\mathfrak{p}}(y_Q) = 0$, and the summation is extended over all points $\mathfrak{p}$ of K.)

Corollary (Mahler's Theorem). *Let S be any finite set of points of K and let $x \in L \setminus K$. The set of points $Q \in C_K$ for which $\nu_{\mathfrak{p}}((x_Q)_\infty) = 0$ for $\mathfrak{p} \in S$ is finite.*

(One must put $y = x^{-1}$ and remember that any infinite set of points of C_K has unbounded height. If the condition $\nu_{\mathfrak{p}}((x_Q)_\infty) = 0$ for $\mathfrak{p} \notin S$ were satisfied for the points of such a set, then $\sum_{\mathfrak{p} \in S} \nu_{\mathfrak{p}}((x_Q^{-1})_0) \longrightarrow \infty$, contrary to (34).)

We shall prove Theorem 4 in the next two sections and here note only that the assertion of the theorem is trivially false when K is an algebraic number field (although the corollary remains true). For let K be the field of rational numbers and consider the function y^{-1} on a curve C in Weierstrass normal form $y^2 = x^3 + Ax + B$. Let $Q \in C_K$ be any point of infinite order. By a result of Lutz [7] some prime p divides the denominator of y_Q to a positive power. There is an integer $d \neq 0$ such that dQ belongs to the Lutz subgroup on C relative to p. Put

$$Q_n = p^n dQ.$$

It follows from the properties of the Lutz subgroup that $\nu_p((y_{Q_n}^{-1})_0)$ increases indefinitely as $n \longrightarrow \infty$ (as a linear function of n). Hence (34) cannot remain true.

§12. Auxiliary results

We shall fix some $v \in L \setminus K$ and some derivation ∂ of K. Without loss of generality we may suppose that there is a $t \in K$ for which $\partial t = 1$. We shall also fix the differential $\omega = u\,dv$ of the first kind on C. Since the K/k-trace of C is trivial, $\bar{\omega}$, $\partial\bar{\omega}$ are linearly independent over K. Hence we have

Lemma 19. *The module of Picard-Fuchs equations on C reduced to ω is generated by a single equation with a representative of the type*

$$(\partial_v^2 u + a\partial_v u + bu)\,dv = dw, \quad a, b \in K, \quad w \in L. \tag{35}$$

Lemma 20. *Let $e_p = \nu_p(dt) + 1$ ($t \in K$, $\partial t = 1$). Then for any $a \in K$ we have*

$$\nu_p(\partial a) \geq \nu_p(a) - e. \tag{36}$$

Proof. In K the derivation ∂ coincides with d/dt, and so

$$\nu_p(\partial a) = \nu_p(da) - \nu_p(dt) \geqslant \nu_p(a) - \nu_p(dt) - 1 = \nu_p(a) - e_p.$$

This proves the lemma.

Finally we shall need the following partial formulation of the "decomposition theorem" of A. Weil.

A function defined for all $P \in C_K$ and taking values in the group of divisors of K/k is called a *distribution*. The distributions f, g are called *equivalent* if the quotients $f^{-1}(P)g(P)$ and $f(P)g^{-1}(P)$ divide some fixed divisor of K/k for all $P \in C_K$. The distribution f is called *integral* if all its values are integral divisors of K/k.

The decomposition theorem asserts that there is a set of distributions $A(P)$ in 1-1 correspondence with the divisors A of L/K and with the following properties:

(a) If A is an integral divisor then $A(P)$ is an integral distribution.

(b) $(A_1A_2)(P) = A_1(P)A_2(P)$.

(c) If g.c.d. $(A_1, A_2) = 1$, then g.c.d. $(A_1(P), A_2(P)) = 1$ for all P (where of course the first g.c.d. is in the sense of divisors on L/k and the second of divisors on K/k).

(d) $(v)(P) \sim (v_P)$ for any $v \in L$. (Of course (v_P) denotes the divisor of $v_P \in K$ in K/k. Further, we assign to v_P the unit divisor if $v_P = 0$ or ∞. Here $\sim$ is the symbol of equivalence for distributions.)

§13. Proof of Theorem 4

It is easy to see that it is enough to prove Theorem 4 for any finite extension of K. Hence we may suppose that the zeros of y are in C_K.

Let O be some fixed point of C_K and v a local parameter there. We shall give C_K the group law in which O is zero, so, as usual $P_1 + P_2 = P_3$ if the divisor $P_1 P_2/OP_3$ is principal.

We define the finite set S of points of K/k by the two following conditions (the notation being the same as in Lemma 19 and the decomposition theorem).

(1) S contains all zeros and poles of a and factors of the divisor of the differential dt.

(2) If $\mathfrak{p} \notin S$ and $\nu_{\mathfrak{p}}(O(Q)) \neq 0$, then

$$\nu_{\mathfrak{p}}(x_Q) = \nu_O(x)\nu_{\mathfrak{p}}(O(Q)), \tag{37}$$

where x is any one of u, v, $\partial_v u$, w (w being normalized by the condition $w_O = 0$).

The existence of such a set follows at once from the decomposition theorem, which also implies the following replacement of (37) for $\mathfrak{p} \in S$.

(3) There is a constant c such that

$$|\nu_{\mathfrak{p}}(x_Q) - \nu_O(x)\nu_{\mathfrak{p}}(O(Q))| < c, \tag{38}$$

for all $\mathfrak{p} \in S$ and all $Q \in C_K$ with $\nu_{\mathfrak{p}}(O(Q)) \neq 0$, where x is one of u, au, v, $\partial_v u$, w.

We shall now consider the function $\mu(Q) = \mu(O, Q)$ belonging to (35). By Lemma 16

$$\mu(Q) = w_Q + au_Q \partial v_Q + (\partial_v u)_Q \, \partial v_Q + \partial \, (u_Q \, \partial v_Q).$$

for almost all $Q \in C_K$. We now make some estimations. Suppose first that $\mathfrak{p} \notin S$, $\nu_{\mathfrak{p}}(O(Q)) \neq 0$. Then

$$\nu_{\mathfrak{p}}(w_Q) = \nu_O(w)\,\nu_{\mathfrak{p}}(O(Q)) \geqslant \nu_{\mathfrak{p}}(O(Q)),$$
$$\nu_{\mathfrak{p}}(au_Q \partial v_Q) \geqslant (\nu_O(u) + \nu_O(v))\,\nu_{\mathfrak{p}}(O(Q)) - 1 = \nu_{\mathfrak{p}}(O(Q)) - 1$$

(on using 37), Lemma 20 and the fact that $\nu_O(v) = 1$, $\nu_O(u) = \nu_O(\omega/dv) = -\nu_O(dv) = 0$). Similarly on using Lemma 20 again and noting that $\partial_v u \in o_Q$ we have:

$$\nu_{\mathfrak{p}}((\partial_v u)_Q \, \partial v_Q) \geqslant \nu_{\mathfrak{p}}(O(Q)) - 1,$$
$$\nu_{\mathfrak{p}}(\partial\,(u_Q \partial v_Q)) \geqslant \nu_{\mathfrak{p}}(O(Q)) - 2.$$

Suppose now that $\mathfrak{p} \in S$, $\nu_{\mathfrak{p}}(O(Q)) \neq 0$. By the same arguments but using (38) instead of (37) we have

$$\nu_{\mathfrak{p}}(w_Q) \geqslant \nu_{\mathfrak{p}}(O(Q)) - c,$$
$$\nu_{\mathfrak{p}}(au_Q \, \partial v_Q) \geqslant \nu_{\mathfrak{p}}(O(Q)) - 2c - e,$$
$$\nu_{\mathfrak{p}}((\partial_v u)_Q \, \partial v_Q) \geqslant \nu_{\mathfrak{p}}(O(Q)) - 2 - e,$$
$$\nu_{\mathfrak{p}}(\partial\,(u_Q \partial v_Q)) \geqslant \nu_{\mathfrak{p}}(O(Q)) - 2c - 2e,$$

where $e = \max_{p\in S}(\nu_p(dt) + 1)$. On combining all these estimates we get the following estimate for the order of the divisor of zeros of $\mu(Q) \in K$:

$$\deg \mu(Q) \geqslant \sum_{p\notin S} \varepsilon(\nu_p(O(Q)) - 2) + \sum_{p\in S} \varepsilon(\nu_p(O(Q)) - d), \quad d = 2c + 2e.$$

By the Mordell-Weil theorem and the kernel theorem $\deg \mu(Q)$ is bounded above on the set of $Q \in C_K$ of infinite order. Hence there is a constant h_0 such that

$$\sum_{p\notin S} \varepsilon(\nu_p(O(Q)) - 2) + \sum_{p\in S} \varepsilon(\nu_p(O(Q)) - d) < h_0 \tag{39}$$

for all $Q \in C_K$. In essence this is the strongest formulation of the theorem of finitude for elliptic curves which can be obtained in this way. The transition to (34) already weakens our result somewhat.

Let us take h_0 and S in such a way, that (34) remains true for all factors O_i of the divisor of zeros of y.

Let $(y)_0 = \prod_i O_i^{r_i}$. From the obvious inequality

$$\varepsilon(x_1 + x_2) \leqslant \varepsilon(x_1) + \varepsilon(x_2)$$

we have

$$\sum_{p\notin S} \varepsilon(x_p(O_i^{r_i}(Q)) - 2r) + \sum_{p\in S} \varepsilon(\nu_p(O_i^{r_i}(Q)) - dr) < h_0 r_i, \tag{40}$$

where $r = \max r_i$. Now we use the coprimeness of $O_i(Q)$ and $O_j(Q)$ for $i \neq j$ and note that $\sum_i \epsilon(x_i - s) = \epsilon(\sum_i x_i - s)$ if there is at most one x_i distinct from zero and $s > 0$. On summing (40) with respect to i we thus obtain

$$\sum_{p\notin S} \varepsilon(\nu_p((y)_0(Q)) - 2r) + \sum_{p\in S} \varepsilon(\nu_p((y_0)(Q)) - dr) < h_0 \sum_i r_i. \tag{41}$$

Since there are only finitely many elements in the second sum we may replace it by

$$\sum_{p\in S} \varepsilon(\nu_p((y)_0(Q)) - 2r),$$

on replacing the constant on the right-hand side by a greater one. Finally (possibly again at the price of increasing the constant) we may replace the divisor $(y)_0(Q)$ by $(y_Q)_0$. Since clearly

$$\varepsilon(\nu_p((y_Q)_0) - 2r) = \varepsilon(\nu_p((y_Q)) - 2r),$$

we thus obtain (34).

BIBLIOGRAPHY

[1] I. R. Šafarevič, *Principal homogeneous spaces defined over a function field*, Trudy Mat. Inst. Steklov. 64 (1961), 316–346.

[2] S. Lang, *Introduction to algebraic geometry*, 1959.

[3] ———, *Abelian varieties*, Interscience Tracts in Pure and Applied Mathematics, No. 7, Interscience, New York, 1959. MR 21 #4959.

[4] ———, *Diophantine geometry*, Interscience Tracts in Pure and Applied Mathematics, No. 11, Interscience, New York, 1962. MR 26 #119.

[5] ———, *Integral points on curves*, Inst. Hautes Études Sci. Publ. Math. No. 6 (1960), 27–43. MR 24 #A86.

[6] S. Lang and A. Néron, *Rational points of abelian varieties over function fields*, Amer. J. Math. 81 (1959), 95–118. MR 21 #1311.

[7] E. Lutz, *Sur l'équation $y^2 = x^3 - Ax - B$ dans les corps p-adiques*, J. Reine Angew. Math. 177 (1937), 238–247.

[8] Ju. Manin, *Algebraic curves over fields with differentiation*, Izv. Akad. Nauk SSSR Ser. Mat. 22 (1958), 737–756. (Russian) MR 21 #2652.

[9] ———, *On the Hasse-Witt matrix of an algebraic curve*, ibid. 25 (1961), 153–172. (Russian) MR 23 #A1638.

[10] ———, *Diophantine equations over functional fields*, Dokl. Akad. Nauk SSSR 139 (1961), 806–809 = Soviet Math. Dokl. 2 (1961), 1009–1012. MR 24 #A2576.

[11] L. J. Mordell, *On the rational solutions of the indeterminate equation of the third and fourth degrees*, Proc. Cambridge Philos. Soc. 21 (1922), 179–192.

[12] E. Picard and G. Simart, *Théorie des fonctions algébraiques de deux variables independents*, Paris, 1897–1906.

[13] J. Igusa, *Abstract vanishing cycle theory*, Proc. Japan Acad 34 (1958), 589–594. MR 21 #50.

[14] ———, *Class number of a definite quaternion with prime discriminant*, Proc. Nat. Acad. Sci. U.S.A. 44 (1958), 312–314. MR 20 #5183.

Translated by:

J. W. S. Cassels

Изв. Акад. Наук СССР
Сер. Матем. Том 53(1989), № 2

Math. USSR Izvestiya
Vol. 34(1990), No. 2

LETTER TO THE EDITOR

YU. I. MANIN

Professor R. F. Coleman has sent me his preprint [2] devoted to an up-to-date account of my proof [1] of Mordell's function field conjecture. In [2] an error in [1] is discovered and corrected. Below I explain what local changes must be done in the language of [1].

a) In §6.2 $B_\sigma r$ (respectively $B_\sigma r^1$) must be replaced everywhere by $B_\sigma r^\sigma$ (respectively $B_\sigma r^{1\sigma}$).

b) In the formulation of the kernel theorem (§5.1, Theorem 2) insert the following restriction of "general position" type: "... under the condition that the matrix consisting of periods of A and all their derivatives with respect to the parameters has maximum rank $2n$". Add the same restriction to Corollary 2 in §8.1.

c) Add the following text to the end of §9: "... if the Jacobian of C satisfies the kernel theorem.

"Otherwise we denote by $\tilde\partial : K \to K$ the mapping with the property $\partial x_P = (\tilde\partial x)_P$ for all x and almost all P. Obviously this is a derivation. We consider a smooth relatively proper model of C over a smooth affine base S (a model of K), $f: \tilde C \to S$. Let S be small enough that $\Omega^1_S = O_S\, dt$ for some $t \in \Gamma(O_S)$ and $\partial = d/dt$. Then $\tilde\partial$ induces a meromorphic vector field on $\tilde C$, the lift of ∂. If it is holomorphic over a nonempty open subset of S, then C is determined by K because the Kodaira-Spencer class vanishes.

"We now assume that $\tilde\partial$ has a pole along the multivalued section Y of the morphism f. Consider a finite affine covering of $\tilde C$ with the following properties: Y intersects every one of its elements U, and Ω^1_U is freely generated by dt and dv for some v regular on U. Then $\tilde\partial = \partial/\partial t + z\partial/\partial v$, where $z = \tilde\partial v$, and z has a pole on $Y \cap U$. Now from the formula $\tilde\partial(v_P) = z_P$ for almost all P, considered as sections of f, it follows that almost all P do not intersect $Y \cap U$. Since this is true for all U, we obtain a contradiction with Siegel's functional theorem on integral points."

This completes the corrections; now we explain their logic. After $B_\sigma r$ is replaced by $B_\sigma r^\sigma$, the existence of r^0 asserted in §6.2 is not obvious. The condition of general position circumvents this difficulty, since it implies uniqueness of decomposition of Abel arguments of a point with respect to the periods. The kernel theorem is used twice in the proof of the Mordell conjecture, in §9 and in §10, but in §10 only the case of general position appears in the construction itself. In §9 the case where general position may be violated can be examined by means of the supplementary argument presented in c) above.

Whether the kernel theorem is true without the assumption of general position is unclear at this time. Another variant is proved in [2], but Coleman was not able to use it for applications to the Mordell conjecture.

Received 25/OCT/88

BIBLIOGRAPHY

1. Yu. I. Manin, *Rational points of algebraic curves over function fields*, Izv. Akad. Nauk SSSR Ser. Mat. **27** (1963), 1395–1440; English transl. in Amer. Math. Soc. Transl. (2) **50** (1966).

2. R. F. Coleman, *Manin's proof of the Mordell conjecture*, Preprint, 1988.

Translated by B. SILVER

Reprinted from *Math. USSR-Sb.* **6** (1968).

CORRESPONDENCES, MOTIFS AND MONOIDAL TRANSFORMATIONS

Ju. I. MANIN

Let V be a nonsingular cubic hypersurface in $\mathbf{P}^4$ over a finite field of q elements. In their paper [1] Bombieri and Swinnerton-Dyer proved that the Riemann hypothesis holds for the zeta function of V. In particular, letting ν_n denote the number of rational points on V over the field of q^n elements, we have

$$\nu_n = 1 + q^n - q^n \sum_{i=1}^{10} \omega_i^n + q^{2n} + q^{3n}, \qquad (1)$$

where ω_i are algebraic integers such that

$$|\omega_i| = q^{1/2}. \qquad (2)$$

The idea of this paper originated in a simple observation, which permits an easy proof of a more general result. Namely, the cubic hypersurfaces V are *unirational*. This property together with standard results of the theory of étale cohomology is found to be completely sufficient to prove

Theorem. *Nonsingular three-dimensional projective unirational varieties V over a finite field satisfy the Riemann-Weil hypothesis.*

Sketch of the Proof. We consider a rational map $f\colon \mathbf{P}^3 \dashrightarrow V$ of finite degree. By virtue of Abhyankar's results there exists a commutative diagram of the form

$$\begin{array}{ccc} W & & \\ {\scriptstyle g}\downarrow & \searrow{\scriptstyle h} & \\ \mathbf{P}^3 & \xrightarrow{f} & V, \end{array} \qquad (3)$$

in which g is a birational morphism which splits into a sequence of monoidal transformations with nonsingular centers, and h is a morphism of finite degree (at a generic point). Now the calculation of the zeta function of V can be carried out in two steps.

First Step. We compute the zeta function of the variety W. For simplicity we shall assume that the centers of the monoidal transformations into which g splits are geometrically irreducible (this can be obtained by a finite extension of the field of constants, which does not affect our result). Then every such center is either a point or a curve. Let $X' \to X$ be a monoidal transformation with such a center. The inverse image of the center is either the projective plane P^2 or a fibering over a curve with fiber P^1, whence, using obvious notation, we find:

$$\nu_n(X') = \nu_n(X) + q^n + q^{2n} \quad \text{(center = a point)},$$
$$\nu_n(X') = \nu_n(X) + q^n - q^n \sum_{i=1}^{2g} \eta_i^n + q^{2n} \quad \text{(center = a curve)}, \tag{4}$$

where in the second formula η_i are the characteristic roots of the Frobenius morphism of the blown-up curve: $|\eta_i| = q^{1/2}$.

Iterating formula (4), we can lift from P^3 to $\mathbb{F}$. Comparing the formula obtained for $\nu_n(\mathbb{F})$ with the Lefschetz formula, we discover that the characteristic roots of the Frobenius morphism on the cohomology space of $\mathbb{F}$ have the form $1, q, q^2, q^3$ or $q\eta$, where $|\eta| = q^{1/2}$ and η are algebraic integers.

Second Step. The morphism h defines a homomorphism of cohomology spaces $h^*: H^*(V) \to H^*(\mathbb{F})$, compatible with the action of the morphism. Hence the characteristic roots on $H^*(V)$ constitute a part of the roots on $H^*(\mathbb{F})$ only if h^* has no kernel. But this follows from Poincaré duality, which ensures the existence of the Gysin homomorphism $h_*: H^*(\mathbb{F}) \to H^*(V)$ with the property $h_*h^* = \deg h$ (we neglect the details, which are connected with the necessity of tracing a twisting via the Tate module $T_l(G_m)$). This proves the theorem.

There are two deficiencies in the argument given:

a) It remains unclear which characteristic roots correspond to the cohomology space of a given dimension.

But it would not be hard to deal with this matter; more essential is the following shortcoming of the argument.

b) It is restricted to finite ground fields, since it is based on a direct calculation of the points, as in the proof of Bombieri and Swinnerton-Dyer.

Obviously it would be possible to avoid this restriction, having calculated the ring $H^*(X')$ in terms of the ring $H^*(X)$ and the properties of the center of the monoidal transformation $X' \to X$, and it is this calculation that makes up one of the main results of this paper, with the essential improvement that we are not occupied with the cohomology ring of the variety X', but rather with a much more interesting object.

This object—the *motif of the variety* X'—was introduced by Grothendieck. Grothendieck's goal consisted of the construction of a universal cohomology theory for algebraic varieties. We describe this idea rather explicitly.

Let $V(k)$ be the category of projective nonsingular algebraic varieties over a field k. For any variety X of $V(k)$ we denote by $C(X)$ the Q-algebra of classes of cycles on X up to numerical equivalence. We shall call an arbitrary element of $C(X \times Y)$ a *correspondence* between the varieties X and Y. The classical concept of the composition of the correspondences $f \in C(X \times Y)$ and $g \in C(Y \times Z)$ permits us to construct $g \circ f \in C(X \times Z)$. Correspondences possess all the formal properties of morphisms, which gives us a *new category* $CV(k)$, whose objects are the same as those of $V(k)$ and whose morphisms are correspondences. It is well known that in the classical case $k = C$, correspondences induce homomorphisms of the cohomology spaces $H^*(X, C)$. In fact, this property follows formally from the functoriality of graded cohomology rings, the Künneth formula and Poincaré duality.

Cohomology theories on $V(k)$ which satisfy these conditions will be called *geometric*. There exist many such theories: étale cohomology with l-adic coefficients for different l, the classical cohomologies induced by different embeddings of the ground field k into C (if char $k = 0$ and k is "not too large"). But, whether or not this is true of k, it is possible to consider any geometric cohomology theory on $V(k)$ as a functor on $CV(k)$.

More precisely, we consider a functor $V(k)^0 \to CV(k)$ which associates to each variety X, considered as an object in $CV(k)$, and to each morphism $f\colon X \to Y$ the class of the graph of f in the ring $C(Y \times X)$. Then for any geometric cohomology theory $H\colon V(k)^0 \to \mathcal{H}$ there exists a functor $CV(k) \to \mathcal{H}$ which makes the diagram

$$\begin{array}{ccc} & & \mathcal{H} \\ & \overset{H}{\nearrow} & \uparrow \\ V(k)^0 & & \\ & \searrow & \\ & & CV(k) \end{array} \qquad (5)$$

of functors commutative.

We remark that the category of values $\mathcal{H}$ of every geometric cohomology theory is abelian [in the given examples $\mathcal{H}$ is simply the category of graded linear spaces, and in the general case the *abelian-ness* (or even semisimplicity) of $\mathcal{H}$ should be included in the definition]. Meanwhile, the category $CV(k)$ is far from abelian. In particular, it contains many morphisms which are projectors that do not correspond to the direct decomposition of an object. It is possible formally to add the kernels and images of all the projectors to $CV(k)$; this gives us a category which we will denote by $\widetilde{CV}(k)$. The cohomology functor $CV(k) \to \mathcal{H}$ is developed via this category $\widetilde{CV}(k)$. Hence for technical reasons it is easier to consider as the basic category not $\widetilde{CV}(k)$, but rather a modified category.

Namely, we denote by $CV^0(k)$ the category with the same objects as $CV(k)$, but with a smaller number of morphisms: in this category $\mathrm{Hom}(X, Y)$ will consist only of correspondences of codimension equal to $\dim X$ (the same as the codimension of the graph of any morphism of varieties $Y \to X$). Adding the kernels and images of all the projectors in $CV^0(k)$ to it, we will obtain the *category of motifs* $\widetilde{CV}^0(k)$. By the same token, every motif is represented by a direct sum of "pieces of varieties", i.e. of objects of the form (X, p), where X is an n-dimensional variety, $p \in C^n(X \times X)$, $p^2 = p$ (in the sense of composition of correspondences). Any geometric cohomology theory extends to a functor of motifs.

All these definitions are entirely formal. A very interesting part of the theory of motifs is entirely conjectural; a set of conjectures on algebraic cycles, whole validity is claimed by Grothendieck, implies, because of the basic hypothesis on it, that the functor $h\colon V(k)^0 \to \widetilde{CV}^0(k)$ (category of motifs) constructed above *is by itself a geometric cohomology theory*. Consequently this theory is universal in the obvious sense of the word.

In connection with this conjecture, Grothendieck calls h the *functor of motivic cohomology*. Let

X be some variety, $h(X)$ its motif. Since for any geometric cohomology theory H the object $H(X)$ depends only on $h(X)$, Grothendieck call $H(X)$ the *realization of the motif* $h(X)$. For example, there are l-adic realizations, Hodge realizations, etc. For the reasoning behind the use of the word cf. Herbert Read's remark ([6], French p. 16): "Cézanne's method of painting was first to choose his 'motif'—a landscape, a person to be portrayed, a still-life; then to bring into being his visual apprehension of this *motif*; and in this process to lose nothing of the vital intensity that the *motif* possessed in its actual existence." (In order to preserve the vital intensity of a motif, realization should obviously be a functor. *Ju. M.*)[1)]

However, nowhere in this paper do we touch upon either the conjectural part of the theory of motifs or upon its connection with cohomology theories. Technically speaking, the basic new results of the paper are connected with calculations (in the categories $CV(k)$ of correspondences and $\widetilde{CV^0}(k)$ of motifs), which were suggested by Grothendieck's proof of the Riemann-Roch Theorem. As is known (see [8] and to a lesser extent [2]), in the course of the proof of this theorem it turns out to be necessary to study the ring $C(X')$, where $X' \to X$ is a monoidal transformation. We shall prove that all the results obtained by this investigation have an interpretation in the language of correspondences, and that in a finite calculation they permit us to compute the motif $h(X')$. Since, moreover, for any morphism $Y \to X$ of finite degree the motif $h(X)$ is a distinguished direct summand of the motif $h(Y)$ (Example 1, §3), this also permits us to compute the motives of three-dimensional unirational varieties.

We now give a more detailed survey of the contents of the paper.

The first section is devoted to a description of "global intersection theories", on the basis of which it is possible to construct the category of correspondences. The situation is that taking the ring of classes of cycles modulo *numerical* equivalence for $C(X)$ is quite optional; other functors are also suitable, for example, K-theory. K-correspondences deserve a special study which is not carried out here.

In the second and third sections we introduce the category of correspondences $CV(k)$. The culmination of our study of it is in the simple but important "principle of identities", which is the basis of the majority of the subsequent computations.

In the fourth section we study the phenomena connected with the existence of a grading on the ring $C(X)$.

The fifth section presents Grothendieck's construction[2)] of the *pseudo-abelian completion* of an additive category. It is used to pass from the category of correspondences to motifs.

This completes the introductory part of the paper; the rest of it is devoted to the computation of the structure of some motifs. In order to state some of the results, we remark that the category of motifs possesses the operation of *tensor product*, which is induced from a direct product in $V(k)$.

---

1) *Translator's note.* In fact, the author gives a Russian version of the French translation of this quotation. I have given the original English version.

2) Grothendieck has informed me that this construction was first introduced by Karoubi in the context of Banach categories.

In the sixth section we introduce the fundamental Tate motif L.[1] It is defined by the expansion of a motif of a projective space:

$$h(\mathbf{P}^r) = 1 \oplus \mathbf{L} \oplus \dots \oplus \mathbf{L}^r,$$

where $\mathbf{L}^i = \mathbf{L} \otimes \cdots \otimes \mathbf{L}$ (i factors). (The name of L is related to the fact that the group $H^2_{et}(\mathbf{L}, \mathbf{Z}_l)$ is dual to the Tate module $T_l(G_m) = \varprojlim \mu_{l^n}$.) The operation of "twisting" with the help of the motif L, introduced in §8, plays the same large role in the theory of motifs as twisting with the help of the Tate module $T_l(G_m)$ does in the theory of étale cohomology.

In § 7 we calculate the motif of a fiber space $h(\mathbf{P}(\mathcal{E}))$, where $\mathcal{E}$ is a locally free sheaf of rank r on the variety X:

$$h(\mathbf{P}(\mathcal{E})) = h(X) \otimes h(\mathbf{P}^r) = \bigoplus_{i=0}^{r} h(X) \otimes \mathbf{L}^i.$$

The ninth section contains the main result of the paper. Let $X' \to X$ be a monoidal transformation of X with center in a subvariety Y of codimension r. Then

$$h(X') = h(X) \oplus (\bigoplus_{i=1}^{r-1} h(Y) \otimes \mathbf{L}^i).$$

This is the motivic variant of the identities for the number of points (4). (Of course, the equalities of (4) follow formally from this, the same as does a more general Lefschetz formula, relating not only to the Frobenius morphism.) And what is more, we compute the "multiplication morphism" $h(X') \otimes h(X') \to h(X')$ induced by the diagonal $X' \to X' \times X'$. This permits us to describe the cohomology ring $H^*(X')$ for any geometric cohomology theory H (at least, if we disregard torsion). However, as was stated, in this paper we do not take up cohomology.

In §10 we obtain a subsidiary result on *motifs of curves* X. Such a motif has the form $h(X) = 1 \oplus X^+ \oplus \mathbf{L}$, and the category consisting of "one-dimensional motivic cohomology of curves" X^+ is equivalent to the category of Jacobian varieties. In essence, this assertion is derived from the work of the Italian algebraic geometers and from Abel. Grothendieck has a generalization of this result, identifying the category of "one-dimensional motifs" with the category of abelian varieties.

The stated assertion about the category of motifs of curves is then used in §11 to prove the motivic variant of formula (1). As in article [1], to a three-dimensional unirational variety X we associate an abelian variety A_X (up to isogeny). This correspondence is a functor; in particular, the representation of the Frobenius morphism on $H^3(X)$ is computed via its characteristic roots on A_X. We remark that in the case of a cubic hypersurface X, Bombieri and Swinnerton-Dyer attach a beautiful geometric interpretation to A_X. It is found that A_X is the Albanese variety of the "space of lines" on X. This result should undoubtedly be valid over an arbitrary field. A more precise analysis of the morphisms g and h in diagram (3) for the case of a cubic hypersurface X

1) Grothendieck has indicated to me that it would be more reasonable to call Tate's motif "the motif dual to L". The convention adopted here is harmless except that in the "twisting" of §8 we get the signs opposite to those in "twisting" by Tate's module of G_m in étale cohomology.

probably permits the establishment of this assertion.

In conclusion we pose a question connected with the as yet unsolved problem of the existence of a unirational three-dimensional variety which is not rational, a question which seems interesting in its own right.

Let k be an algebraically closed field. We let $\mathfrak{AB}$ denote the category of abelian varieties over k, and by $\mathfrak{AB}_Q$ the category of abelian varieties up to isogeny. We construct the Grothendieck groups $K(\mathfrak{AB})$ and $K(\mathfrak{AB}_Q)$ and denote by K' and K'' the factor groups of $K(\mathfrak{AB})$ and $K(\mathfrak{AB}_Q)$ modulo the subgroups generated by the classes of Jacobian varieties.

Question. *Are the groups K' and K'' trivial?*

Some birational invariants lie in the group K''.

We shall call the three-dimensional (nonsingular projective) varieties X, Y *weakly birationally equivalent* if there exists a sequence of varieties $X_0 = X, X_1, \cdots, X_{2n} = Y$ and birational morphisms f_i, each of which splits into the product of monoidal transformations with nonsingular centers:

$$\begin{array}{ccccccccc} & X_1 & & X_3 & & X_{2n-1} & \\ {}^{f_1}\swarrow & & \searrow^{f_2}\ {}^{f_3}\swarrow & & \searrow^{f_4}\ \swarrow & & \searrow^{f_{2n}} \\ X = X_0 & & X_2 & & \cdots & & X_{2n} = Y. \end{array}$$

Into correspondence with the unirational variety X we put an element of the group K'', which is the class of the abelian variety A_X described above. It follows from our results that this element depends only the the weak birational equivalence class of X.

I learned about the theory of motifs from lectures which Grothendieck gave at the Insitut des Hautes Études Scientifiques (Paris) in the spring of 1967. The calculation of the Chow ring of a monoidal transformation was set forth in his seminar on the Riemann-Roch Theorem, of which I was an auditor. This paper is no more than the fruit of assiduous meditation on these beautiful ideas.

§ 1. Global intersection theory

We shall consider the category $V(k)$ of smooth projective varieties defined over a field k.

Throughout this paper we shall fix a "global intersection theory" on $V(k)$ in Grothendieck's sense [3]. By definition, it consists of the following data:

a) A contravariant functor $X \to C(X)$ on $V(k)$ with values in the category of commutative Λ-algebras, where Λ is a fixed ring.

For any morphism $\phi: X \to Y$ the corresponding homomorphism of Λ-algebras is denoted by the symbol $\phi^*: C(Y) \to C(X)$. The identity of the ring is denoted by the element $1_X \in C(X)$.

b) The construction of a covariant functor on C, preserving only the Λ-module structure.

The homomorphism of Λ-modules $C(X) \to C(Y)$ corresponding to the morphism $\phi: X \to Y$ is denoted by ϕ_*.

c) For any two varieties $X, Y \in V(k)$ we are given a Λ-algebra homomorphism $C(X) \otimes_\Lambda C(Y) \to C(X \times Y)$.

The image of the element $x \otimes y$ *under this mapping will be denoted by* $x \times y$.

d) For any irreducible variety X an augmentation homomorphism $C(X) \to \Lambda$ is given, which is an isomorphism for $X = \mathrm{Spec}\, k$.

$\mathrm{Spec}\, k$ is a final object of the category $V(k)$ and will often be denoted by the letter e.

The data described should satisfy certain axioms, the most important of which are the following ones.

Multiplication Axiom. *Let* $X \in V(k)$, *and let* $\delta_X\colon X \to X \times X$ *be the diagonal morphism. Then the composition homomorphism of* Λ*-algebras*

$$C(X) \underset{\Lambda}{\otimes} C(X) \to C(X \times X) \xrightarrow{\delta_X^*} C(X)$$

coincides with the homomorphism of multiplication: $\delta_X^*(x \times y) = xy$.

Projection Formula. *Let* $\phi\colon X \to Y$ *be a morphism, and take* $x \in C(X)$, $y \in C(Y)$. *Then*

$$\varphi_*(x\varphi^*(y)) = \varphi_*(x)\, y.$$

For a complete list of the remaining axioms and their elementary consequences, which we shall tacitly use, we refer the reader to Gorthendieck's note [3]. The following functors satisfy all the necessary requirements, to which we shall apply the formal part of the theory of correspondences:

1. $C(X) = K(X)$, $\Lambda = \mathbf{Z}$ (see the report of Borel and Serre [2] and Grothendieck's seminar [8]).

2. $C(X)$ is the ring of classes of cycles on X under an adequate equivalence relation (see [7]). (In particular, the Chow ring of classes of cycles modulo rational equivalence $A(X)$.) We shall also consider theories obtained by a change of the ground ring.

3. $C(X) = H^{ev}(X)$ is the ring of cohomology classes of even dimension with coefficients in a $\mathbf{Q}$-algebra Λ for $k = \mathbf{C}$.

In the abstract case we have to consider étale cohomology

$$H^{ev}(X) = \sum_i H^{2i}_{et}(X, Q_l)(i),$$

twisting properly (see [9]).

4. $C(X)$ is the ring of even-dimensional cohomology classes of type (p, p) in the sense of Hodge (over $\mathbf{C}$) or the ring of invariant classes in $\bigoplus H^{2p}_{et}(X)(p)$ relative to the action of the Galois group $\mathrm{Gal}(\bar{k}/k)$ for any field k (cf. [9]).

§2. C-correspondences

Definition. Any element of the ring $C(X \times Y)$ is called a *C-correspondence* between the varieties $X, Y \in V(k)$.

Let $f \in C(X \times Y)$, $g \in C(Y \times Z)$. The correspondence

$$g \circ f = p_{13*}(p_{12}^*(f)\, p_{23}^*(g)) \in C(X \times Z).$$

is called the *composition of the correspondences* f and g.

(Here and in what follows notation of the type p_{13}: $X \times Y \times Z \to X \times Z$ means the projection morphism onto the product of the first and third factors.)

We shall call C-correspondences simply correspondences if the choice of the global intersection theory is clear from the context.

Lemma. a) *Let* $\Delta_X = \delta_{X*}(1_X) \in C(X \times X)$ *be the "diagonal class". Then for any correspondences* $f \in C(X \times Y)$, $g \in C(Y \times X)$ *we have*

$$f \circ \Delta_X = f, \quad \Delta_X \circ g = g.$$

b) *Composition of correspondences is associative.*

Proof. Leaving the first assertion as an excerise to the reader, we give a proof of associativity. Let X_i $(i = 1, \cdots, 4)$ be varieties and $f_{i,i+1} \in C(X_i \times X_{i+1})$.

In the formulas below, p_{13} is the projection with $X_1 \times X_2 \times X_3$; p_{14} the projection with $X_1 \times X_3 \times X_4$; p'_{14} the projection with $X_1 \times X_2 \times X_3 \times X_4$. Several times we shall use identities of the type $p^*(x) = x \times 1_Y$, where p: $X \times Y \to X$ is the projection.

By definition we have

$$\begin{aligned} f_{34}\circ(f_{23}\circ f_{12}) &= p_{14*}\{(p_{13*}[(f_{12} \times 1_{X_3})(1_{X_1} \times f_{23})] \times 1_{X_4})(1_{X_1} \times f_{34})\} \\ &= p_{14*}(p_{13} \times \mathrm{id}_{X_4})_*\{[(f_{12} \times 1_{X_3})(1_{X_1} \times f_{23}) \times 1_{X_4}](p_{13} \times \mathrm{id}_{X_4})^*(1_{X_1} \times f_{34})\} \\ &= p'_{14*}[(f_{12} \times 1_{X_3} \times 1_{X_4})(1_{X_1} \times f_{23} \times 1_{X_4})(1_{X_1} \times 1_{X_2} \times f_{34})]. \end{aligned}$$

(In passing from the first line to the second we first replaced $p_{13*}[\cdots] \times 1_{X_4}$ by $(p_{13} \times \mathrm{id}_{X_4})_*([\cdots] \times 1_{X_4})$, and then we applied the projection formula to the morphism $p_{13} \times \mathrm{id}_{X_4}$.)

Applying analogous transformations to the composition $(f_{34} \circ f_{23}) \circ f_{12}$, we can reduce it to precisely the same form.

The lemma is proved. It permits us to make the following

Definition. The category of C-correspondences $CV(k)$ is defined by the following data.

a) $\mathcal{O}\mathcal{B}\, CV(k) = \mathcal{O}\mathcal{B}\, V(k)$.

(The variety X as an object of the category of C-correspondences will be denoted by the symbol $\bar{X}$.)

b) For any two objects $\bar{X}, \bar{Y}$ in $CV(k)$,

$$\mathrm{Hom}(\bar{X}, \bar{Y}) = C(X \times Y).$$

c) Composition of morphisms in $CV(k)$ is composition of correspondences.

Every morphism of the category $V(k)$ defines a morphism in the opposite direction in the category of correspondences.

In fact, let ϕ: $Y \to X$ be a morphism of varieties. Its *graph* will be the morphism

$$\Gamma_\phi = (\phi \times \mathrm{id}_Y) \circ \delta_Y : Y \xrightarrow{\delta_Y} Y \times Y \xrightarrow{\phi \times \mathrm{id}_Y} X \times Y.$$

Definition. $c(\phi) = \Gamma_{\phi *}(1_Y) \in \mathrm{Hom}(\bar{X}, \bar{Y}) = C(X \times Y)$.

Proposition. *The collection of mappings*

$$X \rightsquigarrow \bar{X}, \quad \varphi \rightsquigarrow c(\varphi)$$

is a contravariant functor on $V(k)$ with values in $CV(k)$.

Proof. We restrict ourselves to verifying that for a sequence of morphisms $Z \xrightarrow{\phi} Y \xrightarrow{\psi} X$ in $V(k)$ we have $c(\psi\phi) = c(\phi) \circ c(\psi)$. The remaining properties are easily checked.

First of all, we have a commutative diagram

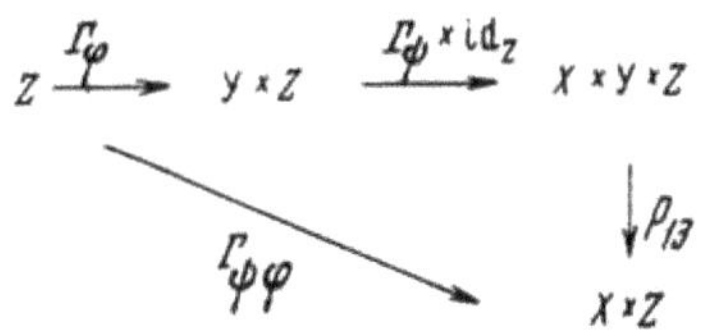

From this it follows that

$$c(\psi\varphi) = (\Gamma_{\psi\varphi})_*(1_Z) = p_{13*}((\Gamma_\psi \times \mathrm{id}_Z)_* \Gamma_{\varphi *}(1_Z)).$$

On the other hand, by definition

$$c(\varphi) \circ c(\psi) = p_{13*}((1_X \times \Gamma_{\varphi *}(1_Z))(\Gamma_{\psi *}(1_Y) \times 1_Z)).$$

Therefore it suffices to prove that the arguments of p_{13*} on the right sides of these formulas are the same.

We have

$$(1_X \times \Gamma_{\varphi *}(1_Z))(\Gamma_{\psi *}(1_Y) \times 1_Z) = (1_X \times \Gamma_{\varphi *}(1_Z))(\Gamma_\psi \times \mathrm{id}_Z)_*(1_{Y\times Z})$$

(projection formula)

$$= (\Gamma_\psi \times \mathrm{id}_Z)_*[(\Gamma_\psi \times \mathrm{id}_Z)^*(1_X \times \Gamma_{\varphi *}(1_Z))] = (\Gamma_\psi \times \mathrm{id}_Z)_*[(\Gamma_\psi \times \mathrm{id}_Z)^* p_{23}^* \Gamma_{\varphi *}(1_Z)]$$

$$= (\Gamma_\psi \times \mathrm{id}_Z)_* \Gamma_{\varphi *}(1_Z),$$

because $p_{23}(\Gamma_\psi \times \mathrm{id}_Z) = \mathrm{id}_{Y\times Z}$.

The proof is complete.

The meaning of the result just proved is that we "converted" $V(k)$ into an additive category. More precisely, the category $CV(k)$ is Λ-additive: all the groups of morphisms $\mathrm{Hom}(\bar{X}, \bar{Y})$ are Λ-modules and composition of morphisms is Λ-linear with respect to all the arguments.

For any correspondence $f \in \mathrm{Hom}(\bar{X}, \bar{Y})$ we denote by $f^t \in \mathrm{Hom}(\bar{Y}, \bar{X})$ the image of f under the homomorphism of "transposing the factors" $C(X \times Y) \to C(Y \times X)$. We have $(f \circ g)^t = g^t \circ f^t$, so that the category $CV(k)$ is equivalent to the dual category.

The sum of varieties in $CV(k)$ corresponds to their direct sum $\overline{X \sqcup Y} = \bar{X} \oplus \bar{Y}$).

The operation in $CV(k)$ corresponding to the direct product in $V(k)$ is naturally called the *tensor product.*

Definition. a) $\bar{X} \otimes \bar{Y} = \overline{X \times Y}$.

b) Let $f_1 \in \mathrm{Hom}(\bar{X}_1, \bar{Y}_1)$, $f_2 \in \mathrm{Hom}(\bar{X}_2, \bar{Y}_2)$. We set

$$f_1 \otimes f_2 = s_{23*}(p_{13}^*(f_1)\, p_{34}^*(f_2)) \in \mathrm{Hom}(\overline{X}_1 \otimes \overline{X}_2, \overline{Y}_1 \otimes \overline{Y}_2) = C(X_1 \times X_2 \times Y_1 \times Y_2),$$

where s_{23} is the morphism which switches the second and third factors:

$$s_{23}: X_1 \times Y_1 \times X_2 \times Y_2 \to X_1 \times X_2 \times Y_1 \times Y_2.$$

Lemma. $(f_1 \otimes f_2) \circ (g_1 \otimes g_2) = (f_1 \circ g_1) \otimes (f_2 \circ g_2)$.

The proof is left to the reader as an exercise.

§3. Functorial properties of correspondences

We put $\bar{X}(\bar{T}) = \mathrm{Hom}(\bar{T}, \bar{X})$ in the category $CV(k)$.

For any correspondence $f \in \mathrm{Hom}(\bar{Y}, \bar{X})$ and any object $\bar{T}$ we define the natural homomorphism of modules of "$\bar{T}$-points" f_T: $\bar{Y}(\bar{T}) \to \bar{X}(\bar{T})$ by $f_T(g) = f \circ g$ for each element $g \in \bar{Y}(\bar{T}) = C(T \times Y)$.

Special case. We put $T = e = \mathrm{Spec}\, k$ and $g \in \bar{Y}(\bar{T}) = C(e \times Y) = C(Y)$. Then $f_e(g) = p_{2*}(f p_1^*(g))$, where p_1, p_2 are projections of $Y \times X$. In particular, the mappings $\bar{Y} \rightsquigarrow C(Y)$, $f \rightsquigarrow f_e$: $C(Y) \to C(X)$ permit us to consider C as a functor on the category $CV(k)$ (and not only on $V(k)$).

Now we remark that we can construct another mapping between the same target and source as f_T. Namely, every correspondence $f \in \mathrm{Hom}(\bar{Y}, \bar{X})$ defines a correspondence

$$\mathrm{id}_T \otimes f \in \mathrm{Hom}(\overline{T} \otimes \overline{Y}, \overline{T} \otimes \overline{X})$$

and a homomorphism of Λ-modules

$$(\mathrm{id}_T \otimes f)_e : (\overline{T} \otimes \overline{Y})(e) \to (\overline{T} \otimes \overline{X})(e),$$

where $(\bar{T} \otimes \bar{Y})(e) = C(e \times T \times Y) = C(T \times Y) = \bar{Y}(\bar{T})$ and analogously $(\bar{T} \otimes \bar{X})(e) = \bar{X}(\bar{T})$.

In fact the constructed mappings coincide.

Lemma. $f_T = (\mathrm{id}_T \otimes f)_e$.

Proof. Let $g \in \bar{Y}(\bar{T})$. By definition,

$$f_T(g) = f \circ g = p_{13*}(p_{12}^*(g)\, p_{23}^*(f)), \quad (\mathrm{id}_T \otimes f)_e(g) = p_{34*}[(\mathrm{id}_T \otimes f)\, p_{12}(g)].$$

All the morphisms here and below are contained in the diagram

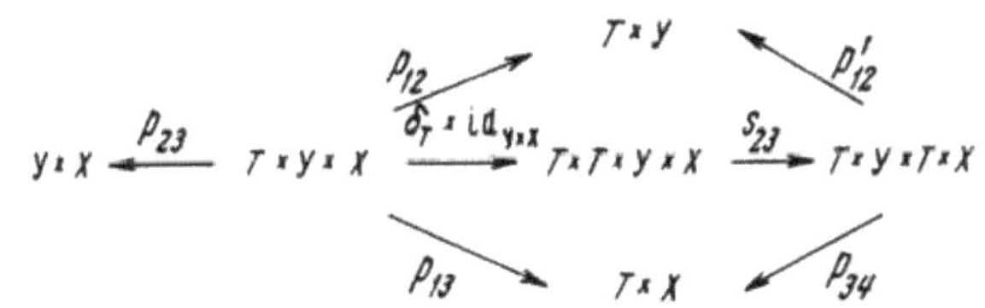

Using the obvious commutativity relations

$$p_{34}s_{23}(\delta_T\times \mathrm{id}_{Y\times X}) = p_{13},\quad p'_{12}s_{23}(\delta_T\times \mathrm{id}_{Y\times X}) = p_{12}$$

and the projection formula, we find that

$$\begin{aligned} p_{34*}[(\mathrm{id}_T\otimes f)\,p'^{*}_{12}(g)] &= p_{34*}[s_{23*}(\delta_T\times \mathrm{id}_{Y\times X})_*(1_X\times f)\,p'^{*}_{12}(g)] \\ &= p_{34*}s_{23*}(\delta_T\times \mathrm{id}_{Y\times X})_*[(1_X\times f)(\delta_T\times \mathrm{id}_{Y\times X})^* s_{23}^* p'^{*}_{12}(g)] \\ &= p_{13*}[(1_X\times f)\,p^*_{12}(g)] = p_{13*}[p^*_{12}(g)\,p^*_{23}(f)], \end{aligned}$$

which proves what we required.

The lemma just proved will be used to establish some identities among correspondences.

We will say that *the homomorphism* h*:* $C(X)\to C(Y)$ *of* Λ*-modules is represented by the correspondence* $f\in \mathrm{Hom}(\bar{X},\bar{Y})$ if $h = f_e$.

Obviously, if h_1 and h_2 are represented by the correspondences f_1 and f_2, then $h_1\circ h_2$ is represented by the correspondence $f_1\circ f_2$. We shall show that for any morphism of varieties ϕ: $X\to Y$ the homomorphisms ϕ^* and ϕ_* are represented by correspondences. The same one is related to the homomorphism $C(X)\to C(X)$ of multiplication by any element of the ring $C(X)$.

Lemma. *Let* ϕ*:* $X\to Y$ *be a morphism of varieties,* $x\in C(X)$*. Then the homomorphism* ϕ^* *is represented by the correspondence* $c(\phi)$*, the homomorphism* ϕ_* *is represented by the correspondence* $c(\phi)^t$*, and the homomorphism of multiplication by* x *is represented by the correspondence* $\iota_x = \delta_{X*}(x)\in C(X\times X)$.

Proof. Using the identities $p_1\Gamma\phi = \phi$ and $p_2\Gamma\phi = \mathrm{id}_Y$ ($X\times Y$ is the source of the projections), for all $y\in C(Y)$ we find that

$$c(\varphi)_e(y) = p_{2*}(\Gamma_{\varphi*}(1_X)\,p_1^*(y)) = p_{2*}\Gamma_{\varphi*}(1_X\Gamma_\varphi^* p_1^*(y)) = \varphi^*(y).$$

Analogously, for all $x\in C(X)$

$$c(\varphi)^t_e(x) = p_{2*}(\Gamma_{\varphi*}(1_X)^t p_1^*(x)) = p_{1*}(\Gamma_{\varphi*}(1_X)\,p_2^*(x)) = p_{1*}\Gamma_{\varphi*}(1_X\Gamma_\varphi^* p_2^*(x)) = \varphi_*(x).$$

Finally, for $x'\in C(X)$

$$c(x)_e(x') = p_{2*}(\delta_{X*}(x)\,p_1^*(x')) = p_{2*}\delta_{X*}(x\delta_X^* p_1^*(x')) = xx',$$

because $p_2\delta_X = p_1\delta_X = \mathrm{id}_X$. This proves the lemma.

An immediate application of the definitions shows that in the previous notation we have the

Lemma. *For any morphism ϕ: $X \to Y$ and variety T*

$$\mathrm{id}_T \otimes c(\varphi) = c(\mathrm{id}_T \times \varphi), \quad \mathrm{id}_T \otimes c(\varphi)^t = c(\mathrm{id}_T \times \varphi)^t, \quad \mathrm{id}_T \otimes c_x = c_{1_T \times x},$$

where $x \in C(X)$.

Collecting all our results, we obtain the fundamental

Corollary. *For any morphism ϕ: $X \to Y$ and variety T we have*

$$c(\varphi)_T = (\mathrm{id}_T \times \varphi)^* : C(T \times Y) \to C(T \times X); \quad c(\varphi)^t_T = (\mathrm{id}_T \times \varphi)_* : C(T \times X) \to C(T \times Y),$$

$$(c_x)_T = \textit{multiplication by } 1_T \times x : C(T \times X) \to C(T \times X) \quad \textit{for all} \quad x \in C(X).$$

We can now state a general principle for obtaining identities between correspondences.

Suppose that a diagram D of objects and morphisms of the category $V(k)$ is given. Furthermore, let an identity of the form I: $\sum_{i=1}^{r} a_i F_i = 0$ hold, where $a_i \in \Lambda$, and the F_i are homomorphisms of the Λ-module $C(X)$, $X \in \mathcal{O}\mathcal{B}D$, such that every homomorphism F_i is the composition of a finite number of homomorphisms of the form ϕ^*, ϕ_* ($\phi \in \mathrm{Mor}\, D$) and of multiplications by fixed elements of the ring $C(X)$, $X \in \mathcal{O}\mathcal{B}D$.

For any variety T we denote by I_T the identity obtained from I by replacing all the objects X by $T \times X$ in I, all the morphisms ϕ by $\mathrm{id}_T \times \phi$ and all the multiplications by x by multiplications by the element $1_T \times x$.

We shall call an identity I *universal* if all I_T are valid identities.

Identity Principle. *In the preceding notation the following two assertions are equivalent.*

a) *I is a universal identity.*

b) *If in I the homomorphisms ϕ^*, ϕ_* and "multiplication by x" are respectively replaced by $c(\phi)$, $c(\phi)^t$ and c_x, then a valid identity between correspondences is obtained.*

The proof is immediately obtained from the preceding corollary: one must take into account that an identity holds between correspondences if and only if it holds between the modules of $\overline{T}$-points for all T.

We illustrate this principle by some examples.

Example 1. Let C be the Chow theory ($C(X)$ = the ring of classes of cycles on X up to rational equivalence). We consider a diagram D: ϕ: $X \to Y$, where ϕ is a morphism having finite degree d. Then, by the definition of ϕ_*, we have $\phi_*(1_X) = d1_Y$, from which by the projection formula we find an identity I: $\phi_*\phi^* = d \cdot \mathrm{id}_Y$, because $\phi_*\phi^*(y) = \phi_*(1_X\phi^*(y)) = \phi_*(1_X)y = dy$. The identity I is universal, since for all T the morphism $\mathrm{id}_T \times \phi$: $T \times X \to T \times Y$ has degree d. Therefore the identity $c(\phi)^t \circ c(\phi) = d \cdot \mathrm{id}_Y$ between correspondences holds.

We replace the ground ring Z by a ring in which d is invertible, and we put $p = (1/d)c(\phi) \circ c(\phi)^t \in \mathrm{Hom}(\overline{X}, \overline{X})$. Then

$$p^2 = \frac{1}{d^2} c(\varphi) \circ (c(\varphi)^t \circ c(\varphi)) \circ c(\varphi)^t = p,$$

i.e. p is a projector. $CV(k)$ is a nonabelian category, and, generally speaking, p does not have an image in it. However, when we define the category of motifs below by formally adding the images and the kernels of such projectors, we see that the image of p is isomorphic to $\bar{Y}$, so that $\bar{Y}$ is an "isolated direct summand" on $\bar{X}$. This is a parody of Poincaré's "complete reducibility theorem".

Example 2. We consider a diagram D:

$$\begin{array}{ccc} Y' & \xrightarrow{\varepsilon} & X' \\ \psi\downarrow & & \downarrow\varphi, \\ Y & \xrightarrow{\iota} & X \end{array}$$

in which ι is a closed embedding of a variety Y of codimension r, ϕ is a monoidal transformation with center at Y, $Y' = \phi^{-1}(Y)$, ψ is the natural projection of Y' onto Y, and ϵ the canonical embedding of Y' into X'. Grothendieck considers such a diagram in the proof of the Riemann-Roch Theorem (see [2], [8]) and proves some identities, only two of which we consider as examples (the others are examined analogously):

$$\forall y \in C(Y): \iota^*\iota_*(y) = yn, \quad n = \begin{cases} \lambda_{-1} N, & \text{if } C \text{ is } K\text{-theory}, \\ c_r(N), & \text{if } C \text{ is the Chow theory}/\mathbf{Q}, \end{cases}$$

$$\forall y \in C(Y): \varphi^*\iota_*(y) = \varepsilon_*(\psi^*(y)\,\Phi), \quad \Phi = \begin{cases} \lambda_{-1} F, & \text{if } C \text{ is } K\text{-theory}, \\ c_{r-1}(F), & \text{if } C \text{ is the Chow theory}/\mathbf{Q}. \end{cases}$$

(Here N is the conormal bundle of Y in X, $c_r(N)$ its rth Chern class, F a sheaf on Y' which is the kernel of the canonical epimorphism $\psi^*(N) \to \mathcal{O}_{Y'}(1) \to 0$, and $c_{r-1}(F)$ its $(r-1)$th Chern class.)

Both these identities are universal, since the diagram which is obtained from D by multiplying all the objects by T and all the morphisms by id_T is again a diagram of the same type, associated with a monoidal transformation of the variety $T \times X$ with center at $T \times Y$. Moreover, the elements n and Φ corresponding to such a monoidal transformation have the form $1_T \times n$ and $1_T \times \Phi$. From this we find two identities between correspondences:

$$c(\iota) \circ c(\iota)^t = c_n, \quad c(\varphi) \circ c(\iota)^t = c(\varepsilon)^t \circ c_\Phi \circ c(\psi).$$

As we mentioned in the Introduction, every "geometric" cohomology theory on $V(k)$ can also be considered as a functor on the category of correspondences. From this it follows that Grothendieck's identities also hold in any such cohomology theory (modulo torsion). This gives an answer to a question of Grothendieck, posed in [8] (Exposé XIV, pp. 10–11).

§4. Graded correspondences

In this section we assume that the intersection theory C under consideration is a functor with values in the category of commutative Λ-algebras, *graded by degrees* ≥ 0 such that $C^i(X) = 0$ for $i > \dim X$ (examples 2–4 of §4 will satisfy all the requirements, but K-theory won't). We impose the following conditions on the grading (cf. [3]).

e) The homomorphisms ϕ^* are homogeneous of degree zero.

f) If X and Y are equidimensional varieties of dimensions n and m respectively, then for any morphism ϕ: $X \to Y$ the homomorphism of graded groups ϕ_*: $C(X) \to C(Y)$ is homogeneous of degree $m - n$.

g) $C(X) \underset{\Lambda}{\otimes} C(Y) \to C(X \times Y)$ is a homogeneous homorphism of degree zero.

h) For any irreducible variety X the augmentation homomorphism ϵ: $C(X) \to \Lambda$ takes all the $C^i(X)$, $i \geq 1$, into zero and induces an isomorphism of $C^0(X)$ with Λ.

We shall show that the groups of morphisms in the category of C-correspondences also have a natural grading.

For any $X, Y \in V(k)$, X equidimensional of dimension n, we put

$$\mathrm{Hom}^i(\overline{X}, \overline{Y}) = C^{i+n}(X \times Y).$$

We shall call the elements of $\mathrm{Hom}^i(\bar{X}, \bar{Y})$ *homogeneous correspondences of degree* i. (In the general case we split X into the union of equidimensional components and define the degree relative to each component. We shall usually assume tacitly that X and Y are equidimensional or even irreducible.) We note that the degree i of a correspondence can be negative.

Lemma. *The degree of a composition of homogeneous correspondences equals the sum of the degrees of these correspondences.*

Proof. Let $\dim X = n$, $\dim Y = m$, $f \in C^{i+n}(X \times Y)$ and $g \in C^{j+m}(Y \times Z)$. Then by property e) we have $p_{12}^*(f) p_{23}^*(g) \in C^{i+j+n+m}(X \times Y \times Z)$, and by property f), $g \circ f \in C^{i+j+n}(X \times Z)$, which proves the lemma.

Examples. 1) Let ϕ: $X \to Y$ be a morphism; the assumptions on X and Y are the same as above. Then the degree of $c(\phi)$ equals zero and the degree of $c(\phi)^t$ equals $n - m$.

2) Let $x \in C^i(X)$. Then the degree of c_x equals i.

Since the identity correspondence has degree zero, we can consider a new category $CV^0(k)$, whose objects are the varieties of $V(k)$ and whose morphisms are correspondences of degree zero.

Like $CV(k)$, the category $CV^0(k)$ is not abelian. For example, as we shall see later, there exist many morphisms-projectors, which do not correspond to any decomposition of an object into a direct sum. To Grothendieck we owe the idea of "formally" adding the kernels and images of all projectors. The meaning of this partial "abelianization" of the category of C-correspondences consists in the fact that we detect a series of interesting direct decompositions of "varieties", immediately transferring the cohomology groups to them.

Grothendieck's construction is studied in the following section. Since it is completely formal and relative to an arbitrary additive category, we give only sketches of the proofs.

§5. Pseudo-abelian categories and motifs

Definition. An additive category $\mathfrak{D}$ is called *pseudo-abelian* if the following condition holds in it:

(P) For any projector $p \in \mathrm{Hom}(X, X)$, $X \in \mathfrak{OB}\mathfrak{D}$, there exists a kernel $\mathrm{Ker}\, p$, and the canonical homomorphism $\mathrm{Ker}\, p \oplus \mathrm{Ker}(\mathrm{id}_X - p) \to X$ is an isomorphism.

Definition. Let $\mathfrak{D}$ be an additive category. Its *pseudo-abelian completion* is the category $\tilde{\mathfrak{D}}$, defined by the following data: $\mathcal{OB}\tilde{\mathfrak{D}}$ are pairs (X, p), where $x \in \mathcal{OB}\mathfrak{D}$ and $p \in \mathrm{Hom}_{\mathfrak{D}}(X, X)$ is an arbitrary projector.

$$\mathrm{Hom}_{\tilde{\mathfrak{D}}}((X, p), (Y, q)) = \left\{ \begin{matrix} \text{the group of } f \in \mathrm{Hom}_{\mathfrak{D}}(X, Y) \\ \text{for which } fp = qf \end{matrix} \right\} \Big/ \left\{ \begin{matrix} \text{the subgroup of } f \\ \text{for which } fp = qf = 0 \end{matrix} \right\}.$$

Composition of morphisms in $\tilde{\mathfrak{D}}$ is induced by composition in $\mathfrak{D}$. We shall let $\tilde{f}$ denote the class of the morphism f in $\mathrm{Hom}((X, p), (Y, q))$.

The name of $\tilde{\mathfrak{D}}$ is justified by the following formal lemma.

Lemma. a) *The category* $\tilde{\mathfrak{D}}$ *is pseudo-abelian.*

b) *The mapping* $X \rightsquigarrow \tilde{X} = (X, \mathrm{id})$, $f \rightsquigarrow \tilde{f}$ *can be extended uniquely to a functor* $G: \mathfrak{D} \to \tilde{\mathfrak{D}}$, *which is completely faithful and possesses the following universal property:*

For every additive functor $F: \mathfrak{D} \to \mathfrak{E}$, *where* $\mathfrak{E}$ *is a pseudo-abelian category, there exists an additive functor* $\tilde{F}: \tilde{\mathfrak{D}} \to \mathfrak{E}$ *such that the functors* F *and* $\tilde{F}G$ *are equivalent.*

The proof is omitted.

We need two simple statements about projectors in additive categories.

Lemma. *Let* $X, Y \in \mathfrak{D}$ *and let* $Y \underset{a'}{\overset{a}{\rightleftarrows}} X$ *be two morphisms such that* $a'a = \mathrm{id}_Y$. *Then the morphism* $aa': X \to X$ *is a projector and the morphisms* a, a' *induce mutually reciprocal isomorphisms* $\tilde{Y} \underset{a'}{\overset{a}{\rightleftarrows}} (X, aa')$ *in the pseudo-abelian completion* $\tilde{\mathfrak{D}}$.

The proof reduces to a formal verification.

Proposition. *In the additive category* $\mathfrak{D}$ *let the sequence*

$$Y \xrightarrow{a} X \xrightarrow{b} Z \tag{6}$$

be such that the morphism a *has a right inverse* a' *and for all* $T \in \mathcal{OB}\mathfrak{D}$ *the corresponding sequence of groups of* T*-points*

$$0 \to Y(T) \xrightarrow{a_T} X(T) \xrightarrow{b_T} Z(T) \to 0 \tag{7}$$

is exact. Then the sequence (6) *is exact in the pseudo-abelian completion* $\tilde{\mathfrak{D}}$ *and splits, i.e. it is isomorphic to an exact sequence of the form*

$$\tilde{Y} \xrightarrow{(\mathrm{id}_Y, 0)} \tilde{Y} \oplus \tilde{Z} \xrightarrow{p_1} \tilde{Z}.$$

Sketch of the Proof. By the lemma, $\tilde{X} = (X, aa') \oplus (X, 1 - aa')$, and moreover, $(X, aa') \simeq \tilde{Y}$. Therefore it suffices to verify that the morphism b induces an isomorphism $\tilde{b}: (X, 1 - aa') \xrightarrow{\sim} Z$.

First, a simple check shows that b in fact induces a morphism $\tilde{b}$, because $ba = 0$.

To construct an inverse morphism to $\tilde{b}$ we put $T = Z$ in (7) and let $c' \in X(Z)$ denote some preimage of the element $\mathrm{id}_Z \in Z(Z)$. Then we put $c = (1 - aa')c_*'$.

It is easy to check that $bc = \mathrm{id}_Z$ and $(1 - aa')c = c$.

From the last equality it follows that c induces a morphism $\tilde{c}$: $\tilde{Z} \to (X, 1 - aa')$.

The morphisms $\tilde{b}$ and $\tilde{c}$ are mutual inverses. Similarly, $\tilde{b}\tilde{c} = \mathrm{id}_{\tilde{Z}}$, because $bc = \mathrm{id}_Z$. It is a little less obvious that $\tilde{c}\tilde{b} = \mathrm{id}_{(X,1-aa')}$. This equality is equivalent to $(1 - aa')cb = 1 - aa'$ or $cb = 1 - aa'$, because $(1 - aa')c = c$. From the exactness of sequence (7) we find for $T = X$ that $1 - cb = ad$ for some $d \in Y(X)$. Multiplying on the left by a', we obtain $d = a'(1 - cb)$, and hence $1 - cb = aa'(1 - cb) = a'$, since $aa'c = 0$.

The proposition is proved.

We complete this section with two definitions relative to the category of correspondences.

Definition. As in the preceding section, let C denote a graded intersection theory. We call the pseudo-abelian completion of the category $CV^0(k)$ of correspondences of degree zero the *category of C-motifs*.

The objects of the category $\widetilde{CV}^0(k)$ will often be called simply *motifs*. In particular, the object $\tilde{X} = (X, \mathrm{id})$ will be called the *motif of the variety X*. Every C-motif has the form (X, p), where X is a k-variety, $p \in \mathrm{Hom}^0(\bar{X}, \bar{X})$, $p^2 = p$.

The operation of the tensor product carries over to motifs:

Definition. $(X, p) \otimes (Y, q) = (X \otimes Y, p \otimes q)$.

That this is the correct definition is left to the reader, as are the definition of $f_1 \otimes f_2$ for morphisms f_1, f_2 of motifs and the proofs of the properties of the tensor product, analogous to those stated at the end of §2.

§6. The Tate motif

Let C be a graded intersection theory.

The variety X will be called *C-special* if it is connected, its set of k-points is not empty, and for any morphism ϕ: $e = \mathrm{Spec}\, k \to X$ the image $\phi_*(e) \in C(X)$ is one and the same (independent of ϕ).

Example. In Chow theory and in the theories of Examples 3–4 (§1) the varieties P_k^n and the various products of them are special. If the field k is algebraically closed, and $C(X)$ is the ring of classes of X relative to *numerical* equivalence, then all connected varieties are C-special.

Definition. Let X be a C-special variety of dimension n. We define the projectors p_0^X, $p_n^X \in \mathrm{Hom}^0(\bar{X}, \bar{X})$ by the formulas

$$p_0^X = e_X \times 1_X, \quad p_n^X = 1_X \times e_X.$$

Proposition. a) *The motif (X, p_0^X) is isomorphic to the motif e.*

b) *The motif (X, p_n^X) depends only on the dimension of the variety X. Moreover, in $\widetilde{CV}^0(k)$ we have*

$$\mathrm{Hom}\,((X, p_n^X), (Y, p_m^Y)) = \begin{cases} \Lambda & \text{for } n = m, \\ 0 & \text{for } n \neq m \end{cases}$$

(*equality means a canonical isomorphism*).

Proof. a) Let $\phi: e \to X$ be a k-point, and $\psi: X \to e$ the canonical morphism. Then $c(\phi)$ and $c(\psi)$ induce mutually inverse isomorphisms of motifs:

$$\widetilde{c(\varphi)}: (X, p_0^X) \to \tilde{e}\ ;\ \widetilde{c(\psi)}: \tilde{e} \to (X, p_0^X).$$

The verification reduces to a trivial use of the definitions.

b) We compute $\mathrm{Hom}((X, p_n^X), (Y, p_m^Y))$. Let an element of this set be induced by the morphism $f \in \mathrm{Hom}^0(\bar{X}, \bar{Y}) = C^n(X \times Y)$. Then $p_m^Y \circ f = f \circ p_n^X$. We have

$$p_m^Y \circ f = p_{13*}((f \times 1_Y) \circ (1_X \times 1_Y \times e_Y)) = p_{1*}(f) \times e_Y,$$
$$f \circ p_n^X = p_{13*}((1_X \times e_X \times 1_Y) \circ (1_X \times f)) = p_{13*}(1_X \times ((e_X \times 1_Y) \circ f)).$$

By properties f) and h) of intersection theory, $p_{1*}(f) \in C^{n-m}(X)$ and $(e_X \times 1_Y) \circ f \in C^{2n}(X \times Y)$. Hence for $m > n$ we have $p_m^Y \circ f = 0$ for all f, and for $m < n$ we have $f \circ p_n^X = 0$ for all f.

Therefore nontrivial morphisms between such motifs can only exist for $n = m$.

In this case $p_{1*} \in C^0(X)$. We put $\deg f = \epsilon(p_{1*}(f)) \in \Lambda$. Obviously the mapping $f \to \deg f$ effects an embedding $\mathrm{Hom}((X, p_n^X), (Y, p_n^Y)) \to \Lambda$; moreover, it is an epimorphism since $f = e_X \times 1_Y$ induces a morphism of motifs such that $\deg f = 1$. Finally, $f = e_X \times 1_Y$ and $g = 1_X \times e_Y$ define mutually inverse isomorphisms of (X, p_n^X) and (Y, p_n^Y).

The proposition is proved.

Definition. $\mathbf{L} = (\mathbf{P}^1, p_1^{\mathbf{P}^1})$ is called the *Tate* (*C*)*-motif*.

In order to represent the geometric meaning of the Tate motif better, we remark that, in all the intersection theories encountered, $p_1^{\mathbf{P}^1} + p_0^{\mathbf{P}^1} = \mathrm{id}_{\mathbf{P}^1}$, so that $\widetilde{\mathbf{P}}^1 = \tilde{e} \oplus \mathbf{L}$.

Lemma. *If* $\widetilde{\mathbf{P}}^1 = \tilde{e} \oplus \mathbf{L}$, *then* $(X, p_n^X) \simeq \mathbf{L}^n (= \mathbf{L}^{\otimes n})$ *for all n-dimensional C-special varieties X.*

Sketch of the Proof. We put $Y = \mathbf{P}^1 \times \cdots \times \mathbf{P}^1$ (n factors). Then $(X, p_n^X) \simeq (Y, p_n^Y)$. On the other hand, $\widetilde{Y} = (\widetilde{\mathbf{P}}^1)^{\otimes n} = \bigoplus_{i=0}^n \binom{n}{i} \mathbf{L}^{\otimes i}$ (where we naturally assume $\mathbf{L}^0 = \tilde{e}$). It is easy to verify that p_n^Y projects $\widetilde{Y}$ onto the last summand $\mathbf{L}^{\otimes n}$, because $p_n^Y = (p_1^{\mathbf{P}^1})^{\otimes n}$

The motif of a projective space. We shall assume that C is one of the "geometric" intersection theories (§ 1, Examples 2–4), so that $C(\mathbf{P}^n) = \Lambda[x]$, where $x^{n+1} = 0$, x being the "class of a hyperplane" in $\mathbf{P}^n$. Then it is not hard to establish that

$$\mathbf{P}^n = \tilde{e} \oplus \mathbf{L} \oplus \ldots \oplus \mathbf{L}^n. \tag{8}$$

Namely, we put $p_i = x^{n-i} \times x^i \in C^n(\mathbf{P}^n \times \mathbf{P}^n)$. It is immediate that p_i are pairwise orthogonal projectors and that $\sum_{i=0}^n p_i = \mathrm{id}$. Therefore $\widetilde{\mathbf{P}}^n$ decomposes into a direct sum of the images of all the p_i. But $p_0 = p_0^{\mathbf{P}^n}$ and $p_n = p_n^{\mathbf{P}^n}$, so that the first and last summands in this sum are respectively isomorphic to $\tilde{e}$ and to $\mathbf{L}^n$. For the remaining summands it is possible to show for example that the standard linear embedding $f: \mathbf{P}^i \to \mathbf{P}^n$ ($i < n$) induces an isomorphism of motifs

$$\widetilde{c(f)}: (\mathbf{P}^n, p_i) \simeq (\mathbf{P}^i, p_i).$$

We leave the details of the proof to the reader.

We compare formula (8) with the classical make-up of complex projective space. The Tate motif is analogous to a "two-dimensional cell" (or rather to its cohomology, since motifs are a contravariant functor).

§7. The motif of the fiber space $P(\mathcal{E})$

Let X be a variety, $\mathcal{E}$ a locally free sheaf of rank $r+1$ on X. We recall that $P(\mathcal{E})$ denotes the "projectivized fibering" Proj $S(\mathcal{E})$, where $S(\mathcal{E})$ is the sheaf of symmetric algebras over $\mathcal{E}$ generated by $\mathcal{E}$.

There is a canonical projection $\psi: P(\mathcal{E}) \to X$. All its fibers are projective spaces. $C(P(\mathcal{E}))$ is a $C(X)$-algebra relative to the homomorphism ψ^*.

Let C be an intersection theory in which the theory of Chern classes holds with the usual properties and in which we also have the following theorem.

Theorem. a) *The $C(X)$-module $C(P(\mathcal{E}))$ is free of rank $r+1$ and the elements $1, x, \cdots, x^r$ (x is the class of a "hyperplane") form a basis for this module. Moreover, x satisfies the relation*

$$\sum_{i=0}^{r+1} (-1)^i \gamma_i x^{r+1-i} = 0,$$

where γ_i is the ith Chern class of the sheaf $\mathcal{E}$.

b)
$$\psi_*(x^k) = \begin{cases} 0 & \text{for } 0 < k < r, \\ 1_X & \text{for } k = r \end{cases}$$

(*see* [3], [8]).

We shall show that the universality of this result permits us immediately to reduce it to the language of correspondences, and, in particular, to prove that

$$\mathbf{P}(\mathcal{E}) = \widetilde{X} \otimes \widetilde{\mathbf{P}}^r = \bigoplus_{i=0}^{r} X \otimes \mathbf{L}^i.$$

(By the same token, the motif $\mathbf{P}(\mathcal{E})$ depends only on the rank of the sheaf $\mathcal{E}$ and is constructed just as if $\mathcal{E}$ were free! The nontriviality of $\mathcal{E}$ is manifested, however, in the multiplicative structure of the motif $\widetilde{P(\mathcal{E})}$; see below.)

Definition. a) We define the correspondences $f_i \in \mathrm{Hom}\,(\overline{P(\mathcal{E})}, \overline{P(\mathcal{E})})$, $i = 0, \cdots, r$, by a downward induction on i, setting $f_r = c(\psi) \circ c(\psi)^t$ and, if $f_r, \cdots, f_{i+1}$ are already defined,

$$f_i = f_r \circ c_{x^{r-i}} \circ \left(\Delta_{P(\mathcal{E})} - \sum_{k=i+1}^{r} c_{x^k} \circ f_k \right).$$

b) We put $p_{r-i} = c_{x^i} \circ f_i$.

Proposition. *The p_i are pairwise orthogonal projectors of degree zero, and* $\sum_{i=1}^{r} p_i = \mathrm{id}_{P(\mathcal{E})}$.

First of all, after using Grothendieck's theorem and the results of §3, by a downward induction on i we shall immediately obtain, for all $a_i \in C(X)$,

$$(p_{r-i})_* \left(\sum_{k=0}^{r} \psi^*(a_k) x^k \right) = \psi^*(a_i) x^i.$$

Therefore the homomorphisms $(p_{r-i})_*: C(\mathbf{P}(\mathcal{E})) \to C(\mathbf{P}(\mathcal{E}))$ form a complete orthonormal system of projectors. This result is universal in view of the functoriality of base change. The identity principle then gives the required fact.

Now let $\mathcal{E}'$ and $\mathcal{E}''$ be two locally free sheaves on X of the same rank $r+1$. We construct canonical isomorphisms of motifs $(\mathbf{P}(\mathcal{E}'), p_i') \xrightarrow{\sim} (\mathbf{P}(\mathcal{E}''), p_i'')$, where p_i' and p_i'' are constructed with respect to $\mathcal{E}'$ and $\mathcal{E}''$ just as was p_i with respect to $\mathcal{E}$.

For this we modify the previous construction somewhat by introducing a correspondence be-between $\mathbf{P}(\mathcal{E}')$ and $\mathbf{P}(\mathcal{E}'')$, which represents an isomorphism of groups

$$C(\mathbf{P}(\mathcal{E}')) \to C(\mathbf{P}(\mathcal{E}'')): \sum_{i=0}^{r} (\psi')^*(a_i)(x')^i \leadsto \sum_{i=0}^{r} (\psi'')^*(a_i)(x'')^i.$$

Definition. We define the correspondences $g_i \in \operatorname{Hom}(\overline{\mathbf{P}(\mathcal{E})}, \overline{X})$, $i = 0, \cdots, r$, by a downward induction on i, setting $g_r = c(\psi)^t$ and, if $g_r, \cdots, g_{i+1}$ are already defined,

$$g_i = c(\psi)^t \circ c_{x^{r-i}} \circ \left(\Delta_{\mathbf{P}(\mathcal{E})} - \sum_{k=i+1}^{r} c_{x^k} \circ f_k \right)$$

Obviously, $c(\psi) \circ g_i = f_i$.

The correspondence g_i represents a homomorphism of "ith coefficients":

$$C(\mathbf{P}(\mathcal{E})) \to C(X): \sum_{k=0}^{r+1} \psi^*(a_k) x^k \leadsto a_i.$$

From these homomorphisms it is now easy to complete the construction.

Definition. $h_i' = c_{(x'')^i} \circ c(\psi'') \circ g_i' \in \operatorname{Hom}(\overline{\mathbf{P}(\mathcal{E}')}, \overline{\mathbf{P}(\mathcal{E}'')})$. The morphisms $h_i'' \in \operatorname{Hom}(\overline{\mathbf{P}(\mathcal{E}'')}, \overline{\mathbf{P}(\mathcal{E}')})$ are constructed analogously.

More precisely, just as above, we have proved the

Theorem. *The correspondences h_i', h_i'' induce mutually inverse isomorphisms*

$$(\mathbf{P}(\mathcal{E}'), p_i') \simeq (\mathbf{P}(\mathcal{E}''), p_i'')$$

for all $i = 0, \cdots, r$.

Corollary. $\widetilde{\mathbf{P}(\mathcal{E})} = \oplus_{i=0}^{r} \widetilde{X} \otimes \mathbf{L}^i$.

Proof. By the theorem the calculations can be carried out for the trivial sheaf $\mathcal{E} = \mathcal{O}_X^{r+1}$. Then a comparison with the computations of the preceding section shows that

$$(\widetilde{\mathbf{P}(\mathcal{E})}, p_i) = (\widetilde{X} \otimes \mathbf{P}^r, \mathrm{id}_X \otimes p_i^{\mathbf{P}^r}) = \widetilde{X} \otimes \mathbf{L}^i.$$

In particular, it follows from this that $\operatorname{Ker} p_0 = \bigoplus_{i=1}^r \widetilde{X} \otimes \mathbf{L}^i$. We shall need this result in a succeeding section to compute the motif of a monoidal transformation.

Thus we have completely determined the "additive" structure of the motif $\widetilde{\mathbf{P}(\mathcal{E})}$. However, it still has a "multiplicative" structure, as does every motif of a variety.

Definition. The multiplicative structure on the motif U is defined by the assignment of a "multiplication" morphism $U \otimes U \to U$.

The multiplicative structure on the motif of a variety X, which is given by the multiplication $\widetilde{c(\delta_X)}: \widetilde{X} \otimes \widetilde{X} \to \widetilde{X}$, is called *canonical*.

In the following section we encounter noncanonical multiplications on motifs. Here we only note that Grothendieck's theorem also permits us to compute the canonical multiplication on $\widetilde{\mathbf{P}(\mathcal{E})} = \bigoplus_{i=0}^r \widetilde{X} \otimes \mathbf{L}^i$ without particular difficulty in terms of a) the canonical multiplication on $\widetilde{X}$, and b) the characteristic classes of the sheaf $\mathcal{E}$.

We shall not give the details; their reconstruction would be a good exercise for the reader.

Finally, we remark that the result of this section deserves a generalization. Let G be a connected linear group over k, $H \subset G$ a subgroup containing a Borel subgroup. Let P be a principal fiber bundle over the variety X with structure group G. The analog of Grothendieck's theorem for computing $C(P/H)$ via $C(X)$ and the "characteristic classes" of P would permit us to compute the motif $\widetilde{P/H}$.

§ 8. Twisting

Later we shall need some properties of "twisting" with the help of the motif L.

For integers $i \geq 0$ we define a functor from the category of motifs into itself, by setting, for any motif U and morphism f:

$$U \rightsquigarrow U(i) = U \otimes \mathbf{L}^i; \quad f \rightsquigarrow f(i) = f \otimes \mathrm{id}_{\mathbf{L}^i}.$$

It is easy to check that $(f \circ g)(i) = f(i) \circ g(i)$ and $f(i) \otimes g(j) = (f \otimes g)(i+j)$. Moreover, it is obvious that $(U(i))(j) = U(i+j)$ and $(f(i))(j) = f(i+j)$.

Lemma. *The functors described are completely faithful, i.e. for any $i \geq 0$ the homomorphisms of groups*

$$\operatorname{Hom}(U, V) \to \operatorname{Hom}(U(i), V(i)) : f \rightsquigarrow f(i)$$

are isomorphisms for all U and V.

Proof. It obviously suffices to prove the lemma for $i = 1$. We limit our verification to the case when $U = \widetilde{X}$ and $V = \widetilde{Y}$ are motifs of varieties; the general case follows easily from this.

First we shall show that if $f(1) = 0$, then $f = 0$, too. We have

$$\widetilde{X}(1) = (X \times \mathbf{P}^1, \mathrm{id}_X \otimes p_1), \quad \widetilde{Y}(1) = (Y \times \mathbf{P}^1, \mathrm{id}_Y \otimes p_1).$$

The morphism $f(1)$ is induced by the correspondence $f \otimes \mathrm{id}_{\mathbf{P}^1} \in \operatorname{Hom}^0(\bar{X} \otimes \bar{\mathbf{P}}^1, \bar{Y} \otimes \bar{\mathbf{P}}^1)$. It is zero if the condition $0 = (f \otimes \mathrm{id}_{\mathbf{P}^1}) \circ (\mathrm{id}_X \otimes p_1) = f \otimes p_1$ holds. But from the definition of p_1 and

Grothendieck's theorem permitting the computation of the ring $C(X \times Y \times \mathbf{P}^1 \times \mathbf{P}^1)$ it then follows easily that $f = 0$ (see below).

We shall now show that any element $g \in \mathrm{Hom}(\widetilde{X}(1), \widetilde{Y}(1))$ has the form $f(1)$ for some $f \in \mathrm{Hom}(\widetilde{X}, \widetilde{Y})$.

First of all, by Grothendieck's theorem $C(X \times Y \times \mathbf{P}^1 \times \mathbf{P}^1)$ is a free module over $C(X \times Y)$ with the basis

$$1_{X\times Y} \times 1_{\mathbf{P}^1\times\mathbf{P}^1};\quad 1_{X\times Y}\times p_0;\quad 1_{X\times Y}\times p_1;\quad 1_{X\times Y}\times p_0p_1,$$

where, as above, $p_0 = e_{\mathbf{P}^1} \times 1_{\mathbf{P}^1}$ and $p_1 = 1_{\mathbf{P}^1} \times e_{\mathbf{P}^1}$. From this it follows that any element of $\mathrm{Hom}(\bar{X} \otimes \mathrm{L}, \bar{Y} \otimes \mathrm{L})$ is uniquely represented in the form of a linear combination of elements of the form $f_1 \otimes 1_{\mathbf{P}^1\times\mathbf{P}^1}$; $f_2 \otimes p_0$; $f_3 \otimes p_1$; $f_4 \otimes p_0p_1$ with coefficients in Λ, where $f_i \in \mathrm{Hom}(\bar{X}, \bar{Y})$.

It is easy to check that $1_{\mathbf{P}^1\times\mathbf{P}^1} \circ p_1 = p_1$ and $(p_0p_1) \circ p_1 = p_1 \circ (p_0p_1) = 0$. From this it follows that any element of $\mathrm{Hom}(\widetilde{X} \otimes \mathrm{L}, \widetilde{Y} \otimes \mathrm{L})$ is induced by some morphism of the form $f \otimes p_1 \in \mathrm{Hom}(X \otimes p_1, Y \otimes p_1)$, which completes the proof.

Now we can define for motifs the groups of "homogeneous morphisms of degree i" for any $i \in \mathbf{Z}$.

Definition. We put

$$\mathrm{Hom}^i(U, V) = \varinjlim \mathrm{Hom}(U(n+i), V(n)).$$

(For different n the groups on the right are canonically identified by means of a twisting.)

The composition $\mathrm{Hom}^i(U, V) \times \mathrm{Hom}^j(V, W) \to \mathrm{Hom}^{i+j}(U, W)$ is defined in the following manner.

It is possible to find an n such that the element $f \in \mathrm{Hom}^i(U, V)$ is represented by the element $f_n \in \mathrm{Hom}(U(n+i), V(n))$, and $g \in \mathrm{Hom}^j(V, W)$ by the element $g_n \in \mathrm{Hom}(V(n), W(n-j))$. Then $g \circ f$ is represented by the morphism $g_n \circ f_n$. With the help of the lemma it is again easy to see that the result does not depend on n.

Example. We denote by $d_{ij} \in \mathrm{Hom}^{j-i}(\mathrm{L}^i, \mathrm{L}^j)$ the morphism corresponding to the identity in $\mathrm{Hom}(\mathrm{L}^n, \mathrm{L}^n)$ for large n. Obviously we have $d_{ii} = \mathrm{id}_{\mathrm{L}}$ and $d_{jk} \circ d_{ij} = d_{ik}$ for all $i, k, j \in \mathbf{Z}$.

Proposition. *We define a homomorphism*

$$C^i(X \times Y) = \mathrm{Hom}^i(\bar{X}, \bar{Y}) \to \mathrm{Hom}^i(\widetilde{X}, \widetilde{Y}) = \varinjlim(\widetilde{X} \otimes \mathrm{L}^{n+i}, \widetilde{Y} \otimes \mathrm{L}^n),$$

putting $g \mapsto (g \otimes d_{n+i,n})_{n \geq n_0}$.

Then this is a homomorphism of groups, compatible with composition.

By the same token, we have discovered that the group $\mathrm{Hom}^i(\bar{X}, \widetilde{Y})$ has a simple geometric meaning. Another interpretation of this result consists in the fact that replacing $CV(k)$ by $CV^0(k)$ is not connected with a loss of information: passage to the pseudo-abelian completion and twisting with the help of the Tate motif permits restoration of $CV(k)$ with respect to $CV^0(k)$.

Proof. Compatibility with composition follows immediately from the formula

$$(f \otimes d_{jk}) \circ (g \otimes d_{ij}) = (f \circ g) \otimes d_{ik}.$$

We shall show that the mapping described above is an isomorphism. We separate the case $i > 0$;

the case $i < 0$ is studied completely analogously. It is necessary to prove that the mapping

$$g \rightsquigarrow g \otimes d_{i,0} : C^i(X \times Y) \to \operatorname{Hom}(\widetilde{X} \otimes \mathbf{L}^i, \widetilde{Y})$$

is an isomorphism.

We compute $\operatorname{Hom}(\widetilde{X} \otimes \mathbf{L}^i, \widetilde{Y})$. We have $\mathbf{L}^i = (\mathbf{P}^i, 1_{\mathbf{P}^i} \times x^i)$, whence $\widetilde{X} \otimes \mathbf{L}^i = (X \times \mathbf{P}^i, \mathrm{id}_X \otimes (1_{\mathbf{P}^i} \times x^i))$. Every element

$$f \in \operatorname{Hom}^0(\overline{X} \otimes \overline{\mathbf{P}}^i, Y) = C^{n+i}(X \times \mathbf{P}^i \times Y)$$

is a linear combination over Λ of elements of the form $s_{23}^{\bullet}(g_k \times x^k)$, $g_k \in C^{n+i-k}(X \times Y)$, where s_{23}: $X \times \mathbf{P}^i \times Y \to X \times Y \times \mathbf{P}^i$ is the switch mapping. The correspondence f induces a morphism of motifs if and only if $f \circ (\mathrm{id}_X \otimes (1_{\mathbf{P}^i} \times x^i)) = f$. The composition on the left is computed with the help of the multiplication in $X \times \mathbf{P}^i \times X \times \mathbf{P}^i \times Y$. Applying the permutation $(12345) \to (13524)$ to this composition, it is not difficult to show that

$$s_{23}^{\bullet}(g_k \times x^k) \circ (\mathrm{id}_X \otimes (1_{\mathbf{P}^i} \times x^i)) = s_{23}^{\bullet}(g_k \times x^{i+k}) = \begin{cases} s_{23}^{\bullet}(g_k \times x^i) & \text{for } k = 0, \\ 0 & \text{for } k \neq 0. \end{cases}$$

But $s_{23}^{\bullet}(g \times x^i) = g \otimes d_{i,0}$. From this we obtain the desired assertion.

Remark. The objects we have called motifs are called *effective motifs* by Grothendieck. According to Grothendieck, motifs are objects in some large category which is obtained if we formally add the "negative powers" of the Tate motif $\mathbf{L}$. This is necessary for the formulation of "Poincaré duality", but for our purposes it is not necessary.

§9. The motif of a monoidal transformation

In this section $\Lambda = \mathbf{Q}$, and $C(X)$ is the $\mathbf{Q}$-algebra of classes of cycles on X with respect to rational or numerical equivalence. We shall return to the situation of the second example of §3. We recall the notation introduced there.

Let X be a variety over k, $Y \xrightarrow{i} X$ the closed embedding of a subvariety of codimension r, ϕ: $X' \to X$ a monoidal transformation with center Y, $Y' = \phi^{-1}(Y)$, ψ: $Y' \to Y$ the restriction of the morphism ϕ to Y', and ϵ: $Y' \to X'$ the canonical embedding of Y' into X'.

We need a theorem of Grothendieck on the structure of the ring $C(X')$. We begin by introducing the objects which figure in the statement of the theorem.

We denote by N the conormal bundle to Y in X, and we put $n = c_r(N) \in C(Y)$ (rth Chern class). Analogously $n' = c_1(N') \in C(Y')$, where N' is the conormal bundle to Y' in X'.

Furthermore, we denote by F the sheaf on Y' which is the kernel of the canonical epimorphism $\psi^*(N) \to O_{Y'}(1) \to 0$, and we put $\Phi = c_{r-1}(F) \in C(Y')$ ($(r-1)$th Chern class).

The rings $C(Y)_N$ and $C(Y')_{N'}$. Preserving the additive structure of the ring $C(Y)$, we introduce a new multiplication in it, setting $y_1 * y_2 = y_1 y_2 n$.

We denote the ring thus obtained (generally speaking, without an identity) by $C(Y)_N$.

Analogously $C(Y')_{N'}$ is defined: instead of n we must use the "correcting factor" n'.

The ring $C(X)\oplus C(Y')_{N'}$. The additive group of this ring is the direct sum of the groups $C(X)$ and $C(Y')_{N'}$. The multiplication in the direct summands coincides with that defined above. Finally, by definition $(x, 0)(0, y') = (0, \psi^*\iota^*(x)y')$ for any $x \in C(X)$ and $y' \in (Y')_{N'}$.

A homomorphism α: $C(Y)_N \to C(X)\oplus C(Y')_{N'}$:

$$\alpha(y) = (\iota_*(y), \; -\psi^*(y)\Phi);$$

A homomorphism α': $C(X)\oplus C(Y')_{N'} \to C(Y)_N$:

$$\alpha'(x, y') = -\psi_*(y');$$

A homomorphism β: $C(X)\oplus C(Y')_{N'} \to C(X')$:

$$\beta(x, y') = \varphi^*(x) + \varepsilon_*(y').$$

Theorem (Grothendieck [8], Exposé VII). *The sequence*

$$0 \to C(Y)_N \xrightarrow{\alpha} C(X)\oplus C(Y')_{N'} \xrightarrow{\beta} C(X') \to 0$$

is exact. Moreover, α *and* β *are ring homomorphisms, and* α' *is a group homomorphism right inverse to* α.

As in the investigation of $P(\xi)$, we shall show that an analogous decomposition occurs in the sequence of C-motifs. First we define motivic variants of the homomorphisms α, α', β:

The morphisms a, a', b *of motifs*:

$$a = (c(\iota)^t \otimes d_{r,0}, \; -c_\Phi \circ c(\psi)\otimes d_{r,1}) : \widetilde{Y}\otimes \mathbf{L}^r \to \widetilde{X}\oplus(\widetilde{Y}'\otimes \mathbf{L}),$$
$$a' = (c(\psi)^t \otimes d_{1,0})\,p_2 : \widetilde{X}\oplus(\widetilde{Y}'\otimes \mathbf{L}) \to \widetilde{Y}$$

(here p_2 is the projection onto $\widetilde{Y}'\otimes \mathbf{L}$ in the category of motifs; moreover, we identify $\widetilde{X}\otimes\widetilde{e}$ with $\widetilde{X}$, etc.),

$$b = c(\varphi) + c(\varepsilon)^t \otimes d_{1,0} : \widetilde{X}\oplus(\widetilde{Y}'\otimes \mathbf{L}) \to \widetilde{X}'.$$

(The correction factors $d_{i,j}$ and $\mathbf{L}^k$ are added in order to make the degrees of the correspondences zero: the correspondences which represent α, α', β are morphisms in $\widetilde{CV}(k)$, but not in $\widetilde{CV}^0(k)$. This complicates the situation insignificantly by comparison with $\widetilde{P(\xi)}$. See below for a more detailed discussion.)

We begin with the additive structure.

Theorem on the additive structure of the motif $\widetilde{X}'$. *The sequence*

$$0 \to \widetilde{Y}(r) \xrightarrow{a} \widetilde{X}\oplus\widetilde{Y}'(1) \xrightarrow{b} \widetilde{X}' \to 0$$

is exact and splits in the category of C-motifs.

Proof. The argument will consist of two steps. First, we establish the analogous result in the pseudo-abelian completion of the category $CV(k)$, and not $CV^0(k)$, i.e. admitting correspondences of arbitrary degree. Then there is no need for twisting: we put

$$\bar{a} = (c(\iota)^t, \ -c_{\Phi} \circ c(\psi)) : \bar{Y} \to \bar{X} \oplus \bar{Y}', \quad \bar{a}' = c(\psi)^t p_2 : \bar{X} \oplus \bar{Y}' \to \bar{Y}, \qquad (9)$$
$$\bar{b} = c(\varphi) + c(\varepsilon)^t : \bar{X} \oplus \bar{Y}' \to \bar{X}'$$

and we consider the corresponding sequence in the category $CV(k)$:

$$0 \to \bar{Y} \xrightarrow{\bar{a}} \bar{X} \oplus \bar{Y} \xrightarrow{\bar{b}} \bar{X}' \to 0. \qquad (10)$$

To prove that is is exact and splits in $\widetilde{CV}(k)$, we apply the Proposition of § 5. It is necessary to check that its hypotheses hold here. To this end we remark, first of all, that for any variety T the sequence of groups

$$0 \to \bar{Y}(\bar{T}) \xrightarrow{\bar{a}_T} (\bar{X} \oplus \bar{Y})(\bar{T}) \xrightarrow{\bar{b}_T} \bar{X}'(\bar{T}) \to 0$$

coincides with the sequence written out in the statement of Grothendieck's theorem, for a monoidal transformation of $T \times X$ with center at $T \times Y$ (relative to the imbedding $\mathrm{id}_T \times \iota$). The exactness of this sequence follows from this. Analogously, $\bar{a}'\bar{a} = \mathrm{id}_Y$ by the identity principle. The requirement follows from this.

The second step consists of transferring the sequence (10) into the category $\widetilde{CV}^0(k)$. The obstruction to this is that $\bar{a}, \bar{a}'$ and $\bar{b}$ are not homogeneous morphisms of degree zero. An appropriate way of twisting the objects and morphisms is found before the statement of the theorem. We only explain why it actually gives the proper result. Let us say that in the category $CV(k)$ the commutative diagram

$$\begin{array}{ccc} \bar{X} & \xrightarrow{f} & \bar{Y} \\ & {}_h\nwarrow \quad \nearrow_g & \\ & \bar{Z} & \end{array}$$

is given, where f, g, h are homogeneous correspondences of degrees r, s, t respectively. Then we have a commutative diagram in the category of motifs:

$$\begin{array}{ccc} \tilde{X}(n+r) & \xrightarrow{f \otimes d_{n+r,\, n}} & \tilde{Y}(n) \\ {}_{h \otimes d_{n+s,\, n+r}}\nwarrow & & \nearrow_{g \otimes d_{n+s,\, n}} \\ & \tilde{Z}(n+s) & \end{array}$$

(for any n). In fact, the morphisms in this diagram have degree zero, and the commutativity follows from the proposition at the end of the previous section. From the same proposition we get the exactness of the twisted sequence of T-points.

Analogously we could state a more general principle of the "identity principle" type, which, however, is completely trivial. The proof is finished.

Corollary. *The motif of the image X' of a monoidal transformation of the variety X with center on the subvariety Y of pure codimension r is isomorphic to*

$$\widetilde{X}' = \widetilde{X} \oplus \left(\bigoplus_{i=1}^{r-1} \widetilde{Y} \otimes \mathbf{L}^i \right).$$

Proof. By what we have proved, $\widetilde{X}' = (X \sqcup Y' \times \mathbf{P}^1, \mathrm{id} - a'a)$. But, as is easy to see, the projector $\mathrm{id} - a'a$ is the identity on X, and on $\overline{Y' \times \mathbf{P}^1}$ coincides with the projector $(1 - p_0^{Y'}) \otimes p_0$, where $p_0 = e_{\mathbf{P}^1} \times 1_{\mathbf{P}^1}$, and $p_0^{Y'}$ is defined as in §7, for $Y' = \mathbf{P}(N)$. The requirement follows from this if the structure of the motif $\widetilde{\mathbf{P}(N)}$ is considered.

We now pass to the multiplicative structure of the motif X'.

The motifs $\widetilde{Y}_N(r)$ and $\widetilde{Y}'_{N'}(1)$. These will denote the motifs $\widetilde{Y}(r)$ and $\widetilde{Y}'(1)$ with an additional structure, namely multiplication morphisms

$$(\widetilde{Y} \otimes \mathbf{L}^r) \otimes (\widetilde{Y} \otimes \mathbf{L}^r) \xrightarrow{\widetilde{c(\delta_Y) \otimes d_{2r,r}}} \widetilde{Y} \otimes \mathbf{L}^r \xrightarrow{c_n \otimes \mathrm{id}} \widetilde{Y} \otimes \mathbf{L}^r$$

(we identify $\mathbf{L}^r \otimes \mathbf{L}^r$ with $\mathbf{L}^{2r}$. The first morphism has degree $-r$, and the second, r; therefore multiplication is a morphism of degree zero). The multiplication in $\widetilde{Y}'_{N'}(1)$ is defined analogously:

$$(\widetilde{Y}' \otimes \mathbf{L}) \otimes (\widetilde{Y}' \otimes \mathbf{L}) \xrightarrow{c(\delta_{Y'}) \otimes d_{2,1}} \widetilde{Y}' \otimes \mathbf{L} \xrightarrow{c_{n'} \otimes \mathrm{id}} \widetilde{Y} \otimes \mathbf{L}.$$

The motif $\widetilde{X} \otimes \widetilde{Y}'_{N'}(1)$. We introduce a (commutative) multiplication on this structure so that on $\widetilde{X}$ it coincides with the canonical one, on $\widetilde{Y}'_{N'}(1)$ with the one introduced above, and "crosswise" it is defined by the composition of morphisms

$$\widetilde{X} \otimes (\widetilde{Y}' \otimes \mathbf{L}) \xrightarrow{(c(\psi) = c(i) \otimes d_{0,1}) \otimes \mathrm{id}} (\widetilde{Y}' \otimes \mathbf{L}) \otimes (\widetilde{Y}' \otimes \mathbf{L}) \xrightarrow{c(\delta_{Y'}) \otimes d_{2,1}} \widetilde{Y}' \otimes \mathbf{L}.$$

(On the first arrow id is relative to $\widetilde{Y}' \otimes \mathbf{L}$.) The degree of the first morphism equals 1, and that of the second is -1, so their composition belongs to Hom^0.

Theorem on the multiplicative structure of a monoidal transformation. *The homomorphisms*

$$a : \widetilde{Y}_N(r) \to \widetilde{X} \oplus \widetilde{Y}'_{N'}(1), \quad b : \widetilde{X} \oplus \widetilde{Y}'_{N'}(1) \to \widetilde{X}'$$

are compatible with multiplication (the canonical multiplication on $\widetilde{X}'$).

Proof. As in the proof of the preceding theorem, we establish the corresponding result in the category $CV(k)$. This means that instead of the morphisms a, b we consider the morphisms $\bar{a}$, $\bar{b}$ introduced above and prove their compatibility with the multiplications which coincide with the canonical ones on $\bar{X}$, $\bar{X}'$, and on $\bar{Y}$, $\bar{Y}'_{N'}$ and between $\bar{X}$ and $\bar{Y}'_{N'}$ are defined by the compositions

$$\overline{Y}\otimes\overline{Y}\xrightarrow{c(\delta_{Y})}\overline{Y}\xrightarrow{c_n}\overline{Y},\quad \overline{Y}'\otimes\overline{Y}'\xrightarrow{c(\delta_{Y'})}\overline{Y}'\xrightarrow{c_{n'}}\overline{Y}',$$
$$\overline{X}\otimes\overline{Y}'\xrightarrow{c(\psi)\circ c(\iota)\otimes id}\overline{Y}'\otimes\overline{Y}'\xrightarrow{c(\delta_{Y'})}\overline{Y}'.$$

If the compatibility with such multiplications is proved, then the validity of the corresponding results for motifs will follow from the previous considerations on twisting. More precisely, a typical situation will appear as follows. Let motifs U and V with multiplications $m\colon U\otimes U\to U$ and $n\colon V\otimes V\to V$ be given. Moreover, let a morphism $f\colon U\to V$ which is compatible with the multiplications be given, i.e. such that $f\circ m = n\circ(f\otimes f)$. Let us assume that f, m and n are homogeneous of degrees r, s and t respectively. Instead of U, V and f we consider $U(r+t)$, $V(t)$ and $f\otimes d_{r+t,t}$ respectively and introduce multiplications on $U(r+t)$ and $V(t)$:

$$U\otimes \mathbf{L}^{r+t}\otimes U\otimes \mathbf{L}^{r+t}\xrightarrow{m\otimes d_{2(r+t),\,r+t}} U\otimes \mathbf{L}^{r+t},$$
$$V\otimes \mathbf{L}^{t}\otimes V\otimes \mathbf{L}^{t}\xrightarrow{n\otimes d_{2t,t}} V\otimes \mathbf{L}^{t}.$$

All the new morphisms have degree zero, and a simple diagram chase shows that $f\otimes d_{r+t,t}$ is compatible with the new multiplication.

We now return to the category $CV(k)$ and prove the compatibility with the multiplications in it.

Let U be an object with multiplication. Then for any variety T we can introduce a multiplication on the group $U(\overline{T})$ in the following manner: in order to multiply two elements $f, g\in \mathrm{Hom}(\overline{T}, U)$ it is necessary to apply the multiplication morphism $m\colon U\otimes U\to U$ to a $\overline{T}$-point of the motif

$$U\otimes U\colon\quad \overline{T}\xrightarrow{c(\delta_T)^t}\overline{T}\otimes\overline{T}\xrightarrow{f\otimes g} U\otimes U.$$

In other words, this multiplication is defined by the formula

$$f * g = m_T\,[(f\otimes g)\circ c(\delta_T)^t],$$

and it suffices to prove that $\overline{a}_T$ and $\overline{b}_T$ are compatible with such a multiplication for all T.

On the other hand, applying Grothendieck's theorem to a monoidal transformation of $T\times X$ with center at $T\times Y$, we obtain that $\overline{a}_T$ and $\overline{b}_T$ are in fact compatible with multiplication, but with one defined another way.

Namely, let U be a variety; identifying $\overline{U}(\overline{T})$ with $C(T\times U)$, we obtain a ring structure on $\overline{U}(\overline{T})$: the usual one for $U = X$ or X', or with a multiplication changed by a correction factor for $U = \overline{Y}_N$ or $\overline{Y}'_{N'}$. This multiplication on $\overline{U}(\overline{T})$ will be denoted by a dot. (The multiplication between $C(T\times X)$ and $C(T\times Y)$ is still more superficial; its description is obvious from the definition of the ring $C(X')\oplus C(Y')_{N'}$.)

For the proof of our theorem it is therefore necessary to show that the two multiplications described on the groups $\overline{U}(\overline{T})$ coincide.

The coincidence of the two multiplications on Y, Y', X and X'. Let U be one of these varieties, and let $m\in C(U)$ be the element n, n', 1_X, $1_{X'}$, respectively. Let T be an arbitrary variety, and let $f, g\in \overline{U}(\overline{T}) = \overline{T\times U(e)}$. Then the usual multiplication is obtained if we apply the morphism

$$\overline{T\times U}\otimes\overline{T\times U}\xrightarrow{c(\delta_{T\times U})}\overline{T\times U}\xrightarrow{c_{1_T\times m}}\overline{T\times U}$$

to the elements f and g, i.e.

$$f\quad g = (c_{1_T\times m})_e\, c\,(\delta_{T\times U})_e\,(f\times g)\in \overline{T\times U}\,(e),$$

where $f\times g\in(\overline{T\times U}\otimes\overline{T\times U})(e)$.

We must compare this formula with the formula for $f\bullet g$:

$$f\bullet g=(c_m)_T c\,(\delta_U)_T\,[(f\otimes g)\circ c\,(\delta_T)^t].$$

By the lemma of §3, $(c_{1_T}\times m)_e=(c_m)_T$. Therefore the first factors in the formulas for $f\circ g$ and $f\bullet g$ coincide. Furthermore, $c(\delta_U)_T=c(\mathrm{id}_T\times\delta_U)_e$. Consequently it suffices to prove the identity

$$c\,(\delta_{T\times U})_e\,(f\times g)=c\,(\mathrm{id}_T\times\delta_U)_e\,[(f\otimes g)\circ c\,(\delta_T)^t].$$

By the corollary of §3,

$$c\,(\delta_{T\times U})_e\,(f\times g)=(\delta_{T\times U})^*\,(f\times g),$$
$$c\,(\mathrm{id}_T\times\delta_U)_e\,[(f\otimes g)\circ c\,(\delta_T)^t]=(\mathrm{id}_T\times\delta_U)^*\,[(f\otimes g)\circ c\,(\delta_T)^t].$$

Let $\delta_{3,T}\colon T\to T\times T\times T$ be the threefold diagonal. From the definition it follows easily that $c(\delta_T)^t=\delta_{3,T*}(1_T)$. Therefore

$$(f\otimes g)\circ c\,(\delta_T)^t=(f\otimes g)\circ\delta_{3,T*}\,(1_T).$$

We consider the diagram

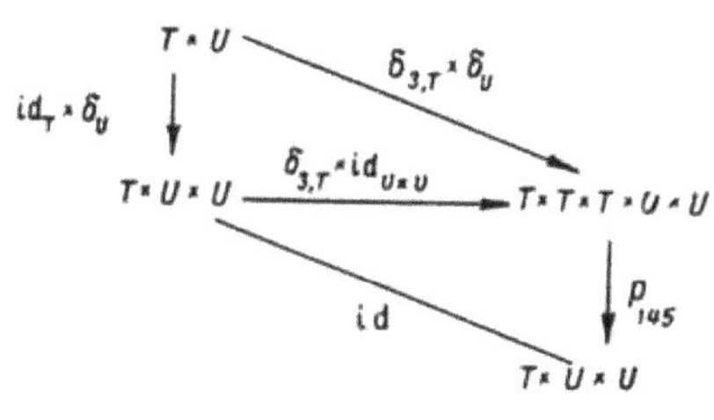

Using it, we find that

$$(f\otimes g)\circ\delta_{3,T*}\,(1_T)=p_{145*}\,((1_T\times(f\otimes g))\,(\delta_{3,T}\times\mathrm{id}_{U\times U})_*\,(1_T\times 1_{U\times U}))$$
$$=(\delta_3\times\mathrm{id}_{U\times U})^*\,(1_T\times(f\otimes g)).$$

Applying $(\mathrm{id}_T\times\delta_U)_*$ to both sides of the equality and continuing to use the commutativity of the diagram, we obtain

$$c\,(\mathrm{id}_T\times\delta_U)_e\,[(f\otimes g)\circ c\,(\delta_T)^t]=(\delta_{3,T}\times\delta_U)^*\,(1_T\times(f\otimes g))$$

$$=(\delta_{3,T}\times\delta_U)^*\,p_{2345}^*\,(f\otimes g)=(\delta_T\times\delta_U)^*\,(f\otimes g)=(\delta_{T\times U})^*\,(f\times g)=c\,(\delta_{T\times U})_e\,(f\times g).$$

which was required.

The coincidence of the two multiplications on $\bar{X}\otimes\bar{Y}'$. *Let* $g\in\bar{X}(\bar{T})$ *and* $f\in\bar{Y}'_{N'}(\bar{T})$. In order to multiply f and g with the help of the ring structure on $\bar{X}\otimes\bar{Y}'_{N'}$ described above, we must apply the multiplication morphism

$$\bar{X}\otimes\bar{Y}'\xrightarrow{c(\psi)\circ c(\iota)\otimes\mathrm{id}}\bar{Y}'\otimes\bar{Y}'\xrightarrow{c(\delta_{Y'})}\bar{Y}'$$

to the T-point

$$\bar{T}\xrightarrow{c(\delta_T)^t}\bar{T}\otimes\bar{T}\xrightarrow{g\otimes f}\bar{X}\otimes\bar{Y}'.$$

In other words,

$$f*g=c(\delta_{Y'})_T\,(c(\psi)\circ c(\iota)\otimes\mathrm{id}_{Y'})_T\,[(g\otimes f)\circ c(\delta_T)^t].$$

As above, the second multiplication is defined by the formula

$$f\cdot g=c(\delta_{T\times Y'})_e\,[(1_T\times\iota\psi)^*g\times f].$$

Transforming as above, we find that

$$(g\otimes f)\circ c(\delta_T)^t=(\delta_{3,T}\times\mathrm{id}_{X\times Y'})^*(1_T\times g\otimes f).$$

Considering the diagram

$$\begin{array}{ccccccc}
T\times Y' & \xrightarrow{\mathrm{id}_T\times\delta_{Y'}} & T\times Y'\times Y' & \xrightarrow{\mathrm{id}_T\times\iota\psi\times\mathrm{id}_{Y'}} & T\times X\times Y' & \xrightarrow{\delta_{3,T}\times\mathrm{id}_{X\times Y'}} & T\times T\times T\times X\times Y' \\
 & & & & & \searrow^{\mathrm{id}} & \downarrow p_{145} \\
 & & & & & & T\times X\times Y',
\end{array}$$

we obtain

$$f*g=(\mathrm{id}_T\times\delta_{Y'})^*(\mathrm{id}_T\times\iota\psi\times\mathrm{id}_{Y'})^*(\delta_{3,T}\times\mathrm{id}_{X\times Y'})^*(1_T\times(g\otimes f)).$$

Getting rid of 1_T with the help of the formula $1_T\times(g\otimes f)=p_{2345}^*(f\otimes g)$, we find that

$$f*g=(\mathrm{id}_T\times\delta_{Y'})^*(\mathrm{id}_T\times\iota\psi\times\mathrm{id}_{Y'})^*(\delta_T\times\mathrm{id}_{X\times Y'})^*(g\otimes f).$$

We now consider the commutative diagram

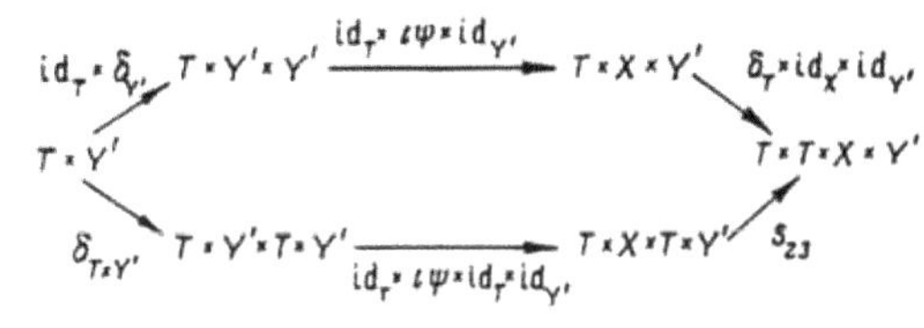

The elements $f \cdot g$ and $f * g$ are respectively obtained from $g \otimes f \in C(T \times T \times X' \times Y')$ by a motion contrary to the arrows on the lower and upper parts of this diagram; therefore they coincide.

The proof is finished.

§10. Motifs of curves

Unless the contrary is explicitly stated, in this section $C(X)$ denotes the algebra of classes of cycles on X modulo numerical equivalence.

Let X be a connected variety of dimension n. The canonical element $e_X \in C^n(X)$ is defined in two cases: if X has a k-point (then e_X is the class of this point) or if $C(X)$ is a $\mathbf{Q}$-algebra (then e_X is the class of a rational 0-cycle of the first degree). We shall constantly assume that one of these conditions holds.

Then, as above, it is possible to consider the projectors $p_0^X = e_X \times 1_X$ and $p_n^X = 1_X \times e_X$. We put $p_+^X = \mathrm{id}_X - p_0^X - p_n^X$ and $X^+ = (X, p_+^X)$. The motif X^+ is the motif of the variety X, "refined" from trivial components of lower and higher dimension.

Proposition. *Let X be a curve. Then for any i, j*

$$\mathrm{Hom}(X^+ \otimes \mathbf{L}^i, \mathbf{L}^j) = 0, \quad \mathrm{Hom}(\mathbf{L}^i, X^+ \otimes \mathbf{L}^j) = 0.$$

Sketch of the Proof. It suffices to consider four cases: in each of the two equalities $i = 0$ or $j = 0$. The verification reduces to an easy computation in the ring $C(X \times \mathbf{P}^k)$, which is generated by linearly independent elements of the form $1_X \times x^i$, $e_X \times x^i$.

We now go into the calculation of $\mathrm{Hom}(X^+, Y^+)$, where X, Y are curves. First we shall assume that C is the *Chow theory*: in particular, $C^1(X) = \mathrm{Pic}(X)$. By J_X we denote the Jacobian variety of the curve X. As is well known, there exists a standard correspondence $F_X \in C^1(X \times J_X) = \mathrm{Hom}(\tilde{X}, \tilde{J}_X)$ which possesses the following property. *For every correspondence $f \in C^1(X \times Y) = \mathrm{Hom}(\tilde{X}, \tilde{Y})$ with the condition*

$$f \cdot (e_X \times 1_Y) = f(1_X \times e_Y) = 0 \tag{11}$$

there exists a uniquely determined morphism $\psi: Y \to J_X$ such that $f = (\mathrm{id}_X \times \psi)^*(F_X)$ (cf. [5], [4]). Moreover, $p_{2*}(F_X) = p_{2*}(F_X \cdot (e_X \times 1_{J_X})) = 0$.

We apply this assertion to $Y = X$, $f = p_+^X$. It defines a morphism of varieties $\phi_X: X \to J_X$ such that $p_+^X = (\mathrm{id}_X \times \phi)^*(F_X)$.

Proposition. *The correspondences $F_X \circ p_+^X \in \mathrm{Hom}(\tilde{X}, \tilde{J}_X)$ and $p_+^X \circ c(\phi_X) \in \mathrm{Hom}(\tilde{J}_X, \tilde{X})$ induce morphisms of motifs $F_X^+: X^+ \to J_X^+$ and $c(\phi)^+: \tilde{J}_X^+ \to X^+$, where $c(\phi)^+ \circ F_X^+ = \mathrm{id}_{X^+}$. In particular, the motif X^+ is an isolated direct summand in $\tilde{J}_X$.*

Proof. We shall show that $c(\phi_X) \circ F_X = p_+^X$. Obviously the proposition follows from this, because p_+^X is a projector. We consider the commutative diagram $(\phi = \phi_X)$

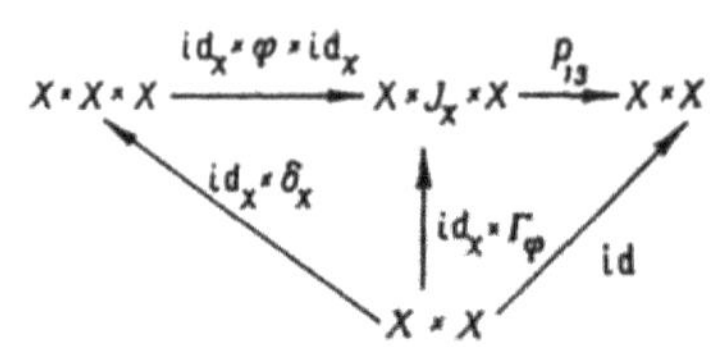

Using it and standard identities, we find that

$$
\begin{aligned}
c(\varphi) \circ F_X &= p_{13*}((1_X \times \Gamma_{\varphi*}(1_X)) \cdot (F_X \times 1_X)) = p_{13*}((\mathrm{id}_X^{-1} \times \Gamma_\varphi)_* (1_{X \times X}) \cdot (F_X \times 1_X)) \\
&= p_{13*}(\mathrm{id}_X \times \Gamma_\varphi)_* (\mathrm{id}_X \times \Gamma_\varphi)^* (F_X \times 1_X) = (\mathrm{id}_X \times \Gamma_\varphi)^* (F_X \times 1_X) \\
&= (\mathrm{id}_X \times \delta_X)^* (\mathrm{id}_X \times \varphi \times \mathrm{id}_X)^* (F_X \times 1_X) = (\mathrm{id}_X \times \delta_X)^* [(\mathrm{id}_X \times \varphi)^* \times 1_X] \\
&= (\mathrm{id}_X \times \delta_X)^* (p_+^X \times 1_X) = p_+^X.
\end{aligned}
$$

This proves the proposition.

Remark. It is possible to make this result considerably more precise, proving that the motif $\tilde{J}_X$ is an "exterior algebra" of the motif X^+. But this is not necessary for our purposes.

Now we again assume that C is a numerical equivalence theory, and denote by F_X, $c(\phi_X)$, F_X^+, $c(\phi_X)^+$, etc., the images of the elements introduced earlier with the same notation in C^1. Then the correspondences f which satisfy condition (11) induce morphisms $f^+: X^+ \to Y^+$, and the mapping $f \leadsto f^+$ is bijective.

Let $\theta_Y \in C^1(J_Y \times J_Y)$ be the class of a "Poincaré divisor" (under the canonical identification of J_Y with $\check{J}_Y$ (cf. [4], [5])). Then with respect to any morphism of abelian varieties $\lambda: J_X \to J_Y$ it is possible to construct a correspondence

$$
t(\lambda) = (\varphi_X \times \varphi_Y)^* (\lambda \times \mathrm{id}_Y)^* (\theta_Y),
$$

satisfying condition (11). The following result is classical.

Proposition. *The mapping* $X^+ \leadsto J_X$, $\lambda \leadsto T(\lambda)$ *induces an equivalence of a subcategory of motifs whose objects are motifs* X^+ *(and whose morphisms are all possible morphisms among them), with the category whose objects are Jacobian varieties, and whose morphisms are "Λ-linearized" morphisms of abelian varieties:* $\Lambda \otimes \mathrm{Hom}(J_X, J_Y)$ *(Λ is an arbitrary aglebra, the ground ring for the chosen intersection theory).*

In particular, if $\Lambda = Q$, Jacobian varieties are considered "up to isogeny". This category is nonabelian, but, as is well known, its pseudo-abelian completion will already be abelian: in fact, it is equivalent to the category of all abelian varieties "up to isogeny", which is semisimple. In the following section we use this in order to associate canonically with every three-dimensional unirational variety some abelian variety playing the role of Weil's "three-dimensional Jacobian".

§ 11. Three-dimensional unirational varieties

Here C is the Chow theory over Q.

Theorem. *Let X be a three-dimensional unirational variety over the field k. Then over a finite extension of k its motif has the form*

$$\widetilde{X}=\widetilde{e}\oplus a\mathbf{L}\oplus U\otimes\mathbf{L}\oplus a\mathbf{L}^2\oplus\mathbf{L}^3, \tag{12}$$

where U is a direct summand of a motif of the form $\oplus Y_i^+$, the Y_i being curves.

Proof. Let $\phi\colon \mathbf{P}^3\to X$ be a rational map of finite degree. By the theorem on the resolution of singularities of mappings (for finite characteristic and dimension 3 the corresponding result was proved by Abhyankar), there exists a commutative diagram of the form

$$\begin{array}{ccc} & X' & \\ {}^{\chi}\swarrow & & \searrow^{\psi} \\ \mathbf{P}^3 & \xrightarrow[\varphi]{} & X \end{array},$$

where ψ is a morphism of finite degree, and χ is a birational morphism which factors into the composition of monoidal transformations with nonsingular geometrically irreducible centers over a finite extension of the field k. These centers can be either zero-dimensional, and then a multiple of the motif $\mathbf{L}\oplus\mathbf{L}^2$ is added to the motif of the previous variety, or one-dimensional, and then a motif of the form $\widetilde{Y}\otimes\mathbf{L}=\mathbf{L}\oplus Y^+\otimes\mathbf{L}\oplus\mathbf{L}^2$ is added. Therefore the motif X' has the form

$$\widetilde{X}'=\widetilde{e}\oplus a\mathbf{L}\oplus(\bigoplus_i Y_i^+)\otimes\mathbf{L}\oplus a\mathbf{L}^2\oplus\mathbf{L}^3. \tag{13}$$

Since the degree of ψ is finite, Example 1 of § 3 shows that $\widetilde{X}=(\widetilde{X}', p)$, where p is the projector $(1/\deg\psi)c(\psi)\circ c(\psi)^t$. But from the first proposition of the preceding section it follows that any projector of the motif $\widetilde{X}'$ is "diagonal" relative to decomposition (13). Its component on $(\oplus Y_i^+)\otimes\mathbf{L}$ has the form $p'\otimes\mathrm{id}_{\mathbf{L}}$, where $p'\in\mathrm{End}\,(\oplus_i Y_i^+)$ in view of the proposition of § 3. The theorem follows from this.

Corollary. *Let X be a three-dimensional unirational variety, U the motif defined by decomposition (12), and A_X a class of isogenous abelian varieties associated to the motif U by the results of the preceding section. Then the mapping $X\leadsto A_X$ extends to a functor from the category of such varieties with correspondences of degree zero as morphisms into the category of abelian varieties up to isogeny.*

In particular, the Lefschetz formula for the number of fixed points of a correspondence holds, in which the part corresponding to three-dimensional cohomology can be calculated by means of the trace in the endomorphism ring of the corresponding abelian varieties. This proves the conjecture stated in the paper [1] for cubic hypersurfaces.

Received 4 JULY 68

BIBLIOGRAPHY

[1] E. Bombieri and H.P.F. Swinnerton-Dyer, *On the local zeta function of a cubic threefold*, Ann. Scuola Norm. Sup. Pisa (3) 21 (1967), 1–29. MR 35 #2894.

[2] A. Borel and J.-P. Serre, *Le théorème de Riemann-Roch*, Bull. Soc. Math. France 86 (1958), 97–136. MR 22 #6817.

[3] A. Grothendieck, "Sur quelques propriétés fondamentales en théorie des intersections," *Séminaire C. Chevalley: anneaux de Chow et applications*, Secrétariat mathématique, Paris, 1958. MR 22 #1572.

[4] S. Lang, *Abelian varieties*, Interscience Tracts in Pure and Appl. Math., no. 7, Interscience, New York, 1959. MR 21 #4959.

[5] D. Mumford, *A remark on Mordell's conjecture*, Amer. J. Math. 87 (1965), 1007–1016. MR 32 #4083.

[6] H. Read, *A concise history of modern painting*, Praeger, New York, 1959; French transl., Paris, 1960.

[7] P. Samuel, *Relations d'équivalence en géométrie algébrique*, Proc. Internat. Congress Math., 1958, Cambridge Univ. Press, New York, 1960, pp. 470–487. MR 22 #6807; MR 22 p. 2546.

[8] *Séminaire de géométrie algébrique*. Fasc. 6 (1966/67), Exposé 7 par P. Berthelot; Exposé 14 par A. Grothendieck, Inst. Hautes Études Sci., Paris, 1968.

[9] J. Tate, *Algebraic cycles and poles of zeta functions*, Arithmetical Algebraic Geometry (Proc. Conf. Purdue Univ., 1963) Harper & Row, New York, 1965, pp. 93–110. MR 37 #1371.

Translated by:
J. S. Joel

Mat. Sbornik
Tom 86 (128) (1971), No. 1

Math. USSR Sbornik
Vol. 15 (1971), No. 1

THREE-DIMENSIONAL QUARTICS AND COUNTEREXAMPLES TO THE LÜROTH PROBLEM

V. A. ISKOVSKIH AND Ju. I. MANIN

Abstract. In this paper we prove that any birational mapping between smooth hypersurfaces of degree four is an isomorphism. Since B. Segre constructed examples of smooth unirational quartics, this leads to a negative resolution of the three-dimensioal Lüroth problem.

Bibliography: 13 items.

Introduction

1. In this paper we shall prove the following

Main Theorem. *Let $\chi: V \longrightarrow V'$ be a birational mapping between two smooth hypersurfaces of degree four (quartics) in* $\mathbf{P}^4$ *over any field k. Then* χ *is an isomorphism.*

Corollary. a. *The birational classification of smooth three-dimensional quartics is the same as the projective classification.*

b. *The group of birational mappings of any smooth three-dimensional quartic V into itself is finite. Consequently V is not rational (i.e. not birationally equivalent to* $\mathbf{P}^3$).

On the other hand, B. Segre [12] found examples of smooth unirational quartics (at least when $\mathrm{Char}\, k \neq 2, 3$). His construction is reproduced in §9 of this paper. Likewise, we will obtain examples of unirational varieties in dimension three which are not rational. This resolves the Lüroth problem negatively, as was expected. We observe that the absence of nontrivial birational endomorphisms signifies the extraordinary "rigidity" of our unirational varieties. Whether or not all smooth quartics are unirational is unknown.

2. The statement of the theorem and many of the fundamental ideas of the proof are due to Fano [3], [4]. Since Fano's work has often been cited and subjected to criticism (see [11]), it is relevent to briefly describe Fano's approach to this and related problems. It consists of three steps.

AMS 1970 *subject classifications.* Primary 14E05, 14J10.

a. **Numerical invariants of the mapping χ.** Suppose a hyperplane section of the variety V' is taken by the mapping χ^{-1} into a surface of degree $4n$ inside V. Generally speaking, the linear system of these surfaces is not complete; it has a system of base points and curves which are perhaps "infinitely close". To each base point or curve B_{i-1} we can ascribe a multiplicity ν_i.

The numbers $(n; \nu_i)$, and also the system of singularities, the base points and curves B_{i-1}, form a fundamental invariant of the mapping χ. It is not completely uniquely determined, since it is connected with the choice of a "resolution of singularities" of the mapping χ, which is not unique: the correct definitions are given in §§2 and 3 of our paper.

b. **Fano's inequalities.** A simple but important lemma of Fano asserts that for $n \geq 2$ at least one of the multiplicities ν_i is large: $\nu_i > 2n$ (if B_{i-1} is a curve).

This is a special case of much more general inequalities which are valid for varieties of any dimension whose anticanonical class is ample (including, for example, $\mathbf{P}^4$). In this form Fano's lemma is stated and proved in §2 below.

c. **Reduction to a contradiction.** It is not hard to establish that χ is a (projective) isomorphism if $n = 1$ (Proposition 4.2). Fano tried to prove that the case $n \geq 2$ is not possible. The essential gaps in his paper begin here. Finally, the idea of the proof consists of verifying the incompatibility of singularities of large multiplicity ν_i on surfaces of small degree $4n$ with the unfixedness (or only positively) of this system of surfaces. In this connection it is necessary to examine a series of cases separately, depending on the position of the singularity B_{i-1} with maximum multiplicity ν_i. In almost all of these cases Fano's arguments do not withstand critical examination.

One of the false reasonings of Fano is the following. On a surface of degree $4n$ if there is a singular point of multiplicity greater than $2n$, the arithmetic genus of this surface becomes negative, which contradicts its birational invariance, since the genus of hyperplane sections of V' equals one. The error here is the following: if we put Fano's arguments correctly into shape, we find that the calculation of later singularities, curves, can give a positive contribution to the arithmetic genus, despite intuition based on the two-dimensional case. For an analogous reason we cannot use the (triple) self-intersection index instead of genus.

In this paper we propose another method of reducing to a contradiction, which is informally described in §4.3, and apply it to the simplest situations in §5. The greatest difficulty in carrying it out occurs in the case when the curve of maximal multiplicity is "infinitely close" to a point on V. Here, in general, Fano's inequalities are found to be insufficient because of possible incidences between the base varieties B_i. Fortunately, however, the more such incidences there exist, the more significantly Fano's inequalities can be strengthened, which in a finite computation again allows us to obtain a contradiction. This is the subtlest part of the proof, and is practically absent in Fano. It is in §§6, 7 and 8 of this paper.

3. A considerable part of the intermediate results of this paper are relevant to

a broader class of objects than quartics: *smooth three-dimensional varieties with an ample anticanonical sheaf.* Fano studied these varieties, and in particular their birational properties, in a long-lived series of papers (see [5] and the bibliography of Roth's book [11]), and we shall call them *Fano varieties.* Among the results announced by him are the following.

The degree d of a Fano variety (by definition, the cube of the anticanonical class) takes some even values from 2 to 64 (the upper bound is extended to 72 if we allow singularities in "general position"). Double coverings with sextic ramification surface have degree 2; quartic in $\mathbf{P}^4$, degree 4; intersections of quadrics and cubics in $\mathbf{P}^5$, degree 6; cubics in $\mathbf{P}^3$, degree 24 (!); $\mathbf{P}^3$, degree 64. Except for $d = 24$, all the varieties of degree $d \geq 16$ are rational; the same is true for $d = 12$.

(Unfortunately, there are no modern accounts of these results of Fano. It would be especially interesting to explain whether or not the theorem on the boundedness of the degree of the anticanonical class carries over to varieties of dimension ≥ 4 on which this class is ample. Fano's method in the three-dimensional case is connected with a larger surplus; it was independently reconstructed by D. Mumford and V. A. Iskovskih, but is unpublished.)

For the study of birational properties of Fano varieties of degree $d \leq 10$ and $d = 14$ Fano attempted to apply his idea relating to the case $d = 4$, which was described above. The essential difference of varieties of degree $d \geq 6$ is that it is known that nontrivial birational mappings exist between them, so step c) above, the "reduction to a contradiction" in subsection 2, must be replaced by another argument. Supposedly, this argument must have the following character. In advance we construct a standard set of simple birational mappings of V into itself (for example, symmetry relative to points on a cubic, etc.). After it is shown that for any nontrivial birational mapping $\chi: V \to V'$ there exists a standard mapping $\phi: V \to V$ such that the composition $\chi \circ \phi: V \to V'$ has smaller "degree" than χ, or in some other sense is "simpler" than χ. The sequence of simplications must terminate in an isomorphism. Such a proof scheme would lead to a theorem on the coincidence of the birational classification with the projective one, and a description of the generators of the group of birational endomorphisms. A good two-dimensional model of this hypothetical theory could be the investigation of del Pezzo surfaces in [10] (see also [9]). It is clear that a very early result of this type is Noether's theorem on the two-dimensional Cremona group; Fano's method also goes back to the ideas of Noether.

For the actual realization of this program we must have, on the one hand, a developed technique for studying the singularities of birational mappings, and, on the other hand, a sufficiently extensive inventory of standard mappings: ideally about one for each singularity of maximum multiplicity whose existence has not been excluded beforehand by considerations like those studied in this paper. Fano's work contains some very interesting material on both these questions, but his techniques are known to be insufficient even for the simplest cases.

The authors also hope to return to Fano varieties of higher degree. The methods of this paper (with simplifications) are also suitable for $d = 2$; we restrict ourselves to quartics here, because it is only among them that there are known to be unirational varieties.

4. In conclusion we give some fragmentary remarks on three-dimensional birational geometry.

a) The methods of this paper are essentially based on the representation of birational mappings in inductive limits of Chow rings relative to birational morphisms of varieties of a given birational class.

Clemens and Griffiths have successfully applied another representation, by means of the three-dimensional (intermediate) Jacobians of Weil, to the study of the three-dimensional cubic (written communication).

These representations are complementary to a known extent. It is possible that a combination of them would also permit progress in the resolution of other long outstanding problems, for example, in clarifying the structure of the three-dimensional Cremona group.

b) The question of a systematic classification of unirational varieties remains completely open. What place in the system of models do the Fano varieties occupy, including the singular ones?

c) The torsion in the three-dimensional integral cohomology group of a smooth projective three-dimensional variety V over $\mathbf{C}$ is a birational invariant. (It is the Brauer-Grothendieck group of the variety V.) Can it be nontrivial for unirational varieties?(1) Any of the logically possible answers would be interesting. We recall that Serre proved triviality of the fundamental group and, hence, that of the Severi torsion of unirational varieties.

§1. Birational morphisms and Chow rings

1.1. Throughout the paper k will denote a fixed algebraically closed field. All of the varieties and mappings between them to be considered will be supposed to be defined over k.

Let V be a smooth projective variety. By the symbol $A(V) = \bigoplus_{i=0}^{\dim V} A^i(V)$ we denote the ring of classes of cycles modulo numerical equivalence, graded by codimension. If V is irreducible, the mapping

$$A^{\dim V}(V) \to \mathbf{Z} : z \rightsquigarrow (z) = \deg z$$

allows us to identify $A^{\dim V}(V)$ with $\mathbf{Z}$. The notation (z) is chosen by analogy with the "index of intersection".

A class $z \in A^i(V)$ is said to be *positive* if it contains positive cycles or zero; the set of them is denoted by $|z|$ by analogy with the notation for a linear system of divisors. An element $z \in A^1(V)$ is said to be *unfixed* if it is positive and the intersection of the supports of all divisors of $|z|$ has codimension ≥ 2. The class of

(1) *Added in proof.* M. Artin and D. Mumford have constructed examples of unirational varieties with nonzero torsion.

the divisors belonging to a linear system is automatically unfixed if the projective dimension of this system is ≥ 2 and it has no fixed component.

These concepts and similar ones are used many times below to prove the nonnegativity of different intersection indices. Example: suppose $z \in A^1(V)$ is unfixed, and that $y \in A^{\dim V-1}(V)$ is the class of an irreducible curve. Then $(zy) \geq 0$ only if y does not belong to the intersection of the supports of all divisors of $|z|$, i.e. is not a base curve of $|z|$, if $|z|$ is a linear system.

1.2. Let ϕ: $V_2 \to V_1$ be a morphism of smooth projective varieties. It defines two mappings: $\phi^*: A(V_1) \to A(V_2)$ (homomorphism of graded rings with identity) and $\phi_*: A(V_2) \to A(V_1)$ (homomorphism of graded groups of degree $\dim V_1 \to \dim V_2$). They are connected by the following "projective formula".

$$\varphi_*(a_2\varphi^*(a_1)) = \varphi_*(a_2)a_1; \quad a_i \in A(V_i).$$

(Cf. the details in the Chevalley seminar [6].)

The homorphisms ϕ^* and ϕ_* are first defined on cycles lying in "sufficiently general position" to the morphism ϕ. It is then established that there are such cycles in each class, and that the result does not depend on the choice of representative. We will recall the concrete constructions where we need them.

The following properties of the homomorphisms ϕ^* and ϕ_* will be constantly use below.

a) Let $\phi: V_2 \to V_2$ be a birational morphism (i.e. a morphism which is a birational mapping) of smooth projective varieties. Then ϕ^* is an inclusion, and ϕ_* is a section of it, i.e. $\phi_* \phi^* = \text{id}$. Therefore

$$A(V_2) = \varphi^* A(V_1) \oplus \operatorname{Ker} \varphi_*.$$

From the projection formula we see that $\operatorname{Ker}\phi_*$ is also a $\phi^*(A(V_1))$-module.

(This result for monoidal transformations is contained in a theorem of Grothendiec [9]; from this it can be derived for arbitrary birational morphisms over which a composition of monoidal transformations dominates. The latter is true for all such morphism in characteristic 0 (Hironaka [7]) and in finite characteristic for $\dim V \leq 3$ (Abhyanka [1]).)

b) Suppose ϕ: $V_2 \to V_1$ is a monoidal transformation with smooth irreducible center, $E \subset V_2$ the inverse image of the center, and $e \in A^1(V_2)$ its class. Then

$$A^1(V_2) = \varphi^* A^1(V_1) \oplus \mathbf{Z}e, \quad \varphi_*(e) = 0.$$

c) Let $c_1(V) \in A^1(V)$ denote the first Chern class; under the hypotheses of b) we let B denote the center of ϕ; then

$$c_1(V_2) = \varphi^*(c_1(V_1)) - \delta e, \quad \delta = \dim V_1 - \dim B - 1.$$

In particular,

$$\varphi_*(c_1(V_2)) = c_1(V_1).$$

From Grothendieck's Riemann-Roch theorem [2] it follows that the last equality holds for all birational morphisms ϕ.

d) Suppose $\dim V_1 = \dim V_2 = 3$. Under the hypotheses of b) and c) we let b denote the class of B in $A(V_1)$. We also set

$$f = \begin{Bmatrix} \text{class of a line on } E \text{ for } \delta = 2 \\ \text{class of a fiber on } E \text{ for } \delta = 1 \end{Bmatrix} \in A^2(V_2).$$

Then the ring $A(V_2)$ is constructed in the following manner (see [9]).

Additive structure.

$$A(V_2) = \varphi^* A(V_1) \oplus \mathbf{Z} e \oplus \mathbf{Z} f; \qquad \varphi_*(e) = \varphi_*(f) = 0.$$

Multiplicative structure. For $\delta = 2$ (i.e. B is a point, E a plane):

$$\begin{aligned} &\phi^* A(V_1) \text{ annihilates } e \text{ and } f; \\ &e^2 = -f;\ (e^3) = -(ef) = 1; \end{aligned}$$

For $\delta = 1$ (i.e. B is a curve, E a ruled surface):

$$\begin{aligned} &\phi^* A^1(V_1) \text{ annihilates } f; \\ &\phi^* A^2(V_1) \text{ annihilates } e \text{ and } f; \\ &\phi^*(a)e = (ab)f,\ \forall a \in A^1(V); \\ &e^2 = -\phi^*(b) + \gamma f; \\ &(ef) = -1,\ (e^3) = -\gamma, \end{aligned}$$

where

$$\gamma = 2\operatorname{genus}(B) - 2 + (c_1(V_1)b).$$

(The formula $(ef) = -1$ also holds for $\dim V \neq 3$ under the same definitions of e and f; we shall use this remark in the following section, where Fano's lemma is proved in greater generality than is strictly necessary for later use.)

§2. Fano's lemma

2.1. Let V and V' be smooth projective varieties, $\chi: V \dashrightarrow V'$ a birational mapping between them. Here and later by a *resolution (of the singularities) of* χ we will mean a commutative diagram with the following properties:

$$\begin{array}{ccc} V_N & \longrightarrow & \\ \downarrow{\scriptstyle \varphi_{N,N-1}} & & \\ \vdots & & \\ V_{i+1} & & \Big\downarrow \psi \\ \downarrow{\scriptstyle \varphi_{i+1,i}} & & \\ V_i & & \\ \vdots & & \\ \downarrow{\scriptstyle \varphi_{1,0}} & & \\ V_0 = V & \xrightarrow{\chi} & V' \end{array}$$

a) *ψ is a birational morphism;*

b) *$\phi_{i+1,i}: V_{i+1} \to V_i$ is a monoidal transformation with smooth irreducible center $B_i \subset V_i$.*

A resolution of χ is not uniquely determined; in the following section we will single out a narrower class of resolutions with the properties needed later; but for the statement and proof of Fano's lemma this is sufficient.

Furthermore we set E_i = the inverse image of B_{i-1} in V_i relative to $\phi_{i,i-1}$, e_i = the class of $E_i \subset A^1(V_i)$, b_i = the class of $B_i \subset A^{\delta_i+1}(V_i)$, where $\delta_i = \dim V - \dim B_i - 1$, and f_i = {the class of a line in a fiber of E_i} $\subset A^{\dim V-1}(V_i)$. Finally, suppose for $N \geq i \geq j \geq 0$

$$\varphi_{i,j} = \varphi_{j+1,j} \circ \dots \circ \varphi_{i,i-1} : V_i \to V_j; \quad \varphi_{i,i} = \mathrm{id}.$$

We shall assume that $A(V_j) \subset A(V_i)$ relative to ϕ^*_{ij} and $A(V') \subset A(V_N)$ relative to ψ^* (cf. 1. 2a)). Formally this means that in computations in the Chow rings $A(V_i)$ and $A(V')$ we have the right to omit the symbols of the mappings ϕ^*_{ij} and ψ^*, leaving only ϕ_{ij*} and ψ_*. This convention will particularly often be used in calculation of intersection indices via the projection formula.

2.2. **Fano's Lemma.** *In the notation of* §2.1 *assume that the following conditions hold*:

a) *The group* $A^1(V)$ *is cyclic, and* $c_1(V) > 0$; *let* $q = [A^1(V): \mathbf{Z}c_1(V)] \geq 1$.

b) *We are given the class of an ample divisor* $l' \in A^1(V')$ *for which the classes* $wl' - mc_1(V')$ *are not positive for all* $m \geq m_0$.

In $A^1(V_N)$ *define the numbers* n *and* $\nu_i \in \mathbf{Z}$ *by the formula*

$$l' = \frac{n}{w} c_1(V) - \sum_{i=1}^{N} \nu_i e_i$$

(n *and* ν_i *exist and are unique, since* $A^1(V_N) = A^1(V) \oplus \bigoplus_{i=1}^{N} \mathbf{Z}e_i$ *according to* 1,2b)). *Then* $n \geq 1$, $\nu_i \geq 0$ *and for each* $n \geq m_0$ *there exists an* i, $1 \leq i \leq N$, *such that*

$$\nu_i > \frac{\delta_i}{w} n. \tag{1}$$

Proof. By the projection formula

$$(e_i c_1(V)^{\dim V-1}) = (\varphi_{N,i*}(e_i) c_1(V)^{\dim V-1}) = 0.$$

Therefore

$$0 < (l' \psi_*(c_1(V)^{\dim V-1})) = (l' c_1(V)^{\dim V-1}) = \frac{n}{w}(c_1(V)^{\dim V}),$$

since the class $c_1(V)$ is ample; consequently $n \geq 1$.

Analogously, applying the projection formula to ϕ_{ij}, ϕ_{ji} or ϕ_{i0}, we find that $(e_j f_i) = 0$ for $i \neq j$, and $(c_1(V) f_i) = 0$, whence

$$(l' f_i) = -\nu_i (e_i f_i) = \nu_i$$

(cf. the end of §1). On the other hand, the class kl' is unfixed on V_N for $k \geq k_0$ (since it is very ample on V'), and the class f_i is positive on V_N (since on V_i it is represented by any fiber of E_i, including those not contained in B_i nor in $\phi_{j,i}(B_j)$ for any $j > i$). Thus $\nu_i = (l' f_i) \geq 0$.

To prove the basic inequality (1) we compare the formulas

$$l' = \frac{n}{w} c_1(V) - \sum_{i=1}^{N} \nu_i e_i, \quad c_1(V_N) = c_1(V) - \sum_{i=1}^{N} \delta_i e_i$$

(cf. 1.2c)). Granting that $\psi_* \phi^* = \mathrm{id}$ and $\psi_*(c_1(V_N)) = c_1(V')$, we find that

$$\psi_*(wl' - nc_1(V_N)) = wl' - nc_1(V') = \sum_{i=1}^{N} (n\delta_i - w\nu_i)\psi_*(e_i). \tag{2}$$

By hypothesis, for $n \geq m_0$ this class is not positive; simultaneously all the e_i, and hence $\psi_*(e_i)$, are positive. Therefore at least one of the coefficients of (2) must be negative; and this is (1).

The lemma is proved.

2.3. **Remarks.** The geometric interpretation of Fano's lemma follows. From the proof we see that n/w is the degree of the class $\phi_{N,0*}(l')$ relative to $c_1(V)$; in particular, if a hyperplane section of V corresponds to $c_1(V)$, then n/w is the usual projective degree of the inverse images under χ of hyperplane sections of V' (in the imbedding l'). On the other hand, in the following section the numbers ν_i will be interpreted as the "multiplicities" of the varieties B_{i-1} in the subscheme of base points of the system $(\phi_{N,i}\circ\psi^{-1})|l'|$, the inverse image of V_i of the system of hyperplane sections of V'. Therefore Fano's lemma asserts that if the "degree" n/w of the birational mapping is large, the multiplicity of at least one component of the base set is great.

We consider two typical examples.

a. $V = V' = \mathbf{P}^2$, $c_1(V) = c_1(V')$ is the class of cubic curves, $w = 3$, l' is the class of a line, and $3l' - nc_1(V')$ is negative for $n \geq 2$. Therefore if the birational mapping $\chi: \mathbf{P}^2 \to \mathbf{P}^2$ is such that χ^{-1} takes lines into curves of degree $n \geq 2$, for any resolution of χ we have $\nu_i > n/3$ for some i. This is a weakening of a famous lemma of Noether, which says that $\nu_i + \nu_j + \nu_k > n$ for three different i, j, k.

b. V and V' are smooth hypersurfaces of degree four in $\mathbf{P}^4$, $c_1(V)$ and $c_1(V')$ classes of hyperplane sections, $w = 1$, $l' = c_1(V')$, and the class $l' - nc_1(V')$ is negative for $n \geq 2$. Thus for any birational mapping χ^{-1} of "degree" $n \geq 2$ and any resolution of it we have Fano's inequalities

$$\exists i,\ \nu_i > \begin{cases} 2n, & \text{if } B_{i-1} \text{ is a point,} \\ n, & \text{if } B_{i-1} \text{ is a curve.} \end{cases}$$

Starting with §4 this example will be the main object of study, and Fano's inequalities (and stronger inequalities which can be extracted from the nonpositivity of the class (2)) will be the principal springboard for the proofs.

We observe that in these examples the condition of nonpositivity of the class $wl' - mc_1(V')$ for $m \geq m_0$ holds for the trivial reason that the group $A^1(V')$ is cyclic and $c_1(V')$ is positive. In other cases another useful criterion of this can be found, which we give here without proof, since it will not be used in the following.

2.4. **Lemma.** *Assume that $H^0(V', \Omega_{V'}^{\otimes n}) = 0$ for $n \geq 2$ and that $D \subset V'$ is a smooth irreducible divisor for which*

$$\dim H^0(D, \mathcal{O}_D(D)) \geqslant 2, \qquad \dim H^0(D, \Omega_D^{\otimes n}) \leqslant 1 \quad \textit{for } n \geqslant 1.$$

Then (class D) $- nc_1(V')$ *is nonpositive for* $n \geq 2$.

§3. Multiplicities and permissible resolutions

3.1. Let $I \subset \mathcal{O}_V$ be a coherent sheaf of ideals on a variety V, and $x \in V$ some point (in the scheme sense). The *order of the point* x *relative to* I (or to subschemes of $\operatorname{Spec} \mathcal{O}_V/I$) is the number (or the symbol ∞)

$$\nu_x(I) = \sup \{\nu \mid m_x^\nu \supset I_x\},$$

where $m_x \subset \mathcal{O}_x$ is the maximal ideal.

This function is semicontinuous in x: $\nu_{x'}(I) \geq \nu_x(I)$ if x' is a specialization of x.

3.2. **Lemma.** *Let $\phi: V_2 \dashrightarrow V_1$ be a monoidal transformation with smooth irreducible center B; x a generic point of B, and $E = \phi^{-1}(B) \subset V_2$. Further, let $D \subset V_1$ be an effective divisor, and $I_D \subset \mathcal{O}_{V_1}$ its ideal. Then $\nu_x(I_D)$ is the multiplicity of the component of E in $\phi^{-1}(D)$.*

Here $\phi^{-1}(D)$ is the full preimage of D; its class is the same as $\phi^*(\text{class}\, D)$. We recall that the difference $\phi^{-1}(D) - \nu_x(I_D)E$ is the proper (or weak) preimage of D; cf. [7] and [13].

This lemma allows us to interpret the coefficients in 2.1 as multiplicities.

3.3. **Definition.** Under the conditions of 2.1 we assume in addition that $A^1(V) = \operatorname{Pic} V$ and $A^1(V') = \operatorname{Pic} V'$, so that for any class $z \in A^1(V)$ $|z|$ is the complete linear system of effective divisors of this class (and analogously for V' and V_i). The *ith base subscheme* $\widetilde{B}_i \subset V_i$ (connected with the resolution of 2.1) will be the scheme-theoretic intersection

$$\widetilde{B}_i = \bigcap_{D \in |l'|} \varphi_{N,i} \psi^{-1}(D).$$

We also choose a basis $\{D^{(k)}\} \subset |l'|$ and set $\widetilde{D}_i^{(k)} = \phi_{N,i}\psi^{-1}(D^{(k)})$ so that $\widetilde{B}_i = \bigcap_k \widetilde{D}_i^{(k)}$.

We recall that the intersection of a family of closed subschemes is defined as the sum of their ideals.

Obviously $\phi_{i,i-1}(\operatorname{supp} \widetilde{B}_i) \subset \operatorname{supp} \widetilde{B}_{i-1}$.

3.4. **Proposition.** *Under the conditions of 2.1 and 3.3 let x_i be a generic point of $\widetilde{B}_i$, and $\widetilde{I}_i$ the ideal of $\widetilde{B}_i$. Then $\nu_{i+1} = \nu_{x_i}(\widetilde{I}_i)$ (cf. [7]).*

In general, for any point $x \in V_i$ we shall call $\nu_x(\widetilde{I}_i)$ the multiplicity of x (relative to χ and l'), intending the order of x in a base subscheme of the inverse image of the system of hyperplane sections of V'.

3.5. **Definition.** A resolution (of the singularities) of the mapping χ described in 2.1 is said to be *permissible* if the following conditions hold:

a. All of the $\widetilde{B}_i$ are nonempty, $B_i \subset \widetilde{B}_i$, and the schemes $\widetilde{B}_i$ and $\widetilde{D}_i^{(k)}$ are normally flat along B_i.

b. $\forall\, i, \forall\, x \in B_i, \nu_x(\widetilde{I}_i) = \max_{y \in V_i} \nu_y(\widetilde{I}_i)$.

c. $\forall\, i, \forall\, x \in V_{i+1}, y = \phi_{i+1,i}(x)$, we have $\nu_x(\widetilde{I}_{i+1}) \geq \nu_y(\widetilde{I}_i)$.

From 3.2 and 3.5 we see that $\nu_1 \geq \nu_2 \geq \cdots \geq \nu_N$ if the resolution is permissible. Hironaka [7] showed that in case Char $k = 0$ any birational mapping has a permissible resolution. The same can be deduced from the results of Abhyankar [1] for Char $k \neq 0$ and dim $V = 3$ (the case dim $V = 2$ was known earlier); see also [13].

3.6. **Proposition.** *Under the conditions of 2.1 suppose* dim $V = 3$ *and that the resolution is permissible. If* E_i *is a ruled surface, none of its fibers belongs to the base subscheme* $\tilde{B}_i$.

In particular, B_i *cannot be a fiber of* E_i.

Proof. The subscheme $\tilde{B}_i$ coincides with the weak preimage of $\tilde{B}_{i-1}$ under the mapping $\phi_{i,i-1}$, which is a monoidal transformation with center $B_{i-1} \subset \tilde{B}_{i-1}$. The inverse image of B_{i-1} in $\tilde{B}_i$ is "that part" of the subscheme $\tilde{B}_i$ whose support is contained in E_i. From the normal flatness of $\tilde{D}_{i-1}^{(k)}$ along B_{i-1} it follows that this inverse image is flat over B_{i-1}; therefore the fibers of E_i cannot be contained in its support, which was asserted.

§4. Plan and start of the proof of the main theorem

4.1. *Beginning here and continuing to the end of the main theorem the following notation and conventions will be fixed:*

V and V' are smooth hypersurfaces of degree four in $\mathbf{P}^4$.

$\chi: V \dashrightarrow V'$ is a birational mapping.

We choose a permissible resolution of χ, to which all the notation of 2.1 is related: ψ, $\phi_{i,j}$, V_i, B_i, E_i, b_i, e_i, etc.

l and l' are the classes of hyperplane sections on V and V' respectively.

The numbers n and ν_i are defined, as in Fano's Lemma 2.2, by the formula

$$l' = nl - \sum_{i=1}^{N} \nu_i e_i, \quad n \geqslant 1 \quad \nu_i \geqslant 0;$$

for $n \geq 2$ they must satisfy Fano's inequality:

$$\exists i, \ \nu_i > \delta_i n, \text{ where } \delta_i = \begin{cases} 2 & \text{if } B_{i-1} \text{ is a point,} \\ 1 & \text{if } B_{i-1} \text{ is a curve.} \end{cases} \tag{3}$$

4.2. **Proposition.** *If* $n = 1$, *then* χ *is an isomorphism.*

Proof. From the definition of n (cf. 2.3) it is clear that the set of divisors $\phi_{N,0*}\psi^{-1}|l'|$ is a linear subsystem of $|nl|$. For $n = 1$ it must coincide with $|l|$, because its (projective) dimension equals the dimension of $|l'|$, i.e. 4, and is the same as the dimension of $|l|$.

But the mapping χ is always reestablished with respect to $\phi_{N,0*}\psi^{-1}|l'|$ up to isomorphism as follows: it is defined outside the base subschemes of this system and in the standard way is given by sections of the sheaf $O_V(n)$, whose zeros belong to $\phi_{N,0*}\psi^{-1}|l'|$. In the given case these are all the sections, which complete the proof.

4.3. *In all of what follows we shall assume that* $n \geq 2$. This assumption leads to a contradiction, after which the main theorem will follow from Proposition 4.2.

To obtain a contradiction the following three assertions are proved in turn. We will call any curve or point B_{i-1} whose multiplicity satisfies Fano's inequality (3) a *maximal singularity* (of the mapping χ in the chosen resolution).

a) $\nu_i < 2n$, *for all* i, *so that a maximal singularity cannot be a point* (proved in §5).

b) *Under the mapping* $\phi_{i-1,0}$ *a maximal singularity* B_{i-1} *cannot remain a curve in* V_0 (proved in §5).

c) *Under the mapping* $\phi_{i-1,0}$ *a maximal singularity* B_{i-1} *cannot become a point in* V_0 (proved in §§6–8).

Although the difficulties in the proofs of these three cases increase sharply, the general idea is everywhere the same: establish that Fano's inequality for the multiplicity ν_i is too strong and is therefore incompatible with the fact that the (image of the) system l' on V_i is unfixed.

The typical method of mounting this argument is the following: we construct a "test" class $y \in A^1(V_N)$ (or $y \in A^2(V_N)$) such that from the geometric considerations of unfixedness and positivity we must have $(l'^2 y) \geq 0$ (or $(l' y) \geq 0$). On the other hand, a direct calculation of this index using n and ν_i instead of Fano's inequalities gives a negative estimate from above for it.

In a pure form this idea works for case a). In case b) it is emended by the simple technique of "isolation" of a maximal singularity (cf. §5). In case c), in order to reach a contradiction it is found to be necessary not only to choose the test class y very carefully, but also to again sharpen Fano's inequalities by returning to the proof of Fano's Lemma in 2.2 and using the nonpositivity of the cycle (2) more effectively.

We observe also that the intersection index $(l'^2 y)$ (or $(l' y)$) is a quadratic (a linear) form of n and ν_i, whose coefficient for n^2 (for n) is proportional to the degree of V, in this case 4. The contribution of the multiplicities ν_i is generally negative, and for $(c_1(V)^3) = 4$ it is hardly sufficient to give the index a negative value. Thus for $(c_1(V)^3) > 4$ such arguments already do not lead to a contradiction, and in their place we must apply the technique of "twisting" the maximal singularity via a suitable set of birational mappings, as in Noether's theorem or for del Pezzo surfaces (cf. [10] and the Introduction).

§5. Maximal singularities

5.1. **Multiplication table.** To compute intersection indices in the ring $A(V_N)$ we shall use the following tables:

Nonzero indices	*Zero indices*
$l^3 = 4;$	$(l^2 e_k) = 0;$
$(le_k^2) = -(lb_{k-1});$	$(le_i e_k) = 0$ for $i \neq k;$
$(e_i e_k^2) = -(e_i b_{k-1});$	$(e_i^2 e_k) = 0$ for $i < k;$
$e_k^3 = \begin{cases} 1, & \text{if } E_k \text{ is a plane;} \\ -\gamma_{k-1} & \text{if } E_k \text{ is ruled;} \end{cases}$	$(e_i e_j e_k) = 0$ for $i < j < k;$

$$\gamma_{k-1} = 2 \operatorname{genus} B_{k-1} - 2 + (lb_{k-1}) - \sum_{i=1}^{k-1} (e_i b_{k-1}) \delta_i.$$

Sketch of the proof. All the formulas (except the first one) follow from the relations in 1.2 d) and the projection formula applied to one of the morphisms ϕ_{ik}. For example,

$(l^2 e_k) = (l^2 \phi_{k,0*}(e_k)) = 0$, because $\phi_{k,0*}(e_k) = 0$;

$(e_i e_j e_k) = (e_i e_j \phi_{k,j*}(e_k)) = 0$ for $i \leq j < k$, because $\phi_{k,j*}(e_k) = 0$;

$(le_k^2) = -(lb_{k-1}) + \gamma(lf_k)$ if E_k is ruled; but $(lf_k) = (l\phi_{k,0*}(f_k)) = 0$, etc.

The coefficient γ_{k-1} in the formula for e_k^3 is computed according to 1.2d) and 1.2c):

$$\gamma_{k-1} = 2 \operatorname{genus} B_{k-1} - 2 + (c_1(V_{k-1}) b_{k-1}); \qquad c_1(V_{k-1}) = c_1(V_0) - \sum_{i=1}^{k-1} \delta_i e_i.$$

Proposition. 5.2. a. *$\nu_i < 2n$ for all i.*

b. *If $\phi_{i-1,0}(B_{i-1})$ is a curve, then $\nu_i \leq n$. Thus a maximal singularity of the mapping χ (cf. 4.3) can only be a curve whose image in V_0 is a point.*

Proof. a) Using the multiplication table,

$$(ll'^2) = 4n^2 - \sum_{i=1}^{N} \nu_i^2 (lb_{i-1}) > 0. \tag{4}$$

Here $(ll'^2) > 0$, because the class $\phi_{N,0*}(l'^2)$ on V is represented by positive curves, the inverse image of plane sections of V'. Moreover, $\nu_1 \geq \nu_2 \geq \cdots \geq \nu_N$, since the resolution is permissible. Therefore, if $\max \nu_i \geq 2n$, already $\nu_1 \geq 2n$. From (4) we see immediately that this is impossible if B_0 is a curve, since then $(lb_0) > 0$.

Now we assume that B_0 is a point. Then we can "decrease" the test class l to $l - e_1$:

$$((l - e_1) l'^2) = 4n^2 - \nu_1^2 - \sum_{i=2}^{N} \nu_i^2 ((l - e_1) b_{i-1}) > 0. \tag{5}$$

Positivity of the index on the left is preserved, because on V_1 $l - e_1$ is represented by the proper preimages of hyperplane sections on V passing through B_0 and hence has an empty base set, and l'^2 is represented by a system of positive curves on V, in which we known that there are not only lines passing through B_0. By the same considerations,

$$((l - e_1) b_{i-1}) = ((l - e_1) \phi_{i-1,0*}(b_{i-1})) \geq 0$$

for all i. Thus from (5) we see that again $\nu_1 < 2n$. This completes the proof of the first part of the proposition.

We now proceed to the second part. Let $B = \phi_{i-1,0}(B_{i-1})$ (set-theoretic image). We assume that B is a curve. Then from the projection formula we see that $(lb_{i-1}) > 0$, and inequality (4) shows that $(lb_{i-1}) \leq 3$, i.e. $\deg B \leq 3$ (the degree in the standard projective imbedding). We reduce the assumption $\nu_i > n$ to a contradiction.

Since B is irreducible, it can be singular only in case it is a plane cubic curve, and then there is one double point on it. We discuss this case separately.

b_1) *B is a nonsingular curve.* We construct an auxiliary monoidal transformation $\bar{\phi}: \bar{V} \to V$ with center B, and we inject $A(V)$ into $A(\bar{V})$ by means of $\bar{\phi}^*$. Let $\bar{e} \subset A(\bar{V})$ be the class of the inverse image of B, and ν the multiplicity of a generic point of B relative to l' (§3.4). By the property of permissibility $\nu \geq \nu_i > n$.

Using the multiplication table in $A(\bar{V})$ we easily find:

$$\deg B = 1: ((l - \bar{e})^2 (nl - \nu\bar{e})) = 3n - 3\nu < 0;$$
$$\deg B = 2: ((2l - \bar{e})(nl - \nu\bar{e})^2) = 8n^2 - 4n\nu - 4\nu^2 < 0;$$
$$\deg B = 3, \textit{ in a plane}: ((3l - \bar{e}))(nl - \nu\bar{e})^2) = 12n^2 - 6n\nu - 6\nu^2 < 0;$$
$$\deg B = 3, \textit{ in a space}: ((2l - \bar{e})(nl - \nu\bar{e})^2) = 8n^2 - 6n\nu - 5\nu^2 < 0.$$

On the other hand, the left side must be nonnegative in every case. In fact, by Lemma 3.2 the system $|nl - \bar{\nu}e|$ is unfixed on $\bar{V}$, since it is represented by the inverse image of the system $\phi_{N,0}\psi^{-1}(|l'|)$ on V. On the other hand, the test system $|al - \bar{e}|$ ($a = 1, 2, 3$) in all four cases generally has an empty base subscheme on $\bar{V}$. This follows from the fact that the intersection of hyperplanes (quadrics, cylinders, quadrics) passing through B cut out B as a *subscheme* (not just set-theoretically). This is a simple verification which we leave to the reader: in the case of a plane curve B it is based on the fact that B is the intersection of a plane and a two-dimensional cone with base B as schemes.

b_2) *B is a singular curve.* Let $\bar{B}_0 \in V$ be its double point, $\bar{\phi}_1: \bar{V}_1 \to V$ a monoidal transformation with center $\bar{B}_0$, $\bar{E}_1$ the inverse image of $\bar{B}_0$ in $\bar{V}_1$, and $\bar{B}_1$ the proper inverse image of B in $\bar{V}_1$. Then $\bar{B}_1$ is a nonsingular curve, and we can construct another monoidal transformation $\bar{\phi}_2: \bar{V}_2 \to \bar{V}_1$ with center $\bar{B}_1$. Let $\bar{E}_2 = \bar{\phi}_2^{-1}(\bar{B}_1)$.

We inject $A(V)$ and $A(\bar{V}_1)$ into $A(\bar{V}_2)$ by means of $\bar{\phi}_2^* \circ \bar{\phi}_1^*$ and $\bar{\phi}_2^*$ respectively. Let $\bar{e}_i$ be the classes of $\bar{E}_i$ in $A(\bar{V}_2)$. Furthermore, let $\bar{\nu}_1$ be the multiplicity of $\bar{B}_0$ with respect to l', and $\bar{\nu}_2$ the multiplicity of a generic point of B (and $\bar{B}_1$) relative to l'; as above, $\bar{\nu}_1 \geq \bar{\nu}_2 \geq \nu_i > n$.

According to the multiplication table in $A(\bar{V}_2)$,

$$((3l - 2\bar{e}_1 - \bar{e}_2)(nl - \bar{\nu}_1\bar{e}_1 - \bar{\nu}_2\bar{e}_2)^2) = 12n^2 - 6\bar{\nu}_2^2 - 6n\bar{\nu}_2 - 2(\bar{\nu}_1 - \bar{\nu}_2)^2 < 0.$$

On the other hand, the intersection index on the left must be nonnegative: as above, for this it suffices to establish that $|3l - 2\bar{e}_1 - \bar{e}_2|$ has no base curves on $\bar{V}_2$; this is checked with no trouble by a direct computation.

The proof is complete.

Remark. As was already noted, the proofs of the two assertions of 5.2 are somewhat different: in the first we used a fixed resolution of χ, and in the second we used auxiliary monoidal transformations of V, which "isolate" the chosen singularity of large multiplicity without taking the remaining singularities into account. Both ways are close, and many computations can be carried out in two variations. However, they are connected with somewhat different intuitive representations, and the authors found it useful to demonstrate both types of reasoning in a comparatively simple example.

5.3. We now obtain the first results on the blowing down of maximal singularities to a point in V.

We shall use the following terminology: let $x \in V_i$ be a point (in the scheme sense). We shall call $j \leq i$ a *jump index* (or simply a *jump*) for x if $\phi_{i,j}(x) \in E_j$. In other words, j is a jump index for x if $\phi_{j,j-1}$ *is not* a local isomorphism in any neighborhood of the point $\phi_{i,j}(x)$. An index j is called a jump for an irreducible subvariety $C \subset V_i$ if it is a jump for a *generic point* of C.

5.4. **Proposition.** *Let $B_{i-1} \subset V_{i-1}$ be a maximal singularity, i.e. B_{i-1} is a curve and $\nu_i > n$. Then the following assertions hold:*

a. *Let k be the smallest jump index for B_{i-1} such that $\phi_{i-1,k}(B_{i-1})$ is a curve. Then E_k is a plane, and $\phi_{i-1,k}(B_{i-1}) \subset E_k$ is a line on it.*

b. *If $j > k$ is a jump index for B_{i-1}, then E_j is a ruled surface with base $B_{j-1} \cong \mathbf{P}^1$, and $\phi_{i-1,j}(B_{i-1})$ is a section on it. In particular, the mapping $B_{i-1} \to \phi_{i-1,j}(B_{i-1})$ is an isomorphism for all $i-1 \geq j \geq k$, and $B_{i-1} \cong \mathbf{P}^1$.*

Proof. a. By Proposition 5.2b, $\phi_{i-1,0}(B_{i-1})$ is a point. Therefore after the jump $V_k \to V_{k-1}$ the curve B_{i-1} must become a point; this is only possible if $\phi_{i-1,k}(B_{i-1})$ lies on a plane E_k or is the fiber of a ruled surface E_k. The second case is excluded since our chosen resolution is permissible.

Thus E_k is a plane. For short we put $B_{i-1,k} = \phi_{i-1,k}(B_{i-1})$. Let $m > k$ be the preceding jump for B_{i-1}, or $m = i$ if there is no such jump. Then B_{m-1} coincides with the proper preimage of $B_{i-1,k}$ on V_{m-1}. Obviously $\nu_{m-1} \geq \nu_{i-1} > n$. If the degree of the curve $B_{i-1,k}$ on E_k equals d, then a sufficiently general line on E_k intersects $B_{i-1,k}$ at d different points, in a neighborhood of which $\phi_{m-1,k}$ is biregular. Therefore the class $f_k - df_m$ will represent the proper preimage of almost any line of E_k on V_m. Consequently $(l'(f_k - df_m)) \geq 0$, but

$$(l'(f_k - df_m)) = \nu_k - d\nu_m.$$

Hence $\nu_k > d\nu_m > dn$, which contradicts Proposition 5.2a for $d \geq 2$.

Thus $\phi_{i-1,k}(B_{i-1})$ is a line on the plane E_k.

b. Now let $j > k$ be any jump for B_{i-1}. The divisor E_j cannot be a plane in view of the definition of k. Further, let d be the degree of the projection of $\phi_{i-1,j}(B_{i-1})$ onto the base B_{j-1} of a ruled surface E_j. Then, letting $m > j$ denote the preceding jump (or i), as above, we find that, on the one hand, $(l'(f_j - df_m)) \geq 0$, and, on the other hand,

$$(l'(f_i - df_m)) = \nu_i - d\nu_m \Rightarrow d \leqslant 1$$

by Proposition 5.2a.

Thus $\phi_{i-1,j}(B_{i-1})$ is a section of E_j, so that under each jump the curve B_{i-1} is mapped isomorphically onto a fiber of the image. In principle it could acquire singularities between jumps, but this is also impossible, since before then, as it is blown down, it turns into $\mathbf{P}^1$. This completes the proof.

§6. A sharpening of Fano's inequalities

6.1. Beginning here we shall assume that all the $\nu_i < 2n$ (cf. 5.2a), and use the information on maximal curves obtained in 5.4.

Attempts to construct a test class, giving a contradiction with Fano's inequalities, become more ineffective as the maximal curve B_{i-1} makes more jumps. However, the existence of these jumps is connected with the presence of incidences in the set of base curves B_{i-1} and exceptional divisors E_i. Each incidence point leads to the existence of an additional component on the full preimage of E_i in V_N. Regrouping the terms in the Fano cycle (2) in order to isolate the coefficients of the positive parts (possibly fewer) of the classes e_i allows us to obtain a stronger inequality than (1); however, this new inequality is already not for the individual multiplicities ν_i, but for linear combination of them. The coefficients of this combination are included in the definition of the test class. After this a contradiction is reached by the former scheme.

This section is devoted to the construction of the coefficients and a proof of the strong inequality, §7 to the construction of the test class and a proof of the inequality $(l'^2y) \geq 0$, and §8 to the inequality $(l'^2y) < 0$.

6.2. **Definition.** a) $M = \{E_i | \nu_i > n\}$ is the set of *maximal divisors* (it is nonempty by Fano's lemma).

b) The *order relation* on M is

$$E_i > E_j \Leftrightarrow i > j \text{ and } \phi_{i,j}(E_i) \subset E_j$$

(in other words, j is a jump for B_{i-1}).

c) The *oriented graph* $\Gamma(M)$: its set of vertices is M; an arrow (oriented edge) goes from E_i to E_j if $i > j$ and B_{i-1} is contained in the proper preimage of E_j on V_{i-1}.

6.3. Here are some simple remarks about the graph $\Gamma(M)$. We shall write $E_{i,j}$ (E_i "lowered" onto V_j) instead of $\phi_{i,j}(E_i)$ for $i > j$, and E_i^j (E_i "raised" to V_j) for the *proper* preimage of E_i on V_j for $j > i$.

We assume that $E_i > E_j$, but that there exists no index k such that $E_i > E_k > E_j$. Then in $\Gamma(M)$ an arrow goes from E_i to E_j. (In fact, from the assumption it follows that the mapping $\phi_{i-1,j}: V_{i-1} \to V_j$ is an isomorphism in a neighborhood of a generic point of B_{i-1}, and since $\phi_{i-1,j}(B_{i-1}) \subset E_j$, also $B_{i-1} \subset E_j^{i-1}$.) Such arrows in $\Gamma(M)$ will be called *short* and the remaining ones, *long*.

If an arrow goes from E_i to E_j, automatically $E_i > E_j$. Inserting between E_i and

E_j a maximal sequence $E_i > E_{i_1} > \cdots > E_{i_k} > E_j$, we will obtain that for any long arrow in $\Gamma(M)$ there exists a connected sequence of short arrows ("path of arrows") with the same beginning and end.

The connected components of the graph $\Gamma(M)$ are not necessarily trees. Here is some information about them.

From Proposition 5.2 it follows that $\phi_{i,0}(E_i)$ is a point for any $E_i \in M$. Therefore all the minimal elements of M are planes. If E_i is a minimal element of M, the point B_{i-1} does not undergo any jumps: a neighborhood of it is locally isomorphically mapped into V. From this it is clear that for each minimal element E_i the set of all the elements $E_j \in M$ with $E_j \geq E_i$ forms a complete set of vertices of some connected component of the graph $\Gamma(M)$.

We set $I(M) = \{i \mid \nu_i > n\}$. We shall write $i \succ j$ if an arrow goes from E_i to E_j in $\Gamma(M)$. Obviously, $i \succ j \Rightarrow E_i > E_j \Rightarrow i > j$, but not conversely: the relation $i \succ j$ is not necessarily transitive.

6.4. **A new basis in** $A^1(V)$. For all i, $1 \leq i \leq N$, we set

$$c_i = e_i - \sum_{j \succ i} e_j \in A^1(V_N) \tag{6}$$

(we shall assume that the sum Σ equals zero if there are no indices j with $j \succ i$, for example if $i \notin M$).

The transition matrix from (e_i) to (c_j) obviously has one on the diagonal and zeros below the diagonal, so that (together with l) the classes (c_i) form a new basis in $A^1(V_N)$.

6.5. **Lemma.** *All the classes c_i are positive.*

Proof. We set $c_{i,k} = \phi_{N,k*}(c_i)$. Obviously $c_{i,i} = e_i$ is positive, and $c_{i,N} = c_i$. We check the positivity of all $c_{i,k}$ by induction from k to $k+1$ for all $k \geq i$. We have

$$c_{i,k+1} = \begin{cases} c_{i,k}, & \text{if } k+1 \text{ is not } \succ i; \\ c_{i,k} - e_{k+1}, & \text{if } k+1 \succ i. \end{cases}$$

In the first case positivity is of course preserved. In the second case in the (unique) positive divisor representing $c_{i,k}$ on V_k we include $E_i^k \supset B_k$ as components (by definition of $\succ$), so that E_{k+1} is necessarily included in the full preimage of this divisor on V_{k+1} (Lemma 3.2). Consequently $c_{i,k+1}$ is positive. The lemma is proved.

Now we invert the matrix of the substitution (6); more precisely, we solve the reciprocal problem. The coefficients of the inverse matrix are combinatorial invariants of the graph $\Gamma(M)$.

6.6. **Lemma-Definition.** *For all i and j, $1 \leq i \leq j \leq N$, set*

$$r_{ij} = \begin{cases} \text{number of paths of arrows from } E_i \text{ to } E_j \text{ in } \Gamma(M) \text{ if } i > j,\ (i, j) \in I(M); \\ 1 \text{ if } i = j; \\ 0 \text{ otherwise.} \end{cases}$$

Then the following two equalities are equivalent:

$$\sum_{i=1}^{N} n_i e_i = \sum_{i=1}^{N} m_i c_i \Leftrightarrow \forall i,\ m_i = \sum_{j \leqslant i} r_{ij} n_j. \tag{7}$$

Proof. First of all we observe that

$$\sum_{i=1}^{N} n_i e_i = \sum_{i=1}^{N} m_i c_i \Leftrightarrow m_i = n_i + \sum_{j \prec i} m_j \tag{8}$$

(pay attention to the different directions of the arrow $\succ$ in (6) and (8)!!). For the proof we should substitute (6) in the left side of (8), transfer the nondiagonal terms to the left, change the order of summation and rename indices:

$$\sum_{l=1}^{N} n_l e_l = \sum_{i=1}^{N} m_l \Big(e_l - \sum_{j \succ l} e_j\Big) \Leftrightarrow \sum_{l=1}^{N} \Big(n_l e_l + \sum_{j \succ l} m_l e_j\Big) = \sum_{l=1}^{N} m_l e_l$$

$$\Leftrightarrow \sum_{j=1}^{N} n_j e_j + \sum_{j=1}^{N} e_j \sum_{l \prec j} m_l = \sum_{j=1}^{N} m_j e_j.$$

Now, repeatedly substituting the right side of (8) for m_i, we express (m_i) by means of (n_i):

$$m_i = n_i + \sum_{i_1 \prec i} m_{i_1} = n_i + \sum_{i_1 \prec i} \Big(n_{i_1} + \sum_{i_2 \prec i_1} m_{i_2}\Big)$$

$$= \ldots = n_i + \sum_{i_1 \prec i} n_{i_1} + \sum_{i_2 \prec i_1 \prec i} n_{i_2} + \sum_{i_3 \prec \ldots \prec i} n_{i_3} + \ldots .$$

Since each sequence of indices $i_k \prec i_{k-1} \prec \cdots \prec i$ is simply a path of arrows from E_i to E_{i_k} in the graph $\Gamma(M)$, it is clear that in the last multiple sum the term n_j with index j is included precisely r_{ij} times. This proves the implication to the right in (7); the implication to the left follows from the invertibility of (r_{ij}). This proves the lemma.

6.7. **Strengthened Fano's inequality.** *There exists an index $a \in I(M)$ such that*

$$\sum_{j \leqslant a} r_{aj} \nu_j > \sum_{j \leqslant a} r_{aj} \delta_j n, \tag{9}$$

where we recall that $\delta_j = 2$ if E_j is a plane, and $\delta_j = 1$ otherwise.

Proof. We set $n_i = \delta_i n - \nu_i$ and we write the class (2) from the proof of Fano's lemma in the basis $(\psi_*(c_i))$, using (7):

$$\sum_{i=1}^{N} (n\delta_i - \nu_i)\, \psi_*(e_i) = \sum_{i=1}^{N} \Big(\sum_{j \leqslant i} r_{ij}(n\delta_j - \nu_j)\Big)\, \psi_*(c_i).$$

Since this cycle is nonpositive, and all the $\psi_*(c_1)$ are positive (Lemma 6.5), at least one of the coefficients of the $\psi_*(c_i)$ is negative. Let a be the corresponding index. It is only necessary to check that $a \in I(M)$. But if $a \notin I(M)$, then $r_{aj} = 0$ for all $j < a$, and $r_{aa} = 1$, so that the coefficient of $\psi_*(c_a)$ equals $n\delta_a - \nu_a \geq 0$, since $\nu_a \leq n$.

Inequality (9) is proved.

Now we choose once for all an index $a \in I(M)$ with the property (9), and we shall work only with the part of the graph $\Gamma(M)$ which lies below (in the sense of $>$) the "critical vertex" E_a. We fix all the new objects connected with this choice of a:

6.8. **Definitions.** a) $S = \{E_i | E_i \leq E_a \text{ in } M\}$.

b) $\Gamma(S)$ is the oriented subgraph of $\Gamma(M)$ whose set of vertices equals S, and two vertices are joined by an arrow if they are in $\Gamma(M)$.

c) $I_1(S) = \{i | E_i \in S,\ E_i \text{ is a ruled surface}\}$;
$I_2(S) = \{i | E_i \in S,\ E_i \text{ is a plane}\}$;
$I(S) = I_1(S) \cup I_2(S)$.

d) $r_j = r_{aj}$ in the old notation is the number of paths of arrows from the highest vertex E_a of the graph $\Gamma(S)$ to the vertex $E_j \in S$ ($r_j = 0$ if $E_j \notin S$; $r_a = 1$).

These definitions will be used in the following section for the construction and investigation of the test class. We conclude this section with the statement of a quadratic inequality which follows from (9) which will only be necessary to estimate the index of intersection with the test class:

6.9. **Quadratic inequality.**

$$\sum_{i \in I(S)} r_i \nu_i^2 > \frac{\left(\sum_{i \in I(S)} r_i \delta_i\right)^2}{\sum_{i \in I(S)} r_i} n^2. \tag{10}$$

(This follows from a standard estimate of the minimum of a positive quadratic form of the variables ν_j in the halfspace (9).)

§7. The test class

We retain the notation of the preceding section.

7.1. **Definition.** The element

$$y = \Big(\sum_{i \in I_2(S)} r_i\Big) l - \sum_{i \in I_1(S)} r_i e_i \in A^1(V_N)$$

is called the *test class*.

7.2. **Positivity Lemma.** *Let $C \subset V_i$ (for some i) be an irreducible curve, $z =$ class $(C) \in A^2(V_N)$. Assume that for all $j \leq i$ for which E_j is a ruled surface, $\phi_{ij}(C)$ is not a fiber of E_j. Then $(yz) \geq 0$.*

Proof. We shall denote by C_j (for all $j \leq i$) the set-theoretic image $\phi_{ij}(C) \subset V_j$ and consider the cases when C_0 is a curve or a point separately.

Case 1. *C_0 is a curve.*

Let $i_1 = \min\{i | i \in I(S)\}$. The graph $\Gamma(S)$ is connected, since it consists of all $E_i \leq E_a$. The argument in §6.3 then shows that E_{i_1} is a plane, one of the minimal elements of M, and some neighborhood of the point $B_{i_1 - 1}$ is mapped isomorphically into $V_0 = V$. From this it follows that the class $l - e_{i_1}$ is representated by V_{i_1} by the inverse image of a system of hyperplane sections on V passing through the point $\phi_{i_1 0}(E_{i_1})$. Since, moreover, $\phi_{N,i_1}.(z)$ is positive by assumption, we have, first of all, $((l - e_{i_1})z) \geq 0$. Furthermore, if $i_2 \succ i_1$, then

$$((e_{i_1} - e_{i_2})z) \geqslant 0.$$

In fact, $e_{i_1} - e_{i_2}$ is represented on V_{i_2} by a positive cycle. For $i \geq i_2$ the index is nonnegative, since $\phi_{i,i_2}(C)$ does not lie on this cycle (otherwise C_0 would not be a curve). But if $i_2 > i$, then $((e_{i_1} - e_{i_2})z) = (e_{i_1} z) \geq 0$ by the projection formula; and if $i_1 > i$, then $((e_{i_1} - e_{i_2})z) = 0$.

Finally, for any $i_k \in S$, $i_k \neq i_1$, we have $E_{i_k} > E_{i_1}$ (cf. 6.3), so that there exists a path of arrows in $\Gamma(S)$: $i_k \succ i_{k-1} \succ \cdots \succ i_1$. Consequently

$$((l - e_{i_k})z) = ((l - e_{i_1})z) + ((e_{i_1} - e_{i_2})z) + \ \ldots\ + ((e_{i_{k-1}} - e_{i_k})z) \geqslant 0.$$

Finally we find

$$(yz) = \Big(\Big(\sum_{j \in I_2(S)} r_j(l - e_j)\Big)\, z\Big) = \sum_{j \in I_2(S)} r_j((l - e_j)z) \geqslant 0.$$

For the proof of the inequality $(l'^2 y) \geq 0$ (Corollary 7.7 below) this is sufficient.

Case 2. *C_0 is a point.* This case is noticeably more difficult, but both the result and the technique of analyzing it will be used in §8 in the proof of the reverse inequality $(l'^2 y) < 0$.

If $(e_j z) = 0$ for all $j \in I_2(S)$, then $(yz) = 0$. Therefore it is further possible to restrict ourselves by the assumption that $\exists\, j \in I_2(S)$, $(e_j z) \neq 0$. We set

$$f = \min\{j \in I_2(S) \mid (e_j z) \neq 0\}$$

and prove some auxiliary lemmas.

7.3. **Lemma.** $C_f \subset E_f$.

Proof. Since C_0 is a point, there exists an index j such that C_j is still a curve, but C_{j-1} is already a point. Then $C_j \subset E_j$ and E_j is a plane, because, by assumption, C_j cannot be a fiber of a ruled surface.

If $j \neq f$, then $j < f$, since otherwise $(e_f z) = 0$ by the projection formula. Moreover, in this case $E_f > E_j$, since $E_f \cap C_f \neq \varnothing$ and

$$\varnothing \neq \varphi_{f,j}(E_f \cap C_f) \subset C_j \subset E_j,$$

so that $\phi_{f,j}(E_f) \subset E_j$, because E_f is a plane.

Consequently $j \in I_2(S)$ (S consists of *all* vertices $\leq E_a$). But this contradicts the choice of f, because $-(e_j z)$ is the product of the degree of the projection $C \to C_j$ with the degree of the curve C_j on the plane E_j, and hence $(e_j z) \neq 0$.

The lemma is proved.

This lemma selects the unique positive term in the expression for (yz):

$$(yz) = -r_f(e_f z) - \sum_{j > f} r_j (e_j z),$$

where $(e_f z) = -$ (degree of C_f) $\times$ (degree of the projection $C \to C_f$).

Now it is necessary to verify that all the remaining (negative) terms cannot compensate for the first one. First we explain their position in the graph $\Gamma(S)$.

7.4. **Lemma.** *In the same notation let $k > f$, $k \in I_2(S)$ and $(e_k z) \neq 0$. Then one of two cases occurs:*

a) *If the morphism $\phi_{k,f}: V_k \to V_f$ is an isomorphism in a neighborhood of a generic point of the curve C_k, then $k \succ f$.*

b) *Otherwise let $j(k) = \max\{j \mid j$ is a jump for $C_k\}$. Then $j(k) \in I_1(S)$, and*

$$k \succ\!\!- j(k) > f; \qquad (b_{j(k)-1} e_f) < 0; \qquad C_{j(k)} \subset E_{j(k)}.$$

Proof. In case a) we have

$$(C_{k-1} \subset E_f^{k-1};\ (e_k z) \neq 0) \Rightarrow B_{k-1} \subset C_{k-1} \Rightarrow B_{k-1} \subset E_f^{k-1} \Rightarrow k \succ\!\!- f.$$

In case b), first of all, obviously $k > j(k) > f$, since there is a jump along the path from k to f, and k cannot be a jump since E_k is a plane. Since $j(k)$ is the first jump, the same argument as in case a) shows that $k \succ j(k)$. According to the definition of a jump $C_{j(k)} \subset E_{j(k)}$; but C_f is still a curve, so $E_{j(k)}$ must be a ruled surface. Together with E_k it belongs to S, because $E_k > E_{j(k)}$.

It remains to verify that $(b_{j(k)-1}\, e_f) < 0$. But from what has been said it is clear that $B_{j(k)-1} = C_{j(k)-1}$. Therefore the image of $B_{j(k)-1}$ on E_f coincides with C_f, and the assertion follows from Lemma 7.3. This proves Lemma 7.4.

7.5. **Lemma.** *In the same notation*

$$r_f \geqslant \sum_{k \in J} r_k, \tag{11}$$

where $J = \{k \in I_2(S) \mid k \neq f,\ (e_k z) \neq 0\}$.

Proof. For each $k \in J$ we choose a particular path in $\Gamma(S)$ in the following manner.

In case a) of Lemma 7.4, $k \succ\!\!- f$; we take the path along one arrow joining k with f.

In case b) of Lemma 7.4, $k \succ j(k)$; we select a path passing from E_k to $E_{j(k)}$ along one arrow, and then from $E_{j(k)}$ to E_f along an arbitrary path consisting of a sequence of short arrows. It is important to observe that along the path from $E_{j(k)}$ to E_f there will be no planes besides E_f; otherwise $C_{j(k)}$ would be contracted to a point before it reached E_f.

We denote the chosen path from E_k to E_f by γ_k.

Now we construct a mapping

$$\coprod_{k \in J} (\text{set of paths from } E_a \text{ to } E_k) \xrightarrow{\Phi} (\text{set of paths from } E_a \text{ to } E_f), \tag{12}$$

which puts a path γ from E_a to E_k in correspondence with its composition with the path γ_k from E_k to E_f.

Obviously the number of elements in the set on the left side of (12) equals $\Sigma_{k \in J} r_k$, and the number on the right is r_f. Therefore (11) will follow from (12) if we can show that Φ is an inclusion.

The injectivity of Φ in turn follows from the fact that none of the paths γ_k is contained in another: in fact, all of them begin on different planes, and later do not pass through planes, except for the very end.

This proves the lemma.

7.6. **End of the proof of the positivity lemma.** First of all we have (case 2)

$$(yz) = \left(\left(\left(\sum r_j\right) l - \sum r_j e_j\right) z\right) = -r_l(e_l z) - \sum_{k\in J}' r_k(e_k z).$$

By Lemma 7.3, $(e_l z) < 0$. Therefore, replacing r_l on the right by $\Sigma_{k\in J} r_k$, by (11), we obtain an estimate from below for (yz). Thus

$$(yz) \geqslant \sum_{k\in J} r_k((-e_l - e_k)z).$$

Now, as in case 1, it suffices to verify that $((-e_l - e_k)z) \geq 0$ for all $k \in J$.

Let $\deg C_l$ be the degree of C_l on $E_l \cong \mathbf{P}^2$, d the degree of the projection $\phi_{k-1,l}$: $C_{k-1} \to C_l$, and m the multiplicity of the point B_{k-1} on C_{k-1}. Then by the projection formula we have

$$-(e_l z) - (e_k z) = (d \deg C_l - m) \times (\text{degree of the projection } C \to C_{k-1}).$$

Therefore it suffices to establish that $m \leq d \deg C_l$. This is a simple algebro-geometric fact: for $d = 1$ it reduces to the assertion that the multiplicity of a singular point of a plane curve does not exceed the degree of the curve. In the general case it is established by considering the normalizations of C_{k-1} and C_l.

The positivity lemma is proved.

7.7. **Corollary.** $(l'^2 y) \geq 0$.

Proof. The class l'^2 on V_N is represented by proper preimages of plane sections of V'. But almost all the plane sections do not intersect the singularities of ψ^{-1}. Their inverse images on V_N are irreducible and almost all are not contained in inverse images of E_i. From this the assertion follows via the projection formula and the simple case 1 of the positivity lemma.

§8. The last inequality

We complete the proof of the main theorem by proving that the strengthened Fano's inequality (10) contradicts Corollary 7.7.

8.1. **Lemma.** $(l'^2 y) < 0$.

Proof. By the multiplication table (§5.1),

$$(l'^2 y) = \left(\left(nl - \sum_{i=1}^{N} \nu_i e_i\right)^2 \left(\left(\sum_{j\in I_1(S)} r_j\right) l - \sum_{j\in I_1(S)} r_j l_j\right)\right)$$

$$= 4n^2\left(\sum_{j\in I_1(S)} r_j\right) - \sum_{j\in I_1(S)} r_j \nu_j^2 - \sum_{j=1}^{N} (b_{j-1} y)\nu_j^2. \qquad (13)$$

From the permissibility of the resolution and the difficult case 2 of Lemma 7.2 it follows that $(b_{j-1}y) \geq 0$ for all j. Therefore, keeping only the terms corresponding to $j \in I_1(S)$ in the very last sum, we are reduced to an estimate from above on (l'^2y).

We consider one such term and apply Lemma 7.4 to the class $z = b_{j-1}$ in order to describe the structure of the nonzero factors of this term:

$$(b_{j-1}y) = -\sum_{k\in I_1(S)} r_k(b_{j-1}e_k), \quad j\in I_1(S). \tag{14}$$

Let $f(j-1)$ be the index for B_{j-1} which corresponds to the index f of Lemma 7.3 for C. Then $(b_{j-1}e_{f(j-1)}) = -1$ by Proposition 5.4. Moreover, if $k \in I_2(S)$, $k \neq f(j-1)$, but $(b_{j-1}e_k) \neq 0$, then necessarily $(b_{j-1}e_k) = 1$: in fact, this index is positive, and it cannot be ≥ 2, since otherwise in blowing down E_k the image of the curve B_{j-1} would acquire a point of multiplicity $(b_{j-1}e_k)$, contradicting Proposition 5.4b. Thus (14) has the form

$$(b_{j-1}y) = r_{f(j-1)} - \sum_{k\in J} r_k, \tag{15}$$

where $J = \{k \in I_2(S) | k \neq f(j-1), (b_{j-1}e_k) \neq 0\}$. We shall now prove that

$$r_{f(j-1)} \geqslant r_j + \sum_{k\in I} r_k. \tag{16}$$

For this we act as in the proof of Lemma 7.5 with little change.

We divide the set of indices J in the sum (16) into groups (some or all of which may be empty). We observe that $f(j-1) < k \leq j-1$ for any $k \in J$.

Zero group: those k for which $V_k \to V_{f(j-1)}$ is an isomorphism in a neighborhood of a generic point of $B_{j-1,k}$. For them $k \succ f(j-1)$, according to Lemma 7.4.

Group with cardinality $j(k)$: those k for which $j(k)$ is fixed to be the first jump of $B_{j-1,k}$ along the path down to $E_{f(j-1)}$. For them $j(k) \in I_1(S)$ and $k \succ j(k)$ according to Lemma 7.4.

Now let $k \in (j) \cup J$. For each such k we choose a path γ_k from E_k to $E_{f(j-1)}$ in the following manner.

If $k = j$, then γ_j is any path of short arrows. It exists by the definition of $f(j-1)$. In it there are no planes until $E_{f(j-1)}$, since otherwise B_{j-1} would be contracted to a point before reaching $E_{f(j-1)}$.

If k belongs to the zero group, then $k \succ f(j-1)$ and γ_k is a path along this arrow.

If k belongs to a group with cardinality $j(k)$, then $k \succ j(k)$, and γ_k is a path along this arrow until $E_{j(k)}$ and then any path from $E_{j(k)}$ to $E_{f(j-1)}$ along short arrows. It exists, since $j(k)$ is one of the jumps for B_{j-1}, so that from Proposition 5.4b) we see that $E_{j(k)} > E_{f(j-1)}$.

Now we put any path from the vertex E_a in S to one of the E_k, $k \in (j) \cup I$, into correspondence with its composition with γ_k. From this (16) will follow, as in Lemma 7.5, if we only check that none of the paths γ_k is contained in another. But all of them, except for γ_j, begin at different planes, and do not pass through any other planes until $E_{f(j-1)}$. As for γ_j, it starts with the ruled surface E_j, but since $j > k > j(k)$ for all $k \in I$, this beginning cannot belong to any of the paths γ_k.

Thus inequality (16) is established.

Substituting (16) in (15), we find that $(b_{j-1}y) \geq r_j \quad \forall\, j \in I_1(S)$. Therefore

$$\sum_{j=1}^{N} (b_{j-1}y)\, v_j^2 \geqslant \sum_{j\in I_1(S)} (b_{j-1}y)\, v_j^2 \geqslant \sum_{j\in I_1(S)} r_j v_j^2.$$

Substituting this in (13), we will obtain

$$(l'^2 y) \leqslant 4n^2 \Big(\sum_{j\in I_2(S)} r_j \Big) - \sum_{j\in I_2(S)} r_j v_j^2 - \sum_{j\in I_1(S)} r_j v_j^2 = 4\, n^2 \Big(\sum_{j\in I_2(S)} r_j \Big) - \sum_{j\in I(S)} r_j v_j^2.$$

Now using (10), we find

$$(l'^2 y) < 4\, n^2 \Big(\sum_{j\in I_2(S)} r_j \Big) - \frac{\Big(\sum\limits_{j\in I_1(S)} r_j - 2 \sum\limits_{j\in I_2(S)} r_j \Big)^2}{\sum\limits_{j\in I_1(S)} r_j + \sum\limits_{j\in I_2(S)} r_j}\, n^2,$$

or writing Σ_k in place of $\Sigma_{j\in I_k(S)} r_j$,

$$\frac{\Sigma_1 + \Sigma_2}{n}(l'^2 y) < 4\,(\Sigma_1 + \Sigma_2)\Sigma_2 - (\Sigma_1 + 2\,\Sigma_2)^2 = -\,\Sigma_1^2 \leqslant 0.$$

Herewith the proofs of Lemma 8.1 and the main theorem are complete.

§9. Examples of unirational quartics

In this section we present a general construction of smooth unirational quartics, due to B. Segre [12], and examples showing that the conditions of this construction are actually realized.

9.1. **Construction.** Let $V \subset \mathbf{P}^4$ be a smooth hypersurface of degree four over k. We will call a line $L \subset \mathbf{P}^4$ a *triple tangent* to V at a point $x \in V$ if either $x \in L \subset V$ or the intersection cycle $L \cap V$ has the form $3x + y$ (possibly $y = x$). In the second case the point y will be called a *fourth point of intersection* of L with V.

Let $x \in V$, and let $C(x)$ be the intersection of V with the tangent hyperplane to V at x (inside $\mathbf{P}^4$). A point x will be called *good* if the surface $C(x)$ is irreducible and reduced, the point x has multiplicity exactly 2 on it, and the quadratic tangent cone to $C(x)$ at the point x (inside $\mathbf{P}^4$) is also irreducible and reduced. Under these conditions the set of triple tangents to V at the point x coincides with the set of generators of the quadratic tangent cone to $C(x)$ at x. Moreover, among these triple tangents, only a finite number can lie in V: otherwise the tangent cone would lie in $C(x)$, despite the irreducibility and reducedness of $C(x)$.

It is known that there exists a dense open set of good points $V_0 \subset V$: a somewhat stronger result is established in Lefschetz' theory.

Now let $TV_0 \to V_0$ be the projectivized tangent bundle to V_0: its fiber over $x \in V_0$ is the set of tangent directions to V_0 at x. We denote by $W_0 \subset TV_0$ the closed subset consisting of the directions which correspond to triple tangent lines in $\mathbf{P}^4 \supset V$. Let $\pi : W_0 \to V_0$ be the canonical projection. The fiber of π over any point $x \in V_0$

is isomorphic to the base of the quadratic tangent cone to $C(x)$, i.e. to $\mathbf{P}^1$. Since on each fiber only a finite number of points correspond to triple tangents contained in V, there exists a rational map $\rho: W_0 \to V$ which puts a point $y \in W_0$ into correspondence with the fourth point of intersection of the triple tangent corresponding to y in $\mathbf{P}^4 \supset V$ with V.

A further idea of Segre consists of constructing inside W_0 a three-dimensional rational variety such that the restriction of ρ onto it has finite degree. Since π fibers W_0 by rational curves, as a first approximation we can take $\pi^{-1}(F_0)$, where $F_0 \subset V_0$ is some rational surface, provided in general that it exists. We impose a somewhat stronger requirement on V.

Condition a). *There exists a hyperplane section $F \subset V$ (necessarily not smooth) which is a rational surface and for which the intersection $F_0 = F \cap V_0$ is not empty.*

We put $U_0 = \pi^{-1}(F_0)$. The mapping $\pi: U_0 \to F_0$ fibers U_0 by smooth rational curves, so that to prove the rationality of U_0 it suffices to find a rational section of the morphism π over F_0. But a double section automatically exists: in fact, suppose the point $x \in F_0$ is smooth on F; then the tangent plane to F_0 at x intersects the quadratic tangent cone to $C(x)$ at x in two generators (or in one, with multiplicity two). These two generators give two points in the fiber of U_0 over x. Let $S \subset U_0$ be the closure of the geometric locus of these pairs of points.

Condition b). *The variety $S \subset U_0$ is divided into two components, $S = S_0 \cup S_1$, such that S_0 (and S_1) is a rational section of the morphism $\pi: U_0 \to F_0$.*

If this condition holds, U_0 is rational, since on a general fiber of π (in the sense of schemes) there is a rational point, so that it is a trivial form of $\mathbf{P}^1$ over the field of rational functions on F_0.

Condition c). *The rational mapping ρ, restricted to S_0, $\rho: S_0 \dashrightarrow F$, is surjective at a generic point.*

A sufficiently general point of S_0 corresponds to a triple tangent to F, so the fourth point of its intersection with V lies in F, so that ρ maps S_0 (rationally) into F. If this mapping is already surjective at a generic point, then, as we shall now prove, $\rho: U_0 \to V$ will also be surjective at a generic point. The rationality of U_0 is already ensured, so V will be unirational.

If $\rho: U_0 \to V$ were not surjective at a generic point, the closure of the image under ρ of the domain of definition of ρ would be an irreducible surface. But $S_0 \subset U_\rho$; consequently, in view of Condition c), this closure could only be F. But suppose $x \in F_0$ is any point which is smooth on F. Then $C(x)$ does not coincide with F, so that almost all the generators of the quadratic tangent cone to $C(x)$ at x do not lie in the hyperplane cutting out F. Consequently their four points of intersection with V do not belong to F, contrary to the assumption.

9.2. **Example.** The construction of an example begins with the choice of a surface F: here it is essential that Conditions b) and c) indeed express the properties of the triple tangents to F and therefore do not depend on the inclusion of F into V. The example of F given by Segre possesses the required properties only in case $\mathrm{Char}\, k \neq 2, 3$. The authors have not tried to remove these restrictions; it is possible that the same method also allows construction of examples for the two exceptional characteristics.

Surface F. It is given by the homogeneous equation in $\mathbf{P}^3$:

$$T_1^4 - 6T_1^2T_2^2 + T_2^4 + T_3^4 + T_3^3T_4 = 0.$$

It has a unique singular point (0001); this point is a triple point; therefore F is rational. The discriminant of the equation defining the two triple tangents at a generic point of F coincides with the Hessian of the form giving F up to a quadratic multiple. In the given case this Hessian equals $[36(T_1^2 + T_2^2)T_3^2]^2$; therefore the surface of triple tangents S splits. Condition b) and the part of Condition a) which does not involve V follow from this. Condition c) is verified by a direct calculation.

Hypersurface V. It is given by the homogeneous equation in $\mathbf{P}^4$:

$$T_0^4 + T_0T_4^3 + T_1^4 - 6\,T_1^2T_2^2 + T_2^4 + T_3^4 + T_3^3T_4 = 0.$$

Here F is a section of V by the hyperplane $T_0 = 0$. The smoothness is verified immediately. To check the remaining part of Condition a) it suffices to find a good point on V, lying on F and different from (00001). For example, such a point is (01113).

The degree of the constructed mapping $\mathbf{P}^3 \to V$ equals 24.

Received 2/JUNE/71

BIBLIOGRAPHY

[1] S. Abhyankar, *Resolution of singularities of embedded algebraic surfaces*, Pure and Appl. Math., vol. 24, Academic Press, New York, 1966. MR 36 #164.

[2] A. Borel and J.-P. Serre, *Le théorème de Riemann-Roch*, Bull. Soc. Math. France 86 (1958), 97–136. MR 22 #6817.

[3] G. Fano, *Sopra alcune varietà algebriche a tre dimensioni aventi tutti i generi nulli*, Atti Accad. Torino 43 (1907/08), 973–977.

[4] ———, *Osservazioni sopra alcune varietà non razionali aventi tutti i generi nulli*, Atti Accad. Torino 50 (1915), 1067–1071.

[5] ———, *Nuove ricerche sulle varietà algebriche a tre dimensioni a curve-sezioni canoniche*, Pont. Acad. Sci. Comment. 11 (1947), 635–720. MR 12, 355.

[6] A. Grothendieck, *Sur quelques propriétés fondamentales en théorie des intersections*, Séminaire Chevalley 2e année: 1958. Anneaux de Chow et applications, Exposé 4, Secrétrariat mathématique, Paris, 1958. MR 22 #1572.

[7] H. Hironaka, *Resolution of singularities of an algebraic variety over a field of characteristic zero*. I, II, Ann. of Math. (2) 79 (1964), 109–326. MR 33 #7333.

[8] V. A. Iskovskih, *Rational surfaces with a pencil of rational curves*, Mat. Sb. 74 (116) (1967), 608–638 = Math. USSR Sb. 3 (1967), 563–588. MR 36 #3786.

[9] Ju. I. Manin, *Correspondences, motifs and monoidal transformations*, Mat. Sb. 77 (119) (1968), 475–507 = Math. USSR SB. 6 (1968), 439–470. MR 41 #3482.

[10] Ju. I. Manin, *Rational surfaces over perfect fields.* II, Mat. Sb. 72 (114) (1967), 161–192 = Math. USSR Sb. 1 (1967), 141–168. MR 37 #1374.

[11] L. Roth, *Algebraic threefolds, with special regard to problems of rationality*, Ergebnisse der Mathematik und ihrer Grenzgebiete, Heft 6, Springer-Verlag, Berlin 1955. MR 17, 897.

[12] B. Segre. *Variazione continua ed omotopia in geometria algebrica,* Ann. Mat. Pura Appl. (4) 50 (1960), 149–186. MR 22 #12432.

[13] O. Zariski, *Reduction of the singularites of algebraic three dimensional varieties,* Ann. of Math. (2) 45 (1944), 472–542. MR 6, 102.

Translated by:
J. S. Joel

Uspekhi Mat. Nauk 39:6 (1984), 47-73 Russian Math. Surveys 39:6 (1984), 51-83

New directions in geometry[1]

Yu.I. Manin

CONTENTS

Introduction 51

§1. A-varieties and A-divisors 55

§2. Riemann-Roch theorems 60

§3. Problems and perspectives of A-geometry 64

§4. Superspaces 67

§5. Schubert supercells 73

§6. The geometry of supergravitation 77

References 82

Introduction

Twenty five years ago André Weil published a short article "De la métaphysique aux mathématiques" [37]. The mathematicians of the 18th century, he explained, had the habit of talking about the metaphysics of the analysis of infinitesimals or the theory of equations. By this they understood a certain system of obscure analogies that are difficult to state, but nevertheless play at times a decisive role in scientific discovery.

Weil's inimitable style should be quoted directly. "Rien n'est plus fécond, tous les mathématiciens le savent, que ces obscures analogies, ces troubles reflets d'une théorie à une autre, ces furtives caresses, ces brouilleries inexplicables; rien aussi ne donne plus de plaisir au chercheur. Un jour vient où l'illusion se dissipe; le pressentiment se change en certitude; les théories jumelles révèlent leur source commune avant de disparaître; comme l'enseigne la *Gítá* on atteint à la connaissance et à l'indifférence en même temps. La métaphysique est devenue mathématique, prête à former la matière d'un traité dont la beauté froide ne saurait plus nous émouvoir."

At the 25th Arbeitstagung I thought it appropriate to present variations on this theme. Here it is a question of what analogies. The archetype of an m-dimensional geometrical object is the number space $\mathbf{R}^m$ or, from the time of Descartes, the ring of polynomial functions $\mathbf{R}[x_1, \ldots, x_m]$.

Instead of this, let us consider the ring $\mathbf{Z}[x_1, \ldots, x_m; \xi_1, \ldots, \xi_n]$, where $\mathbf{Z}$ is the set of integers and the ξ_i are "odd" variables that anticommute with

[1] From the contents of a lecture given at the 25th Arbeitstagung, Bonn 1984.

one another and commute with the "even" x_k. It is useful to assume that this ring is a certain geometrical object of dimension $1+m+n$ or, better, $(1; m|n)$: here 1 refers to the "arithmetic" direction of $\mathbf{Z}$, m to the usual geometrical directions of $(x_1, \ldots, x_m)$, and n to the "odd directions" that correspond to the coordinates ξ_i.

From the rise in the 1950's and 1960's of the ideology of ringed spaces it has been difficult even to say precisely what geometrical object we have in mind. This is now easy to introduce—it is the "affine superscheme" Spec $\mathbf{Z}[x_i, \xi_k]$, an object of the category of topological spaces locally ringed by a sheaf of $\mathbf{Z}_2$-graded supercommutative rings.

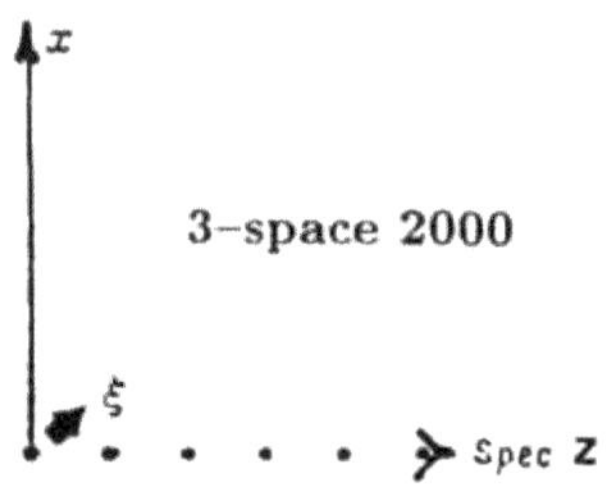

The diagram shows a system of coordinates of the "three-dimensional space 2000": the usual x-axis is complemented by the system of prime numbers and a black arrow corresponding to the odd direction ξ.

In this lecture I intend to present some recent results, the metaphysics of which can be designated by the thesis: "all three types of geometrical dimension have equal rights"

The resemblance between Spec $\mathbf{Z}$ and Spec $k[x]$, or more generally, between fields of algebraic numbers and fields of algebraic functions, is a heuristic principle, which was the source of nearly all the most significant discoveries of Diophantine geometry in the last century. It was this resemblance that Weil wrote about in the article we quoted above. The three theories—of Riemann surfaces, algebraic numbers, and algebraic curves over finite fields—he likened to a trilingual inscription that tells about the same thing in different ways with indecipherable and ruined fragments in different places in the parallel texts.

In this lecture I touch on just one aspect of this resemblance: the idea that schemes of finite type over $\mathbf{Z}$ must be compactified by adding to them a conditional closed fibre over the point at infinity of Spec $\mathbf{Z}$. In two remarkable papers [1], [2], Arakelov was the first to show conclusively that in this way an arithmetical direction becomes completely analogous to a geometrical direction not only in itself, but also in interaction with them. Faltings [10], [11] developed the idea of Arakelov and made certain (for me) the existence of an arithmetical geometry, or A-geometry, in which we construct analogues of all the main ideas and theorems of modern algebraic geometry. The guiding principle in the construction of arithmetical compactifications is now the following idea:

the Einstein-Kähler geometry = ∞-adic arithmetic.

The aim of §§1-3 of the lecture is to collect the fragments of our knowledge that testify to the use of this.

Beginning with §4 we are concerned with odd directions. Algebraic geometers are quite used to representing the spectrum of the algebra of dual numbers Spec $\mathbf{R}[\varepsilon]$, $\varepsilon^2 = 0$, by an arrow, in order to take account of the natural analogous visualization of Spec $\mathbf{R}[\xi]$. However, there is a difference between these arrows, and a very important one. The black arrow of Spec $\mathbf{R}[\varepsilon]$ is not a manifold—it is only an infinitesimal part of a manifold. Algebraically this is expressed, for example, by the fact that the module of the differentials $\Omega^1\mathbf{R}[\varepsilon]/\mathbf{R}$ is not free over $\mathbf{R}[\varepsilon]$: from $\varepsilon^2 = 0$ it follows that $\varepsilon d\varepsilon = 0$. However, the odd arrow of Spec $\mathbf{R}[\varepsilon]$ is a manifold: the formula $\xi d\xi + d\xi \cdot \xi = 0$ for the (even) differential d is satisfied automatically, and $\Omega^1\mathbf{R}[\xi]/\mathbf{R}$ is free.

This example, for all its simplicity, shows that odd nilpotent objects in the structure sheaf actually deserve the name of coordinates. The most remarkable result achieved so far in supergeometry is probably the extension of the Killing-Cartan classification theory to the case of finite-dimensional simple Lie supergroups [15], [16].

Lie supergroups, acting on supermanifolds, "transpose even and odd coordinates", which gives additional grounds for talking about their equal status.

In §§5 and 6 we present some recent results on the geometry of homogeneous supermanifolds due to Weintraub, Penkov, Skornyakov, Voronov, and the author. They refer to the construction of Schubert supercells and the theory of deformations and show that in quite meaningful constructions supergeometry is a natural extension of even geometry.

It is still more fascinating to present the following radical idea:

even geometry = the collective effect of infinite-dimensional odd geometry.

This simplest algebraic model shows how this can arise: the ring homomorphism of formal series of infinitely many variables (x_i even, ξ_j odd)

$$(1)\qquad \mathbf{R}[[x_1, x_2, \ldots]] \to \mathbf{R}[[\ldots\xi_{-1}, \xi_0, \xi_1, \ldots]], \qquad x_i \mapsto \sum_{n=-\infty}^{\infty} \xi_n \xi_{n+i}$$

is an embedding.

A more refined version of this construction was recently discovered in papers on representations of Kac-Moody algebras [17], [7]. It establishes an isomorphism between two realizations of the Lie algebra gl(∞) in the algebras of differential operators $\mathscr{D}_{ev} = \mathrm{Diff}^{\infty}(\mathbf{R}[x_1, x_2, \ldots])$ and $\mathscr{D}_{odd} = \mathrm{Diff}(\mathbf{R}[[\ldots\xi_{-1}, \xi_0, \xi_1, \ldots]])$

$$(2)\qquad \xi_i \frac{\partial}{\partial \xi_j} - \delta_{ij} \leftrightarrow Z_{ij}\left(x, \frac{\partial}{\partial x}\right),$$

where the Z_{ij} are found from the formal series

$$\sum_{i,j} Z_{ij}p^iq^{-j} = qp^{-1}(1-qp^{-1})\left[\exp\left(\sum_1^{\infty} x_i\,(p^i-q^i)\right)\exp\left(\sum_1^{\infty}\frac{\partial}{\partial x_i}\frac{q^{-i}-p^{-i}}{i}\right)-1\right].$$

The isomorphism (2) is established by comparing two natural representations: $\mathcal{D}_{\mathrm{ev}}$ on $\mathbf{R}[x_1, x_2, \ldots]$ and $\mathcal{D}_{\mathrm{odd}}$ on $F = \mathbf{R}[\xi_i, \partial/\partial\xi_j]/I$, where I is the left ideal generated by the ξ_i, $i < 0$, and the $\partial/\partial\xi_j$, $j \geqslant 0$. It is, therefore, convenient to represent the class 1 mod I, the generator of the cyclic $\mathbf{R}[\xi_i, \partial/\partial\xi_j]$-module F, in the form of the infinite outer product $\bigwedge_{n=-\infty}^{-1} \xi_i$, and the whole module F as a variant of the outer algebra of countably many variables, generated by monomials of the form $\bigwedge_{i\in J} \xi_i$, where $J \subset \mathbf{Z}$, card $(\mathbf{Z}_- \triangleleft J) < \infty$. The isomorphism of representations

$$F \simeq \mathbf{R}[x_1, x_2, \ldots] \tag{3}$$

then appears as a development of the simple idea (1).

The study of geometries with odd coordinates was begun by physicists and to a significant degree it continues to be left to them [12], [13], [34]. In particular, the mathematical foundations of supergeometry have been successively developed by Berezin [5], who was quick to see the role and fundamental place of this extension of our geometrical intuition.

The odd coordinates model the internal degrees of freedom of fundamental particles of matter—leptons and quarks. These particles have half-integer spin and are subject to the Fermi-Dirac statistic. Those fundamental particles that carry over the interaction between particles of matter (photons, gluons, $W^{\pm}$, Z, ...) turn out to be bosons. The map (1) is a toy model of the phenomenon well known to physicists—the rise of boson collective degrees of freedom in a condensate of fermion pairs. The formulae (2) and (3) essentially first appeared in physical papers on the theory of dual strings.

The idea that fermion degrees of freedom are primary in relation to boson degrees of freedom has been expressed repeatedly in different forms. It still awaits a mathematical formulation. Our usual four space-time coordinates ($x_0 = ct$, x_1, x_2, x_3) possibly reflect only the collective effect, a coherent state of the world of quantum fields at low temperature, which makes possible the existence of life in its usual form.

In the expectation of a future theory, which will lead to a deeper understanding of the fundamental degrees of freedom, physicists are now discussing models of quantum field theory that unite the fundamental interactions in a single Lagrangian. The most promising constructions of this kind are concerned with models of simple and extended supergravitation [34], whose Lagrangians include the action of Hilbert-Einstein gravitation, the action of gauge fields and fields of matter in a single supersymmetric and general covariant combination.

In §6 we describe the geometry of simple supergravitation from a new point of view, which is a superspace in the form of a distorted flag supermanifold that preserves part of the Schubert cells.

Serious differences between the geometry of supergravitation and the usual extension of Riemannian geometry to the supercase force us to believe that an analogue of a super-Kähler structure must be non-trivial. Therefore, so far I have not risked making a proposal about how an A-geometry with odd coordinates could be constructed.

A comparison of the progress in understanding the geometry of the arithmetical direction and the odd directions shows that the analogues that have been established between them and even dimensions have a different character. If we attempt to reconstruct the ruined parts of this triple language, we can pose two questions.

a) *How do we compactify a supermanifold in odd directions?*

Apparently, the lack of this procedure hinders the construction of a cohomology theory of supermanifolds in which the Schubert supercells have classes that depend not only on their substructure.

Leites put forward the conjecture that in a suitable category there must be an object "purely odd projective superspace": the factor of the complement Spec $k[\xi_1, \ldots, \xi_n] \backslash$Spec k under the action of the multiplicative group $(t, \xi_i) \mapsto (t\xi_i)$. In the usual sense of the word, the corresponding space is empty.

b) *What is "the group that intermixes the arithmetical direction with the (even) geometrical directions"?*

Of course, there are no groups in the naive sense of the word, but there may be a "category of representations of this group". It is also possible that there are rings of correspondences (or the category of representations of them) between the $\mathbf{Z}$-direction and the x-direction. A recent paper of Mazur and Wiles [27], which establishes that the p-adic zeta-function of Kubota and Leopoldt is the divisor of some modular p-adic zeta-function defined purely in characteristic p, testifies to the use of the last proposition.

In preparing this lecture I have taken the advice of many people. I would especially like to thank Igor Rostislavovich Shafarevich, who for three decades has taught his students about arithmetico-geometrical analogies. I have also had invaluable contact with A.A. Veilinson, whose geometrical intuition has often helped me to see things in a proper light.

§1. A-varieties and A-divisors

1. A-varieties.

Let K be an algebraic number field of finite degree, R its ring of integers, and $S = S_f \cup S_\infty$ the set of all points of K, Archimedian and non-Archimedean. For $v \in S$ we denote by K_v the completion of K with respect to v. The corresponding system of normalizations $| \ |_v : K \to \mathbf{R}^*$ satisfies the identity $\prod_v |a|_v = 1$ for all $a \in K$. (We put $|a|_v = |a|$ if $K_v = \mathbf{R}$, $|a|_v = |a|^2$ if $K_v = \mathbf{C}$.) Obviously, $R = \{a \in K : |a|_v \leqslant 1 \text{ for all } v \in S_f\}$.

We define an *A-variety* as a structure

$$(1) \qquad X = (X_f;\ \omega_v,\ v \in S_\infty).$$

Here X_f is a scheme of finite type, and the morphism $\pi : X_f \to \operatorname{Spec} R$ is surjective and flat, and a generic geometrical fibre of it is a complete non-singular variety. Furthermore, ω_v are Kähler forms on the complex varieties $X_v = (X_f \underset{\mathbf{R}}{\otimes} K_v)(\mathbf{C})$, where $\overline{\omega}_v = \omega_v$ if $K_v = \mathbf{R}$, and for $K_v = \mathbf{C}$ the metrics corresponding to the two embeddings $K_v \to \mathbf{C}$ are complex conjugate.

We denote the volume form corresponding to ω_v by $\operatorname{vol}_v = \omega_v^{\dim X_v}$.

The simplest example of an A-variety is the A-curve $X_f = \operatorname{Spec} R$; here the ω_v do not make sense, and the vol_v give the volume of all points $v \in S_\infty$.

We emphasize that the definition (1) must be regarded very much as preparatory. First of all, we cannot restrict ourselves to schemes that are proper over R; for improper schemes we probably need to impose on ω_v the condition of logarithmic growth at infinity (see [8]). Moreover, a special role is played by forms ω_v that satisfy the Kähler-Einstein equation. (See the explanations in §5.)

2. Invertible *A*-sheaves.

An invertible A-sheaf on the A-variety (1) is a structure

$$(2) \qquad \mathcal{L} = (\mathcal{L}_f;\ h_v,\ v \in S_\infty).$$

Here $\mathcal{L}_f$ is an invertible sheaf on X_f, and h_v is a Hermitian metric on $\mathcal{L}_v = \mathcal{L}_f \underset{\mathbf{R}}{\otimes} K_v$ with the obvious reality conditions and the following property:

(3) *the curvature form F_v of the connection corresponding to h_v is ω_v-harmonic.*

We recall that if s is a local holomorphic section of $\mathcal{L}_v$, then $F_v = \bar{\partial}\partial \log h_v(s, s)$ in the domain of definition of s. For an arithmetic curve (3) is assumed to be vacuous.

The tensor product $\mathcal{L} \otimes \mathcal{L}'$ of two invertible A-sheaves is defined in the obvious way. The group of classes of invertible A-sheaves with respect to the tensor product is denoted by $\operatorname{Pic}_A X$. Its zero element is the class of the structure A-sheaf

$$\mathcal{O}_X = (\mathcal{O}_{X_f};\ h_v \mid h_v(1, 1) = 1 \quad \text{for all} \quad v \in S_\infty).$$

For what follows it is important to observe that for a given ω_v the metric h_v satisfying (3) is unique up to multiplication by a constant.

3. **Sections.**
Let $\mathcal{L}$ be the invertible sheaf (2). We put $H^i(X, \mathcal{L}) = H^i(X_f, \mathcal{L}_f)$. All these are R-modules of finite type, and the usual Riemann-Roch-Grothendieck theorem for schemes [35] says much about their structure. An essentially new object, specific for A-geometry, is the Euler A-characteristic $\chi_A(\mathcal{L})$. If there were a canonical projection $\rho : X \to$ "A-point", then $\chi_A(\mathcal{L})$ should be defined as $\mathbf{R}\rho_*(\mathcal{L})$. In the absence of such a construction only ad hoc definitions of $\chi_A(\mathcal{L})$ are known in various special cases, which are given in §2. The general idea is this: if $H^i(X, \mathcal{L}) = 0$ for $i > 0$, so that $\chi_A(\mathcal{L})$ must be described in terms of $H^0(X, \mathcal{L})$, then $\chi_A(\mathcal{L})$ is the volume of a basic parallelepiped of the image $H^0(X, \mathcal{L}) \to \bigoplus_{v \in S_\infty} H^0(X_v, \mathcal{L}_v) = H^0_\infty(X, \mathcal{L})$ with respect to some volume form on H^0_∞. The proper definition of this volume form is a problem that has not been solved in the general case. A paper of Faltings [10] compels us to suppose that it must be constructed by means of the canonical metric on the bundle with fibre $\bigotimes_i \det H^i(X_v, \mathcal{L}_v)^{(-1)^i}$ on $\mathrm{Pic}_0 X_v$ and an inductive definition on the Néron-Severi group.

If $\chi_A(\mathcal{L})$ has been properly defined, then it must be calculated by means of a certain Riemann-Roch A-theorem for which we need to know the divisors of sections of $\mathcal{L}$ and, more generally, the characteristic classes.

4. ***A*-divisors.**
We define an *A-divisor* on the A-variety (1) as a structure

$$D = (D_f;\ r_v \mid r_v \in \mathbf{R},\ v \in S_\infty),$$

where D_f is a Cartier divisor on X_f. Instead of this, it is convenient to have the symbolic notation

$$D = D_f + \sum_{v \in S_\infty} r_v X(v),$$

where $X(v)$ are conceptual "closed fibres" of the A-variety X over $v \in S_\infty$. (They must not be confused with X_v.) The A-divisors form a group $\mathrm{Div}_A X$.

We define the A-divisor of a section $s \in H^0(X, \mathcal{L})$ as the element

$$\mathrm{div}\, s = \mathrm{div}_f\, s - \sum_{v \in S_\infty} \left(\int_{X_v} \log |s|_v \cdot \mathrm{vol}_v \right) X(v), \tag{4}$$

$$|s|_v^{[\mathbf{C}:K_v]} = h_v(s, s),$$

where $\mathrm{div}_f\, s$ is the Cartier divisor of s. If g is a rational function on X_f, represented as the ratio of two sections of $\mathcal{L}$, then the definition of a principal A-divisor, which follows from (4), does not depend on $\mathcal{L}$:

$$\mathrm{div}\, g = \mathrm{div}_f\, g - \sum_{v \in S_\infty} \left(\int_{X_v} \log |g|_v\, \mathrm{vol}_v \right) X(v). \tag{5}$$

Finally, by the same formula (4) we can define the divisor of a meromorphic section.

For any A-divisor D it is not difficult to define the A-sheaf $\mathcal{O}(D)$ with a section whose divisor is D. Namely, for $D = D_f$ we put

$$\mathcal{O}(D_f) = (\mathcal{O}_{X_f}(D_f);\ h_v), \tag{6}$$

where h_v is the unique metric on $\mathcal{O}_{X_f}(D_f) \otimes K_v$ that satisfies (3) and the normalization condition

$$\int\limits_{X_v} \log |1_{D_f}|_v \cdot \mathrm{vol}_v = 0, \qquad v \in S_\infty, \tag{7}$$

1_{D_f} being the meromorphic section of $\mathcal{O}_{X_f}(D_f)$ whose divisor is D_f.
(4) explains the meaning of (7): $\operatorname{div} 1_D = \operatorname{div}_f 1_D = D_f$, which justifies (6).

Next, we put

$$\mathcal{O}\Big(\sum_{v\in S_\infty} r_v X(v)\Big) = \Big(\mathcal{O}_{X_f};\ h_v\,|\,|1|_v = \exp\Big(-r_v\Big(\int\limits_{X_v} \mathrm{vol}_v\Big)^{-1}\Big)\Big). \tag{8}$$

Again by (4),

$$\operatorname{div} 1 = \sum_{v\in S_\infty} r_v X(v),$$

which shows that (8) is well-defined.

As in the purely geometrical case, we can now construct the exact sequence

$$\begin{array}{ccccccccc} 0 & \to & \mathrm{Div}_A^0 X & \to & \mathrm{Div}_A X & \to & \mathrm{Pic}_A X & \to & 0 \\ & & & & \cup\!\!\!\!\!\!\! & & \cup\!\!\!\!\!\!\! & & \\ & & & & D & \mapsto & \text{class } \mathcal{O}(D), & & \end{array}$$

where $\mathrm{Div}_A^0 X$ is the group of principal A-divisors.

5. Green's functions.

From the definition of 4. it is clear that the main information about the Archimedean component of A-divisors on X is carried by the functions

$$G_v(D_v, x) = |1_{D_v}|_v(x), \qquad x \in X_v. \tag{9}$$

On a compact Kähler manifold (X_v, ω_v) they are uniquely defined by the following conditions:

a) $G_v(D_v, x)$ is real-analytic for $x \notin \operatorname{supp} D_v$, and close to $\operatorname{supp} D_v$ the function $G_v(D_v, x)/|g(x)|_v$ is real-analytic, where g_v is the local equation of D_v.

(b) the (1, 1)-form $\bar\partial\partial \log G_v(D_v, x)$ is ω_v-harmonic outside D_v; the corresponding flow is a linear combination of a harmonic form and the flow D_v.

(c)
$$\int\limits_{X_v} \log G_v(D_v, x) \cdot \mathrm{vol}_v = 0.$$

Also:

(d)
$$G_v(D_v' + D_v'', x) = G_v(D_v', x)\, G_v(D_v'', x).$$

(e) $G_v(\operatorname{div} g_v, x) = c_g \cdot |g_v(x)|_v$ for any meromorphic function g_v, where c_g is determined from c).

In the last chapter of Lang's book [22] he explains how to calculate Green's functions on Abelian varieties and curves by means of theta-functions and differentials of the third kind, respectively. The Kähler metric must be taken as flat in the first case and induced by the flat metric of the Jacobian in the second case. These metrics are used in the proof of the Riemann-Roch-Arakelov-Faltings A-theorem, which we state in §2.

Since the function (9) is not constant (apart from trivial cases), the "closed fibres" $X(v)$, $v \in S_\infty$, must be represented as analogous to infinitely degenerate reductions. In fact, for $v \in S_f$ the degeneracy of $X(v)$ manifests itself in the fact that a meromorphic function or section of an invertible sheaf can have different orders on different components of the fibre $X(v)$, and so instead of $|1_{D_f}|_v$ for degenerate $v \in S_f$ we must consider $|1_{D_f}|_{v_i}$, where v_i are the normalizations corresponding to these components. In turn, these data can be replaced by the function $|1_{D_f}|(x)$, where $x \in X(R_v)$, by putting $|1_{D_f}|(x) = |1_{D_f}|_{v_i}$ if the reduction of x is in the i-th component of $X(v)$.

This analogy suggests that an A-divisor concentrated on $X(v)$, $v \in S_\infty$, should be regarded not as a constant r_v, but possibly a volume form on X_v, and we must take the integral sign in the definitions (4) and (5).

To state this more precise definition, it is useful to compare the ∞-adic situation with the Mumford-Schottky p-adic curves. But for now we take a deeper version of (3)–(5).

6. The intersection index.

Let $\sigma : \operatorname{Spec} R \to X_f$ be a section of the structure morphism $\pi : X_f \to \operatorname{Spec} R$. We regard the image of σ as a closed A-curve Y lying in X. We define the intersection index of Y with the A-divisor D for which the intersection of $\operatorname{supp} Y \cap \operatorname{supp} D$ with the common fibre is empty:

$$\langle Y, D\rangle = \sum_{v \in S_f} (Y, D_f)_v \log q_v + \sum_{v \in S_\infty} \langle Y, D\rangle_v. \tag{10}$$

Here $(Y, D_f)_v$ for $v \in S_f$ is the sum of the local intersection indices of Y and D at closed points X_f over v, and q_v is the number of residue classes. Furthermore, for $v \in S_\infty$

$$\langle Y, D_f\rangle_v = -\log G_v(D_{f,v}, y_v),$$

$$\left\langle Y, \sum_{v \in S_\infty} r_v X(v)\right\rangle = \sum_{v \in S_\infty} r_v \left(\int_{X_v} \mathrm{vol}_v\right)^{-1}.$$

Comparing (10) and (4), we obtain an equivalent definition: let $\pi^*(1_D)$ be a section of the A-sheaf $\sigma^*(\mathcal{O}(D))$ on the A-curve Y induced by the sheaf $\mathcal{O}(D)$ on X: then

$$\langle Y, D\rangle = -\log \prod_v |\pi^*(1_D)|_v. \tag{11}$$

From the product formula it follows that the right-hand side of (11) does not change if we replace $\sigma^*(1_D)$ by another non-zero section $\sigma^*(\mathcal{O}(D))$. This motivates the following general definition of the degree of an invertible A-sheaf $\mathcal{L}$ on the A-curve $Y = \operatorname{Spec} R$:

$$\deg \mathcal{L} = -\log \prod_v |s|_v, \tag{12}$$

where s is a non-zero meromorphic section of $\mathcal{L}$.

§2. Riemann-Roch theorems

1. The geometrical Riemann-Roch theorem.
Let us recall the statement of the simplest version of the Riemann-Roch-Hirzebruch theorem for smooth projective varieties over a field. Let X be such a variety of dimension d, $\mathcal{L}$ an invertible sheaf on it, $\chi(\mathcal{L}) = \sum_i (-1)^i \dim H^i(\mathcal{L})$, $c_1(\mathcal{L})$ the Chern class, and $\mathrm{td}_i(X)$ the Todd classes of X. Then

$$\chi(\mathcal{L}) = \left\langle \sum_{i=0}^{d} \frac{1}{i!} c_1(\mathcal{L})^i \, \mathrm{td}_{d-i}(X) \right\rangle, \tag{1}$$

where $\langle\,\rangle$ on the right-hand side of (1) denotes the intersection index, calculated in the Chow ring or in some cohomology ring with coefficients of characteristic zero. In particular,

$$\chi(\mathcal{L}^n) = \sum_{i=0}^{d} \frac{n^i}{i!} \langle c_1(\mathcal{L})^i \, \mathrm{td}_{d-i}(X) \rangle. \tag{2}$$

In this section we describe three special cases of a hypothetical Riemann-Roch theorem for A-varieties: for a projective A-space, an A-curve, and an A-surface. As we have already said in §1.3, the first problem, which is not yet solved in the general case, is to give a regular definition of $\chi_A(\mathcal{L})$.

2. Projective A-space.
Let us consider an A-variety, for simplicity over Spec $\mathbf{Z}$, $\mathbf{P}^d = (\mathbf{P}^d_{\mathbf{Z}}, \omega)$, where ω is a Kähler form on $\mathbf{P}^d(\mathbf{C})$. We realize $\mathbf{P}^d_{\mathbf{Z}}$ in the form Proj $S(T_{\mathbf{Z}})$, where $T_{\mathbf{Z}}$ is a free $\mathbf{Z}$-module of rank $d+1$, and put $T = \mathbf{R} \otimes T_{\mathbf{Z}}$. On the sheaf $\mathcal{O}_f(n)$ there is a canonical Hermitian metric whose curvature form is a multiple of ω, therefore, we may denote the corresponding A-sheaf by the symbol $\mathcal{O}(n)$. Since $H^0(\mathbf{P}^d, \mathcal{O}(n)) = S^n(T_{\mathbf{Z}})$, $H^i(\mathbf{P}^d, \mathcal{O}(n)) = 0$ for $i > 0$, $n \geqslant 0$, we must choose some volume form $w_n \in \Lambda_{\mathbf{R}}^{\binom{n+d}{d}}(S^n(T))$ and then put

$$\chi_A(\mathcal{O}(n)) = \log \left| \frac{w_n}{v_n} \right|,$$

where $v_n \in \Lambda_{\mathbf{Z}}^{\binom{n+d}{d}}(S^n(T_{\mathbf{Z}}))$ is one of the two generators of the corresponding cyclic group.

The simplest choice of w_n that we can imagine consists in the following. We consider the isomorphism

$$\varphi_n:\ \Lambda_{\mathbf{Z}}^{\binom{n+d}{d}}(S^n(T_{\mathbf{Z}})) \to [\Lambda_{\mathbf{Z}}^{d+1}(T_{\mathbf{Z}})]^{\otimes \frac{n}{d+1}\binom{n+d}{d}},$$

which takes v_n into $v_1^{\otimes \frac{n}{d+1}\binom{n+d}{d}}$. (Even if $\frac{n}{d+1}\binom{n+d}{d} \notin \mathbf{Z}$, the isomorphism $\varphi_n^{\otimes(d+1)}$ is well-defined: for what follows this is sufficient.) Next we choose some w_1 and put $w_n = (\varphi_n \otimes \mathrm{id}_{\mathbf{R}})^{-1}\left(w_1^{\otimes \frac{n}{d+1}\binom{n+d}{d}}\right)$. We then obtain

$$\chi_A(\mathcal{O}(n)) = \log\left|\frac{w_1}{v_1}\right|^{\otimes \frac{n}{d+1}\binom{n+d}{d}} = \frac{n}{d+1}\binom{n+d}{d}\chi_A(\mathcal{O}(1)).$$

By § 1 (8) multiplication of the A-sheaf $\mathcal{L}$ by the A-sheaf $\mathcal{O}\left(\frac{r}{n}\mathbf{P}(\infty)\right)$ expands the metric on $\mathcal{L}_\infty$ $\exp\left(-\frac{r}{n \operatorname{vol} \mathbf{P}_\infty}\right)$ times, so that if we postulate the corresponding action on w_1 we find that

$$\chi_A\left(\mathcal{O}(1) \otimes \mathcal{O}\left(\frac{r}{n}\mathbf{P}(\infty)\right)\right) = \chi_A(\mathcal{O}(1)) + \frac{r(d+1)}{n \operatorname{vol} \mathbf{P}_\infty}$$

and finally,

$$\text{(3)}\quad \chi_A(\mathcal{O}(n) \otimes \mathcal{O}(r\mathbf{P}(\infty))) = \frac{n}{d+1}\binom{n+d}{d}\chi_A(\mathcal{O}(1)) + \binom{n+d}{d}\frac{r}{\operatorname{vol} \mathbf{P}_\infty}.$$

Comparison of (3) with (1) and (2) shows that $\mathbf{P}^d$ behaves like a $(d+1)$-dimensional variety and makes it possible to guess the "Todd A-classes" $td^A_{d+1-i}(\mathbf{P}^i_{\mathbf{Z}})$.

3. The A-curve.

Let $X = \operatorname{Spec} R$, $n = [R : \mathbf{Z}]$, $r_1 = \operatorname{card}\{v \in S_\infty | K_v = \mathbf{R}\}$, $n = r_1 + 2r_2$. Let $\mathcal{L} = (\mathcal{L}_f, h_v)$ denote an invertible A-sheaf over X, and $L = H^0(X, \mathcal{L}_f)$. According to § 1.3,

$$\chi_A(\mathcal{L}) = \operatorname{vol}\left(\left(\bigoplus_{v \in S_\infty} L \otimes K_v\right)/L\right).$$

Following Weil [36] and Szpiro [33], we choose the volume form on $\bigotimes_{v \in S_\infty} L \otimes K_v$ implied in this definition as

$$w = \frac{2^n}{2^{r_1}\pi^{r_2}} \prod_{v \in S_\infty} w_v,$$

where w_v is the volume element corresponding to the Euclidean metric on L_v associated with the Hermitian metric h_v. The success of this choice is motivated by the following results, which are close to the corresponding facts for curves over finite fields.

The Riemann-Roch theorem is

$$\chi_A(\mathcal{L}) = \deg \mathcal{L} + \chi_A(\mathcal{O}_X),$$

where $\deg \mathcal{L}$ is defined by § 1, (12).

The Euler A-characteristic of the structure sheaf is

$$\chi_A(\mathcal{O}_X) = r_2 \log \frac{\pi}{2} - \frac{1}{2}\log \Delta_K,$$

where Δ_K is the absolute value of the discriminant of K.

We put $H^0_A(\mathcal{L}) = \{s \in \mathcal{L} \mid |s|_v \leqslant 1 \textit{ for all } v\}$. Then $H^0_A(\mathcal{O}) = \{0\} \cup$ *roots of unity in* R, an analogue of constants in the functional case. From Minkowski's theorem we can deduce that if $\chi_A(\mathcal{L}) \geqslant 0$, then $H^0_A(\mathcal{L}) \neq \{0\}$.

4. The A-surface.

Following Arakelov and Faltings, we consider the A-surface $X = (X_f, \omega_v)$, where $X_f \xrightarrow{\pi} \operatorname{Spec} R$ is a semistable family of curves over $\operatorname{Spec} R$ with a smooth irreducible common fibre of genus $g > 0$, and

$$\omega_v = \mathrm{vol}_v = -\frac{1}{2\pi i}\sum_{k=1}^{g} \nu_{k,v} \wedge \bar{\nu}_{k,v};$$

here $(\nu_{1,v}, \ldots, \nu_{g,v})$ is a basis of differentials of the first kind on X_v, orthonormal with respect to the scalar product

$$\langle \nu, \nu' \rangle = -\frac{1}{2\pi i}\int_{X_v(\mathbf{C})} \nu \wedge \bar{\nu}'.$$

We denote by Ω_f the relative dualizing sheaf of the morphism π. Then $\Omega_v = \Omega^1 X_v$. The canonical A-sheaf $\Omega = (\Omega_f, h_v)$ is *uniquely* determined by the following condition, which normalizes the metrics h_v: for any point $x \in X_v$ the map $\mathrm{res}_x : \Omega^1 X_v \otimes \mathcal{O}_{X_v}(x) \to \mathbf{C}$ is an isometry of the geometrical fibre of this sheaf at the point x with $\mathbf{C}$.

5. The arithmetical Euler characteristic.

The problem of defining $\chi_A(\mathcal{L})$ for an invertible A-sheaf $\mathcal{L}$ on X was solved by Faltings in the following way [10].

The main step is the definition of Hermitian metrics on the spaces $\det_{\bar{K}_v} H^0(\mathcal{L}_v) \otimes \det^{-1}_{\bar{K}_v} H^1(\mathcal{L}_v) \otimes \bar{K}_v$, $v \in S_\infty$, which are regarded as volume forms on the virtual space $H^0(\mathcal{L}_v) - H^1(\mathcal{L}_v)$. The elegance of the construction consists in the fact that it is applicable to the case when $H^0(\mathcal{L}_v) = H^1(\mathcal{L}_v) = \{0\}$: then the corresponding determinant space is $\bar{K}_v = \mathbf{C}$.

This case ($\deg \mathcal{L}_v = g-1$) is the basis of an induction on the (usual) degree of $\mathcal{L}$: any sheaf $\mathcal{L}$ can be represented in the form $\mathcal{L}_0(D)$, where $H^0(\mathcal{L}_{0,v}) = H^1(\mathcal{L}_{0,v}) = 0$, and D is a horizontal A-divisor, which can be taken to be a sum of sections, extending the base if necessary. After this we can simultaneously define $\chi_A(\mathcal{L}_0(D))$ and prove the Riemann-Roch formula by induction on $\deg D_v$, provided that we can establish that $\chi_A(\mathcal{L})$ is independent of the representation of $\mathcal{L}$ in the form $\mathcal{L}_0(D)$, which is by no means unique.

With this aim we need to go over to a variety of modules. We consider $\mathrm{Pic}_{g-1}X_v = M_v$. Let $\mathcal{E}_v$ denote the universal sheaf over $X_v \times M_v$. Let $\pi_2 : X_v \times M_v \to M_v$ be the projection. Then on M_v there is an invertible sheaf $\det R\pi_{2*}\mathcal{E}_v$, and its geometrical fibre at $y \in M_v$ corresponding to $\mathcal{L}_v(y) = \mathcal{E}_v|(X_v \times \{y\})$ is canonically identified with $\det H^0(\mathcal{L}_v(y)) \otimes \det^{-1} H^1(\mathcal{L}_v(y)) \otimes \bar{K}_v$.

Over the set $U = \{y \in M_v | H^0(\mathcal{L}_v(y)) = H^1(\mathcal{L}_v(y)) = 0\}$ the sheaf $\det R\pi_{2*}\mathcal{E}$ has a canonical unit section. On the other hand, $M_v \setminus U$ is a θ-divisor, and a simple argument shows that it is the identification $\det R\pi_{2*}\mathcal{E} = \mathcal{O}_{M_v}(-\theta)$ under which the distinguished section goes into 1. Thus, the choice of A-structure on $\mathcal{O}_{M_v}(-\theta)$ gives a simultaneous normalization of all the $\chi_A(\mathcal{L}_v)$. For this normalization to be compatible with the inductive definition of $\chi_A(\mathcal{L}_0(D))$ it turns out to be sufficient to choose a Kähler structure on M_v that becomes $-\frac{1}{2\pi i}\sum_{j=1}^{g}\sum_{k=1}^{g} \pi_k^* \nu_{j,v} \wedge \pi_k^* \bar{\nu}_{j,v}$ under a lifting to X_v^g.

We can now state the Riemann-Roch theorem for A-surfaces according to Arakelov and Faltings.

6. ***Theorem.*** a) $\chi_A(\mathcal{O}(D)) = \frac{1}{2}\langle D, D-K\rangle + \chi_A(\Omega)$, *where* $\Omega = \mathcal{O}(K)$, *and* $\langle\ \rangle$ *is the intersection index defined in* §1.6.

b)
$$\chi_A(\Omega) = \frac{1}{12}(\langle K, K\rangle + \delta), \text{ where } \delta = \sum_{v \in S_\infty} \delta_v(X_v)$$

and where $\delta_v(X_v) = \log \operatorname{card}$ (*singular points of* $X(v)$) *for* $v \in S_f$; *finally*, δ_v *for* $v \in S_\infty$ *is a certain real-analytic function on the space of modules of Riemann surfaces of genus g, measuring the closeness of* X_v *to the boundary.* ∎

Of course, b), which is the arithmetical formula of Noether, follows in the geometrical case from the Riemann-Roch-Grothendieck theorem for the morphism π.

The structure of $\delta_v(X_v)$ for $v \in S_\infty$ is also compatible with the representation of $X(v)$ as a degenerate fibre (see §1.5).

Faltings proves the second part of the theorem by means of an argument that is to some extent analogous to the introduction of Pic_{g-1} in the proof of the first part. In essence he uses canonical A-structures on schemes of modules and standard A-sheaves on them, the Riemann-Roch-Grothendieck theorem for their arithmetically finite part, and additional arguments from Kähler geometry, which "compactify" this information.

In the next section we discuss what work needs to be done to put A-geometry on a firm basis.

§3. Problems and perspectives of A-geometry

1. The problem of defining basic categories.

In the definition of A-manifolds adopted in § 1.1 we have not imposed any conditions on the Kähler forms ω_v. In the proof of the Arakelov-Faltings theorems, however, there occur distinguished Kähler structures. We give definitions of these structures in a more general context.

Let $\mathcal{E}_v$ be a locally free sheaf on a compact Kähler manifold (X_v, ω_v). Let h_v be a Hermitian metric on $\mathcal{E}_v$. We choose a holomorphic local coordinate system (z^α) on X_v and a basis of local holomorphic sections (s_i) of $\mathcal{E}_v$. We put $h_{ij} = h_v(s_i, s_j)$. The curvature tensor of the canonical connection corresponding to h_v has the form

$$F_{ij\alpha\beta} = -\partial^2/\partial z^\alpha \partial \bar{z}^\beta (h_{ij}) + h^{cb} \partial/\partial z^\alpha (h_{ib}) \cdot \partial/\partial \bar{z}^\beta (h_{aj}),$$

where $(h^{ab}) = (h_{ij})^{-1}$. Let $\omega_v = \sqrt{-1} g_{\alpha\beta}\, dz^\alpha \wedge d\bar{z}^\beta$ and $(g^{\gamma\delta}) = (g_{\alpha\beta})^{-1}$. Then $(\mathcal{E}_v, h_v)$ is called a Hermite-Einstein sheaf if

$$g^{\alpha\beta} F_{ij\alpha\beta} = \lambda h_{ij},$$

where λ is a constant. (It can be calculated explicitly: putting $n = \dim X_v$, we have $\lambda = (2\pi n \int_{X_v} c_1(\mathcal{E}_v)\omega_v^{n-1}) / (\mathrm{rk}\, \mathcal{E} \int_{X_v} \omega_v^n)$.)

On the holomorphic tangent sheaf $\mathcal{T}X_v$ there is a Hermitian metric $g_v = 2g_{\alpha\beta} dz^\alpha d\bar{z}^\beta$. Then (X_v, ω_v) is called a Kähler-Einstein manifold if $(\mathcal{T}X_v, g_v)$ is a Hermite-Einstein sheaf.

The problem of the existence and uniqueness of a Hermite-Einstein structure on a sheaf $\mathcal{E}_v$ was considered by Kobayashi [19] and Donaldson [9]. Kobayashi proved that the existence of such a structure h_v implies the semistability of $\mathcal{E}_v$, and that a semistable Hermite-Einstein sheaf $(\mathcal{E}_v, h_v)$ is isomorphic to the direct sum $\oplus (\mathcal{E}_v^{(i)}, h_v^{(i)})$, where the $\mathcal{E}_v^{(i)}$ are already stable. (Stability is defined as the condition that the function $\mu(\mathcal{F}) = \deg_{\omega_v} \mathcal{F} / \mathrm{rk}\, \mathcal{F}$ is monotonic on the subsheaves $\mathcal{F} \subset \mathcal{E}_v$, where $\deg_{\omega_v} \mathcal{F} = \int_{X_v} c_1(\mathcal{F}) \omega_v^{n-1}$.)

On the other hand, Donaldson showed that on projective algebraic surfaces (X_v, ω_v) every stable sheaf has a unique Hermite-Einstein structure (up to multiplication of h_v by a constant). The same is true for algebraic curves and, presumably, in the general case.

The existence and uniqueness of Kähler-Einstein metrics on Kähler manifolds was studied by Yau [38] and Aubin [3]. From Yau's results it follows, in particular, that for $c_1(X_v) = 0$ in every cohomology class of Kähler metrics there is a unique Kähler-Einstein metric. Aubin proved that if there is a form in the class $c_1(X_v)$ that corresponds to a negative definite metric, then on X_v there is a unique Kähler-Einstein structure with constant $\lambda = -1$.

The existing fragments of A-geometry apparently point to the special role of those A-manifolds for which the structures (X_v, ω_v), $v \in S_\infty$, are Hermite-Einstein manifolds. This condition appears as a regular analogue of the minimality condition for X_f over Spec R in the sense of birational geometry. Next, on a given A-manifold locally free A-sheaves should apparently be defined as structures $\mathcal{E} = (\mathcal{E}_f; h_v, v \in S_\infty)$, where $\mathcal{E}_f$ is a locally free sheaf on X_f, and $(\mathcal{E}_v, h_v)$ are Hermite-Einstein sheaves on (X_v, ω_v). In the case $\text{rk}\, \mathcal{E}_f = 1$ this leads to the definition of §1.2.

The categorical aspects of these definitions are needed in clearing up. Since the Hermite-Einstein property is apparently a proper definition only for locally free and semistable sheaves, its proper generalization must refer to "complete complexes of A-sheaves", a concept that still has to be defined by analogy with [35].

Since for any $v \in S$ there are selected volume forms on A-sheaves $\mathcal{E}_v$ (for $v \in S_f$ they are induced by the integral structure), complexes of A-sheaves have invariants of "torsion" type, which must be replaced by the Euler characteristic. Apparently, for $v \in S_\infty$ the corresponding component χ_A is defined as the analytic torsion of the Dolbeault complex according to Quillen [30].

It is not clear how we should define morphisms of A-manifolds. The theory of hyperbolicity of Kobayashi [18] must have a relation to this problem. For it is based on the study of the morphisms $D \to X_v$ and $X_v \to D$, where $D = \{z \in \mathbf{C} : |z| \leqslant 1\}$ is the unit disc, which is an analogue of Spec $\mathbf{Z}_p$, so that the theory of metrics of Carathéodory and Kobayashi must be parallel to the theory of $\mathbf{Z}_p$-models of $\mathbf{Q}_p$-manifolds.

2. The problem of canonical A-structures on varieties of modules and A-varieties of modules.

As the papers of Arakelov and Faltings show, on varieties of modules of curves and invertible sheaves on curves there are selected structures of A-varieties, the use of which leads to good formulations of the basic theorems. (For a space of modules of curves having a boundary, this principle must be regarded as heuristic, since there is no certainty in the basic definitions.)

From this point of view we should study varieties of modules of stable bundles on curves whose degree is prime to the rank. The first unsolved problem here is to extend the Riemann-Roch-Arakelov-Faltings theorem to A-sheaves of any rank over A-surfaces. In particular, for sheaves $\mathcal{E}$ of rank at least 2 there appears the second Chern A-class $c_2^A(\mathcal{E})$, which should be an interesting arithmetic invariant.

Once the category of A-varieties has been properly defined, we can hope for more, namely, the existence of object-modules in this category. Of course, the first problems here are again connected with purely "infinite" effects, that is, with problems of Kähler geometry. In this respect a paper by Koiso [20] deserves serious attention; this shows, in particular, that on the base of a normal and stable family of Kähler-Einstein structures there is a canonical Kähler structure. Unfortunately, in the majority of cases it is not known whether it is itself a Kähler-Einstein structure.

3. The problem of intersection theory on A-varieties.
In an important paper [4] Veilinson constructed a theory of regulators and laid the foundations of the technique necessary for the construction of intersection theory on A-varieties of any dimension. We describe a small part of his results, which emphasize the role of arithmetical K-theory as a cohomology theory.

Let X_f be a regular flat projective scheme over $\mathbf{Z}$, where $\dim X_f = d+1$. We fix two cycles z_i of pure codimension l_i in X_f. Suppose that $l_0+l_1 = d+1$ and that $\mathrm{supp}(z_0 \cap z_1) \cap X_{f,\mathbf{Q}} = \emptyset$. We describe Veilinson's construction of the A-intersection index $\langle z_0, z_1\rangle$. For simplicity we assume that both cycles are $\mathbf{Q}$-homologous to zero on a common fibre, which allows us to disregard the metric ω_∞ on X_∞: in this case the intersection index does not depend on ω_∞. We put

$$\langle z_0, z_1\rangle = \langle z_0, z_1\rangle_\infty + \sum_{p\in S_f} \log p \cdot \langle z_0, z_1\rangle_p$$

and define the fibrewise intersection indices $\langle\ \rangle_v$ by means of some cohomology theories.

For the construction of $\langle\ \rangle_\infty$ we need the Deligne-Veilinson cohomology $H^j_{\mathscr{D}}$. If X_∞ is a smooth complex manifold, then for any ring of coefficients $B \subset \mathbf{R}$ and $B(j) = (2\pi\sqrt{-1})^j B$ this cohomology is defined by the formula

$$H^k_{\mathscr{D}}(X_\infty, B(j)) = R^k\Gamma(X_\infty, B(j)_{\mathscr{D}}),$$

$$B(j)_{\mathscr{D}} = \mathrm{Cone}(F^j \oplus B(j) \xrightarrow{\alpha} \Omega^\cdot)[-1],$$

where $F^j = \Omega^{\geqslant j}$ is the complex of holomorphic forms $\Omega^\cdot$, truncated on the left, and $\alpha = \alpha_1 - \alpha_2$, where the α_i are the natural embeddings. In the non-compact case we use forms with logarithmic singularities. Standard cohomology methods also make it possible to construct $\mathscr{D}$-cohomology of simplicial schemes and relative $\mathscr{D}$-cohomology, to define classes of algebraic cycles, to construct multiplication, and to prove the Poincaré duality theorem.

In the notation of the beginning of this subsection we put $U_i = X_\infty - \mathrm{supp}\, z_{i,\infty}$. If the classes $z_{i,\infty}$ in $H^{2l_i}_{\mathscr{D}}(X_\infty, \mathbf{R}(l_i))$ are zero, then from the Mayer-Vietoris exact sequence we find that the classes $cl_{\mathscr{D}}\, z_i \in H^{2l_i}_{\mathscr{D}}(X, U_i; \mathbf{R}(l_i))$ have the form $\partial\zeta_i$, where $\zeta_i \in H^{2l_i-1}_{\mathscr{D}}(U_i, \mathbf{R}(l_i))$. After this we construct the class $\zeta_0 \cup \zeta_1 \in H^{2d}_{\mathscr{D}}(U_0 \cap U_1, \mathbf{R}(d+1))$ and its image, the "intersection index"

$$(z_0 \cap z_1)_\infty = \partial(\zeta_0 \cup \zeta_1) \in H^{2d+1}_{\mathscr{D}}(X, \mathbf{R}(d+1)).$$

We denote by $\pi : X_\infty \to \operatorname{Spec} \mathbf{R}$ a structure morphism. Finally, the A-intersection index of z_0 and z_1 at the arithmetic infinity is

$$\langle z_0, z_1\rangle_\infty = \pi_* (z_0 \cap z_1)_\infty \in H^1_{\mathscr{D}} (\operatorname{Spec} \mathbf{R}, \mathbf{R}(1)) = \mathbf{C}/\mathbf{R}(1) = \mathbf{R}.$$

To define $\langle z_1, z_2\rangle_p$ by a similar formula, Veilinson uses the K-cohomology:

$$H^j_K (X_f, \mathbf{Q}(i)) = K^{(i)}_{2i-j} (X_f) \otimes \mathbf{Q} = \{a \in K_{2i-j} (X_f) \otimes \mathbf{Q} \mid \psi^p (a) = p^i a\},$$

where ψ^p is the Adams operation, and also the corresponding versions of the relative theory. As above, by putting $U_i = X_f \setminus \operatorname{supp} z_i$ we can construct the "linkage index" class

$$(z_0 \cap z_1)_f \in H^{2d+2}_K (X_f, U_0 \cup U_1; \mathbf{Q}(d+1)).$$

Let $S' = \operatorname{supp} \pi_f (X_f \setminus (U_0 \cup U_1))$, where $\pi_f : X_f \to \operatorname{Spec} \mathbf{Z}$ is a structure morphism. This is a finite set of prime numbers. Now $\langle z_0, z_1\rangle_p = 0$ for $p \notin S'$, and for $p \in S'$ the number $\langle z_0, z_1\rangle_p$ is a version of the direct image $\pi_{f*} (z_0 \cap z_1)_f$, localized with respect to p.

4. The problem of the arithmetical Euler characteristic and Riemann-Roch A-theorems.

I can add a little to what was said above. It is possibly worth while to draw attention once more to the fact that the volume forms needed to define $\chi_A (\mathscr{L})$ in full generality must be closely connected with the analytic torsion of Dolbeault complexes, so that the Riemann-Roch A-theorem in Grothendieck's form may require a theory of relative torsion (in families).

Furthermore, independently of a hypothetical general theory, problems of A-geometry for flag manifolds $(G/P)_{\mathbf{Z}}$ may turn out to be very interesting and accessible by direct methods. A recent paper by Bombieri and Vaaler [6] should certainly be considered from this point of view. In particular, it suggests that Minkowski's classical "geometry of numbers" must be interpreted in A-geometry as the theory of the ∞-component of characteristic classes.

§4. Superspaces

1. Examples of supermanifolds.

Manifolds (for example, smooth or complex-analytic) can be specified by local charts and transition functions between them. Before we define supermanifolds formally, we describe some examples, specified in this way.

a) *m/n-dimensional affine superspace.* It is defined by a global coordinate system $(x_1, \ldots, x_m; \xi_1, \ldots, \xi_n)$, where the x_i commute, and the ξ_i anticommute. In the category of superschemes over a commutative ring A the ring of superfunctions on it is a Grassmann algebra with generators ξ_i over the polynomial ring $A[x_1, \ldots, x_m]$. In the category of smooth supermanifolds the ring of superfunctions is $C^\infty(x_1, \ldots, x_m)[\xi_1, \ldots, \xi_n]$.

b) *m|n-dimensional projective superspace.* It is defined by an atlas of $m+1$ charts U_i; each of the U_i is an $m|n$-dimensional superspace. By introducing a system of homogeneous coordinates $(X_0, X_1, \ldots, X_m; \Xi_1, \ldots, \Xi_n)$ we can describe the coordinates $(x^i_j, j \neq i | \xi^i_1, \ldots, \xi^i_n)$ on U_i by putting $x^i_j = X_j/X_i$, $\xi^i_j = \Xi_j/X_i$. From this it is easy to restore the transition functions.

c) *The super-Grassmannian of $d_0|d_1$-dimensional linear subspaces of a $(d_0+c_0|d_1+c_1)$-dimensional linear space.* The following atlas of a Grassmannian is called standard.

We consider $(d_0+d_1) \times (d_0+c_0+d_1+c_1)$-matrices split into four blocks, so that the one in the top left is a $d_0 \times (d_0+c_0)$-matrix. For each subset I of columns of such a matrix containing d_0 columns of the left half and d_1 columns of the right half we construct the matrix

$$Z_I = \begin{array}{c} \\ d_0 \\ d_1 \end{array}\!\!\begin{array}{c} \begin{array}{cccc} c_0 & d_0 & d_1 & c_1 \end{array} \\ \left(\begin{array}{c|c|c|c} x_I & \begin{matrix}1 & & \\ & \ddots & \\ & & 1\end{matrix} & 0 & \xi_I \\ \hline \xi_I & 0 & \begin{matrix}1 & & \\ & \ddots & \\ & & 1\end{matrix} & x_I \end{array}\right) \\ \underbrace{\hphantom{xxxxxxxxxxxx}}_{I} \end{array} \tag{1}$$

In this matrix the columns of I form a unit submatrix, in the blocks of the main diagonal in the remaining places there are the even variables x^{qb}_I, and in the blocks of the opposite diagonal there are the odd variables ξ^{qb}_I. The variables (x^{qb}_I, ξ^{qb}_I) are independent coordinates on the I-th chart U_I. Let B_{IJ} be the submatrix of Z_I formed by the columns with the numbers J. The rules for a change of coordinates are given by the equations $Z_J = B^{-1}_{IJ} Z_I$.

If we put here $d_0|d_1 = 1|0$, $d_0+c_0|d_1+c_1 = m+1|n$, we obtain the projective superspace of the previous section. If we put $d_1 = c_1 = 0$, we obtain a standard description of the Grassmannian, covered by large cells.

We now turn to an accurate definition of the category of superspaces, some objects of which were introduced above. As a preliminary we briefly describe the basic constructions of superalgebra. For further details, see [5], [23], [24].

2. Superalgebra.

The main laws of composition that are important for geometry are naturally divided into additive and multiplicative ones. In superalgebra all additive groups are endowed with the $\mathbf{Z}_2$-grading, with which all multiplications are compatible. We write $A = A_0 \oplus A_1$ and $\tilde{a} = \varepsilon$ if $a \in A_\varepsilon$; then $\widetilde{ab} = \tilde{a}+\tilde{b}$. Elements of A_0 are called even and elements of A_1 odd. A characteristic peculiarity of a superalgebra is that in the definitions and axioms and in polynomial identities of basic structures there appear the additional factors ± 1 (compared with ordinary algebra). Let us give some basic examples.

Let $A = A_0 \oplus A_1$ be an associative ring. The supercommutator of homogeneous elements $a, b \in A$ is defined by the formula $[a, b] = ab - (-1)^{\tilde{a}\tilde{b}} ba$. The ring A is supercommutative if $[a, b] = 0$ for all a and b. When 2 is invertible (which we always assume), then $a^2 = [a, a]/2 = 0$ for all $a \in A_1$. Supercommutators satisfy two identities:

$[a, b] = -(-1)^{\tilde{a}\tilde{b}}[b, a]$ and $[a, [b, c]] + (-1)^{\tilde{a}(\tilde{b}+\tilde{c})}[b, [c, a]] + (-1)^{\tilde{c}(\tilde{a}+\tilde{b})}[c, [a, b]] = 0.$

These two identities (plus bilinearity) are the defining relations for Lie superalgebras. Ring morphisms preserve the $\mathbf{Z}_2$-grading.

Over supercommutative rings, just as over ordinary commutative rings, the concepts of a left, right, and bimodule are equivlaent: left and right multiplications are connected by the formula as $= (-1)^{\tilde{a}\tilde{s}} sa$. A new construction is the change of parity functor Π: $(\Pi S)_0 = S_1$, $(\Pi S)_1 = S_0$; right multiplications of A by S and ΠS coincide. A free A-module of rank $p|q$ is a module isomorphic to $A^{p|q} = A^p \oplus (\Pi A)^q$. The tensor algebra of A-modules is distinguished from the usual tensor algebra by the introduction of a sign in the canonical isomorphism

$$\varphi : S \otimes T \simeq T \otimes S : \varphi(s \otimes t) = (-1)^{\tilde{s}\tilde{t}} t \otimes s.$$

In the category of A-modules there are internal Hom's; the usual homomorphisms are Hom_0. Elements of Hom_1 are also called odd morphisms: they satisfy the identity $f(as) = (-1)^{\tilde{a}} af(s)$.

Morphisms between free A-modules can be described by matrices; we must remember that the transition from row-coordinates to columns is also associated with changes of signs. Matrices are often written in the standard format, like (1), when in the implied bases of modules the first places are occupied by even elements and the last by odd elements. GL is defined as the group of invertible matrices that represent even morphisms.

The superdeterminant, or Berezinian, $\mathrm{Ber} : \mathrm{GL}(p|q, A) \to A_0^{\bullet}$ was devised by Berezin. It is a rational function of the elements of the matrix and in the standard format it is given by the formula

$$\mathrm{Ber}\begin{pmatrix} B_1 & B_2 \\ B_3 & B_4 \end{pmatrix} = \det(B_1 - B_2 B_4^{-1} B_3) \det B_4^{-1}.$$

The kernel of Ber is denoted by SL. Because Ber is rational, super-Grassmannians do not have Plücker coordinates.

The Berezinian of a free module Ber S is defined as a free module of rank $1|0$ or $0|1$ (depending on the parity of q, where $\mathrm{rk}\, S = p|q$), generated by the $D(s_1, \ldots, s_{p+q})$, where (s_i) is a basis of S, with the relations $D(f(s_1), \ldots, \ldots f(s_{p+q})) = \mathrm{Ber}\, f \cdot D(s_1, \ldots, s_{p+q})$. It replaces the concept of the maximal exterior degree.

Binlinear forms with symmetry conditions on the A-module T are divided into four main types: OSp (even symmetric), SpO (even antisymmetric), ΠO (odd symmetric), ΠSp (odd antisymmetric). The form b^{Π} on ΠT is defined by the formula $b^{\Pi}(\Pi t_1, \Pi t_2) = (-1)^{\tilde{t}_1} b(t_1, t_2)$. This construction preserves the parity, but changes the symmetry, so that it is often sufficient to consider forms of the types OSp and ΠSp.

The functor of automorphisms of non-degenerate forms defines supergroups of the corresponding types. Apart from SL, OSp, and ΠSp, in algebra there is one more series of supergroups of classical type: these are the groups Q of automorphisms of an odd involution $p : T \to T$, $\tilde{p} = 1$, $p^2 = \mathrm{id}$. Their Lie superalgebras, after some reduction (like factorization by the centre) form the classical part of the classification of Kac [15] of simple finite-dimensional Lie superalgebras. There is still a continuous family of exceptional superalgebras and the Cartan superalgebras of vector fields on the spectra of a Grassmann algebra.

Superdifferentiation $X : A \to A$ satisfies the Leibniz formula $X(ab) = (Xa)b + (-1)^{\widetilde{X}\tilde{a}} aXb$. There are two natural modules of relative differentials of an A-algebra B: $\Omega^1_{ev} B/A$ and $\Omega^1_{odd} B/A$, that classify even and odd differentiations, respectively. Henceforth we shall use $\Omega^1_{odd} B/A$, because the corresponding de Rham algebra turns out to be supercommutative, whereas for $\Omega^1_{ev} B/A$ it is superanticommutative.

3. Supergeometry.

The most general geometrical definition of a "space with even and odd coordinates" is now the concept of a superspace. A superspace is a pair $(M, \mathcal{O}_M)$, where M is a topological space and $\mathcal{O}_M$ a sheaf of local supercommutative rings on it. Morphisms of superspaces are morphisms of locally ringed spaces.

The objects of the basic geometrical categories are differentiable and analytic manifolds, analytic spaces, and schemes—they are all superspaces with $\mathcal{O}_M = (\mathcal{O}_M)_0$. We call such superspaces purely even.

In the general case we put $J_M = \mathcal{O}_M \cdot \mathcal{O}_{M,1}$, $\mathrm{Gr}_i\mathcal{O}_M = J^i_M/J^{i+1}_M$. Furthermore, $M_{rd} = (M, \mathrm{Gr}_0\,\mathcal{O}_M)$, $\mathrm{Gr}\,M = (M, \bigoplus_{i=0}^{\infty} \mathrm{Gr}_i\,\mathcal{O}_M)$. The structure sheaf $\mathrm{Gr}\,M$ has the natural $\mathbf{Z}$-grading; when we regard $\mathrm{Gr}\,M$ as a superspace, we have in mind the corresponding $(\mathbf{Z}/2\mathbf{Z})$-grading.

By means of these constructions we can define the very simple and important class of superspaces. A supermanifold, differentiable, analytic, or algebraic, is a superspace $(M, \mathcal{O}_M)$ such that M_{rd} is a purely even manifold of the corresponding class, and the sheaf $\mathcal{O}_M$ is locally isomorphic to $\mathrm{Gr}\,\mathcal{O}_M$, which, in turn, is isomorphic to the Grassmann algebra of the locally free sheaf over $\mathcal{O}_M/J_M$ of finite rank J_M/J^2_M. (We observe that since the rank J_M/J^2_M is purely odd, this Grassmann algebra is symmetric.)

Supermanifolds can be described in terms of local coordinate systems $(x_1, \ldots, x_m; \xi_1, \ldots, \xi_n)$: the sheaf $\mathcal{O}_M$ consists of functions $\sum_\alpha f_\alpha(x)\xi^\alpha$ of the corresponding class. A typical feature of supergeometric constructions is their invariance under local changes of coordinates that interchange even and odd coordinates, of the type $y = x + \xi_1\xi_2$, $\eta_1 = (1 + x^2)\xi_1$, $\eta_2 = \xi_2$.

Under such a change of coordinates the description of the growth of $f(x)$ becomes $f(y - \frac{1}{1+x^2}\eta_1\eta_2) = f(y) - \frac{f'(y)}{1+x^2}\eta_1\eta_2$. The occurence of derivatives in the formulae for coordinate changes plays a special role in the Lagrangian formalism of supersymmetric field theory. It shows that in the case of continuous supergeometry the natural sheaf must contain generalized functions; the possibilities revealed here have not been studied.

Among the most important superspaces, which are not necessarily supermanifolds, there are the analytic superspaces ($\mathcal{O}_{M,0}$ is the structure sheaf of a purely even analytic space, and $\mathcal{O}_{M,1}$ is coherent over $\mathcal{O}_{M,0}$) and the superschemes (whose definition is obvious).

We regard the concept of a locally free sheaf of $\mathcal{O}_M$-modules as the replacement of a vector bundle over a superspace. In the standard way we define the tangent sheaf $\mathcal{T}M$, the cotangent sheaf $\Omega^1 M = \Omega^1_{\text{odd}} M$, and their relative versions. For supermanifolds we always construct such sheaves over the field of constants; they are locally free; their rank is called the (super)dimension of the space.

The reader will now have no difficulty in converting the descriptions of §4.1 into the definitions of the corresponding objects of algebraic, complex-analytic, or smooth categories. We observe that, as in the purely even case, Grassmannians are endowed with the tautological sheaf: over U_I it is generated by the rows of the matrix Z_I. Over a projective superspace this sheaf is denoted by $\mathcal{O}(-1)$.

4. Methods of constructing superspaces.

a) Let $(M, \mathcal{O}_M)$ be a purely even locally ringed space, and $\mathcal{E}$ a locally free sheaf of modules of rank $0|q$ over $\mathcal{O}_M$. Then $M, \mathcal{O}_{M,\bullet} = S(\mathcal{E})$ is a superspace. In the differentiable (C^∞-) category any supermanifold is splittable, that is, it is obtained in this way from an ordinary manifold and a bundle over it. In the analytic and algebraic categories this is not so, for example, Grassmann supermanifolds are analytically unsplittable, except when the tautological sheaf has purely even or purely odd rank (see §4.7 below).

b) As in purely even analytic and algebraic geometry, the most important superspaces are defined by their point functors. Such are the superalgebraic groups GL, SL, OSp, ΠSp; the flag supermanifolds $F(d_1, \ldots, d_k; T)$, where $d_1 < d_2 < \ldots < d_k < \dim T$ are the superdimensions of the components of the flag. One should expect that the main theorems about representable functors and spaces of modules of purely even geometry have analogues in supergeometry, although systematic work in this direction has hardly started. We give two results due to Weintraub, which concern the local theory of Kodaira-Spencer deformations. The basic definitions of this theory go over to superspaces without substantial changes. The infinitesimal deformations form the ring $\mathbf{C}[x, \xi]/(x^2, x\xi)$.

Let M be a complex compact supermanifold and $\mathcal{T}M$ its tangent sheaf.

5. ***Theorem.*** a) *Suppose that* $\dim H^1(M, \mathcal{T}M) = a|b$. *If* $H^2(M, \mathcal{T}M) = 0$, *then there is a local deformation of M in the category of analytic supermanifolds with base* $B = \mathbf{C}^{a|b}$ *for which the Kodaira-Spencer morphism* $\rho\colon T_0B \to H^1(M, \mathcal{T}M)$ *is an isomorphism.*

b) *Any deformation of M over a supermanifold for which the Kodaira-Spencer map ρ is epimorphic is complete: in particular, if ρ is an isomorphism, then it is a versal deformation.* ∎

Let us give one example. Let M be a compact analytic manifold of dimension $1|1$. It is completely determined by the Riemann surface $M_0 = M_{\mathrm{rd}}$ and the invertible sheaf $\Pi J_M = \mathcal{L}$ on M_0. Suppose that $g > 1$, where g is the genus of M_0, $\deg \mathcal{L} = 0$, and $\mathcal{L}$ is not isomorphic to $\mathcal{O}_{M_0}$. In this case, $\operatorname{rk} H^1(M, \mathcal{T}M) = 4g-3|4g-4$. The even part $4g-3$ of this dimension refers to the classical manifold Z of deformations of the pair $(M_0, \mathcal{L})$: it is fibered into Jacobians over the coarse manifold of modules of curves of genus g. The theorem above shows that outside the zero section Z is endowed with the canonical structure of a supermanifold of odd dimension $4g-4$. This structure deserves further study.

6. ***Theorem.*** *Let M be a closed compact subsupermanifold of a complex supermanifold M′. Let* $\mathcal{N}$ *denote the sheaf normal to M in M′.*

a) *Suppose that* $\dim H^0(M, \mathcal{N}) = a|b$. *If* $H^1(M, \mathcal{N}) = 0$, *then there is a versal local deformation of M into M′ whose base is a supermanifold of dimension* $a|b$.

b) *The deformation of M into M′ with base B is complete if and only if the Kodaira-Spencer map* $\rho\colon T_0B \to H^0(M, \mathcal{N})$ *is surjective.* ∎

7. *Example.* We turn once more to the definition of a super-Grassmannian and illustrate some of our constructions by this example. The super-Grassmannian $G = G(1|1;\ \mathbf{C}^{2|2})$ is covered by four affine superspaces of dimension $2|2$; the matrices Z_I for it are:

$$\begin{pmatrix} x_1 & 1 & \xi_1 & 0 \\ \eta_1 & 0 & y_1 & 1 \end{pmatrix}, \quad \begin{pmatrix} x_2 & 1 & 0 & \xi_2 \\ \eta_2 & 0 & 1 & y_2 \end{pmatrix}, \quad \begin{pmatrix} 1 & x_3 & \xi_3 & 0 \\ 0 & \eta_3 & y_3 & 1 \end{pmatrix}, \quad \begin{pmatrix} 1 & x_4 & 0 & \xi_4 \\ 0 & \eta_4 & 1 & y_4 \end{pmatrix}.$$

The transition functions can be calculated by using the prescriptions in §4.1, for example,

$$\begin{pmatrix} x_1 & \xi_1 \\ \eta_1 & y_1 \end{pmatrix} = \begin{pmatrix} x_4 & \xi_4 \\ \eta_4 & y_4 \end{pmatrix}^{-1} = \begin{pmatrix} x_4^{-1} + x_4^{-2} y_4^{-1} \xi_4 \eta_4 & -x_4^{-1} y_4^{-1} \xi_4 \\ -x_4^{-1} y_4^{-1} \eta_4 & y_4^{-1} - x_4^{-1} y_4^{-2} \xi_4 \eta_4 \end{pmatrix}.$$

From this it follows immediately that $G_{\mathrm{rd}} = \mathbf{P}^1 \times \mathbf{P}^1$. If we take a little trouble, we can define a canonical isomorphism of $\mathcal{O}_{G_{\mathrm{rd}}}$-modules

$$(\mathcal{O}_{G,1})^2 = \Omega^2(\mathbf{P}^1 \times \mathbf{P}^1).$$

Therefore, the obstruction ω to the splitting of G lies in the group

$$H^1(\mathcal{T}(\mathbf{P}^1 \times \mathbf{P}^1) \otimes \Omega^2(\mathbf{P}^1 \times \mathbf{P}^1)) = H^1(\mathbf{P}^1, \Omega^1\mathbf{P}^1) \oplus H^1(\mathbf{P}^1, \Omega^1\mathbf{P}^1) = \mathbf{C} \oplus \mathbf{C}.$$

A direct calculation of the Čech cocycle in the standard covering shows that this obstruction is equal to (1, 1), so G is unsplittable.

(We remark that projective superspace is splittable:
$\mathbf{P}^{m|n} = (\mathbf{P}^m, S(\Pi\, \mathcal{O}^n_{\mathbf{P}^m}(-1)))$.)

Moreover, G is not a projective variety: the image of the Picard group $H^1(G, \mathcal{O}^{\bullet}_{G,0}) \to \mathrm{Pic}(\mathbf{P}^1 \times \mathbf{P}^1)$ consists of classes of sheaves $\mathcal{O}(a, -a)$, $a \in \mathbf{Z}$, because the obstruction to the extendability of the sheaf $\mathcal{O}(a, b)$ from G_{rd} to G is proportional to $(a+b)\omega$. From this it is not difficult to deduce that all the Grassmannians $G(a|b; \mathbf{C}^{m|n})$ with $0 < a < m$, $0 < b < n$ are non-projective, since G admits a closed embedding into them.

Restricted use of the projective technique in algebraic supergeometry forces us from the very start to revise many proofs in which invertible sheaves are usually used.

For example, the dualizing sheaf on a smooth complex supermanifold X is isomorphic to Ber $\Omega^1_{ev}X$. This conjecture was made by Deligne, and a proof, not using hypotheses about projectiveness, was obtained by Penkov, who showed in [29] that working with $\mathcal{D}$-modules on supermanifolds provides in this case an effective method of reducing the situation to the purely even case.

§5. Schubert supercells

1. Statement of the problem.

Let G be a semisimple algebraic group, and $B \subset G$ a Borel subgroup of it. The orbits of G on $G/B \times G/B$ form a finite stratification of this manifold, and the strata Y_w are naturally numbered by elements w of the Weyl group $W = W(G)$ (the Bruhat decomposition). We call the manifolds $X_w(b) = (\{b\} \times G/B) \cap Y_w \subset G/B$ Schubert cells. Similarly, by means of the orbits of G on $G/B \times G/P$ we can define Schubert cells for any parabolic subgroup P. The geometry of Schubert cells plays a large role in the theory of characteristic classes and representation theory.

Our aim in this section is to introduce Schubert supercells for homogeneous supermanifolds of classical type and to establish analogues of classical results for them.

Let T be a linear superspace of dimension $m|n$ over a field of characteristic not equal to 2. We consider the following algebraic supergroups with fundamental representation space T: a) $G = \mathrm{SL}(T)$; b) $G = \mathrm{OSp}(T)$, the group of automorphisms of a non-degenerate even symmetric form $b: T \to T^*$; c) $G = \Pi\mathrm{Sp}(T)$, the group of automorphisms of a non-degenerate odd antisymmetric form $b: T \to T^*$; d) $G = Q(T)$, the group of automorphisms of an odd involution $p: T \to T$, $p^2 = \mathrm{id}$. In the cases $\Pi\mathrm{Sp}$ and Q we obviously have $m = n$.

As an analogue of the classical manifolds G/B we must consider the supermanifolds F of complete flags in T that are compatible with b or p in the cases G = OSp, ΠSp, Q (for precise definitions, see below). We mention some differences between this and the classical situation.

Firstly, stabilizers of complete flags B are not, generally speaking, maximal soluble subsupergroups. However, in the theory of the highest weight [16] and in the theory of Schubert supercells they play exactly the same role as the classical Borel subgroups. Secondly, not all subgroups B are conjugate to each other: F splits into connected components. Thirdly, the stratification of $F \times F$ that we wish to construct is not purely set-theoretical: it is a partition of $F \times F$ into the union of locally closed subsuperschemes. The flag model is well suited to this construction.

The most urgent unsolved problem is to construct a cohomology theory in which classes of Schubert cells are the free generators. As we mentioned in the Introduction, this possibly requires a certain "compactification with respect to odd directions".

We now describe some of the details.

2. Connected components of flag supermanifolds.

These connected components are naturally numbered (with one exception) by sets ${}^{G}I$, which we now define. We put

$$^{SL}I = \{(\delta_1, \ldots, \delta_{m+n}) | \delta_i = 1|0 \text{ or } 0|1, \sum_{i=1}^{m+n} \delta_i = m|n\}.$$

Also, let

$$^{OSp}I = \{(\delta_1, \ldots, \delta_{m+n}) \in {}^{SL}I | \delta_i = \delta_{m+n+1-i} \quad (i = 1, \ldots, m+n)\},$$

$$^{\Pi Sp}I = \{(\delta_1, \ldots, \delta_{2m}) \in {}^{SL}I | \delta_i = \delta^c_{2m+1-i} \quad (i = 1, \ldots, 2m)\},$$

where $(p|q)^c = q|p$. Finally, ${}^{Q}I = \{ \underbrace{(1|1, \ldots, 1|1)}_{m} \}$ (a one-element set).

The flag f: $0 = \mathscr{S}_0 \subset \mathscr{S}_1 \subset \ldots \subset \mathscr{S}_{m+n} = T$ has the type $I \in {}^{G}I$ if $\delta_i(I) = \operatorname{rk} \mathscr{S}_i/\mathscr{S}_{i+1}$. For the groups G = OSp, ΠSp, and Q the flag f is said to be G-invariant if

$$b(\mathscr{S}_i) = \mathscr{S}^{\perp}_{m+n-i} \text{ for OSp, } \Pi\text{Sp}; \quad p(\mathscr{S}_i) = \mathscr{S}_i \text{ for } Q.$$

The functor ${}^{G}F_I$ on the category of superschemes over the ground field sets up a correspondence between the superscheme S and the set of flags of the type $I \in {}^{G}I$ in the sheaf $\mathcal{O}_S \otimes T$ that are G-invariant for $G \neq$ SL. (A sequence of subsheaves of $\mathcal{O}_S$-modules $\ldots \subset \mathscr{S}_i \subset \ldots \subset \mathcal{O}_S \otimes T$ is called a flag if all the embeddings $\mathscr{S}_i \subset \mathscr{S}_j$ are locally direct, and in the definition of G-invariance there appear the forms $\mathrm{id}_{\mathcal{O}_S} \otimes b$ and the involution $\mathrm{id}_{\mathcal{O}_S} \otimes p$.)

3. ***Theorem.*** *The functor ^GF_I is an irreducible supermanifold in all cases except $G = \mathrm{OSp}(2r|2s)$, $r \geqslant 1$, where ^GF_I consists of two components of the same dimension. Moreover:*

$$\dim {}^{\mathrm{SL}}F_I = \left(\frac{m(m-1)}{2} + \frac{n(n-1)}{2} \,\middle|\, mn\right);$$

$$\dim {}^{\mathrm{OSp}}F_I = \begin{cases} (r^2+s^2|(2r+1)s) & \text{for } m=2r+1,\ n=2s>0; \\ (r^2-r+s^2|2rs) & \text{for } m=2r,\ n=2s>0; \end{cases}$$

$$\dim {}^{Q}F_I = \left(\frac{m(m-1)}{2} \,\middle|\, \frac{m(m-1)}{2}\right);$$

$$\dim {}^{\Pi\mathrm{Sp}}F_I = \left(rs + \frac{r(r-1)}{2} + \frac{s(s+1)}{2} \,\middle|\, rs - \frac{r(r+1)}{2} + \frac{s(s-1)}{2}\right),$$

where $r|s = \sum_{i=1}^{m} \delta_i(I)$. ■

Of course, the functors of incomplete flags and the standard morphisms "projections over subflags" are also representable. The same is true in the relative version, for flags in a given locally free sheaf over a superscheme. It is convenient to prove the theorem in the same way: to construct directly schemes of relative Grassmannians, as in §4, and to carry out an induction on the length of a flag. The main part of these constructions is described in the book [24].

We now put $^GF = \coprod_{I\in {}^GI} {}^GF_I$ and denote by $\mathscr{S}_i \subset \mathcal{O}_F \otimes T$ the components of the tautological flag. On $^GF \times {}^GF$ there are two natural flags: $\{p_1^*(\mathscr{S}_i)\}$ and $\{p_2^*(\mathscr{S}_j)\}$, where $p_{1,2}$ are the projections on ^GF. In the classical theory each G-orbit corresponds to those points of $^GF \times {}^GF$ over which the type of the mutual disposition of the flags $\{p_1^*(\mathscr{S}_i)\}$ and $\{p_2^*(\mathscr{S}_j)\}$ is fixed. We can imitate this definition in supergeometry, replacing points by morphisms.

We define the type of the mutual disposition of two complete flags $(\mathscr{S}_i)$ and $(\mathscr{S}'_j)$ in T as the matrix $d_{ij} = \mathrm{rk}\,(\mathscr{S}_i + \mathscr{S}'_j)$.

We define the Weyl groups ^GW, together with their natural action on ^GI:

$$^{\mathrm{SL}}W = S_{m+n};\quad {}^QW = S_m;$$

$$^GW = \{g \in S_{m+n} |\, g\,({}^GI) \subset {}^GI\} \text{ for } G = \mathrm{OSp}, \Pi\mathrm{Sp}.$$

We observe that, generally speaking, ^GW is not the same as the Weyl group G_{rd}: for example, for SL the latter is equal to $S_m \times S_n$, not S_{m+n}. As will be clear from what follows, in the theory of Schubert supercells this "large" group, containing the odd reflections, plays the required role.

4. ***Lemma.*** *The types of the mutual disposition of pairs of complete G-invariant flags are in a natural bijection with the triples $(I, J; w)$, where $I, J \in {}^GI$, $w \in {}^GW$, $J = w(I)$.* ■

The proof is a purely combinatorial construction.

5. **Schubert supermanifolds.**

In the notation of §5.3, for $w \in {}^GW$, $I, J \in {}^GI$ we now put

$$|Y_{w,IJ}| = \{x \in {}^GF_I \times {}^GF_J |\, \mathrm{rk}\,(p_1^*\mathscr{S}_i(x) + p_2^*\mathscr{S}_j(x)) = d_{ij,w},$$

where $(d_{ij,w})$ is the type of mutual disposition that corresponds to the triple $(I, J;\ w)$ by the lemma of §5.4. (If $J \neq w(I)$, it is convenient to assume that $|Y_{w,IJ}| = \emptyset$.) Next we put

$$|Y_w| = \coprod_{I,J} |Y_{w,IJ}| \subset {}^G F \times {}^G F.$$

6. **Theorem.** *On each set $|Y_w|$ there is the canonical structure of a locally closed subsuperscheme $Y_w \subset {}^G F \times {}^G F$, defined by the fact that $\coprod_w Y_w$ is a flattening partition for the family of sheaves $\mathcal{S}_{ij} = p_1^*(\mathcal{S}_i) + p_2^*(\mathcal{S}_j)$ on ${}^G F \times {}^G F$* ∎

We recall that this condition has the following meaning: any morphism $q : X \to {}^G F \times {}^G F$ for which all the sheaves $q^* (\mathcal{S}_{ij})$ are flat proceeds uniquely by means of a certain morphism q': $X \to \coprod_w Y_w$. The proof of the existence of a flattening partition goes just as in the purely even case.

7. **Superlength.**

In the classical theory, the dimension of the Schubert cell corresponding to an element $w \in W$ is equal to the minimal length of a factorization of w into the product of basic reflections. An analogue of this assertion in supergeometry can be stated in terms of the following data.

We define basic reflections in ${}^G W$ as the following elements:

$\sigma_i = (i, i+1)$ for $G = \mathrm{SL}, Q$; $\sigma_i = (i, i+1)\,(m+n+1-i, m+n-i)$, $i+1 \leqslant \left[\frac{m+n}{2}\right]$; $\tau_l = (l, m+n+1-l)$, $l = \left[\frac{m+n}{2}\right]$ for $G = \mathrm{OSp}, \Pi\mathrm{Sp}$.

We define the superlength ${}^G l(w)$, $w \in {}^G W$, by the following inductive rule: it is a vector of superdimension $({}^G l_{IJ}(w) | I,J \in {}^G I)$:

a) For a basic reflection $\sigma \in {}^G W$ we put ${}^G l_{IJ}(\sigma) = 0$ for $J \neq \sigma(I)$. In the remaining cases:

σ \ G	SL	OSp	ΠSp	Q
σ_i	$1\|0\ (I=J)$ $0\|1\ (I \neq J)$	$1\|0\ (I=J)$; $0\|1\ (I \neq J)$	$1\|0\ (I=J)$; $0\|1\ (I \neq J)$	$1\|1$
τ_l	—	$1\|0\ (\delta_l(I)=1\|0,$ $m\|n=2r+1\|2s)$; $1\|1\ (\delta_l(I)=0\|1,$ $m\|n=2r+1\|2s)$; $0\|0\ (\delta_l(I)=1\|0,$ $m\|n=2r\|2s)$; $1\|0\ (\delta_l(I)=0\|1,$ $m\|n=2r\|2s)$	$0\|1\ (\delta_l(I)=1\|0)$; $0\|0\ (\delta_l(I)=0\|1)$	—

b) Let $w = \sigma^k \dots \sigma^1$ be an incontractible decomposition of w into a product of basic reflections. We put $I_i = \sigma^i \dots \sigma^1(I)$ and then:

$$^{G}l_{IJ}(w) = \sum_{i=0}^{k-1} {}^{G}l_{I_i, I_{i+1}}(\sigma^{i+1}), \quad \text{if} \quad J = w(I);$$

$$^{G}l_{IJ}(w) = 0, \quad \text{if} \quad J \neq w(I).$$

8. ***Theorem.*** *The projection $Y_{w,IJ} \to {}^{G}F_I$ is surjective, and the ${}^{G}F_I$-local superspace $Y_{w,IJ}$ is relative affine of dimension ${}^{G}l_{IJ}(w)$.* ∎

In other words, the dimension of the Schubert supercell $Y_w(b)$ is equal to the superlength of w.

From the geometrical proof of the theorem there follow some purely combinatorial facts: ${}^{G}l_{IJ}(w)$ does not depend on the choice of the incontractible decomposition of w; ${}^{G}l_{IJ}(w) = {}^{G}l_{IJ}(w^{-1})$ for $G \neq \Pi\mathrm{Sp}$; finally, for $G = \Pi\mathrm{Sp}$:

$$^{G}l_{IJ}(w) + \dim {}^{G}F_I = {}^{G}l_{JI}(w^{-1}) + \dim {}^{G}F_J$$

if $J = w(I)$.

§6. The geometry of supergravitation

1. **Minkowski space and Schubert cells.**
The aim of this section is to describe a model of pure supergravitation from the point of view presented in the paper [26], which considers the geometry of supergravitation as a distortion of the geometry of a certain flag superspace.

To explain the essence of our approach we recall the usual scheme of presenting the general theory of relativity. Space-time without a gravitational field is the Minkowski space of the special theory of relativity $\mathbf{R}^4$, with the metric which in an inertial system of coordinates has the form $dx_0^2 - \sum_{i=1}^{3} dx_i^2$. A gravitational field distorts the picture: space-time becomes a smooth four-dimensional manifold M^4 with a pseudo-Riemannian metric $g_{ab}dx^a dx^b$. The dynamics are described by the Lagrangian $R \cdot \mathrm{vol}_g$, where R is the scalar curvature and vol_g the volume form of the metric g. (We do not consider the contribution of matter.)

Models of supergravitation in superspace, which have been considered in many recent papers [12], [13], [34], are described by certain geometrical structures on a smooth supermanifold $M^{m|n}$, by means of which the super-Lagrangian is defined—a section of the sheaf $\mathrm{Ber}\, M = \mathrm{Ber}\, \Omega^1_{ev} M$. Physically interesting are many cases with $m \neq 4$, for example, the case $m|n = 11|32$ is now regarded as the most fundamental.

A question that is far from clear is: what are these original structures on $M^{m|n}$, that is, what are the kinematics of supergravitation? The simplest assumption, that we should consider a supermetric, turns out to be inadequate.

The most universal method consists in taking Cartan's approach: the geometry is described by an affine connection on $M^{m|n}$ Then the centre of gravity of the problem is transferred to the guessing of connections and their extension in terms of "free fields" Considerable success has been achieved in this direction, but the lack of geometrical guiding principles hinders the work very much.

The point of view presented here, which is effective for $m|n = 4|4$ (simple gravitation) and $m|n = 4|4N$, consists in the interpretation of connections as integrability conditions that define certain families of submanifolds of $M^{m|n}$. The geometry of these families is a distorted version of the geometry of Schubert supercells.

From this point of view we consider the simplest example: the Plücker-Klein-Penrose model of Minkowski space.

Let T be a four-dimensional complex space (Penrose twistors). Let $G = G(2; T)$ be the Grassmannian of planes of T, $\mathcal{S} = \mathcal{S}_l$ the tautological sheaf on G, and $\mathcal{S}_r = (T \otimes \mathcal{O}_G/\mathcal{S}_l)^*$. There is a canonical isomorphism $\Omega^1G = \mathcal{S}_l \otimes \mathcal{S}_r$, and the subsheaf $\Lambda^2\mathcal{S}_l \otimes \Lambda^2\mathcal{S}_r \subset S^2(\Omega^1G)$ can be interpreted as a holomorphic conformal metric on G. We choose a large cell $U \subset G$. The complement $G \setminus U$ is a divisor, and the sheaves $\Lambda^2\mathcal{S}_l$ and $\Lambda^2\mathcal{S}_r$ have sections ε_l and ε_r over U with a pole of the first order on $G \setminus U$. The metric of $\varepsilon_l \otimes \varepsilon_r \in \Gamma(U, S^2\Omega^1G)$ is uniquely determined up to multiplication by a constant. We consider the real structure ρ on $T \oplus T^*$ that interchanges T and T^*. The involution ρ acts on **C**-points of $G(2; T)$ in view of the canonical identification $G(2; T) = G(2; T^*)$. Suppose that ρ is compatible with $(U, \varepsilon_l, \varepsilon_r)$ in the sense that $U^\rho = U$ and $\varepsilon_l^\rho = \varepsilon_r$. Then it is not difficult to verify the following facts:

a) Real (that is, ρ-invariant) points of the large cell U form a space $\mathbf{R}^4$. The metric of $\varepsilon_l \otimes \varepsilon_r$ restricted to it is the Lorentz metric.

b) Three-dimensional Schubert manifolds in the Grassmannian G that intersect $U(\mathbf{R})$ specify light cones of the metric in $U(\mathbf{R})$.

c) Two-dimensional Schubert manifolds do not intersect $U(\mathbf{R})$. In $U(\mathbf{C})$ they form two connected families of complex planes. These families play a large role in the theory of Yang-Mills equations: if the connection (on some bundle over U) is integrable along the members of one of these families, it is called an (anti)autodual solution of these equations.

d) One-dimensional Schubert manifolds that intersect $U(\mathbf{R})$ specify null-geodesics of the metric in $U(\mathbf{R})$.

To sum up, we can say the following: the kinematics of the general theory of relativity in M^4 are defined by specifying a distorted system of Schubert 3- or 1-manifolds (which specifies a conformal metric) and violating the conformal invariance by a choice of the spinor metrics of ε_l, ε_r (which fixes the metric g_{ab}). Below we give a similar description of supergravitation. In §§6.2 and 6.3 we describe a flag model of the Minkowski superspace $M^{4|4N}$.

In §6.4 we explain that the distortion of $M^{4|4N}$ consists in postulating the preservation of the two families of $0|2N$-dimensional (purely odd!) Schubert supermanifolds. Finally, in §§6.5 and 6.6 we introduce dynamics by finding the Lagrangian of simple supergravitation and its expression in terms of the Ogievetskii-Sokachëv prepotential [28].

2. **Minkowski superspace.**
We fix an integer $N \geqslant 1$ and denote by T a linear complex superspace of dimension $4|N$. We put $M = F(2|0, 2|N; T)$: an S-point of M is a flag of type $\mathcal{S}^{2|0} \subset \mathcal{S}^{2|N}$ in $\mathcal{O}_S \otimes T$. Next, we introduce the left and right superspaces of Grassmannians

$$M_l = G(2|0; T), \quad M_r = G(2|N; T) = G(2|0; T^*).$$

Let $\mathcal{S}^{2|0} \subset \mathcal{S}^{2|N}$ be the tautological flag in $\mathcal{O}_M \otimes T$, and $\tilde{\mathcal{S}}^{2|0} \subset \tilde{\mathcal{S}}^{2|N}$ the flag orthogonal to it in $\mathcal{O}_M \otimes T^*$. We put $\mathcal{F}_l = \mathcal{S}^{2|N}/\mathcal{S}^{2|0}$, $\mathcal{F}_r = \tilde{\mathcal{S}}^{2|N}/\tilde{\mathcal{S}}^{2|0} = \mathcal{F}_l^*$. There are the canonical projections $\pi_{l,r}: M \to M_{l,r}$. We put $\mathcal{T}_l M = \mathcal{T}M/M_r$, $\mathcal{T}_r M = \mathcal{T}M/M_l$. (We are working in the category of complex-analytic supermanifolds.) Since M is obviously a relative Grassmannian over $M_{l,r}$ a standard argument leads to the canonical isomorphisms

$$\mathcal{T}_l M = (\mathcal{S}^{2|0})^* \otimes \mathcal{F}_l, \quad \mathcal{T}_r M = \mathcal{F}_l^* \otimes (\tilde{\mathcal{S}}^{2|0})^*.$$

By convolution with respect to $\mathcal{F}_l$ we then obtain the natural bilinear map

$$(1) \qquad a: \mathcal{F}_l M \otimes \mathcal{F}_r M \to (\mathcal{S}^{2|0})^* \otimes (\tilde{\mathcal{S}}^{2|0})^*.$$

On the other hand, the $\mathcal{T}_{l,r} M$ as relative tangent sheaves are integrable distributions, that is, they are locally direct subsheaves of Lie superalgebras in $\mathcal{T}M$ of rank $0|2N$. Through each point of M_{rd} there pass exactly two closed subsupermanifolds of dimension $0|2N$ touching $\mathcal{T}_l M$ and $\mathcal{T}_r M$, respectively: these are the Schubert cells of which we spoke. The supercommutator of $\mathcal{T}_l M$ and $\mathcal{T}_r M$ determines the Frobenius form

$$(2) \qquad b: \mathcal{T}_l M \otimes \mathcal{T}_r M \to \mathcal{T}M/(\mathcal{T}_l M + \mathcal{T}_r M) = \mathcal{T}_0 M.$$

The geometrical picture that we wish to preserve under distortion is summed up in the following proposition.

3. ***Proposition.*** a) *The sum $\mathcal{T}_l M + \mathcal{T}_r M$ in $\mathcal{T}M$ is direct, and $\mathcal{T}_l M + \mathcal{T}_r M$ is a locally direct subsheaf of $\mathcal{T}M$ of rank $0|4N$.*
b) *There is a unique isomorphism $\mathcal{T}_0 M = (\mathcal{S}^{2|0})^* \otimes (\mathcal{S}^{2|0})^*$ under which the forms* (1) *and* (2) *coincide.* ■

As in §6.1, we must introduce a real superstructure ρ on $T \oplus T^*$ (in the superalgebra $(ab)^\rho = (-1)^{\tilde{a}\tilde{b}} a^\rho b^\rho$; for further details, see [24]). Under the condition $T^\rho = T^*$ we have $(\mathcal{T}_l M)^\rho = \mathcal{T}_r M$, $(\mathcal{S}^{2|0})^\rho = \tilde{\mathcal{S}}^{2|0}$. We can verify that over a ρ-invariant large cell in M the set of standard sections $\mathcal{T}_l M$, $\mathcal{T}_r M$, and $\mathcal{T}_0 M$ generates the Poincaré superalgebra described in works on physics (see [34], for example).

4. Distorted superspace.

We define the distorted superspace of N-extended supergravitation as the complex supermanifold $M^{4|4N}$ with the following structures:

a) Integrable complex distributions $\mathcal{T}_r M, \mathcal{T}_l M \subset \mathcal{T} M$ of rank $0|2N$, whose sum is direct.

b) Locally free sheaves $\mathcal{S}_l, \mathcal{S}_r$ of rank $2|0$ and $\mathcal{F}_l = \mathcal{F}_r^*$ of rank $0|N$ and the structure isomorphisms $\mathcal{T}_l M = \mathcal{S}_l^* \otimes \mathcal{F}_l$, $\mathcal{T}_r M = \mathcal{F}_r \otimes \mathcal{S}_r^*$.

c) A real structure ρ on M having on M_{rd} a four-dimensional manifold of fixed points and its extensions to $\mathcal{S}_l \oplus \mathcal{S}_r$ and $\mathcal{F}_l \oplus \mathcal{F}_r$, which interchange left and right sheaves.

d) The volume forms $v_{l,r} \in \Gamma(M_{l,r}, \operatorname{Ber} M_{l,r})$ with the condition $v_l^\rho = v_r$. The choice of these volume forms corresponds to that of the spinor metrics ε_l and ε_r in §6.1.

On this set of data we impose the following axiom. We put

$$\mathcal{T}_0 M = \mathcal{T} M/(\mathcal{T}_l M \oplus \mathcal{T}_r M).$$

Then the Frobenius form

$$\varphi: \mathcal{T}_l M \otimes \mathcal{T}_r M \to \mathcal{T}_0 M$$

must coincide with the form of the convolution with respect to $\mathcal{F}_l$ under a suitable identification $\mathcal{T}_0 M = \mathcal{S}_l^* \otimes \mathcal{S}_r^*$, as in b) of the Proposition in §6.3.

5. The Lagrangian.

Let M be the superspace of N-extended supergravitation. Then from the data above we can construct the canonical isomorphism

$$\text{(3)} \qquad \operatorname{Ber} M = [\pi_l^* \operatorname{Ber} M_l \otimes \pi_r^* \operatorname{Ber} M_r]^{(2-N)/(4-N)}.$$

(Here and below we define $\operatorname{Ber} M$ as $\operatorname{Ber}{}^*(\Omega^1_{odd} M)$, in contrast to §6.1.)

Therefore, the volume forms v_l and v_r allow us to define the section (for $N \neq 4$)

$$\text{(4)} \qquad w = (\pi_l^* v_l \otimes \pi_r^* v_r)^{(N-2)/(N-4)} \in \Gamma(M, \operatorname{Ber} M).$$

For $N = 1$ it turns out to be the regular Lagrangian of simple supergravitation. The case $N = 2$ is not automatically given by (4): in this case, w gives trivial equations of motion. It is possible, however, that if we regard N as a formal variable and take the term of (4) that is linear in $N-2$, we obtain the Lagrangian proposed by Sokachëv. In any case, for $N > 1$ we must consider new connections, that is, presumably, the integrability of generalized Schubert cells of other dimensions.

Undoubtedly, in the flat version we must consider more complicated flag manifolds. For example, in a recent paper by Gal'perin, Ivanov, Kalitsin, Ogievetskii, and Sokachëv, devoted to the case $N = 2$, there appears implicitly the flag supermanifold $F(2|0, 2|1, 2|2; T)$ or, in the distorted version, the projectivization $P(\mathcal{F}_l) = P(\mathcal{F}_r) \to M$.

Similarly, for the richest structure $N = 8$, the $11|32$-dimensional flag supermanifold $F(2|0, 2|1, 2|8; T)$ or, in the distorted version, the projectivization $P(\mathcal{F}_l) \to M$ is the fundamental so-called dimensional reduction, or the generalized Kaluza-Klein model.

6. Prepotential.

In conclusion we give some calculations in coordinates, which make it possible to identify our geometrical picture with the one proposed in [28]. We put $N = 1$. In M_l we choose a local coordinate system (x_l^a, θ_l^α). Let $x_r^a = (x_l^a)^\rho$, $\theta_r^{\dot\alpha} = (\theta_l^\alpha)^\rho$. Suppose that the following conditions are satisfied: the functions $(x_l^a)_{rd}$ are ρ-invariant, and the functions

$$\left(x^a = \frac{1}{2}(x_l^a + x_r^a),\ \theta_l^\alpha,\ \theta_r^{\dot\alpha}\right)$$

form a local coordinate system on M. These coordinates (x_l, θ_l) on M_l, (x_r, θ_r) on M_r, and (x, θ_l, θ_r) on M are called adapted.

We now put

$$H^a = \frac{1}{2i}(x_l^a - x_r^a). \tag{5}$$

These four real and nilpotent superfunctions on M are called the Ogievetskii-Sokachëv prepotential. If we work locally and identify $\mathcal{F}_{l,r}$ with $\Pi\mathcal{O}_M$, then the set H^a completely determines the geometry of the situation, except for the forms $v_{l,r}$ for which we must have in addition:

$$v_{l,r} = \Phi_{l,r}^3 D^*(d\theta_l^\alpha, dx_l^a). \tag{6}$$

Calculations carried out in detail in [26] show that the Lagrangian (3) can be expressed in terms of (5) and (6) by the formula of Wess and Zumino

$$w = \frac{1}{8}\,\mathrm{Ber}\,(E_B^A)\, D^*(d\theta_l^\alpha, d\theta_r^{\dot\beta}, dx^a), \tag{7}$$

where E_B^A is the transition matrix from the frame $(\partial/\partial x^a, \partial/\partial\theta_l^\alpha, \partial/\partial\theta_r^{\dot\alpha}) = (\partial_a, \partial_\alpha, \partial_{\dot\alpha})$ to the frame $\left(\frac{i}{2}[\tilde\Delta_\alpha, \tilde\Delta_{\dot\alpha}], \tilde\Delta_\alpha, \tilde\Delta_{\dot\alpha}\right)$.

This latter frame of vector fields is defined in three steps.

Step 1. $\Delta_\alpha = \partial_\alpha + X_\alpha^a\partial_a$ and $\Delta_{\dot\alpha} = -\partial_{\dot\alpha} - X^a{}_{\dot\alpha}\partial_a$ by definition form a local basis of $\mathcal{F}_l M$ and $\mathcal{F}_r M$, respectively. From this condition we find the coefficients of X:

$$X_\alpha^a = i\left[\left(1 - i\,\frac{\partial H}{\partial x}\right)^{-1}\right]_b^a \partial_\alpha H^b, \quad X_{\dot\alpha}^a = -i\left[\left(1 + i\,\frac{\partial H}{\partial x}\right)^{-1}\right]_b^a \partial_{\dot\alpha} H^b.$$

Step 2. From $v_{l,r}$ we construct the spinor metrics of $\varepsilon_{l,r} \in \mathrm{Ber}\mathcal{S}_{l,r}$ by the formulae

$$\varepsilon_l = (\pi_l^* v_l)^{1/3} \otimes (\pi_r^* v_r)^{-2/3}, \quad \varepsilon_r = (\pi_l^* v_l)^{-2/3} \otimes (\pi_r^* v_r)^{1/3}.$$

Step 3. The vector fields $\widetilde{\Delta}_\alpha = F_\alpha \Delta$, $\widetilde{\Delta}_{\dot\alpha} = F^{\rho} \Delta_{\dot\alpha}$ are constructed in such a way that $D^*(\widetilde{\Delta}_\alpha) = \varepsilon_l$, $D^*(\widetilde{\Delta}_{\dot\alpha}) = \varepsilon_r$.

The structure frame $\left(\frac{i}{2}[\widetilde{\Delta}_\alpha, \widetilde{\Delta}_{\dot\alpha}], \widetilde{\Delta}_\alpha, \widetilde{\Delta}_{\dot\alpha}\right)$ satisfies all the relations imposed in the Cartan formulation of simple supergravitation.

References

[1] S.Yu. Arakelov, An intersection theory for divisors on an arithmetic surface, Izv. Akad. Nauk Ser. Mat. **38** (1974), 1179-1192. MR **57** # 12505. = Math. USSR-Izv. **8** (1974), 1167-1180.

[2] ———, Theory of intersections on an arithmetic surface, Proc. Internat. Congress Mathematicians, Vancouver 1974, vol. 1, 405-408. MR 57 # 6031.

[3] T. Aubin, Équations du type de Monge-Ampère sur les variétés kählериennes compactes, C.R. Acad. Sci. Paris **283** (1976), A119-A121. MR **55** # 6496.

[4] A.A. Veilinson, Higher regulators and values of L-functions, in: *Sovremennye problemy matematiki* (Modern problems of mathematics), vol. 24, VINITI, Moscow 1984.

[5] F.A. Berezin, *Vvedenie v algebry i analiz s antikommutiruyushchimi peremennymi* (Introduction to algebra and analysis with anticommuting variables), Moscow State Univ., Moscow 1983.

[6] E. Bombieri and J. Vaaler, On Siegel's lemma, Invent. Math. **73** (1983), 11-32.

[7] E. Date, M. Jimbo, M. Kashiwara, and T. Miwa, Solitons, τ-functions, and Euclidean Lie algebras, in: Mathématique et Physique, Sém. ENS 1979-1982, Birkhäuser, Boston 1983, 261-280.

[8] P. Deligne, Preuve des conjectures de Tate et Shafarevitch (d'après G. Faltings), Sém. Bourbaki **616** (1983).

[9] S. Donaldson, Stable holomorphic bundles and curvature, Preprint, Oxford 1983.

[10] G. Faltings, Calculus on arithmetic surfaces, Preprint 1983.

[11] ———, Endlichkeitssätze für Abelsche Varietäten über Zahlkörpern, Invent. Math. **73** (1983), 349-366.

[12] S.J. Gates, M.T. Grisaru, M.Roček, and W. Siegel, Superspace or one thousand and one lessons in supersymmetry, Benjamin, London 1983.

[13] Yu.I. Manin (ed.), *Geometricheskie idei fizike: sbornik perevodov* (Geometrical ideas in physics: a collection of translations), Mir, Moscow 1983.

[14] K. Iwasawa, Analogies between number fields and function fields, in: Some recent advances in the basic sciences, Yeshiva Univ., New York 1969, vol. 2, 203-208. MR **41** # 172.

[15] V.G. Kac, Lie superalgebras, Adv. in Math. **26** (1977), 8-96. MR **58** # 5803.

[16] ———, Representations of classical Lie superalgebras, Lecture Notes in Math. **676** (1978), 597-626. MR **80f**:17006.

[17] ——— and D.A. Kazhdan, Structure of representations with highest weight of infinite-dimensional Lie algebras, Adv. in Math. **34** (1979), 97-108. MR **81d**:17004.

[18] S. Kobayashi, Hyperbolic manifolds and holomorphic mappings, Marcel Dekker, New York 1970. MR **43** # 3503.

[19] ——— Curvature and stability of vector bundles, Proc. Jap. Acad. Sci. **58**:4 (1982), 158-162.

[20] N. Koiso, Einstein metrics and complex structures, Invent. Math. **73** (1983), 71-106. MR **85d**:58018.

[21] B. Kostant, Graded manifolds, graded Lie theory, and prequantization, Lecture Notes in Math. **570** (1977), 177-306. MR **58** # 28326.

[22] S. Lang, Fundamentals of Diophantine geometry, Springer-Verlag, Berlin-Heidelberg-New York 1983.

[23] D.A. Leites, Introduction to the theory of supermanifolds, Uspekhi Mat. Nauk **35**:1 (1980), 3-57. MR **81j**:58003.
= Russian Math. Surveys **35**:1 (1980), 1-64.

[24] Yu.I. Manin, *Kalibrovochnye polya i golomorfnaya geometriya* (Gauge fields and holomorphic geometry), Nauka, Moscow 1984.

[25] ———, Flag superspaces and supersymmetric Yang-Mills equations, in: Arithmetic and geometry. Papers dedicated to I.R. Shafarevich on the occasion of his sixtieth birthday, Birkhäuser, Boston 1983, vol. 2, 175-188.

[26] ———, The geometry of supergravitation and Schubert supercells, Zap. Nauchn. Sem. LOMI **133** (1984), 160-176.

[27] B. Mazur and A. Wiles, Analogies between function fields and number fields, Amer. J. Math. **105** (1983), 507-521. MR **84g**:12003.

[28] V.I. Ogievetskii and E.S. Sokachëv, A gravitational axial superfield and the formalism of differential geometry, Yader. Fizika (Nuclear Physics) **31** (1980), 821-839. Phys. Abstracts 80-57164.

[29] I.B. Penkov, *D*-modules on supermanifolds, Invent. Math. **71** (1983), 501-512. MR **85b**:32015.

[30] D. Quillen, Arithmetic surfaces and analytic torsion, Arbeitstagung, Bonn 1983.

[31] I.R. Shafarevich, Algebraic number fields, Proc. Internat. Congress Mathematicians, Stockholm 1962, 163-176. MR **34** # 2569.

[32] ———, Minimal models of two-dimensional schemes, Tata Inst. of Fundamental Research, Bombay **37** (1965).

[33] L. Szpiro, La conjecture de Mordell (d'après G. Faltings), Sém. Bourbaki **619** (1983).

[34] S.W. Hawking and M. Roček (ed.), Superspace and supergravity, Cambridge Univ. Press, Cambridge 1981.

[35] P. Berthelot, A. Grothendieck, and L. Illusie (ed.), Théorie des intersections et théorème de Riemann-Roch, Lecture Notes in Math. **225** (1971).

[36] A. Weil, Sur l'analogie entre les corps de nombres algébriques et les corps de fonctions algébriques, in: Collected Papers, Springer-Verlag, Berlin-Heidelberg-New York 1980, vol. 1, 236-240. MR **80k**:01067a.

[37] ———, De la metaphysique aux mathématiques, in: Collected Papers, Springer-Verlag, Berlin-Heidelberg-New York 1980, vol. 2, 408-412. MR **80k**:01067b.

[38] S.T. Yau, Calabi's conjecture and some new results in algebraic geometry, Proc. Nat. Acad. Sci. USA **74** (1977), 1798-1799. MR **56** # 9467.

Translated by E.J.F. Primrose

Mathematical Institute of the USSR
Academy of Sciences

Received by the Editors 21 July 1984

Advanced Studies in Pure Mathematics 17, 1989
Algebraic Number Theory — in honor of K. Iwasawa
pp. 289-308

Arrangements of Hyperplanes, Higher Braid Groups and Higher Bruhat Orders

Yu. I. Manin and V. V. Schechtman

Dedicated to Kenkichi Iwasawa on his seventieth birthday

Introduction

Let z_i be coordinate functions on C^n. Consider the arrangement of hyperplanes $D_{ij}: z_i - z_j = 0$ in C^n and let $U = C^n - \bigcup D_{ij}$ be its complement. The fundamental group of U is called the (colored) braid group. This group and the topology of U has been studied in many papers.

The family $\{D_{ij}\}$ is a special example of hyperplane arrangements which we call discriminantal ones. This article is devoted to the study of the topological and combinatorial properties of the discriminantal arrangements.

Among the vast literature on arrangements of hyperplanes we can mention Cartier's Bourbaki report [1] and an important paper [5]. Recently their study was stimulated by the theory of multidimensional hypergeometric functions (cf. [8–10]) and certain models of quantum and statistic physics (see [6], [7] and the bibliography therein).

In section 1 of this paper we recall some results on the hyperplane arrangements, define discriminantal arrangements (considered previously in [6], [7] and [10]), define higher braid groups and calculate their nilpotent completions.

In section 2 we introduce posets $B(n, k)$. Their definition is motivated by combinatorics of the discriminantal arrangements. The poset $B(n, 1)$ is essentially the symmetric group S_n with its weak Bruhat order. We prove some fundamental properties of $B(n, k)$ including the higher analogs of the Coxeter relations.

The results of section 2 were previously announced in [6], [7].

Actually, the construction of section 2 defines on S_n a canonical structure of $(n-1)$-category, whose "1-coskelton" is the category associated to the weak Bruhat order. This $(n-1)$-category is introduced in the section 3. Its structure is closely related to the combinatorial structure of the convex closure of a general orbit of S_n in $\boldsymbol{R}^n$.

Received October 16, 1987.

In conclusion we would like to thank V. A. Hinich, who explained to us the notion of n-category used in section 3, and O. V. Ogievetsky who helped us to prove the main theorem of section 2.

§ 1. Discriminantal arrangements and higher braid groups

1. Preliminaries on hyperplane arrangements. Let F be a field, $D_1, \cdots, D_n$ a finite family of affine hyperplanes in F^m. The main combinatorial invariant of this arrangement is a ranked poset Z, consisting of affine subspaces $D_{i_1} \cap \cdots \cap D_{i_k}$ ordered by inverse inclusion and endowed with codimension as a rank function. Let $U = F^m - \bigcup D_i$. Then for $F = C$ the following topological invariants of U are determined by Z.

a. *The nilpotent completion of* $\pi_1(U)$. Let G be a group, $G = \Gamma_0 G \supset \Gamma_1 G = [G, G] \supset \cdots \supset \Gamma_{j+1} G = [\Gamma_j G, G] \supset \cdots$ its lower central series. Using the central extension

$$1 \longrightarrow \Gamma_j G / \Gamma_{j+1} G \longrightarrow G / \Gamma_{j+1} G \longrightarrow G / \Gamma_j G \longrightarrow 1$$

one can inductively define a Lie Q-algebra $[G/\Gamma_j G] \otimes Q$ and the projective limit of these algebras $[G]_Q^{\wedge}$ which we shall call the nilpotent completion of G.

For an arrangement of hyperplanes $D = (D_1, \cdots, D_n)$ denote by $L(D)$ the graded Lie Q-algebra generated by degree 1 elements $h_1, \cdots, h_n$ satisfying the following relations:

$$R_S : [h_i, \sum_{j \in S} h_j] = 0, \qquad \forall i \in S$$

for all subsets $S \subset \{1, \cdots, n\}$ such that

$$\operatorname{codim} \bigcap_{j \in S} D_j = 2, \quad \operatorname{codim} \bigcap_{j \in S'} D_j > 2 \qquad \text{if } S' \text{ strictly contains } S.$$

Let $L(D)^{\wedge}$ be the completion of $L(D)$ with respect to the lower central series.

1.1. Proposition ([2], [3]). *For* $U = C^m - \bigcup D_j$ *we have*

$$[\pi_1(U)]_Q^{\wedge} \cong L(D)^{\wedge}. \qquad \blacksquare$$

b. *Cohomology of* U. Consider a grassmannian Q-algebra Λ generated by symbols e_i, $i = 1, \cdots, n$. If $S = (i_1, \cdots, i_k)$, $1 \leqslant i_1 < \cdots < i_k \leqslant n$ put $e_S = e_{i_1} \cdots e_{i_k}$. Define a Q-linear differential ∂ in Λ by $\partial 1 = 0$, $\partial e_i = 1$ and $\partial(e_S e_T) = \partial e_S \cdot e_T + (-1)^{|S|} e_S \partial e_T$. Clearly $\partial^2 = 0$.

We say that S is dependent if $\operatorname{codim} \bigcap_{i \in S} D_i > |S|$.

Let $f_i = 0$ be an equation of $D_i \subset C^m$. In 1.2, 1.3 and 1.4 below we assume that $\bigcap_{i=1}^{n} D_i \neq \varnothing$.

1.2. Proposition ([5]). *Denote by Δ the ideal in A generated by $\{\partial e_S \mid S$ dependent$\}$. Then the map*

$$\text{the de Rham class of } \frac{1}{2\pi i} d\log f_i \mapsto e_i \bmod \Delta$$

defines an isomorphism of graded rings $H^*(U, Q) \cong A/\Delta$. ■

c. *The Betti numbers of U.* Define the Möbius function of the poset Z by

$$\mu(C^m)=1, \qquad \mu(X)=-\sum_{\substack{Y\supset X\\ Y\in Z}} \mu(Y).$$

1.3. Proposition ([5]). *We have*

$$\dim H^i(U, Q)=\Big|\sum_{\substack{X\in Z\\ \operatorname{codim} X=i}} \mu(X)\Big|. \qquad ■$$

d. *Real arrangements.* Assume that all $D_i \subset C^m$ are complexifications of real hyperplanes $D_i^R \subset R^m$. Put $u=|\pi_0(R^m-\bigcup D_i^R)|$.

1.4. Proposition ([11]). *We have*

$$u=\sum_{i\geqslant 0} \dim H^i(U, Q)=\sum_{i\geqslant 0}\Big|\sum_{\substack{X\in Z\\ \operatorname{codim} X=i}} \mu(X)\Big|. \qquad ■$$

2. Discriminantal arrangements. Let now $H_1^0, \cdots, H_n^0 \subset F^k$ be a family of affine hyperplanes in general position. This means in particular that codim $(H_{i_1}^0 \cap \cdots \cap H_{i_a}^0)=a$ for all $1\leqslant i_1<\cdots<i_a\leqslant n$ (we agree that codim $Y>k$ in F^k means that $Y=\varnothing$).

Let $U(n, k)$ be the manifold of arrangements $H_1, \cdots, H_n$ enjoying two properties: a) H_i is parallel to H_i^0 for all $1\leqslant i\leqslant n$; b) $H_1, \cdots, H_n$ are in general position.

Clearly, $U(n, k)$ is a subset of the space F^n of all parallel transports of H_i^0. Moreover, $F^n-U(n, k)$ is a union of hyperplanes in F^n which we will now describe.

Denote by $C(n, a)$ the set of subsets in $(1, \cdots, n)$ of cardinality a. For $K\in C(n, a)$ put

(1) $D_K=$ the set of $(H_1, \cdots, H_n)$ in F^n such that $\bigcap_{i\in K} H_i\neq\varnothing$.

Clearly $D_K=F^n$ if $|K|\leqslant k$ and

(2) $$\operatorname{codim} D_K=|K|-k \qquad \text{for } K\geqslant k+1.$$

In particular, D_J for $J \in C(n, k+1)$ are pairwise distinct hyperplanes in F^n. One easily sees (cf. Proposition 4 below) that

$$U(n, k) = F^n - \sum_{J \in C(n, k+1)} D_J. \tag{3}$$

We shall call the set of hyperplanes (D_J) a *discriminantal arrangement* in F^n Strictly speaking, it depends on $(H_1^0, \cdots, H_n^0)$ and not only on n, k. However, we shall be concerned mostly with its combinatorial invariants which are constant on an open Zariski dense subset of all n-arrangements in F^n Stating properties of such invariants we shall tacitly assume that our arrangements (H_i^0) are general in this sense. Note that the discriminantal arrangement is almost never general!

3. Definition. The higher braid group $T(n, k)$ is the fundamental group $\pi_1(U(n, k))$ (for $F = C$). ■

If $k = 1$ we get the ordinary braid group.

Remark. It would be natural to consider also the manifold $V(n, k)$ of arbitrary families of n hyperplanes in C^k in general position. It coincides with $U(n, k)$ for $k = 1$ but leads to a different generalization of the braid group for $k > 1$. Unfortunately, $V(n, k)$ is not in general a complement of hyperplanes.

In the next Proposition we state some properties of the poset $Z(n, k)$ generated by hyperplanes D_J. Unfortunately we were unable to obtain its complete combinatorial description.

4. Proposition. a) *The map $K \mapsto D_K$ (cf. (1)) defines an injection of the lattice of subsets $\coprod_{k+1 \leqslant a \leqslant n} C(n, a)$ ordered by inverse inclusion, with operation $\wedge = \cup$ and rank function $a - k$ into the poset (geometric lattice [5]) $Z(n, k)$.*

b) *The codimension one elements in $Z(n, k)$ are D_J for all $J \in C(n, k+1)$. They are pairwise distinct.*

c) *The codimension two elements in $Z(n, k)$ are D_K for $K \in C(n, k+2)$ and $D_{J_1} \cap D_{J_2}$ for $J_i \in C(n, k+1)$, $|J_1 \cup J_2| \geqslant k+3$. They are pairwise distinct. Moreover, $D_K \subset D_J$ for $J \in C(n, k+1)$ iff $J \supset K$, and $D_{J_1} \cap D_{J_2} \subset D_J$ iff either $J = J_1$, or $J = J_2$.*

d) *$Z(n, k)$ has a unique minimal element of codimension $n-k$, namely $D_{\{1, \cdots, n\}}$.*

Proof. Everything follows from two simple remarks. First, for arbitrary subsets K_1, K_2 with $|K_i| \geqslant k+1$ we have $K_1 \subset K_2 \Leftrightarrow D_{K_1} \supset D_{K_2}$. Second, if $|J| = k+1$ and $D_J \supset D_{K_1} \cap \cdots \cap D_{K_a}$ then $J \subset \bigcup_{i=1}^{a} K_i$.

In order to prove the second assertion suppose that $J \not\subset \bigcup K_i$. Choose $j \in J - \bigcup K_i$. Then the condition $(H_1, \cdots, H_n) \in D_{K_1} \cap \cdots \cap D_{K_a}$ implies no restrictions on the position of H_j: the other hyperplanes being fixed, this one can be freely moved. Contrariwise, the condition $(H_1, \cdots, H_n) \in D_J$ fixes the position of H_j unambiguously once the position of all H_i, $i \in J - \{j\}$ is known since H_j must pass through the point $\bigcap_{i \in J - \{j\}} H_i$. Therefore in this case $D_J \not\supset D_{K_1} \cap \cdots \cap D_{K_a}$.

Let now $a = |K| \geqslant k+1$. Then we can find $J_1, \cdots, J_{a-k}$ with $|J_i| = k+1$, $|J_{i+1} - \bigcup_{b=1}^{i} J_b| = 1$ and $K = \bigcup J_i$. Therefore $D_K = \bigcap D_{J_i}$ and codim $D_K = a-k$. Then rest of our assertions follow from this and two remarks stated in the beginning.

The minimal element of $Z(n, k)$ consists of all arrangements intersecting in a point (if $n \geqslant k+1$). This point can be chosen arbitrarily in F^k. ∎

5. Theorem. *The nilpotent completion* $[T(n, k)]_Q^{\wedge}$ *of the higher braid group is generated by elements* h_J, $J \in C(n, k+1)$ *subjected to the relations*

$$(4) \qquad [h_{J_1}, h_{J_2}] = 0, \qquad \text{if } |J_1 \cup J_2| \geqslant k+3,$$

$$(5) \qquad [h_J, \sum_{I \subset K} h_I] = 0, \qquad \text{if } K \in C(n, k+2),\ J \subset K. \qquad ∎$$

This follows immediately from Propositions 1.1 and 4 a), b).

6. Higher braid groups of real arrangements. If the initial arrangement $(H_1^0, \cdots, H_n^0)$ is the complexification of a real one, then its discriminantal arrangement also is the complexification of a real one. In this case one can use the results of Randell [4] to compute the fundamental group $T(n, k)$ itself. We get the following information.

For each pair $a = (J_1, J_2)$, $J_i \in C(n, k+1)$, $|J_1 \cup J_2| \geqslant k+3$, introduce generators $\alpha_1(a)$, $\alpha_2(a)$. For each set $K \in C(n, k+2)$ introduce generators $\alpha_1(K), \cdots, \alpha_{k+2}(K)$. Then $\pi_1(U(n, k))$ is isomorphic to the group $P(n, k)$ generated by these elements subjected to the relations

$$(4') \qquad \alpha_1(a)\alpha_2(a) = \alpha_2(a)\alpha_1(a),$$

$$(5') \qquad \begin{aligned} \alpha_1(K)\alpha_2(K) \cdots \alpha_{k+2}(K) &= \alpha_{k+2}(K)\alpha_1(K)\alpha_2(K) \cdots \alpha_{k+1}(K) \\ &= \cdots = \alpha_{k+2}(K)\alpha_{k+1}(K) \cdots \alpha_1(K), \end{aligned}$$

$$(6) \qquad \text{some equalities among } \alpha_i(a),\ \alpha_j(K).$$

The relations (4′) and (5′) correspond to (4) and (5) respectively. However, if one uses Randell's prescriptions, the relations (6) cannot be written in pure combinatorial terms, since they require taking into account some inequalities, i.e. certain characteristics of the real part of the picture.

7. Topology of $U(n, k)$ for $n \leqslant k+3$. If $n=k+1$, we have clearly $U(k+1, k)=C^{k+1}-C^k$ and $T(k+1, k)=Z$.

$U(k+2, k)$ is a complement of a union of $k+1$ hyperplanes in C^{k+2} passing through a common axis C^k. We leave to the reader a calculation of its cohomology and fundamental group. Proposition 4 suffices also for calculation of the topology of $U(k+3, k)$.

The following table describes the structure of $Z(k+3, k)$. We use the following notation: $(ij)=D_J$, $J=(1, \cdots, k+3)-(i, j)$; $(i)=D_K$, $K=(1, \cdots, k+3)-(i)$; $(ij, lm)=D_{J_1} \cap D_{J_2}$, $(i, j) \cap (l, m)=\varnothing$.

codim	Inclusion diagram	Number of elements
3	$\mathbf{1}=C^k$	1
2	$(i)\cdots$ $\quad$ $(ij, lm)\cdots$	$(k+3)+\frac{1}{8}k(k+1)(k+2)(k+3)$
1	$(ij)\cdots$	$\frac{1}{2}(k+2)(k+3)$
0	$\mathbf{0}=C^{k+3}$	1

Hence, by induction on codim

$$\mu(\mathbf{0})=1;\ \mu((ij))=-1;\ \mu((i))=-\sum_{j \neq i} \mu((ij))-1=k+1;$$

$$\mu((ij, lm))=-\mu((ij))-\mu((lm))-\mu(\mathbf{0})=1;$$

$$\mu(\mathbf{1})=-(k+1)(k+3)-\tfrac{1}{8}k(k+1)(k+2)(k+3)+\tfrac{1}{2}(k+2)(k+3)-1$$
$$=-\tfrac{1}{8}k(k+3)(k^2+3k+6)-1.$$

Put $h^i=\dim H^i(U(k+3, k))$. We have $h^0=1$, $h^i=0$ for $i>3$ and, using Proposition 1.4

$$h^1=\tfrac{1}{2}(k+2)(k+3);\qquad h^2=\tfrac{1}{8}(k+1)(k+3)(k^2+2k+8);$$
$$h^3=\tfrac{1}{8}k(k+3)(k^2+3k+6)+1.$$

The sum of these Betti numbers equals to the number of isotopy classes of arrangements of $k+3$ ordered hyperplanes in general position in $\boldsymbol{R}^k$. It is 62 for $k=2$, 140 for $k=3$.

§ 2. Higher Bruhat orders

1. Notation. Let $(I, <)$ be a finite totally ordered set. Denote by $C(I, k)$ the set of all subsets of I of cardinality k.

Usually I will be $(1, \cdots, n)=\underline{n}$ or $(2, \cdots, n)=\underline{n}-(1)$. There is a lexicographic total order on $C(I, k)$: if $J=(j_1, j_2, \cdots, j_k), j_i<j_{i+1}$, and $J'=(j'_1, j'_2, \cdots, j'_k), j'_i<j'_{i+1}$, then $J\leqslant J'$ means that either $j_1<j'_1$, or $j_1=j'_1$ but $j_2<j'_2$ etc.

For $K\in C(I, k+1)$, we shall call a *K-packet* the set $P(K)=\{J\,|\,J\in C(I, k), J\subset K\}$. If $K=(i_1, \cdots, i_{k+1}), i_j<i_{j+1}$, then $P(K)$ consists of the sets $K^\wedge_a=K-(i_a)$, $a=1, \cdots, k+1$. We have lexicographically $K^\wedge_{k+1}<K^\wedge_k<\cdots<K^\wedge_1$.

Below we shall consider various total orders on $C(I, k)$ which we shall denote ρ, ρ', σ etc. The notation $\rho=J_1J_2\cdots J_N$, $N=\binom{n}{k}$ means that $J_i\rho J_j$ for all $i<j$.

2. Definition. a. A total order on $C(I, k)$ is called *admissible* if on each packet it induces either a lexicographic order or the inverse lexicographic one.

We denote by $A(I, k)$ the set of all admissible total orders on $C(I, k)$.

b. Two total orders $\rho, \rho'\in A(I, k)$ are called *elementarily equivalent*, if they differ by an interchange of two neighbours which do not belong to a common packet.

We denote by $B(I, k)$ the quotient of $A(I, k)$ by the corresponding equivalence relation. The natural projection is $\pi: A(I, k)\to B(I, k)$.

c. An *inversion* in the order $\rho\in A(I, k)$ is an element $K\in C(I, k+1)$ such that ρ induces on $P(K)$ the antilexicographic order. We denote by $\mathrm{Inv}(\rho)\subset C(I, k+1)$ the set of all inversions in ρ and by $\mathrm{inv}(\rho)$ its cardinality. Clearly, for $r=\pi(\rho)=\pi(\rho')$ we can set $\mathrm{Inv}(r)=\mathrm{Inv}(\rho)=\mathrm{Inv}(\rho')$. ∎

Remarks. 1) The lexicographic order $\rho_{\min}$ and the antilexicographic order $\rho_{\max}$ are clearly admissible. We have $\mathrm{Inv}(\rho_{\min})=\varnothing$, $\mathrm{Inv}(\rho_{\max})=C(I, k+1)$. In general, if ρ is admissible, then the inverse order ρ^t also is admissible, and we have $\mathrm{Inv}(\rho^t)=C(I, k+1)-\mathrm{Inv}(\rho)$. If ρ and ρ' are elementarily equivalent then the same is true for ρ^t, ρ'^t, hence t acts upon $B(I, k)$.

2) Suppose that for some $K\in C(I, k+1)$ members of the packet $P(K)$ form a *chain* with respect to an admissible order ρ, i.e. any element of $C(I, k)$ lying between two elements of $P(K)$ belongs to $P(K)$. Define $p_K(\rho)$ as an order in which this chain is reversed while all the rest elements conserve their positions. Evidently $p_K(\rho)$ is admissible, and

$$\mathrm{Inv}\,(p_K(\rho))=\begin{cases}\mathrm{Inv}\,(\rho)\cup\{K\}, & \text{if } K\notin \mathrm{Inv}\,(\rho),\\ \mathrm{Inv}\,(\rho)-\{K\}, & \text{if } K\in \mathrm{Inv}\,(\rho).\end{cases}$$

One easily sees that if ρ and ρ' are elementarily equivalent and if $p_K(\rho)$, $p_K(\rho')$ are both defined then they also are elementarily equivalent.

For $r \in B(I, k)$ we put

$$N(r)=\{K\in C(I, k+1)\,|\,P(K) \text{ forms a chain for some } \rho_0\in r\},$$
$$p_K(r)=\pi(p_K(\rho_0)) \text{ for such } K\in N(r) \text{ and } \rho_0.$$

Clearly, $N(r^t)=N(r)$.

3) Let $f\colon I\to J$ be a strictly increasing map of totally ordered finite sets. It induces maps $f_*\colon C(I, k)\to C(J, k)$ which are strictly increasing with respect to lexicographic orders and map packets into packets. Therefore each admissible order on $C(J, k)$ induces an admissible order on $C(I, k)$, whence we get a map $f^*\colon A(J, k)\to A(I, k)$. It is compatible with elementary equivalencies so that we have a map $f^{\cdot}\colon B(J, k)\to B(I, k)$. In this way our constructions are functorial.

4) For $k=1$ we have $C(I, 1)\simeq I$, $A(I, 1)\simeq$ the set of all total orderings of I, i.e. the symmetric group of permutations of I, $B(I, 1)=A(I, 1)$; $p_K(\rho)$ is obtained from ρ by transposing two neighbours forming K. (Case $k=n-2$ is described below, cf. Lemma 7).

Thus the following theorem which is the main result of this section is an extension to the case $k\geqslant 2$ of principal properties of the weak Bruhat order on the symmetric group (see e.g. [12]).

3. Theorem. a) *The relation*

$$r\leqslant r' \Longleftrightarrow \exists K_i\in C(n, k+1),\ K_i\in N(p_{K_{i-1}}\cdots p_{K_1}(r))-\mathrm{Inv}\,(p_{K_{i-1}}\cdots p_{K_1}(r)),$$
$$r'=p_{K_m}\cdots p_{K_1}(r)$$

is a partial order on $B(n, k)$.

b) *This partial order defines on* $B(n, k)$ *the structure of a ranked poset, with rank function inv, a unique minimal element* $r_{\min}=\pi(\rho_{\min})$ *and a unique maximal element* $r_{\max}=\pi(\rho_{\max})$.

c) *The map*

$$\{r_{\min}<p_{K_1}(r_{\min})<\cdots<p_{K_N}\cdots p_{K_1}(r_{\min})\}\longmapsto\rho=K_1\cdots K_N$$

defines a bijection

$$\{\textit{the set of maxmial chains in } B(n, k)\}\xrightarrow{\sim} A(n, k+1).$$

d) *Every element* $r\in B(n, k)$ *is uniquely defined by the set* $\mathrm{Inv}\,(r)$. ■

Before we start proving this theorem we digress to give a geometric motivation of our combinatorial notions in terms of the discriminantal hyperplane arrangements.

Choose a real hyperplane arrangement $H_1^0, \cdots, H_n^0 \subset \boldsymbol{R}^{k-1}$. The components of the corresponding discriminantal arrangement $D_J \subset \boldsymbol{R}^n$ are numbered by $J \in C(n, k)$ (cf. (1)). Below we shall consider only the real part of this picture, so that D_J divides $\boldsymbol{R}^n$ into two parts.

Choose in $\boldsymbol{R}^n$ a generic plane and orient it. The arrangement D_J intersects it in a family of lines. The intersection points of these lines are in a bijection with the set

$$C(n, k+1) \cup \text{ the set of unordered pairs } J_1, J_2 \in C(n, k) \text{ such that } |J_1 \cup J_2| \geqslant k+2.$$

This follows from section 1, Proposition 4.

Draw in our plane P a closed path in the positive direction intersecting each line $D_J \cap P$ twice and containing inside it all intersection points of lines. Then starting with some initial point we shall intersect hyperplanes D_J in some order $D_{J_1}, \cdots, D_{J_R}$, $R=\binom{n}{k}$ and then again $D_{J_1}, \cdots, D_{J_R}$.

One can show that if the initial point and the numbering of H_i^0 are properly chosen, the order $J_1, \cdots, J_R$ will be the lexicographic one.

Now we shall fix the initial point and the endpoint of the first half of our path (after intersecting each D_J once) and deform the path in such a way that at each moment t the deformed path Γ_t(intersects at most one intersection point of lines $P \cap D_{J_i}$. At these critical moments the intersection order of J's will enjoy the following changes (a two-dimensional picture will make it evident to the reader).

When Γ_t intersects $P \cap D_{J_i} \cap D_{J_j}$, $|J_i \cup J_j| \geqslant k+2$, then J_i and J_j which were neighbours just before the critical moment change places.

When Γ_t intersects $P \cap D_K$ then the members of the packet $P(K)$ which formed a chain just before the critical moment become intersected in reverse order.

In this way we get elementary equivalencies and inversions.

Now we return to the proof of the Theorem 3 which requires a number of lemmas. The rest of this section is devoted to this proof.

4. Lemma. *In each class $r \in B(n, k)$ there exists an order $\rho \in A(n, k)$ such that all elements of $C(n, k)$ containing 1 form a chain with respect to this order.*

Proof. We shall say that elements $J, J' \in C(n, k)$ commute if they do not belong to a packet, i.e. if $|J \cup J'| \geqslant k+2$ (cf. (4), section 1).

Let ρ be an order from $A(n, k)$, S a subset of $C(n, k)$. We shall denote by $\bar{S}$ the minimal chain containing S. Choose an element $J \in \bar{S}-S$. We shall say that J can be pushed out to the left (resp. to the right) from S, if J commutes with all $J' \in S$ such that $J'\rho J$ (resp. $J\rho J'$).

Now take an arbitrary class $r \in B(n, k)$ represented by an order ρ. Let S consist of all elements $J_1\rho J_2 \cdots \rho J_N$, $N=\binom{n-1}{k-1}$, containing 1.

We affirm that any element $J \in \bar{S}-S$ can be pushed out from S either to the left or to the right (or both). In effect, let $J=(j_1, \cdots, j_k)$, $K=\{1\} \cup J$. Any element of S not commuting with J must be of the form $(1, j_1, \cdots, \hat{j}_p, \cdots, j_k)=K^{\wedge}_{p+1}$ for some p, $1 \leqslant p \leqslant k$. Moreover, $J=K^{\wedge}_1$. Since ρ is admissible, J must be either the maximal or the minimal element of the packet $\{K^{\wedge}_i\}$ with respect to ρ. Therefore J can be pushed out of S.

Let now J', J'' be such a couple of elements of $\bar{S}-S$ that $J'\rho J''$, J' cannot be pushed out to the left and J'' cannot be pushed out to the right. We affirm that in this case J' and J'' must commute. In effect, if they do not commute, we have $J'=K^{\wedge}_p$, $J''=K^{\wedge}_q$ for some $K \in C(n, k+1)$, where $K=(i_1 \cdots, i_{k+1})$, $1<i_1, 1 \leqslant p, q \leqslant k+1$. By assumption, all sets $\{1\} \cup K^{\wedge}_{s,p}=(1, i_1, \cdots, \hat{i}_s, \cdots, \hat{i}_p, \cdots, i_{k+1})$, $1 \leqslant s \leqslant k+1$, $s \neq p$, lie to the left of J' with respect to ρ. In particular, for $J=\{1\} \cup K^{\wedge}_{p,q}$ we have $J\rho J'\rho J''$. Since J clearly does not commute with J'', J'' cannot be pushed out of S to the left. But by assumption it cannot be pushed out to the right either which contradicts our previous result.

Now we shall apply elementary equivalencies to ρ trying to push out all elements of $\bar{S}-S$ so that in the end S forms a chain. If all elements of $\bar{S}-S$ can be pushed out to the left then it suffices to interchange them in turn with all elements of S lying to the left of them. Otherwise let $J \in \bar{S}-S$ be the maximal of all elements that cannot be pushed out to the left. Then it can be pushed out to the right from S by changing places with all its right neighbours, belonging to S or not. The proof concludes by induction on the number of elements which cannot be pushed out to the left. ∎

An order ρ whose existence is asserted in Lemma 4 will be called a *good* one. We shall write it in the standard form

$$\rho=J'_1 \cdots J'_a;\ J_1 \cdots J_N;\ J'_{a+1} \cdots J'_M, \tag{7}$$

where $1 \in J_i$, $1 \notin J'_i$, $N=\binom{n-1}{k-1}$, $M=\binom{n-1}{k}$. Put $J_t=\{1\} \cup L_t$, $\rho'=L_1 \cdots L_N \in A(\underline{n}-\{1\}, k-1)$, $\sigma=J'_1 \cdots J'_a J'_{a+1} \cdots J'_M \in A(\underline{n}-\{1\}, k)$. Then we can rewrite (7) in the form

(8) $$\rho=\rho_1 1*\rho'\rho_2, \qquad \sigma=\rho_1\rho_2.$$

We shall call ρ_1, ρ_2, $1*\rho'$ the left, right and middle part of ρ respectively. Left or right part may be empty.

5. Lemma. a) $\mathrm{Inv}(\rho')=\{J_1', \cdots, J_a'\}$ = *the set of elements of the left part of* ρ; $1*\mathrm{Inv}(\rho')=\{\{1\}\cup J_1', \cdots, \{1\}\cup J_a'\}$ = *the set of inversions of* ρ, *containing* 1.

b) *Any packet from* $C(n, k)$ *either is disjoint with the middle part of* ρ, *or* k *of its members lie in the middle part. In the latter case the exceptional member lies in the left part of* ρ, *iff the whole packet belongs to* $\mathrm{Inv}(\rho)$.

Corollary. *Left, right and middle parts of* ρ *as sets depend only on* $r=\pi(\rho)$.

Proof. Let $K\in C(n, k+1)$. If $1\notin K$ then $P(K)$ is disjoint with the middle part of ρ. If $1\in K$ then only $K_1^\wedge$ does not contain 1. If ρ induces on $P(K)$ the lexicographic order then $K_1^\wedge$ belongs to the right part, otherwise to the left one. The rest of the packet belongs to the middle part. If we delete 1 from them we obtain a packet in $C(\underline{n}-\{1\}, k-1)$, which belongs to $\mathrm{Inv}(\rho')$ exactly when $K_1^\wedge$ is in the left part of ρ. ∎

6. Lemma. *Let* $\rho\in A(n, k)$, $r=\pi(\rho)$, $K\in C(n, k+1)$, $\bar{P}$ *is the minimal* ρ-*chain containing the packet* $P=P(K)$. *Then the following properties are equivalent.*

a) $K\in N(r)$, *i.e.* P *forms a chain with respect to an appropriate representative of* r.

b) *Every element* $L\in\bar{P}-P$ *can be pushed out either to the left or to the right.*

Proof. We shall consider only the case $K\notin\mathrm{Inv}(r)$; applying it to r^t we shall get the rest. Implication a)⇒b) results from the following observation. Suppose that P forms a chain with respect to an order $\rho'\in r$. Then in the series of elementary equivalencies connecting ρ to ρ' every element $L\in\bar{P}-P$ must change places with all elements of P lying to the same side of L where it is eventually pushed out.

We shall prove b)⇒a) by induction on $n+k$. The first case $n=k=1$ is trivial.

First note that it suffices to prove b)⇒a) for an arbitrary representative ρ_0 of r. In effect, if b) is true for some ρ then it is true for any ρ' elementarily equivalent to ρ.

Take for ρ_0 a good order (7). Consider two cases.

1) $1 \in K$, i.e. $K=(1, i_2, \cdots, i_{k+1})$. Then $\{K_{k+1}, \cdots, K_2\} \subset \{J_1, \cdots, J_N\}$ and since the induced order on P is the lexicographic one, we have $K_1^{\wedge}=(i_2, \cdots, i_{k+1})=J_j'$ for some $j \geqslant a+1$. By the inductive assumption for $(n-1, k-1)$ it follows from b) that the order ρ_0 can be changed to an equivalent one in such a way that $\{K_{k+1}, \cdots, K_2\}$ form a chain. We shall assume that this is true for ρ_0.

Suppose first that $j>a+1$. We affirm that in this case $J_{a+1}', \cdots, J_{j-1}'$ commute with J_j' so that one can put J_j' to the $(a+1)$-th place.

In effect, if J_t' does not commute with $J_j'=K_1$ for some $a+1 \leqslant t \leqslant j-1$, then for some $s \geqslant 2$ we have $i_s \notin J_t'$. But $1 \notin J_t'$, hence J_t' cannot be pushed out of P neither to the left nor to the right, which contradicts our assumption.

Let now $K_1^{\wedge}=J_{a+1}'$. Then among all sets $J_1, \cdots, J_N$ only the elements of P do not commute with $K_1^{\wedge}$. But P already forms a chain and $K_1^{\wedge}$ can be moved to the left to extend this chain.

2) $1 \notin K$. We shall show that in this case all members of P belong simultaneously either to the left or to the right part of ρ_0. Therefore we can apply the inductive assumption for the case $(n-1, k)$.

In effect, suppose that $\{K_{k+1}, \cdots, K_{t+1}\} \subset \{J_1', \cdots, J_a'\}$ and $\{K_t, \cdots, K_1\} \subset \{J_{a+1}', \cdots, J_N'\}$. Let $1 \leqslant p \leqslant t$, $t+1 \leqslant q \leqslant k+1$. Then $K_{p,q}^{\wedge} \cup \{1\} \in \{J_1, \cdots, J_N\}$ does not commute neither with $K_p^{\wedge}$ nor with $K_q^{\wedge}$ and therefore cannot be pushed out of P, contrary to our assumption. ■

7. Lemma. a) $A(n, n-1)=B(n, n-1)=\{K_n \cdots K_1, K_1 \cdots K_n\}$ *where* $K=(1, \cdots, n)$.

b) *The poset* $B(n, n-2)$ *is described by the following picture*:

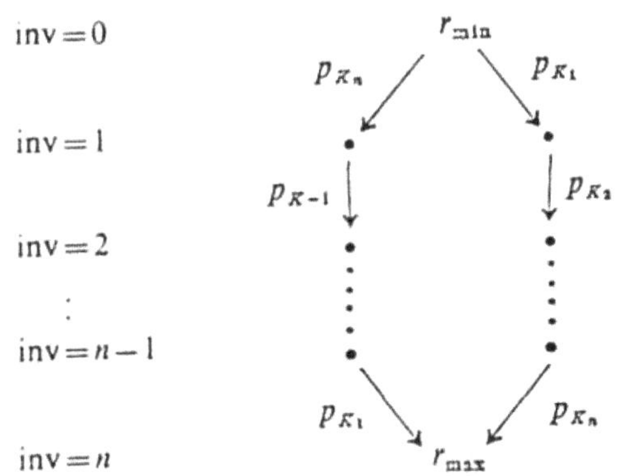

Proof. The first statement is evident.

No less evident is the structure of $B(3, 1)$:

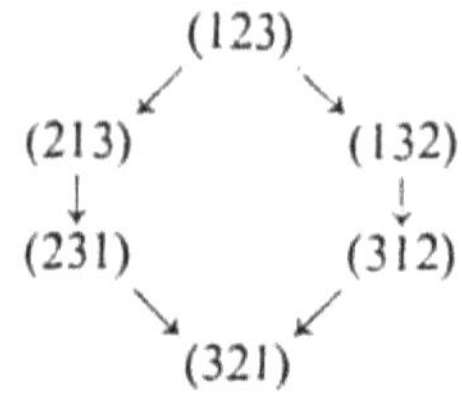

Now we shall prove by induction the following statement which is a reformulation of the second part of our Lemma.

$(B)_{n-1}$. For any $r \in B(n-1, n-3)$ we have either $\mathrm{Inv}(r)=\{K^{\wedge}_1, \cdots, K^{\wedge}_i\}$ or $\mathrm{Inv}(r)=\{K^{\wedge}_{n-1}, \cdots, K^{\wedge}_{n-i}\}$ for an appropriate i, where $K=\{1, \cdots, n-1\}$. In the first case we have $N(r)=\{K^{\wedge}_i, K^{\wedge}_{i+1}\}$, in the second $N(r)=\{K^{\wedge}_{n-i}, K^{\wedge}_{n-i-1}\}$ with exception of border effects: for $i=0$, where $r=r_{\min}$, and for $i=n-1$, where $r=r_{\max}$, we have $N(r)=\{K^{\wedge}_1, K^{\wedge}_{n-1}\}$.

In order to deduce $(B)_n$ from $(B)_{n-1}$ take an arbitrary $r \in B(n, n-2)$ with a good representative (7). Then $M=n-1$ and since $\bigcup J'_j=\underline{n}-\{1\}$, sets J'_j form a packet. Consider three possibilities.

1) $a=0$. From lemma 5 it follows that $\mathrm{Inv}(\rho')=\varnothing$ (in notation (8)). By inductive assumption, ρ' is equivalent to the lexicographic order on $C(\underline{n}-\{1\}, k-1)$ and $N(\pi(\rho'))=\{(\underline{n}-\{1\})^{\wedge}_1, (\underline{n}-\{1\})^{\wedge}_{n-1}\}$. Therefore

$$\underline{n}^{\wedge}_1 \in N(r) \subset \{\underline{n}^{\wedge}_2, \underline{n}^{\wedge}_n, \underline{n}^{\wedge}_1\},$$

where the first two sets in the right-hand side emerge after adding $\{1\}$ to the members of ρ' and the third one corresponds to the packet (J'_j). This packet is ordered by ρ either lexicographically or antilexicographically. In the first case $r=r_{\min}$, the packet $P(\underline{n}^{\wedge}_n)$ can be moved to form a chain in an order representing r, but the packet $P(\underline{n}^{\wedge}_2)$ cannot be moved in this way since its member $(\underline{n}^{\wedge}_2)^{\wedge}_1=(\underline{n}-\{1\})^{\wedge}_1$ is among $(J_1, \cdots, J_v)$ and cannot be put to the right part. Hence $N(\pi(\rho))=\{\underline{n}^{\wedge}_1, \underline{n}^{\wedge}_n\}$ in accord with $(B)_n$.

In the second case $r=p_{\underline{n}-\{1\}}(r_{\min})$ and for similar reasons one can form a chain from the packet $P(\underline{n}^{\wedge}_2)$ but not from the packet $\underline{n}^{\wedge}_n$ since $(\underline{n}^{\wedge}_n)^{\wedge}_1=(\underline{n}-\{1\})^{\wedge}_{n-1}$ is ρ-maximal. Therefore $N(r)=\{\underline{n}^{\wedge}_1, \underline{n}^{\wedge}_2\}$ as should be by $(B)_n$.

2) $0<a<n-1$. Since here $\mathrm{Inv}(\rho')=\{J'_1, \cdots, J'_a\}$, ρ' is neither minimal, nor maximal one. And since the packet $\{J'_i\}$ does not belong to one part of ρ, it is not contained in $N(r)$ (cf. the end of the proof of Lemma 6).

By inductive assumption $(B)_{n-1}$, one of two cases can occur:

$$\mathrm{Inv}(\rho')=\{(\underline{n}-\{1\})^{\wedge}_1, \cdots, (\underline{n}-\{1\})^{\wedge}_a\},\ N(\pi(\rho'))=\{(\underline{n}-\{1\})^{\wedge}_a, (\underline{n}-\{1\})^{\wedge}_{a+1}\}$$

or

$$\operatorname{Inv}(\rho')=\{(\underline{n}-\{1\})^{\wedge}_{n-1},\cdots,(\underline{n}-\{1\})^{\wedge}_{n-a}\},$$
$$N(\pi(\rho'))=\{(\underline{n}-\{1\})^{\wedge}_{n-a},(\underline{n}-\{1\})^{\wedge}_{n-a-1}\}.$$

To get $\operatorname{Inv}(r)$, $N(r)$ one must add $\{1\}$ and include the packet $P(\underline{n}-\{1\})$ into $\operatorname{Inv}(r)$ if necessary. This should be done iff $J'_b=(\underline{n}-\{1\})^{\wedge}_b$ for $1\leqslant b\leqslant a$, i.e. in the first case. Hence we have respectively

$$\operatorname{Inv}(r)=\{\underline{n}^{\wedge}_1,\underline{n}^{\wedge}_2,\cdots,\underline{n}^{\wedge}_{a+1}\},\qquad N(r)=\{\underline{n}^{\wedge}_{a+1},\underline{n}^{\wedge}_{a+2}\}$$

or

$$\operatorname{Inv}(r)=\{\underline{n}^{\wedge}_n,\cdots,\underline{n}^{\wedge}_{n-a+1}\},\qquad N(r)=\{\underline{n}^{\wedge}_{n-a+1},\underline{n}^{\wedge}_{n-a}\}$$

in accordance with $(B)_n$.

3) $a=n-1$. This case is treated similarly to $a=0$. ■

8. Lemma. *The following properties of $r\in B(n,k)$ are equivalent:*

a) *r is a maximal (resp. a minimal) element of $B(n,k)$.*

b) $r=\pi(\rho_{\max})$ *(resp.* $\pi(\rho_{\min})$*).*

c) $\operatorname{Inv}(r)=C(n,k+1)$ *(resp.* $\operatorname{Inv}(r)=\varnothing$*).*

Let $r_{\max}=p_{K_R}\cdots p_{K_1}(r_{\min})$, $R=\binom{n}{k+1}$. Then $K_1\cdots K_R\in A(n,k+1)$.

Corollary. *Any two elements of $A(n,k)$ are connected by a series of elementary equivalencies and operations p_K.*

Proof. Clearly b)$\Rightarrow$c)$\Rightarrow$a). We shall show that a)$\Rightarrow$b). In order to do that we shall prove by induction on $n+k$ the conjunction of two statements: if $\operatorname{Inv}(r)\neq\varnothing$, r is not minimal: otherwise r contains the lexicographical order. The first case is evident. The inductive step depends on the value of a in a good representative ρ of r (see (7), (8)).

$a=0$. Here $\operatorname{Inv}(\rho')=\varnothing$. Hence if $\operatorname{Inv}(r)\neq\varnothing$, all inversions of ρ lie in $\rho_2\in A(\underline{n}-\{1\},k)$. By inductive assumption then ρ_2 is not minimal, therefore r is not minimal. If $\operatorname{Inv}(r)=\varnothing$, both ρ' and ρ_2 are equivalent to lexicographical orders. hence r also is.

$a>0$. In this case $\operatorname{Inv}(r)\neq\varnothing$. Let J' be the maximal element of ρ_1 and $K=\{1\}\cup J'$. We shall show that $K\in N(r)$. Since $K\in\operatorname{Inv}(r)$, it follows that $p_K(r)<r$, hence r is not minimal.

First we prove that $J'\in N(\rho')$, i.e. that $K^{\wedge}_2,\cdots,K^{\wedge}_{k+1}$ can be gathered to form a chain in $1*\rho'$. In order to show this we shall apply Lemma 6 and check that any element L lying between two elements of this packet (but not belonging to it) can be pushed out. Assume the contrary and let L lie between $K_p^\wedge$ and $K_q^\wedge$ not commuting with them. Then $L = K_p^\wedge \cap K_q^\wedge \cup \{l\}$ for some $l \in L$. Therefore the set $M = K \cup L$ contains only $k+2$ elements and the situation which we want to exclude must be realizable already in $A(k+2, k)$. But in this situation we would have that the maximal element of ρ_1 is not contained in $N(r)$ while this is impossible by Lemma 7.

Thus we may assume that $K_2^\wedge, \cdots, K_{k+1}^\wedge$ form a chain in $1*\rho'$. Clearly $K_1^\wedge$ can be moved to this chain since all sets lying in between commute with $K_1^\wedge$.

We have established that a), b), c) are equivalent. In order to prove the last statement we must choose among $K_1, \cdots, K_R$ all members of a packet $P(L)$, $L \in C(n, k+2)$, and to show that the order induced on them is either lexicographic one, or inverse. But if we shall look in all our constructions only at elements and subsets of L, we shall reduce our task to that in $B(k+2, k)$ which is dealt with in Lemma 7.

Up to now we have proved statements a) and b) of Theorem 3. We can now conclude.

9. Proof of 3c). Our map is clearly injective. In view of Corollary to Lemma 8, in order to prove surjectivity it suffices to establish that in each equality

$$r_{max} = p_{K_R} \cdots p_{K_1}(r_{min})$$

one can change places the neighbouring $p_{K_t}, p_{K_{t-1}}$ if K_t, K_{t-1} commute, or reverse the order of the members of a packet if they are applied consecutively, without breaking the validity of such an equality. Note that if only the right hand side makes sense it necessarily coincides with r_{max} in view of Lemma 8c).

Suppose that K_t, K_{t-1} commute. Then $P(K_t) \cap P(K_{t-1}) = \varnothing$. From Lemma 6 it follows that if one can make chain first out of $P(K_{t-1})$ and second out of $P(K_t)$ applying only elementary equivalencies, one can make this also in reverse order.

Finally, let us show that if $p_{L_1^\wedge} \cdots p_{L_{k+2}^\wedge}(r)$ makes sense for some $L \in C(I, k+2)$, then $p_{L_{k+2}^\wedge} \cdots p_{L_1^\wedge}(r)$ also makes sense and gives the same result.

Both operators act nontrivially only upon $C(L, k) \subset C(n, k)$ and coincide in view of Lemma 7. Any element $J \in C(n, k) - C(L, k)$ does not commute with at most three elements of the set $\{L_{p,q}^\wedge\}$: if $|L \cap J| \leqslant k-2$, then J commutes with all $L_{p,q}^\wedge$, and if $|L \cap J| = k-1$ and $L - J = \{l_p, l_q, l_r\}$, then J does not commute with $L_{p,q}^\wedge, L_{p,r}^\wedge, L_{q,r}^\wedge$. From Lemma 7 it follows that in each of the packets $P(L_p)$, $P(L_q)$, $P(L_r)$ the set J lies at the same side of both elements with which it does not commute. Therefore the same will be true for the packet with reverse order. Hence if one of the expressions we work with make sense, so is the other one. ■

10. Proof of 3d). By induction on $n+k$ we shall show that if $\mathrm{Inv}(\rho)=\mathrm{Inv}(\sigma)$ then $\rho\sim\sigma$. Let $\rho=\rho_1 1*\rho'\rho_2$, $\sigma=\sigma_1 1*\sigma'\sigma_2$ From Lemma 5 it follows that ρ_1 and σ_1 coincide as sets and $\mathrm{Inv}(\rho_1\rho_2)=\mathrm{Inv}(\sigma_1\sigma_2)$, $\mathrm{Inv}(\rho')=\mathrm{Inv}(\sigma')$. Hence by inductive assumption $\pi(\rho_1\rho_2)=\pi(\sigma_1\sigma_2)$, $\pi(\rho')=\pi(\sigma')$.

Clearly, it follows that $\pi(\rho)=\pi(\sigma)$. ■

11. Question. In [12], [13] a nice combinatorial description of the set $A(n, 2)$ is given in terms of Young tableaux. Is there a generalization to $A(n, k)$, $k>2$?

§ 3. $(n-1)$-category S_n

1. Definition. An n-spheric set A consists of sets $A_0, A_1, \cdots, A_n$ and maps

$$s_k, t_k: A_k \rightrightarrows A_{k-1};\ 1\leqslant k\leqslant n,$$
$$i_k: A_k \longrightarrow A_{k+1};\ 0\leqslant k\leqslant n-1$$

such that

a) $s_{k-1}s_k=s_{k-1}t_k$; $t_{k-1}s_k=t_{k-1}t_k$.

b) $s_{k+1}i_k=t_{k+1}i_k=\mathrm{id}_{A_k}$. ■

We shall sometimes omit the subscript k and write s, t, i. The elements of A_k are called k-cells of A. A k-cell is called degenerate if it is contained in $i(A_{k-1})$.

A geometric realization of the n-spheric set A is the n-dimensional CW-complex in which to a nondegenerate k-cell there corresponds a k-dimensional disk $D(k)$ represented as the union of an open k-ball and two open k'-balls for all $0\leqslant k'<k$. Example:

$D(2)$: 

Let A be an n-spheric set. For $k'<k$ we set

$$s_{kk'}=s_{k'+1}s_{k'+2}\cdots s_k,\ t_{kk'}=t_{k'+1}\cdots t_k: A_k\longrightarrow A_{k'},$$
$$i_{k'k}=i_{k-1}\cdots i_{k'+1}i_{k'}: A_{k'}\longrightarrow A_k.$$

2. Definition. A (small) n-category is an n-spheric set C endowed with a family of multiplication maps

$$\mu_{pq}: \{(f, g) \in C_p \times C_p \mid s_{pq}(f) = t_{pq}(g)\} \longrightarrow C_p, \qquad 0 \leqslant q < p.$$

We shall denote $\mu_{pq}(f, g)$ by $f \circ_q g$.

These maps must satisfy the following axioms:

(AO) Let $f, g \in C_p$, $q < p$, $s_{pq}(f) = t_{pq}(g)$. Then $s_p(f \circ_q g) = s_p(f) \circ_q s_p(g)$, $t_p(f \circ_q g) = t_p(f) \circ_q t_p(g)$ if $q < p-1$, $s_p(f \circ_q g) = s_p(g)$, $t_p(f \circ_q g) = t_p(f)$ if $q = p-1$.

(Ass 1) $(f \circ_q g) \circ_q h = f \circ_q (g \circ_q h)$.

(Ass 2) Let $p < q$; $f, f', g, g' \in C_r$ and $t_q(f) = s_q(f')$, $t_q(g) = s_q(g')$, $t_p(f) = s_p(g)(= t_p(f') = s_p(g'))$. Then $(f \circ_q f') \circ_p (g \circ_q g') = (f \circ_p g) \circ_q (f' \circ_p g')$.

(Id) Let $f \in C_p$, $q < p$. Then $f \circ_q i_{qp} s_{pq}(f) = i_{qp} t_{pq}(f) \circ_q f = f$. ■

Elements of C_p are called p-morphisms, 0-morphisms are called objects.

For $n = 1$ we get the usual definition of a category.

Let C be an n-category and $m < n$. Then we can define an m-category $\tau_{\leqslant m} C$ as follows:

$$(\tau_{\leqslant m} C)_p = C_p \qquad \text{for } p < m,$$

$$(\tau_{\leqslant m} C)_m = \operatorname{Coker}(s, t; C_{m+1} \rightrightarrows C_m).$$

We shall show now how the construction of section 2 allows us to define for each $n \geqslant 1$ an $(n-1)$-category $\boldsymbol{S}_n$.

Let $I = \{1, 2, \cdots, n\}$, $\mathscr{P} \subset C(I, k)$ an arbitrary subset.

3. Definition (cf. sec. 2, $n^{\circ}2$). a) A total order ρ on $\mathscr{P}$ is called admissible if for each packet P it induces on $P \cap \mathscr{P}$ either the lexicographic order or the reverse one.

We denote by $A(n, k; \mathscr{P})$ the set of all admissible total orders on $\mathscr{P}$.

b) Two total orders $\rho, \rho' \in A(n, k; \mathscr{P})$ are called elementarily equivalent, if they differ by a transposition of two neighbours which do not belong to a common packet.

We denote by $B(n, k; \mathscr{P})$ the quotient of $A(n, k; \mathscr{P})$ by the induced equivalence relation and by π the natural projection.

c) An inversion in the order $\rho \in A(n, k; \mathscr{P})$ is an element $K \in C(I, k+1)$ such that $P(K) \subset \mathscr{P}$ and induces on $P(K)$ the antilexicographic order. We denote by $\operatorname{Inv}(\rho) \subset C(I, k+1)$ the set of all inversions of ρ. Clearly $\operatorname{Inv}(\rho) = \operatorname{Inv}(\rho')$ if $\pi(\rho) = \pi(\rho')$ so that one can define $\operatorname{Inv}(r)$ for any $r \in B(n, k; \mathscr{P})$. As in section 2 we define $p_K(\rho)$.

4. Definition. a) Let $r, r' \in B(n, k; \mathscr{P})$ and $\operatorname{Inv}(r') \supset \operatorname{Inv}(r)$. Set $\mathscr{P}' = \operatorname{Inv}(r') - \operatorname{Inv}(r)$. Call an *arrow* from r to r' an element $f \in B(n, k+1; \mathscr{P}')$ such that if $f \sim K_1 \cdots K_M$, we have $r' = p_{K_1} \cdots p_{K_M}(r)$ (cf.

sec. 2, Theorem 3a)). We shall write simply $f;\, r \to r'$ and denote by $\mathrm{Ar}(r, r')$ the set of arrows from r to r'.

b) Composition: if $f: r \to r'$, $f \sim K_1 \cdots K_M$; $g: r' \to r''$, $g \sim K_1' \cdots K_N'$, we put $g \circ f: r \to r''$, $g \circ f \sim K_1' \cdots K_N' K_1 \cdots K_M$.

c) Concatenation: let $f_0, f_1: r \to r'$; $g_0, g_1: r' \to r''$; $h_0: f_0 \to f_1$, $h_0 \sim K_1 \cdots K_M$; $h_1: g_0 \to g_1$, $h_1 \sim K_1' \cdots K_N'$. Then $h = h_1 * h_0 : \sim K_1 \cdots K_M K_1' \cdots K_N' \sim K_1' \cdots K_N' K_1 \cdots K_M$ is an arrow from $g_0 f_0$ to $g_1 f_1$. ■

5. Now we can define $\boldsymbol{S}_n$. Set $\boldsymbol{S}_{n.0} = S_n$. For $1 \leqslant p \leqslant n-1$, a p-morphism of $\boldsymbol{S}_n$ consists of the following data:

a) A family of subsets $\mathscr{P}_i \subset C(I, i+1)$, $i = 0, 1, \cdots, p$, such that $\mathscr{P}_0$ $C(I, 1) = S_n$.

b) A family of pairs $r_0^i, r_1^i \in B(n, i+1; \mathscr{P}_i)$, $i = 0, 1, \cdots, p-1$, and an element $r^p \in B(n, p+1; \mathscr{P}_p)$ such that

$$r_0^i, r_1^i \in \mathrm{Ar}(r_0^{i-1}, r_1^{i-1}),\ i = 1, \cdots, p-1, \quad r^p \in \mathrm{Ar}(r_0^{p-1}, r_1^{p-1}).$$

Thus $\mathscr{P}_j = \mathrm{Inv}(r_1^{i-1}) - \mathrm{Inv}(r_0^{i-1})$ for $i \geqslant 1$.

For $r = (r_0^0, r_1^0; r_0^1, r_1^1; \cdots; r_0^{p-1}, r_1^{p-1}; r^p) \in \boldsymbol{S}_p$ set

$$s(r) = (r_0^0, r_1^0; \cdots; r_0^{p-2}, r_1^{p-2}; r_0^{p-1}),$$
$$t(r) = (r_0^0, r_1^0; \cdots; r_0^{p-2}, r_1^{p-2}; r_1^{p-1}),$$
$$i(r) = (r_0^0, r_1^0; \cdots; r_0^{p-1}, r_1^{p-1}; r^p, r^p; \mathrm{Id}),$$

where Id means the identity, an only element of $B(n, p+2; \varnothing)$.

This data define on $\boldsymbol{S}_n = \{\boldsymbol{S}_{n.0}, \cdots, \boldsymbol{S}_{n.n-1}\}$ a structure of a spheric set. Now we shall describe multiplication.

Let $r, r' \in \boldsymbol{S}_p$, $t_q(r) = s_q(r')$. If $q = p-1$, we have

$$r = (r_0^0, r_1^0; \cdots; r_0^{p-1}, r_1^{p-1}: r^p),$$
$$r' = (r_0^0, r_1^0; \cdots; r_0^{p-2}, r_1^{p-2}; r_1^{p-1}, r_2^{p-1}; r'^p)$$

and we set

$$r' \circ_{p-1} r = (r_0^0, r_1^0; \cdots; r_0^{p-2}, r_1^{p-2}; r_1^{p-1}, r_2^{p-1}; r'^p \circ r^p).$$

If $q < p-1$, we have

$$r = (r_0^0, r_1^0; \cdots; r_0^{q-1}, r_1^{q-1}; r_0^q, r_1^q; \cdots; r^p),$$
$$r' = (r_0^0, r_1^0; \cdots; r_0^{q-1}, r_1^{q-1}; r_1^q, r_2^q; r'^{q+1}_0, r'^{q+1}_1; \cdots; r'^p)$$

and we set

$$r' \circ_q r = (r_0^0, r_1^0; \cdots; r_0^{q-1}, r_1^{q-1}; r_0^q, r_2^q; r_0'^{q+1} \circ r_0^{q+1},$$
$$r_1'^{q+1} \circ r_1'^{q+1}; r_0'^{q+2} * r_0'^{q+2}, r_1'^{q+2} * r_1'^{q+2}; \cdots; r'^p * r^p).$$

It is straightforward to check the axioms of the $(n-1)$-category. The truncated category $\tau_{\leqslant 1}\boldsymbol{S}_n$ is a category with objects S_n in which $\mathrm{Hom}(x, y)$ is either empty or consists of one element. The latter occurs iff $x \geqslant y$ with respect to the weak Bruhat order.

6. **Example.** 2-category $\boldsymbol{S}_3$:

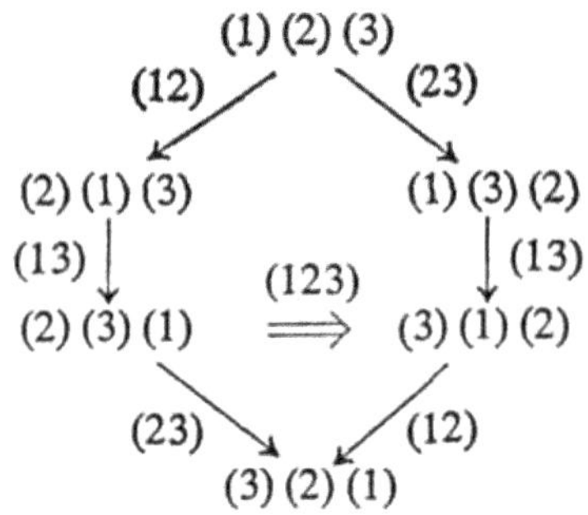

7. **Convex hull.** Let $x = (x_1, \cdots, x_n)$ be a point with pairwise distinct coordinates, M_n the convex hull of points $(x_{\sigma(1)}, \cdots, x_{\sigma(n)})$, $\sigma \in S_n$. We shall be interested only in combinatorial structure of M_n which does not depend on a choice of x. The $(n-1)$-polytope M_n in a sense may be considered as a "geometric realization" of $\boldsymbol{S}_n$. More precisely, the set of vertices of M_n is (bijective to) $S_n = \boldsymbol{S}_{n,0}$. The set of 1-faces is bijective to the set of indecomposable 1-morphisms of $\boldsymbol{S}_n$ (this is well-known in the theory of the weak Bruhat order). In general, each p-face of M_n is a product $M_{p_1} \times \cdots \times M_{p_k}$, where $\sum (p_i - 1) = p$. We shall call *indecomposable* the faces M_{p+1}. E.g., M_4 has eight indecomposable 2-faces (hexagons M_3) and six decomposable ones (quadrangles $M_2 \times M_2$).

Conjecture. The set of indecomposable p-faces of M_n is naturally bijective to the set of indecomposable p-morphisms of $\boldsymbol{S}_n$.

This is true for $p = 0, 1$ and also for $p = n$ (one morphism) and $p = n - 1$ (2n morphisms). The faces $M_{p_1} \times \cdots \times M_{p_k}$ correspond to the products of indecomposable morphisms with commuting factors.

References

[1] P. Cartier, Les arrangements d'hyperplans: un chapitre de géométrie combinatoire, Sém. Bourbaki, 1980/81, Springer Lecture Notes in Math., vol. **801** (1981), p. 1–22.

[2] T. Kohno, Poincaré series of the Malcev completion of generalized pure braid groups, Preprint IHES M/85/36, 1985.

[3] ——, Série de Poincaré-Koszul associée aux groupes de tresses, Invent. Math., **82:1** (1985), 57–76.

[4] R. Randell, The fundamental group of the complement of a union of complex hyperplanes, Invent. Math., **69:1** (1982), 103–108. Correction: **80:3** (1985), 467–468.

[5] L. Orlik and P. Solomon, Combinatorics and topology of complements of hyperplanes, Invent. Math., **56:2** (1980), 167–189.

[6] Yu. I. Manin and V. V. Schechtman, On the higher Bruhat orders related to the symmetric group, Funkc. Analiz i ego pril., **20:2** (1986), 74–75 (in Russian).

[7] ——, Arrangements of real hyperplanes and Zamolodchikov's equations, In: Group-theoretical methods in physics, Moscow, Nauka, vol. **1** (1986), 316–324 (in Russian).

[8] I. M. Gelfand and A. V. Zelevinsky, Algebraic and combinatorial aspects of the general theory of hypergeometric functions, Funkc. Analiz i ego pril., **20:3** (1986), 17–34 (in Russian).

[9] V. A. Vasiliev, I. M. Gelfand and A. V. Zelevinsky, General hypergeometric functions on complex Grassmannians, Funkc. Analiz i ego pril., **21:1** (1987), 23–38 (in Russian).

[10] A. N. Varchenko, Combinatorics and topology of arrangements of affine hyperplanes in a real space, Funkc. Analiz i ego pril., **21:1** (1987), 11–22 (in Russian).

[11] T. Zaslavsky, Facing up to arrangements: face-count formulas for partitions of space by hyperplanes, Amer. Math. Soc. Memoirs, vol. 154, 1975.

[12] P. Edelman and C. Greene, Balanced tableaux, Adv. Mah., **63** (1987), 42–99.

[13] R. Stanley, On the number of reduced decompositions of elements of Coxeter groups, European J. Comb., **5** (1984), 359–372.

Steklov Math. Inst.
Ul. Vavilova 42
117966, GSP-1
Moscow, USSR

PART II. MODULAR FORMS AND DIOPHANTINE EQUATIONS

Izv. Akad. Nauk SSSR
Ser. Mat. Tom 33 (1969), No. 3

Math. USSR – Izvestija
Vol. 3 (1969), No. 3

THE p-TORSION OF ELLIPTIC CURVES IS UNIFORMLY BOUNDED

Ju. I. MANIN

It is shown that the p-component of the torsion of elliptic curves over a numerical field is bounded.

Let k be a field of algebraic numbers of finite degree, X an elliptic curve (a one-dimensional abelian variety) over k, and Φ its group of k-points of finite order.

In his survey ([3], § 22), Cassels observed:

"The following conjecture is part of the folklore:

Conjecture. *For given k (and in particular for $k = Q$), the order of Φ is bounded.*"

In this note we prove the corresponding statement for the p-component of the group Φ:

Theorem 0. *Let p be a fixed prime. There exists a constant c (depending only on p and k) such that the order of the p-torsion group of the k-points of an elliptic curve defined over k does not exceed c.*

(In fact, the result is somewhat stronger; see the corollary to Theorem 2, and also Theorem 3.)

The proof runs as follows. If the curve X has a cyclic subgroup D of period p^n defined over k, then the quotient group X/D is also defined over k. The absolute variants of the curves X and X/D are connected by a modular equation whose genus increases without limit for $n \to \infty$. In particular, the Mordell conjecture would imply that k has only a finite number of invariants corresponding to curves on which there is a k-point of order p^n, provided the corresponding module equation has genus > 1. (This observation was made independently by Shimura; see Cassels, loc. cit.).

The basic idea of this paper is that the finiteness of the number of solutions to the modular equations *for sufficiently large n* can be proved by using an idea of Dem'janenko [4]. Generalizing his theorem, we show in the first section of the paper that if the jacobian variety of a curve Y over k contains a k-simple abelian variety A (up to isogeny) with multiplicity greater than $\mathrm{rk}\, A(k)/\mathrm{rk}\, \mathrm{End}_k A$, then the set $Y(k)$ is finite.

The condition is verified in the second section for modular equations of sufficiently high degree. The well-known symmetry properties of modular curves are used. The idea is applicable to many other projective systems of curves of "modular type" and shows that, beginning at some point (depending on the field of constants), the curves of such systems have a finite number of rational points.

To complete the proof of the theorem on boundedness of torsion, we verify it for curves with a fixed absolute invariant, i.e., for the forms of a given curve. A simple proof of a somewhat stronger result, using the Serre theory of l-adic representations, is the subject of the last section.

§ 1. Generalization of the theorem of Dem'janenko

Unless otherwise stated, all varieties and morphisms are defined over a fixed field k of algebraic numbers. For any variety X, the set of its k-points will be denoted by $X(k)$.

Let D be an abelian group. By its *rank* (rk D) is meant the number of generators of its free part.

Let X be a normal projective variety and A an abelian variety. Denote by $A(X)$ the group of k-morphisms $X \to A$ taking a fixed point $x \in X(k)$ to zero.

In this section, we prove the following result.

Theorem 1. *Assume that the rank of the Néron-Severi group of X is 1. If* rk $A(X) >$ rk $A(k)$, *then the set $X(k)$ is finite.*

This has been proved by Dem'janenko [4] for the case that X and A are curves.

We shall be using the following particular case of the theorem.

Corollary. *Let X be a curve and A a k-simple abelian variety contained in the jacobian of X with multiplicity $m(X, A)$ (to within isogeny). If*

$$m(X, A) > \operatorname{rk} A(k) / \operatorname{rk} \operatorname{End}_k A,$$

then the set $X(k)$ is finite.

Indeed, under these conditions,

$$\operatorname{rk} A(X) = m(X, A) \operatorname{rk} \operatorname{End}_k A > \operatorname{rk} A(k).$$

We begin with an elementary lemma on euclidean spaces.

Let E be a euclidean space. For any vector $x \in E$ the symbol $h(x)$ will denote the square of its length.

Definition. The system of vectors $x_1, \ldots, x_r \in E$ is called an (ε, N)-system if for all $1 \leq i$, $j \leq r$ we have

$$\left.\begin{aligned} &|h(x_i) / h(x_1) - 1| < \varepsilon, \\ &|h(x_i + x_j) / h(x_1) - 2| < \varepsilon, \quad i \neq j, \end{aligned}\right\} \tag{1}$$

$$h(x_i) > N. \tag{2}$$

Lemma 1. *Let $D \subset E$ be a subgroup. The following two conditions are equivalent:*

a) *D contains a sublattice (discrete subgroup) of rank r.*

b) *For any $\varepsilon > 0$ and $N > 0$, D contains an (ε, N)-system of r vectors.*

Proof. b) $\Rightarrow$ a). It follows from (1) that for fixed N and sufficiently small ε the scalar products (x_i, x_j), $i \neq j$, can be taken arbitrarily close to zero. Therefore the vectors of the system are pairwise "almost" orthogonal and so linearly independent.

a) $\Rightarrow$ b). Let D contain a sublattice of rank r. In the euclidean subspace spanned by it, choose an orthonormal basis $\xi_1, \ldots, \xi_r$.

There exists an $\eta > 0$ such that any system of vectors (ξ_i') with $h(\xi_i - \xi_i') < \eta$ is an $(\epsilon, 1/2)$-system. Now choose a similarity coefficient $\sqrt{R} > 2N$ so large that any sphere of radius $\sqrt{R}\eta$ contains a point of the sublattice. Then there exist elements x_i within it for which

$$h(\xi_i - x_i / \sqrt{R}) < \eta, \quad i = 1, \ldots, r.$$

Clearly, these elements form an (ϵ, N)-system. This proves the lemma.

Proof of Theorem 1. Let $r = \operatorname{rk} A(X)$ and $f_1, \ldots, f_r \in A(X)$ be a free system of elements. We shall show that if the set $X(k)$ is infinite, there exist points $x \in X(k)$ for which $f_1(x), \ldots, f_r(x)$ are independent in $A(k)$, and therefore $\operatorname{rk} A(k) \geq \operatorname{rk} A(X)$ in contradiction to the hypothesis of the theorem. For this purpose we use the properties of (logarithmic) height. All requisite facts concerning Weil height are given in [7], and concerning Tate height on abelian varieties in [8] and [9].

Take a fixed symmetric ample pencil L on the abelian variety A with respect to which the Tate height on the group $A(k)$ is defined. In order to define it on the group $A(X)$, we must also take a fixed ample invertible pencil L_0 on X.

First step. Set $\deg M = (M \cdot L_0^{(d-1)})$ (intersection index) for any invertible pencil M on X, where $d = \dim X$.

The Tate height $\hat{H}_L$ of a morphism $f \in A(X)$ satisfies the inequality

$$|\hat{H}_L(f) - \deg f^*(L)| < c \tag{3}$$

for some constant c (see [7], Chap. III, § 3 and [8]). It induces a euclidean space structure on $A(X) \otimes \mathbf{R}$. Setting $r = \operatorname{rk} A(X)$ and taking fixed ϵ and $N > 0$, choose an (ϵ, N)-system $f_1, \ldots, f_r \in A(X)$.

Second step. For each pencil $L_i = f_i^*(L)$, $L_{ij} = (f_i + f_j)^* L$ on X choose (from an equivalence class of Weil heights) a height h_i [resp., h_{ij}] on the set $X(k)$. Since the rank of the Néron-Severi group of X is unity, the pencils L_i and L_{ij} are commensurate modulo algebraic equivalence. From the criterion for quasi-equivalence of heights (Lang [7], Chap. IV, §) it follows that for $h_1(x) \to \infty$ $(x \in X(k))$ we have

$$\begin{aligned} \lim h_i(x) / h_1(x) &= \deg L_i / \deg L_1, \\ \lim h_{ij}(x) / h_1(x) &= \deg L_{ij} / \deg L_1. \end{aligned} \tag{4}$$

On the other hand, from the functorial property of heights and the definition of Tate height it follows that under the same conditions,

$$\begin{aligned} \lim h_i(x)/h_1(x) &= \lim \hat{h}_L(f_i(x))/\hat{h}_L(f_1(x)), \\ \lim h_{ij}(x)/h_1(x) &= \lim \hat{h}_L((f_i + f_j)(x))/\hat{h}_L(f_1(x)), \end{aligned} \tag{5}$$

where $\hat{h}_L$ is the (numerical) Tate height on $A(k)$.

Third step. Comparing (3), (4), and (5), we see that if N, and then $h_1(x)$, are chosen sufficiently large, we have

$$|\hat{h}_L(f_i(x))/\hat{h}_L(f_1(x)) - \hat{H}_L(f_i)/\hat{H}_L(f_1)| < \varepsilon,$$
$$|\hat{h}_L(f_i(x) + f_j(x))/\hat{h}_L(f_1(x)) - \hat{H}_L(f_i + f_j)/\hat{H}_L(f_1)| < \varepsilon.$$

It follows that the models of the points $f_i(x)$ form a $(2\epsilon, N)$-system in the euclidean space $A(k) \otimes \mathbf{R}$ (with norm $\hat{h}_L$) and are therefore linearly independent.

This completes the proof.

§ 2. Modular curves

Let K denote the algebraic closure of a field of functions of one independent variable over k.

Let X be an elliptic curve over K with absolute invariant transcendental over k; p a prime; and $\xi = (\ldots x_n \ldots) \in T_p(X)$ a nonzero element of the Tate module.

Let j_n $(n \geq 0)$ be the absolute invariant of the curve $X/(x_n)$ (the quotient modulo the subgroup generated by x_n) and for all $n \geq m \geq 0$ set

$$K_{m,n} = k(j_m, j_n) \subset K.$$

The following list of properties of the field $K_{m,n}$ can easily be derived from Igusa's results ([6], § 6; see also [1]):

a) *The field $K_{m,n}$ to within isomorphism over k does not depend on the choice of X and ξ. More precisely, if K'_m is defined in terms of X', ξ', then the mapping $j_m \sim\to j'_m$, $j_n \sim\to j'_n$ extends to an isomorphism $K_{m,n} \to K'_{m,n}$ over k.*

This follows basically from the fact that Gal $(K/k(j_0))$ is transitive on $T_p(X)$.

b) *The mapping $j_m \sim\to j_{m+a}$, $j_n \sim\to j_{n+a}$ extends to an isomorphism $K_{m,n} \to K_{m+a,n+a}$ $(a \geq 0)$.*

Indeed, let $X' = X/(x_a)$, $\xi' = \operatorname{Im} \xi \in T_p(X')$. Then the field $K_{m+a,n+a}$ coincides with the field $K'_{m,n}$ constructed for X', ξ'.

c) *$j_k \in K_{m,n}$ for any $m \leq k \leq n$.*

Indeed, in the last paragraph of [6] Igusa shows that the field $K_{n,n}$ is the minimum field of definition over $K(j_0)$ for the cycle consisting of the point x_n and all its conjugates. Therefore $K_{0,k} \subset K_{n,n}$ for all $k \leq n$, since $x_k = p^{n-k}x_n$.

As in b), the "shift" proves the statement for any m.

d) *The field k is algebraically closed in $K_{m,n}$.*

This was proved by Igusa [6] for $K_{0,n}$ and is obtained for all m by means of the shift.

Let Y_n be the complete normal model of the field $K_{0,n}$ over k. The main result in this section is the following.

Theorem 2. *If n is sufficiently large (the bound depends on p and k), the set $Y_n(k)$ is finite.*

Corollary. *If n is sufficiently large, there exist only a finite number of elements $j \in k$ such that there is defined over k an elliptic curve with invariant j and a cyclic subgroup of period p^n defined over k (i.e., Gal $(\bar{k}/k)$-invariant).*

By the corollary of Theorem 1 it suffices to prove the following result.

Lemma 2. *There exists a simple abelian variety A defined over k whose multiplicity in the jacobian of the curve Y_n increases without limit for $n \to \infty$.*

Proof. Take $a > 0$ such that the genus of the curve Y_a is positive; we show that A can be chosen from the simple components of the jacobian of Y_a.

Consider the n inclusions

$$f_i\colon K_{0,a}\to K_{0,na},\qquad i=0,1,\ldots,n-1;$$
$$f_i(j_k)=j_{k+ai},\qquad k=0,\ldots,a.$$

They give rise to mappings of the jacobians $J(Y_a)\to J(Y_{na})$. If the dimension of the abelian subvariety generated by the images of these mappings increases without limit as $n\to\infty$, then the multiplicity of one of the components of $J(Y_a)$ in $J(Y_{na})$ will also increase without limit.

Assume the converse: that the dimension of the sum of the images is bounded. Then also bounded is the dimension of the space of differentials in $K_{0,na}$ generated by the images of the spaces of differentials of the first kind in $K_{n,a}$ with respect to the inclusion f_i. Since this space increases (because $K_{0,na}\subset K_{0,(n+1)a}\subset\ldots$), it ultimately becomes stabilized. Therefore the subfield L generated by the ratios of these differentials is also stabilized.

When the curve Y_a is not hyperelliptic, we immediately arrive at a contradiction. Indeed, in this case the ratios of the differentials of the first kind of the subfield $f_i(K_{0,a})$ generate it, and these subfields together generate the entire field $K_{0,na}$, which increases without limit.

Now let Y_a be hyperelliptic. Then the field $f_i(K_{n,a})$ is generated over the field of ratios of its differentials, and so over L, by a quadratic root. Therefore

$$[K_{0,na}:L(j_0)]\leqslant[K_{0,na}:L]\leqslant 2^n,$$

so that $[K_{0,na}:k(j_n)]\leq b2^n$, where $b=[L(j_0):k(j_0)]$ is fixed. But this is impossible, since ([6, 1])

$$[K_{0,na}:k(j_0)]=p^{na}(1+1/p)>b2^n$$

for large n (for $p=2$ we have $a>1$: the genus of Y_1 is zero). This proves the lemma.

§ 3. Torsion of curves with fixed invariant

To complete the proof of Theorem 0 we must show that the p-torsion of curves with given invariant $j\in k$ is bounded. In fact, *all torsion is bounded*: it suffices to verify this for complete local fields, and over such fields the number of curves with a fixed invariant is finite. (Note that in [5] a somewhat stronger result is obtained. The method used there is based on an explicit estimate of the deviation of the Tate height from the Weil height and merits special analysis).

The aim of this section is to prove a similar result for cyclic p-subgroups rational over k.

Theorem 3. *Let X be an elliptic curve over k that has no complex multiplication. Then on the set of its k-forms the order of a maximal cyclic p-subgroup rational over k is bounded.*

Proof. Let $\{X_h\}$ be a set (of representatives of all classes) of forms of X. For each form X_h choose an isomorphism over $\bar{k}$: $f_h:\bar{X}_h\to\bar{X}$ ($\bar{X}=X_k\otimes\bar{k}$, etc.). Let P_h^n be the set of those primitive points of period p^n in $X_h(\bar{k})$ that generate a cyclic subgroup rational over k. Put

$$P^n=\bigcup_h f_h(P_h^n)\subset X(\bar{k}).$$

Suppose the order of the rational p-subgroups on the set $\{X_h\}$ is unbounded. Then P^n *is nonempty for all* n. Multiplication by p^k defines a mapping $P^{n+k}\to P^n$, with respect to which the sets P^n form a projective system. Since the projective limit of nonempty finite sets is nonempty ([2], Chap. I, Appendix, Theorem 1), the Tate module $T_p(X)$ contains an element $\xi=(\ldots x_n\ldots)$ such that

$$x_n = f_{h_n}(x_n'), \quad x_n' \in P_{h_n}^n \subset X_{h_n}(\bar{k}).$$

For any Galois group element $s \in \mathrm{Gal}(\bar{k}/k)$ we have

$$x_n^s = a_n(s)\,x_n, \quad a_n(s) \in \mathbf{Z}/p^n\mathbf{Z},$$

because $f_h^s = \pm f_h$ and x_n' and $x_n'^s$ generate the same cyclic subgroup. Moreover, $a_{n+1}(s) \equiv a_n(s) \bmod p^n$. Hence there exists a p-adic number $a(s)$ such that $\xi^s = a(s)\xi$. This means that the element $\xi \in T_p(X)$ generates a $\mathrm{Gal}(\bar{k}/k)$-invariant submodule in $T_p(X)$. By Serre's results [10], this is impossible if X has no complex multiplication, because the representation of the corresponding p-adic Lie algebra is irreducible ([10], Chap. IV, § 2).

Received 8 DEC 68

BIBLIOGRAPHY

[1] A. Borel et al., *Seminar on complex multiplication*. Lecture Notes in Math., no. 21, Springer-Verlag, Berlin and New York, 1966. MR 34 #1278.

[2] N. Bourbaki, *Topologie générale*, 3éme éd., Actualités Sci. Indust., no. 1143, Hermann, Paris, 1960. MR 25 #4021.

[3] J. W. S. Cassels, *Diophantine equations with special reference to elliptic curves*. J. London. Math. Soc. 41 (1966), 193-291. MR 33 #7299.

[4] V. A. Dem'janenko, *Rational points of a class of algebraic curves*. Izv. Akad. Nauk SSSR Ser. Mat. 30 (1966), 1373-1396; English transl., Amer. Math. Soc. Transl. (2) 66 (1968), 246-272. MR 34 #5816.

[5] ———, *An estimate of the remainder term in Tate's formula*, Mat. Zametki 3 (1968), 271-278 = Math. Notes 3 (1968), 173-177. MR 37 #2751.

[6] J. Igusa, *Fibre systems of Jacobian varieties*. III, *Fibre systems of elliptic curves*, Amer. J. Math. 81 (1959), 453-476. MR 21 #3422.

[7] S. Lang, *Diophantine geometry*, Interscience Tracts in Pure and Appl. Math., no. XI, Interscience, New York, 1962. MR 26 #119.

[8] Ju. I. Manin, *The Tate height of points on an Abelian variety, its variants and applications*. Izv. Akad. Nauk SSSR Ser. Mat. 28 (1964), 1363-1390; English transl., Amer. Math. Soc. Transl. (2) 59 (1966), 82-110. MR 30 #3886.

[9] D. Mumford, *A remark on Mordell's conjecture*, Amer. J. Math. 87 (1965), 1007-1016. MR 32 #4083.

[10] J.-P. Serre, *Abelian l-adic representations and elliptic curves*, Benjamin, New York, 1968.

Actes, Congrès intern. math., 1970. Tome 1, p. 401 à 411.

LE GROUPE DE BRAUER-GROTHENDIECK EN GÉOMÉTRIE DIOPHANTIENNE

par Y. I. MANIN

Introduction.

Soit k un corps, $[k: \mathbf{Q}] < \infty$, $\mathscr{V}$ une classe de variétés algébriques sur k. Soit $V \in \mathscr{V}$ et soit k_v un complété de k.

La première condition nécessaire pour l'existence d'un k-point sur V est que, pour tout v, $V(k_v)$ ne soit pas vide. Si pour tout $V \in \mathscr{V}$ cette condition est suffisante, on dit que le principe de Hasse est vérifié par $\mathscr{V}$.

Si $\mathscr{V}$ est la classe des espaces homogènes principaux sur une variété abélienne V_0, Cassels [1] et Tate ont donné une obstruction au principe de Hasse et cette obstruction est effectivement calculable. Plus précisément, soit $V \in \mathscr{V}$ et soit $b \in H^1(G, V_0(\bar{k}))$ la classe de cohomologie correspondante, où $G = \mathrm{Gal}(\bar{k}/k)$. Si tous les $V(k_v)$ sont non vides, b appartient au groupe de Chafarevitch-Tate $Ш(V_0)$. Soit $\hat{V}_0$ la variété abélienne duale de V_0 et soit $Ш(V_0) \times Ш(\hat{V}_0) \to \mathbf{Q}/\mathbf{Z}$ l'accouplement canonique de Cassels-Tate. Pour que $V(k)$ soit non vide, il faut que la classe b de V appartienne au noyau (à gauche) de cet accouplement.

Le premier but de cet exposé est de démontrer que l'utilisation du groupe de Brauer-Grothendieck $\mathrm{Br}(V)$ de la variété V permet de construire une obstruction parfaitement générale au principe de Hasse. En calculant cette obstruction pour différentes classes $\mathscr{V}$, nous obtenons tout d'abord une démonstration simple et unifiée d'une série de résultats déjà connus. On obtient ainsi :

a) La forme de Cassels-Tate et la « suite duale » de Cassels [1, 2].

b) Le théorème de Voskresensky sur le groupe Ш pour les tores [14].

c) Les contre-exemples de Swinnerton-Dyer [12], de Mordell [9] et de Cassels et de Guy [3] au principe de Hasse pour certaines surfaces cubiques.

La construction de l'obstruction est liée à l'existence d'un accouplement général $\mathrm{Br}(V) \times Z(V) \to \mathrm{Br}(k)$, où $Z(V)$ est le groupe des cycles de dimension zéro sur V. En appliquant cet accouplement lorsque $V(k)$ est non vide, on peut obtenir des résultats qui vont plus loin :

d) Des minorations pour le « rang » de la surface cubique V (nombre de points de $V(k)$ à partir desquels on peut construire tous les autres par la méthode « des sécantes et des tangentes » ; voir [7]).

e) Un certain accouplement de $\mathrm{Br}(V)$ et du noyau d'Albanese dans le groupe de Chow de dimension zéro, cet accouplement étant à valeurs dans $\mathrm{Br}(k)$. Comme l'a démontré récemment Mumford [10], le deuxième groupe est de dimension infinie si

$k = C$, dim $V = 2$ et $p_g(V) > 0$. L'étude de ce groupe par voie arithmétique peut présenter un intérêt.

Construction de l'obstruction.

Soit V une variété sur k et soit $\mathrm{Br}(V)$ son groupe de Brauer-Grothendieck (voir Grothendieck [5]). A tout élément $a \in \mathrm{Br}(V)$ et à tout point géométrique $x \in V(K)$, $K \supset k$, est attachée une spécialisation $a(x) \in \mathrm{Br}(K)$. (Si a est la classe d'une algèbre d'Azumaya A sur V, $a(x)$ est la classe de la fibre géométrique $A(x)$).

Supposons que k soit un corps de nombres et que $V(k_v)$ soit non vide pour tout v. Soit $V(A)$ l'espace des k-adèles de V.

Fixons un sous-groupe B de $\mathrm{Br}(V)$.

DÉFINITION. — *a) Les adèles (x_v), $(y_v) \in V(A)$ sont B-équivalents si*

$$\forall a \in B, \quad \forall v, \quad a(x_v) = a(y_v) \in \mathrm{Br}(k_v).$$

b) On désignera par E l'espace quotient de $V(A)$ par la relation de B-équivalence.

c) Soit $X \in E$; on définit un caractère $i_X : B \to Q/Z$ par la formule

$$i_X(a) = \sum_v \mathrm{inv}_v(a(x_v)),$$

où (x_v) adèle quelconque de la classe X et où $\mathrm{inv}_v : \mathrm{Br}(k_v) \hookrightarrow Q/Z$ est le plongement classique (c'est un isomorphisme si v n'est pas archimédienne).

1. THÉORÈME DE L'OBSTRUCTION : $V(k) \subset \bigcup_{i_X = 0} X \subset V(A)$.

La démonstration est triviale : si l'adèle $(x_v) \in X$ est un adèle principal et correspond au point $x \in V(k)$, alors $a(x)$ appartient à $\mathrm{Br}(k)$ pour tout $a \in B$ et, par conséquent,

$$\sum_v \mathrm{inv}_v a(x) = 0.$$

Si l'on choisit convenablement V et B, *il se peut que $i_X \neq 0$ pour tout X* : alors $V(k)$ est vide, bien que $V(A)$ soit non vide.

Avant de passer aux exemples, donnons quelques propriétés du *groupe de Brauer-Grothendieck* $\mathrm{Br}(V)$.

Les k-morphismes canoniques $V \to k$ et $V \otimes \bar{k} \to V$ induisent des homomorphismes $\mathrm{Br}(k) \to \mathrm{Br}(V)$ et $\mathrm{Br}(V) \to \mathrm{Br}(V \otimes \bar{k})$. Désignons par $\mathrm{Br}_0(V)$ l'image du premier homomorphisme et par $\mathrm{Br}_1(V)$ le noyau du deuxième.

La filtration

$$\mathrm{Br}_0(V) \subset \mathrm{Br}_1(V) \subset \mathrm{Br}_2(V) = \mathrm{Br}(V)$$

joue un rôle essentiel dans l'étude du groupe de Brauer. Il est évident que, quels que soient B et X, on a $B \cap \mathrm{Br}_0(V) \subset \mathrm{Ker}\, i_X$. Ici, seuls les sous-groupes B contenus dans $\mathrm{Br}_1(V)$ nous intéressent. C'est pourquoi il faut savoir calculer $\mathrm{Br}_1(V)/\mathrm{Br}_0(V)$. Ce groupe est décrit, dans une large mesure, par le résultat suivant, dans lequel on suppose que V est une varité projective lisse :

Soit V_0 la variété d'Albanese de V, $\hat{V}_0$ la variété duale et NS le groupe de Néron-Severi.

2. Théorème. — *Si $H^3(G, \bar{k}^*) = 0$, on a une suite exacte*

$$\mathrm{Pic}(V\otimes \bar{k})^G \to NS(V\otimes \bar{k})^G \to H^1(G, \hat{V}_0(\bar{k})) \xrightarrow{\varphi} \mathrm{Br}_1(V)/\mathrm{Br}_0(V) \xrightarrow{\psi} H^1(G, NS(V\otimes \bar{k})) \to H^2(G, \hat{V}_0(\bar{k})).$$

Cas particuliers

a) Si $H^1(G, NS(V\otimes \bar{k})) = 0$ (par exemple, si V est une courbe, ou, plus généralement, si $NS(V\otimes \bar{k})$ est libre et G opère dessus de façon triviale), il existe un épimorphisme à noyau fini

$$H^1(G, \hat{V}_0(\bar{k})) \xrightarrow{\varphi} \mathrm{Br}_1(V)/\mathrm{Br}_0(V) \to 0.$$

b) Si $V_0 = \{0\}$, alors $\mathrm{Br}_1(V)/\mathrm{Br}_0(V) \cong H^1(G, NS(V\otimes \bar{k}))$.

Pour la démonstration, nous aurons besoin de deux lemmes.

3. Lemme. — $\mathrm{Br}_1(V)/\mathrm{Br}_0(V) \cong H^1(G, \mathrm{Pic}(V\otimes \bar{k})$.

Démonstration. — Soit $\bar{k}(\xi)$ le corps des fonctions rationnelles sur $V\otimes \bar{k}$. De la lissité de V, on déduit facilement, par passage à la fibre générique, la suite exacte (cf. Grothendieck [5], II)

$$0 \to \mathrm{Br}_1(V) \to H^2(G, \bar{k}(\xi)^*) \to H^2(G, \mathrm{Div}(V\otimes \bar{k})).$$

On peut l'inclure dans le diagramme commutatif:

$$\begin{array}{ccccccc}
 & & \mathrm{Br}(k) & \xrightarrow{\sim} & \mathrm{Br}(k) & & \\
 & & \downarrow & & \downarrow & & \\
0 & \to & \mathrm{Br}_1(V) & \to & H^2(G, \bar{k}(\xi)^*) & \to & H^2(G, \mathrm{Div}(V\otimes \bar{k})) \\
 & & & & \downarrow & & \downarrow\wr \\
0 & \to & H^1(G, \mathrm{Pic}(V\otimes \bar{k})) & \to & H^2(G, \bar{k}(\xi)^*/\bar{k}^*) & \to & H^2(G, \mathrm{Div}(\xi\otimes \bar{k})) \\
 & & & & \downarrow & & \\
 & & & & H^3(G, \bar{k}^*) = 0 & &
\end{array}$$

La ligne du bas de ce diagramme provient de la suite exacte de G-modules

$$0 \to \bar{k}(\xi)^*/\bar{k}^* \to \mathrm{Div}(V\otimes \bar{k}) \to \mathrm{Pic}(V\otimes \bar{k}) \to 0,$$

tandis que la colonne provient de la suite $0 \to \bar{k}^* \to \bar{k}(\xi)^* \to \bar{k}(\xi)^*/\bar{k}^* \to 0$.

En utilisant la partie gauche de ce diagramme, on obtient facilement le résultat cherché. (Ce raisonnement a été utilisé par Lichtenbaum [6] dans le cas où V est une courbe et où, par conséquent, $\mathrm{Br}_1(V) = \mathrm{Br}(V)$).

Remarque. — La condition $H^3(G, \bar{k}^*) = 0$ peut être remplacée par la suivante: $V(k)$ est dense pour la topologie de Zariski.

4. Lemme. — $\mathrm{Pic}^0(V\otimes \bar{k}) = \hat{V}_0(\bar{k})$ *en tant que G-modules.*

Démonstration. — Pic^0 est le groupe des classes des diviseurs qui sont algébriquement équivalents à zéro. L'application canonique $\alpha: V\times V \to V_0$ induit un plongement

$$\alpha^*: \mathrm{Pic}^0(V_0) \to \mathrm{Pic}^0(V\times V);$$

en outre, l'application $L \mapsto p_1^*L \otimes p_2^*L^{-1}$ (où $p_i : V \times V \to V$ est la $i^{\text{ème}}$ projection) définit un plongement

$$\beta^* : \mathrm{Pic}^0(V) \to \mathrm{Pic}^0(V \times V).$$

On voit facilement que α^* et β^* ont même image, formée des classes des faisceaux inversibles sur $V \times V$ qui sont triviaux sur la diagonale. Ceci donne des isomorphismes $\hat{V}_0(k) \to \mathrm{Pic}^0(V_0)$; on a une construction analogue sur le corps $\bar{k}$.

Démonstration du théorème 2. — On l'obtient en considérant la suite exacte de cohomologie associée à la suite de G-modules

$$0 \to \mathrm{Pic}^0(V \otimes \bar{k}) \to \mathrm{Pic}(V \otimes \bar{k}) \to NS(V \otimes \bar{k}) \to 0$$

et en appliquant les lemmes 3 et 4.

Applications. — Nous nous limiterons à deux exemples :

5. La surface de Swinnerton-Dyer [12]. — Cette surface V est donnée par l'équation homogène sur $\boldsymbol{Q}$:

$$T_3(T_0 + T_3)(T_0 + 2T_3) = \prod_{i=1}^{3} (T_0 + \theta^{(i)}T_1 + \theta^{(i)2}T_2) \tag{1}$$

où les $\theta^{(i)}$ sont les trois racines du polynôme $\theta^3 + 7(\theta + 1)^2 = 0$. On vérifie facilement que $V(\boldsymbol{Q}_p)$ et $V(\boldsymbol{R})$ sont non vides. Soit $K = \boldsymbol{Q}(\theta)$. C'est une extension cubique normale ; $V \otimes K$ est birationnellement équivalente à $\boldsymbol{P}_K^2$. On a donc $\mathrm{Br}(V) = \mathrm{Br}_1(V)$ et $V_0 = 0$. Il découle du théorème 2 que

$$\mathrm{Br}_1(V)/\mathrm{Br}_0(V) = H^1(\mathrm{Gal}\,K/k, \mathrm{Pic}\,(V \otimes K)) \cong \boldsymbol{Z}_3 \times \boldsymbol{Z}_3.$$

Étant donné que $G_0 = \mathrm{Gal}(K/k)$ est cyclique, il est commode d'identifier $\mathrm{Br}(V)$ au sous-groupe de $\mathrm{Br}(\boldsymbol{Q}(\xi))$ que voici :

$$\begin{aligned} &\mathrm{Ker}\,[H^2(G_0, K(\xi)^*) \to H^2(G_0, \mathrm{Div}\,(V \otimes K))] \\ &\qquad \cong \mathrm{Ker}\,[H^0(G_0, K(\xi)^*) \to H^0(G_0, \mathrm{Div}\,(V \otimes K))] \subset k(\xi)^*/N_{K/k}(K(\xi)^*), \end{aligned}$$

où ξ est un point générique de V.

Désignons par B le sous-groupe de $\mathrm{Br}_1(V)$ engendré par les deux éléments de $k(\xi)^*/N_{K/k}(K(\xi)^*)$ représentés par les fonctions

$$f_1 = \frac{T_0 + T_3}{T_3}, \quad f_2 = \frac{T_0 + 2T_3}{T_3}.$$

(Leurs diviseurs sont des normes, comme on le voit sur l'équation (1) ; c'est pourquoi les cocycles fonctionnels correspondants se décomposent dans la cohomologie à coefficients dans $\mathrm{Div}\,(V)$. Un calcul facile montre que $\mathrm{Br}(V)$ est engendré par $\mathrm{Br}_0(V)$ et B).

Une étude locale élémentaire permet d'établir le fait suivant :

Proposition. — *a) Soit $x \in V(\boldsymbol{R})$ ou $V(\boldsymbol{Q}_p)$, avec $p \neq 7$, $T_3(T_0 + T_3)(T_0 + 2T_3)(x) \neq 0$. Alors*

$$\mathrm{inv}_v(f_1(x)) = \mathrm{inv}_v(f_2(x)) = 0.$$

b) Soit $x \in V(\boldsymbol{Q}_7)$. Alors, ou bien $\mathrm{inv}_7(f_1(x)) \neq 0$, ou bien $\mathrm{inv}_7(f_2(x)) \neq 0$.

Le résultat relatif à $\boldsymbol{Q}_7$ provient de ce que 7 est complètement ramifié dans K, et $f_1(x)$ et $f_2(x)$ doivent être des unités locales, tandis que $f_2(x) - f_1(x) = 1$. Mais les normes des unités locales sont congrues à $\pm 1 \pmod 7$ et leur différence ne peut donc être égale à 1.

COROLLAIRE. — *L'ensemble $V(\boldsymbol{Q})$ est vide.*

En effet, il découle clairement de la proposition précédente que, pour tout adèle $(x_v) \in V(A)$, on a $\sum_v \mathrm{inv}_v(f_1(x)) \neq 0$ ou $\sum_v \mathrm{inv}_v(f_2(x)) \neq 0$.

On peut interpréter de la même façon les contre-exemples de Mordell [9] qui constituent plusieurs séries infinies.

L'exemple plus fin de Cassels et Guy [3] rentre aussi dans notre schéma général mais non sans quelques difficultés. Pour les détails, voir [7].

VARIÉTÉS ABÉLIENNES. — Soit V un espace homogène principal sur une variété abélienne V_0 et soit $b \in H^1(G, V_0(\bar{k}))$ sa classe de cohomologie, où $G = \mathrm{Gal}(\bar{k}/k)$. Si tous les $V(k_v)$ sont non vides, alors $b \in Ш(V_0)$.

Choisissons maintenant le sous-groupe B de $\mathrm{Br}(V)$. L'obstruction la plus économique est obtenue si l'ensemble $E = V(A) \bmod B$ est composé d'une seule classe, autrement dit si, pour tout $a \in B$, la spécialisation $a(x_v)$ ne dépend pas du choix de x_v dans $V(k_v)$. Pour cela, il suffit que $a \otimes k_v \in \mathrm{Br}_0(V \otimes k_v)$ pour tout v; dans ce cas, l'obstruction est alors l'homomorphisme $i: B \to \boldsymbol{Q}/\boldsymbol{Z}$ défini par

$$i(a) = \sum_v \mathrm{inv}_v(a(x_v)).$$

En particulier, le théorème 2 montre qu'on peut prendre pour B le groupe suivant :

$$B = \rho^{-1} \circ \varphi(Ш(\hat{V}_0)), \tag{2}$$

où $\rho: \mathrm{Br}(V) \to \mathrm{Br}(V)/\mathrm{Br}_0(V)$ est l'épimorphisme canonique.

Dans ces conditions, on a

6. THÉORÈME. — *Soit $a \in B$, $\rho(a) = \varphi(a')$, $a' \in Ш(\hat{V}_0)$, et soit $b \in Ш(V_0)$ la classe de l'espace V. On a alors*

$$i(a) = \langle b, a' \rangle \in \boldsymbol{Q}/\boldsymbol{Z},$$

où $\langle\, ,\rangle$ est la forme de Cassels-Tate $Ш(V_0) \times Ш(\hat{V}_0) \to \boldsymbol{Q}/\boldsymbol{Z}$.

La démonstration s'obtient par une simple comparaison de deux définitions (cf. la proposition 8 *c*) ci-dessous, qui contient l'idée centrale des calculs).

Remarques.

a) Regardons ce que l'on obtient si on construit l'obstruction à l'aide d'un sous-groupe plus grand que le groupe (2). Ceci a une raison d'être si l'obstruction précédente est nulle, ou, autrement dit, si (classe V) $\in Ш(V_0)$, avec les notations de Cassels [2]. Posons alors $B' = \rho^{-1} \circ \varphi(H^1(G, \hat{V}_0(\bar{k}))$. Il est évident que pour toute classe $X \in V(A) \bmod B'$, nous avons $B \in \mathrm{Ker}\, i_X$, où B est défini par (2). Chaque classe X définit donc un caractère

$$i'_X : B'/B = H^1(G, \hat{V}_0(\bar{k}))/Ш(\hat{V}_0) \to \boldsymbol{Q}/\boldsymbol{Z}.$$

Si, comme le suppose Cassels, Ш est fini, alors $Ш(V_0) = \{0\}$ et $V \cong V_0$; l'ensemble $V_0(A)$ mod B' possède une structure de groupe naturelle. On peut alors montrer que i' fournit une dualité topologique

$$V_0(A) \bmod B' \times H^1(G, \hat{V}_0(\bar{k}))/Ш(\hat{V}_0) \to \mathbf{Q}/\mathbf{Z},$$

et donner une description plus transparente de la B'-équivalence.

Ceci constitue l'étape essentielle dans la construction d'une suite exacte duale à

$$0 \to H^1(G, \hat{V}_0(\bar{k}))/Ш(\hat{V}_0) \to \sum_v H^1(G_v, \hat{V}_0(k_v)) \to Б(\hat{V}_0) \to 0.$$

En fait, Cassels [2] considère une suite à 4 termes, avec $Ш(\hat{V}_0)$ à gauche, et il obtient la suite duale

$$0 \leftarrow Ш(V_0) \leftarrow \theta \leftarrow \prod_v V_0'(k_v) \leftarrow V_0(k)^{\sim} \leftarrow 0 \tag{3}$$

où $V_0'(k_v)$ est le quotient de $V_0(k_v)$ par sa composante neutre et $V_0(k)^{\sim}$ le complété de $V_0(k)$ pour la topologie induite.

b) Soit V_0 un tore et $\bar{V}$ sa fermeture projective lisse. Voskressensky a construit la suite exacte suivante ([14], théorème 6):

$$0 \leftarrow Ш(V_0) \leftarrow H^1(G, \mathrm{Pic}(\bar{V} \otimes \bar{k}) \leftarrow \prod_v V_0(k_v) \leftarrow V_0(k)^{\sim} \leftarrow 0. \tag{4}$$

L'analogie avec (3) est presque complète puisque, en raison du théorème 2, on a

$$H^1(G, \mathrm{Pic}(\bar{V} \otimes \bar{k})) \cong \mathrm{Br}_1(\bar{V})/\mathrm{Br}_0(\bar{V})$$

(dans (3), le groupe θ est effectivement dual de $\mathrm{Br}_1(V_0)/\mathrm{Br}_0(V_0) \ldots$).

c) Plusieurs propriétés de la forme de Cassels-Tate résultent directement de notre définition. Toutefois, l'additivité par rapport au premier argument (voir théorème 6) n'est pas évidente. C'est pourquoi nous allons reformuler la « composante locale » de notre construction de manière à rendre évidente l'existence d'un accouplement complétant les homomorphismes i_X.

Produit scalaire.

Soit k un corps parfait, V une variété sur k, $Z(V)$ le groupe des cycles de V de dimension zéro. Les points fermés $x \in V$ forment une base de ce groupe.

7. Définition. — *L'accouplement*

$$\mathrm{Br}(V) \times Z(V) \to \mathrm{Br}(k)$$

est défini par la formule

$$(a, \Sigma n_i(x_i)) = \Sigma n_i \operatorname{cor}_{k(x_i)/k} a(x_i) \tag{5}$$

pour $a \in \mathrm{Br}(V)$, $\Sigma n_i x_i \in Z(V)$, où cor est l'homomorphisme de corestriction

$$\mathrm{Br}(k(x_i)) \to \mathrm{Br}(k).$$

Cette définition a été proposée dans [8]. Lichtenbaum [6], et avant lui D. K. Faddeev [4], ont étudié une construction analogue lorsque V est une courbe; dans le cas $\dim V \geqslant 2$, on observe de nouveaux phénomènes qui sont intéressants.

Pour formuler les principales propriétés de l'accouplement ci-dessus, considérons la filtration

$$Z_1(V) \subset Z_a(V) \subset Z_0(V) \subset Z(V).$$

Le groupe $Z_0(V)$ est formé des cycles de degré zéro :

$$z = \Sigma n_i x_i \in Z_0(V) \Leftrightarrow \deg z = \Sigma n_i [k(x_i):k] = 0.$$

Pour définir $Z_a(V)$, on utilise le fait que l'application canonique d'Albanese $\alpha : V \times V \to V_0$ induit l'homomorphisme « de sommation » $\beta : Z_0(V) \to V_0(k)$. Son noyau est $Z_a(V)$.

Enfin, $Z_1(V)$ est le groupe des cycles rationnellement équivalents à zéro. Il est engendré par les cycles de la forme $z(0) - z(\infty)$, où $\{ z(t), t \} \subset V \times \boldsymbol{P}^1$ est un système plat quelconque de cycles de V de dimension zéro, de base $\boldsymbol{P}^1$.

8. PROPOSITION.

a) Pour tout $a \in \mathrm{Br}_0(V)$ *et tout* $z \in Z_0(V)$, *on a* $(a, z) = 0$.

b) Pour tout $a \in \mathrm{Br}(V)$ *et tout* $z \in Z_1(V)$, *on a* $(a, z) = 0$.

c) Soit $\varphi : H^1(G, \hat{V}_0(\bar{k})) \to \mathrm{Br}_1(V)/\mathrm{Br}_0(V)$ *et soit* $\rho : \mathrm{Br}_1(V) \to \mathrm{Br}_1(V)/\mathrm{Br}_0(V)$ *les applications canoniques. Si* $\rho(a) = \varphi(a')$, $z \in Z_0(V)$ *et si* $V(k)$ *n'est pas vide, on a*

$$(a, z) = (a', \beta(z))_T, \tag{6}$$

où $(\, , \,)_T$ *est l'accouplement de Tate* 13 : $H^1(G, \hat{V}_0(\bar{k})) \times V_0(k) \to \mathrm{Br}(k)$. *En particulier,* $\rho^{-1} \circ \varphi(H^1(G, V_0(\bar{k})))$ *et* $Z_a(V)$ *sont orthogonaux.*

(Il est probable que la condition « $V(k)$ est non vide » n'est pas indispensable).

De cette proposition découle immédiatement le théorème suivant :

9. THÉORÈME. — *Le produit scalaire défini ci-dessus induit des accouplements*

$$\mathrm{Br}(V)/\mathrm{Br}(V_0) \times Z_0(V)/Z_1(V) \to \mathrm{Br}(k), \tag{7}$$

$$\mathrm{Br}_1'(V)/\mathrm{Br}_0(V) \times Z_0(V)/Z_1(V) \to \mathrm{Br}(k), \tag{8}$$

où $\mathrm{Br}_1'(V) = \rho^{-1} \circ \varphi(H^1(G, \hat{V}_0(\bar{k}))) \subset \mathrm{Br}_1(V)$;

$$\mathrm{Br}(V)/\mathrm{Br}_1'(V) \times Z_a(V)/Z_1(V) \to \mathrm{Br}(k). \tag{9}$$

Commentaires. — Il n'y a rien à dire de spécial sur l'accouplement (7); il est intervenu dans la construction de l'obstruction au principe de Hasse. L'accouplement (8) pourrait être considéré comme une « explication » de la construction de Tate, si celle-ci n'était pas déjà classique.

L'accouplement (9) est vraisemblablement nouveau. Les deux groupes qui y figurent ne peuvent être non triviaux que si $\dim V \geqslant 2$, et ils ont encore été très peu étudiés (même pour les variétés abéliennes). Voilà ce que l'on sait à leur sujet :

Supposons pour simplifier que $H^1(G, NS(V \otimes \bar{k})) = 0$ (pour des V abéliennes et, plus généralement, pour des variétés sans torsion, cette condition peut être réalisée après extension finie du corps de base).

Alors $\mathrm{Br}'_1(V) = \mathrm{Br}_1(V)$ (théorème 2). D'après la définition de Br_1, le morphisme de projection $V \otimes \bar{k} \to V$ induit un plongement

$$\mathrm{Br}(V)/\mathrm{Br}_1(V) \hookrightarrow (\mathrm{Br}(V \otimes \bar{k}))^G,$$

où G agit sur $\mathrm{Br}(V \otimes \bar{k})$ par le deuxième facteur. D'après Grothendieck [5], il existe un isomorphisme canonique

$$\mathrm{Br}(V \otimes \bar{k}) = H^2(V, \mu_\infty)/(\boldsymbol{Q}/\boldsymbol{Z} \otimes NS(V \otimes \bar{k})),$$

où $H^2(V, \mu_\infty) = \varinjlim H^2(V, \mu_n)$. En particulier, $\mathrm{Br}(V \otimes \bar{k})$ est un groupe divisible de corang fini $B_2 - \rho$.

Selon une conjecture de Tate, le groupe $(\mathrm{Br}(V \otimes \bar{k}))^G$ doit être fini si $[k:\boldsymbol{Q}] < \infty$. On connait peu de choses sur l'action de G sur $H^2(V, \mu_\infty)$ dans le cas des corps locaux. Lorsque $V = X \times Y$, où X et Y sont des courbes elliptiques, on peut construire, en utilisant la théorie de Serre [11], des exemples intéressants de V pour lesquels le corang de $(\mathrm{Br}(V \otimes \bar{k}))^G$ est non nul. Toutefois, j'ignore dans quelle mesure $\mathrm{Br}(V)/\mathrm{Br}_1(V)$ peut être différent de $(\mathrm{Br}(V \otimes \bar{k}))^G$.

Considérons maintenant le groupe $Z_a(V)/Z_1(V)$. Le seul résultat connu concernant ce groupe est le théorème suivant de Mumford [10]: *si* $k = \boldsymbol{C}$ *et* $\dim H^0(V, \Omega_V^2) > 0$, *alors* $Z_a(V)/Z_1(V)$ *est de dimension infinie.* Malheureusement, la méthode de Severi-Mumford ne donne presque pas de renseignements supplémentaires sur ce groupe; en particulier, j'ignore dans quelle mesure il peut être non trivial sur un corps dénombrable.

C'est pourquoi il est logique d'essayer d'utiliser l'accouplement (9) pour l'étude de ce groupe.

Il est à noter que les deux approches sont liées, d'une certaine façon, à l'existence de cycles transcendants de dimension deux sur V (par l'intermédiaire des inégalités $h^{2,0} > 0$ ou $B_2 - \rho > 0$). N'y a-t-il pas d'explication de ce fait?

Autre question: peut-on construire un élément infiniment divisible dans $Ш(V_0)$, où V_0 est une variété abélienne convenable, en utilisant l'obstruction sur $\mathrm{Br}(V)/\mathrm{Br}_1(V)$? En tout cas, la structure du noyau d'Albanese doit jouer un rôle dans l'arithmétique des variétés abéliennes.

10. Démonstration de la proposition 8

a) Soit $a \in \mathrm{Br}_0(V)$ l'image de $a' \in \mathrm{Br}(k)$. Alors, pour tout point fermé $x \in V$, $a(x)$ est l'image de $\mathrm{res}_{k/k(x)} a'$. On a donc

$$\mathrm{cor}_{k(x)/k}\, a(x) = [k(x): k]a$$

et par suite

$$(a, z) = \deg(z) \cdot a \quad \text{pour tout} \quad z \in Z(V).$$

b) Soit $C \subset V \times \boldsymbol{P}^1$ le graphe d'un système plat irréductible $z(t)$ de cycles de dimension zéro sur V. La compatibilité de la corestriction et de la spécialisation montre que pour tout élément $a \in \mathrm{Br}(V)$ et tout point $t \in \boldsymbol{P}^1(k)$, on a:

$$(a, z(t)) = (\mathrm{cor}_{C/\boldsymbol{P}^1}(p_1^*(a)\,|_C), t),$$

où $p_1 : V \times \boldsymbol{P}^1 \to V$ est la projection et l'homomorphisme $\mathrm{cor}_{C/\boldsymbol{P}^1} : \mathrm{Br}(C) \to \mathrm{Br}(\boldsymbol{P}^1)$ est induit par la corestriction habituelle aux points génériques. Mais

$$\mathrm{Br}(\boldsymbol{P}^1) = \mathrm{Br}_0(\boldsymbol{P}^1) = \mathrm{Br}(k),$$

et $(a, z(t))$ ne dépend donc pas de t.

c) La démonstration de la formule (6) repose sur une comparaison détaillée des définitions.

Puisque $V(k)$, par hypothèse, est non vide, il existe une application $\alpha_0 : V \to V_0$ telle que l'application d'Albanese $\alpha : V \times V \to V_0$ soit donnée par $\alpha(x, y) = \alpha_0(x) - \alpha_0(y)$. Identifions $Z(V)$ avec $Z(V \otimes \bar{k})^G$. Alors, l'application $\beta : Z_0(V) \to V(k)$ est donnée par la formule

$$\beta\Big(\sum_{x \in V(\bar{k})} n_x(x)\Big) = \Sigma n_x \alpha_0(x).$$

Soit $a' \in H^1(G, \hat{V}_0(\bar{k}))$; choisissons un cocycle $\{ a_s \}$ représentant a' dans une extension normale finie K/k où a' se décompose. Soit en outre :

$$z = \sum_{x \in V(\bar{k})} n_x(x) \in Z_0(V).$$

Nous garderons ces notations jusqu'à la fin de la démonstration.

Calcul de $(a', \beta(z))_T$.

Suivant la méthode de Tate [13], calculons d'abord $\delta a'$, où δ provient de la suite exacte (où S désigne l'homomorphisme « somme ») :

$$0 \to Z_a(\hat{V}_0 \otimes K) \to Z_0(\hat{V}_0 \otimes K) \xrightarrow{S} \hat{V}_0(K) \to 0.$$

Pour image inverse de a_s dans $Z_0(V_0 \otimes K)$, prenons le cycle $(a_s) - (0)$. La classe de cohomologie $\delta a'$ est alors représentée par le cocycle

$$\{ (sa_t) - (a_{st}) + (a_s) - (0) \} \in Z^2(G, Z_a(\hat{V}_0 \otimes K)).$$

Soit $D \subset \hat{V}_0 \times V$ un diviseur de Poincaré. Pour tout point géométrique $x \in \hat{V}_0(K)$, notons $D(x) \in \mathrm{Div}\,(V_0 \otimes K)$ sa fibre géométrique en x. Si $s, t \in G$, le diviseur

$$D(sa_t) - D(a_{st}) + D(a_s) - D(0) \tag{10}$$

est un diviseur principal d'après le théorème du carré. Soit $g_{s,t}$ la fonction rationnelle correspondante sur $V_0 \otimes K$.

Ici se terminent les opérations pour le premier terme a'.

Il reste encore à choisir une image inverse pour S du point $\beta(z)$. Nous prenons pour cela le cycle

$$\sum_{x \in V(\bar{k})} n_x(\alpha_0(x)) \in Z_0(V_0 \otimes \bar{k})^G.$$

Maintenant, par définition, la classe de cohomologie $(a', \beta(z))_T$ appartenant à $H^2(G, K^*)$ est représentée par le cocycle

$$\Big\{ \prod_{x \in V(\bar{k})} g_{s,t}(\alpha_0(x))^{n_x} \Big\} \in Z^2(G, K^*). \tag{11}$$

(Le diviseur de Poincaré D doit être choisi tel que Supp $(\Sigma s D_t)$ ne contienne pas de points $\alpha_0(x)$ avec $n_x \neq 0$, ce que l'on peut toujours réaliser).

Calcul de (a, z).

Il faut d'abord calculer l'élément

$$\varphi(a') \in \mathrm{Br}_1(V)/\mathrm{Br}_0(V).$$

L'homomorphisme composé $\hat{V}_0(K) \to \mathrm{Pic}^0(V_0 \otimes K) \xrightarrow{\alpha_0^*} \mathrm{Pic}^0(V \otimes K) \hookrightarrow \mathrm{Pic}(V \otimes K)$ transforme le cocycle $\{ a_s \}$ en

$$\{ \mathrm{Cl}_{V\otimes K}(\alpha_0^*(D(a_s)) \} \in Z^1(G, \mathrm{Pic}(V \otimes K)).$$

De plus (voir la démonstration du lemme 3), l'homomorphisme bord associé à la suite

$$0 \to K(\xi)^*/K^* \to \mathrm{Div}(V \otimes K) \to \mathrm{Pic}(V \otimes K) \to 0,$$

transforme ce cocycle en le cocycle $\{ f_{s,t} \bmod K^* \} \in Z^2(G, K(v)^*/K)$, où

$$\mathrm{div}(f_{s,t}) = \alpha_0^*(D(sa_t) - D(a_{st}) + D(a_s)). \tag{12}$$

La classe de ce cocycle est égale à l'image de $\varphi(a')$ dans $H^2(G, K(\xi)^*/K)$. La définition 7 montre clairement que (a, z) est représenté par le cocycle

$$\{ \prod_{x \in V(k)} f_{s,t}(x)^{n_x} \} \in Z^2(G, K^*). \tag{13}$$

Mais le diviseur D peut être choisi de telle sorte que $D(0) = 0$. On voit alors, en comparant les formules (10) et (12), que, après une normalisation convenable, on a

$$f_{s,t} = \alpha_0^*(g_{s,t}),$$

et donc

$$f_{s,t}(x) = g_{s,t}(\alpha_0(x)).$$

Ceci démontre l'égalité de (11) et (13) et termine la démonstration de la proposition 8.

BIBLIOGRAPHIE

[1] J. W. S. Cassels. — Arithmetic on curves of genus 1. IV. Proof of the Hauptvermutung, *Journ. für die reine u. angew. Math.*, Bd. 211, H.1/2 (1962), pp. 95-112.

[2] J. W. S. Cassels. — Arithmetic on curves of genus 1. VII. The dual exact sequence, *Journ. für die reine u. angew. Math.*, Bd. 216, H.3/4 (1964), pp. 150-158.

[3] J. W. S. Cassels, M. J. T. Guy. — On the Hasse principle for cubic surfaces, *Mathematika*, 13 (1966), pp. 111-120.

[4] Д. К. Фаддеев. — К теории алгебр над полем алгебраических функций одной переменной, Вестник ЛГУ, № 7, сер. мат., в. 2 (1957), 45-51.

[5] A. Grothendieck. — Le groupe de Brauer, I, II, III, *Séminaire Bourbaki*, 290 (1965), 297 (1965), *IHES notes* (1966).

[6] S. Lichtenbaum. — Duality theorems for Curves over p-adic fiels, *Inventiones Math.*, vol. 7, f. 2 (1969), pp. 120-136.

[7] Ю. И. Манин. — Кубические формы : алгебра, геометрия, арифметика (в печати).

[8] —. — Кубические гиперповерхности Ш. Лупы Муфанг и эквивалентность Брауэра. Мат. сборник, т. 79, в. 2 (1969), 155-170.

[9] L. Mordell. — On the conjecture for the rational points on a cubic surface, *J. Lond. Math. Soc.*, 40 (1965), pp. 149-158.

[10] D. MUMFORD. — Rational equivalence of O-cycles on surfaces, *Journ. Math. Kyoto Univ.*, vol. 9, N. 2 (1969), pp. 195-204.
[11] J. P. SERRE. — *Abelian l-adic representations and elliptic curves*, New York, Benjamin, 1968.
[12] H. P. F. SWINNERTON-DYER. — Two special cubic surfaces, *Mathematika*, 9 (1962), pp. 54-56.
[13] J. TATE. — WC-groups over p-adic fields, *Seminaire Bourbaki*, exp. 156 (1957).
[14] В. Е. ВОСКРЕСЕНСКИЙ. — Бирациональные свойства линейных алгебраических групп, Известия АН СССР, сер. мат.

Steklov Mathematical Institute
Academy of Sciences of USSR
ul Vavilova 42,
Moscow V 333 (U. R. S. S.)

Izv. Akad. Nauk SSSR
Ser. Mat. Tom 36 (1972), No. 1

Math. USSR Izvestija
Vol. 6 (1972), No. 1

PARABOLIC POINTS AND ZETA-FUNCTIONS OF MODULAR CURVES

Ju. I. MANIN

Abstract. In this paper we obtain explicit formulas for the values at the center of the critical strip of Dirichlet series connected with weight 2 parabolic forms of the group $\Gamma_0(N)$. In particular, these formulas allow us to verify the Birch–Swinnerton-Dyer conjecture on the order of a zero for uniformizable elliptic curves over certain Γ-extensions. We also give applications to noncommutative reciprocity laws.

Introduction

Let X be an elliptic curve over the field $\mathbf{Q}$, N its conductor, ω a Néron differential, and $L(X, s)$ the canonical Dirichlet series. Further, let X_N be the standard modular curve over $\mathbf{Q}$ parametrized by the group $\Gamma_0(N)$. Weil [16] conjectured that there exists a morphism $\psi : X_N \to X$ over $\mathbf{Q}$ such that the differential $\psi^*(\omega)$ lifted to the uppper halfplane H has the same Fourier coefficients as the Dirichlet series $L(X, s)$ (see the precise formulation in §5.2 of this paper). We call such a morphism ψ a *Weil uniformization* of the curve X.

In this paper we show that the existence of a Weil uniformization for the curve X allows us to give explicit formulas for the values of $L(X, 1)$, and also $L(X \otimes K, 1)$, for all possible abelian extensions $K \supset \mathbf{Q}$. These explicit formulas have the structure predicted by the Birch–Swinnerton-Dyer conjectures. Comparison with Mazur's theory [10] of elliptic curves over Γ-extensions of $\mathbf{Q}$ also shows a good agreement with the Birch–Swinnerton-Dyer conjecture. In particular, Mazur's "anomalous prime numbers" appear in an analytic context.

The general idea for obtaining explicit formulas for $L(X \otimes K, 1)$ consists in the following. Let $\Phi(z)dz$ be the preimage of ω on H, and let $\{0, i\infty\}$ be the path on $X_N(\mathbf{C})$ which is the image of the imaginary semiaxis on H. From the classical integral representation for $L(X, s)$, we find, after a suitable normalization of ψ, that

$$L(X, 1) = \int_0^{i\infty} \Phi(z)\,dz = \int_{\{0, i\infty\}} \psi^*(\omega) = \int_{\psi_*\{0, i\infty\}} \omega.$$

Since $\psi_*(i\infty)$ is the zero point on $X(\mathbf{C})$, it follows that $L(X, 1)$ is the Abel-Jacobi

AMS (MOS) subject classifications (1970). Primary 14G10; Secondary 14H99.

argument of the image of $0 \in H$, i. e. of some parabolic point on X_N. If the images of 0 and $i\infty$ coincided, then to compute $L(X, 1)$ we would have to integrate ω over some closed path in $X(\mathbf{C})$ (and even $X(\mathbf{R})$), so that the number $L(X, 1)$ would be an integral multiple of the minimal real period of the differential ω. In general this is not the case. Nevertheless, we were able to find a technical device which allows us to reduce to integration over closed paths in $X(\mathbf{C})$. Hecke operators are used to do this in the case of ground field $\mathbf{Q}$, and in the case of an abelian extension $K \supset \mathbf{Q}$ we use the expansion of the series $L(X \otimes K, s)$ with respect to the characters χ of the Galois group and the Hecke-Weil lemma on Mellin tranforms of the series $L_\chi(X, s)$. This device and the resulting formulas make up the conceptual center of the paper; they are presented in §§3–5.

The homology classes in the group $H_1(X_N(\mathbf{C}), \mathbf{Z})$ over which we must integrate $\psi^*(\omega)$ to compute $L(X \otimes K, s)$ are *fundamental arithmetic invariants of the curves* X and X_N. Hence the paper begins by studying them: in §1 we prove a new theorem on the structure of the first homology group of a curve uniformized by any subgroup G of the modular group, and in §2 we specialize this theorem to the case $G = \Gamma_0(N)$.

§§6 and 7 contain applications of these results. Namely, §6 is devoted to comparing them with the Birch–Swinnerton-Dyer conjecture. To make this comparison, we must have independent information about the rank of $X(K)$ and the order of the Tate-Šafarevič group of the curve $X \otimes K$. Mazur's theory [10], [11] (see also [9]) obtains several results of this type, and they actually lend themselves to detailed comparison with our formulas.

In §7 we give new exact formulas for the coefficients of parabolic forms relative to the group $\Gamma_0(N)$, which call to mind the noncommutative reciprocity law or the Eichler relations, where, however, we have an indefinite rather than a definite quaternion quadratic form.

Finally, §8 contains tables of arithmetic invariants and a discussion of them.

I am grateful to A. N. Andrianov, whose conversations with me stimulated some new ideas for this paper.

I am also grateful to M. Z. Rozenfel'd, who computed vast tables of the functions $x^\pm$ for the group $\Gamma_0(11)$ on the computer "System 4" of the Institute for Control Problems of the Academy of Sciences of the USSR, and to V. Drinfel'd, who composed similar tables for the groups $\Gamma_0(N)$ with $N = 14, 17, 19$ and who kindly agreed to their publication in this article.

After completing this work, I learned that Professor Birch (England) has also obtained some results close to ours, and that Professor Mazur (USA) and Professor Swinnerton-Dyer (England) have independently examined the functions $x^\pm$.

§1. Homology of modular curves

1. 1. General information. For the duration of the paper we use the following notation: $H = \{z \in \mathbf{C} | \operatorname{Im} z > 0\}$ is the upper halfplane; $\overline{H} = H \cup \mathbf{Q} \cup (i\infty)$ is its compactification with the usual topology;

$$\Gamma = \left\{ z \rightsquigarrow \frac{az+b}{cz+d} \,\middle|\, \begin{pmatrix} a & b \\ c & d \end{pmatrix} \in SL(2, \mathbf{Z}) \right\}$$

is the group of automorphisms of $\overline{H}$, which we identify with $PL(2, \mathbf{Z}) = SL(2, \mathbf{Z})/(\pm 1)$ and whose elements are written as the corresponding matrices.

Let $G \subset \Gamma$ be a subgroup of finite index. The topological space $X_G(\mathbf{C}) = G \backslash \overline{H}$ has a natural structure of a smooth compact complex space of topological dimension 2. By $\phi : \overline{H} \to X_G(\mathbf{C})$ we always designate the natural projection mapping.

Let $i \in H$, $i^2 = -1$; $\rho = e^{\pi i/3} \in H$. Points in the set $\phi(\Gamma i \cup \Gamma \rho) \subset X_G(\mathbf{C})$ are called elliptic points, and points in $\phi(\mathbf{Q} \cup (i\infty)) = \phi(\Gamma(i\infty)) \subset X_G(\mathbf{C})$ are called parabolic points. The map ϕ is unramified outside these points. Both sets are finite.

1.2. **The classes $\{\alpha, \beta\}_G$.** Let $\alpha, \beta \in \overline{H}$ be two points such that $\phi(\alpha) = \phi(\beta) \in X_G(\mathbf{C})$, or, equivalently, $G\alpha = G\beta$. Then any path from α to β on $\overline{H}$ becomes a closed path on $X_G(\mathbf{C})$ whose homology class depends only on α and β. This homology class will always be denoted by the symbol $\{\alpha, \beta\}_G \in H_1(X_G(\mathbf{C}), \mathbf{Z})$.

More generally, integration allows us to associate a homology class with *real* coefficients to any pair of points $\alpha, \beta \in \overline{H}$ even if $\phi(\alpha) \neq \phi(\beta)$. We consider the differentials of the first kind $\omega \in H^0(X_G(\mathbf{C}), \Omega^1)$. Any class $\gamma \in H_1(X_G(\mathbf{C}), \mathbf{Z})$ determines a functional on the space of these differentials: $\omega \rightsquigarrow \int_\gamma \omega$. The group of such functionals forms a lattice of maximal rank in the dual space of $H^0(X_G(\mathbf{C}), \Omega^1)$. Extending this map by $\mathbf{R}$-linearity, we obtain an $\mathbf{R}$-isomorphism

$$H_1(X_G(\mathbf{C}), \mathbf{R}) \simeq \mathrm{Hom}_{\mathbf{C}}(H^0(X_G(\mathbf{C}), \Omega^1), \mathbf{C}).$$

Consequently for any two points $\alpha, \beta \in \overline{H}$ the functional $\omega \rightsquigarrow \int_\alpha^\beta \phi^*(\omega)$ can be identified with a real first homology class, which we shall denote by $\{\alpha, \beta\}_G$ in the general case. Obviously, if $\phi(\alpha) = \phi(\beta)$ this notation coincides with the earlier notation. We shall sometimes write $\{\alpha, \beta\}$ instead of $\{\alpha, \beta\}_G$ and $\int_{\{\alpha,\beta\}} \omega$ instead of $\int_\alpha^\beta \phi^*(\omega)$.

1.3. **First properties of the classes $\{\alpha, \beta\}$.** Obviously, $\{\alpha, \alpha\} = 0$, $\{\alpha, \beta\} = -\{\beta, \alpha\}$. The following properties are also obtained immediately from the definition:

a) $\{\alpha, \beta\} + \{\beta, \gamma\} + \{\gamma, \alpha\} = 0$.

b) $\{g\alpha, g\beta\}_G = \{\alpha, \beta\}_G$ for all $g \in G$.

c) If the genus of $X_G(\mathbf{C})$ is nonzero, then $\{\alpha, \beta\}_G \in H_1(X_G(\mathbf{C}), \mathbf{Z})$ if and only if $\beta \in G\alpha$ or, equivalently, $\phi(\alpha) = \phi(\beta)$.

(Sufficiency follows from the definition, and necessity follows from the Abel-Jacobi inversion theorem.)

The following fact requires a somewhat more detailed discussion.

1.4. **Proposition.** *Let $\alpha \in \overline{H}$. The map*

$$G \to H_1(X_G(\mathbf{C}), \mathbf{Z}) : g \rightsquigarrow \{\alpha, g\alpha\}_G$$

is a surjective group homomorphism which does not depend on the choice of α. The kernel of this homomorphism is generated by the commutator, by the elliptic elements, and by the parabolic elements of the group G.

Proof. The fact that the map $g \rightsquigarrow \{\alpha, g\alpha\}$ is a homomorphism is easily obtained formally by applying 1.3 a) and b):

$$\{\alpha, gh\alpha\} = \{\alpha, g\alpha\} + \{g\alpha, gh\alpha\} = \{\alpha, g\alpha\} + \{\alpha, h\alpha\}.$$

To prove the remaining assertions, we must use a direct geometric interpretation of this homomorphism.

We first suppose that $\phi(\alpha) \in X_G(\mathbf{C})$ is neither an elliptic nor a parabolic point. Let H^0 be the complement of $\Gamma i \cup \Gamma\rho$ in H, and let $X_G^0(\mathbf{C})$ be the complement of the set of elliptic and parabolic points. The map $\phi\colon H^0 \to X_G^0(\mathbf{C})$ is an unramified covering with Galois group G. Consequently any point $\alpha \in H^0$ determines a surjective homomorphism $\pi_1(X_G^0(\mathbf{C}), \phi(\alpha)) \to G$. Its explicit description is as follows: suppose we are given a closed path on $X_G^0(\mathbf{C})$ starting at $\phi(\alpha)$. We lift it to a path in H^0 starting at α. The endpoint of this path has the form $g\alpha$ for a uniquely determined element $g \in G$. This element is what we associate to the class of the original path on $X_G^0(\mathbf{C})$.

It is clear from this description that the composite map

$$\pi_1(X_G^0(\mathbf{C}), \varphi(\alpha)) \to G \to H_1(X_G(\mathbf{C}), \mathbf{Z})$$

(the second row takes g to $\{\alpha, g\alpha\}_G$) coincides with the canonical homorphism of the fundamental group of the surface $X_G^0(\mathbf{C})$ into the homology group of the compactification $X_G(\mathbf{C})$. This immediately implies that the map is surjective and does not depend on the choice of the point α. Further, the structure of the group π_1 is well known; using this, we easily observe that the kernel of the homomorphism $\pi_1 \to H_1$ is generated in π_1 by the commutator and the circuits around the elliptic and parabolic points, which contract in compactification. But the images of these circuits in G make up precisely the elliptic and parabolic elements. This completes the discussion of the case $\alpha \in H^0$.

Finally, let $\alpha \in \Gamma(i, \rho, i\infty)$ and $g \in G$. We choose a point $\alpha_0 \in H^0$ so close to α that there exist open neighborhoods $U_1 \supset (\alpha, \alpha_0)$, $U_2 \supset (g\alpha, g\alpha_0)$ in $\overline{H}$ such that the union of their images $\phi(U_1) \cup \phi(U_2)$ in $X_G(\mathbf{C})$ is simply connected. We choose a path from α to $g\alpha$ and one from α_0 to $g\alpha_0$, which coincide outside $U_1 \cup U_2$. Since their images on $X_G(\mathbf{C})$ coincide outside $\phi(U_1) \cup \phi(U_2)$, it follows that the homology classes of these paths are identical, so that $\{\alpha, g\alpha\}_G = \{\alpha_0, g\alpha_0\}_G$. This completes the proof, because all the required properties have been proved for the classes $\{\alpha_0, g\alpha_0\}_G$.

1.5. **Distinguished classes.** Let $J = G\backslash\Gamma$ be the set of right cosets. We define the map

$$\xi : J \to H_1(X_G(\mathbf{C}), \mathbf{R})$$

as follows: if $j \in J$ and g is any representative of the class j, then

$$\xi(J) = \{g(0), g(i\infty)\}_G. \tag{1}$$

Obviously this class does not depend on the choice of the representative g (see 1.3 b)). We have thereby defined a finite family of homology classes $\xi(J)$; we shall call its elements distinguished classes. We note that, in general, they are not integral.

1.6. Proposition. a) *Any class in* $H_1(X_G(\mathbf{C}), \mathbf{Z})$ *can be represented as a sum of distinguished classes. In particular, the distinguished classes generate* $H_1(X_G(\mathbf{C}), \mathbf{R})$ *as an* $\mathbf{R}$*-space.*

b) *The representation of any class* $h \in H_1(X_G(\mathbf{C}), \mathbf{Z})$ *as a sum* $\Sigma m_k\{\alpha_k, \beta_k\}$ *of distinguished classes can be chosen so that* $\Sigma m_k(\phi(\beta_k) - \phi(\alpha_k)) = 0$ *(as a zero-dimensional cycle on* $X_G(\mathbf{C})$*).*

Proof. By Proposition 1.4, any class in $H_1(X_G(\mathbf{C}), \mathbf{Z})$ has the form $\{0, g(0)\}$, where $g \in G$. If $g(0) = i\infty$, then this class is distinguished and $\phi(i\infty) - \phi(0) = 0$. Otherwise, let $g(0) = b/a$ be a rational number in lowest terms, $a > 0$. Also let $b > 0$; the case $b < 0$ is treated similarly. We expand b/a as a continued fraction and consider the successive convergents in lowest terms:

$$\frac{b}{a} = \frac{b_n}{a_n}, \frac{b_{n-1}}{a_{n-1}}, \ldots, \frac{b_0}{a_0} = \frac{b_0}{1}, \frac{b_{-1}}{a_{-1}} = \frac{1}{0}, \frac{b_{-2}}{a_{-2}} = \frac{0}{1}$$

(the last two "fractions" are added formally).

It is well known that $b_k a_{k-1} - b_{k-1} a_k = (-1)^{k-1}$, so that

$$g_k = \begin{pmatrix} b_k & (-1)^{k-1} b_{k-1} \\ a_k & (-1)^{k-1} a_{k-1} \end{pmatrix} \in \Gamma.$$

Hence the classes

$$\left\{\frac{b_{k-1}}{a_{k-1}}, \frac{b_k}{a_k}\right\} = \{g_k(0), g_k(i\infty)\}$$

are distinguished. Finally, by 1.3 a),

$$\left\{0, \frac{b}{a}\right\} = \sum_{k=-1}^{n} \left\{\frac{b_{k-1}}{a_{k-1}}, \frac{b_k}{a_k}\right\} = \sum_{k=-1}^{n} \xi(Gg_k). \tag{2}$$

This representation obviously has the required properties, so the proof is finished.

1.7. Relations between distinguished classes. The group Γ acts on the right on the set of right cosets $J = G\backslash\Gamma$. In this group we consider the two special elements

$$s = \begin{pmatrix} 0 & -1 \\ 1 & 0 \end{pmatrix} \text{ and } t = \begin{pmatrix} 1 & -1 \\ 1 & 0 \end{pmatrix};\ s^2 = t^3 = \mathrm{id}.$$

The element s takes $(0, i\infty)$ to $(i\infty, 0)$ and successive application of t takes $(0, i\infty)$ to $(i\infty, 1)$ and then to $(1, 0)$. Using the definition of $\xi(j)$ (see (1)) and property 1.3 a), we find two types of relations between distinguished classes:

$$\xi(j) + \xi(js) = 0, \quad \xi(j) = 0, \quad \text{if} \quad j = js; \tag{3}$$

$$\xi(j) + \xi(jt) + \xi(jt^2) = 0, \quad \xi(j) = 0, \quad \text{if} \quad j = jt \tag{4}$$

(the second relation in each group follows because there is no torsion). We show that in some sense this system of relations is complete. In order to formulate the result precisely, we introduce some new notation.

1. 8. a) **Algebraic formulation.** We let C designate the abelian group generated by the symbols (j) for all $j \in G\backslash\Gamma$, with the relations

$$(j)+(js)=0, \quad (j)=0, \quad \text{if} \quad j=js. \tag{3'}$$

We call the elements of the group C *G-chains*. Let $g \in \Gamma$ belong to the class $j \in J$, $g = \begin{pmatrix} a & b \\ c & d \end{pmatrix}$. By the *boundary* of the chain (j) we mean the difference between the two parabolic points: $\phi(g(i\infty)) - \phi(g(0)) = Ga/c - Gb/d$. We consider this difference as an element of the free abelian group generated by the set of parabolic points $G\backslash(Q \cup (i\infty))$. Since s interchanges 0 and $i\infty$, it follows from (3') that the boundary operator extends to the entire group C by linearity. We designate its kernel by Z; the elements of the kernel are called *G-cycles*.

Finally, let B be the subgroup of C generated by elements (j) for all $j \in J$ with the condition $j = jt$, and by the elements $(j) + (jt) + (jt^2)$ for the remaining j. We easily see that $B \subset Z$; the elements of B are called *G-boundaries*.

The map $\xi : G\backslash\Gamma \to H_1(X_G(\mathbf{C}), \mathbf{R})$, defined in 1.5, extends to a homomorphism ξ : $C \to H_1(X_G(\mathbf{C}), \mathbf{R})$ because $\xi(j) + \xi(js) = 0$. Here $\xi(Z)$ coincides with the integral homology subgroup by Proposition 1.6 b), and $\xi(B) = 0$ by (4).

Thus we obtain a surjective homomorphism

$$Z/B \to H_1(X_G(\mathbf{C}), \mathbf{Z}). \tag{5}$$

b) **Geometric formulation.** The groups C, Z and B can be realized as the subgroups of 1-chains (1-cycles, 1-boundaries) of some cell complex $K(G)$, which we shall call a *parabolic complex*. Here is its description.

0-cells are the elements of the set of parabolic points $G\backslash(Q \cup i\infty)$.

1-cells are in one-to-one correspondence with the set of orbits of the group (id, s), which acts on the right on $G\backslash\Gamma$. Every such cell (j, js) joins two 0-cells in the boundary (j) (or (js)): Ga/c and Gb/d, if $\begin{pmatrix} a & b \\ c & d \end{pmatrix}$ belongs to the class j. If these 0-cells coincide, then the corresponding 1-cell is a loop. We choose the orientation arbitrarily.

2-cells are of two sorts: two-sided and triangular. Let $j \in J$, $j = jt$. Then the 1-cell corresponding to (j, js), as described above, is a loop: if $g \in \Gamma$ belongs to the class j, then $Gg(0) = Ggt(0) = G(i\infty)$. We glue this loop by a 2-cell: we call such cells two sided.

Finally, let $j \in J$, $j \neq jt$. Then the 1-simplices (j, js), (jt, jts) and (jt^2, jt^2s), form a triangle; we glue it with a 2-cell; we call such 2-cells triangular.

It is now clear that there exists a map $C \to C_1(K(G))$ which takes a G-chain (j) to a $K(G)$-chain: "the simplex (j, js), oriented from Gb/d to Ga/c" (if $\begin{pmatrix} a & b \\ c & d \end{pmatrix}$ belongs to the class j). We easily see that this map induces an isomorphism $Z/B \xrightarrow{\sim} H_1(K(G), \mathbf{Z})$, which, together with (5), gives us a surjective map

$$H_1(K(G), \mathbf{Z}) \to H_1(X_G(\mathbf{C}), \mathbf{Z}). \tag{6}$$

The following theorem is the fundamental result of this section. It gives a representation of the group $H_1(X_G(\mathbf{C}), \mathbf{Z})$ by generators and relations which is convenient for computation and is functorial in G.

1.9. **Theorem.** *The maps* (5) *and* (6) *are isomorphisms.*

Proof. We construct a complex L of the space $X_G(\mathbf{C})$ and represent $H_1(X_G(\mathbf{C}), \mathbf{Z})$ as the factor group $Z_1(L)/B_1(L)$ of the 1-cycles of L by the 1-boundaries.

We then imbed the group Z in $Z_1(L)$ in such a way that $z \bmod B_1(L) = \xi(z)$ for all $z \in Z$.

Finally, we show that the boundary of any 2-cell of the complex L belongs to Z (under the above imbedding $Z \subset Z_1(L)$) and coincides with one of the generators of the group B of the form (j) (for $j = jt$) or $(j) + (jt) + (jt^2)$.

Obviously all these results and the surjectivity of ξ give us the isomorphism $Z/B \simeq Z_1(L)/B_1(L) = H_1(X_G(\mathbf{C}), \mathbf{Z})$, which we are trying to establish.

We realize this program in several steps.

a) *Preparation for constructing the complex L.* Let $\alpha, \beta \in \overline{H}$ be two points. We let $\langle\alpha, \beta\rangle$ designate the segment joining them along the geodesic oriented from α to β. (We recall that the geodesics are semicircles and lines orthogonal to the real axis.)

The triangles, quadrilaterals, etc. which we refer to will be the figures on $\overline{H}$ formed by geodesic segments joining the vertices of these figures, and also their ϕ-images on $X_G(\mathbf{C})$.

We let E'' designate the interior of the triangle with vertices $(0, 1, i\infty)$, and we let E' designate the union of the interior of the quadrilateral with vertices $(i, \rho, 1+i, i\infty)$ and the side $\langle i, \rho\rangle$, except for the vertex i. Each of the quadrilaterals E', tE' and t^2E' is a fundamental region for the entire group Γ. In addition, all the 1-simplices in Figure 1—the half-sides and half-medians of the triangle E''—imbed homomorphically into $X_\Gamma(\mathbf{C})$. (These are both classical assertions.)

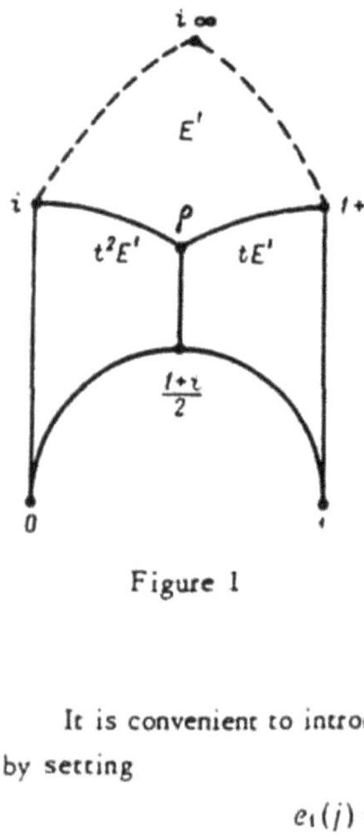

Figure 1

b) *Description of L.*

0-cells. These are all the parabolic points and all the i-elliptic points on $X_G(\mathbf{C})$, the images of the vertices and the midpoints of the sides of the triangles gE'', $g \in \Gamma$.

1-cells. These are the images of the half-sides of the triangles gE'', $g \in \Gamma$, oriented "from the vertex to the midpoint," i. e. from the parabolic point to the i-elliptic point.

It is convenient to introduce a family of 1-cells indexed by the classes $j \in G\backslash\Gamma$ by setting

$$e_1(j) = \text{image } \langle g(i\infty), g(i)\rangle \text{ in } X_G(\mathbf{C}) \tag{7}$$

for any representative $g \in \Gamma$ of the class j.

Since Γ is transitive on half-sides, it follows that any 1-cell has the form $e_1(j)$ for some $j \in J$. On the other hand, the stationary subgroup of i in Γ is equal to (id, s); hence $e_1(j) = e_1(k)$ is only possible for $k, j \in J$ if $k = j$ or $k = js$; otherwise $e_1(j)$ and $e_1(k)$ have different endpoints. Moreover, for $j \neq js$ we have $e_1(j) \neq e_1(js)$. In fact, in this case the images of the classical fundamental region $(-\bar{\rho}, i\infty, \rho)$ and its s-translation on $X_G(\mathbf{C})$ are disjoint.

2-cells. They are of two types: two-sided and triangular.

The two-sided $e_2'(j)$ are indexed by the classes $j \in J$ for which $jt = j$: by definition, the cell $e_2'(j)$ is $\phi(gE')$, where $g \in \Gamma$ is any representative of the class j. (The interior of E' maps to $e_2'(j)$, and the half-median $g\langle\rho, i\rangle$ becomes a cut from the center to the boundary of this cell.) We orient $e_2'(j)$ in the usual way. Then

$$\partial e_2'(j) = e_1(j) - e_1(js). \tag{8}$$

In fact, $\partial e_2'(j)$ consists of the images of the paths $\langle g(i\infty), g(i)\rangle$ and $\langle g(1+i), g(i\infty)\rangle$, where g is a representative of j. According to (7), the first of them is $e_1(j)$, while the second is equal to the image of

$$\langle gt^2(1+i), gt^2(i\infty)\rangle = \langle g(i), g(0)\rangle = \langle gs(i), gs(i\infty)\rangle,$$

i. e. $-e_1(js)$. Obviously $e_2'(j) \neq e_2'(k)$, if $j \neq k$.

The family of triangles $e_2''(j)$ is indexed by the classes $j \in J$ for which $jt \neq j$. By definition, the cell $e_2''(j)$ is $\phi(gE'')$, where $g \in \Gamma$ is any representative of the class j. (This is the cell: $E'' = E' \cup tE' \cup t^2E'$, and all the classes j, jt and jt^2 are distinct.) In the usual orientation induced by the complex structure, we have

$$\partial e_2''(j) = \sum_{a=0}^{2} (e_1(jt^a) - e_1(jt^a s)) \tag{9}$$

(this follows from the analogous formula for the boundary of E'' on $\bar{H}$, which the reader can easily verify). Obviously $e_2''(j) = e_2''(k) \Leftrightarrow j = kt^a$ for some a.

We easily see that L is a complex of the space $X_G(\mathbf{C})$.

c) *Conclusion of the proof of Theorem* 1.9. Following the plan announced at the beginning of the proof, we construct the groups C_1, Z_1 and B_1 of 1-chains, 1-cycles, and 1-boundaries of the complex L, and we define the imbedding $C \to C_1$ as follows: the G-chain (j) corresponds to the L-chain $e_1(js) - e_1(j)$. We easily see that this definition is correct and commutes with the boundary homomorphism (the group of linear combinations of parabolic points is naturally imbedded in the group of 0-chains of L).

We show that the kernel of this homomorphism $C \to C_1$ is trivial. We consider a nonzero G-chain $\Sigma n_j(j)$. Using the relations $(j) + (js) = 0$ and $(j) = 0$ for $j = js$, we may assume this expression normalized so that $n_j n_{js} = 0$ for all j. This G-chain corresponds to the L-chain $\Sigma n_j(e_1(js) - e_1(j))$. If $n_j \neq 0$, then $j \neq js$, and hence, as noted above, $e_1(j) \neq e_1(js)$. In addition, if $n_j n_k \neq 0$, then $j \neq k, ks$, so that all the simplices $e_1(j)$, $e_1(js)$, $e_1(k)$ and $e_1(ks)$ are distinct. It hence follows that $\Sigma n_j(e_1(js) - e_1(j)) \neq 0$.

We now assume that $Z \subset Z_1$ using the above imbedding. Then $(j) \bmod B_1(L) = \xi(j)$, because the chain $e_1(js) - e_1(j)$ belongs to the homology class $\{g(0), g(i\infty)\}$

by (7), where $g \in \Gamma$ is any representative of the class j. It hence follows by linearity that $z \bmod B_1(L) = \xi(z)$ for all $z \in Z$.

Finally, it is clear from (8) and (9) that all the generators of $B_1(L)$–the boundaries of 2-cells–belong to $B \subset Z$ and have the form $-(j)$ for $j = jt$ and $-(j)-(jt)-(jt^2)$ for $j \neq jt$. Hence $Z \cap B_1(L) = B$.

This completes the proof of the theorem.

Remarks. a) The parabolic complex $K(G)$ constructed in 1.8 b) has the same 1-homology as L, but is more economic than L; of the 0-cells of L only the parabolic points are left, since a pair of 1-cells of L with a common i-elliptic vertex corresponds to a single 1-cell in $K(G)$.

b) The construction of the G-complex in 1.8 a) is formally applicable to any subgroup $G \subset \Gamma$, for example to the unit subgroup. Taking into account the possible interest in studying the limits $\varprojlim H_1(X_G(\mathbf{C}))$ and $\varinjlim H^1(X_G(\mathbf{C}))$ over systems of subgroups $(G_i) \subset \Gamma$, we mention the following algebraic situation: in the case $G = \{e\}$, the system of equations (3) admits an explicit parametric solution. In order to construct it, we recall that Γ is the free product of its subgroups $\mathbf{Z}_2$ and $\mathbf{Z}_3$, which are generated by s and t, respectively. Consequently any element of Γ can be uniquely represented as a word $es^{\alpha_0}t^{\beta_0}\dots s^{\alpha_n}t^{\beta_n}$, where $\alpha_i = 0$ or 1 and $\beta_i = 0, 1$ or 2; in addition, α_i and β_i can only be zero at the ends of a word.

By t-words we mean the word et^2, and also all words with $\beta_n = 1$; by t^2-words we mean all words with $\beta_n = 2$, except for et^2; by s-words we mean all words with $\beta_n = 0$ and $\alpha_n = 1$. Every t^2-word can be uniquely represented in the form $g(st^2)^m$, where $m \geq 0$ and g is a t-word. Every s-word, except for es, can be uniquely represented in the form $g(st^2)^m s$, where $m \geq 0$ and g is a t-word.

We introduce a family of independent variables $U(g)$ indexed by all t-words g of the group Γ.

1.10. Proposition. *The infinite system of equations* (3) *in the unknowns* $\xi(g)$, $g \in \Gamma$, *has the following general solution:*

$$\xi(e) = -U(t) - U(t^2), \quad \xi(s) = U(t) + U(t^2),$$

$$\xi(st^2) = -U(t) - U(t^2) - U(st), \quad \xi(h) = U(h),$$

$$\xi(h(st^2)^m) = \sum_{i=0}^{m-1} (-1)^{m-i} U(h(st^2)^i st) + (-1)^m U(h),$$

$$\xi(h(st^2)^m s) = -\xi(h(st^2)^m)$$

for any t-word h.

Proof. The relations $\xi(gs) = -\xi(g)$ and $\xi(gt^2) = -\xi(gt) - \xi(g)$ allow us inductively to express all the ξ indexed by s-words and t^2-words in terms of the ξ indexed by t-words. It can be immediately verified that $\xi(g)$ can be chosen independently for all the t-words g, and that the above formulas are obtained as a result of induction.

(Passing to a nontrivial subgroup $G \subset \Gamma$ naturally imposes additional relations on the parameters $U(g)$.)

§2. The curves X_N

2.1 From now on, we shall work with the subgroups $G \subset \Gamma$ of the form

$$G = \Gamma_0(N) = \left\{ \begin{pmatrix} a & b \\ c & d \end{pmatrix} \in \Gamma \,\middle|\, c \equiv 0 \bmod N \right\}$$

For the most part we keep the notation of §1, but we write $X_N(\mathbf{C})$, $\{\alpha, \beta\}_N$ etc. instead of $X_{\Gamma_0(N)}(\mathbf{C})$, $\{\alpha, \beta\}_{\Gamma_0(N)}$. Some of the results become trivial or require slight modifications in the case when genus $X_N(\mathbf{C}) = 0$; we usually exclude this case without explicit mention.

The basic purpose of this section is to specialize the results of §1 to the case of the groups $\Gamma_0(N)$ and "explicitly" compute the groups $H_1(X_N(\mathbf{C}), \mathbf{Z})$. However, we begin by describing the special properties of the Riemann surfaces $X_N(\mathbf{C})$ which we need later.

The principal property is the existence of a special smooth projective curve X_N defined over $\mathbf{Q}$ for which the space $X_N(\mathbf{C}) = \Gamma_0(N) \backslash \overline{H}$ is canonically identified with the set of $\mathbf{C}$-points of X_N (also $X_N(\mathbf{C})$ in the traditional notation).

We enumerate some features of the $\mathbf{Q}$-structure and the induced $\mathbf{R}$-structure on X_N.

Let $j : H \to C$ be the classical modular invariant; this is a holomorphic function on H; we define j_N by the condition $j_N(z) = (Nz)$. Let $\mathbf{Q}(j, j_N)$ be the field of rational functions generated by j and j_N. It has transcendence degree 1, and $\mathbf{Q}$ is algebraically closed in it. The curve X_N is a smooth projective model of this field.

We further set $Y_N = \operatorname{Spec} \mathbf{Q}[j, j_N]$; this is an affine model of the field. The map $H \to Y_N(\mathbf{C}) : z \rightsquigarrow (j(z), j(Nz))$ extends to a map $\phi : \overline{H} \to X_N(\mathbf{C})$ which, in turn, induces an isomorphism $\Gamma_0(N) \backslash \overline{H} \xrightarrow{\sim} X_N(\mathbf{C})$.

Let $f \in \mathbf{C}(j, j_N)$ be a rational function on $X_N \otimes \mathbf{C}$. Its lifting $\phi^*(f)$ to H expands in a Fourier series $\Sigma a_n e^{2\pi i n z}$ with a finite number of coefficients $a_n \neq 0$ for $n < 0$. This function is defined over $\mathbf{Q}$, i. e. it belongs to $\mathbf{Q}(j, j_N)$, if and only if $a_n \in \mathbf{Q}$ for all n. Analogously, the differential on $X_N \otimes \mathbf{C}$ with Fourier expansion on H

$$\sum b_n e^{2\pi i (n-1) z} d(e^{2\pi i z}) = 2\pi i \sum b_n e^{2\pi i n z} dz$$

is defined over $\mathbf{Q}$ if and only if $b_n \in \mathbf{Q}$ for all n.

The local ring of the point $\phi(i\infty)$ consists of all functions with Fourier coefficients $a_n = 0$ for $n < 0$, so that, algebraically, $e^{2\pi i z}$ is a preferred formal parameter of this ring in the $\mathbf{Q}$-structure. In particular, $f(\phi(i\infty)) = \phi^*(f(i\infty)) = a_0$. Hence the values at the point $\phi(i\infty)$ of all functions defined over $\mathbf{Q}$ belong to $\mathbf{Q}$. This means that $\phi(i\infty) \in X_N(\mathbf{Q})$.

The map $z \rightsquigarrow -1/Nz$ belongs to the normalizer of the group $\Gamma_0(N)$, and hence induces an involution on $\Gamma_0(N)\backslash\bar{H}$. We easily see that this involution interchanges j and j_N, and thus comes from the canonical involution of the curve X_N over $\mathbf{Q}$. This involution takes $\phi(i\infty)$ to $\phi(0)$; hence also $\phi(0) \in X_N(\mathbf{Q})$.

The parabolic points on $X_N(\mathbf{C})$ other than $\phi_N(0)$ and $\phi_N(i\infty)$ do not necessarily belong to $X_N(\mathbf{Q})$ (or even $X_N(\mathbf{R})$).

In fact, complex conjugation acts naturally on $X_N(\mathbf{C})$. Denoting this action by a bar, we have

$$\overline{\varphi(z)} = \varphi(-\bar{z}).$$

In other words, reflection of H relative to the imaginary axis becomes complex conjugation on $X_N(\mathbf{C})$. This follows from the formula $\overline{e^{2\pi i z}} = e^{-2\pi i \bar{z}}$ and from the fact that $e^{2\pi i z}$, as an analytic local parameter, is defined over $\mathbf{R}$. Hence to construct a non-real parabolic point it suffices to find a rational number $\alpha \in \mathbf{Q}$ such that $-\alpha \notin \Gamma_0(N)\alpha$. Such numbers always exist if N is divisible by the square of a prime number ≥ 3, as is clear from the classical description given below of parabolic points.

Parabolic points. The parabolic points of $X_N(\mathbf{C})$ are in one-to-one correspondence with the classes $\Gamma_0(N)\backslash\mathbf{Q} \cup (i\infty)$. In order to describe them we introduce the set $\Pi(N)$, which consists of pairs of the form $[\delta;\, a \bmod (\delta, N\delta^{-1})]$. Here δ runs through all positive divisors of N, and the second coordinate of the pair runs through any invertible class of residues modulo the greatest common divisor of δ and $N\delta^{-1}$. If $(\delta, N\delta^{-1}) = 1$ we sometimes put simply 1 in place of the second coordinate.

2.2. **Proposition.** *Let* $\delta | N$, $u, v \in \mathbf{Z}$; $(u, v\delta) = (v, N\delta)^{-1} = 1$. *The map* $\mathbf{Q} \cup (i\infty) \to \Pi(N)$ *of the form*

$$\frac{u}{v\delta} \rightsquigarrow [\delta;\, uv \bmod (\delta, N\delta^{-1})], \qquad i\infty \rightsquigarrow [N;\, 1]$$

gives an isomorphism of the set of parabolic points on X_N *with* $\Pi(N)$.

Proof. The substitution

$$\begin{pmatrix} 1 & 1 \\ N & N+1 \end{pmatrix} \in \Gamma_0(N)$$

takes $i\infty$ into $\mathbf{Q}$, so that it suffices to examine the action of $\Gamma_0(N)$ on $\mathbf{Q}$. The substitution $\begin{pmatrix} a & b \\ Nc & d \end{pmatrix}$ takes

$$\frac{u}{v\delta} \text{ to } \frac{au + bv\delta}{Ncu + dv\delta}.$$

This fraction is irreducible, and $\delta = (Ncu + dv\delta, N)$, because $(dv, N\delta^{-1}) = 1$. Finally,

$$(au + bv\delta)(N\delta^{-1}cu + dv) \equiv aduv \equiv uv \bmod (\delta, N\delta^{-1}),$$

because $ad - Nbc = 1 \Rightarrow ad \equiv 1 \bmod (\delta, N\delta^{-1})$. Consequently every class $\Gamma_0(N)u/v\delta$ corresponds to the same element in $\Pi(N)$. The induced map $\Gamma_0(N)\backslash\mathbf{Q} \cup (i\infty) \to \Pi(N)$ is obviously surjective; the fractions u/δ, $(u, \delta) = 1$, cover all pairs with first coordinate δ. Finally, this map can either be checked to be injective directly, or else we can refer to the fact that both sets consist of the same number of elements $\sum_{\delta|N}\phi((\delta, N\delta^{-1}))$. The proposition is proved.

Remark. The stationary subgroup of the point u/δ in $\Gamma_0(N)$ is generated by the element

$$\begin{pmatrix} u & u' \\ \delta & \delta' \end{pmatrix}\begin{pmatrix} 1 & e \\ 0 & 1 \end{pmatrix}\begin{pmatrix} \delta' & -u' \\ -\delta & u \end{pmatrix},$$

where $e = N\delta^{-1}/(\delta, N\delta^{-1})$ and $u\delta' - u'\delta = 1$.

2.3. **The set** $\Gamma_0(N)\backslash\Gamma$. We define the set $\mathbf{P}^1(\mathbf{Z}/(N))$, "the projective line over $\mathbf{Z}/(N)$" using homogeneous coordinates.

Let $\tilde{c} = c \bmod N$, $\tilde{d} = d \bmod N$ be two residue classes mod N which are represented by the relative prime integers c and d. We call two such pairs $(\tilde{c}, \tilde{d})$ and $(\tilde{e}, \tilde{f})$ equivalent if there exists an invertible residue class $u \in (\mathbf{Z}/(N))^*$ such that $(u\tilde{c}, u\tilde{d}) = (\tilde{e}, \tilde{f})$. We designate the equivalence class of the pair $(\tilde{c}, \tilde{d})$ by the symbol $\tilde{c} : \tilde{d}$. By definition, the set of these classes is $\mathbf{P}^1(\mathbf{Z}/(N))$.

The group Γ acts on the right on $\mathbf{P}^1(\mathbf{Z}/(N))$ by the formula

$$(\tilde{e}:\tilde{f})\begin{pmatrix} a & b \\ c & d \end{pmatrix} = (\tilde{a}\tilde{e} + \tilde{c}\tilde{f}) : (\tilde{b}\tilde{e} + \tilde{d}\tilde{f}).$$

2.4. **Proposition.** *The map* $\Gamma \to \mathbf{P}^1(\mathbf{Z}/(N))$, *which associates the matrix* $\left(\begin{smallmatrix} a & b \\ c & d \end{smallmatrix}\right)$ *to the point* $\tilde{c} : \tilde{d}$ *is constant on the cosets* $\Gamma_0(N)g$ *and induces an isomorphism of right* Γ*-sets*

$$\Gamma_0(N)\backslash\Gamma \stackrel{\sim}{\to} \mathbf{P}^1(\mathbf{Z}/(N)).$$

Proof. We immediately verify that the map

$$\begin{pmatrix} a & b \\ c & d \end{pmatrix} \rightsquigarrow \tilde{c}:\tilde{d}$$

is constant on right cosets and commutes with the action of Γ. In addition, the group Γ acts transitively on both sets $\Gamma_0(N)\backslash\Gamma$ and $\mathbf{P}^1(\mathbf{Z}/(N))$, the unit class goes to the point $(\tilde{0} : \tilde{1})$, and the stationary subgroups of these two elements coincide: they equal $\Gamma_0(N)$. This completes the proof.

From now on, we shall often identify $\Gamma_0(N)\backslash\Gamma$ with $\mathbf{P}^1(\mathbf{Z}/(N))$ by means of the above isomorphism. We translate the structures in §1 connected with $\Gamma_0(N)\backslash\Gamma$ to the language of $\mathbf{P}^1(\mathbf{Z}/(N))$. See 1.5 for the definition of the map ξ: we recall that

$$s = \begin{pmatrix} 0 & -1 \\ 1 & 0 \end{pmatrix} \quad \text{and} \quad t = \begin{pmatrix} 1 & -1 \\ 1 & 0 \end{pmatrix}$$

2.5. Corollary. a) *The function* $\xi: \mathbf{P}^1(\mathbf{Z}/(N)) \to H_1(X_N(\mathbf{C}), \mathbf{R})$ *has the form*

$$\xi(\tilde{c}:\tilde{d}) = \left\{\frac{b}{d}, \frac{a}{c}\right\}_N, \tag{10}$$

where a, b, c, d *are any integers with the conditions* $ad - bc = 1$, $\tilde{c} = c \bmod N$, $\tilde{d} = d \bmod N$ *(by definition,* $a/0 = i\infty$*).*

b) *The action of the elements* s *and* t *on* $\mathbf{P}^1(\mathbf{Z}/(N))$ *is described by the formulas*

$$(\tilde{c}:\tilde{d})s = -\tilde{d}:\tilde{c}, \quad (\tilde{c}:\tilde{d})t = (\tilde{c}-\tilde{d}):\tilde{c}. \tag{11}$$

c) *Complex conjugation (see the end of 2.1) acts on the distinguished classes* $\xi(\tilde{c}:\tilde{d})$ *by the formulas*

$$\overline{\xi(\tilde{c}:\tilde{d})} = -\xi(\tilde{d}:\tilde{c}). \tag{12}$$

All these facts are verified directly from the definitions.

In 1.8 we define the "boundary" of any element in $\Gamma_0(N)\backslash\Gamma$: this is an element of the free abelian group generated by the parabolic points of $X_N(\mathbf{C})$. Identifying the set of parabolic points with $\Pi(N)$ as in 2.2, we describe the boundary map:

2.6. Corollary. *The boundary of the "simplex"* $\tilde{c}:\tilde{d}$ *equals*

$$\partial(\tilde{c}:\tilde{d}) = \left[\delta_1; \frac{c}{\delta_1} d^{-1} \bmod (\delta_1, N\delta_1^{-1})\right] - \left[\delta_2; -c^{-1}\frac{d}{\delta_2} \bmod (\delta_2, N\delta_2^{-1})\right], \tag{13}$$

where $\delta_1 = (c, N)$, $\delta_2 = (d, N)$.

Proof. It is clear from (10) and the definition of the boundary that the boundary is equal to the difference between the classes $\Gamma_0(N)a/c$ and $\Gamma_0(N)b/d$. By Proposition 2.2, the point a/c corresponds to the pair $[\delta_1; a(c/\delta_1) \bmod (\delta_1, N\delta_1^{-1})]$, and $ad - bc = 1$, so that $ad \equiv 1 \bmod (\delta_1, N\delta_1^{-1})$ and $a \equiv d^{-1}$. The second pair is computed analogously, and this proves the corollary.

2.7. Theorem. a) *Construct the maximal torsion-free abelian group generated by the symbols* $(\tilde{c}:\tilde{d})$, *one for each point* $\tilde{c}:\tilde{d} \in \mathbf{P}^1(\mathbf{Z}/(N))$, *with the relations*

$$(\tilde{c}:\tilde{d}) + (-\tilde{d}:\tilde{c}) = 0, \tag{14}$$

$$(\tilde{c}:\tilde{d}) + ((\tilde{c}-\tilde{d}):\tilde{c}) + (-\tilde{d}:(\tilde{c}-\tilde{d})) = 0. \tag{15}$$

Further, let $H(N)$ *designate the subgroup in it which is the kernel of the boundary homomorphsim (13). Then the map* $\xi: (\tilde{c}:\tilde{d}) \rightsquigarrow \{b/d, a/c\}_N$, *as in (10), induces an isomorphism*

$$\xi: H(N) \simeq H_1(X_N(\mathbf{C}), \mathbf{Z}).$$

b) *Let*

$$\frac{b}{a} = \frac{b_n}{a_n}, \frac{b_{n-1}}{a_{n-1}}, \ldots, \frac{b_1}{a_1}, \frac{b_0}{a_0} = \frac{b_0}{1}$$

be the successive convergents of the rational number $b/a > 0$. Then

$$\left\{0, \frac{b}{a}\right\}_N = \sum_{k=1}^{n} \xi((-1)^{k-1}\tilde{a}_k : \tilde{a}_{k-1}). \tag{16}$$

2.8. **Special case.** Let $N = p$, a prime number. In this case there are two parabolic points $[1;\,1]$ and $[p;\,1]$ in the notation of 2.2. The points of $\mathbf{P}^1(\mathbf{Z}/(p))$ have the form $\tilde{c} : 1$ or $\tilde{1} : \tilde{0}$. The simplex in $K(\Gamma_0(N))$ corresponding to the pair $\tilde{0} : \tilde{1}$ and $\tilde{1} : \tilde{0}$, joins $[1;\,1]$ and $[p;\,1]$; all the other simplices of the parabolic complex are loops which begin and end at $[1;\,1]$. Hence, introducing the affine coordinate system $\tilde{c} : \tilde{1} = \tilde{c}$, $\tilde{1} : \tilde{0} = \infty$, in $\mathbf{P}^1(\mathbf{Z}/(p))$, we find that the map $\xi : \mathbf{Z}/(p) \cup (\infty) \to H_1(X_p(\mathbf{C}), \mathbf{Z})$, whose definition is provisionally completed by the conditions $\xi(0) = \xi(\infty) = 0$, is the *universal function* satisfying the functional equations:

$$\left.\begin{aligned} \xi(\tilde{c}) + \xi(-\tilde{c}^{-1}) &= 0, \\ \xi(\tilde{c}) + \xi(1 - \tilde{c}^{-1}) + \xi\left(\frac{1}{1-\tilde{c}}\right) &= 0, \\ \xi(0) = \xi(\infty) &= 0. \end{aligned}\right\} \tag{17}$$

(Universality holds in the class of such functions with values in torsion-free abelian groups.) We note that $\xi(\tilde{c}) = \{0, 1/c\}$ by (10) if $\tilde{c} \neq 0$.

§3. Arguments of parabolic points

In this section we give explicit expressions for the integrals $\int_{\{\alpha,\beta\}} \omega$ in the case when the class $\{\alpha, \beta\}$ is not necessarily integral. To formulate and prove our results, we need some elementary facts about Hecke operators. We give them in the limited context in which we need them.

3.1. **Hecke operators and parabolic forms.** Let $a, b, c, d \in \mathbf{R}$, $ad - bc > 0$. For any function Φ on H we set

$$\Phi \Big| \begin{pmatrix} a & b \\ c & d \end{pmatrix}(z) = \Phi\left(\frac{az+b}{cz+d}\right) \frac{d}{dz}\left(\frac{az+b}{cz+d}\right).$$

This defines a right action of the group $PL(2, \mathbf{R})$ on the space of functions on H. This action extends by linearity to the entire group ring:

$$\Phi \Big| \left(\sum a_i g_i\right) = \sum a_i \Phi \Big| g_i, \qquad a_i \in \mathbf{C}, \quad g_i \in PL.$$

The following special elements of this ring are called Hecke operators:

$$T_m = \sum_{\substack{d/m \\ b=0,\dots,d-1}} \begin{pmatrix} md^{-1} & b \\ 0 & d \end{pmatrix}, \quad m \in \mathbf{Z}, \quad m > 0. \tag{18}$$

They satisfy the following relations on the space of functions on H with period 1: $T_m T_n = T_{mn}$ for $(m, n) = 1$, and $T_p T_{p^r} = T_{p^{r+1}} + pT_{p^{r-1}}$.

Let $G \subset \Gamma$ be a subgroup of finite index. A function $\Phi(z)$ on H is called a G-parabolic form if there exists a differential of the first kind ω on the surface $X_G(\mathbf{C})$ such that $\phi^*(\omega) = \Phi(z)dz$, where $\phi : H \to X_G(\mathbf{C})$ is the canonical projection.

3.2. **Proposition.** *If* $(m, N) = 1$, *then* T_m *takes the space of* $\Gamma_0(N)$-*parabolic forms* P_N *into itself.*

Thus the operators $\{T_m | (m, N) = 1\}$ generate a commutative operator alegbra on the space of $\Gamma_0(N)$-parabolic forms P_N. They are Hermitian relative to the Peterson scalar product. It is also worthwhile to keep in mind that the $\mathbf{Q}$-subspace $\phi_N^*(H^0(X_N, \Omega))$ is invariant relative to T_m : this is clear either from the direct description of the action of T_m on the Fourier coefficients (Atkin and Lehner [1], formula (3.1)), or else from the invariant definition of T_m using correspondences on $X_N \times X_N$. In particular, P_N has a basis of eigenfunctions for the Hecke algebra all of whose Fourier coefficients are algebraic.

The theory of Hecke operators with indices not prime to N is more complicated. We shall only indicate the operators U_p, p prime:

$$U_p = \sum_{b=0}^{p-1} \begin{pmatrix} 1 & b \\ 0 & p \end{pmatrix} = T_p - \begin{pmatrix} p & 0 \\ 0 & 1 \end{pmatrix}. \tag{19}$$

In the article by Atkin and Lehner [1] it is shown that $U_p(P_N) \subset P_N$ if $p \mid N$, and that the U_p commute with all the T_m, $p \nmid m$.

We shall henceforth assume N fixed; $\{\alpha, \beta\}_N$ denotes the element in $H_1(X_N(\mathbf{C}), \mathbf{R})$, defined in 1.2. In addition to the general properties of the classes $\{\alpha, \beta\}$ we note that $\{\alpha + m, \beta + n\}_N = \{\alpha, \beta\}_N$ for all $m, n \in \mathbf{Z}$. This follows because the parabolic element $\left(\begin{smallmatrix} 1 & m \\ 0 & 1 \end{smallmatrix}\right)$ belongs to $\Gamma_0(N)$ if we use Proposition 1.4: the class $\{\alpha, \alpha + m\}_N = \{\alpha, \left(\begin{smallmatrix} 1 & m \\ 0 & 1 \end{smallmatrix}\right)\alpha\}$ is equal to zero as the image of a parabolic element under the homomorphism $\Gamma_0(N) \to H_1(X_N(\mathbf{C}), \mathbf{Z})$.

3.3. **Theorem.** *Let the* $\Gamma_0(N)$-*parabolic form* $\Phi = \phi^*(\omega)/dz$ *be an eigenfunction for the Hecke operator* T_m, $(m, N) = 1$: $\Phi | T_m = c_m \Phi$. *Then*

$$\Big(\sum_{d/m} d - c_m\Big) \int_0^{i\infty} \Phi\, dz = \sum_{\substack{d/m \\ b \bmod d}} \int_{\{0, \frac{b}{d}\}} \omega. \tag{20}$$

Comment. Since $(m, N) = 1$, we have $b/d \in \Gamma_0(N)(0)$ for all $d|m$ by Proposition 2.2. Hence $\{0, b/d\} \in H_1(X_N(\mathbf{C}), \mathbf{Z})$, so that the right side of (20) consists of integral linear combinations of the fundamental periods of the differential ω with respect to some integral homology basis. The coefficients of these linear combinations are computed using the theory in §§1 and 2 (see, in particular, formula (16)). In addition,

$\Sigma_{d|m} d - c_m \neq 0$ for sufficiently large m by the well-known growth estimates for the coefficients of parabolic forms.

Thus we may assume that the expressions (20) give us an explicit form for the arguments of the point $\phi(0)$ under the Abel-Jacobi map of the curve $X_N(\mathbf{C})$ (with origin $\phi(i\infty)$) relative to the basis of differentials of the first kind for which the corresponding parabolic forms are eigenfunctions for the Hecke operators.

Another point of view on formulas (20) emerges if we consider m variable and fix Φ. Then, under the assumption $\int_0^{i\infty} \Phi dz \neq 0$, (20) and (16) give expressions for the eigenvalues c_m of the operators T_m on Φ in terms of the expansion in continued fractions of all numbers of the form b/d, $d|m$, $0 \leq b \leq d-1$.

Both points of view lead to interesting number-theoretic results, which we shall examine in greater detail below in §§6 and 7.

3.4. **Proof of Theorem 3.3.** For any element $g \in PL(2, \mathbf{R})$ and function Φ on H we have

$$\int_\alpha^\beta (\Phi \,|\, g) dz = \int_\alpha^\beta \Phi(gz) d(gz) = \int_{g(\alpha)}^{g(\beta)} \Phi(z) dz.$$

Using this, we obtain the following formulas for the action of the Hecke operators (18):

$$\int_\alpha^\beta (\Phi \,|\, T_m) dz = \sum_{d/m} \sum_{b=0}^{d-1} \int_{\frac{m}{d^2}\alpha + \frac{b}{d}}^{\frac{m}{d^2}\beta + \frac{b}{d}} \Phi dz. \tag{21}$$

We susbtitute $\alpha = 0$, $\beta = i\infty$ here and use the fact that $\Phi | T_m = c_m \Phi$:

$$c_m \int_0^{i\infty} \Phi dz = \sum_{d/m} \sum_{b=0}^{d-1} \left(\int_{\frac{b}{d}}^{0} + \int_0^{i\infty} \right) \Phi dz,$$

so that

$$\left(\sum_{d/m} d - c_m \right) \int_0^{i\infty} \Phi dz = \sum_{d/m} \sum_{b=0}^{d-1} \int_0^{\frac{b}{d}} \Phi dz = \sum_{\substack{d/m \\ b \bmod d}} \int_{\{0, \frac{b}{d}\}} \omega,$$

as was to be proved.

More generally, this same device of "closing the path of integration" allows us to compute the arguments of any parabolic point.

3.5. **Theorem.** *Under the conditions of Theorem 3.3, let $\alpha \in \mathbf{Q}$. Then*

$$\left(\sum_{d/m} d - c_m \right) \int_\alpha^{i\beta} \Phi dz = \sum_{\substack{d/m \\ b \bmod d}} \int_{\{\alpha, \frac{m}{d^2}\alpha + \frac{b}{d}\}_N} \omega. \tag{22}$$

For every N there exist infinitely many values of m, $(m, N) = 1$, such that $\{\alpha, m\alpha/d^2 + b/d\}_N \in H_1(X_N(\mathbf{C}), \mathbf{Z})$ for all b. Hence if Φ is an eigenfunction for *all the Hecke operators* T_m, $(m, N) = 1$, *then* m *can be chosen so that the right side of* (22) *contains the periods of* ω *over integral homology classes.*

Proof. (22) follows immediately from (21), as in the previous theorem.

To prove the second assertion, we set $\alpha = u/v\delta$, where $\delta | N$ and $(u, v\delta) = (v, N\delta^{-1}) = 1$, and take for m any prime number l with the conditions $l \nmid N$ and $l \equiv 1 \bmod (\delta, N\delta^{-1})$.

According to 1.3 c), integrality of the class $\{\alpha, l\alpha/d^2 + b/d\}$ is equivalent to the inclusion $l\alpha/d^2 + b/d \in \Gamma_0(N)\alpha$. Taking into account that $d = 1$ or l and using Proposition 2.2, we conclude that we must verify that, in the irreducible representation of any of the fractions

$$l\alpha = \frac{lu}{v\delta}, \qquad \frac{1}{l}\alpha + \frac{b}{l} = \frac{u + bv\delta}{lv\delta}$$

the product (numerator) $\times$ (denominator) δ^{-1} is congruent to $uv \bmod (\delta, N\delta^{-1})$. We must consider separately the cases when this fraction is reducible (then the greatest common divisor of the numerator and denominator equals l) and when it is irreducible. In both cases the required congruence follows from $l \equiv 1 \bmod (\delta, N\delta^{-1})$.

We note that all $l \nmid N$ are suitable for square-free N, and that all $l \equiv 1 \bmod N$.

The theorem is proved.

The case when the eigenvalues of T_m on Φ are rational is especially interesting: it then follows from (21) and (22) that the corresponding arguments of all the parabolic points of $X_N(\mathbf{C})$ are rational linear combinations of the fundamental periods. Here is the algebraic-geometric formulation of this fact:

3.6. Corollary. *Let* $\psi: X_N \to X$ *be a morphism of curves over* $\mathbf{Q}$, *and let the space* $\psi^*(H^0(X, \Omega_X))$ *be invariant relative to the Hecke operators* T_m, $(m, N) = 1$, *and have a basis of eigenvectors for* T_m *with rational eigenvalues at least for some sufficiently large* m. *Then for any two parabolic points* $x, y \in X_N(\mathbf{C})$ *the divisor class* $\psi(x) - \psi(y)$ *on* $X \otimes \mathbf{C}$ *has finite order.* (1)

3.7. Special case. Let $\psi : X_N \to X$ be a morphism of X_N onto an elliptic curve X over $\mathbf{Q}$. We call this morphism a *Weil uniformization* for X (in the weak sense) if $\psi \circ \phi(i\infty)$ is zero on X, and the one-dimensional subspace $(\psi \circ \phi)^* H^0(X, \Omega^1)$ is invariant relative to the operators T_m, $(m, N) = 1$, with rational eigenvalues.

The following assertions are easily deduced from the above:

a) *If the curve* X *has a weak uniformization* $\psi : X_N \to X$, *then there exists another weak uniformization for which the images of all the parabolic points coincide with zero on* X.

In fact, it suffices to take the composition of ψ with the multiplication $X \xrightarrow{n} X$ for suitable n and then use 3.6.

---

(1) *Added in proof*. V. Drinfel'd has shown me that Theorem 3.5 easily implies that such classes even have finite order on X_N.

b) *Let* $\psi : X_N \to X$ *be a weak uniformization, let* ω *be a differential of the first kind on* X, *and let* $\Phi(z)dz$ *be its preimage on* H. *Further, let* γ^+ *be a generator of the subgroup of real (invariant relative to conjugation) classes in* $H_1(X_N(\mathbf{C}), \mathbf{Z})$, $W^+ = \int_{\gamma^+} \omega$, *and let* t *be the maximal period of a point of finite order in* $X(\mathbf{Q})$. *Then*

$$\int_0^{i\infty} \Phi\, dz = \frac{s}{t} W^+, \quad s \in \mathbf{Z}. \tag{23}$$

In fact,

$$\int_0^{i\infty} \Phi\, dz = \int_{\psi\circ\varphi(0)}^{\psi\circ\varphi(i\infty)} \omega.$$

The second integral is taken over the image of the imaginary axis, which lies entirely in $X(\mathbf{R})$ and joins the point of finite order $\psi \circ \phi(0) \in X(\mathbf{Q})$ with zero $\psi \circ \phi(i\infty) \in X(\mathbf{Q})$. This implies the assertion.

3.8. Finally, we give a somewhat strengthened result from the second chapter of [9], where the device of closing the path was first introduced in a somewhat different context. Here we are not required to apply the Hecke operators, but, on the other hand, the integrand contains parabolic forms of a special type arising in the Hecke-Weil theory and in the study of zeta-functions of modular curves over abelian extensions of the field $\mathbf{Q}$.

Suppose that $\Phi = \Sigma a_n e^{2\pi i n z}$ is a parabolic form relative to $\Gamma_0(N)$, $m > 1$, $m \in \mathbf{Z}$; let $\chi : \mathbf{Z} \to \mathbf{C}$ be a primitive Dirichlet character mod m. We set

$$g(\chi) = \sum_{b \bmod m} \chi(b) e^{2\pi i \frac{b}{m}} \text{ (Gaussin sum) and } \Phi_\chi = \sum_{n=1}^{\infty} \chi(n) a_n e^{2\pi i n z};$$

finally, let $\Phi dz = \phi^*(\omega)$.

3.9. **Theorem.** *Let* $\delta = (m, N)$, *and let* $(\delta, N\delta^{-1}) = 1$. *Then*

$$\int_0^{i\infty} \Phi_\chi\, dz = \frac{g(\chi)}{m} \sum_{b \bmod m} \bar{\chi}(b) \int_{\{-\frac{b}{m}, \frac{1}{\delta}\}_N} \omega \tag{24}$$

and $\{-b/m, 1/\delta\}_N \in H_1(X_N(\mathbf{C}), \mathbf{Z})$ *for all* b mod m, $\bar{\chi}(b) \neq 0$.

Proof. In fact, by a well-known lemma (see Weil [16], Ogg [12] or Manin [9]),

$$\Phi_\chi(z) = \frac{g(\chi)}{m} \sum_{b \bmod m} \bar{\chi}(-b) \Phi\left(z + \frac{b}{m}\right),$$

so that

$$\int_0^{i\infty} \Phi_\chi\, dz = \frac{g(\chi)}{m} \sum_{b \bmod m} \bar{\chi}(b) \int_{-\frac{b}{m}}^{i\infty} \Phi(z)\, dz = \frac{g(\chi)}{m} \sum_{b \bmod m} \bar{\chi}(b) \int_{-\frac{b}{m}}^{\alpha} \Phi(z)\, dz$$

for any $\alpha \in \overline{H}$, because $\Sigma_{b \bmod m} \overline{\chi}(b) = 0$ since the character χ is primitive.

In particular, if $(\delta, N\delta^{-1}) = 1$, then for any b with $\overline{\chi}(b) \neq 0$, i. e. $(b, m) = 1$, we have $b/m \in \Gamma_0(N)\delta^{-1}$ by Proposition 2.2. Hence $\{b/m, 1/\delta\}_N \in H_1(X_N(\mathbf{C}), \mathbf{Z})$, which explains the choice $\alpha = 1/\delta$. The theorem is proved.

In [9] we examine the case $\delta = 1$.

§4. L-series at the center of the critical strip

4.1. Let ω be a differential of the first kind on $X_N \otimes \mathbf{C}$. As above, we set $\Phi_\omega = \phi_N^*(\omega)/dz = -2\pi i \sum a_n e^{2\pi i n z}$ and define the Dirichlet series L_ω by the formula

$$L_\omega(s) = \sum_{n=1}^{\infty} a_n n^{-s}. \tag{25}$$

It is well known that $L_\omega(s)$ has an analytic continuation onto the entire plane given by the formula

$$L_\omega(s) = -\frac{1}{2\pi i}\frac{(2\pi)^s}{\Gamma(s)}\int_0^\infty \Phi_\omega(iy)\, y^{s-1}\, dy$$

and, in particular,

$$L_\omega(1) = \int_0^{i\infty} \Phi_\omega(z)\, dz = \int_{\{0, i\infty\}_N} \omega \tag{26}$$

(see, for example, Manin [9], Lemma 9.2).

This allows us to interpret the fundamental results of the last section in terms of explicit formulas for the values of the series L_ω at the point $s = 1$, the center of their critical strip; and we gather these formulas together here for more convenient reference.

4.2. **Theorem.** a) *Under the conditions of* 3.1, *suppose that* Φ_ω *is an eigen-form for the Hecke operator* T_m, $(m, N) = 1$, *and that* $a_1 = 1$. *Then*

$$\Bigl(\sum_{d/m} d - a_m\Bigr) L_\omega(1) = \sum_{\substack{d/m \\ b \bmod d}} \int_{\{0, \frac{b}{d}\}_N} \omega. \tag{27}$$

b) *Under the conditions of* 3.1, *let* χ *be a primitive character* mod $m > 1$, *let* $g(\chi)$ *be the Gaussian sum, and let* $L_{\omega,\chi}(s) = \Sigma_{n=1}^{\infty} \chi(n) a_n n^{-s}$. *If* $(m, N) = \delta$, $(\delta, N\delta^{-1}) = 1$, *then*

$$L_{\omega,\chi}(1) = \frac{g(\chi)}{m} \sum_{b \bmod m} \overline{\chi}(b) \int_{\{-\frac{b}{m}, \frac{1}{\delta}\}_N} \omega. \tag{28}$$

Proof. Formula (27) follows from (20) and (26) if we take into account that $\Phi_\omega | T_m = a_m \Phi_\omega$ for $a_1 = 1$. Formula (28) coincides with (24).

4.3. It is interesting to compare formulas (27) and (28) with the Birch–Swinnerton-Dyer conjecture (see [8], [9] and [15]). In the next two sections we do this for Weil uniformized elliptic curves over $\mathbf{Q}$ and for some abelian extensions of $\mathbf{Q}$. Here we shall limit ourselves to a remark on the behavior of the curve X_N itself.

Let

$$\Phi_{\omega(k)} = -2\pi i \sum_{n=1}^{\infty} a_n(k) e^{2\pi i n z}$$

be the family in P_N consisting of the eigenfunctions for the Hecke operators with the proper multiplicities ($k = 1, \cdots$, genus X_N; $a_1(k) = 1$). Then the product $\prod_{k=1}^{\text{genus}} L_{\omega(k)}(s)$ coincides with the Hasse-Weil series of the curve X_N corresponding to the one-dimensional cohomology of X_N, to within the Euler factors for $p | N$. It appears that the precise form of these exceptional factors is not known: Serre suggests characterizing them in terms of the l-adic representations connected with X_N, but these are not sufficiently well known. On the other hand, Atkin and Lehner [1] introduced the useful notions of "new forms" in P_N and the canonical partition of P_N into two terms. The first term is generated by the new forms; there a one-dimensional subspace corresponds to every weight of the Hecke algebra. The second term is generated by the "old forms," which are constructed in a natural way from the new forms in P_d for $d | N$.

A more detailed examination of this construction and its translation in $H_1(X_N(\mathbf{C}), \mathbf{Z})$ (or, rather, $H_1(X_N, \mathbf{Q})$) should allow us to conjecture the correct form of the L-function of the curve X_N and compute the exact value of $L(1)$ using (27).

§5. Weil uniformization

5.1. Let X be an elliptic curve on $\mathbf{Q}$. The following notation will be fixed for the duration of this and the next sections: N is the conductor of X; ω is a Néron differential on X; $L(X, s) = \sum_{n=1}^{\infty} a_n n^{-s}$ is the canonical Dirichlet series connected with X, the fundamental part of the Hasse-Weil zeta-function of this curve, and $L(X \otimes K, s)$ is the analogous series for X over K if $K \supset \mathbf{Q}$ is any finite extension. We emphasize that the Euler factors of L at the points of degeneracy of X are assumed to be normalized in the way that is now generally accepted, as described, for example, in Weil's article [16] and in the author's survey article [9].

5.2. Definition. A *Weil uniformization* (in the strong sense) of the curve X is a morphism of curves $\psi: X_N \to X$ over $\mathbf{Q}$ with the following properties:

a) $\psi \circ \phi: \overline{H} \to X_N(\mathbf{C}) \to X(\mathbf{C})$ takes $i\infty$ to the zero point of $X(\mathbf{Q})$.

b) $(\psi \circ \phi)^* \omega = \Phi(z) dz = -2\pi i \sum_{n=1}^{\infty} a_n e^{2\pi i n z} dz$, where (a_n) are the coefficients of the Dirichlet series $L(X, s)$.

c) The form $\Phi(z)$ is an eigenfunction of all the Hecke operators T_m (see (18)) for $(m, N) = 1$, of all the operators U_p for the primes $p | N$ (see (19)), and of the standard involution operator $\begin{pmatrix} 0 & 1 \\ -N & 0 \end{pmatrix}$ of the curve X_N.

It probably follows from a) and b) that $\Phi(z)$ is a "new form" in the sense of Atkin-Lehner (see [1]). Then properties c) are automatically fulfilled. In any case, conditions a), b) and c) are not independent (see, for example, Cartier [4]).

Weil conjecture. *Any elliptic curve over* $\mathbf{Q}$ *admits a Weil uniformization in the strong sense.*

For discussions of this conjecture, see, in particular, the articles by Weil [16], by the author [9], §10, and by Cartier [4]. In [16] and [9] it is shown that, for curves admitting a Weil uniformization, the L-series over $\mathbf{Q}$ and over any abelian extension of $\mathbf{Q}$ have analytic continuations onto the entire plane, as in §4. In this section we give formulas for the values of L-series at one (the center of their critical strip), and we deduce from them that the uniformization ψ is unique.

5.3. Let the curve X, the differential ω and the strong uniformization ψ be fixed. We introduce two fundamental number-theoretic functions connected with (X, ω, ψ). We let γ^+ and γ^- designate the generators of the groups of real classes and of purely imaginary classes, respectively, in $H_1(X(\mathbf{C}), \mathbf{Z})$ and we set $W^{\pm} = \int_{\gamma^{\pm}}\omega$.

5.4. **Definition** The functions $x^{\pm}: \mathbf{Q} \cup (i\infty) \to \mathbf{Q}$ are defined by the equations

$$\psi_*\{-\alpha, 0\}_N \pm \psi_*\{\alpha, 0\}_N = x^{\pm}(\alpha)\gamma^{\pm}$$

(the signs are taken either all plus or all minus).

We recall that $\{\alpha, 0\}_N \in H_1(X_N(\mathbf{C}), \mathbf{R})$. $x^{\pm}(\alpha)$ is rational by the results of 3.6 and 3.7. If the denominator of α is relatively prime to N, then we even have $x^{\pm}(\alpha) \in Z$. The functions x^+ and x^- have period 1; x^+ is even, and x^- is odd. We have normalized the signs in a way that is not entirely natural because of a desire to remain compatible with the notation in [9].

Using these functions, we can explicitly distinguish irrationality in the formulas for $L(1)$.

5.5. **Theorem.** *For all* m, $(m, N) = 1$, *we have*

$$\left(\sum_{d/m} d - a_m\right) L(X, 1) = -\frac{W^+}{2}\left(\sum_{\substack{d/m \\ b \bmod d}} x^+\left(\frac{b}{d}\right)\right). \tag{29}$$

Proof. Formula (29) follows from (27) with $\psi^*(\omega)$ in place of ω, if we take into account that the differential ω is defined over $\mathbf{R}$, so that $L(X, 1) \in \mathbf{R}$ and

$$\overline{\int_{\{0, \frac{b}{d}\}} \psi^*(\omega)} = \int_{\{0, -\frac{b}{d}\}} \psi^*(\omega).$$

Hence the real part of the sum on the right in (27), which is equal to half of its sum with it complex conjugate, has the form

$$\frac{1}{2}\sum_{\substack{d/m \\ b \bmod d}} \int_{\{0, \frac{b}{d}\} + \{0, -\frac{b}{d}\}} \psi^*(\omega) = -\frac{1}{2}\sum_{\substack{d/m \\ b \bmod d}} x^+\left(\frac{b}{d}\right)\int_{\gamma^+}\omega = -\frac{W^+}{2}\sum_{\substack{d/m \\ b \bmod d}} x^+\left(\frac{b}{d}\right).$$

as was to be proved.

To formulate the next result, we consider two abelian extensions $K \subset K'$ of the field $\mathbb{Q}$. Let the discriminants D and D' of these fields be relatively prime to N. Further, let r_1 be the number of real points of the field K, r_2 the number of purely imaginary points, and $r = r_1 + r_2$; let r_1', r_2' and r' have the analogous meaning for K'. We let M designate the set of Dirichlet characters belonging to K' but not to K; let m_χ be the conductor of the charater χ; sign $\chi = +$ if $\chi(-1) = 1$ and $-$ if $\chi(-1) = -1$.

5.6. **Theorem.** *In the above notation we have*

$$L(X \otimes K', s)/L(X \otimes K, s)|_{s=1}$$

$$= \pm \left|\frac{D}{D'}\right|^{\frac{1}{2}} (W^+)^{r'-r} (W^-)^{r_2'-r_2} \prod_{\chi \in M} \frac{1}{2} \left(\sum_{b \bmod m_\chi} \chi(b) x^{\operatorname{sign} \chi}\left(\frac{b}{m_\chi}\right)\right). \tag{30}$$

Proof. We set $L_\chi(X, s) = \sum_{n=1}^{\infty} \chi(n) a_n n^{-s}$ and use the formula

$$L(X \otimes K, s) = \prod L_\chi(X, s), \tag{31}$$

where χ runs through the Dirichlet characters associated with the field K (here we use the fact that D and N are relatively prime: otherwise the product in the right side of (31) may differ from the canonical Dirichlet series for $X \otimes K$ by a finite number of Euler factors; see [9], Lemma 7.3). Dividing the formulas (31) corresponding to K' and K by one another and substituting in the right side of expression (28), we find

$$L(X \otimes K', s)/L(X \otimes K, s)|_{s=1} = \prod_{\chi \in M} L_\chi(X, 1)$$

$$= \prod_{\chi \in M} \frac{g(\chi)}{m_\chi} \left(\sum_{b \bmod m_\chi} \bar{\chi}(b) \int_{\{-\frac{b}{m}, 0\}} \psi^*(\omega)\right). \tag{32}$$

Further, we know that

$$|g(\chi)| = m_\chi^{\frac{1}{2}} \text{ and } \prod_{\chi \in M} \frac{g(\chi)}{m_\chi} = \pm \left|\frac{D}{D'}\right|^{\frac{1}{2}}$$

by the Hasse-Artin formula. Finally, the inner sum in (32) transforms as in the previous theorem, giving

$$\sum_{b \bmod m_\chi} \bar{\chi}(b) \int_{\{-\frac{b}{m}, 0\}} \psi^*(\omega) = \frac{W^{\operatorname{sign} \chi}}{2} \sum_{b \bmod m_\chi} \bar{\chi}(b) x^{\operatorname{sign} \chi}\left(\frac{b}{m_\chi}\right). \tag{33}$$

To complete the transisition from (32) to (30), it remains to note that the number of even characters in M equals $r' - r$, and the number of odd characters equals $r_2' - r$. The theorem is proved.

We note that, since the fields K and K' are normal over $\mathbb{Q}$, it follows that only the following three combinations of the numbers r and r' are possible: either $r_2 = r_2' = 0$, or $r_2 = 0$, $r_2' = \frac{1}{2} r'$ or else $r_2 = \frac{1}{2} r$, $r_2' = \frac{1}{2} r'$

5.7. **Theorem.** *If a Weil uniformization (more precisely, a pair (X, ω)) exists for the curve X, then it is unique.*

Proof. The values of $L(X, 1)$ and $L_\chi(X, 1)$ are uniquely defined and are characterized by formulas (29) and (33), in which the numbers $x^{\pm}(\alpha)$ can be computed from any uniformization. We consider all characters χ with *prime* conductor $l \nmid 2N$. Then (30) allows us to compute the sum $\Sigma_{b \bmod l} x^+(b/l)$, and (33) allows us to compute all the sums $\Sigma_{b \bmod l} \chi(b) x^{\pm}(b/l)$ with nonprincipal characters mod l in terms which do not depend on the choice of uniformization. Consequently the numbers $x^{\pm}(b/l)$ for $l \nmid 2N$ do not depend on the choice of uniformization ψ. These numbers determine the values of the homomorphism of homology groups $\psi_* : H_1(X_N(\mathbf{C}), \mathbf{Z}) \to H_1(X(\mathbf{C}), \mathbf{Z})$ on the homology classes of the form $\{0, b/l\}_N$, as is clear from Definition 5.4. Suppose that these classes generate the entire group $H_1(X_N(\mathbf{C}), \mathbf{Z})$. Then it follows from the above that the homomorphism ψ_* is uniquely determined. But ψ is also determined uniquely from ψ_* if we require, as in 5.2 a), that the distinguished points of the curves X_N and X corresponds to each other. Hence it remains to prove the following fact:

5.8. **Lemma.** *The homology classes $\{0, b/l\}_N \in H_1(X_N(\mathbf{C}), \mathbf{Z})$ generate the entire homology group when $l \nmid 2N$ runs through the prime numbers and b runs through a complete system of residues* mod l.

Proof. We use a method of Weil [16]. Let $\left(\begin{smallmatrix} a & b \\ Nc & d \end{smallmatrix}\right) \in \Gamma_0(N)$ be any element. We have

$$\begin{pmatrix} a & b \\ Nc & d \end{pmatrix} \begin{pmatrix} 1 & x \\ 0 & 1 \end{pmatrix} = \begin{pmatrix} * & * \\ * & Ncx + d \end{pmatrix}.$$

By Dirichlet's theorem, x can always be chosen so that the number $Ncx + d$ is a prime $\neq 2$; obviously it does not divide $2N$. Hence $\Gamma_0(N)$ is generated by the elements and by the matrices having a prime $l \nmid 2N$ in the lower right-hand corner.

Setting $\alpha = 0$ in Proposition 1.4, we now immediately obtain the assertion of the lemma, and with it the uniqueness theorem.

Remark. Another variant of the uniqueness theorem (with a completely different proof) is contained in Cartier's report [4].

§ 6. The Birch–Swinnerton-Dyer conjecture and Mazur's theory

We keep the notation of the last section; in particular, X is an elliptic curve over $\mathbf{Q}$, N is its conductor, and $\psi : X_N \to X$ is a Weil uniformization in the strong sense.

In this section we compare formulas (29) and (30) with the Birch–Swinnerton-Dyer conjecture. Its complete formulation in the form convenient for us is given in [9]. Here we limit ourselves to several special cases with which our results may be directly compared.

All of these special cases, along with the conditional results whose proofs use the Birch–Swinnerton-Dyer conjecture and the conditional formulas of this type, are marked in the text with an asterisk.

6. 1. Conditional formula for $L(X, 1)$:

$$L(X, 1) = \begin{cases} 0, & \text{if rank } X(\mathbf{Q}) > 0, \\ W^+ \dfrac{[Ш]}{[\widetilde{X}(Q)]^2} [\pi_0(X(\mathbf{R}))] \prod_{p/N} [\pi_0(X \bmod p)]. \end{cases} \qquad (34)^*$$

In this and the other formulas, the following notation is used: $[G]$ is the number of elements in the set G; Ш is the Šafarevič-Tate group of the curve X; $\widetilde{X}(\mathbf{Q})$ is the group of $\mathbf{Q}$-points of finite order; $\pi_0(X(\mathbf{R}))$ is the group of connected components of the real points of X; and $\pi_0(X \bmod p)$ is the group of connected components of the closed fiber of the Néron model of the curve X over the point p.

In comparing $(34)^*$ with (23) and (29), the following circumstances deserve mention:

a) The general structure of the formulas is the same: $L(X, 1)$ is the product of W^+ by a rational number. However, in our formula the denominator of this number divides $[\widetilde{X}(\mathbf{Q})]$ and even the maximal period of the points of finite order in $X(\mathbf{Q})$. This evidently means that the local factors in $(34)^*$ must strongly cancel with $[\widetilde{X}(\mathbf{Q})]^2$. Ligozat [8] computed these factors for all twelve curves X_N of genus 1. According to his computations, the product of the local factors is always exactly equal to $[\widetilde{X}(\mathbf{Q})]$, and the hypothetical value of $[Ш]$ equals 1.

b) The condition $L(X, 1) = 0$ expresses a simple topological property of the uniformization map $\psi \circ \phi : H \to X(\mathbf{C})$. This is the property that $\psi \circ \phi$ takes the imaginary semiaxis to a closed path which is homotopic to zero in $X(\mathbf{R})$. In particular, in this case there are branch points of $\psi \circ \phi$ on the imaginary semiaxis, namely the zeros of the form $(\psi \circ \phi)^* \omega$. Do they have any relation to the points of infinite order in $X(\mathbf{Q})$ whose existence is predicted by the Birch–Swinnerton-Dyer conjecture?

6. 2. Conditional formulas for $L(X \otimes K', s)/L(X \otimes K, s)|_{s=1}$. With the conditions and notation of Theorem 5. 6, we further suppose that $X(K') = X(K)$: the group of rational points of X does not increase when going from K to K'. Then the Birch–Swinnerton-Dyer conjecture leads to the following expression (for the details, see [9]):

$$L(X \otimes K', s)/L(X \otimes K, s)|_{s=1} = \left|\frac{D}{D'}\right|^{\frac{1}{2}} (W^+)^{r'-r} \, |W^-|^{r_2-r} [\pi_0(X(\mathbf{R}))]^{r'-r} \times \frac{\prod_{v'/N} [\pi_0(X \bmod v')]}{\prod_{v/N} [\pi_0(X \bmod v)]} \frac{[Ш']}{[Ш]}. \qquad (35)^*$$

The notation here is similar to that used in $(34)^*$, except that $[Ш]$ denotes the order of the Šafarevič-Tate group of the curve $X \otimes K$ and $[Ш']$ denotes the same for the curve $X \otimes K'$, and so on. In the case when $\operatorname{rk} X(K') > \operatorname{rk} X(K)$, we must have zero on the right in $(35)^*$, while if the index $[X(K') : X(K)]$ is finite, then the right side is multiplied by a rational number, which we do not write out explicitly here.

Comparing (30) and (35)* again shows a good structural agreement of the formulas and allows us to derive a hypothetical formula for the ratio of the orders of the Šafarevič-Tate groups under the above conditions:

$$\frac{[Ш']}{[Ш]}=\frac{\prod\limits_{v/N}[\pi_0(X \bmod v)]}{[\pi_0(X(R))]^{r'-r}\prod\limits_{v'/N}[\pi_0(X \bmod v')]}\cdot\prod_{\chi\in M}\frac{1}{2}\Big(\sum_{b \bmod m_\chi}\chi(b)x^{\operatorname{sign}\chi}\Big(\frac{b}{m_\chi}\Big)\Big). \qquad (36)^*$$

The left side must be the square of a rational number by Cassels' theorem. We can independently prove the following assertion about the right side being a square.

6. 3. **Proposition.** *Let* $\overline{M}=\{\chi\in M\mid \chi$ *is not real*$\}$. *Then*

$$\prod_{\chi\in\overline{M}}\frac{1}{2}\Big(\sum_{b \bmod m_\chi}\chi(b)x^{\operatorname{sign}\chi}\Big(\frac{b}{m_\chi}\Big)\Big)=\Delta S^2, \qquad (37)$$

where Δ *and* S *are integers and* Δ *consists only of ramified primes in the value field of the characters* $\chi\in\overline{M}$.

Proof. According to 5. 2 c), the standard involution $z\rightsquigarrow -1/Nz$ also induces an involution on the curve X. Let this involution act on the homology of X by multiplication by $-C=+1$ (the sign is chosen to agree with Weil's notation in [16]).

We now compute its action directly. It acts on the group $\Gamma_0(N)$ by matrix conjugation $\begin{pmatrix}0&1\\-N&0\end{pmatrix}$ and hence takes $\begin{pmatrix}a&b\\Nc&d\end{pmatrix}$ to $\begin{pmatrix}a&c\\Nb&d\end{pmatrix}^{-1}$. Thus the class

$$\Big\{-\frac{b}{a},0\Big\}=\Big\{-\frac{b}{a},\begin{pmatrix}a&b\\Nc&d\end{pmatrix}\Big(-\frac{b}{a}\Big)\Big\}$$

goes to the class

$$-\Big\{-\frac{c}{a},0\Big\}=-\Big\{-\frac{c}{a},\begin{pmatrix}a&c\\Nb&d\end{pmatrix}\Big(-\frac{c}{a}\Big)\Big\}$$

(we are using Proposition 1. 4). But $c\equiv -N^{-1}b^{-1} \bmod a$, because $ad-Nbc=1$.

Turning now to Definition 5. 4, we find from this a functional equation for the functions $x^{\pm}$:

$$x^{\pm}\Big(\frac{b}{a}\Big)=Cx^{\pm}\Big(\frac{-N^{-1}b^{-1} \bmod a}{a}\Big),$$

so that

$$\begin{aligned}\sum_b{}'\chi(b)x^{\pm}\Big(\frac{b}{m}\Big)&=\sum_b{}'\chi(-N^{-1}b^{-1} \bmod m)x^{\pm}\Big(\frac{-N^{-1}b^{-1} \bmod m}{m}\Big)\\&=C\overline{\chi}(-N)\sum_b{}'\overline{\chi}(b)x^{\pm}\Big(\frac{b}{m}\Big)\end{aligned} \qquad (38)$$

(here $m = m_\chi$, b runs through the residue classes mod m, and $x^{\pm}$ is chosen corresponding to sign χ). Thus, combining the sums in (37) corresponding to complex conjugation characters and bringing all factors outside, we obtain

$$\prod_{\chi \in \overline{M}} \frac{1}{2}\left(\sum_{b \bmod m_\chi} \chi(b) x^{\pm}\left(\frac{b}{m_\chi}\right)\right) = \varepsilon T^2, \tag{39}$$

where ϵ is a root of unity and T is an integer in the value field of the characters $\chi \in \overline{M}$ (T is integral because $(m_\chi, N) = 1$ by assumption, because $x^{\pm}(b/m) \in \mathbf{Z}$, and finally because the functions $b \rightsquigarrow \chi(b) x^{\operatorname{sign} \chi}(b/m_\chi)$ are even functions, so that the sums $\Sigma_{b \bmod m_\chi}$ contain each term twice).

On the other hand, $\Pi_{\chi \in \overline{M}}$ is an ordinary integer, because all of the conjugates over $\mathbf{Q}$ appear with every sum under the product sign. Formula (39) shows that adjoining a square root of this number (divided by ϵ) to $\mathbf{Q}$ does not take us outside the value field of the characters χ. Consequently, only primes which ramify in this value field appear with odd exponent in the left side of (32). (This argument was mentioned to me by A. N. Andrianov.)

The proposition is proved.

Remark. Of course, formula (38) is also applicable to real characters χ; it is trivial in the case $C\chi(-N) = 1$, and in the case $C\chi(-N) = -1$ it shows that $\Sigma_{b \bmod m} \chi(b) x^{\pm}(b/m) = 0$. This argument was used earlier in a somewhat different form to actually construct forms of the curve X over quadratic extensions (corresponding to real χ) whose L-series vanishes at $s = 1$ (see, for example, the appendix to Birch [2]).

6.4. We now compare the behavior of the right sides of (30) and $(35)^*$ over cyclotomic Γ-extensions K of the field $\mathbf{Q}$. In this case we have Mazur's results [10], [11], [9] concerning the behavior of the groups $X(K)$ and Ш$(X \otimes K)$ obtained using Iwasawa's theory of Γ-modules. Our formulas agree very well with the conditional interpretation of Mazur's theory in the language of L-functions, and they also allow us to make some predictions in the cases when the Γ-module technique has so far been insufficient.

We introduce the following notation. Let l be an odd prime, $l \nmid 2N$ (the case $l = 2$ differs in inessential details, and the case $l \mid N$ requires separate consideration, which we shall not go into here). Let $G = \{\epsilon \in \mathbf{Z}_l \mid \epsilon^{l-1} = 1\}$. The group G acts on the field $\mathbf{Q}(\zeta_n): \zeta_n = e^{2\pi i/l^n}$, $\zeta_n \rightsquigarrow \zeta_n^g$, $g \in G$. We set $K_n = \mathbf{Q}(\zeta_{n+1})^G$, $K_\infty \bigcup_{n=1}^{\infty} K_n$. Obviously $\mathbf{Q} = K_0 \subset K_1 \subset \cdots \subset K_\infty$. The Galois group $\Gamma = \operatorname{Gal} K_\infty/\mathbf{Q}$ is canonically isomorphic to $(1 + l\mathbf{Z}_l)^* \overset{\log}{\cong} \mathbf{Z}_l$. We set $\Gamma_n = \Gamma^{l^n}$; then $\operatorname{Gal}(K_n/\mathbf{Q}) \cong \Gamma/\Gamma_n \cong \mathbf{Z}_l/l^n$.

The extension $K_{n-1}/\mathbf{Q}$ corresponds to the primitive Dirichlet characters modulo $1, \ldots, l^n$ whose values are roots of 1 of order l^k. All of these characters are even. Let M_{n-1} be the subset of them which are associated with the field K_{n-1} but not with the field K_{n-2}.

We set

$$\Lambda_{n-1} = \prod_{\chi \in M_{n-1}} \frac{1}{2}\left(\sum_{b \bmod l^n} \chi(b)\, x^+\left(\frac{b}{l^n}\right)\right). \tag{40}$$

According to (30), Λ_{n-1} is the nontrivial additional factor in the value of L at one which appears in going from K_{n-2} to K_{n-1}. If $X(K_{n-2}) = X(K_{n-1})$, then, by (36)*, this same number must also be the nontrivial factor in the expression for the order of $Ш_{n-1}$. As has already been noted, $\Lambda_n \in \mathbf{Z}$.

We further set $\lambda = L(X, 1)/W^+$; this is a rational number (possibly zero). We recall that (a_n) are the coefficients of the canonical series $L(X, s)$.

6.5. **Theorem.** a) $\Lambda_1 \equiv 0 \bmod l$ *if and only if either* $a_l \equiv 1 \bmod l$, *or* l *divides the numerator of* λ *(all primes divide zero).*

b) *For* $n \geq 2$ *we have* $\Lambda_n \equiv 0 \bmod l$ *if and only if either* $a_l \equiv 1 \bmod l$, *or* $a_l \equiv 0 \bmod l$, *or* l *divides the numerator of* λ.

In particular, for all other l *we have* $\Lambda_n \neq 0$ *for all* n.

Proof. Let l_n be a prime divisor of l in $Q(\zeta_n)$. If $m_\chi = l^n$, then $\chi(b)^{l^{n-1}} = 1$ for all b, $(b, l) = 1$, so that $\chi(b) \equiv 1 \bmod l_{n-1}$. Consequently

$$\Lambda_{n-1} \equiv \prod_{\chi \bmod l} \frac{1}{2}\left(\sum_{\substack{b \bmod l^n \\ (b,l)=1}} x^+\left(\frac{b}{l^n}\right)\right) \bmod l \equiv \left(\frac{1}{2}\sum_{(b,l)=1} x^+\left(\frac{b}{l^n}\right)\right)^{\varphi(l^{n-1})} \bmod l. \tag{41}$$

Hence we must clarify the behavior of $\frac{1}{2}\sum_{(b,l)=1} x^+(b/l^n) \bmod l$. To do this we set $m = l^n$, $n \geq 2$, in formula (29):

$$\frac{1}{2}\sum_{\substack{b \bmod l^i \\ i \leq n}} x^+\left(\frac{b}{l^i}\right) = \left(a_{l^n} - \sum_{i=0}^{n} l^i\right)\lambda. \tag*{$(42)_n$}$$

Subtracting formula $(42)_{n-1}$ from $(42)_n$, we find

$$\frac{1}{2}\sum_{b \bmod l^n} x^+\left(\frac{b}{l^n}\right) = (a_{l^n} - a_{l^{n-1}} - l^n)\lambda. \tag*{$(43)_n$}$$

In the sum on the left the obstacle is the residue classes $b \equiv 0 \bmod l$; in order to remove them, we again subtract $(43)_{n-1}$ from $(43)_n$:

$$\frac{1}{2}\sum_{\substack{b \bmod l^n \\ (b,l)=1}} x^+\left(\frac{b}{l^n}\right) = (a_{l^n} - 2a_{l^{n-1}} + a_{l^{n-2}} - l^n + l^{n-1})\lambda. \tag{44}$$

We now consider the cases $n = 2$ and $n > 2$ separately.

The case $n = 2$. Since $a_{l^2} = a_l^2 - l$, we find from (44) that

$$\frac{1}{2}\sum_{(b,l)=1} x^+\left(\frac{b}{l^2}\right) = (a_l - 1 + l)(a_l - 1 - l)\lambda.$$

It is immediately clear from this that if $l\mid\lambda$ or $l\mid(a_l-1)$, then $l\mid\Lambda_1$. Conversely, if $l\mid\Lambda_1$ but $l\nmid(a_l-1)$, then l must divide the numerator of λ. This proves assertion a).

The case $n>2$. We have

$$a_{l^n}=a_la_{l^{n-1}}-la_{l^{n-2}}\equiv a_la_{l^{n-1}}\bmod l,$$

so that $a_{l^n}\equiv a_l^n\bmod l$. Hence the coefficient of λ in (44) is congruent to $a_l^{n-2}(a_l-1)^2\bmod l$. It is hence clear that if $a_l(a_l-1)\equiv 0\bmod l$ and l does not appear in the denominator of λ, or if $l\mid\lambda$, then we have $l\mid\Lambda_{n-1}$. We now suppose that $a_l(a_l-1)\equiv 0\bmod l$, but that l appears in the denominator of λ. The number $(a_l-1)\lambda$ is an integer by $(43)_1$; $a_l-l-1\neq 0$, because this is the number of points on the reduction of X mod l; finally, a_1-1-l is divisible by no higher than the first power of l for $l>2$ by the Weil estimate $|a_l|<2\sqrt{l}$. Consequently the denominator of λ is not divisible by l^2 and $a_l\equiv 1\bmod l$. On the other hand, if we twice use the formula $a_{l^n}=a_la_{l^{n-1}}-la_{l^{n-2}}$, we easily obtain

$$a_{l^n}-2a_{l^{n-1}}+a_{l^{n-2}}=(a_l-1)^2a_{l^{n-2}}-l(a_la_{l^{n-3}}+a_{l^{n-2}}-2a_{l^{n-3}}).$$

Hence the coefficient of λ in (43) is divisible by l^2, so that the left side of (43) and Λ_{n-1} are divisible by l.

Conversely, if l divides Λ_{n-1} but does not divide the numerator of λ, then it is clear from (44) that l divides

$$a_{l^n}-2a_{l^{n-1}}+a_{l^{n-2}}\equiv a_l^{n-1}(a_l-1)\bmod l.$$

The theorem is proved.

We derive several conditional corollaries from the theorem, using the Birch–Swinnerton-Dyer conjecture, and we compare them with Mazur's unconditional results.

We recall that $l\nmid 2N$ and that primes l for which $a_l\equiv 0\bmod l$, are called *supersingular for* X (Deuring), while those for which $a_l\equiv 1\bmod l$, are called *anomalous* for X (Mazur).

6.6. Corollary*. *If* $[X(\mathbf{Q})]<\infty$, *and if* l *does not divide the numerator of* $\lambda=L(X,1)/W^+$ *and is neither anomalous nor supersingular for* X, *then the group* $X(K_\infty)$ *is finite, and the* l-*component of the groups* Ш$(X\otimes K_n)$ *has bounded order as* $n\to\infty$.

In fact, according to Theorem 6.5, under the conditions of the corollary we have $L(X\otimes K_n,1)\neq 0$ for all n, so that $\mathrm{rk}\,X(K_n)=0$. On the other hand, as Mazur showed, the group $\tilde{X}(K_\infty)$ is finite. Finally, the ratio of local factors in $(36)^*$ becomes 1 as $n\to\infty$ to within a 2-component, and by Theorem 6.5 the new factors in $\Pi_\chi(\tfrac{1}{2}\Sigma)$ are not divisible by l.

This corollary is conditional, but this very result is proved precisely in Mazur's theory under the assumption that l does not divide

$$[\text{Ш}(X)]\prod_{p/N}[\pi_0(X\bmod p)]$$

instead of the numerator of λ: this agrees well with $(34)^*$.

6.7. **Corollary***. *If* $[X(\mathbf{Q})] < \infty$ *and* l *either divides the numerator of* λ *or is a supersingular or anomalous prime for* X, *then*

$$\mathrm{rk}\, X(K_n) + [Ш(X \otimes K_n)^{(l)}] \to \infty \quad as \quad n \to \infty.$$

In fact, if the quotients $L(X \otimes K_n, s)/L(X \otimes K_{n-1}, s)|_{s=1}$ equal zero for infinitely many values of n, then $\mathrm{rk}\, X(K_n) \to \infty$; otherwise

$$\frac{[Ш(X \otimes K_n)]}{[Ш(X \otimes K_{n-1})]} \equiv 0 \mod l$$

for all $n \geq n_0$ by Theorem 6.5 and formula (36)*, if we take into account the stabilization of the local factors.

The parallel unconditional result in Mazur's theory was only proved for anomalous primes, and asserts the following: *if* $a_l \equiv 1 \bmod l$ *then either* $\mathrm{rk}\, X(K_\infty) > 0$ *(and this rank is necessarily finite)*, or $[Ш(X \otimes K_n)^{(l)}] \to \infty$, *or else both hold together.*

Thus, in this place Mazur's theory partially overlaps Corollary 6.7*, but partially complements it: combining both results, we find for $a_l \equiv 1 \bmod l$ we *must** have $[Ш(X \otimes K_n)^{(l)}] \to \infty$.

The supersingular primes have so far resisted the Γ-module technique; hence it might be interesting to note a partial result relating to them:

6.8. **Corollary***. *If* $[X(\mathbf{Q})] < \infty$, *and if* l *is supersingular and does not divide the numerator of* λ, *then the group* $X(K_1)$ *is still finite.*

In fact, Theorem 6.5 b) shows that $L(X \otimes K_1, 1) \neq 0$. For anomalous numbers l with the condition $L(X, 1) \neq 0$ there are no apparent reasons why $L(X \otimes K_1, 1)$ cannot vanish, but the author does not know any examples where it does vanish.

6.9. **Corollary***. *If* $\mathrm{rk}\, X(\mathbf{Q}) > 0$, *then for all* $l \nmid 2N$

$$\mathrm{rk}\, X(K_n) + [Ш(X \otimes K_n)^{(l)}] \to \infty.$$

The reasoning is the same as in the proof of Corollary 6.7, since in this case we must have $\lambda = 0$, so that $\Lambda_n \equiv 0 \bmod l$ for all n and l.

Mazur [10] conjectured that the rank of $X(K_n)$ remains bounded (at least for non-supersingular l).

The parallel conjecture under our conditions is the following:

6.10. **Conjecture.** $\Lambda_n \neq 0$ *for all* $n \geq n_0(X, l)$.

I am unable to prove this result in any case except those which are included in Theorem 6.5. Possibly investigating Λ_n p-adically for $p \mid N$ could give useful information.

In certain special circumstances we can prove that the numbers Λ_n are divisible by certain special primes. In order to formulate the result precisely, we introduce the following

6.11. **Definition.** An isogeny $\chi : X \to Y$ of elliptic curves over $\mathbf{Q}$ is called *admissible* if, for any prime p, it induces a separable morphism of the connected components of the closed fibers of the Néron models of the curves X and Y over p.

The following two properties of admissible isogenies can be proved without difficulty:

a) *Let $\chi : X \to Y$ be an admissible isogeny, and let ω_Y be a Néron differential. Then $\chi^*(\omega_Y)$ is a Néron differential on X.*

In fact, $\chi^*(\omega_Y)$ is regular on the Néron model of X, and a divisor of zeros can only include the components of the fibers where the isogeny χ is inseparable.

b) *If an admissible isogeny $\chi : X \to Y$ exists, then the conductors and the canonical L-series of the curves X and Y coincide.*

We need only verify that the divisors of the conductor and the Euler factors of the L-series coincide at the points of degenerate reduction, and this is proved directly from the definitions.

Let $\chi : X \to Y$ be an isogeny, and let $\gamma_X^{\pm}$ and $\gamma_Y^{\pm}$ be the generators of the real (+) and imaginary (−) homology classes of the curves X and Y, respectively. We shall say that χ has type (q^+, q^-) if $\chi_*(\gamma_X^{\pm}) = q^{\pm}\gamma_Y^{\pm}$.

6.12. **Proposition.** *Let $\chi : X \to Y$ be an admissible isogeny of curves of type (q^+, q^-), and let N be their common conductor.*

a) *If $\psi : X_N \to X$ is a Weil uniformization of the curve X, then its composition with χ is a Weil uniformization of the curve Y.*

b) *Let $x_Y^{\pm}$ be the functions associated to the curve Y according to Definition 5.4. If the denominator of the number $\alpha \in \mathbf{Q}$ is relatively prime to N, then*

$$x_Y^+(\alpha) \equiv 0 \bmod q^+, \quad x_Y^-(\alpha) \equiv 0 \bmod q^-. \tag{45}$$

Proof. Assertion a) is obtained from properties 6.11 a), b) and the definition of uniformization in 5.2: we need only choose the Néron differentials on X and Y compatibly. Assertion b) follows from Definition 5.4 and the definition of type (q^+, q^-).

6.13. **Corollary.** *Let $K' \supset K$ be abelian extensions of the field $\mathbf{Q}$ with discriminant relatively prime to N, and let M, r, r_i, etc. be defined as they were before Theorem 5.6. Then*

$$\prod_{\chi\in M} \frac{1}{2}\left(\sum_{b \bmod m_\chi} \chi(b)\, x_Y^{\operatorname{sign}\chi}\left(\frac{b}{m_\chi}\right)\right) \equiv 0 \bmod (q^+)^{r'-r}(q^-)^{r_2'-r_1}. \tag{46}$$

Using formula $(36)^*$, we can derive from this conditional corollaries concerning the behavior of $Ш(X \otimes K')$ and $X(K')$, if we only ensure no cancellation of q^+ and q^- with the local factors in the right side of $(36)^*$. For example, this result is obtained for the curve Y by precisely the same reasoning as in Corollary 6.7.

6.14. **Corollary*.** *Let $\mathbf{Q} \subset K_1 \subset \cdots \subset K_n \subset \cdots$ be a Γ-extension corresponding to the prime $l \nmid 2N$. Then either the rank of $Y(K_n)$ increases without bound, or else the order of $[Ш(Y \otimes K_n)]$ is divisible by $(q^+)^{[K_n : \mathbf{Q}] - \text{const}}$ as $n \to \infty$.*

(The constant in the exponent of q^+ appears because the group $X(K_n)$ can grow in the first few steps of the Γ-extension, and also the contribution of the local factors in $(36)^*$ does not manage to stablize to 1.)

This result can be compared with Proposition 9.1 in Mazur's article [10].

6.15. **Examples and remarks.** a) The existence of admissible isogenies is a rather exceptional phenomenon. (The multiplication $X \xrightarrow{n} X$ is not admissible for $n > 1$!) If the kernel of $\chi : X \to Y$ is cyclic and is generated by a rational point x of order q, then for $(q, N) = 1$ admissibility follows from Lutz's theorem that the "coordinates of x are integers," i.e. the reduction of the kernel of χ does not become trivial. However, the case of common divisors of q and N requires special investigation.

Here is the data on the existence of admissible isogenies of curves X_N of genus one; $q^- = 1$, so that we only give q^+:

N	11	14	15	17	19	20	21	27	49
q^+	5	3	4	4	3	2	4	3	2

Corollary 6.14^* for the curve X_{11} was proved by Mazur [10] for $l = 5$; in this case the rank of $Y(K_n)$ equals zero for all n.

b) Assertion (45) relates to the behavior of the functions $x_Y^\pm$ constructed for the *image* Y of an admissible isogeny. However, observing lengthy tables of the functions $x^\pm$ for the curve X_{11} compelled us also to suggest some regularity in the behavior mod $q^\pm$ of the functions $x_X^\pm$ constructed for the *domain* X of the admissible isogeny. More precisely, the following assertion is fulfilled in the tables ($X = X_{11}$, $q^+ = 5$):

$$\textit{the residue class } x_{11}^+\left(\frac{b}{a}\right) \bmod 5 \textit{ depends only on } a \tag{47}$$

(*for all* $a \not\equiv 0 \bmod 11$ *and* $(b, a) = 1$).

The analogous property is observed for $N = 17, 19, 27$; see §8.

Although this assertion seems to have the same nature as property (45), I have not been able to prove it. The assertion is rather striking, since in the computation of $x_{11}^+(b/a)$ the denominators of the convergents to b/a are operated on modulo 11, and not modulo 5.

A natural generalization of the conjecture (47) is the conjecture that *the residue classes* $x_X^\pm(a/b)$ mod $q^\pm$ *are constant with respect to* b.[1] We note that the congruence (46) would also follow from this assertion, which is weaker than (45), because $\Sigma_b \chi(b) = 0$ for any nonprincipal character χ.

Evidence for (47) is noted in the commentary on the tables in §8.

§7. Noncommutative reciprocity law

As the basic result of this section, in 7.3 we fomulate a special case of Theorem

(1) *Added in proof.* Drinfel'd has proved an assertion of this type with another interpretation of the numbers $q^\pm$. Swinnerton-Dyer has obtained an analogous fact.

7.9 on the coefficients of parabolic forms, which can be derived from formula (20). The features of a noncommutative reciprocity law emerge in this special case.

We begin by formulating the necessary concepts.

7.1. **Admissible solutions.** Let $d > 1$ be an integer. A solution of the equation $d = \Delta\Delta' + \delta\delta'$ is an ordered quadruple of numbers $(\Delta, \Delta', \delta, \delta')$ satisfying this equation. A solution is called *admissible* if it consists of integers satisfying the addition conditions

$$(\Delta, \delta) = (\Delta', \delta') = 1, \qquad \Delta > \delta > 0, \tag{48}$$

and also

$$\text{either } \Delta' > \delta' > 0, \tag{49}$$

$$\text{or else } \delta' = 0,\ \Delta = d,\ \Delta' = 1,\ \ 0 \leqslant \delta < \frac{d}{2}. \tag{50}$$

We call the solutions (50) *boundary* solutions.

The *set* of admissible solutions of the equation $d = \Delta\Delta' + \delta\delta'$ determines a finite *family* of pairs (Δ, δ) which appear in these solutions. We shall later need to sum functions of pairs of integers over the terms of this family. Hence for practical purposes it suffices to think of this family of pairs (Δ, δ) as the set of different pairs, each equipped with a multiplicity.

We shall also call such pairs (Δ, δ) d-admissible.

7.2. Let X be an elliptic curve over $\mathbf{Q}$ with conductor N which has a Weil uniformization in the strong sense. Let $L(X, s)$ be its canonical Dirichlet series, and let a_n be the nth coefficient.

If Δ is an integer, we set $\widetilde{\Delta} = \Delta \bmod N$. If $(\Delta, \delta) = 1$, then $\widetilde{\Delta} : \widetilde{\delta}$ denotes a point of the projective line $\mathbf{P}^1(\mathbf{Z}/(N))$, as in 2.3.

With this notation we have the following

7.3. **Fundamental Theorem.** *Suppose that $L(X, 1) \neq 0$. Then there exists a function $y : \mathbf{P}^1(\mathbf{Z}/(N)) \to \mathbf{Q}$ depending only on X such that the following holds for any prime $l \nmid 2N$:*

$$1 - a_l + l = \sum_{l = \Delta\Delta' + \delta\delta'} y(\widetilde{\Delta} : \widetilde{\delta}), \tag{51}$$

where the summation on the right is over the family of all l-admissible pairs (Δ, δ).

Remarks. a) The function y can be expressed explicitly in terms of the function x^+ for the curve X (see formula (74) in 7.10).

b) The left side of (51) is the number of $\mathbf{Z}/(l)$-points on the reduction of X mod l, and the right side is some sum over the solutions of the equation $l = \Delta\Delta' + \delta\delta'$ taken mod N. The general form of this symmetry:

(an equation depending on N, taken mod l)

$\updownarrow$

(an equation depending on l, taken mod N)

brings to mind a reciprocity law. It relates explicitly to noncommutative extensions, since a_l in (51) is the trace of the Frobenius automorphism of the fields obtained by adjoining to $\mathbb{Q}$ the points of finite order on the curve X (see Shimura [14]).

Another point of view regarding equation (51) is that it gives information on the representations of l by the indefinite quadratic form $\Delta\Delta' + \delta\delta'$. Eichler [5] gave a general technique for obtaining such formulas for representations by positive forms (using theta-functions). It seems that our result has another nature.

7.4. The plan of proof for Theorem 7.3 and its generalization is as follows. Formula (20) gives an expression for $1 - a_l + l$ in terms of integrals over the homology classes $\{0, b/l\}_N, 0 \leq b \leq l-1$. Formula (16) allows us to represent each class $\{0, b/l\}_N$ as a sum of distinguished classes $\xi(\tilde{c} : \tilde{d})$ whose arguments are (up to sign) the ratios of the denominators mod N of the successive convergents of b/l. Finally, a lemma of Heilbronn [6] allows us to go from continued fractions to solutions of the equation.

We begin the proof by giving Heilbronn's lemma.

7.5. **Formal continued fractions.** Following Heilbronn [6], we introduce the polynomials $Q_i \in \mathbb{Z}[T_1, \cdots, T_n, \cdots], i \geq -1$, by the inductive formulas

$$Q_{-1} = 0, \qquad Q_0 = 1, \qquad Q_n = T_n Q_{n-1} + Q_{n-2} \text{ for } n \geq 1.$$

Obviously, $Q_n \in \mathbb{Z}[T_1, \cdots, T_n]$, so that we may write Q_n (and its particular values) as a polynomial in n arguments for $n \geq 1$. We shall also apply this same notation for $n = 0, -1$, but then we do not pay attention to any arguments.

It is easy to verify that

$$Q_n(T_1, \ldots, T_n) = Q_n(T_n, \ldots, T_1). \tag{52}$$

The following formula gives the connection with continued fractions:

$$\frac{Q_{n-1}(T_2, \ldots, T_n)}{Q_n(T_1, \ldots, T_n)} = \cfrac{1}{T_1 + \cfrac{1}{T_2 + \cdots \cfrac{1}{T_n}}}. \tag{53}$$

It remains valid for $n = 0$ if we take the right side equal to zero in this case.

The successive convergents to (53) are defined by the formulas

$$\frac{Q_{n-1}(T_2, \ldots, T_n)}{Q_n(T_1, \ldots, T_n)}, \ldots, \frac{Q_{m-1}(T_2, \ldots, T_m)}{Q_m(T_1, \ldots, T_m)}, \ldots, \frac{Q_0}{Q_1}, \frac{Q_{-1}}{Q_0} = \frac{0}{1}. \tag{54}$$

The index of a convergent is the index of its denominator. We have

$$\begin{aligned} &Q_{m-1}(T_2, \ldots, T_m)Q_{m-1}(T_1, \ldots, T_{m-1}) \\ &- Q_m(T_1, \ldots, T_m)Q_{m-2}(T_2, \ldots, T_{m-1}) = (-1)^{m-1}. \end{aligned} \tag{55}$$

The connection with the equation $d = \Delta\Delta' + \delta\delta'$ will be established in 7.7 using the fundamental formula

$$Q_n(T_1, \ldots, T_n) = Q_m(T_1, \ldots, T_m)\, Q_{n-m}(T_{m+1}, \ldots, T_n) + Q_{m-1}(T_1, \ldots, T_{m-1})\, Q_{n-m-1}(T_{m+2}, \ldots, T_n), \tag{56}$$

which makes sense and remains valid for all $0 \le m \le n$. It is proved by induction, decreasing m from the obvious cases $m = n$ and $m = n - 1$.

7.6. **Expansion of rational numbers in continued fractions.** Let $0 < \alpha < ½$ be a rational number. It uniquely determines an integer $n = n(\alpha) \ge 1$ and positive integers $c_1, \cdots, c_n$ such that $c_1 \ge 2$, $c_n \ge 2$ and

$$\alpha = \frac{Q_{n-1}(c_2, \ldots, c_n)}{Q_n(c_1, \ldots, c_n)} = \cfrac{1}{c_1 + \cdots + \cfrac{1}{c_n}}. \tag{57}$$

The number n is called the length of the (continued fraction) expansion of α, and the numbers $c_1, \cdots, c_n$ are called the partial quotients of α. Substituting $c_1, \cdots, c_n$ for $T_1, \cdots, T_n$ in (54), we obtain the successive convergents of α, and also their numerators and denominators.

7.7. **Heilbronn's Lemma.** *Let $d > 2$ be an integer. The following two families of ordered pairs of integers coincide:*

a) *The pairs of neighboring denominators (from larger to smaller) in the sequence of convergents of all possible rational numbers of the form b/d, $(b, d) = 1$, $1 \le b < d/2$.*

b) *The pairs (Δ, δ) taken for all possible admissible solutions of the equation $d = \Delta\Delta' + \delta\delta'$.*

(Coincidence of families means coincidence of sets and multiplicities: see 7.1.)

Proof. The first family of pairs is indexed by the set consisting of elements of the form

$$\left[\text{the fraction } \alpha = \frac{b}{d}\left((b, d) = 1 \text{ and } 1 \leqslant b < \frac{d}{2}\right); \text{the integer } 1 \leqslant m \leqslant n(\alpha)\right]. \tag{58}$$

This element corresponds to the pair [mth denominator, $(m - 1)$th denominator of the convergents of α] in the family a).

The second family of pairs is indexed by the set

$$\text{admissible solutions of the equation } d = \Delta\Delta' + \delta\delta'. \tag{59}$$

We shall construct mutually inverse maps of the sets (58) $\leftrightarrow$ (59) which preserve the pairs in the families a) and b).

The map (58) $\to$ (59). Let $(c_1, \cdots, c_n)$ be the partial quotients for α, and let $1 \le m \le n = n(\alpha)$. We set

$$\Delta = Q_m(c_1, \ldots, c_m), \quad \delta = Q_{m-1}(c_1, \ldots, c_{m-1}),$$
$$\Delta' = Q_{n-m}(c_{m+1}, \ldots, c_n), \quad \delta' = Q_{n-m-1}(c_{m+1}, \ldots, c_n). \tag{60}$$

It is clear from (53), (56) and (57) that $d = \Delta\Delta' + \delta\delta'$. Since the c_i are positive, it follows from the recursion relations for Q_n that $\Delta > \delta > 0$ and $\Delta' > \delta' \geq 0$. From (55) we have $(\Delta, \delta) = (\Delta', \delta') = 1$. It remains to verify that the admissible boundary solutions are obtained for $\delta' = 0$. But if $\delta' = 0$, then $m = n$, since $\Delta = d$ and $\Delta' = 1$; finally, $1 \leq \delta \Delta d/2$, because $c_n \geq 2$ (apply the recursion relation $\Delta = c_m \delta + Q_{m-2}$).

The map (59) $\to$ (58). Let $(\Delta, \Delta', \delta, \delta')$ be an admissible solution. If it is a boundary solution, we set

$$m = n = \text{the length of the expansion of } \delta/\Delta = \delta/d;$$

we define the numbers $c_1, \cdots, c_n$ by the formula

$$\frac{\delta}{d} = \frac{1}{c_n + \cdots + \cfrac{1}{c_1}}, \quad c_n \geqslant 2, \quad c_1 \geqslant 2, \tag{61}$$

and the number α by the formula

$$\alpha = \frac{1}{c_1 + \cdots + \cfrac{1}{c_n}}. \tag{62}$$

The denominator of α equals d; this follows from (61), (62) and (52). The numerator of α does not exceed $d/2$; this follows because $\delta/d < \frac{1}{2} \Rightarrow c_n \geq 2$, if we use the recursion relations for Q_n together with (53).

The nonboundary solutions give the pairs (58) with $m < n(\alpha)$. If $(\Delta, \Delta', \delta, \delta')$ is not a boundary solution, we define $c_1, \cdots, c_n$ and m by the formulas

$$\frac{\delta}{\Delta} = \frac{1}{c_m + \cdots + \cfrac{1}{c_1}}, \quad c_1 \geqslant 2, \tag{63}$$

$$\frac{\delta'}{\Delta'} = \frac{1}{c_{m+1} + \cdots + \cfrac{1}{c_n}}. \quad c_n \geqslant 2, \tag{64}$$

and we set

$$\sigma = \frac{1}{c_1 + \cdots + \cfrac{1}{c_n}}.$$

The denominator of α obviously equals d; this follows from (63), (64) and (56). In addition, $0 < \alpha < \frac{1}{2}$, since $c_1 \geq 2$.

It is automatically verified that these set maps are mutually inverse and preserve the pairs which interest us. The lemma is proved.

Remark. It is clear from the proof that if the pair (Δ, δ) corresponds to the pair [mth denominator, $(m-1)$th denominator], then

$$m = \text{the length of the expansion of } \delta/\Delta = n(\delta/\Delta) \tag{65}$$

(see (63)).

7.8. We now proceed to formulate a theorem which contains Theorem 7.3 as a special case. Let $\Phi(z) = \sum_{n=1}^{\infty} a_n e^{2\pi i n z}$ be a $\Gamma_0(N)$-parabolic form which is an eigenfunction relative to all Hecke operators T_m with $(m, N) = 1$. We suppose that $a_1 = 1$; then $\Phi | T_m = a_m \Phi$ for all m, $(m, N) = 1$.

Further, for any point $\tilde{c} : \tilde{d} \in \mathbf{P}^1/(N))$ set

$$\eta(\tilde{c}:\tilde{d}) = \xi(\tilde{c}:\tilde{d}) - \xi(\tilde{d}:\tilde{c}), \tag{66}$$

where the classes $\xi(\tilde{c} : \tilde{d}) \in H_1(X_N(\mathbf{C}), \mathbf{R})$ are defined in 2.5. Finally, if

$$\int_0^{i\infty} \Phi(z)\, dz = \int_{\{0, i\infty\}} \omega \neq 0,$$

where ω is the differential of the first kind on $X_N(\mathbf{C})$ corresponding to Φ, then we define the function $y : \mathbf{P}^1(\mathbf{Z}/(N)) \to \mathbf{C}$ by the formula

$$y(\tilde{c}:\tilde{d}) = \int_{\eta(\tilde{c}:\tilde{d})} \omega \Big/ \int_{\{0, i\infty\}} \omega. \tag{67}$$

We note that $\eta(-\tilde{c} : \tilde{d}) = \eta(\tilde{c} : \tilde{d})$, so that y is an even function.

7.9. **Theorem.** *With the notation and assumptions of the last subsection, for any m, $(m, 2N) = 1$, we have*

$$\sum_{d/m} d - a_m = \sum_{\substack{d/m \\ d>1}} \tau\left(\frac{m}{d}\right) \sum_{d = \Delta\Delta' + \delta\delta'} y(\tilde{\Delta} : \tilde{\delta}), \tag{68}$$

where $\tau(m)$ is the number of divisors of m, and the (Δ, δ) in the inner sum runs through all d-admissible pairs.

Proof. According to formula (20) of Theorem 3.3,

$$\left(\sum_{d/m} d - a_m\right) \int_{\{0, i\infty\}} \omega = \sum_{d/m} \sum_{b \bmod d} \int_{\left\{0, \frac{b}{d}\right\}} \omega.$$

Each irreducible fraction b/d on the right obviously appears $\tau(m/d)$ times in the form $b\delta/d\delta$ for all possible $\delta/(m/d)$. The contribution from the sum with $d = 1$ equals zero, since $\{0, b\} = 0$ for $b \in \mathbf{Z}$. Consequently

$$\left(\sum_{d/m} d - a_m\right) \int_{\{0, i\infty\}} \omega = \sum_{\substack{d/m \\ d>1}} \tau\left(\frac{m}{d}\right) \sum_{\substack{b \bmod d \\ (b,d)=1}} \int_{\left\{0, \frac{b}{d}\right\}} \omega. \tag{69}$$

The inner sum on the right is the integral of ω over the class $\sum_{(b,d)=1} \{0, b/d\}$. Using the fact that

$$\left\{0, \frac{d-b}{d}\right\}_N = \left\{0, -\frac{b}{d}\right\}_N,$$

we represent this class in the form

$$\sum_{\substack{1\leq b<\frac{d}{2}\\(b,d)=1}}\left(\left\{0,\frac{b}{d}\right\}_N+\left\{0,-\frac{b}{d}\right\}_N\right). \tag{70}$$

Let $d = d_n, \cdots, d_0$ be the successive denominators of the convergents of b/d. According to formula (16) of Theorem 2.7,

$$\left\{0,\frac{b}{d}\right\}=\sum_{k=1}^{n}\xi((-1)^{k-1}\widetilde{d}_k:\widetilde{d}_{k-1}). \tag{71}$$

Since the class $\{0, b/d\}$ is complex conjugate to $\{0, -b/d\}$, it follows from (12) that

$$\left\{0,-\frac{b}{d}\right\}=-\sum_{k=1}^{n}\xi((-1)^{k-1}\widetilde{d}_{k-1}:\widetilde{d}_k). \tag{72}$$

Combining (70), (71), (72) and (66), we find, after summing over b, $1 \leq b < d/2$, $(b, d) = 1$, and applying Heilbronn's lemma, that

$$\sum_{\substack{(b,d)=1\\1\leq b\leq d}}\left\{c,\frac{b}{d}\right\}=\sum_{d=\Delta\Delta'+\delta\delta'}\eta(\widetilde{\Delta}:\widetilde{\delta}), \tag{73}$$

where the sum on the right is taken over admissible solutions. The sign $(-1)^{k-1}$ in (72) disappears because η is even.

It is now clear that, combining (69) and (73), we obtain formula (68): we need only divide through both sides of (69) by $\int_{\{0,i\infty\}}\omega$ and recall the definition (67) of the function y. The theorem is proved.

7.10. **Remarks.** a) Under the conditions of Theorem 7.3, it is not hard to express the function y in terms of x^+. Namely, if $ad - bc = 1$, then

$$y(\widetilde{c}:\widetilde{d})=2\,\frac{x^+\left(\frac{a}{c}\right)-x^+\left(\frac{b}{d}\right)}{x^+(i\infty)}. \tag{74}$$

In fact, then

$$\xi(\widetilde{c}:\widetilde{d})=\left\{\frac{b}{d},\frac{a}{c}\right\}=\left\{\frac{b}{d},0\right\}-\left\{\frac{a}{c},0\right\}.$$

Using the definition of x^+, we find

$$\eta(\widetilde{c}:\widetilde{d})=\left(x^+\left(\frac{b}{d}\right)-x^+\left(\frac{a}{c}\right)\right)\gamma^+.$$

Since, in addition, $\{0, i\infty\} = -\frac{1}{2}x^+(i\infty)\,y^+$, it hence follows that (74) holds.

b) It is natural to write abelian reciprocity laws in the form $\prod_l S_l = 1$, where the S_l are certain symbols and l runs through the prime numbers and ∞. We can formally derive an analogous relation from (51). Let $L(X, s) = \prod_l L_l(X, s)$, where L_l are the local factors of the L-series. We suppose that $\prod_l L_l(X, 1) = L(X, 1)$ *in the sense of* some type of (nonabsolute) convergence. For $l \nmid N$ we have $L_l(X, 1) = (1 - a_l + l)/l$. Hence from (51) we find

$$\prod_{l/2N} L_l(X, 1) \prod_{l \nmid 2N} \left(\frac{1}{l} \sum_{l=\Delta\Delta'+\delta\delta'} y(\bar{\Delta}:\bar{\delta}) \right) L(X, 1)^{-1} = 1. \tag{75}$$

Here it is natural to associate the factor $L(X, 1)^{-1}$ to the point at infinity in the field $\mathbf{Q}$.

c) It is interesting to connect formula (51) with the Sato-Tate conjecture on the distribution of (a_l) as $l \to \infty$ (see, for example, Serre [13]). The right side of (51) could possibly be treated by an independent statistical investigation.

In fact, the sums $\sum_k \phi(d_k, d_{k-1})$, where (d_k) are the successive denominators of the convergents to α, have been studied before. Lévy's book [7] *contain facts on the* distribution *almost everywhere* of such sums, for irrational α as well (some natural conditions ensuring convergence are imposed on ϕ). We are interested in the mean of such sums over all *rational* α with fixed denominator l and in the distribution of this mean when $l \to \infty$. It was to solve such a problem that Heilbronn [6] proved Lemma 7.7: he was interested in the function $\phi \equiv 1$, and he obtained the principal term of its asymptotic behavior. In our case the principal term is known in advance: it is $l + 1$, and a_l is a "random error."

A natural approach to studying the sums (51) is to expand the function y in terms of *some elementary functions*. For example, *for N prime it would suffice to study the* distribution of the sums over admissible solutions of the form

$$\sum_{l=\Delta\Delta'+\delta\delta'} \chi(\Delta)\bar{\chi}(\delta),$$

where χ is any multiplicative character mod N.

d) We would like to note a similarity between the considerations of this section and the constructions in Chapters V and VI of Venkov's book [3]. Comparing these results may lead to a better understanding of them.

e) The condition $\int_0^{i\infty} \Phi(z)dz \neq 0$ is only used to go from continued fractions to the equation $l = \Delta\Delta' + \delta\delta'$. If we do not insist on this, Theorem 3.5 allows us to give explicit formulas for the coefficients of any parabolic forms.

§8. Tables, their computation and use

The basic content of this section is tables of the functions $x^{\pm}(\alpha)$ for $\alpha \in \mathbf{Q}$, constructed for curves X_N of genus 1 for $N = 11$, 17, 19 and 27. Before proceeding to a discussion of the method used to compute these tables and the possibilities for using them, we shall bring together in one place and recall all the notation needed here, which was introduced in various places in the article.

8. 1. Notation and definitions. The integer $N > 0$ is fixed, H is the complex upper halfplane, $\overline{H} = H \cup \mathbf{Q} \cup (i\infty)$ and $X_N(\mathbf{C}) = \Gamma_0(N)\backslash\overline{H}$. For any $\alpha, \beta \in \overline{H}$ the symbol $\{\alpha, \beta\}_N \in H_1(X_N(\mathbf{C}), \mathbf{R})$ designates the homology class of the path on $X_N(\mathbf{C})$ which is the image of a path on $\overline{H}$ from α to β. Further, $\mathbf{P}^1(\mathbf{Z}/(N)) = \{$classes of pairs $\widetilde{c} : \widetilde{d} \,|\, \widetilde{c} = c \bmod N,\ \widetilde{d} = d \bmod N,\ (c, d) = 1\}$. The function $\xi_N : \mathbf{P}^1(\mathbf{Z}/(N)) \to H_1(X_N(\mathbf{C}), \mathbf{R})$ is defined by the equation

$$\xi_N(\widetilde{c}:\widetilde{d}) = \left\{\frac{b}{d}, \frac{a}{c}\right\}_N \quad \text{for any} \quad ad - bc = 1.$$

Now let the genus of $X_N(\mathbf{C})$ equal 1. Then the subgroup of classes in $H_1(X_N(\mathbf{C}), \mathbf{Z})$ invariant (anti-invariant) relative to conjugation is infinite cyclic; let γ^+ (γ^-) be any generator of this group.

After choosing γ^+ and γ^-, the functions $x_N^{\pm} : \mathbf{Q} \cup (i\infty) \to \mathbf{Q}$ are defined by the equations

$$\{-\alpha, 0\}_N \pm \{\alpha, 0\}_N = x^{\pm}(\alpha)\gamma^{\pm}.$$

The fundamental functions to be tabulated, which we first introduce here, are $\xi_N^{\pm} : \mathbf{P}^1(\mathbf{Z}/(N)) \to \mathbf{Q}$, which are defined by the formulas

$$\xi_N(\widetilde{c}:\widetilde{d}) \mp \xi_N(\widetilde{d}:\widetilde{c}) = \xi_N^{\pm}(\widetilde{c}:\widetilde{d})\gamma^{\pm}. \tag{76}$$

If we have at our disposal a table of the functions $\xi_N^{\pm}$ (their domain of definition consists of $N\prod_{p|N}(1 + 1/p)$ points, and their range is the rational numbers with rather small numerators and denominators), it is not hard to compute an arbitrarily long table of the functions $x_N^{\pm}$ by using the formula

$$x_N^{\pm}\left(\frac{b}{a}\right) = \mp \sum_{k=1}^{n} \xi^{\pm}((-1)^{k-1}\widetilde{a}_k : \widetilde{a}_{k-1}), \tag{77}$$

where $a_n = a, a_{n-1}, \cdots, a_0 = 1$ are the denominators of the successive convergents of b/a (equation (77) is derived from the definitions and formula (16)). We further recall that the $x^{\pm}$ have period 1 and that x^+ is even and x^- is odd, so that we may limit ourselves to the arguments $0 < b/a \le \frac{1}{2}$.

If in addition $\xi_N^+(\widetilde{0} : \widetilde{1}) \neq 0$ (this holds for all N for which the genus of X_N equals one), then we can tabulate the function $y_N : \mathbf{P}^1(\mathbf{Z}/(N)) \to \mathbf{Q}$, which is defined by the formula

$$y_N(\widetilde{c}:\widetilde{d}) = 2\,\frac{\xi_N^+(\widetilde{c}:\widetilde{d})}{\xi_N^+(\widetilde{0}:\widetilde{1})} \tag{78}$$

(and so it is proportional to ξ_N^+. Nevertheless, it is instructive to tabulate it separately from ξ_N^+, because it is used for different purposes).

8. 2. Use of the tables. a) The fundamental function $\xi_N^{\pm}$ in the tables is necessary for computing $x_N^{\pm}$.

b) The functions $x_N^{\pm}$ are used to compute $L(X_N \otimes K, 1)$ over abelian extensions $K \subset \mathbf{Q}$, and also to compute the individual factors of these components corresponding to the different Dirichlet characters. Using the Birch–Swinnerton-Dyer conjecture, we can then make hypothetical estimates from below for the rank of the group $X(K)$ (which, of course, must be verified independently). Thus, for example, we can collect experimental data on questions which remain unsolved in Mazur's theory (the presence of jumps in the rank in a Γ-extension tower; the behavior of the rank for supersingular l, etc.).

c) The function y_N is used to compute the coefficients of $L(X, s)$ in any quantity using formula (51).

In addition, the tables can simply be looked over with the idea of trying to observe *anything curious.*

8.3. Computation of the tables. The compilation of the tables of the functions $\xi_N^{\pm}$ is in the first place based on Theorem 2.7 (and formula (12), which is necessary to choose γ^+ and γ^-). This theorem alone is sufficient to compute $\xi^{\pm}(\tilde{c} : \tilde{d})$ for all $\tilde{c}$ and $\tilde{d}$ which are not divisors of zero in $\mathbf{Z}/(N)$. In particular, if N is prime, then we obtain in this way all values of $\xi^{\pm}$, except for the values at the points $1 : 0$ and $0 : 1$ (here and later we shall omit the tilde over the numbers, since no ambiguity can arise if N is fixed). To compute the missing values of $\xi^{\pm}$ we must then use Theorems 3.3 and 3.5 (if N is prime Theorem 3.3 suffices) for any prime value of m, $m \nmid N$. For this purpose we must know in advance several coefficients of the canonical L-series.

The general plan is as follows.

a) To compile a list of the points $\mathbf{P}^1(\mathbf{Z}/(N))$.

b) To solve the system of equations (14) and (15), i.e. to find integral linear expressions for the symbols $(\tilde{c} : \tilde{d})$ in terms of independent parameters. The general number of parameters equals

$$2(\text{genus of } X_N) + (\text{number of parabolic points on } X_N) - 1.$$

The parameters must be chosen so that only 2(genus of X_N) parameters appear in the expression for $(\tilde{c} : \tilde{d})$ with $\tilde{c}$ and $\tilde{d}$ not divisors of zero, i.e. 2 parameters appear in the case genus $X_N = 1$, which is the case we shall work with.

For N not very large (less than a hundred), the system (14)–(15) can easily be solved by hand if we successively examine the 3-equations of (15) and the 2-equations of (14) which "link" these 3-equations. Each time, if we solve the next 3-equation in which at least one of the unknowns has already been found using the previous 2-equation, either we obtain a new free parameter or else we obtain a relation among the old parameters; if we use a reasonable procedure, the latter possibility rarely occurs.

c) To choose γ^+ and γ^- from among the linear combinations of the symbols $(\tilde{c} : \tilde{d})$ with $\tilde{c}$ and $\tilde{d}$ not divisors of zero, and to compute $\xi^{\pm}(\tilde{c} : \tilde{d})$ for these points $\tilde{c} : \tilde{d}$.

d) To compile a rather large table of the functions $x^{\pm}(b/a)$ for $(a, N) = (b, N) = 1$, using formula (16). The "extra" parameters which may appear in the separate terms of the sum

(16) because $(a_k, N) > 1$ automatically cancel out in the sum, so that the result is expressed in terms of y^+ and y^-.

e) To compute several coefficients of $L(X, s)$, which are necessary in order to apply Theorems 3. 3 and 3. 5.

f) Using Theorems 3. 3 and 3. 5, to compute $\xi^{\pm}(\tilde{c} : \tilde{d})$ for the missing values of $\tilde{c} : \tilde{d}$. Here formulas (20) and (22) must be considered as equations for the classes $\{0, i\infty\}_N$ and $\{a, i\infty\}_N$, respectively. The coefficients of these classes are computed using e), and the classes in the right side of (20) and (22) are computed as in d).

The curve X_{11}

Equation: $y^2 = t(t^3 - 20t^2 + 56t - 44)$.

$$\gamma^{\pm} = \left\{-\frac{1}{3}, 0\right\} \pm \left\{\frac{1}{3}, 0\right\}.$$

$\mathbf{P}^1(\mathbf{Z}/(11))$	1 : 0	0 : 1	1 : 1	2 : 1	3 : 1	4 : 1	5 : 1	6 : 1	7 : 1	8 : 1	9 : 1	10 : 1
y	−2	2	0	10	5	−5	−10	−10	−5	+5	10	0
ξ^+	$\frac{2}{5}$	$-\frac{2}{5}$	0	−2	−1	1	2	2	1	−1	−2	0
ξ^-	0	0	0	0	1	1	0	0	−1	−1	0	0

a	$\frac{1}{2}$	$\frac{1}{3}$	$\frac{1}{4}$	$\frac{1}{5}$	$\frac{2}{5}$	$\frac{1}{6}$	$\frac{1}{7}$	$\frac{2}{7}$	$\frac{3}{7}$	$\frac{1}{8}$	$\frac{3}{8}$	$\frac{1}{9}$	$\frac{2}{9}$	$\frac{4}{9}$	$\frac{1}{10}$	$\frac{3}{10}$	$\frac{1}{12}$	$\frac{5}{12}$
$x^+(\alpha)$	2	1	−1	−2	3	−2	−1	−1	4	1	1	2	−3	2	0	0	0	5
$x^-(\alpha)$	0	1	1	0	1	0	−1	1	0	−1	1	0	1	0	0	2	0	1

a	$\frac{1}{13}$	$\frac{2}{13}$	$\frac{3}{13}$	$\frac{4}{13}$	$\frac{5}{13}$	$\frac{6}{13}$	$\frac{1}{14}$	$\frac{3}{14}$	$\frac{5}{14}$	$\frac{1}{15}$	$\frac{2}{15}$	$\frac{4}{15}$	$\frac{7}{15}$	$\frac{1}{16}$	$\frac{3}{16}$	$\frac{5}{16}$	$\frac{7}{16}$	$\frac{1}{17}$
$x^+(\alpha)$	2	−3	−3	2	2	2	1	−4	1	−1	−1	−1	4	−2	−2	3	3	−2
$x^-(\alpha)$	0	−1	1	2	2	0	1	0	1	1	−1	1	0	0	0	1	−1	0

a	$\frac{2}{17}$	$\frac{3}{17}$	$\frac{4}{17}$	$\frac{5}{17}$	$\frac{6}{17}$	$\frac{7}{17}$	$\frac{8}{17}$	$\frac{1}{18}$	$\frac{5}{18}$	$\frac{7}{18}$	$\frac{1}{19}$	$\frac{2}{19}$	$\frac{3}{19}$	$\frac{4}{19}$	$\frac{5}{19}$	$\frac{6}{19}$	$\frac{7}{19}$	$\frac{8}{19}$
$x^+(\alpha)$	3	−2	−2	−2	3	3	3	−1	−1	4	1	1	−4	−4	1	1	1	6
$x^-(\alpha)$	−1	0	2	2	1	1	−1	−1	1	2	−1	1	0	0	1	1	1	0

a	$\frac{9}{19}$	$\frac{1}{20}$	$\frac{3}{20}$	$\frac{7}{20}$	$\frac{9}{20}$	$\frac{1}{21}$	$\frac{2}{21}$	$\frac{4}{21}$	$\frac{5}{21}$	$\frac{8}{21}$	$\frac{10}{21}$	$\frac{1}{23}$	$\frac{2}{23}$	$\frac{3}{23}$	$\frac{4}{23}$	$\frac{5}{23}$	$\frac{6}{23}$	$\frac{7}{23}$
$x^+(\alpha)$	1	2	−3	2	2	0	0	0	0	0	0	0	0	0	0	−5	0	0
$x^-(\alpha)$	−1	0	−1	2	0	0	0	0	2	2	0	0	0	−2	0	1	2	2

α	$\frac{8}{23}$	$\frac{9}{23}$	$\frac{10}{23}$	$\frac{11}{23}$	$\frac{1}{24}$	$\frac{5}{24}$	$\frac{7}{24}$	$\frac{11}{24}$	$\frac{1}{25}$	$\frac{2}{25}$	$\frac{3}{25}$	$\frac{4}{25}$	$\frac{6}{25}$	$\frac{7}{25}$	$\frac{8}{25}$	$\frac{9}{25}$
$x^+(\alpha)$	0	5	5	0	2	−3	−3	2	1	1	1	−4	1	1	1	1
$x^-(\alpha)$	2	1	−1	0	0	−1	1	0	1	−1	−1	0	1	1	1	1

α	$\frac{11}{25}$	$\frac{12}{25}$	$\frac{1}{26}$	$\frac{3}{26}$	$\frac{5}{26}$	$\frac{7}{26}$	$\frac{9}{26}$	$\frac{11}{26}$	$\frac{1}{27}$	$\frac{2}{27}$	$\frac{4}{27}$	$\frac{5}{27}$	$\frac{7}{27}$	$\frac{8}{27}$	$\frac{10}{27}$	$\frac{11}{27}$
$x^+(\alpha)$	1	1	−1	4	−1	−1	−1	4	−2	3	−2	−2	−2	−2	3	3
$x^-(\alpha)$	−1	1	1	0	−1	1	1	0	0	1	−2	0	2	2	1	1

α	$\frac{13}{27}$	$\frac{1}{28}$	$\frac{3}{28}$	$\frac{5}{28}$	$\frac{9}{28}$	$\frac{11}{28}$	$\frac{13}{28}$	$\frac{1}{29}$	$\frac{2}{29}$	$\frac{3}{29}$	$\frac{4}{29}$	$\frac{5}{29}$	$\frac{6}{29}$	$\frac{7}{29}$	$\frac{8}{29}$	$\frac{9}{29}$
$x^+(\alpha)$	3	−2	3	−2	3	3	3	−1	−1	−1	−1	−1	−1	−1	−1	4
$x^-(\alpha)$	1	0	1	0	1	1	1	−1	1	1	−1	1	−1	1	1	2

α	$\frac{10}{29}$	$\frac{11}{29}$	$\frac{12}{29}$	$\frac{13}{29}$	$\frac{14}{29}$	$\frac{1}{30}$	$\frac{7}{30}$	$\frac{11}{30}$	$\frac{13}{30}$
$x^+(\alpha)$	−1	−1	4	4	4	1	−4	1	6
$x^-(\alpha)$	1	1	2	0	0	−1	2	1	0

Here we only publish the part of the tables compiled for $\Gamma_0(11)$. The values of $x^{\pm}$ were computed for all 11-integral rational numbers with denominators ≤ 83, and also with denominators 19^2 and 29^2 (19 and 29 are supersingular primes for X_{11}).

In these tables $x^+(b/a) \bmod 5$ always depends only on a. In addition, in the tables $|x^+| \leq 9$ and $|x^-| \leq 5$.

The curve X_{17}

Equation: $y^2 + xy = x^3 - 4x^2 + 4x - 15$.

$$\gamma^{\mp} = \left\{-\frac{1}{3}, 0\right\} \pm \left\{\frac{1}{3}, 0\right\}.$$

$P^1(Z(17))$	1 : 0	0 : 1	1 : 1	2 : 1	3 : 1	4 : 1	5 : 1	6 : 1	7 : 1	8 : 1	9 : 1	10 : 1	11 : 1	12 : 1	13 : 1	14 : 1	13 : 1	16 : 1
y	−1	1	0	4	2	0	2	−2	−2	−4	−4	−2	−2	2	0	2	4	0
ξ^+	$\frac{1}{2}$	$-\frac{1}{2}$	0	−2	−1	0	−1	1	1	2	2	1	1	−1	0	−1	−2	0
ξ^-	0	0	0	0	1	0	1	1	1	0	0	−1	−1	−1	0	−1	0	0

α	$\frac{1}{2}$	$\frac{1}{3}$	$\frac{1}{4}$	$\frac{1}{5}$	$\frac{2}{5}$	$\frac{1}{6}$	$\frac{1}{7}$	$\frac{2}{7}$	$\frac{3}{7}$	$\frac{1}{8}$	$\frac{3}{8}$	$\frac{1}{9}$	$\frac{2}{9}$	$\frac{4}{9}$	$\frac{1}{10}$	$\frac{3}{10}$
$x^+(\alpha)$	2	1	0	1	1	−1	−1	−1	3	−2	2	−2	2	2	−1	−1
$x^-(\alpha)$	0	1	0	1	1	1	1	1	1	0	0	0	0	0	−1	1

α	$\frac{1}{11}$	$\frac{2}{11}$	$\frac{3}{11}$	$\frac{4}{11}$	$\frac{5}{11}$	$\frac{1}{12}$	$\frac{5}{12}$	$\frac{1}{13}$	$\frac{2}{13}$	$\frac{3}{13}$	$\frac{4}{13}$	$\frac{5}{13}$	$\frac{6}{13}$	$\frac{1}{14}$	$\frac{3}{14}$	$\frac{5}{14}$
$x^+(\alpha)$	−1	−1	−1	3	3	1	1	0	0	0	0	0	4	1	1	1
$x^-(\alpha)$	−1	1	1	1	1	−1	1	0	0	0	2	0	0	−1	1	1

α	$\frac{1}{15}$	$\frac{2}{15}$	$\frac{4}{15}$	$\frac{7}{15}$	$\frac{1}{16}$	$\frac{3}{16}$	$\frac{5}{16}$	$\frac{7}{16}$
$x^+(\alpha)$	2	−2	−2	2	0	0	0	4
$x^-(\alpha)$	0	0	0	0	0	2	2	0

In these tables $x^+(b/a)$ mod 4 depends only on a.

The curve X_{19}

Equation: $y^2 = t(t^3 - 16t^2 + 64t + 76)$.

$$\gamma^{\pm} = \left\{-\frac{1}{4}, 0\right\} \pm \left\{\frac{1}{4}, 0\right\}.$$

$P^1(Z/(19))$	1 : 0	0 : 1	1 : 1	2 : 1	3 : 1	4 : 1	5 : 1	6 : 1	7 : 1	8 : 1	9 : 1	10 : 1
y	−2	2	0	6	6	3	−3	−6	0	0	−6	−6
ξ^+	$\frac{2}{3}$	$-\frac{2}{3}$	0	−2	−2	−1	1	2	0	0	2	2
ξ^-	0	0	0	0	0	1	1	0	0	0	0	0

$P^1(Z/(19))$	11 : 1	12 : 1	13 : 1	14 : 1	15 : 1	16 : 1	17 : 1	18 : 1
y	0	0	−6	−3	3	6	6	0
ξ^+	0	0	2	1	−1	−2	−2	0
ξ^-	0	0	0	−1	−1	0	0	0

α	$\frac{1}{2}$	$\frac{1}{3}$	$\frac{1}{4}$	$\frac{1}{5}$	$\frac{2}{5}$	$\frac{1}{6}$	$\frac{1}{7}$	$\frac{2}{7}$	$\frac{3}{7}$	$\frac{1}{8}$	$\frac{3}{8}$	$\frac{1}{9}$	$\frac{2}{9}$	$\frac{4}{9}$	$\frac{1}{10}$	$\frac{3}{10}$	$\frac{1}{11}$
$x^+(\alpha)$	2	2	1	−1	2	−2	0	3	0	0	0	−2	1	1	−2	4	0
$x^-(\alpha)$	0	0	1	1	0	0	0	1	0	0	0	0	1	1	0	0	0

α	$\frac{2}{11}$	$\frac{3}{11}$	$\frac{4}{11}$	$\frac{5}{11}$	$\frac{1}{12}$	$\frac{5}{12}$	$\frac{1}{13}$	$\frac{2}{13}$	$\frac{3}{13}$	$\frac{4}{13}$	$\frac{5}{13}$	$\frac{6}{13}$	$\frac{1}{14}$	$\frac{3}{14}$	$\frac{5}{14}$	$\frac{1}{15}$	$\frac{2}{15}$
$x^+(\alpha)$	−3	3	0	3	0	0	−2	−2	1	4	1	4	−1	−1	2	1	−2
$x^-(\alpha)$	1	1	0	1	0	0	0	0	1	0	1	0	−1	1	0	−1	0

α	$\frac{4}{15}$	$\frac{7}{15}$	$\frac{1}{16}$	$\frac{3}{16}$	$\frac{5}{16}$	$\frac{7}{16}$	$\frac{1}{17}$	$\frac{2}{17}$	$\frac{3}{17}$	$\frac{4}{17}$	$\frac{5}{17}$	$\frac{6}{17}$	$\frac{7}{17}$	$\frac{8}{17}$	$\frac{1}{18}$	$\frac{5}{18}$	$\frac{7}{18}$
$x^+(\alpha)$	1	4	2	−1	2	−1	2	−1	−4	−1	5	2	2	2	0	3	3
$x^-(\alpha)$	1	0	0	1	0	1	0	1	0	1	1	0	0	0	0	−1	−1

In these tables $x^+(b/a)$ mod 3 depends only on a.

The curve X_{27}

Equation: $y^2 = 4x^3 + 1$ (curve with complex multiplication).

$$\gamma^{\pm} = \left\{-\frac{1}{4}, 0\right\} \pm \left\{\frac{1}{4}, 0\right\}.$$

$\mathbf{P}^1(\mathbf{Z}/(27))$	1 : 0	0 : 1	1 : 1	2 : 1	3 : 1	4 : 1	5 : 1	6 : 1	7 : 1	8 : 1	9 : 1	10 : 1	11 : 1	12 : 1
y	−2	2	0	6	3	3	3	0	−3	0	−1	0	−3	−3
ξ^+	$\frac{2}{3}$	$-\frac{2}{3}$	0	−2	−1	−1	−1	0	1	0	$\frac{1}{3}$	0	1	1
ξ^-	0	0	0	0	$\frac{1}{3}$	1	1	$\frac{2}{3}$	1	0	$\frac{1}{3}$	0	1	$\frac{1}{3}$

$\mathbf{P}^1(\mathbf{Z}/(27))$	13 : 1	14 : 1	15 : 1	16 : 1	17 : 1	18 : 1	19 : 1	20 : 1	21 : 1	22 : 1	23 : 1
y	−6	−6	−3	−3	0	−1	0	−3	0	3	3
ξ^+	2	2	1	1	0	$\frac{1}{3}$	0	1	0	−1	−1
ξ^-	0	0	$-\frac{1}{3}$	−1	0	$-\frac{1}{3}$	0	−1	$-\frac{2}{3}$	−1	−1

$\mathbf{P}^1(\mathbf{Z}/(27))$	24 : 1	25 : 1	26 : 1	1 : 3	2 : 3	4 : 3	5 : 3	7 : 3	8 : 3	1 : 9	2 : 9
y	3	6	0	−3	3	0	0	3	−3	1	1
ξ^+	−1	−2	0	1	−1	0	0	−1	1	$-\frac{3}{3}$	$-\frac{1}{3}$
ξ^-	$-\frac{1}{3}$	0	0	$\frac{1}{3}$	$-\frac{1}{3}$	$-\frac{2}{3}$	$\frac{2}{3}$	$\frac{1}{3}$	$-\frac{1}{3}$	$\frac{1}{3}$	$-\frac{1}{3}$

α	$\frac{1}{2}$	$\frac{1}{4}$	$\frac{1}{5}$	$\frac{2}{5}$	$\frac{1}{7}$	$\frac{2}{7}$	$\frac{3}{7}$	$\frac{1}{8}$	$\frac{3}{8}$	$\frac{1}{10}$	$\frac{3}{10}$	$\frac{1}{11}$	$\frac{2}{11}$	$\frac{3}{11}$	$\frac{4}{11}$	$\frac{5}{11}$	$\frac{1}{13}$
$x^{+}(\alpha)$	2	1	1	1	−1	2	2	0	0	0	0	−1	−1	2	2	2	−2
$x^{-}(\alpha)$	0	1	1	1	1	0	0	0	0	0	0	1	1	0	0	0	0

α	$\frac{2}{13}$	$\frac{3}{13}$	$\frac{4}{13}$	$\frac{5}{13}$	$\frac{6}{13}$	$\frac{1}{14}$	$\frac{3}{14}$	$\frac{5}{14}$	$\frac{1}{16}$	$\frac{3}{16}$	$\frac{5}{16}$	$\frac{7}{16}$	$\frac{1}{17}$	$\frac{2}{17}$	$\frac{3}{17}$	$\frac{4}{17}$	$\frac{5}{17}$	$\frac{6}{17}$
$x^{+}(\alpha)$	1	1	1	1	1	−2	1	1	−1	−1	2	2	0	0	0	0	0	0
$x^{-}(\alpha)$	1	1	1	1	1	0	1	1	−1	1	0	0	0	0	0	0	0	0

α	$\frac{7}{17}$	$\frac{8}{17}$	$\frac{1}{19}$	$\frac{2}{19}$	$\frac{3}{19}$	$\frac{4}{19}$	$\frac{5}{19}$	$\frac{6}{19}$	$\frac{7}{19}$	$\frac{8}{19}$	$\frac{9}{19}$	$\frac{1}{20}$	$\frac{3}{20}$	$\frac{7}{20}$	$\frac{9}{20}$	$\frac{1}{22}$
$x^{+}(\alpha)$	3	3	0	0	0	0	3	0	0	3	3	−1	−1	2	2	1
$x^{-}(\alpha)$	1	1	0	0	0	0	1	0	0	1	1	−1	1	0	0	−1

α	$\frac{3}{22}$	$\frac{5}{22}$	$\frac{7}{22}$	$\frac{9}{22}$	$\frac{1}{23}$	$\frac{2}{23}$	$\frac{3}{23}$	$\frac{4}{23}$	$\frac{5}{23}$	$\frac{6}{23}$	$\frac{7}{23}$	$\frac{8}{23}$	$\frac{9}{23}$	$\frac{10}{23}$	$\frac{11}{23}$
$x^{+}(\alpha)$	−2	1	1	1	1	−2	−2	1	1	1	1	1	1	1	4
$x^{-}(\alpha)$	0	1	1	1	−1	0	0	1	1	1	1	1	1	1	0

α	$\frac{1}{25}$	$\frac{2}{25}$	$\frac{3}{25}$	$\frac{4}{25}$	$\frac{6}{25}$	$\frac{7}{25}$	$\frac{8}{25}$	$\frac{9}{25}$	$\frac{11}{25}$	$\frac{12}{25}$	$\frac{1}{26}$	$\frac{3}{26}$	$\frac{5}{26}$	$\frac{7}{26}$	$\frac{9}{26}$	$\frac{11}{26}$
$x^{+}(\alpha)$	2	−1	−1	−1	−1	2	2	2	2	2	0	0	0	3	0	3
$x^{-}(\alpha)$	0	1	1	1	1	1	0	0	0	0	0	0	2	1	0	1

In these tables $x^{+}(b/a)$ mod 3 depends only on a.

Received 10/OCT/71

BIBLIOGRAPHY

1. A. Atkin and J. Lehner, *Hecke operators on* $\Gamma_0(m)$, Math. Ann. 185 (1970), 134–160.

2. B. J. Birch, *Diophantine analysis and modular functions*, Internat. Colloq. Algebraic Geometry (Tata Inst. Fund. Res., Bombay, 1968), Oxford Univ. Press, London, 1969, pp. 35–42. MR 41 #3478.

3. B. A. Venkov, *Elementary number theory*, ONTI, Moscow, 1937. (Russian)

4. P. Cartier, *Groupes formels, fonctions automorphes et fonctions zeta des courbes elliptiques*, 1970 (preprint).

5. M. Eichler, *Quaternäre quadratische Formen und die Riemannsche Vermutung für die Kongruenzzeta-funktion*, Arch. Math. 5 (1954), 355–366. MR 16, 116.

6. H. Heilbronn, *On the average length of a class of finite continued fractions*, Number Theory and Analysis (Papers in Honor of Edmund Landau), Plenum, New York, 1969, pp. 87–96. MR 41 #3406.

7. P. Lévy, *Théorie de l'addition des variables aléatoires*, Gauthier-Villars, Paris, 1937.

8. G. Ligozat, *Fonctions L des courbes modulaires*, Séminaire Délange-Pisot-Poitou, 1969/70, Exposé 9, Secrétariat mathématique, Paris, 1970.

9. Ju. I. Manin, *Cyclotomic fields and modular curves*, Uspehi Mat. Nauk 26 (1971), no. 6 (162), 7–71 = Russian Math. Surveys 26 (1971), no. 6 (to appear).

10. B. Mazur, *Rational points of abelian varieties with values in towers of number fields*, 1965 (preprint).

11. ——, *Arithmétique des courbes elliptiques sur les corps cyclotomiques (notes de J. F. Boutot)*, Orsay, 1970.

12. A. Ogg, *Modular forms and Dirichlet series*, Benjamin, New York, 1969. MR 41 #1648.

13. J.-P. Serre, *Abelian l-adic representations and elliptic curves*, Benjamin, New York, 1968. MR 41 #8422.

14. G. Shimura, *A reciprocity law in non-solvable extensions*, J. Reine Angew. Math. 221 (1966), 209–220. MR 32 #5637.

15. P. Swinnerton-Dyer, *The conjectures of Birch and Swinnerton-Dyer, and of Tate*, Proc. Conf. Local Fields (Driebergen, 1966), Springer, Berlin, 1967, pp. 132–157. MR 37 #6287.

16. A. Weil, *Über die Bestimmung Dirichletscher Reihen durch Funktionalgleichungen*, Math. Ann. 168 (1967), 149–156. MR 34 #7473.

Translated by N. KOBLITZ

Mat. Sbornik
Tom 89 (131) (1972), No. 2

Math. USSR Sbornik
Vol. 18 (1972), No. 2

HEIGHT ON FAMILIES OF ABELIAN VARIETIES

Ju. G. ZARHIN AND Ju. I. MANIN

Abstract. Let X be an abelian variety imbedded in projective space, and let L be an induced invertible sheaf on X. In this paper explicit bounds are determined for the difference $\hat{h}_L - h_L$, where $\hat{h}_L$ is the Néron-Tate height and h_L is the Weil height.

Bibliography: 5 titles.

Let X be an abelian variety imbedded in projective space and L an induced invertible sheaf on X. We shall assume that X is defined over a field of algebraic numbers. Then the Néron-Tate height function $\hat{h}_L$ is defined on the set of algebraic points of X. This function plays an important role in the arithmetic of X. On the other hand, the choice of coordinate system in the space in which X is imbedded permits the construction of another function—the Weil height h_L. The difference $\hat{h}_L - h_L$ is bounded in modulus. The main objective of this paper is to give explicit bounds for this difference. They depend on the coordinate system and on the variety X itself; therefore a natural formulation of the problem must assume the simultaneous assignment of a coordinate system in the space enveloping X and in the space of parameters defining the variation of X. Both of these conditions are fulfilled in D. Mumford's theory [5]. We also formulate our results for Mumford's families.

The problem of studying the variation of height in families of abelian varieties was mentioned by S. Lang [2]. V. A. Dem'janenko considered it for elliptic curves. In [4] (§11) Ju. I. Manin observed that effective estimates of $\max |\hat{h}_L - h_L|$ occur in the construction of a conditional algorithm for determining a basis of the group of rational points of an elliptic curve.

The main results of this paper are stated in §3 and proved in §§4–6. The first two sections introduce the required definitions and notation.

§1. Height

Let k be a finite extension of the field $\mathbf{Q}$ of rational numbers. Objects of one of the following types are called *places* v of k: a) prime ideals of the ring of integers

---

AMS (MOS) subject classifications (1970). Primary 14K05.

(non-archimedian places); b) imbeddings of k in $\mathbf{R}$; c) pairs of complex conjugate imbeddings of k in $\mathbf{C}$. A place v uniquely determines a valuation $|\ |_v: k \to \mathbf{R}$ satisfying the following conditions:

$|X|_v = |X'|$, where X' is the image of X under an imbedding of k into $\mathbf{R}$ or $\mathbf{C}$, if v is an archimedian place;

$|p|_v = p^{-1}$ if v is a non-archimedian place and its ideal divides the prime $p \in Z$. We denote by k_v the completion of k in the v-topology and set $n_v = [k_v : \mathbf{Q}_v]$.

1.1. **Definition.** The *height of the point* $x = (x_0 : \cdots : x_n) \in \mathbf{P}^n(k)$ in projective space with a particular system of coordinates is the number

$$h(x) = \frac{1}{[k:\mathbf{Q}]} \sum_v n_v \max_i (\log |x_i|_v).$$

We refer the reader to [3] for the proof of the elementary properties of height. Let us collect some of these properties here for reference.

a. $h(x)$ does not change when k is replaced by another field containing the projective coordinates of x. In particular, it is possible to assume that the base field coincides with the algebraic closure of $\mathbf{Q}$.

b. Let $(x_0'', \cdots, x_m'') = (x_0', \cdots, x_n')A$, where $A = (a_{ij})$ is an $(n+1, m+1)$-matrix with algebraic entries. Let $\nu \le n+1$ be the maximal number of nonzero elements in a column of A. Then in an obvious notation

$$h(x'') \leqslant h(x') + h(\ldots a_{ij} \ldots) + \log \nu.$$

c. Under the Segre mapping $\mathbf{P}^m \times \mathbf{P}^n \to \mathbf{P}^{(m+1)(n+1)-1}$, heights add:

$$h(\ldots x_i y_j \ldots) = h(\ldots x_i \ldots) + h(\ldots y_j \ldots).$$

d. If $k > 0$, then $h(x_1^k : x_2^k : \cdots) = kh(x_1 : x_2 : \cdots)$.

1.2. Now let X be a variety over the field k, L an invertible sheaf on X, and $s_i \in \Gamma(X, L)$ a finite nontrivial family of sections of L. Let $s = (s_1, \cdots, s_n)$ and define a function $h_{L,s}: X(k) \to \mathbf{R}$ by

$$h_{L,s}(x) = h(s_0(x) : \ldots : s_n(x)).$$

We shall call such functions *Weil heights*.

Finally, let X be an abelian variety and L a symmetric very ample sheaf on X. Then there exists a unique function $\hat{h}_L$ $X(k) \to \mathbf{R}$ satisfying the following two conditions:

a. $\hat{h}_L$ is quadratic (as a function on an abelian group).

b. $|h_{L,s} - \hat{h}_L|$ is bounded (for any choice of basis $s = (s_i)$).

The function $\hat{h}_L$ is called the *Néron-Tate height* on X (relative to L).

From the definition it is clear that $\hat{h}_L(x) = \lim_{n\to\infty} (1/4^n) h_{L,s}(2^n x)$.

§2. Mumford coordinates

In the sequel k denotes the algebraic closure of $\mathbf{Q}$, X an abelian variety over k, and L a totally symmetric very ample sheaf on X. Mumford [5] showed that it is possible "almost canonically" to choose a basis in the space $\Gamma(X, L)$. We now describe that part of his results that we need in the following.

Let $g = \dim X$ and let $d = (d_1, \cdots, d_g)$ be a finite sequence of g positive integers with $d_{i+1} | d_i$.

2.1. **The group $\mathfrak{G}(d)$ and its representation.** Let

$$K(d) = \bigoplus_{i=1}^{g} \frac{1}{d_i} \mathbf{Z}/\mathbf{Z}.$$

For any abelian group K the symbol $K^{\wedge}$ will denote the dual $\mathrm{Hom}(K, k^*)$ of K. We let $\mathfrak{G}(d) = k^* \times K(d) \times K(d)^{\wedge}$ and introduce a group structure on $\mathfrak{G}(d)$ by setting

$$(a, x, l)(a', x', l') = (aa'l'(x), x+x', l+l').$$

Next we consider the space $V(d)$ of functions on $K(d)$ with values in k, and for any $(a, x, l) \in \mathfrak{G}(d)$ we define a linear operator $U_{(a,x,l)}: V(d) \to V(d)$ by

$$(U_{(a,x,l)}f)(y) = al(y)f(x+y), \quad y \in K(d), \ f \in V(d).$$

2.2. **The group $\mathfrak{G}(L)$ and its representation.** Under the conditions laid down at the beginning of this section we denote by $\mathfrak{G}(L)$ the group of those automorphisms of the pair (X, L) that act as shifts on a basis of X and linearly on fibers of L (over O_X). The group k^* is naturally imbedded in $\mathfrak{G}(L)$: the imbedding is effected by those automorphisms that act as the identity on a basis.

The group $\mathfrak{G}(L)$ has a natural representation in the k-space $\Gamma(X, L)$.

The following proposition summarizes the first results of [5].

2.3. **Proposition.** a) *For some d there exists an isomorphism $\mathfrak{G}(L) \xrightarrow{\sim} \mathfrak{G}(d)$ acting as the identity on subgroups of k^*. There are only finitely many such isomorphisms. The vector $d = (d_1, \cdots, d_g)$ is called the type of the sheaf L.*

b) *Suppose an isomorphism $\mathfrak{G}(L) \xrightarrow{\sim} \mathfrak{G}(d)$ has been chosen. Then there exists a unique (up to multiplication by a nonzero constant) isomorphism $\Gamma(X, L) \xrightarrow{\sim} V(d)$ under which the natural action of $\mathfrak{G}(L)$ on $\Gamma(X, L)$ passes into the action of $\mathfrak{G}(d)$ on $V(d)$ described above.*

A choice of isomorphisms a) and b) is called an *assignment of a Tate structure* on the pair (X, L).

2.4. **Canonical basis in $\Gamma(X, L)$.** Let a Tate structure be assigned on (X, L). We shall tacitly identify $\Gamma(X, L)$ with $V(d)$, and $\mathfrak{G}(L)$ with $\mathfrak{G}(d)$, relative to this structure. For $a \in K(d)$ we set

$$\delta_a(x) = \begin{cases} 1 & \text{for } x = a \\ 0 & \text{for } x \neq a, \ x \in K(d); \end{cases} \qquad \delta_a \in V(d) = \Gamma(X, L).$$

The family $\delta = (\delta_a)_{a \in K(d)}$ in $\Gamma(X, L)$ forms the basic setting in which we shall compute the Weil height.

2.5. **Tate constants.** For any point $x \in K(d)$ we set

$$(\ldots : q_L(x) : \ldots) = (\ldots : \delta_x^{(0)} : \ldots),$$

where $\delta_x^{(0)}$ is the "value" of the section $\delta_x \in \Gamma(X, L)$ at the zero of X. The entire collection $(\delta_x^{(0)})$ is uniquely defined up to a factor from k^*. Thus $(\cdots : q_L(x) : \cdots)$ gives projective coordinates of the zero of X in the δ-basis. The undetermined common factor we shall tacitly assume chosen so that all formulas below turn out to be valid in the form in which they are written.

2.6. **Multiplication and doubling formulas.** Mumford introduces the notion of compatibility of two Tate structures on (X, L) and (X, L^2), respectively. We do not need the definition, only the following properties of this notion:

a. Compatible pairs of Tate structures on (X, L) and (X, L^2) exist.

b. Any compatible pair of Tate structures on (X, L) and (X, L^2) can be extended by choosing a Tate structure on (X, L^4) so that the pair of structures on (X, L^2) and (X, L^4) also turns out to be compatible.

c. Suppose pairwise compatible Tate structures have been chosen on (X, L), (X, L^2) and (X, L^4). We identify $\Gamma(X, L^i)$ with $V(id)$ as above (here $id = (id_1, \cdots, id_g)$). We denote by $\delta^{(i)} = (\delta_a^{(i)})$, $a \in K(id)$, the delta basis in $\Gamma(X, L^i)$ (cf. §2.4). Tensor multiplication $\Gamma(X, L) \otimes \Gamma(X, L) \to \Gamma(X, L^2)$ is converted into the law of composition $V(d) \otimes V(d) \to V(2d)$, which we denote by $*$, following Mumford.

Mumford establishes an explicit formula for the multiplication $*$, from which it is not hard to derive (vide infra) the following result: for any $a, b \in K(d)$

$$\delta_a^{(1)} * \delta_b^{(1)} = \sum_{\eta \in K(2)} q_{L^2}\left(\frac{a-b}{2} + \eta\right) \delta^{(2)}_{\frac{a+b}{2}+\eta}. \tag{1}$$

Here $K(2) = K(2, \cdots, 2) \subset K(2d) = \bigoplus_{i=1}^g (2d_i)^{-1}\mathbf{Z}/\mathbf{Z}$ are elements of period 2 in $K(2d)$. The elements $(a-b)/2$ and $(a+b)/2$ are chosen so that $(a-b)/2 + (a+b)/2 = a$. The remaining choice is arbitrary, which does not affect the right side because of the summation with respect to η. Similar stipulations will be applied to the remaining formulas.

Let $2\colon X \to X$ be the morphism $x \leadsto 2x$. We identify 2^*L with L^4 and denote by $[2]\colon V(d) \to V(4d)$ the mapping induced by the doubling $2^*\colon \Gamma(X, L) \to \Gamma(X, 2^*L) \cong \Gamma(X, L^4)$. Then

$$[2]\,\delta_a^{(1)} = \sum_{\eta \in K(2)} \delta^{(4)}_{\frac{a}{2}+\eta}. \tag{2}$$

2.7. **Derivation of formulas (1) and (2).** The multiplication formula was given in [5] in the following form (p. 330): let $f, f' \in V(d)$; then for $x \in K(2d)$

$$(f * f')(x) = \sum_{y \in x+K(d)} f(x+y) f'(x-y) q_{L^2}(y).$$

Applying this to $\delta_a^{(1)}$ and $\delta_b^{(1)}$, we find

$$(\delta_a^{(1)} * \delta_b^{(1)})(x) = \sum_{y \in x+K(d)} \delta_a^{(1)}(x+y)\delta_b^{(1)}(x-y) q_{L^2}(y)$$

$$= \begin{cases} 0, & \text{if} \quad 2x \neq a+b, \\ \sum_{\eta \in K(2)} q_{L^2}\left(\frac{a-b}{2} + \eta\right), & \text{if} \quad 2x = a+b. \end{cases}$$

Hence (1) follows.

The doubling formula ([5], p. 330) was given in the form ($f \in V(d)$, $x \in K(4d)$)

$$([2]f)(x) = \begin{cases} 0, & \text{if} \quad x \notin K(2d), \\ f(2x), & \text{if} \quad x \in K(2d). \end{cases}$$

Therefore

$$[2]\delta_a^{(1)}(x) = \begin{cases} 0, & \text{if} \quad 2x \neq a, \\ 1, & \text{if} \quad 2x = a. \end{cases}$$

Obviously this coincides with (2).

§3. The main results

3.1. We retain all notation and conventions introduced earlier. In particular the abelian variety X over a field k of algebraic numbers is fixed as well as the totally symmetric ample invertible sheaf L of type $d = (d_1, \cdots, d_g)$ on X. In addition, we assume that $K(4) = \bigoplus_{i=1}^{d} \frac{1}{4}\mathbf{Z}/\mathbf{Z} \subset K(d)$; hence it follows that L is very ample ([5], p. 340).

For any (finite) abelian group H we set, as above, $H^{\wedge} = \mathrm{Hom}(H, k^*)$. Next we fix a Tate structure on (X, L) and consider the constants

$$\theta_L(x, l) = \sum_{\eta \in K(2)} l(\eta) q_L(x+\eta), \quad x \in K(d), \ l \in K(2)^{\wedge}.$$

Finally, we set

$$h(X, L) = h(\ \ldots\ \theta_L(x, l)\ \ldots)_{\substack{x \in K(d) \\ l \in K(2)^{\wedge}}}. \tag{3}$$

(The point on which the h on the right operates is, according to Mumford, the "modulus" of the abelian variety X, relative to which X can be uniquely recovered. Therefore the number $h(X, L)$ should be regarded as the "height of the variety X", provided with rigidity by L and the Tate structure. It differs from $h(\cdots q_L(x) \cdots)$ by no more than $g \log 2$ and has been introduced only to improve some estimates.)

When in the following we have occasion to consider, besides (X, L), pairs (X, L^2), (X, L^4), etc., it will be assumed that compatible Tate structures have been fixed on all these pairs. The canonical coordinates $\delta_a^{(i)}$, the constants $q_{L^i}(x)$, etc. can be constructed with the help of these structures.

The main result of this paper is the following theorem.

3.2. **Theorem.** *Let $h_{L,\delta}$ be the Weil height on X relative to the canonical basis $\delta = (\delta_a)$, $a \in K(d)$, and $\hat{h}_L$ the Néron-Tate height. Then*

$$-C^- \leqslant \hat{h}_L - h_{L,\delta} \leqslant C^+, \tag{4}$$

where

$$C^- = \left(\frac{2^{2g-1}}{3} + 1\right) h(X, L) + \left(2^{2g-2} + \frac{67}{12}\right) g \log 2, \tag{5}$$

$$C^+ = (2^{2g} - 1) h(X, L) + \left(2^{2g+1} - \frac{1}{3}\right) g \log 2. \tag{6}$$

3.3. **Example.** Let $\dim X = g = 1$, let L be a totally symmetric sheaf on X of type (4) and suppose $|K| = 4$. As Mumford shows ([5], p. 352), the curve X is given in canonical coordinates by the equations

$$\delta_1^2 + \delta_3^2 = 2\Lambda\delta_0\delta_2, \quad \delta_0^2 + \delta_2^2 = 2\Lambda\delta_1\delta_3.$$

Here $\Lambda \in k$ is a parameter that is connected with the coordinates $q_L(0), \cdots, q_L(3)$ by the following relations:

$$q_L(1) = q_L(3), \quad q_L(1)^2 = \Lambda q_L(0) q_L(2), \quad q_L(0)^2 + q_L(2)^2 = 2\Lambda q_L(1)^2.$$

Expressing $q_L(i) q_L(1)^{-1}$ in terms of Λ and using definition (3), we easily obtain from this that

$$h(X, L) = h\left(\sqrt{\Lambda \pm \sqrt{\Lambda^2 - \Lambda^{-2}}} \pm 1\right)$$

(the four combinations of signs $\pm$ give four projective coordinates of a point; the choice of sign in front of the outer root is arbitrary, since it does not change the heights). From (4), (5) and (6) we finally obtain

$$-\frac{5}{3} h(X, L) - \frac{79}{12} \log 2 \leqslant \hat{h}_L - h_{L,\delta} \leqslant 3h(X, L) + \frac{23}{3} \log 2.$$

Compare this estimate with Dem'janenko's inequalities [1], which are written in other coordinates but have exactly the same structure.

The proof of Theorem 3.2 is obtained by comparing a series of estimates having a certain independent interest. We shall select three intermediate results. Their proofs are contained in subsequent sections.

3.4. **Proposition.** *Under the conditions of §3.1 we have*

$$-A^- \leqslant h_{L^4,\delta(4)} - 4h_{L,\delta(1)} \leqslant A^+, \tag{7}$$

where

$$A^- = 2h(X, L^2) + h(X, L^4) + 6g\log 2, \tag{8}$$

$$A^+ = 2(2^{2g} - 1)h(X, L^2) + (2^{2g} - 1)h(X, L^4) + 6g\log 2. \tag{9}$$

3.5. **Proposition.** *Let* $(h \circ 2)(x) = h(2x)$. *Under the same conditions*

$$-B^- \leqslant h_{L,\delta(1)} \circ 2 - 4h_{L,\delta(1)} \leqslant B^+, \tag{10}$$

where

$$B^- = \frac{1}{2}(2^{2g} + 3)h(X, L^2) + h(X, L^4) + \frac{1}{2}h(X, L^8) + \frac{17}{2}g\log 2, \tag{11}$$

$$B^+ = 2(2^{2g} - 1)h(X, L^2) + (2^{2g} - 1)h(X, L^4) + 5g\log 2. \tag{12}$$

3.6. **Proposition.**

$$-\frac{5}{2}g\log 2 \leqslant h(X, L^2) - h(X, L) \leqslant \frac{3}{2}g\log 2. \tag{13}$$

3.7. **Derivation of Theorem 3.2.** We recall that $\hat{h}_L(x) = \lim 4^{-n}h_{L,\delta(1)}(2^n x)$. From (10) we find that $h_{L,\delta(1)} \circ 2 \leq 4h_{L,\delta(1)} + B^+$, and hence, by induction on n, that

$$h_{L,\delta(1)} \circ 2^n \leqslant 2^{2n}h_{L,\delta(1)} + (2^{2(n-1)} + \dots + 1)B^+ = 2^{2n}h_{L,\delta(1)} + \frac{2^{2n}-1}{3}B^+,$$

from which we finally have

$$\hat{h}_L \leqslant h_{L,\delta(1)} + \frac{1}{3}B^+$$

and similarly

$$\hat{h}_L \geqslant h_{L,\delta(1)} - \frac{1}{3}B^-.$$

To estimate the $h(X, L^i)$, $i = 2, 4, 8$, occuring in $B^\pm$ in terms of $h(X, L)$ we apply (13). It yields

$$B^- \leqslant (2^{2g-1} + 3)h(X, L) + \left(3 \cdot 2^{2g-2} + \frac{67}{4}\right)g\log 2,$$

$$B^+ \leqslant 3(2^{2g} - 1)h(X, L) + (3 \cdot 2^{2g+1} - 1)g\log 2.$$

Theorem 3.2 follows from this.

3.8. **Remark.** According to Theorem 3.2 and Proposition 3.6,

$$|\hat{h}_{L^{4^n}} - h_{L^{4^n},\delta(4^n)}| \leqslant C_1 h(X, L^{4^n}) + C_3 \leqslant C_1 h(X, L) + C_2 n + C_3$$

for some constants C_1, C_2 and C_3 depending only on dim X. Again taking account of the equality $\hat{h}_{L^{4^n}} = 4^n\hat{h}_L$, we obtain

$$|\hat{h}_L - 4^{-n}h_{L^{4^n},\delta(4^n)}| = O(n4^{-n}) \quad \text{for} \quad n \to \infty;$$

in particular

$$\hat{h}_L = \lim_{n\to\infty} 4^{-n}h_{L^{4^n},\delta(4^n)}.$$

In other words, it is possible to approximate Néron-Tate height with arbitrary precision by Weil height computed in Mumford coordinates for a sufficiently large power of L and then normalized to reduce it to unit degree.

§4. Proof of Proposition 3.4

4.1. Estimation of A^-. We recall first of all that

$$\theta_L(x, l) = \sum_{\eta\in K(2)} l(\eta)\, q_L(x+\eta), \quad x\in K(d),\ l\in K(2)^{\wedge}.$$

From the inversion formula $q_L(x+\eta) = 2^{-g}\sum_{l\in K(2)^{\wedge}} l(\eta)\theta_L(x, l)$ it is clear that

$$h(X, L) = h(\ \dots\ \theta_L(x, l)\ \dots) \leqslant h(\dots\ q_L(x)\ \dots) + g\log 2,$$
$$h(X, L) \geqslant h(\ \dots\ q_L(x)\ \dots) - g\log 2.$$

We shall frequently make use of these inequalities later. From (1) we now find that

$$2h_{L,\delta(1)} = h_{L^2,\delta(1)*\delta(1)} \leqslant h_{L^2,\delta(2)} + h(\ \dots\ q_{L^2}(x)\ \dots) + g\log 2$$
$$\leqslant h_{L^2,\delta(2)} + h(X, L^2) + 2g\log 2.$$

Applying this inequality to L^2 in place of L, we obtain

$$2h_{L^2,\delta(2)} \leqslant h_{L^4,\delta(4)} + h(X, L^4) + 2g\log 2.$$

Finally, substituting this estimate for $h_{L^2,\delta(2)}$ into the right side of the preceding inequality, we obtain

$$4h_{L,\delta(1)} \leqslant h_{L^4,\delta(4)} + 2h(X, L^2) + h(X, L^4) + 6g\log 2.$$

This is relation (8).

4.2. Estimation of A^+. The estimate from above is somewhat more difficult. We first pass to another basis in $\Gamma(X, L^2)$. Let a run through some system of representatives of the classes $K(2d)/K(2)$, and let l run through the characters of the group $K(2)$. We set $\Delta^{(2)}_{a,l} = \sum_{\eta\in K(2)} l(\eta)\delta^{(2)}_{a+\eta}$; the inversion formulas have the form

$$\delta^{(2)}_{a+\eta} = 2^{-g}\sum_{l\in K(2)^{\wedge}} l(\eta)\,\Delta^{(2)}_{a,l}.$$

We need the following lemma of Mumford ([5], p. 339).

4.3. **Lemma.** a) *If $b, a\in K(2d)$ with $b-a\in K(d)$, then*

$$\theta_{L^2}(b, l)\quad \Delta^{(2)}_{a,l} = \sum_{\eta\in K(2)} l(\eta)\,\delta^{(1)}_{a+b+\eta} * \delta^{(1)}_{a-b+\eta}.$$

b) *For any $l\in K(2)^{\wedge}$ there exists an element b in each class of $K(2d)$ mod $K(d)$ such that*

$$\theta_{L^2}(b, l) = \sum_{\eta\in K(2)} l(\eta)\, q_{L^2}(b+\eta) \neq 0.$$

We shall derive an estimate for A^+ from this lemma.

For each pair (a, l) by which the $\Delta^{(2)}_{a,l}$ are enumerated we select an element $b \equiv a \bmod K(d)$, depending only on $a \bmod K(d)$, such that $\theta_{L^2}(b, l) \neq 0$, and we write $r_{a,l} = \theta_{L^2}(b, l)^{-1}$. Then

$$\Delta^{(2)}_{a,l} = r_{a,l} \sum_{\eta \in K(2)} l(\eta)\, \delta^{(1)}_{a+b+\eta} * \delta^{(1)}_{a-b-\eta},$$

whence

$$h_{L^2,\Delta^{(2)}} \leqslant 2h_{L,\delta^{(1)}} + h(\ \dots\ r_{a,l}\ \dots) + g\log 2. \tag{14}$$

To estimate $h(\cdots r_{a,l} \cdots) = h(\cdots \theta_{L^2}(b, l)^{-1} \cdots)$ we note that among the $r_{a,l}$ there are no more than $[K(2d)\colon K(d)]|K(2)^{\wedge}| = 2^{2g}$ different nonzero numbers. Therefore, multiplying all coordinates by their product we obtain

$$h(\dots\ r_{a,l}\ \dots) = h(\dots\ \theta_{L^2}(b, l)^{-1}\ \dots)$$
$$\leqslant h(\dots \text{ (products of } \theta_{L^2}(b, l) \text{ in } 2^{2g} - 1 \text{ pieces) } \dots) \leqslant (2^{2g} - 1)\, h(X, L^2).$$

Substituting in (14), we find

$$h_{L^2,\Delta^{(2)}} \leqslant 2h_{L,\delta^{(1)}} + (2^{2g} - 1)\, h(X, L^2) + g\log 2. \tag{15}$$

But $h_{L^2,\delta^{(2)}} \leq h_{L^2,\Delta^{(2)}} + g \log 2$, whence

$$h_{L^2,\delta^{(2)}} \leqslant 2h_{L,\delta^{(1)}} + (2^{2g} - 1)\, h(X, L^2) + 2g\log 2. \tag{16}$$

We apply this inequality to L^2 in place of L:

$$h_{L^4,\delta^{(4)}} \leqslant 2h_{L^2,\delta^{(2)}} + (2^{2g} - 1)\, h(X, L^4) + 2g\log 2.$$

Finally we have

$$h_{L^4,\delta^{(4)}} \leqslant 4h_{L,\delta^{(1)}} + 2(2^{2g} - 1)\, h(X, L^2) + (2^{2g} - 1)\, h(X, L) + 6g\log 2.$$

This is relation (9).

§5. Proof of Proposition 3.5

5.1. Estimation of B^+. From (2) we find

$$h_{L,\delta^{(1)}} \circ 2 = h_{L,[2]\delta^{(1)}} \leqslant h_{L^4,\Delta^{(4)}},$$

where $\Delta^{(4)} = (\Delta^{(4)}_{a,l})$ with $\Delta^{(4)}_{a,l} = \Sigma_{\eta \in K(2)}\, l(\eta)\delta^{(4)}_{a+\eta}$ as in §4.2 (because $[2]\delta^{(1)}_a = \Delta^{(4)}_{a/2,0}$). Applying (15), written for L^2 in place of L, we obtain

$$h_{L,\delta^{(1)}} \circ 2 \leqslant 2h_{L^2,\delta^{(2)}} + (2^{2g} - 1)\, h(X, L^4) + g\log 2.$$

Substituting (16) in the right side, we finally find

$$h_{L,\delta^{(1)}} \circ 2 \leqslant 4h_{L,\delta^{(1)}} + 2(2^{2g} - 1)\, h(X, L^2) + (2^{2g} - 1)\, h(X, L^4) + 5g\log 2.$$

This is relation (12).

Estimation of B^-. We begin with the following lemma.

5.2. Lemma.

$$\Delta_{a,l}^{(4)} * \Delta_{a,l}^{(4)} = \sum_{\zeta\in K(2)} l(\zeta)\,\theta_{L^8}\left(\frac{\zeta}{2},0\right)[2]\,\delta_{2a+\zeta}^{(2)}. \tag{17}$$

(Note that $\theta_{L^8}(\zeta/2, 0)$ does not depend on the choice of $\zeta/2$.)

Proof. First of all, by definition

$$\Delta_{a,l}^{(4)} * \Delta_{a,l}^{(4)} = \sum_{\xi,\eta\in K(2)} l(\xi+\eta)\,\delta_{a+\xi}^{(4)} * \delta_{a+\eta}^{(4)} = \sum_{\zeta\in K(2)} l(\zeta) \sum_{\xi\in K(2)} \delta_{a+\xi}^{(4)} * \delta_{a+\xi+\zeta}^{(4)}.$$

Now, applying Lemma 4.3a) to L^4 in place of L and to $l = 0$, we find

$$\sum_{\xi\in K(2)} \delta_{a+\xi}^{(4)} * \delta_{a+\xi+\zeta}^{(4)} = \sum_{\xi\in K(2)} \delta^{(4)}_{\left(a+\frac{\zeta}{2}\right)+\frac{\zeta}{2}+\xi} * \delta^{(4)}_{\left(a+\frac{\zeta}{2}\right)-\frac{\zeta}{2}+\xi}$$

$$= \theta_{L^8}\left(\frac{\zeta}{2},0\right)\Delta^{(8)}_{a+\frac{\zeta}{2},\,0} = \theta_{L^8}\left(\frac{\zeta}{2},0\right)[2]\,\delta^{(2)}_{2a+\zeta}$$

by virtue of (2). Hence (17) follows.

We now proceed to estimate B^-. From (17) we obtain

$$2h_{L^4,\Delta^{(4)}} \leqslant h_{L^2,\delta^{(2)}} \circ 2 + h(X, L^8) + g\log 2.$$

Substituting (16), written for a doubled point, in the right side, we find

$$2h_{L^4,\Delta^{(4)}} \leqslant 2h_{L,\delta^{(1)}} \circ 2 + (2^{2g}-1)\,h(X, L^2) + h(X, L^8) + 3g\log 2.$$

In addition, according to (7)

$$4h_{L,\delta^{(1)}} \leqslant h_{L^4,\delta^{(4)}} + A^- \leqslant h_{L^4,\Delta^{(4)}} + A^- + g\log 2.$$

Comparing these two inequalities, we obtain

$$4h_{L,\delta^{(1)}} \leqslant h_{L,\delta^{(1)}} \circ 2 + \frac{1}{2}(2^{2g}+3)\,h(X, L^2) + h(X, L^4)$$

$$+ \frac{1}{2}h(X, L^8) + \frac{17}{2}g\log 2.$$

This yields (11).

§6. Proof of Proposition 3.6

6.1. Estimation from below. According to formula (A′) on page 333 of [5], for any $u, v \in K(2d)$ satisfying $u + v \in K(d)$ we have

$$q_L(u+v)\,q_L(u-v) = \sum_{\eta\in K(2)} q_{L^2}(u+\eta)\,q_{L^2}(v+\eta). \tag{18}$$

Setting $v = 0$ in (18) we obtain $q_L(u)^2 = \sum_{\eta\in K(2)} q_{L^2}(u+\eta)q_{L^2}(\eta)$, whence

$$h(\ldots q_L(u)\ldots)\leqslant h(\ldots q_{L^2}(u)\ldots)+\frac{1}{2}g\log 2,$$

and, as in §4.1,

$$h(X,L)\leqslant h(X,L^2)+\frac{5}{2}g\log 2.$$

6.2. **Estimation from above.** First we show that for any $x\in K(2d)$ and $l\in K(2)^{\frown}$

$$\theta_{L^2}(x,l)^2=\sum_{\eta\in K(2)} l(\eta)q_L(2x-\eta)q_L(\eta). \tag{19}$$

In fact, using the definition of $\theta_{L^2}(x,l)$ and (18), we find

$$\theta_{L^2}(x,l)^2=\sum_{\xi,\eta\in K(2)} l(\xi+\eta)q_{L^2}(x+\xi)q_{L^2}(x+\eta)$$

$$=\sum_{\zeta\in K(2)} l(\zeta)\sum_{\eta\in K(2)} q_{L^2}(x+\eta)q_{L^2}(x+\zeta+\eta)=\sum_{\zeta\in K(2)} l(\zeta)q_L(2x+\zeta)q_L(\zeta).$$

Now from (19) we find

$$2h(X,L^2)=h(\ldots\theta_{L^2}(x,l)^2\ldots)\leqslant 2h(\ldots q_L(x)\ldots)+g\log 2$$

and finally, since $h(\cdots q_L(x)\cdots)\leq h(X,L)+g\log 2$, we have

$$h(X,L^2)\leqslant h(X,L)+\frac{3}{2}g\log 2.$$

Received 30/MAY/72

BIBLIOGRAPHY

1. V. A. Dem'janenko, *An estimate of the remainder term in Tate's formula*, Mat. Zametki 3 (1968), 271–278 = Math. Notes 3 (1968), 173–177. MR 37 #2751.

2. S. Lang, *Diophantine geometry*, Interscience Tracts in Pure and Appl. Math., no. 11, Interscience, New York, 1962. MR 26 #119.

3. Ju. I. Manin, *The Mordell-Weyl theorem*, Appendix to the Russian transl. of D. Mumford, *Abelian varieties*, "Mir", Moscow, 1971.

4. ———, *Cyclotomic fields and modular curves*, Uspehi Mat. Nauk 26 (1971), no. 6 (162), 7–71 = Russian Math. Surveys 26 (1971), no. 6, 7–78.

5. D. Mumford, *On the equations defining abelian varieties*. I, Invent. Math. 1 (1966), 287–354. MR 34 #4269.

Translated by B. SILVER

Reprinted from *J. Reine Angew. Math.* **262/263** (1973).

Periods of p-adic Schottky groups

By *Yu. Manin* and *V. Drinfeld* at Moscow

Dedicated to Professor Helmut Hasse on his 75th birthday

Introduction

Let K be a finite extension of the p-adic field Q_p, and

$$G = \mathrm{PGL}\,(2, K) = \mathrm{GL}\,(2, K)/K^*.$$

An element $g \in G$ is called *hyperbolic*, if it is represented by a matrix whose characteristic roots have different absolute values. A subgroup $\Gamma < G$ is called *Schottky group*, if it is finitely generated and every $g \in \Gamma$, $g \neq e$ is hyperbolic. A Schottky group is *free* (Ihara) and *discrete*. Let Σ be the closure of the set of fixed points of elements of Γ in $P^1_{\bar{K}}$. In fact, $\Sigma < P^1(K)$, and $\Omega = P^1(\bar{K})\setminus\Sigma$ is precisely the set where Γ acts discontinuously.

In a beautiful and important recent work [3] D. Mumford proved that the holomorphic quotient space $\Gamma\setminus\Omega$ has a natural structure of an algebraic curve X_Γ over K of genus $n = \mathrm{rank}\,\Gamma$. This result generalizes a theorem of Tate ($n = 1$), but Mumford's method is quite different. Mumford constructs his quotients in a very geometric manner and in fact in a much more general setting, over complete local rings of any dimension.

In this paper we reconsider the question and introduce the automorphic functions relative to Γ. The principal effect of this approach[1]) is that it allows the analytic construction of the jacobian J_Γ of X_Γ (together with its canonical polarization) and of the canonical map $X_\Gamma \to J_\Gamma$.

So this note can be looked at as a very natural morphism from Mumford [3] to Mumford [4]. Our restriction by p-adic case is most probably inessential but significantly simplifies the exposition.

§ 1. Preliminaries

We collect here some facts on the trees of PGL (2) and Γ needed later. For the proofs we refer to Serre [5], Mumford [3] or to the account [6].

Let V be a two-dimensional vector space over K, P^1_K the space of lines in V, $G = \mathrm{PGL}\,(2, K) = \mathrm{Aut}_K P^1_K$. Denote by $O < K$ the ring of integers, $m < O$ the maximal ideal. Let M_1, $M_2 < V$ be two free O-submodules of rank two. They are called *equivalent*, if $M_1 = a M_2$ for some $a \in K$. The equivalence class of M is noted $\{M\}$. The set Δ^0 of all equivalence classes $\{M\}$ is the set of vertices of a *graph* Δ: if $M_1 < M_2$, $M_1/M_2 \cong O/m$,

[1]) *Added in proof*: As Mumford informed us, analogous results were obtained independently by Jerry Myers.

then $\{M_1\}$ and $\{M_2\}$ are connected by an *edge*. This graph Δ is in fact an infinite but locally finite *tree*, on which G acts. Its ends are naturally identified with $P^1(K)$.

Now let $\Gamma < G$ be a Schottky group. Every element $g \in \Gamma$, $g \neq e$, has two fixed points in $P^1(K)$. There is a unique *line* (path without back-tracking and infinite in both directions) in Δ, whose ends are fixed by g. This line is called the *axis* of g, and g acts on it by the translation to the distance $\operatorname{ord}_K q$, where g is conjugate to $\begin{pmatrix} q & 0 \\ 0 & 1 \end{pmatrix}$ in PGL (2, K).

Call $\Delta_\Gamma < \Delta$ the least subtree of Δ containing the axes of all elements of Γ (this corresponds to Mumford's Δ'_Γ on page 163 of [3]). The set of its ends in $P^1(K)$ is precisely Σ.

The graph $\Gamma\backslash\Delta_\Gamma$ is finite. It has a nice geometric interpretation as the graph of the closed fibre of the minimal smooth model over O of the curve $X_\Gamma = \Gamma\backslash\Omega$ (see [3]).

§ 2. Divisors and automorphic functions

We fix once and for all a Schottky group $\Gamma < G$ of rank n; Ω, Σ etc. are constructed for Γ.

Call a *K-divisor on* Ω a function $\Omega(\bar{K}) \to Z$: $z \mapsto n_z$ with the following properties:

a) $n_{z_1} = n_{z_2}$, if z_1 and z_2 are conjugate over K.

b) There is a finite extension $L < K$ such that $n_z \neq 0 \to z \in \Omega(L)$.

c) The set $\{z \mid n_z \neq 0\}$ has no limit points in Ω.

We denote such a divisor $\Sigma n_z z = d$. The set supp d is defined as $\{z \mid n_z \neq 0\}$. A divisor d is called *finite* if supp d is finite.

The group Γ acts on the group $\mathfrak{D}$ of K-divisors.

Lemma 1. a) *The group of Γ-invariant K-divisors consists exactly of the divisors of the form* $\sum_{g\in\Gamma} g(d)$, *where d is a finite K-divisor on* Ω.

b) $\Sigma g(d) = 0$ *if and only if there exist* $g_i \in \Gamma$, $i = 1, \ldots, k$ *and finite K-divisors* d_i *such that* $d = \sum_{i=1}^{k} (1 - g_i) d_i$.

The proof is not difficult and we omit it (see [6]). The only noteworthy point is the existence of a domain $S < \Omega$ such that $\Omega = \bigcup_{g\in\Gamma} gS$ and such that the set $S(L)$ is compact for any finite $L > K$. In fact one can take for S a finite union of affinoid subdomains of Ω. This can be proved by means of the graph theoretic interpretation of Ω in § 1 (see [6]).

Corollary. *There exists a "degree" homomorphism*:

$$\mathfrak{D}^\Gamma \xrightarrow{\deg} Z: \sum_{g\in\Gamma} g\left(\sum_{i=1}^{K} m_i z_i\right) \mapsto \sum_{i=1}^{K} m_i.$$

We will also speak on the degree of a finite divisor. This can cause no confusion.

Weierstrass products. Let $\sum_{g\in\Gamma} g(d)$ be a Γ-invariant divisor of degree 0. We will construct a meromorphic function on Ω with this divisor. As d is finite of degree zero there exists a K-rational function on P^1, w_d with divisor d. Take a point $z_0 \in \Omega\setminus \bigcup_{g\in\Gamma} g(\operatorname{supp} d)$ and consider the formal product

$$W_{d,z_0} = \prod_{h\in\Gamma} \frac{w_d(hz)}{w_d(hz_0)}. \tag{1}$$

Proposition 1. *For any $z \in \Omega$ the product* (1) *can be written as the product of a finite number of terms having zeroes or poles at z and a convergent infinite factor. The convergence is uniform outside any Γ-invariant neighbourhood of* $\operatorname{supp}(d) \cup \Sigma$.

Proof. Let $z, z_0 \in \Omega$. We show first that $\delta(hz, hz_0) \to 0$ for any $\mathrm{PGL}(2, O) = \mathrm{Aut}\{M\}$-invariant distance δ in $P^1(\bar{K})$, $M < V$. Take the line in Δ with ends z, z_0 and consider the vertex Λ on it closest to Δ_Γ. Also take a vertex Λ_1 of Δ_Γ. Now the distance in Δ from Λ_1 to $h\Lambda$ tends to infinity with h approximately as the length of h in a reduced representation relative to some fixed free base of Γ (this is easily deduced from [5]).

It follows that the paths from Λ_1 to hz, hz_0 begin to diverge at a greater and greater distance from Λ_1. An easy geometric argument then shows that the same is true for any $\Lambda_2 \in \Delta^0$. This means that $\delta(hz, hz_0) \to 0$ for any $\mathrm{Aut}\{M\}$-invariant distance δ, where $\{M\} = \Lambda$.

Now we see that $w_d(hz)/w_d(hz_0) \to 1$ as h changes. In fact this is the quotient of values of a rational function in a pair of variable points which become closer and closer always staying outside a preassigned neighbourhood of the zeroes and poles of the function.

Since K is nonarchimedean we conclude that (1) converges at all points $z \in \Omega \setminus \bigcup_g g(\operatorname{supp} d)$. A slightly closer look at the picture also shows uniform convergence.

Automorphy factors. We maintain

Proposition 2. *In the previous notation, for all $g \in \Gamma$ we have:*

$$W_{d,z_0}(gz) = \mu_d(g) W_{d,z_0}(z), \tag{2}$$

where $\mu_d(g) \in K^*$ are multiplicative in d and g and are independent of z_0.

Proof. We have

$$W_{d,z_0}(gz) = \prod_{h \in \Gamma} \frac{w_d(hgz)}{w_d(hz_0)} = \prod_{h \in \Gamma} \frac{w_d(hz)}{w_d(hg^{-1}z_0)} .$$

Hence

$$\mu_d(g) = \prod_{h \in \Gamma} \frac{w_d(hz_0)}{w_d(hg^{-1}z_0)} . \tag{3}$$

Changing z_0 only multiplies W by a constant; so from (2) one sees that $\mu_d(g)$ does not depend on z_0. Multiplicativity in d and g is also clear. Finally $\mu_d(g)$ is obviously contained in K^* if $z_0 \in \Omega(K)$. If z_0 has to be chosen in $\Omega(L)$ for some $L > K$, then $\mu_d(g) \in L^*$ is $\mathrm{Gal}\, \bar{K}/K$-invariant (being independent of z_0), and hence is still in K^*.

The case $\sum_{g \in \Gamma} g(d) = 0$ is particularly interesting since in this case $W_{d,z_0}(z)$ is a unit in Ω. Lemma 1 shows that it suffices to consider $d = (e - 1)z_1$, where $e \in \Gamma$, $z_1 \in \Omega$.

Theorem 1. *Put $H = \Gamma/[\Gamma, \Gamma]$. Let $g, e \in \Gamma$ and let χ, ε be the classes of g, e in H. Put*

$$\langle \chi, \varepsilon \rangle = \mu_{(e-1)z_1}(g).$$

Then $\langle \chi, \varepsilon \rangle$ depends only on χ, ε (not on g, e and z_1). The mapping

$$\langle\ ,\ \rangle : H \times H \to K^*$$

is a symmetric bimultiplicative pairing.

Proof. Take $w_{(e-1)z_1}(z) = \frac{z - ez_1}{z - z_1}$, where z is an affine coordinate function on P^1 (we identify a point with its z-coordinate). For any $z_0 \in \Omega \setminus (\Gamma \infty \cup \Gamma z_1)$ we get from (3)

$$\mu_{(e-1)z_1}(g) = \prod_{h \in \Gamma} \frac{hz_0 - ez_1}{hz_0 - z_1} \frac{hg^{-1}z_0 - z_1}{hg^{-1}z_0 - ez_1}. \tag{4}$$

Using the projective invariance of a cross-ratio we can rewrite (4) as follows:

$$\begin{aligned}\mu_{(e-1)z_1}(g) &= \prod_{h \in \Gamma} \frac{h^{-1}ez_1 - z_0}{h^{-1}ez_1 - g^{-1}z_0} \frac{h^{-1}z_1 - g^{-1}z_0}{h^{-1}z_1 - z_0} \\ &= \prod_{h \in \Gamma} \frac{w_{g^{-1}z_0 - z_0}(h_{z_1}^{-1})}{w_{g^{-1}z_0 - z_0}(h_{ez_1}^{-1})} = \mu_{(g^{-1}-1)z_0}(e^{-1}).\end{aligned} \tag{5}$$

Now the left hand side of (5) does not depend on z_0 and is multiplicative in g and the right hand side does not depend on z_1 and is multiplicative in e. Since they take values in K^* they depend in fact only on χ, ε and rewriting (5) in the form $\langle \chi, \varepsilon \rangle = \langle \varepsilon^{-1}, \chi^{-1} \rangle$ we get the symmetry.

Later we will show that this scalar product $\langle \, , \rangle$ is essentially the Riemann form of the Jacobian of X_Γ, H being the part of homology group of X_Γ killed by the covering $\Omega \to X_\Gamma$.

The formula (4) is not very convenient to work with since it involves the points z_0, z_1 although the result does not depend on them.

We will now give a more useful expression.

Let the images of $g, e \in \Gamma$ in H be independent. Choose a system of representatives of double cosets $(e^m) \setminus \Gamma / (g^n)$ and denote it $C(e \mid g)$. It is easy to see that every element of Γ can be uniquely written in the form $e^m h g^n$, where $m, m \in \mathbb{Z}$, $h \in C(e \mid g)$.

Similarly, let $g \in \Gamma$ be an element whose image in H is not divisible and take a system $C_0(g \mid g)$ of representatives of cosets $(g^m) \setminus \Gamma / (g^n)$ excluding the identity coset. Then every element of Γ either belongs to (g^n) or can be uniquely written in the form $g^m h g^n$, where $m, n \in \mathbb{Z}$, $h \in C_0(g \mid g)$.

In the following theorem we keep these conditions on g, e (resp. g). Finally for any hyperbolic g we set

$$z_g^+ = \lim_{n \to \infty} g^n z_0, \qquad z_g^- = \lim_{n \to -\infty} g^n z_0.$$

These are attractive and repulsive fixed points of g respectively; any z_0 will do which is not a fixed point of g itself.

Theorem 2. a) *Put* $\chi, \varepsilon = g, e \bmod [\Gamma, \Gamma]$ *respectively. Then*

$$\langle \chi, \varepsilon \rangle = \prod_{h \in C(e|g)} \frac{w(z_e^+ - z_e^-; hz_g^+)}{w(z_e^+ - z_e^-; hz_g^-)} \tag{6}$$

(*where we write* $w(d; z)$ *instead of* $w_d(z)$ *for typographical reasons*).

b) *Put* $\chi = g \bmod [\Gamma, \Gamma]$. *Then*

$$\langle \chi, \chi \rangle = q \prod_{h \in C_0(g|g)} \frac{w(z_g^+ - z_g^-; hz_g^+)}{w(z_g^+ - z_g^-; hz_g^-)} \tag{7}$$

where g *is represented by a matrix conjugate to* $\begin{pmatrix} q & 0 \\ 0 & 1 \end{pmatrix}$, $|q| < 1$.

A particular case. If the coordinate system on P^1 is taken so that $g = \begin{pmatrix} q & 0 \\ 0 & 1 \end{pmatrix}$, then (7) becomes

$$\langle \chi, \chi \rangle = q \prod \frac{bc}{ad}, \qquad \begin{pmatrix} a & b \\ c & d \end{pmatrix} \in C_0(g \mid g).$$

Proof. To calculate $\langle\chi, \varepsilon\rangle$ rewrite (4) in the form

$$\langle\chi, \varepsilon\rangle = \prod_{h\in\Gamma} \frac{ez_1 - hz_0}{ez_1 - hg^{-1}z_0} \frac{z_1 - hg^{-1}z_0}{z_1 - hz_0}.$$

Then we collect together all members h of a coset $(e^{-m})\backslash\Gamma/(g^n)$ and reorder the product:

$$\langle\chi, \varepsilon\rangle = \prod_{h\in C(e|g)} \prod_{m=-\infty}^{\infty} \prod_{n=-\infty}^{\infty} \frac{ez_1 - e^{-m}hg^n z_0}{ez_1 - e^{-m}hg^{n-1}z_0} \frac{z_1 - e^{-m}hg^{n-1}z_0}{z_1 - e^{-m}hg^n z_0}.$$

Now the inner product $\prod_n$ is easily calculated. Fixing m, h denote the n-th member of this product by $\frac{\alpha_n}{\alpha_{n-1}} \frac{\beta_{n-1}}{\beta_n}$ in the obvious notation. Then

$$\prod_{n=-N}^{N} \frac{\alpha_n}{\alpha_{n-1}} = \frac{\alpha_N}{\alpha_{-N-1}}, \qquad \prod_{n=-N}^{N} \frac{\beta_{n-1}}{\beta_n} = \frac{\beta_{-N-1}}{\beta_N}.$$

So putting $\alpha_\infty = \lim_{N\to\infty} \alpha_N$ etc. and using the invariance of the cross-ratio we get

$$\prod_{n=-\infty}^{\infty} = \frac{\alpha_\infty}{\alpha_{-\infty}} \frac{\beta_{-\infty}}{\beta_\infty} = \frac{ez_1 - e^{-m}hz_g^+}{ez_1 - e^{-m}hz_g^-} \frac{z_1 - e^{-m}hz_g^-}{z_1 - e^{-m}hz_g^+}$$

$$= \frac{e^{m+1}z_1 - hz_g^+}{e^{m+1}z_1 - hz_g^-} \frac{e^m z_1 - hz_g^-}{e^m z_1 - hz_g^+} = \frac{\gamma_{m+1}}{\delta_{m+1}} \frac{\delta_m}{\gamma_m}.$$

Hence the same trick can be applied to the product $\prod_m$, and we get finally

$$\prod_m \prod_n = \frac{\gamma_\infty}{\gamma_{-\infty}} \frac{\delta_{-\infty}}{\delta_\infty} = \frac{z_e^+ - hz_g^+}{z_e^- - hz_g^+} \frac{z_e^- - hz_g^-}{z_e^+ - hz_g^-}$$

which is the general member of (6).

The proof of (7) is completely similar. The only difference concerns the coset (g^n) for which our computation first gives $\frac{w(z_g^+ - z_g^-; gz_1)}{w(z_g^+ - z_g^-; z_1)}$. Now calculating this in the coordinate system with $g = \begin{pmatrix} q & 0 \\ 0 & 1 \end{pmatrix}$ we get simply q. ∎

We finish this section with a final formula of this type which is proved in the same way:

Theorem 3. *We have*

$$W_{(g-1)z_1, z_0}(z) = \prod_{h\in C(|g)} \frac{w(hz_g^+ - hz_g^-; z)}{w(hz_g^+ - hz_g^-; z_0)}$$

where $C(|g)$ is a set of representatives of $\Gamma/(g^n)$. Hence this function has no zeroes and poles in Ω, does not depend on z_1, and is multiplicative in g. ∎

The last remark immediately follows from

$$(g_1g_2 - 1)z_1 = (g_1 - 1)(g_2z_1) + (g_2 - 1)z_1.$$

Corollary. *The formula*

$$\omega_g = d \log W_{(g-1)z_1, z_0} \tag{8}$$

defines a Γ-invariant holomorphic differential on Ω, which depends additively on $g \in \Gamma$. ∎

§ 3. Analytic objects associated with a Schottky group

Here we will construct a number of K-analytic objects associated with a Schottky group $\Gamma < G$ which are to be algebrized later.

a) *The analytic curve* X_{an}. This is the functor on finite extensions of K:

$$L \mapsto \Gamma\backslash\Omega(L) = X_{an}(L).$$

b) *The analytic group* $\mathfrak{D}_{an}$ *of divisors of degree zero.* This is the functor $L \mapsto \mathfrak{D}_{an}(L) =$ group of Γ-invariant L-divisors on Ω of degree zero (see 2).

c) *The analytic principal divisor group* $\mathfrak{D}^0_{an}$.

We call the divisor $D \in \mathfrak{D}_{an}(L)$ *principal* if there exists a finite L-divisor d such that

$$D = \sum_{g\in\Gamma} g(d) \quad \text{and} \quad \mu_d(g) = 1 \qquad \text{for all } g \in \Gamma.$$

That is, $W_{d,z_0}(z)$ is Γ-invariant and so can be considered as a meromorphic function on X_{an}.

$\mathfrak{D}^0_{an}$ is thus a subfunctor of $\mathfrak{D}_{an}$.

d) *The analytic Jacobian* J_{an} *of the curve* X_{an}.

First construct the algebraic torus $T = \operatorname{Spec} K[H]$ with character group $H = \Gamma/[\Gamma, \Gamma]$. A K-point of T is a homomorphism $H \to K^*$. In particular, the map $\chi \mapsto \langle\chi, \varepsilon\rangle$ for variable $\chi \in H$ and fixed $\varepsilon \in H$ is a point of T. Denote by B the group of all such points. It is defined together with a homomorphism

$$\varphi^{-1}: H \to B,\ \varphi^{-1}(\varepsilon) = \langle\cdot, \varepsilon\rangle.$$

We will show later that B is a discrete subgroup of $T(K)$ of rank n and that φ^{-1} is an isomorphism the inverse of which is a *polarization* in the sense of Mumford [4].

Finally, *the analytic Jacobian* J_{an} is the functor $L \mapsto T(L)/B$.

e) *The analytic Jacobi maps.* They are the functor morphisms

$$j_1: \mathfrak{D}_{an}/\mathfrak{D}^0_{an} \to J_{an},$$
$$j: X_{an} \to J_{an}.$$

Definition of j_1:

$$j_1\Big(\sum_{g\in\Gamma} g(d) \bmod \mathfrak{D}^0_{an}\Big) = \mu_d(\cdot) \bmod B \in T/B$$

where $\mu_d(\cdot)$ is the homomorphism

$$H \to K^*: g \bmod [\Gamma, \Gamma] \mapsto \mu_d(g)$$

considered as a point of T. To show that j_1 is well defined note that if $\sum_{g\in\Gamma} g(d) \in \mathfrak{D}^0_{an}(K(d))$ then by definition of $\mathfrak{D}^0_{an}$ there exists a finite divisor d' such that $\sum_{g\in\Gamma} g(d) = \sum_{g\in\Gamma} g(d')$ and $\mu_{d'}(\cdot) = 1$. Hence by Lemma 1 we have $d - d' = \sum_{i=1}^{K} (1 - g_i)d_i$. So the point of T

$$\mu_d(\cdot) = \mu_d(\cdot)\mu_{d'}(\cdot)^{-1} = \mu_{d-d'}(\cdot) = \prod_{i=1}^{K} \mu_{(1-g_i)d_i}(\cdot)$$

belongs to B, in view of the definition of B and Theorem 1.

Definition of j. To define j choose a point $z_0 \in \Omega(K)$ and for $z_1 \in \Omega$ set

$$j(\Gamma z_1) = \mu_{z_1 - z_0}(\cdot) \bmod B.$$

The map is well defined because

$$\frac{\mu_{gz_1,-z_0}(\cdot)}{\mu_{z_1,-z_0}(\cdot)} = \mu_{(g-1)z_1}(\cdot) \in B.$$

If $\Omega(K)$ is empty then one can define j only over some finite extensions of K where X_{an} has a rational point.

f) *The analytic differentials of the first kind.*

These are differentials ω_g for $g \in \Gamma$ defined by (8). Now we can give a different expression for them. Namely, if $\chi = g \bmod [\Gamma, \Gamma]$, then

$$\omega_g = \bar{j}^*(\chi^{-1} d\chi) \tag{9}$$

where $\bar{j}: \Omega \to T$ is defined by $\bar{j}(z) = \mu_{z-z_0}(\cdot)$ as in e). We prove more:

$$W_{(g-1)z_1,z_0} = \bar{j}^*(\chi).$$

In fact, this is equivalent to

$$W_{(g-1)z_1,z_0}(z) = \chi(\bar{j}(z)) = \mu_{z-z_0}(g)$$

which is easily checked using (1) and (3).

The rest of this paper is devoted to explaining how all these objects can be algebrized. For X_{an} the construction is due to Mumford [3]. The main point is to reduce the algebraization of J_{an} to Mumford [4] or Gerritzen [1] by showing, that (T, B, φ) is a torus with periods and polarization.

Having completed this in § 4 we will make some concluding remarks in § 5.

§ 4. Polarization of J_{an}

We will prove the following

Theorem 4. *Let $\chi \in H$, $\chi \neq 0$; then $|\langle \chi, \chi \rangle| < 1$.*

Corollary 1. *The form*

$$H \times H \to \mathbb{Z} : (\chi, \varepsilon) \mapsto \operatorname{ord} \langle \chi, \varepsilon \rangle$$

is positive definite. ∎

Corollary 2. *The subgroup $B < T(K)$ is discrete and free of rank n; $\varphi^{-1}: H \to B$ is an isomorphism and $\varphi: B \to H$ is a polarization.*

(All of this follows from Theorem 4 more or less by definition).

To prove Theorem 4 we will explicitly calculate the form ord $\langle\ ,\ \rangle$ in terms of the graph Δ_Γ (see § 1). First introduce some notions.

Denote by $C_1(\Gamma \backslash \Delta_\Gamma)$ the group of integral 1-chains of Γ. If e is an oriented edge of $\Gamma \backslash \Delta_\Gamma$, then $-e$ is the same edge with reverse orientation. We define a scalar product on $C_1(\Gamma \backslash \Delta_\Gamma)$, setting

$$[e, e] = 1,\ [e_1, e_2] = 0, \text{ if } e_1 \neq \pm e_2$$

for edges e, e_1, e_2. Then $[\ ,\ ]$ is a positive definite symmetric form.

Now let $g \in \Gamma$, $\chi = g \bmod [\Gamma, \Gamma] \in H$, $\chi \neq 1$. Take the axis X_g of g in Δ_Γ (see § 1) and orient it from z_g^+ to z_g^-. Take a fundamental domain of (g^n) acting on X_g and denote $e_1, \ldots, e_r$ its edges with the induced orientation. Let $\pi: \Delta_\Gamma \to \Gamma \backslash \Delta_\Gamma$ be the natural projection and put

$$c_\chi = \sum_{i=1}^{r} \pi(e_i) \in C_1(\Gamma \backslash \Delta_\Gamma).$$

The chain c_χ is a cycle whose homology class in $H_1(\Gamma\backslash\Delta_\Gamma) \cong H$ coincides with the image of $g \in \Gamma$ under the natural map $\pi_1(\Gamma\backslash\Delta_\Gamma) \cong \Gamma \to H_1(\Gamma\backslash\Delta_\Gamma) = H$. So c_χ depends only on χ and is additive in χ.

Theorem 4 will be an immediate consequence of the following theorem, which is closely related to some results of Grothendieck [2], Exposé IX, 1. 2. 3. 7 and 1. 2. 5.

Theorem 5. *For all $\chi, \varepsilon \in H$ we have*

$$\operatorname{ord}\langle \chi, \varepsilon\rangle = [c_\chi, c_\varepsilon]$$

Proof. Since both parts are bilinear and symmetric in χ, ε it suffices to treat the case when $\chi = \varepsilon \neq 1$ and χ is not divisible in H. So we are in a position to apply Theorem 2, which gives:

$$\langle \chi, \chi\rangle = q \prod_{h \in C_0(g|g)} \frac{w(z_g^+ - z_g^-; hz_g^+)}{w(z_g^+ - z_g^-; hz_g^-)}.$$

Now ord $q = r =$ the length of the fundamental domain of g on its axis X_g. To calculate the ord of a crossratio we use the following identity:

$$\operatorname{ord} \frac{w(a_1 - a_2; z_1)}{w(a_1 - a_2; z_2)} = [\text{axis}(z_1, z_2), \text{axis}(a_1, a_2)]$$

= the number of edges which these axis have in common (with the sign "minus" if orientations are different).

(We omit the proof which is not difficult: compare Mumford [3], propositions 1. 9 and 1. 22 or see [6]).

To apply it note that

$$X_g \text{ (as a chain)} = \sum_{n \in \mathbb{Z}} \sum_{i=1}^{r} g^n e_i,$$

$$hX_g \text{ (as a chain)} = \sum_{n \in \mathbb{Z}} \sum_{i=1}^{r} hg^n e_i.$$

They are infinite chains of Δ_Γ, but all indices [,] will be finite, so we shall use freely the same notation as for $\Gamma\backslash\Delta_\Gamma$. Hence

$$\operatorname{ord}\langle \chi, \chi\rangle = r + \sum_{h \in C_0(g|g)} \sum_{m,n} \sum_{i,j=1}^{r} [g^n e_i, hg^m e_j] = r + \sum_{h \in C_0(g|g)} \sum_{m,n} \sum_{i,j=1}^{r} [e_i, g^{-n}hg^m e_j]$$

$$= r + \sum_{h \in \Gamma\backslash(g^n)} \sum_{i,j=1}^{r} [e_i, he_j] = \sum_{h \in \Gamma} \sum_{i,j=1}^{r} [e_i, he_j].$$

To get from the second equation to the third note that $\{g^{-n}hg^m\} = \Gamma\backslash\{g^n\}$ by definition of $C_0(g|g)$. The last equality follows from $\sum_{n=-\infty}^{\infty} \sum_{i,j=1}^{r} [e_i, g^n e_j] = r$. Finally we must show that

$$\sum_{h \in \Gamma} \sum_{i,j=1}^{r} [e_i, he_j] = [c_\chi, c_\chi] = \sum_{i,j=1}^{r} [\pi(e_i), \pi(e_j)].$$

In fact

$$[\pi(e_i), \pi(e_j)] = \begin{cases} [e_i, he_j] & \text{if there exists a (necessarily unique) } h \in \Gamma \text{ such that } he_i = \pm e_j; \\ 0 & \text{otherwise.} \end{cases}$$

This completes the proof. ■

§ 5. The final steps

We will be rather sketchy here because the general method of proving the algebricity of the analytic objects of § 3 is now clear, although some necessary background material has never been given in full detail for rigid analytic spaces.

From Mumford [3], [4] and Gerritzen [1] we know now that X_{an} and J_{an} "are" in fact an algebraic curve X of genus n and an abelian variety J of dimension n over K respectively. So the holomorphic map $j: X_{an} \to J_{an}$ is algebraic by GAGA. As $X(L) = X_{an}(L)$ for any finite $L < K$, we have $\mathfrak{D}_{an}(L) = \mathfrak{D}_{alg}(L)$ (the group of L-divisors of degree zero on the algebraic curve X). Moreover $\mathfrak{D}^0_{an} < \mathfrak{D}^0_{alg}$ (principal divisors on X), because in § 2 and § 3 we have constructed for each $D \in \mathfrak{D}^0_{an}$ a meromorphic Γ-invariant function on Ω with divisor D which gives a rational function on X.

On the other hand take a divisor $D \in \mathfrak{D}^0_{alg} < \mathfrak{D}_{an}$ and put $D = \sum_{g \in \Gamma} g(d)$. Then we have $\mu_d(\cdot) \equiv 1 \bmod B$. In fact, $\mu_{z_1-z_0}(\cdot) \bmod B$ is $j(\Gamma z_1)$, so that $\mu_d(\cdot)$ is the sum of j-images of points in d, taken in the sense of the group law of J.

But it is well known that such a sum is always zero for algebraic principal D. It follows that we can change d without changing D so that $\mu_d(g) = 1$, that is $D \in \mathfrak{D}^0_{an}$. Hence $\mathfrak{D}^0_{an} = \mathfrak{D}^0_{alg}$. This means that the variety

$$J = \mathfrak{D}_{an}/\mathfrak{D}^0_{an} = \mathfrak{D}_{alg}/\mathfrak{D}^0_{alg}$$

coincides with the algebraic jacobian of X and j is the usual map of X into its jacobian. This is our main result.

There are two minor points to clarify.

Firstly, it follows that invariant functions $W_{d,z_0}(z)$ for principal D in fact represent *all* rational functions on X. This corresponds to the Jacobi-Tate representation of functions on an elliptic curve as quotients of theta-functions. We cannot prove this in an elementary way as in Tate's theory for the lack of something like Laurent series: Ω is too complicated for $n \geqq 2$!

Secondly, the map (8) $g \mapsto \omega_g$ in fact defines a lattice of rank n in $\Gamma(X, \Omega^1_x)$ which generates the whole space of differentials of the first kind on X. To see this look at (9): the B-invariant differentials $\chi^{-1} d\chi$ (for all characters χ of the torus T) form a lattice in

$$\Gamma(J, \Omega^1_J) \xrightarrow{j^*} \Gamma(X, \Omega^1_x).$$

Bibliography

[1] *L. Gerritzen*, On non-archimedean representations of abelian varieties, Math. Ann. **196** (1972), 323—346.

[2] *A. Grothendieck*, Modèles de Néron et monodromie, Lecture Notes in Math. **288** (1972).

[3] *D. Mumford*, An analytic construction of degenerating curves over complete local rings, Compositio Math. **24** (1972), 129—174.

[4] *D. Mumford*, An analytic construction of degenerating abelian varieties over complete rings, Compositio Math. **24** (1972), 239—272.

[5] *J.-P. Serre*, Groupes discrets, Collège de France, Paris 1968/69.

[6] *Yu. I. Manin*, p-adic automorphic functions (in russian), "Itogi Nauki", Moscow 1973.

Steklov Mathematical Institute, Academy of Science USSR, Moscow State University

Eingegangen 23. März 1973

PERIODS OF PARABOLIC FORMS* AND p-ADIC HECKE SERIES

Ju. I. MANIN

To Igor Rostislavovič Šafarevič on his fiftieth birthday

Abstract. The author proves an algebraicity theorem for the periods of parabolic forms of any weight for the full modular group, gives explicit formulas for the coefficients of the forms, and constructs p-adic analogs of their Mellin transforms in the manner of Iwasawa and Mazur.

Bibliography: 13 items.

§1. Introduction. Basic results

The concentrated attention of mathematicians to number theory, while never lessening, has in recent years taken on new forms. Elementary questions about congruences and equations have found themselves becoming interwoven in an intricate and rich complex of constructions drawn from abstract harmonic analysis, topology, highly technical ramifications of homological algebra, algebraic geometry, measure theory, logic, and so on—corresponding to the spirit of Gödel's theorem on the incompleteness of the techniques of elementary arithmetic and on our capabilities of recognizing even those truths which we are in a position to "prove" (see for example [2], [9]).

A new "synthetic" number theory, taking in the legacy of the "analytic" theory, is possibly taking shape under our very eyes. (And here, perhaps, all the connotations of the word "synthetic" are appropriate.)

In our canvas, scarcely encompassable at a glance, any points of contact with concrete number-theoretical facts, whether old or new, take on especial significance. They discipline the imagination, and they provide a breathing space and the opportunity to evaluate the stunning beauty of past discoveries. Ramanujan's peculiar congruence modulo 691 serves as a fresh example. This congruence (and its analogs; see [7]) is so far, our only clue to understanding the 11-dimensional étale cohomology of the so-called Sato variety (see [9], [10], [11], [12]).

The present article is based on an analysis of several recent works on modular forms and has the modest aim of making explicit a part of the number-theoretic informa-

*Editors's note. In English one usually speaks of cusp forms.

AMS (MOS) subject classifications (1970). Primary 10D15, 12B30; Secondary 12B40.

tion which they contain implicitly. A subsequent translation of this information back into, say the language of representation theory is certainly possible and nontrivial, but remains a matter for the future.

Let us now describe in more detail the contents of the article.

1.1. Let $\Gamma = SL(2, Z)/(\pm 1)$, let w be an even integer, and let S_{w+2} be the space of parabolic forms with respect to Γ on the upper half-plane of one complex variable (see [8] for the basic notions).

To each $\Phi(z) \in S_{w+2}$ is associated the following collection of invariants:

a) The periods of Φ: the numbers $r_k(\Phi) = \int_0^{i\infty} \Phi(z) z^k\, dz$, $0 \leq k \leq w$.

b) The coefficients of Φ: the numbers λ_n such that $\Phi(z) = \sum_{n=1}^{\infty} \lambda_n e^{2\pi i n z}$.

c) The Hecke series $L_\Phi(s) = \sum_{n=1}^{\infty} \lambda_n n^{-s}$.

In §§2–7 of this article we construct the system of equations which connects the periods and the coefficients of Φ, and show that solving this system of equations allows us to obtain extremely precise information on both the periods and the coefficients. In particular, Ramanujan's congruence is included in a new series of formulas.

In §§8–10 we construct p-adic analytic functions which are related to the $L_\Phi(s)$ just as the Leopoldt-Kubota-Iwasawa functions [1] are related to ordinary Dirichlet L-functions.

The first group of results is obtained by the application to forms of higher weight of the techniques developed in the author's articles [3]–[5] for forms of weight 2 using Shimura's theory [13]. The second part generalizes to forms of arbitrary weight the methods of Mazur and Swinnerton-Dyer, introduced in [6] for forms of weight 2.

We restrict ourselves here to the consideration of the full modular group along the lines explained above.

We now formulate some of the basic results of the article.

1.2. **The Periods Theorem.** *Let*

$$\Phi(z) = \sum_{n=1}^{\infty} \lambda_n e^{2\pi i n z} \in S_{w+2}$$

be a nonzero form which is an eigenform for all the Hecke operators, so that $\Phi | T_n = \lambda_n \Phi$. *Then the ratios*

$$(r_0(\Phi) : r_2(\Phi) : \ldots : r_w(\Phi)), \quad (r_1(\Phi) : \ldots : r_{w-1}(\Phi))$$

are rational over the algebraic number field $Q(\lambda_1, \cdots, \lambda_n, \cdots)$.

Example. Let $w = 10$. Then $S_{12} = C\Delta(z)$, where

$$\Delta(z) = e^{2\pi i z} \prod_{n=1}^{\infty} (1 - e^{2\pi i n z})^{24} = \sum_{n=1}^{\infty} \tau(n) e^{2\pi i n z}.$$

We have

$$(r_0 : r_2 : r_4) = \left(1 : -\frac{691}{2^2 \cdot 3^4 \cdot 5} : \frac{691}{2^3 \cdot 3^2 \cdot 5 \cdot 7}\right), \quad (r_1 : r_3 : r_5) = \left(1 : -\frac{5^2}{2^4 \cdot 3} : \frac{5}{2^2 \cdot 3}\right).$$

The values of the remining periods are determined by the general relation $r_k(\Phi) = (-1)^{k+1} r_{w-k}(\Phi)$ (see §2 below).

As we have said, Theorem 1.2 is proved by constructing an infinite system of homogeneous linear equations for the periods of Φ. This system naturally breaks up into equations for the even ($k \equiv 0\ (2)$) and odd ($k \equiv 1\ (2)$) periods. We check that each of these subsystems has a 1-dimensional space of solutions; furthermore, the coefficients of the equations lie in $Q(\lambda_1, \cdots, \lambda_n, \cdots)$.

A part of these equations (which all $\Phi \in S_{w+2}$ have in common) has been pointed out by Shimura [13]. Another part, which is different for each Φ, is written out here for the first time in terms of the Hecke operators. For $w = 10$ Shimura's equations provide all the values of the periods with the exception of r_0. As w increases, more and more of the information is contained in the remaining equations.

Let us introduce another interpretation of the periods $r_k(\Phi)$ in terms of the Hecke series. Since

$$L_\Phi(s) = \frac{(2\pi)^s}{\Gamma(s)} \int_0^\infty \Phi(iy)\, y^{s-1}\, dy,$$

it is easy to see that

$$r_k(\Phi) = \frac{k!\, i^{k+1}}{(2\pi)^{k+1}} L_\Phi(k+1), \quad 1 \leqslant k+1 \leqslant w+1.$$

Thus the Periods Theorem, roughly speaking, allows us to compute the values of $L_\Phi(s)$ *at the integer points within the critical strip*, which stretches from $\sigma = 0$ to $\sigma = w + 2$.

It would be very interesting to extend these computations to the remaining integer points. We also remark that our methods do not have anything to say on the arithmetic nature of the "common transcendencental multiplier" for the $r_{2k}(\Phi)$ and the $r_{2k+1}(\Phi)$ respectively.

1.3. **The Coefficients Theorem.** *Under the hypotheses of Theorem 1.2, for all $n \geq 2$ we have*

$$(\sigma_{w+1}(n) - \lambda_n) r_0 = \sum_{n=\Delta\Delta'+\delta\delta'}' \sum_{l=1}^{\left[\frac{w-2}{4}\right]}{}' 2 \binom{w}{2l} r_{2l} (\Delta^{2l}\delta^{w-2l} - \Delta^{w-2l}\delta^{2l}).^* \tag{1}$$

Here $\sigma_{w+1}(n) = \sum_{d|n} d^{w+1}$, $r_i = r_i(\Phi)$, and in the outer summation on the right the sum is taken over all integer solutions of the equation $n = \Delta\Delta' + \delta\delta'$ which satisfy

$$\Delta > \delta > 0, \text{ and}$$

$$\text{either } \Delta' > \delta' > 0 \text{ or } \Delta | n,\ \Delta' = n/\Delta,\ \delta' = 0 \text{ and } 0 < \delta/\Delta \leq 1/2. \tag{2}$$

**Translator's note.* Here and below the convention is observed that Σ' denotes a summation in which certain "boundary" terms are to be taken with the coefficient $1/2$.

Furthermore, the terms in which $\delta/\Delta = \frac{1}{2}$ are to be taken with the coefficient $\frac{1}{2}$.

Example. Again for $w = 10$ we find

$$\sigma_{11}(n) - \tau(n) = \sum_{n=\Delta\Delta'+\delta\delta'}' \frac{691}{18}(\Delta^8\delta^2 - \Delta^2\delta^8) + \frac{691}{6}(\Delta^6\delta^4 - \Delta^4\delta^6).$$

In particular, this formula gives a new proof of Ramanujan's congruence

$$\tau(n) \equiv \sigma_{11}(n) \mod 691.$$

The table given in §7 shows that formula (1) also gives other well-known congruences.

The Coefficients Theorem simply displays the explicit form of a certain part of the above-mentioned equations for the periods. To prove the Periods Theorem we need to use, among other things, information on the coefficient of $r_0(\Phi)$ in (1). On the other hand, if $r_0(\Phi) \neq 0$, then (1) can be considered as an explicit formula for the coefficients λ_n of Φ, and hence the name of the theorem. It is interesting to note that to compute λ_n we must sum some polynomial of degree w in Δ and δ over the solutions of the universal equation $n = \Delta\Delta' + \delta\delta'$; only this polynomial depends on Φ. Note further that $\sigma_{w+1}(n) - \lambda_n$ is the nth coefficient of the modular form $E_{w+2}(z) - \Phi(z)$, E_{w+2} being the Eisenstein series of weight $w+2$.

Thus the Coefficients Theorem provides a series of relations which can be written

$$E_{w+2}(z) - \Phi(z) = a_0 + \sum_{\substack{\Delta,\delta \\ \Delta',\delta'}}' F(\Delta, \delta) e^{2\pi i(\Delta\Delta'+\delta\delta')z}, \tag{3}$$

where $F(\Delta, \delta)$ is a homogeneous polynomial of degree w, and the summation is over the integer points of the region defined by (2). The series on the right of (3) can be considered to be a sort of theta function "with spherical polynomials" constructed for the indefinite quadratic form $\Delta\Delta' + \delta\delta'$. The role of this series still needs to be made explicit.

1.4. To formulate the theorem on p-adic Hecke series let us introduce some more notation.

First of all, fix a form $\Phi(z) \in S_{w+2}$ by the conditions that $\Phi | T_n = \lambda_n \Phi$ for all n, $\Phi(z) = \sum_1^\infty \lambda_n e^{2\pi i n z}$. Next, choose a prime p and an integer Δ_0 with $(\Delta_0, p) = 1$, and put $\Delta = \Delta_0 p$ (for $p \geq 3$) or $\Delta = 4\Delta_0$ (if $p = 2$). Furthermore, we fix an imbedding $O \hookrightarrow \overline{Q}_p$ of the field of all algebraic numbers into the algebraic closure of the p-adic field. In the sequel, all identifications of algebraic numbers with p-adic ones will be with respect to this imbedding.

Consider a primitive Dirichlet character χ modulo Δp^m, $m \geq 0$. χ can be uniquely written as a product $\chi_0\chi_1$, where χ_0 is a primitive character modulo Δ, and χ_1 is a primitive character of period p^m. χ_0 is called the *tame* component of χ, and χ_1 the *wild* component. Set

$$G(\chi) = \sum_{b \bmod \Delta p^m} \chi(b) e^{2\pi i \frac{b}{\Delta p^m}}$$

(the Gauss sum).

Consider the Hecke-Dirichlet series

$$L_\Phi(s, \chi) = \sum_{n=1}^{\infty} \chi(n) \lambda_n n^{-s} = \frac{(2\pi)^s}{\Gamma(s)} \int_0^\infty \Phi_\chi(iy) y^{s-1} dy.$$

From the Periods Theorem we will later deduce that the values of $L_\Phi(k, \chi)$ for $k = 1, \cdots, w+1$ and any χ are "almost algebraic" Here we only need the following fact: *there exist two numbers* $\omega^{\pm} \in C$ *such that the values of* $(1/2\pi i\omega^{\pm})L_\Phi(1, \chi)$ *are integers of the field* $Q(\lambda_1, \cdots, \lambda_n, \cdots, G(\chi), \chi(z))$ (where we divide by ω^+ or ω^- according as $\chi(-1) = +1$ or -1). Fix such ω^+ and ω^-

We are now in a position to state the final result. All the notation which we have introduced in this subsection is used explicitly or implicitly in our formulation.

1.5. **Theorem on p-adic Hecke series.** *Suppose that* λ_p *is a p-adic unit, and denote by* ρ *that p-adic root of the equation* $X^2 - \lambda_p X + p^{w+1} = 0$ *which is also a p-adic unit.*

Then there exists a unique power series $g_{\chi_0}(T)$ *with coefficients in the p-adic completion of the ring of integers of the field* $Q(\lambda_1, \cdots, \lambda_n, \cdots)$ *such that for any character* $\chi = \chi_0\chi_1$ *modulo* Δp^m, $m \geq 0$, *which has tame component* χ_0 *we have*

$$\frac{\Delta p^m}{G(\chi)} \frac{1}{2\pi i \cdot \omega^{\pm}} L_\Phi(1, \chi) = \rho^m g_{\chi_0}(\chi_1(1+q)-1), \tag{4}$$

where either $q = p$ *or* $q = 4$ (*for* $p = 2$).

We will prove this theorem by means of Mazur's "p-adic Mellin transform" of modular forms. This beautiful and important construction allows us to hope for adelic analogs of p-adic Hecke series.

Our result is clearly incomplete: the domain of values of the argument of L_Φ in (4) is too narrow. By analogy with the Leopoldt-Kubota-Iwasawa theory [1] one should expect a formula of type (4) for all the $L_\Phi(k, \chi)$, $k \in Z$, or at any rate for $k = 1, \cdots, w+1$.

It is a pleasure for me to be able to thank I. I. Pjateckiĭ-Šapiro and A. Nasybullin for extremely useful discussions, and B. Mazur and H. P. F. Swinnerton-Dyer for their kindness in sending me their article [6] before publication.

§2. The Shimura-Eichler relations

2.1. **Proposition.** *Let* $\Phi \in S_{w+2}$, *and let* $r_k(\Phi)$ *be the periods of* Φ. *Then for any* $k = 0, 1, \cdots, w$ *we have*

$$r_k(\Phi) + (-1)^k r_{w-k}(\Phi) = 0, \tag*{$(5)_k$}$$

$$r_k(\Phi) + (-1)^k \sum_{\substack{0\leqslant i\leqslant k\\ i\equiv 0(2)}} \binom{k}{i} r_{w-k+i}(\Phi) + (-1)^k \sum_{\substack{0\leqslant i\leqslant w-k\\ i\equiv k(2)}} \binom{w-k}{i} r_i(\Phi) = 0, \qquad (6)_k$$

$$\sum_{\substack{1\leqslant i\leqslant k\\ i\equiv 1(2)}} \binom{k}{i} r_{w-k+i}(\Phi) + \sum_{\substack{0\leqslant i\leqslant w-k\\ i\not\equiv k(2)}} \binom{w-k}{i} r_i(\Phi) = 0. \qquad (7)_k$$

Proof. Let $g = \begin{pmatrix} a & b\\ c & d\end{pmatrix}$, with $\det g \neq 0$, and let $F(z)$ be a polynomial of weight $\leq w$. Let

$$F^g(z) = (cz+d)^w\, F\left(\frac{az+b}{cz+d}\right).$$

Since Φ is a parabolic form of weight $w+2$ for Γ, for any $g \in \Gamma$ we have

$$\int_{g(0)}^{g(i\infty)} \Phi(z) F(z)\, dz = \int_0^{i\infty} \Phi(gz) F(gz)\, d(gz) = \int_0^{i\infty} \Phi(z) F^g(z)\, dz. \qquad (8)$$

(These and subsequent integrals are to be taken along geodesic arcs.) Put $s = \begin{pmatrix} 0 & -1\\ 1 & 0\end{pmatrix}$ and $t = \begin{pmatrix} 1 & -1\\ 1 & 0\end{pmatrix}$. These are the standard generators of Γ, and $s^2 = t^3 = 1$. It is easy to verify that

$$\left(\int_0^{i\infty} + \int_{s(0)}^{s(i\infty)}\right) \Phi(z) F(z)\, dz = \left(\int_0^{i\infty} + \int_{t(0)}^{t(i\infty)} + \int_{t^2(0)}^{t^2(i\infty)}\right) \Phi(z) F(z)\, dz = 0. \qquad (9)$$

Indeed, $s(i\infty) = 0$ and $s(0) = i\infty$, so that the first sum vanishes. In the second sum the integral is taken around the perimeter of a triangle with vertices $(0, i\infty, 1)$; replacing this by a nearby triangle lying entirely within the upper half-plane, and estimating the error terms using standard inequalities, we obtain the desired result.

Now to obtain $(5)_k$, $(6)_k$ and $(7)_k$ we must substitute $F(z) = z^k$ in (9), replace all the paths of integration by the path $(0, i\infty)$ using (8), and use the fact that by definition

$$(z^k)^s = (-1)^k z^{w-k}, \quad (z^k)^t = z^{w-k}(z-1)^k, \quad (z^k)^{t^2} = (-1)^k (z-1)^{w-k}.$$

The first part of (9) then gives us $(5)_k$, whereas the second, in view of the fact that

$$z^k + (z^k)^t + (z^k)^{t^2} = z^k + \sum_{i=0}^{k} \binom{k}{i} (-1)^{k+i} z^{w-k+i} + \sum_{i=0}^{w-k} \binom{w-k}{i} (-1)^i z^i, \qquad (10)$$

gives us an equation which only differs from $(6)_k$ in that the summation runs over all i rather than just over those having the same parity as 0 or k. To break up this equation into two, note that S_{w+2} has a basis made up of *real* forms (that is, forms with real coefficients). But if Φ is real, then the periods $r_{2k+1}(\Phi)$ are real, and the $r_{2k}(\Phi)$ pure imaginary. Hence, in the linear relation in the $r_k(\Phi)$ which arises out of (10), we can take real and imaginary parts; this then leads to $(6)_k$ and $(7)_k$, which proves the proposition.

2.2. Definition. Let S^+ be the space of real solutions of the system of equations $(5)_k$, $(6)_k$ for even k, and $(7)_k$ for odd k, and let S^- be the space of real solutions of the system $(5)_k$, $(6)_k$ for odd k, and $(7)_k$ for even k.

We clearly have $S^+ \subset R^{(w+2)/2}$ and $S^- \subset R^{w/2}$, where $R^{(w+2)/2}$ and $R^{w/2}$ are the real vector spaces with basis indexed by the even and odd integers respectively in the interval $[0, w]$.

Note that $(1, 0, \cdots, 0, -1) \in S^+$; indeed, a straightforward check shows that $r_0(\Phi)$ and $r_w(\Phi)$ have the same coefficient (either 1 or 2) in all the equations (5)–(7).

2.3. Proposition (Eichler, Shimura). *Let S^0_{w+2} be the space of all real parabolic forms of weight $w+2$. Then the following assertions are true:*

a) *The map*

$$S^0_{w+2} \to S^- : \Phi \mapsto (r_1(\Phi),\ r_3(\Phi),\ \ldots,\ r_{w-1}(\Phi))$$

is an isomorphism.

b) *The map*

$$S^0_{w+2} \to S^+ : \Phi \mapsto (r_0(\Phi), r_2(\Phi),\ \ldots,\ r_w(\Phi))$$

is an imbedding of S^0_{w+2} as a subspace of codimension 1 in S^+ which does not contain the vector $(1, 0, \cdots, 0, -1)$.

This result can be considered an immediate reformulation of a particular case of a theorem of Shimura's ([13], Theorem 1, §5; compare also §9 of the same paper). We omit the proof, since it does not have any direct relation to what will follow. Apart from that, the proof which Shimura gives is not the most natural one here, since it relies on an independent computation of the dimensions of S^+ and S^- carried out by Eichler. In fact Proposition 2.3 follows from the theorems of de Rham and Dolbeault, applied to the Sato variety–the w-fold fiber product of the "general" elliptic curve. The periods of the forms Φ are the periods of the differentials of the first kind of the highest order on such a variety.

§3. The action of the Hecke operators on the periods

3.1. Let $g = \begin{pmatrix} a & b \\ c & d \end{pmatrix}$ and $\Phi \in S_{w+2}$. Recall that if we set

$$(\Phi | g)(z) = (\det g)^{\frac{w+2}{2}} (cz + d)^{-w-2} \Phi(gz)$$

then the Hecke operator T_n on S_{w+2} is defined by the formula

$$T_n = n^{\frac{w}{2}} \sum_{d|n} \sum_{b \bmod d} \begin{pmatrix} nd^{-1} & b \\ 0 & d \end{pmatrix}. \tag{11}$$

It is well known that S^0_{w+2} is invariant with respect to all the Hecke operators T_n and has a unique decomposition as a direct sum of 1-dimensional invariant subspaces.

3.2. Proposition. *For all $i, k = 0, \cdots, w$ and $n \geq 0$ there exist coefficients $A_{ik}(n) \in Z$, zero if $i \not\equiv k$ (2), such that*

$$\int_0^{i\infty}(\Phi\,|\,T_n)(z)z^k\,dz=\sum_{l=0}^{w}A_{lk}(n)\,r_l(\Phi)$$

for all $\Phi\in S^0_{w+2}$. *In other words, there exist endomorphisms* $T^{\pm}_n$ *of the spaces* $R^{(w+2)/2}$ *and* $R^{w/2}$ *of* §2.2,

$$T_n^+:R^{\frac{w+2}{2}}\to R^{\frac{w+2}{2}},\quad T_n^-:R^{\frac{w}{2}}\to R^{\frac{w}{2}},$$

satisfying the following conditions:

a) *The matrix of* $T^{\pm}_n$ *in the standard basis has integer entries.*

b) *The images of* S^0_{w+2} *in* $R^{(w+2)/2}$ *and* $R^{w/2}$ *(see Proposition 2.3) are invariant under the* T^+_n *and* T^-_n *respectively.*

c) *The restriction of* $T^{\pm}_n$ *to these images coincides with the action of* T_n *on* S^0_{w+2}.

Proof. First of all let $F(z)$ be an arbitrary polynomial of degree $\le w$. In the expression $\int_0^{i\infty}(\Phi|T_n)(z)F(z)\,dz$ substitute the explicit form (11) for T_n. We will carry out the summation over b mod d over the interval $-d/2\le b\le d/2$; the symbol Σ' will indicate that in the summation in question the terms with $b=\pm d/2$ are to be taken with the coefficient ½. We have

$$\int_0^{i\infty}(\Phi\,|\,T_n)(z)\,F(z)\,dz=\int_0^{i\infty}\sum_{d|n}\ \sum_{-\frac{d}{2}\le b\le\frac{d}{2}}{}^{'}\ \Phi\left(\frac{n}{d^2}z+\frac{b}{d}\right)\frac{n^{w+1}}{d^{w+2}}F(z)\,dz.$$

Replacing $nz/d^2+b/d$ by z in each term, we bring this expression to the form

$$\int_0^{i\infty}(\Phi|T_n)(z)\,F(z)\,dz=\sum_{d|n}\frac{n^w}{d^w}\sum_{-\frac{d}{2}\le b\le\frac{d}{2}}{}^{'}\left(\int_0^{i\infty}-\int_0^{b/d}\right)\Phi(z)\,F\left(\frac{d^2}{n}z-\frac{bd}{n}\right)dz.\qquad(12)$$

To deduce Proposition 3.2. from (12), set $F(z)=z^k$ and expand $(d^2z/n-bd/n)^k$ in powers of z. Clearly we obtain on the right an integral linear combination of the periods $r_l(\Phi)$ and of integrals of the form $\int_0^{b/d}\Phi(z)z^l\,dz$ with $l\le k$. Hence it suffices to check that all such integrals are integer linear combinations of the periods of Φ.

In fact, let $b>0$, $(b,d)=1$, and let $b/d=b_m/d_m,\ b_{m-1}/d_{m-1},\cdots,b_0/d_0=0/1$ be the successive convergents to b/d in irreducible form. It is well known that

$$g_k=\begin{pmatrix}b_k & (-1)^{k-1}\,b_{k-1}\\ d_k & (-1)^{k-1}\,d_{k-1}\end{pmatrix}\in SL(2,\,Z),$$

and hence by (8)

$$\int_0^{b/d}\Phi(z)\,z^l\,dz=\sum_{k=1}^{m}\int_{b_{k-1}/d_{k-1}}^{b_k/d_k}\Phi(z)\,z^l\,dz=\sum_{k=1}^{m}\int_{g_k(0)}^{g_k(i\infty)}\Phi(z)\,z^l\,dz$$

$$=\sum_{k=1}^{m}\int_0^{i\infty}\Phi(z)\,(b_kz+(-1)^{k-1}b_{k-1})^l\,(d_kz+(-1)^{k-1}d_{k-1})^{w-l}\,dz.\qquad(13)$$

This completes the proof. We note that the matrices $T_n^{\pm}$ can be computed explicitly without especial difficulty.

§4. The proof of Theorem 1.2 for odd periods

4.1. Let $\Phi \in S_{w+2}^0$, and suppose that $\Phi|T_n = \lambda_n \Phi$ for all $n \geq 1$. Then the vector $(r_1(\Phi), \cdots, r_{w-1}(\Phi)) \in S^-$ is an eigenvector for all the operators T_n^-, with eigenvalues λ_n (Proposition 3.2 c)). But the subspace $S^- \subset R^{w/2}$ is defined by the Eichler-Shimura equations, which have integer coefficients, and the T_n^- also have integer coefficients. Finally, S^-, together with S_{w+2}^0, has a uniquely defined decomposition into one-dimensional subspaces invariant under all the T_n^-. Hence it follows that the coordinates of the eigenvector $(r_1(\Phi), \cdots, r_{w-1}(\Phi))$ are, up to a common multiple, rational over $Q(\lambda_1, \cdots, \lambda_n, \cdots)$, and this proves one half of the Periods Theorem.

4.2. The only obstacle to proving the second half of the Periods Theorem in precisely the same way is the fact that S^+ has dimension one greater than S_{w+2}^0 (Proposition 2.3 b)), and hence it could a priori turn out that all the T_n^+ have an invariant plane in S^+, on which T_n^+ acts by multiplication by λ_n. Then for one exceptional form Φ we could not guarantee the rationality of $(r_0(\Phi): \cdots : r_w(\Phi))$ over $Q(\lambda_1, \cdots, \lambda_n, \cdots)$.

In fact there is no such exceptional form, but to prove this we need more precise information on the form of the T_n^+, so that it will be convenient to first prove Theorem 1.3.

§5. Proof of Theorem 1.3

5.1. Formula (1) will be obtained as the result of a series of transformations of (12). First substitute $F = 1$ in (12), transfer all the $\int_0^{i\infty}$ from right to left and change signs. If we use the fact that $\Phi|T_n = \lambda_n \Phi$, this gives

$$(n\sigma_{w-1}(n) - \lambda_n) \int_0^{i\infty} \Phi(z)\, dz = \sum_{d|n} \frac{n^w}{d^w} \sum_{-\frac{d}{2} \leqslant b \leqslant \frac{d}{2}}{}^{'} \int_0^{b/d} \Phi(z)\, dz.$$

Fix b and d with $(b, d) = 1$ and collect together the terms $\int_0^{b\delta/d\delta}$ with all $\delta|(n/d)$. The coefficient of such an integral will then be

$$\frac{n^w}{d^w} \sum_{\delta | \left(\frac{n}{d}\right)} \delta^{-w} = \sigma_w\left(\frac{n}{d}\right),$$

so that

$$(n\sigma_{w-1}(n) - \lambda_n) \int_0^{i\infty} \Phi(z)\, dz = \sum_{d|n} \sigma_w\left(\frac{n}{d}\right) \sum_{\substack{1 \leqslant b \leqslant \frac{d}{2} \\ (b,d)=1}}{}^{'} \left(\int_0^{b/d} + \int_0^{-b/d} \right) \Phi(z)\, dz. \tag{14}$$

The term on the right with $d=1$ vanishes. The term with $d=2$ is

$$\sigma_w\left(\frac{n}{2}\right)\int_0^{1/2}\Phi(z)\,dz=\sigma_w\left(\frac{n}{2}\right)\int_{\left(\begin{smallmatrix}1&0\\2&1\end{smallmatrix}\right)(0)}^{\left(\begin{smallmatrix}1&0\\2&1\end{smallmatrix}\right)(i\infty)}\Phi(z)\,dz=\sigma_w\left(\frac{n}{2}\right)\int_0^{i\infty}\Phi(z)(2z+1)^w\,dz$$

$$=\sigma_w\left(\frac{n}{2}\right)(1-2^w)\,r_0(\Phi)+\sigma_w\left(\frac{n}{2}\right)\sum_{k=1}^{w-1}2^k\binom{w}{k}r_k(\Phi). \tag{15}$$

We transform the terms with $d\geq 3$ using (13) with $l=0$, and the completely analogous formula with $-b/d$ in place of b/d. In the notation of (13) we get

$$\left(\int_0^{b/d}+\int_0^{-b/d}\right)\Phi(z)\,dz=\sum_{k=1}^{m}\int_0^{i\infty}\Phi(z)\,[(d_kz+(-1)^{k-1}d_{k-1})^w-(d_k+(-1)^{k-1}d_{k-1}z)^w]dz$$

$$=\sum_{k=1}^{m}\sum_{i=0}^{w}\binom{w}{i}d_k^i(-1)^{(k-1)(w-i)}d_{k-1}^{w-i}(r_i(\Phi)-r_{w-i}(\Phi)).$$

On account of $(5)_i$ we may leave on the right-hand side only the terms with even i, and then sum only over $i=2l\leq(w-2)/2$. Taking out the $r_0(\Phi)$ term again, we get

$$\left(\int_0^{b/d}+\int_0^{-b/d}\right)\Phi(z)\,dz=\sum_{k=1}^{m}2\,(d_{k-1}^w-d_k^w)\,r_0(\Phi)$$

$$+\sum_{k=1}^{m}\sum_{l=1}^{\left[\frac{w-2}{4}\right]}2\binom{w}{2l}(d_k^{2l}d_{k-1}^{w-2l}-d_k^{w-2l}d_{k-1}^{2l})\,r_{2l}(\Phi). \tag{16$_{b,d}$}$$

The first sum on the left is obviously equal to $2(1-d^w)r_0(\Phi)$.

Now substitute (15) and $(16)_{b,d}$ into the right-hand side of (14) and take all the terms in $r_0(\Phi)$ over to the left. The coefficient of $r_0(\Phi)$ on the left is then of the form

$$n\sigma_{w-1}(n)-\lambda_n-\sum_{d\mid n}\sigma_w\left(\frac{n}{d}\right)\cdot 2\,(1-d^w)\cdot\frac{1}{2}\varphi(d). \tag{17}$$

(Here $\phi(d)/2$ for $d\geq 3$ is the number of b with $1\leq b\leq d/2$ and $(b,d)=1$; for $d=2$ the same formula automatically holds by (15).) An elementary calculation shows that (17) is equal to $\sigma_{w+1}(n)-\lambda_n$; that is, the coefficient of $r_0(\Phi)$ in (1).

It remains to check that we can also reduce the right-hand side of the formula obtained from (14) to the same form as the right-hand side of (1).

Up to this point in the computation the right-hand side of (14) has the following shape (after taking the terms in $r_0(\Phi)$ over to the left and using the fact that in (15) we need take only the terms with even k, the others being real):

$$\sigma_w\left(\frac{n}{2}\right)\sum_{l=1}^{\left[\frac{w-2}{4}\right]}\binom{w}{2l}(2^{2l}-2^{w-2l})r_{2l}(\Phi)$$

$$+\sum_{\substack{d\mid n\\ d\geqslant 3}}\sigma_w\left(\frac{n}{d}\right)\sum_{\substack{1\leqslant b<\frac{d}{2}\\ (b,d)=1}}\sum_{k=1}^{m\left(\frac{b}{d}\right)}\sum_{l=1}^{\left[\frac{w-2}{4}\right]}2\binom{w}{2l}(d_k^{2l}d_{k-1}^{w-2l}-d_k^{w-2l}d_{k-1}^{2l})r_{2l}(\Phi). \tag{18}$$

Here the first sum has arisen from (15) by the use of $(5)_k$, and the second from $(16)_{b,d}$; in summing over b and d in (18) one should bear in mind that the length $m(b/d)$ of the expression for b/d as a continued fraction and the denominators d_k are functions of (b/d).

To compare (18) with (1), we use the following lemma of Heilbronn ([4], Lemma 7.7):

5.2. **Lemma.** *The family of pairs (d_k, d_{k-1}) of successive denominators of convergents to all rationals b/d with given $d\geq 3$ and $1\leq b<d/2$, $(b,d)=1$, is precisely the family of pairs (Δ, δ) taken from the solutions of the equation $d=\Delta\Delta'+\delta\delta'$, subject to the conditions $\Delta>\delta>0$, $(\Delta,\delta)=1$ and either $\Delta'>\delta'>0$, $(\Delta',\delta')=1$ or $\Delta=d$, $\Delta'=1$, $1\leq\delta'<d/2$, $\delta=0$, $(\Delta',\delta')=1$.*

5.3. The conditions imposed on the solutions of the equation $d=\Delta\Delta'+\delta\delta'$ in Heilbronn's Lemma only differ from the conditions (2) in the addition of the requirement $(\Delta,\delta)=(\Delta',\delta')=1$; that is, in the requirement that the solutions be primitive. That is compensated for by the factors $\sigma_w(n/d)$ in (18).

More precisely, write $\sigma_w(n/d)=\Sigma_{D\mid n/d}D^w$ and take D^w inside the internal summation:

$$D^w(d_k^{2l}d_{k-1}^{w-2l}-d_k^{w-2l}d_{k-1}^{2l})=(Dd_k)^{2l}(Dd_{k-1})^{w-2l}-(Dd_k)^{w-2l}(Dd_{k-1})^{2l}.$$

After this, to any primitive solution of $d=\Delta\Delta'+\delta\delta'$ and to any $D\mid n/d$ we make correspond the (not necessarily primitive) solution of $n=\bar\Delta\bar\Delta'+\bar\delta\bar\delta'$:

$$\bar\Delta=D\Delta,\quad \bar\delta=D\delta,\quad \bar\Delta'=\frac{n}{Dd}\Delta',\quad \bar\delta'=\frac{n}{Dd}\delta'.$$

The whole sum (18) then turns into a sum of the same form as the right-hand side of (1): the individual terms clearly have the same form, and the summation will be over the same region by Heilbronn's Lemma. This completes the proof of Theorem 1.3.

§6. Proof of Theorem 1.2 for even periods

6.1. **Lemma.** *Let $\Phi\in S^0_{w+2}$, and let $\Phi|T_n=\lambda_n\Phi$ for all $n\geq 1$. Furthermore, let $r'\in S^+$ (Definition 2.2) be a vector such that $r'|T_n^+=\lambda_n r'$ for all $n\geq 1$. Then r' is also the vector of periods for some form $\Phi'\in S^0_{w+2}$ (obviously proportional to Φ).*

Proof. According to Theorem 1.3, for any $\Phi\in S^0_{w+2}$ we have

$$\int_0^{i\infty} (\sigma_{w+1}(n)\Phi' - \Phi'|T_n)(z)\,dz = \sum_{l=1}^{\frac{w-2}{2}} A_{0,2l}(n)\, r_{2l}(\Phi'), \quad A_{0,2l}(n) \in Z. \tag{19}$$

Let $r' = (r'_0, r'_2, \cdots, r'_w)$ and $r'|T_n^+ = \lambda_n r'$ for all $n \geq 1$. Since (19) describes the action of T_n^+ on the r'_0 coordinate, we have

$$\sigma_{w+1}(n)\, r'_0 - \lambda_n r'_0 = \sum_{l=1}^{\frac{w-2}{2}} A_{0,2l}(n)\, r'_{2l}. \tag{20}$$

On the other hand, by Proposition 2.3 there exists a form $\Phi' \in S^0_{w+2}$ such that $ir'_2 = r_2(\Phi'), \cdots, ir'_{w-2} = r_{w-2}(\Phi')$. We will show that it then follows that $ir'_0 = r_0(\Phi')$. It will then follow from $(5)_0$ that $ir'_w = r_w(\Phi')$, and the lemma will be proved.

Substitute $ir'_{2l} = r_{2l}(\Phi')$ for $1 \leq l \leq (w-2)/2$ in (19). This gives

$$\int_0^{i\infty} (\sigma_{w+1}(n)\Phi' - \Phi'|T_n)(z)\,dz = i \sum_{l=1}^{\frac{w-2}{2}} A_{0,2l}(n)\, r'_{2l}. \tag{21}$$

Comparing (20) and (21) gives us

$$\int_0^{i\infty} (\sigma_{w+1}(n)\Phi' - \Phi'|T_n)(z)\,dz = \sigma_{w+1}(n)\, ir'_0 - \lambda_n ir'_0. \tag{22}$$

Now let us divide (22) by $\sigma_{w+1}(n)$ and take the limit as $n \to \infty$. It is well known that $\lim_{n\to\infty} \sigma_{w+1}(n)^{-1}\lambda_n = 0$ for any Φ. It then follows that

$$\lim_{n\to\infty} \sigma_{w+1}(n)^{-1} \int_0^{i\infty} \Phi'|T_n\, dz = 0$$

for any Φ'. Finally we obtain

$$\int_0^{i\infty} \Phi'\, dz = ir'_0,$$

which completes the proof of the lemma.

6.2. **Proof of Theorem 1.2. Conclusion.** Now we can repeat the arguments of §4.1 without alteration: the exceptional form referred to in §4.2 does not exist, since a collection $(\lambda_1, \cdots, \lambda_n, \cdots)$ of eigenvalues of the T_n^+ on $R^{(w+2)/2}$ corresponds by 6.1 to a 1-dimensional invariant subspace of S^+ (provided that $\Phi|T_n = \lambda_n\Phi$ for all n and some $\Phi \in S^0_{w+2}$).

§7. Examples

7.1. We give here a table of the even periods for those values of $w+2$ for which $\dim S_{w+2} = 1$; that is, for $w = 16, 18, 20, 22, 26$. The case $w+2 = 12$ was used as an illustration in the Introduction.

$w+2$	r_0	r_2	r_4	r_6	r_8	r_{10}
16	1	$-\frac{3617}{2^2\cdot 3^2\cdot 5\cdot 7\cdot 13}$	$\frac{3617}{2^3\cdot 3^2\cdot 5\cdot 11\cdot 13}$	$-\frac{3617}{2^3\cdot 3^3\cdot 5\cdot 7\cdot 13}$		
18	1	$-\frac{43867}{2^4\cdot 3^3\cdot 5^4}$	$\frac{43867}{2^6\cdot 3^2\cdot 5^2\cdot 7\cdot 13}$	$-\frac{43867}{2^6\cdot 3^2\cdot 5^3\cdot 7\cdot 11}$		
20	1	$-\frac{283\cdot 617}{2^2\cdot 3^4\cdot 5\cdot 7^2\cdot 17}$	$\frac{283\cdot 617}{2^3\cdot 3^5\cdot 5\cdot 7^2\cdot 17}$	$-\frac{283\cdot 617}{2^3\cdot 3^4\cdot 5\cdot 7^2\cdot 13\cdot 17}$	$\frac{283\cdot 617}{2^3\cdot 3^5\cdot 5\cdot 7^2\cdot 11\cdot 17}$	
22	1	$-\frac{131\cdot 593}{3^2\cdot 5^4\cdot 7\cdot 19}$	$\frac{131\cdot 593}{2^4\cdot 3^2\cdot 5^3\cdot 17\cdot 19}$	$-\frac{131\cdot 593}{2^4\cdot 3^4\cdot 5^4\cdot 7\cdot 19}$	$\frac{131\cdot 593}{2^5\cdot 3^4\cdot 5^3\cdot 7\cdot 13\cdot 19}$	
26	1	$-\frac{43\cdot 657931}{2^3\cdot 3^3\cdot 5^3\cdot 7^2\cdot 13\cdot 23}$	$\frac{97\cdot 657931}{2^5\cdot 3^4\cdot 5^3\cdot 7^2\cdot 11\cdot 13\cdot 23}$	$-\frac{29\cdot 657931}{2^3\cdot 3^4\cdot 5^4\cdot 7^2\cdot 13\cdot 19\cdot 23}$	$\frac{657931}{2^4\cdot 3^4\cdot 5^2\cdot 7^2\cdot 13\cdot 17\cdot 23}$	$-\frac{657931}{2^5\cdot 3^4\cdot 5^3\cdot 7^2\cdot 11\cdot 13\cdot 23}$

These values have been obtained by solving the system of the Eichler-Shimura equations for the even periods, to which was added one more equation of the Hecke type (for $n=2$) to compute r_0.

An interesting feature of these solutions is the fact that from them congruences of the Ramanujan type modulo large primes automatically follow. More precisely, let

$$S^0_{w+2} = R\left(1+\sum_{n=2}^{\infty} \lambda_n^{(w+2)} e^{2\pi i n z}\right).$$

Then from our table and the Coefficients Theorem (formula (1)) we find immediately that

$$\lambda_n^{(16)} \equiv \sigma_{15}(n) \mod 3617, \quad \lambda_n^{(18)} \equiv \sigma_{17}(n) \mod 43867,$$
$$\lambda_n^{(20)} \equiv \sigma_{19}(n) \mod 283\cdot 617,$$
$$\lambda_n^{(22)} \equiv \sigma_{21}(n) \mod 131\cdot 593, \quad \lambda_n^{(26)} \equiv \sigma_{25}(n) \mod 657\,931.$$

For a systematic discussion of congruences of this type from the point of view of l-adic representations see the articles of Serre [9] and Swinnerton-Dyer [12].

§8. Mazur's p-adic integrals

8.1. In [6] Mazur constructs a p-adic analog of the Hecke series of a parabolic form Φ of weight 2, starting out from the integral representation

$$\frac{\Gamma(s)}{(2\pi)^s} L_\Phi(s) = \int_0^\infty \Phi(iy)\, y^{s-1}\, dy. \tag{23}$$

In this section we describe this construction. Rewrite the integral (23) in the form $\int_{R_+^*} \chi^{(s)}(y)\mu_\Phi^{(\infty)}$, where R_+^* is the multiplicative group of positive real numbers, $\chi^{(s)}: R_+^* \to C$, given by $\chi^{(s)}(y) = y^s$, is a quasi-character of R_+^*, and $\mu_\Phi^{(\infty)} = \Phi(iy)y^{-1}\,dy$ is a measure on R_+^* which is associated with Φ.

We start with the description of the p-adic objects corresponding to R_+^*; after which we will introduce a class of p-adic measures with respect to which we can integrate, define the Mazur integral and prove that the p-adic L-function is analytic. The only novelty of our exposition is the consideration of measures which are not necessarily bounded; this is essential for the applications.

In the following sections we will construct p-adic measures $\mu_\Phi^{(p)}$ associated to parabolic forms $\Phi \in S_{w+2}$; we will then compute the corresponding archimedean and nonarchimedean integrals and compare them.

8.2. **The analog of** R_+^*. The following notation and conventions will be fixed from now on:

p will be a prime number, $\Delta_0 > 0$ an integer with $(\Delta_0, p) = 1$, and $q = p$ if $p \geq 3$, but $q = 4$ if $p = 2$. Let $\Delta = \Delta_0 q$, and let $Z_\Delta = \varprojlim Z/\Delta p^n$. Z_Δ is a ring with p-adic topology.

The analog of R_+^* will be *the multiplicative group* Z_Δ^*. Note that

$$Z_\Delta^* \cong \mathrm{Gal}\ Q(1^{p^{-\infty}}, 1^{\Delta_0^{-1}})/Q.$$

There is a canonical group isomorphism $Z_\Delta^* \cong (Z/(\Delta))^* \times (1 + qZ_p)^*$. The projection onto the first factor is given by $a \mapsto a \bmod \Delta$. The inclusion $(Z/(\Delta))^* \hookrightarrow Z_\Delta^*$ (which of course cannot be extended to an imbedding of rings) defines the subgroups of "Teichmüller representatives". The subgroup $(1 + qZ_p)^* = \Gamma$ (in Iwasawa's notation) is isomorphic to Z_p; $1 + q$ is a canonical generator for it. We will frequently write elements of Γ in the form $(1 + q)^\alpha$, with $\alpha \in Z_p$. If $\alpha = \lim \alpha_n$ $(\alpha_n \in Z)$ in the p-adic topology, then $(1 + q)^\alpha = \lim (1 + q)^{\alpha_n}$ in the topology of the ring Z_Δ.

8.3. *The analog of* $\chi^{(s)}$. Choose a finite extension K of the field Q_p. Let O be ring of integers of K, and $m \subset O$ the maximal ideal. Consider the group of p-adically continuous homomorphisms

$$\mathrm{X} = \mathrm{Hom}\,(Z_\Delta^*, O^*).$$

The elements of X will be called p-adic K-characters of Z_Δ^*. It is they that will be the analogs of $\chi^{(s)}$ (rather than the more restricted class of objects introduced below).

Any character $\chi \in \mathrm{X}$ can be uniquely represented in the form $\chi = \chi_0 \cdot \chi_1$, where χ_0 is trivial on $(1 + qZ_p)^*$ and χ_1 is trivial on $(Z/(\Delta))^*$. The character χ_0 is the tame component of χ, and χ_1 the wild one. The group of tame characters is isomorphic to $\mathrm{Hom}\,((Z/(\Delta))^*, O^*)$ and is obviously finite.

The group of wild characters can naturally be given the structure of p-adic analytic group, as $(1 + m)^* \subset O^*$. Indeed, let us associate to the wild character χ the element $t = \chi(1 + q) - 1$. Then $t \in m$, since $\lim_{n\to\infty} (1 + q)^{p^n} = 1$ in Z_Δ^*; hence $\lim_{n\to\infty} (1 + t)^{p^n} = 1$ in O^*. Furthermore, χ is uniquely determined by t, since $1 + q$ is a topological generator of $(1 + qZ_p)^*$, and t can be taken to be any element of m.

We have thus introduced the *local coordinate* t in a neighborhood of the identity character of X:

$$m \ni t \mapsto \chi_{(t)} : \chi_{(t)}\,(\varepsilon\,(1 + q)^\alpha) = (1 + t)^\alpha$$

for all $\alpha \in Z_p$.

It is sometimes convenient to use the *local coordinate* s, analogous to the classical argument s of Dirichlet series:

$$O \ni s \mapsto \chi^{(s)} : \chi^{(s)}\,(\varepsilon\,(1 + q)^\alpha) = (1 + q)^{\alpha s} = \exp\,(\alpha s \log\,(1 + q)).$$

The character $\chi^{(s)}$ is defined for those s for which the series exp is p-adically convergent. In this domain $t = (1+q)^s - 1$. However, for instance, in the case $(1+t)^{p^n} = 1$ the required value of s clearly does not exist, so that the s-coordinate parametrizes a smaller neighborhood of unity than t (which covers all wild characters).

8.4. *The analog of* $\mu^{(\infty)}$. As in §8.3, let K be a finite extension of Q_p. The symbol $|\ |$ will denote the p-adic valuation of K for which $|p| = p^{-1}$; ord will denote the p-adic denominator, so that $\operatorname{ord}(p) = 1$.

We will call any finitely additive function of the open-and-closed subsets of Z_Δ^* with values in K a *K-measure* μ on Z_Δ^*.

Set $I_{a,m} = a + p^m \Delta Z_\Delta$ for $a \in Z_\Delta^*$ and $m \geq 0$. For any measure μ we have

$$\forall a, m \quad \mu(I_{a,m}) = \sum_{b \equiv a \bmod p^{m+1}\Delta} \mu(I_{b,m+1}). \tag{24}$$

Conversely, if we fix a measure on the $I_{a,m}$, subject to the relations (24), then it has a unique extension to a measure μ on the whole of Z_Δ^*, because any open-and-closed set is a disjoint finite union of "intervals" $I_{a,m}$. We will make frequent use of this remark.

Example 1. $\mu(I_{a,m}) = \text{const} \cdot p^{-m}$. This is the invariant measure. Note that as $m \to \infty$, while $I_{a,m}$ is contracting to a point, the measure $\mu(I_{a,m})$ grows infinitely in the p-adic topology of the field K in which μ is taking its values: $|p^{-m}| = p^m \to \infty$.

This is the phenomenon which is vital for our theory of integration. It is clear that the K-measure $\mu(I_{a,m})$ cannot tend to zero as $m \to \infty$ as it does for measures with archimedean values; but it could be bounded. We will see in what follows that the class of functions which are important for our purposes can be integrated with respect to any measure which grows slower than the invariant measure. This remark is essential for the consideration of values of p which are "supersingular" for a given form Φ, as was first shown by Nasybullin.

Example 2. Let $\Delta_0 = 1$, let c be an integer with $c > 0$, and let $0 < a < p^n$, with $(a, p) = 1$. Set

$$\mu_c(a + (p^n)) = \left[\frac{ac}{p^n}\right] - \frac{c-1}{2} \in \frac{1}{2} Z \subset Q_p = K, \tag{25}$$

where [ ] is the usual integer part of a number, and let us check that conditions (24) are fulfilled. (Mazur introduced this measure and showed that the Kubota-Leopoldt p-adic L-functions are defined by an integration with respect to it. It would be exceedingly interesting to determine whether analogous measures exist for the ζ-functions of any totally real fields, which have recently been constructed by Serre [11].)

We must prove that

$$\mu_c(a + (p^n)) = \sum_{b=0}^{p-1} \mu_c(a + bp^n + (p^{n+1})).$$

It follows from (25) that if $i/c \leq a/p^n < (i+1)/c$, then $\mu_c(a + (p^n)) = i - (c-1)/2$ for $i = 0, \cdots, c-1$. Fix i and a satisfying these conditions, and an integer b with $0 \leq b \leq p-1$.

One can easily see that

$$\frac{k}{c} \leqslant \frac{a+bp^n}{p^{n+1}} < \frac{k+1}{c}, \quad \text{where } k = \left[\frac{i+bc}{p}\right].$$

Hence the additivity relation is equivalent to the following:

$$i - \frac{c-1}{2} = \sum_{b=0}^{p-1} \left(\left[\frac{i+bc}{p}\right] - \frac{c-1}{2}\right).$$

We will check this by induction on i. For $i = 0$ we get the classical identity

$$\sum_{b=0}^{p-1} \left[\frac{bc}{p}\right] = \frac{(p-1)(c-1)}{2},$$

which is proved by counting the integer points under the diagonal of the rectangle having vertices $(1,1)$, $(p-1,1)$, $(p-1,c)$ and $(1,c)$. The inductive step from i to $i+1$ is based on the remark that on the right-hand side the summands corresponding to b with $i + bc \equiv -1 \bmod p$ increase by 1, and the others do not change.

8.5. *The analog of integration.* Let μ be a K-measure on Z_Δ^*, let R_m be a system of representatives for $(Z_\Delta/(p^m\Delta))^*$ in Z_Δ^*, and let $f : Z_\Delta^* \to K$ be a function. We will denote by

$$S(f; R_m) = \sum_{b \in R_m} f(b)\,\mu(I_{b,m})$$

the corresponding "Riemann sum".

8.6. **Theorem-Definition.** *There exists a unique limit*

$$\lim S(f; R_m) \overset{\text{def}}{=\!=} \int_{Z_\Delta^*} f\mu$$

taken over all R_m as $m \to \infty$, provided that the following conditions hold:

a) *The measure μ is of moderate growth; that is, by definition,*

$$\varepsilon_m = \max_b |\mu(I_{b,m})|\, p^{-m} \to 0 \quad as \quad m \to \infty.$$

b) *The function f satisfies the "Lipschitz condition": there exists a constant C such that*

$$b \equiv b' \bmod \Delta p^m \Rightarrow |f(b) - f(b')| < Cp^{-m} \quad as \quad m \to \infty.$$

In the applications, the measure μ will be bounded, and hence of moderate growth. The function f will usually be (piecewise) analytic, or even locally constant.

Proof. Let R_m and R'_m be two different systems of representatives mod Δp^m. Then

$$|S(f; R_m) - S(f; R'_m)| \leqslant \max_{b \equiv b'(\Delta p^m)} |f(b) - f(b')| \max_b |\mu(I_{b,m})| \leqslant C\varepsilon_m. \tag{26}$$

Similarly,

$$|S(f;\,R_{m+1})-S(f;\,R_m)|\leqslant\max_{b\equiv b'(\Delta p^m)}|f(b)-f(b')|\max_b|\mu(I_{b,m+1})|\leqslant Cp\epsilon_{m+1}\qquad(27)$$

(using (24)). Since $\epsilon_m \to 0$, the set of all sums $S(f, R_m)$ is bounded. It follows from the fact that K is locally compact that it has an accumulation point, and it can then be seen from (26) and (27) that there can only be one limit point for the sums as $m \to \infty$.

Remark. Archimedean Riemann sums have a limit because although the individual summands are small, there are a correspondingly large number of them to compensate. Nonarchimedean Riemann sums have a limit because although the individual terms are large, there are a correspondingly large number of the special form const $\cdot\, p^n$ of them to compensate. If the summands are not too large (moderate growth of μ), and are not too far apart (Lipschitz condition), summing them is "practically the same" as multiplying any one of them by const $\cdot\, p^n$; but $|p^n| \to 0$.

Now let χ be some p-adic K-character of Z_Δ^*, and let μ be a K-measure on Z_Δ^*. We will show that the *p-adic Mellin-Mazur transform*

$$L(\mu,\chi)=\int_{Z_\Delta^*}\mu\chi$$

is an analytic function of χ. The integral exists in any case, if μ is of moderate growth. This is clear from 8.3: $|(1+t)^{ap^n}-1| < \text{const}\cdot p^{-n}$ as $n \to \infty$, $a \in Z_p$.

8.7. **Theorem.** *Let μ be of moderate growth, and let $\chi = \chi_0\chi_{(t)}$, where χ_0 is the tame and $\chi_{(t)}$ the wild component, as in §8.3. Then there exists a unique power series $g(\chi_0, T) \in K[[T]]$, depending only on χ_0, convergent for all $t \in m$, and such that*

$$L(\mu,\chi_0\chi_{(t)})=g(\chi_0,t)\quad \textit{for all}\quad t\in m.$$

Proof. For any $m \geq 1$ we define the following system of representatives $R_m \subset Z_\Delta^*$ for the classes mod Δp^m:

$$R_m=\{\varepsilon(1+q)^k\,|\,\varepsilon\in(Z/(\Delta))^*,\ 0\leqslant k\leqslant p^m-1\}.$$

By definition, $L(\mu, \chi_0\chi_{(t)}) = \lim_{m\to\infty} S(R_m)$, where the Riemann sums have the form

$$S(R_m)=S(\chi_0\chi_{(t)},R_m)=\sum_{k=0}^{p^m-1}a_k^{(m)}(1+t)^k,\qquad(28)$$

$$a_k^{(m)}=\sum_{\varepsilon\in(Z/\Delta)^*}\chi_0(\varepsilon)\,\mu(\varepsilon(1+q)^k+(p^m\Delta)).\qquad(29)$$

Set

$$S_m(T)=\sum_{k=0}^{p^m-1}a_k^{(m)}(1+T)^k\in K[T]\qquad(30)$$

and let us prove the following facts:

a) *In the coefficient-by-coefficient convergence topology, the limit* $\lim_{m\to\infty} S_m(t) \in K[[T]]$ *exists.*

Denote this limit by $g(\chi_0, T)$.

b) *If* $g(\chi_0, T) = \sum_0^\infty b_n T^n$, *then* $|b_n| \leq \text{const}\cdot p^k$ *for* $p^k \leq n \leq p^{k+1}-1$, *so that* $g(\chi_0, t)$ *is convergent for all* $t \in m$.

c) *For fixed* $t \in m$

$$\lim_{m\to\infty}\left(S_m(t) - \sum_{n=0}^{p^m-1} b_n t^n\right) = 0,$$

so that $g(\chi_0, t)$ *converges to* $\lim S_m(t) = L(\mu, \chi_0\chi_{(t)})$.

We begin with the following lemma.

Lemma. *For all* $0 \leq k \leq p^m - 1$,

$$\sum_{i=0}^{p-1} a_{k+ip^m}^{(m+1)} = a_k^{(m)}. \tag{31}$$

Proof of the Lemma. Substituting definition (29) on both sides of (31) and equating the coefficients of $\chi_0(\epsilon)$ on either side, we see that it suffices to check the identity

$$\sum_{i=0}^{p-1} \mu\left(\varepsilon(1+q)^{k+ip^m} + (p^{m+1}\Delta)\right) = \mu\left(\varepsilon(1+q)^k + (p^m\Delta)\right),$$

which follows from the fact that μ is finitely additive.

We now return to the proof of the theorem. By using (30) and the lemma, we find that

$$S_{m+1}(T) = \sum_{k=0}^{p^{m+1}-1} a_k^{(m+1)}(1+T)^k = \sum_{k=0}^{p^m-1}\sum_{i=0}^{p-1} a_{k+ip^m}^{(m+1)}(1+T)^{k+ip^m},$$

$$S_m(T) = \sum_{k=0}^{p^m-1} a_k^{(m)}(1+T)^k = \sum_{k=0}^{p^m-1}\sum_{i=0}^{p-1} a_{k+ip^m}^{(m+1)}(1+T)^k,$$

and hence

$$S_{m+1}(T) - S_m(T) = \sum_{k=0}^{p^m-1}\sum_{i=0}^{p-1} a_{k+ip^m}^{(m+1)}(1+T)^k\left[(1+T)^{ip^m}-1\right] \overset{\text{def}}{=} \sum_{n=0}^{\infty} A_n^{(m)}T^n. \tag{32}$$

To estimate the coefficients of the series (32) we use the following estimate for binomial coefficients: $|\binom{r}{s}| \leq |r/s|$ (in the p-adic norm). Indeed, this inequality is equivalent to $\operatorname{ord}\binom{r}{s} \geq \operatorname{ord}(r) - \operatorname{ord}(s)$. If the right side is nonpositive, the assertion is trivial. In any case

$$\operatorname{ord}\binom{r}{s} = \sum_{i=1}^{\infty}\left(\left[\frac{r}{p^i}\right] - \left[\frac{s}{p^i}\right] - \left[\frac{r-s}{p^i}\right]\right).$$

But for any $\alpha, \beta \geq 0$ we obviously have $[\alpha+\beta] - [\alpha] - [\beta] \geq 0$, and even $[\alpha+\beta] -$

$[\alpha]-[\beta]\geq 1$ if $\alpha+\beta$ is an integer but α and β are not. This last possibility occurs for those terms of the sum for which $\operatorname{ord}(s)+1\leq i\leq \operatorname{ord}(r)$.

Now turn to (32). Taking account of the fact that $|a_{k+ip^m}^{(m+1)}|\leq \epsilon_{m+1}p^{m+1}$, and using (29) and 8.6 a), we get

$$|A_n^{(m)}|\leqslant \varepsilon_{m+1}p^{m+1}\max_{\substack{s\leqslant n\\ 1\leqslant i\leqslant p-1}}\left|\frac{ip^m}{s}\right|.$$

Suppose $p^{m_0-1}\leq n\leq p^{m_0}-1$. Then the maximum on the right is equal to p^{-m+m_0-1}, and hence

$$|A_n^{(m)}|\leqslant \varepsilon_{m+1}p^{m_0}\quad\text{for all}\quad p^{m_0-1}\leqslant n\leqslant p^{m_0}-1,\quad m\geqslant m_0. \tag{33}^{m_0}$$

We then have

$$b_n=\lim_{m\to\infty}a_n^{(m)}=a_n^{(m_0)}+(a_n^{(m_0+1)}-a_n^{(m_0)})+\dots=a_n^{(m_0)}+\sum_{M\geqslant m_0}A_n^{(M)},$$

and the series on the right is convergent by (33) and the fact that $\epsilon_{M+1}\to 0$ as $M\to\infty$. Hence the series $\sum_0^\infty b_nT^n=\lim S_m(T)$ is defined.

Furthermore, if $p^{m_0-1}\leq n\leq p^{m_0}-1$, then

$$|b_n|\leqslant\max(|a_n^{(m_0)}|,\ |A_n^{(m_0)}|,\ |A_n^{(m_0+1)}|,\dots)\leqslant p^{m_0}\max(\varepsilon_{m_0},\ p\varepsilon_{m_0+1},\ p\varepsilon_{m_0+2},\dots)$$

by (33). Hence $\sum_0^\infty b_nt^n$ is convergent for all $|t|<1$.

Finally, by $(33)_k$,

$$\left|\sum_{n=0}^{p^m-1}b_nt^n-S_m(t)\right|=\left|\sum_{n=0}^{p^m-1}(b_n-a_n^{(m)})t^n\right|=\left|\sum_{n=0}^{p^m-1}\Big(\sum_{M\geqslant m}A_n^{(m)}\Big)t^n\right|$$
$$\leqslant\max_{k\leqslant m}(\varepsilon_{m+1}p^k|t|^{p^{k-1}})\to 0\quad\text{as}\quad m\to\infty$$

for $k\leq m$. This completes the proof of Theorem 8.7.

§9. Measures associated to modular forms

9.1. In this section we will fix a form $\Phi\in S^0_{w+2}$ and two nonzero numbers $\omega^+\in Ri$ and $\omega^-\in R$, which will be chosen later. Consider the two functions on the rational numbers $P^\pm\colon Q\to C$:

$$P^+(x)=\frac{i}{\omega^+}\operatorname{Im}\int_x^{i\infty}\Phi(z)\,dz,\quad P^-(x)=\frac{1}{\omega^-}\operatorname{Re}\int_x^{i\infty}\Phi(z)\,dz. \tag{34}$$

We list the properties of these functions which we will need.

9.2. **Proposition.** a) $P^\pm(x+1)=P^\pm(x)$.

b) *P^+ is even, and P^- is odd.*

c) *Suppose that $\Phi|T_n=\lambda_n\Phi$ for all $n\geq 1$, and let the numbers $r_{2i+1}(\Phi)/\omega^+$ and $r_{2i}(\Phi)/\omega^-$ be algebraic integers for all integers $2i$, $2i+1\in[0,w]$ (see Theorem 1.2). Then $P^\pm(x)$ (for all $x\in Q$) takes values which are algebraic integers in some finite extension of Q.*

d) *Let p be a prime and $\Phi|T_p = \lambda_p\Phi$. Then*

$$\lambda_p P^{\pm}(x) = p^{w} P^{\pm}(px) + \sum_{k=0}^{p-1} P^{\pm}\left(\frac{x+k}{p}\right).$$

Proof. a) follows from the fact that $\Phi(z+1) = \Phi(z)$. b) follows since the coefficients of Φ are real:

$$\overline{\int_{x}^{i\infty} \Phi(z)\,dz} = \int_{x}^{i\infty} \Phi(-\bar{z})\,d\bar{z} = -\int_{x}^{i\infty} \Phi(z)\,dz. \tag{35}$$

To prove c) it will be sufficient to establish that

$$\forall x \in Q, \quad \int_{x}^{i\infty} \Phi(z)\,dz \in \sum_{l=0}^{w} Zr_l(\Phi), \tag{36}$$

and then to separate out the real and imaginary parts, using (35) and the fact that the odd periods are real, whereas the even ones are pure imaginary. But (36) has been established on the way to the proof of Proposition 3.2.

Finally, the equation d) just describes the action of the Hecke operator T_p (see (11) in §3):

$$\lambda_p \int_{x}^{i\infty} \Phi(z)\,dz = \int_{x}^{i\infty} \Phi|T_p\,dz = \int_{x}^{i\infty} p^{\frac{w}{2}}\left(p^{\frac{w+2}{2}}\Phi(pz) + \sum_{k=0}^{p-1} p^{-\frac{w+2}{2}}\Phi\left(\frac{z+k}{p}\right)\right)dz$$

$$= p^{w}\int_{px}^{i\infty} \Phi(z)\,dz + \sum_{k=0}^{p-1}\int_{\frac{x+k}{p}}^{i\infty} \Phi(z)\,dz.$$

Separating out real and imaginary parts completes the proof.

9.3. We will now fix Φ, ω^{+} and ω^{-} satisfying 9.2 c). Choose a prime number p and denote by K one of the p-adic completions of the field in which the functions $P^{\pm}$ take their values. We will identify the $P^{\pm}(x)$ with elements of K without especial mention.

The following lemma was pointed out to me by Nasybullin. It is an improvement and a generalization of the original construction of Mazur and the author. In the notation of §8, let Q_Δ be the set of rational numbers whose denominators divide Δp^{n} for all $n \geq 1$.

9.4. Lemma. *Let $R: Q_\Delta \to K$ be a function with the following properties: for some A and $B \in K$ and all $x \in Q_\Delta$*

$$R(x+1) = R(x) \quad \text{and} \quad \sum_{k=0}^{p-1} R\left(\frac{x+k}{p}\right) = AR(x) + BR(px). \tag{37}$$

Furthermore, let ρ denote any root of the equation $\rho^2 = A\rho + Bp$, with $\rho \neq 0$. Then *there exists a $K(\rho)$-measure μ on Z_Δ^* such that for all $m \geq 0$ and all $a \in Z$*

$$\mu(a + (p^m\Delta)) = \rho^{-m} R\left(\frac{a}{p^m\Delta}\right) + B\rho^{-(m+1)} R\left(\frac{a}{p^{m-1}\Delta}\right). \tag{38}$$

Proof. We must check formula (24). We have

$$\sum_{k=0}^{p-1} \mu(a + p^m\Delta k + (p^{m+1}\Delta)) = \sum_{k=0}^{p-1} \rho^{-(m+1)} R\left(\frac{a + p^m\Delta k}{p^{m+1}\Delta}\right)$$
$$+ \sum_{k=0}^{p-1} B\rho^{-(m+2)} R\left(\frac{a + p^m\Delta k}{p^m\Delta}\right).$$

We transform the first term by means of (37), and the second using the periodicity of R. This gives

$$\rho^{-(m+1)}\left(AR\left(\frac{a}{p^m\Delta}\right) + BR\left(\frac{a}{p^{m-1}\Delta}\right)\right) + B\rho^{-(m+2)} pR\left(\frac{a}{p^m\Delta}\right)$$
$$= \rho^{-(m+2)}(A\rho + Bp) R\left(\frac{a}{p^m\Delta}\right) + B\rho^{-(m+1)} R\left(\frac{a}{p^{m-1}\Delta}\right)$$
$$= \rho^{-m} R\left(\frac{a}{p^m\Delta}\right) + B\rho^{-(m+1)} R\left(\frac{a}{p^{m-1}\Delta}\right) = \mu(a + (p^m\Delta)),$$

and this proves that μ is finitely additive.

9.5. We now apply Lemma 9.4 to the functions $P^{\pm}(x)$ to obtain the basic result of this section:

Starting from given Φ, ω^+, ω^- and K as above, we can construct $K(\rho)$-measures $\mu_\Phi^\pm$ on Z_Δ^* such that

$$\mu_\Phi^\pm(a + (p^m\Delta)) = \rho^{-m} P^\pm\left(\frac{a}{p^m\Delta}\right) - p^w \rho^{-(m+1)} P^\pm\left(\frac{a}{p^{m-1}\Delta}\right),$$
$$\rho^2 - \lambda_p\rho + p^{w+1} = 0, \quad \Phi|T_p = \lambda_p\Phi. \tag{39}$$

The measure (39) *is clearly of moderate growth if* $\operatorname{ord} \rho < 1$.

It would be interesting to widen the type of conditions under which (39) turns out to be of moderate growth. Here we only use the integrality of the $P^{\pm}(x)$, but it could be that the $P^{\pm}(a/p^m\Delta)$ are divisible by a power of p that grows with m, and that would compensate for the ρ^{-m}.

§10. p-adic Hecke functions

10.1. In this section we prove a stronger version of Theorem 1.5.

We keep all the notation of §§8 and 9, partially summarized in §1.4. In particular, χ is a primitive Dirichlet character modulo Δp^m, with $m \geq 0$; K is the p-adic field obtained by completing the field of values of $P^{\pm}(x)$ and of χ. The $K(\rho)$-measures $\mu_\Phi^\pm$ are constructed for a form $\Phi \in S_{w+2}^0$ which is an eigenfunction for all the T_n, $\Phi|T_p = \lambda_p\Phi$, and ρ is a root of $\rho^2 - \lambda_p\rho + p^{w+1} = 0$.

Under these conditions the following holds:

10.2. **Theorem.** *If the measure* $\mu_{\Phi}^{\pm}$ *is of moderate growth, then*

$$\frac{\Delta p^m}{\omega^{\pm} G(\chi)} \int_0^{i\infty} \Phi_\chi(z)\,dz = \rho^m \int_{Z_\Delta^*} \mu_\Phi^{\pm} \chi^*, \tag{40}$$

where the left-hand side of (40) *is an algebraic integer, considered as belonging to* K; *on the right of* (40) $\chi^*(x) = \chi^{-1}(-x)$; *the sign* $\pm$ (*on both sides*) *is to be taken according as* $\chi(-1) = \pm 1$.

To deduce Theorem 1.5 from (40), we must note that

$$\int_0^{i\infty} \Phi_\chi(z)\,dz = -\frac{1}{2\pi i} L_\Phi(1, \chi),$$

put χ in the form $\chi_0\chi_1$, where χ_1 is the wild component of χ, write the p-adic character χ_1 in the form $\chi_{(t)}$, where $t = \chi_1(1+q) - 1$ according to §8.3, and finally apply Theorem 8.7 to the right-hand side.

Proof of Theorem 10.2. To compute the left-hand side of (40), we make use of the standard formula

$$\Phi_\chi(z) = \frac{G(\chi)}{\Delta p^m} \sum_{b \bmod \Delta p^m} \chi^*(b)\,\Phi\left(z + \frac{b}{\Delta p^m}\right)$$

(see for example [3], Lemma 9.4). This gives

$$\frac{\Delta p^m}{G(\chi)} \int_0^{i\infty} \Phi_\chi(z)\,dz = \sum_{b \bmod \Delta p^m} \chi^*(b) \int_{b/\Delta p^m}^{i\infty} \Phi(z)\,dz$$

$$= \frac{1}{2} \sum_{b \bmod \Delta p^m} \left(\chi^*(b) \int_{b/\Delta p^m}^{i\infty} + \chi^*(-b) \int_{-b/\Delta p^m}^{i\infty}\right) \Phi(z)\,dz.$$

The formulas (34) and (35) show at once that

$$\frac{\Delta p^m}{G(\chi)\,\omega^{\pm}} \int_0^{i\infty} \Phi_\chi(z)\,dz = \sum_{b \bmod \Delta p^m} \chi^*(b)\,P^{\pm}\left(\frac{b}{\Delta p^m}\right). \tag{41}$$

To compute the right-hand side of (40) we note that χ^* is the constant function on the "intervals" $b + (\Delta p^m)$, so that $\int_{Z_\Delta^*}$ is just the same as the corresponding Riemann sum:

$$\begin{aligned} \rho^m \int_{Z_\Delta^*} \mu_\Phi^{\pm} \chi^* &= \rho^m \sum_{b \bmod \Delta p^m} \chi^*(b)\,\mu_\Phi^{\pm}(b + (\Delta p^m)) \\ &= \sum_{b \bmod \Delta p^m} \chi^*(b)\,P^{\pm}\left(\frac{b}{\Delta p^m}\right) - p^{\varpi}\rho^{-1} \sum_{b \bmod \Delta p^m} \chi^*(b)\,P^{\pm}\left(\frac{b}{\Delta p^{m-1}}\right). \end{aligned} \tag{42}$$

Thus to show that (41) and (42) coincide we must check that

$$\forall m \geqslant 0 \quad \sum_{b \bmod \Delta p^m} \chi^*(b) P^{\pm}\left(\frac{b}{\Delta p^{m-1}}\right) = 0.$$

For this we break the sum up into partial summations, indexed by the classes $b \bmod p^{m-1}$ (this is even possible when $m = 0$!), and note that $P^{\pm}(b/\Delta p^{m-1})$ depends only on such a class, and that

$$\sum_{b' \equiv b \bmod \Delta p^{m-1}} \chi^*(b') = 0,$$

since χ^* is a primitive character. (This is where the primitivity of χ is important.)

Received 23/JUNE/73

BIBLIOGRAPHY

1. K. Iwasawa, *Lectures on p-adic L-functions*, Ann of Math. Studies, no. 74, Princeton Univ. Press, Princeton, N. J., 1972.

2. H. Jacquet and R. P. Langlands, *Automorphic forms on GL(2)*, Lecture Notes in Math., vol. 114, Springer-Verlag, Berlin and New York, 1970.

3. Ju. I. Manin, *Cyclotomic fields and modular curves*, Uspehi Mat. Nauk 26 (1971), no. 6 (162), 7–71 = Russian Math. Surveys 26 (1971), 7–78.

4. ———, *Parabolic points and zeta-functions of modular curves*, Izv. Akad. Nauk SSSR Ser. Mat. 36 (1972), 19–66 = Math. USSR Izv. 6 (1972), 19–64.

5. ———, *Explicit formulas for the eigenvalues of Hecke operators*, Acta Arith. 24 (1973), 239–240.

6. B. Mazur and H. P. F. Swinnerton-Dyer, *On the p-adic L-series of an elliptic curve*, Invent. Math. (to appear).

7. S. Ramanujan, *On certain arithmetical functions*, Proc. Cambridge Philos. Soc. 22 (1916), 159–184.

8. J.-P. Serre, *Cours d'arithmétique*, Collection SUP: "Le Mathématicien", 2, Presses Universitaires de France, Paris, 1970. MR 41 #138.

9. ———, *Une interprétation des congruences relatives à la fonction τ de Ramanujan*, Séminaire Delange-Pisot-Poitou: 1967/68, Théorie des Nombres, fasc. 1, Exposé 14, Secrétariat mathematique, Paris, 1969. MR 39 #5464.

10. ———, *Congruences et formes modulaires*, Séminaire Bourbaki, 24ᵉ année: 1971/72, Exposé 416, Secrétariat mathématique, Paris, 1971.

11. ———, *Formes modulaires et fonctions zeta p-adiques*, Summer School on Modular Functions (Antwerp, 1972), Lecture Notes in Math., vol. 350, Springer-Verlag, Berlin and New York, 1973, pp. 191–268.

12. H. P. F. Swinnerton-Dyer, *On l-adic representations and congruences for coefficients of modular forms*, Summer School on Modular Functions (Antwerp, 1972), Lecture Notes in Math., vol. 350, Springer-Verlag, Berlin and New York, 1973, pp. 1–55.

13. G. Shimura, *Sur les intégrales attachées aux formes automorphes*, J. Math. Soc. Japan 11 (1959), 291–311. MR 22 #11126.

Translated by M. REID

Mat. Sbornik
Tom 93 (135) (1974), No. 4

Math. USSR Sbornik
Vol. 22 (1974), No. 4

THE VALUES OF p-ADIC HECKE SERIES AT INTEGER POINTS OF THE CRITICAL STRIP

Ju. I. MANIN

Abstract. In this paper we prove that the values of p-adic Hecke series at integer points of the critical strip in essence coincide with the classical ones. We derive a p-adic functional equation.

Bibliography: 2 items.

1. **Introduction.** In [1] by means of Mazur's p-adic integral we constructed p-adic analogues of the Mellin transforms of modular forms of arbitrary weight (for the usual modular group $\Gamma = SL(2, Z)/(\pm 1)$). However, that the values of these functions coincide with the classical values (up to elementary factors) was only verified for characters lying in a small neighborhood of the identity (see below for details). The goal of the present note is to extend this result to a wider range of values of the argument, corresponding in classical terminology to neighborhoods of all the integer points of the critical strip. Moreover, we also give a p-adic analogue of the classical functional equation of Hecke functions.

2. **Basic notation.** Throughout this paper we shall fix the following notation. $\Phi(z) = \sum_1^\infty \lambda_n e^{2\pi i n z}$ is a cusp form of weight $w \geq 12$ for the group Γ. We assume that $\lambda_1 = 1$ and $\Phi \mid T_n = \lambda_n \Phi$ for all $n \geq 1$, where the T_n are the Hecke operators.

Furthermore, for any Dirichlet character χ we put

$$L_\Phi(s, \chi) = \sum_{n=1}^{\infty} \chi(n)\lambda_n n^{-s} = \frac{(2\pi)^s}{\Gamma(s)} \int_0^\infty \Phi_\chi(iy)\, y^{s-1}\, dy,$$

where $\Phi_\chi(z) = \sum_1^\infty \chi(n)\lambda_n e^{2\pi i n z}$.

Now we fix a prime number p and an integer Δ_0, $p \nmid \Delta_0$.

We put $q = p$ for $p \geq 3$, $q = 4$ for $p = 2$ and $\Delta = \Delta_0 q$. Furthermore let $Z_\Delta = \varprojlim Z/\Delta p^n$, and let Z_Δ^* be the multiplicative group of the ring Z_Δ. We fix once for all an imbedding of the field of all algebraic numbers into the algebraic closure of the field

AMS (MOS) subject classifications (1970). Primary 10D15, 14G20; Secondary 44A15.

$\mathbf{Q}_p$, and all the identifications of algebraic numbers with p-adic numbers will be made with respect to this imbedding.

Let χ be a primitive Dirichlet character modulo Δp^m, $m \geq 0$. We put

$$G(\chi) = \sum_{b \bmod \Delta p^m} \chi(b) e^{2\pi i \frac{b}{\Delta p^m}}.$$

Furthermore, we denote by ρ a root of the equation $\rho^2 - \lambda_p \rho + p^{w+1} = 0$.

Let K be a finite extension of the field $\mathbf{Q}_p$ containing the values of the character χ, ρ, $G(\chi)$ and all λ_n. Furthermore let $a \mapsto a_p$ be the composition homomorphism $Z_\Delta^* \to Z_p^* \hookrightarrow K^*$.

In [1] it was proved that there exist numbers $\omega^\pm \in \mathbf{C}$ such that all the values

$$\frac{1}{\omega^\pm} \int_0^{i\infty} \Phi_\chi(iy)\, y^{k-1}\, dy$$

are algebraic integers if the superscript on $\omega^\pm$ is the same as in the formula $(-1)^{k-1}\chi(-1) = \pm 1$. Moreover, these numbers are rational over the field $\mathbf{Q}(\chi(Z), G(\chi), \lambda_1, \lambda_2, \cdots)$. In other words, the values $\Gamma(k)(2\pi)^{-k}(\omega^\pm)^{-1} L_\Phi(k, \chi)$ are algebraic, where k is an integer in the critical strip of $L_\Phi(s, \chi)$. In that paper it was proved that these numbers for $k = 1$ are expressed as the values of a certain p-adic integral, which is a p-adic analytic function of $\chi(1+q) - 1$.

In order to formulate the extension of this result to the remaining values of k, we recall the construction of the p-adic measures $\mu_\Phi^\pm$. For all $x \in \mathbf{Q}$ we set

$$P^+(x) = \frac{i}{\omega^+} \operatorname{Im} \int_x^{i\infty} \Phi(z)\, dz, \quad P^-(x) = \frac{1}{\omega^-} \operatorname{Re} \int_x^{i\infty} \Phi(z)\, dz \tag{1}$$

and furthermore

$$\mu_\Phi^\pm(a + (p^m\Delta)) = \rho^{-m} P^\pm\left(\frac{a}{p^m\Delta}\right) - p^w \rho^{-(m+1)} P^\pm\left(\frac{a}{p^{m-1}\Delta}\right). \tag{2}$$

Here $a \in Z$ and $(a, p) = 1$. The measures $\mu_\Phi^\pm$ are K-valued finitely additive functions on open-closed subsets of the groups Z_Δ^* (in the p-adic topology).

In [1] it was proved that if $\operatorname{ord}_p \rho < 1$ and $f: Z_\Delta^* \to K$ is a function that satisfies a Lipschitz condition with exponent 1 (in the p-adic metric), then the integral

$$\int_{Z_\Delta^*} f(a)\, d\mu_\Phi^\pm(a)$$

exists (in [1] we wrote $\int f/\mu$; here this notation is inconvenient).

Now we can state the main results of this note.

3. **Theorem.** *For all* $k = 0, \cdots, w$ *in case* $\operatorname{ord}_p \rho < 1$ *we have*

$$\frac{\Delta p^m}{G(\chi)} \frac{1}{\omega^\pm} \int_0^{i\infty} \Phi_\chi(z)\, z^k\, dz = \frac{p^m}{(-\Delta p^m)^k} \int_{Z_\Delta^*} \chi^{-1}(-a)\, a_p^k\, d\mu_\Phi^\pm. \tag{3}$$

Here the superscripts on $\omega^{\pm}$ *and* $d\mu^{\pm}_{\Phi}$ *are taken as in the formula* $(-1)^k\chi(-1) = \pm 1$, *where* χ *is primitive* mod Δp^m.

The analyticity of the right side of (3) as a function of χ is proved in [1].

4. **Theorem.** *For all* $k = 0, \cdots, w$ *in case* $\mathrm{ord}_p\, \rho < 1$ *we have*

$$\int\limits_{Z_\Delta^*} \chi^{-1}(-a)\, a_p^k d\mu_\Phi^{\pm}(a) = (-1)^{k+1} \int\limits_{Z_\Delta^*} \chi(a)\, a_p^{w-k} d\mu_\Phi^{\pm}(a). \tag{4}$$

In this functional equation (in contrast to (3), where χ *is a Dirichlet character), for* χ *we can take any p-adic K-characters of the group* Z_Δ^*, *described in* [1], §8.3.

5. Before proving these results we will need a number of new notations and computations. First of all, for any $x \in \mathbf{Q}$ and $n \geq 0$ we put

$$Q_n(x) = \int\limits_0^{i\infty} \Phi(z+x)\, z^n dz.$$

The formula

$$Q_n(x) = \int\limits_x^{i\infty} \Phi(z)(z-x)^n\, dz = \sum_{j=0}^{n} (-1)^j \binom{n}{j} x^j \int\limits_x^{i\infty} \Phi(z)\, z^{n-j} dz \tag{5}$$

shows that these functions are well defined (convergence at the lower end of the path of integration).

6. **Lemma.** a) $Q_n(x+1) = Q_n(x)$ *for all* x.

b) $\displaystyle\sum_{k=0}^{p-1} Q_n\left(\frac{x+k}{p}\right) = \lambda_p p^{-n} Q_n(x) - p^{w-2n} Q_n(px)$ *for all* x.

c) $Q_n(x) + Q_n(-x) = \begin{cases} 2\,\mathrm{Re}\, Q_n(x) & \text{for } n \equiv 1\ (2), \\ 2i\,\mathrm{Im}\, Q_n(x) & \text{for } n \equiv 0\ (2). \end{cases}$

Proof. The first assertion immediately follows from the fact that $\Phi(z+1) = \Phi(z)$, and the third because the coefficients of $\Phi(z)$ are real, as in [1], §9.2. The second assertion is obtained analogously by means of the Hecke operator:

$$\lambda_p Q_n(x) = \int\limits_0^{i\infty} (\Phi \mid T_p)(z+x)\, z^n\, dz$$

$$= \int\limits_0^{i\infty} p^{w+1}\Phi(pz+px)\, z^n\, dz + \sum_{k=0}^{p-1}\int\limits_0^{i\infty} p^{-1}\Phi\left(\frac{z+x+k}{p}\right) z^n\, dz.$$

After taking pz as a new variable of integration in the first integral and z/p in the remaining integrals, we get what we wanted.

7. **Corollary.** *The formulas*

$$\mu_n(a + (p^m\Delta)) = \rho^{-m} p^{mn} Q_n\left(\frac{a}{p^m\Delta}\right) - \rho^{-(m+1)} p^{mn-n+w} Q_n\left(\frac{a}{p^{m-1}\Delta}\right) \tag{6}$$

define a complex-valued finitely additive measure on Z_Δ^*.

This follows immediately from Lemma 6 and Lemma 9.4 of [1].

We note that in (6) it is *not assumed* that $n \leq w$. Unfortunately, because the arithmetic character of the integrals $\int_0^{i\infty} \Phi(z)z^n\,dz$, or what is the same the values of $L_\Phi(n+1, \chi_0)$, are unknown for $n > w$, it is necessary to restrict attention to the values for $n \leq w$ if we wish to ensure the algebraicity of the measures. We can only integrate such measures p-adically. With this goal we put

$$Q_n^+(x) = \frac{i}{\omega^+}\operatorname{Im} Q_n(x), \quad Q_n^-(x) = \frac{1}{\omega^-}\operatorname{Re} Q_n(x).$$

From the results of [1] and (5) it follows immediately that for $0 \leq n \leq w$ these are algebraic integers of the field $Q(\lambda_1, \lambda_2, \cdots)$. Therefore we can construct two K-valued measures $\mu_k^\pm$,

$$\mu_k^\mp(a + (p^m\Delta)) = \rho^{-m}p^{mk}Q_k^\pm\left(\frac{a}{p^m\Delta}\right) - \rho^{-m-1}p^{mk-k+w}Q_k^\pm\left(\frac{a}{p^{m-1}\Delta}\right) \tag{7}$$

for all $0 \leq k \leq w$.

The following simple lemma is the key to the proof of Theorem 3.

8. **Lemma.** *If* $0 \leq k \leq w$ *and* $\operatorname{ord}_p \rho < 1$, *then for all* $a \in Z_\Delta^*$

$$d\mu_k^\pm(a) = (-\Delta)^{-k}a_p^k\,d\mu_0^\pm(a),$$

where $\mu_0^\pm = \mu_\Phi^\pm$ *in the notation of* §2 *and* [1].

Proof. We shall carry out the computation for, say μ_k^-. Combining (7) and (5), for $a^{(m)} \in Z$ we obtain

$$\mu_k^-(a^{(m)} + (p^m\Delta)) = \rho^{-m}p^{mk}\sum_{j=0}^{k}(-1)^j\binom{k}{j}\left(\frac{a^{(m)}}{p^m\Delta}\right)^j\frac{1}{\omega^-}\operatorname{Re}\int_{a^{(m)}/p^m\Delta}^{i\infty}\Phi(z)z^{k-j}\,dz$$

$$-\rho^{-m-1}p^{mk-k+w}\sum_{j=0}^{k}(-1)^j\binom{k}{j}\left(\frac{a^{(m)}}{p^{m-1}\Delta}\right)^j\frac{1}{\omega^-}\operatorname{Re}\int_{a^{(m)}/p^{m-1}\Delta}^{i\infty}\Phi(z)z^{k-j}dz.$$

In these let $m \to \infty$, and $a^{(m)} \to a \in Z^*$. As shown in [1] we learn that the quantities

$$\frac{1}{\omega^-}\operatorname{Re}\int_x^{i\infty}\Phi(z)z^{k-l}\,dz$$

are algebraic integers. Then from the condition $\operatorname{ord}_p \rho < 1$ it follows immediately that all the terms of both sums (including the exterior factors) tend p-adically to zero, except those for $j = k$. Therefore as $m \to \infty$ (in the p-adic metric of K) asymptotically we have

$$\mu_k^-(a^{(m)} + (p^m\Delta)) \sim (-1)^k\left(\frac{a^{(m)}}{\Delta}\right)^k\mu_0^-(a^{(m)} + (p^m\Delta)).$$

But $a^{(m)} \to a \in K$, where a_p is the image of a under the canonical homomorphism

$Z_\Delta^* \to Z_p^* \hookrightarrow K$. Thus our asymptotic equality also means that

$$d\mu_k^-(a) = (-\Delta)^{-k} a_p^k \mu_0^-(a).$$

The case μ_k^+ is examined in precisely the same way, and the equality of $\mu_0^\pm$ with $\mu_\Phi^\pm$ follows from the definitions (2), (5) and (7).

9. **Proof of Theorem 3.** We compute the Archimedean and non-Archimedean integrals separately.

The Archimedean integral. As in [1], we have

$$\Phi_\lambda(z) = \frac{G(\chi)}{\Delta p^m} \sum_{a \bmod \Delta p^m} \chi^{-1}(-a)\, \Phi\left(z + \frac{a}{\Delta p^m}\right),$$

if χ is a primitive Dirichlet character modulo Δp^m. From this and from (5) we find that

$$\frac{\Delta p^m}{G(\chi)} \int_0^{i\infty} \Phi_\lambda(z) z^k\, dz = \sum_{a \bmod \Delta p^m} \chi^{-1}(-a)\, Q_k\left(\frac{a}{\Delta p^m}\right)$$

$$= \frac{1}{2} \sum_a \left(\chi^{-1}(-a)\, Q_k\left(\frac{a}{\Delta p^m}\right) + \chi^{-1}(a)\, Q_k\left(\frac{-a}{\Delta p^m}\right)\right)$$

$$= \omega^\pm \sum_a \chi^{-1}(-a)\, \bar{Q}_k^\pm\left(\frac{a}{\Delta p^m}\right). \tag{8}$$

The last equality follows from Lemma 6c) and the definition of $Q_n^\pm$; the superscripts on $\omega^\pm$ and $Q_k^\pm$ mean the same as in the formula $(-1)^k \chi(-1) = \pm 1$.

The non-Archimedean integral. Since χ^{-1} as a p-adic character is constant on $a + (\Delta p^m)$, the integral $\int_{Z_\Delta^*}$ coincides with the corresponding Riemann sum, from which, using (7), we find that

$$p^m \int_{Z_\Delta^*} \chi^{-1}(-a)\, d\mu_k^\pm(a) = p^m \sum_{a \bmod \Delta p^m} \chi^{-1}(-a)\, \mu_k^\pm(a + (\Delta p^m))$$

$$= p^{mk} \sum_{a \bmod \Delta p^m} \chi^{-1}(-a)\, Q_k^\pm\left(\frac{a}{\Delta p^m}\right) - \rho^{-1} p^{mk-k+w} \sum_{a \bmod \Delta p^m} \chi^{-1}(-a)\, Q_k^\pm\left(\frac{a}{\Delta p^{m-1}}\right). \tag{9}$$

The last sum in (9) vanishes because χ is a primitive character and because $Q_k^\pm$ has period 1 (cf. the last paragraph of [1]). Using Lemma 8, we can replace $d\mu_k^\pm(a)$ in the integral by $(-\Delta)^{-k} a^k\, d\mu_0^\pm(a)$. Finally we will obtain

$$\frac{p^m}{(-\Delta p^m)^k} \int_{Z_\Delta^*} \chi^{-1}(-a)\, a_p^k d\mu_0^\pm(a) = \sum_{a \bmod \Delta p^m} \chi^{-1}(-a)\, Q_k^\pm\left(\frac{a}{\Delta p^m}\right). \tag{10}$$

A comparison of (8) and (10) now proves Theorem 3.

In order to prove Theorem 4 we need yet another lemma, which is actually stronger than the theorem.

10. **Lemma.** $d\mu_0^{\pm}(-a^{-1}) = -a_p^w \, d\mu_0^{\pm}(a)$ *if* $\mathrm{ord}_p \, \rho < 1$.

Proof. Let a and $a' \in \mathbf{Z}$ and $aa' \equiv -1 \bmod \Delta p^m$. We compare $\mu_0(a' + (\Delta p^m))$ with $\mu_0(a + (\Delta p^m))$ and show that the quotient of these two quantities coincides asymptotically with $-(\lim a)^w$ as m grows larger.

First of all, $aa' - N\Delta p^m = -1$ for some $N \in Z$. We put

$$g = \begin{pmatrix} N & a' \\ a & \Delta p^m \end{pmatrix} \in \Gamma.$$

We have

$$\int_{a'/\Delta p^m}^{i\infty} \Phi(z)\,dz = \int_{g(0)}^{g\left(-\frac{\Delta p^m}{a}\right)} \Phi(z)\,dz = \int_0^{-\Delta p^m/a} \Phi(z)\,(az + \Delta p^m)^w\,dz,$$

from which as $m \to \infty$ we obtain

$$Q_0^{\pm}(a'/\Delta p^m) \sim a^w \cdot \begin{cases} \dfrac{i}{\omega^+} \operatorname{Im} \displaystyle\int_0^{-\Delta p^m/a} \Phi(z)\,z^w\,dz \\ \dfrac{1}{\omega^-} \operatorname{Re} \displaystyle\int_0^{-\Delta p^m/a} \Phi(z)\,z^w\,dz. \end{cases} \tag{11}$$

Here it is necessary to take into account that the corresponding integrals are algebraic integers up to $(\omega^{\pm})^{-1}$. Setting $h = \begin{pmatrix} 0 & -1 \\ 1 & 0 \end{pmatrix} \in \Gamma$, we have

$$\int_0^{-\Delta p^m/a} \Phi(z)\,z^w\,dz = \int_{h(i\infty)}^{h(a/\Delta p^m)} \Phi(z)\,zd^wz = -\int_{a/\Delta p^m}^{i\infty} \Phi(z)\,dz. \tag{12}$$

Comparing (11) and (12), we get what we needed.

11. **Proof of Theorem 4.** It follows formally from Lemma 10 by changing the variable of integration from a to $-a^{-1}$ that

$$\int_{Z_\Delta^*} \chi^{-1}(-a)\; a_p^k d\mu_0^{\pm}(a) = \int_{Z_\Delta^*} \chi^{-1}(a^{-1})\,(-a_p)^{-k}\,d\mu_0^{\pm}(-a^{-1})$$
$$= (-1)^{k+1} \int_{Z_\Delta^*} \chi(a)\,a_p^{w-k} d\mu_0^{\pm}(a).$$

12. **Questions and remarks.** a) There are all sorts of reasons to expect that p-adic analogues can be constructed for the arbitrary L-functions of Jacquet-Langlands [2], at least over totally real ground fields. The construction of complex-valued p-adic measures for them has been found by I. I. Pjateckiĭ-Šapiro, M. E. Novodvorskiĭ, and independently by the author. However, the question of the algebraicity of the measures is unsolved so far.

b) A reasonable conversion from Δ apparently must reduce to the construction of functions with domains of definition and ranges of values of adèle type, for which our functions are only the p-component.

c) We recall once again that we have not yet found any way of computing $\int_0^{i\infty} \Phi(z)z^k\,dz$ for $k \geq w + 1$ and of clarifying the arithmetic nature of $\omega^{\pm}$.

Received 22/OCT/33

BIBLIOGRAPHY

1. Ju. I. Manin, *Periods of parabolic forms and p-adic Hecke series*, Mat. Sb. 92 (134) (1973), 378–401 = Math. USSR Sb. 21 (1973), 371–393.

2. H. Jacquet and R. P. Langlands, *Automorphic forms on $GL(2)$*, Lecture Notes in Math., vol. 114, Springer-Verlag, Berlin and New York, 1970.

Translated by J. S. JOEL

Mat. Sbornik
Tom 95 (137) (1974), No. 3

Math. USSR Sbornik
Vol. 24 (1974), No. 3

p-ADIC HECKE SERIES OF IMAGINARY QUADRATIC FIELDS

M. M. VIŠIK AND Ju. I. MANIN

Abstract. In this paper a p-adic analytic function of two variables is constructed whose values in some "common" domain coincide with the values of the family of Hecke L-series of an imaginary quadratic field. The functional equation for such a function is obtained. The p-adic Mellin integral transform is the main technique.

Bibliography: 13 items.

Introduction

Kubota and Leopoldt [5] opened a new chapter in the arithmetic of Dirichlet series by showing that the function from $\mathbf{Z}$ to $\mathbf{Q}$

$$k \longmapsto (-1)^{\frac{k}{2}} \frac{(k-1)!}{(2\pi)^k} (1-p^{k-1}) \zeta(k) \tag{1}$$

can be continued to a p-adic analytic function from $\mathbf{Z}_p$ to $\mathbf{Q}_p$ when k varies in any arithmetic progression $\bmod (p-1)$ and remains even and positive (in the case $k \equiv 0 \bmod (p-1)$ the formulation varies slightly). A similar result is also true for L-series with a Dirichlet character.

Most recently it was discovered that this result is by no means isolated: Serre [12] extended the Leopoldt-Kubota theorem to zeta-functions of any totally real fields; Deligne investigated L-series of totally real fields with Hecke characters; and Mazur and one of the authors [10], [8], [9] studied Mellin's transform of modular forms with respect to (the congruence subgroups of) $SL(2, \mathbf{Z})$. It now seems certain that all the "sufficiently arithmetic" zeta-functions have corresponding p-adic analogs or, more preferably, are archimedean components of some large adelic object connected with the l-adic representations of the Galois group. This was demonstrated by Iwasawa [3], [4] and Lichtenbaum [6] for the Leopoldt-Kubota zeta. However it will be necessary to do a lot more work before the meaning and natural sphere of extending these phenomena can be explained.

In the present article the p-adic (and even a partially adelic) variant of L-series of imaginary quadratic fields with Hecke characters is constructed. Two properties

AMS (MOS) subject classifications (1970). Primary 10H10, 12B30; Secondary 10D15, 12B40, 44A15.

appear in this example for the first time: the *imaginary nature of the ground field* (until now one could work only with totally real fields) and the analytic dependence of our functions on *two* arguments—the p-adic analog of s and the weight ν of the Hecke character (which can take only integral values in the archimedean case). Serre conjectured that the functions also depend analytically on the character (cf. [12], p. 263).

We shall formulate a special case of our results in more detail.

The following notation will be used in the entire subsequent presentation. K is an imaginary quadratic field; O is a maximal order of K; and λ is a Hecke character of K of weight ν modulo an integral ideal $\mathfrak{f}$. Recall that the meaning of this is as follows. λ is a homomorphism of the group of fractional ideals of K that are relatively prime to $\mathfrak{f}$ in $\mathbf{C}^*$. If $\alpha \in K$, α is $\mathfrak{f}$-integral, and $\alpha \equiv 1 \bmod \mathfrak{f}$, then $\lambda((\alpha)) = \alpha^\nu$. It is also convenient to assume that $\lambda(\mathfrak{n}) = 0$ if the numerator of the fractional ideal $\mathfrak{n}$ is not relatively prime to $\mathfrak{f}$.

The Hecke L-function with character λ is defined for $\operatorname{Re} s \gg 0$ by the formulas

$$L(s, \lambda) = \sum_{\mathfrak{n}} \lambda(\mathfrak{n}) N(\mathfrak{n})^{-s} = \prod_{\mathfrak{p} \nmid \mathfrak{f}} (1 - \lambda(\mathfrak{p}) N(\mathfrak{p})^{-s})^{-1},$$

where $\mathfrak{n}$ runs through the integral ideals of the field, the $\mathfrak{p}$ are prime, and $N(\cdot)$ is the norm of the ideal (or number) in parentheses. The Hecke series can be continued meromorphically (and for $\nu \geq 1$ holomorphically) to the whole plane of the variable s.

Now let $\mathfrak{p}$ be a prime ideal of the field K with *prime norm* unramified in K, not dividing $2\mathfrak{f}$, and not taking any of a finite number of prohibited values (the details can be found in the main text, §6.3).

We denote by $\Lambda(\mathfrak{f})$ the set of all Hecke characters mod $\mathfrak{f}$ of weight $\nu(\lambda) \geq 3$.

The following fact follows from our results:

There exists a finite partition of $\Lambda(\mathfrak{f})$ into disjoint subsets $\Lambda(\mathfrak{f}) = \bigcup_{i \in I} \Lambda_i(\mathfrak{f})$ such that for any $i \in I$ and any k_0, $0 \leq k_0 \leq p - 1$, the function with algebraic values

$$(k, \nu(\lambda)) \longmapsto \frac{(\sqrt{-1})^k \Gamma(k)}{(2\pi)^k} \frac{\pi^{\nu(\lambda)}}{\omega^{\nu(\lambda)}} \left(1 - \frac{\lambda(\mathfrak{p})}{N(\mathfrak{p})^k}\right) \left(1 - \frac{p^{k-1}}{\lambda(\bar{\mathfrak{p}})}\right) L(k, \lambda) \tag{2}$$

in the domain $k \equiv k_0 \bmod (p-1)$, $1 \leq k \leq \nu(\lambda)$, where $\nu(\lambda)$ is the weight of λ and $\lambda \in \Lambda_i(\mathfrak{f})$, can be continued to a $\mathfrak{p}$-adic analytic function of two arguments.

Here $\omega \in \mathbf{C}$ is a complex number such that the modular invariants

$$g_2(O\omega) = 60 \sum_{a \in O \setminus \{0\}} \frac{1}{(a\omega)^4}, \quad g_3(O\omega) = 140 \sum_{a \in O \setminus \{0\}} \frac{1}{(a\omega)^6}$$

are integral and algebraic; ω is assumed to be chosen once and for all.

Note that when $\lambda \in \Lambda_i(\mathfrak{f})$, $\nu(\lambda)$ varies in a fixed (depending on i) arithmetic progression with respect to some modulus of the form $M(p-1)^2$.

Example. $K = \mathbf{Q}(i)$; $\mathfrak{f} = (1)$; $p \equiv 1 \bmod 4$; $p = \rho\bar{\rho}$ (here and henceforth the bar denotes complex conjugation); $\mathfrak{p} = (\rho)$; $\omega = 2\int_0^1 (1 - x^4)^{-1/2} dx$; $\lambda((\alpha)) = \alpha^\nu$ for all $\alpha \in K$; $\nu \equiv 0 \bmod 4$. Here $\Lambda(\mathfrak{f})$ is partitioned into sets each containing ν in a fixed progression mod $(p-1)^2$, and our result assumes the following form: *the function*

$$(k, \nu) \longmapsto \frac{(\sqrt{-1})^k (k-1)!}{(2\pi)^k} \frac{\pi^\nu}{\omega^\nu} \left(1 - \frac{\rho^\nu}{p^k}\right)\left(1 - \frac{p^{k-1}}{\bar{\rho}^\nu}\right) \sum_{(a,b)\in \mathbf{Z}^2 \setminus \{0,0\}} \frac{(a+bi)^\nu}{(a^2+b^2)^k} \qquad (3)$$

in the domain $k \equiv k_0 \bmod (p-1)$, $\nu \equiv \nu_0 \bmod (p-1)^2$, $1 \le k \le \nu$, $\nu \ge 4$, $\nu \equiv 0 \bmod 4$, *can be continued to a* $\mathfrak{p}$*-adic analytic function of two arguments.* (The series in (3) on the right converges for $\nu/2 < k \le \nu$ and can be continued analytically to the remaining values of k.)

The $\mathfrak{p}$-adic continuity of a function close to the restriction of (3) to the right boundary of our domain $k = \nu$ was also established by Katz.

The similarities among (1), (2), and (3) are obvious. We shall briefly dwell on the differences and undecided questions.

a) The argument k in formula (1) assumes infinitely many values on one side of the critical strip, whereas in (2) and (3) the argument k assumes only integral values lying strictly inside the critical strip, which is situated from 0 to $\nu+1$ and grows as ν grows. The arithmetic nature of the values of Hecke functions outside these limits is unknown. Two entirely different approaches to their computation—the methods of Damerell [1] and one of the authors [8], [9]—do not allow the possibility of falling beyond the limits of the critical strip. Is this result connected with the fact that the remaining values are "essentially nonalgebraic"? Recall that for the Riemann zeta the analogous old problem on the values at the odd integral points is unsolved.

Note also that we have not analyzed the cases $\nu = 1, 2$ and $\nu = 0$. The last case corresponds to the ordinary Dedekind zeta of a field K; therefore it falls outside our investigation. More precisely, the $\mathfrak{p}$-adic candidates for it are included in our construction, since $\nu = 0$ is contained in the $\mathfrak{p}$-adic domain of definition, but the connection with the archimedean zeta has not been established.

b) It is entirely unclear how to handle primes p not splitting in K, to say nothing of the finite number of exceptional prime ideals that are in one sense or another "ramified", where the technique works partially, but more natural formulations have not been found.

c) The connection between our functions and l-adic representations has not been investigated.

We now briefly describe the plan of this paper.

The Mellin-Mazur nonarchimedean integral representation (see [8] and [10]) is the main technique. Nonarchimedean local measures connected with Hecke series are introduced in §1. Unlike [8] and [10], we consider a finite number of prime ideals at the same time.

The dependence of the local measure on the p-adic argument s is investigated in §2.

Nonarchimedean integral representations are given in §3, and the p-adic functional equation of our L-functions is given in §4.

The remaining sections are devoted to the most detailed part of the paper—the proof of the analyticity.

The restriction of local measures to the boundary of the domain under discussion ($s = \nu$) is computed in §5 in terms of the values of the Weierstrass function.

In §§6 and 7 the problem is translated into algebraic language and reduces to computations on Néron's p-adic model of an elliptic curve with period lattice $O\omega$.

In conclusion, note that the success of the entire program is predicated on the proper choice of local measures in §1. Until now the traditional construction of a local measure connected with Dirichlet series of algebraic number fields was as follows. For the series, say, $\sum_{n=1}^{\infty} a_n n^{-s}$, one set $\mu_s(a + (m)) = \sum_{n \equiv a \bmod m} a_n n^{-s}$ and then regularized an unbounded measure μ_0. More detailed definitions connected with the possibility of removing the Euler factors from a given Dirichlet series is included in [8] and [10] in implicit form. Formulas (4) and (5) below were written as a result of an analysis of these constructions, which has a much wider range of applicability.

§1. S-adic measures

Let K, O, λ, and $\mathfrak{f}$ have the same meaning as in the Introduction. Choose once and for all a finite set S of prime ideals of K with the following properties:

$$\mathfrak{p} \in S \Rightarrow \begin{cases} \mathfrak{p} \text{ is unramified and of the first degree in } K; \\ \mathfrak{p},\ \bar{\mathfrak{p}} \nmid \mathfrak{f};\ \bar{\mathfrak{p}} \notin S. \end{cases}$$

In what follows the letter $\mathfrak{m}$ will denote only *complete integral S-ideals.* This means that each prime $\mathfrak{p} \in S$ divides $\mathfrak{m}$ and that no prime ideal $\mathfrak{q} \notin S$ divides $\mathfrak{m}$.

Set $\hat{O}_S = \varprojlim O/\mathfrak{m}$. This ring is canonically isomorphic to

$$\prod_{\substack{p = N(\mathfrak{P}) \\ \mathfrak{P} \in S}} \mathbf{Z}_p.$$

In particular, all residue classes $\hat{O}_S \bmod \mathfrak{m}\hat{O}_S$ are represented by ordinary integers $a \in \mathbf{Z}$. We shall write $\hat{\mathfrak{m}} = \mathfrak{m}\hat{O}_S$ and denote by $a + \hat{\mathfrak{m}}$ the corresponding residue class in $\hat{O}_S$. For $\operatorname{Re} s >> 0$ set

$$L(a + \hat{\mathfrak{m}}; s, \lambda) = \sum_{(\mathfrak{n}, \mathfrak{m}) = 1} \lambda(\mathfrak{n}) N(\mathfrak{n})^{-s} \exp\left(-2\pi i \frac{a}{N(\mathfrak{m})} N(\mathfrak{n})\right) \tag{4}$$

($\mathfrak{n}$ runs through the integral ideals of the field K). The right side has an analytic continuation for all s, which will serve as the definition of the left side in the general case (Lemma 1.3).

Here is the main result of this section:

1.1. **Theorem.** a) *For fixed λ of weight ν and for variables $\mathfrak{m}$ the mapping*

$$a + \hat{\mathfrak{m}} \mapsto \mu(a + \hat{\mathfrak{m}}; s, \lambda) = \frac{\Gamma(s)}{(2\pi)^s} \frac{\pi^\nu}{\omega^\nu} \frac{N(\mathfrak{m})^{s-1}}{\lambda(\bar{\mathfrak{m}})} L(a + \hat{\mathfrak{m}}; s, \lambda) \tag{5}$$

can be uniquely continued to a finitely additive mapping (measure) $U \mapsto \mu(U; s, \lambda)$ of the set of open-closed subsets in $\hat{O}_S$ into the additive group of meromorphic functions of s.

b) *If $\nu \geq 1$ and $1 \leq k \leq \nu$, where the integer k and the character λ of weight ν are fixed, then for all U the measure $\mu(U; k, \lambda)$ takes algebraic values lying in some module of finite type over the ring $\mathbf{Z}[\lambda(\bar{\mathfrak{p}})^{-1}]_{\mathfrak{p}\in S}$. In particular, the S-components of the denominators of these values are bounded.*

Remark. In §7 it will be established that for all $\mathfrak{p} \in S$, excluding a finite set depending only on $\mathfrak{f}$ and ω but not on S, the values of the measure $\mu(U; k; \lambda)$ are $\mathfrak{p}$-integral.

The main part of the proof is included in four lemmas.

1.2. **Lemma** (additivity). *Suppose that $a \in \mathbf{Z}$, $\mathfrak{p} \in S$, and that $\mathfrak{m}$ is a complete S-ideal. Then*

$$\mu(a + \hat{\mathfrak{m}}; s, \lambda) = \sum_{\substack{a' \equiv a \bmod N(\mathfrak{m}) \\ a' \bmod N(\mathfrak{m}\mathfrak{p})}} \mu(a' + \hat{\mathfrak{p}}\hat{\mathfrak{m}}; s, \lambda). \tag{6}$$

Proof. It suffices to verify this equality for $\operatorname{Re} s >> 0$, when the Dirichlet series (4) can be used. The left and right sides of (6) are equal, respectively, to

$$\frac{\Gamma(s)}{(2\pi)^s} \frac{\pi^\nu}{\omega^\nu} \frac{N(\mathfrak{m})^{s-1}}{\lambda(\bar{\mathfrak{m}})} \sum_{(\mathfrak{n},\mathfrak{m})=1} \lambda(\mathfrak{n}) N(\mathfrak{n})^{-s} \exp\left(-2\pi i \frac{a}{N(\mathfrak{m})} N(\mathfrak{n})\right),$$

$$\frac{\Gamma(s)}{2\pi)^s} \frac{\pi^\nu}{\omega^\nu} \frac{N(\mathfrak{m})^{s-1}}{\lambda(\bar{\mathfrak{m}})} \frac{N(\mathfrak{p})^{s-1}}{\lambda(\bar{\mathfrak{p}})} \sum_{(\mathfrak{n},\mathfrak{m})=1} \lambda(\mathfrak{n}) N(\mathfrak{n})^{-s} \sum_{\substack{a' \equiv a \bmod N(\mathfrak{m}) \\ a' \bmod N(\mathfrak{p}\mathfrak{m})}} \exp\left(-2\pi i \frac{a'}{N(\mathfrak{m}\mathfrak{p})} N(\mathfrak{n})\right).$$

To verify that they coincide, we divide both expressions by the factor before the first sum, then fix an integral ideal $\mathfrak{n}_0$, $(\mathfrak{n}_0, \mathfrak{m}\bar{\mathfrak{p}}) = 1$, and select on both sides only the terms corresponding to the ideals $\mathfrak{n} = \mathfrak{n}_0\bar{\mathfrak{p}}^r$ and $\mathfrak{n} = 0$:

$$\sum_{r \geq 0} \lambda(\mathfrak{n}_0\bar{\mathfrak{p}}^r) N(\mathfrak{n}_0\bar{\mathfrak{p}}^r)^{-s} \exp\left(-2\pi i \frac{a}{N(\mathfrak{m})} N(\mathfrak{n}_0\bar{\mathfrak{p}}^r)\right),$$

$$\frac{N(\mathfrak{p})^{s-1}}{\lambda(\bar{\mathfrak{p}})} \sum_{r \geq 0} \lambda(\mathfrak{n}_0\bar{\mathfrak{p}}^r) N(\mathfrak{n}_0\bar{\mathfrak{p}}^r)^{-s} \sum_{\substack{a' \equiv a \bmod N(\mathfrak{m}) \\ a' \bmod N(\mathfrak{p}\mathfrak{m})}} \exp\left(-2\pi i \frac{a'}{N(\mathfrak{m}\mathfrak{p})} N(\mathfrak{n}_0\bar{\mathfrak{p}}^r)\right).$$

Let us show that the term of the second sum with $r = 0$ vanishes, and that the $(r+1)$th term of the second sum (along with the outer factor) is equal to the rth term of the first sum, i.e.

$$\sum_{\substack{a' \equiv a \bmod N(\mathfrak{m}) \\ a' \bmod N(\mathfrak{p}\mathfrak{m})}} \exp\left(-2\pi i \frac{a'}{N(\mathfrak{p}\mathfrak{m})} N(\mathfrak{n}_0)\right) = 0,$$

$$\frac{N(\mathfrak{p})^{s-1}}{\lambda(\bar{\mathfrak{p}})}\lambda(\mathfrak{n}_0\bar{\mathfrak{p}}^{r+1})\,N(\mathfrak{n}_0\bar{\mathfrak{p}}^{r+1})^{-s}\sum_{\substack{a'\equiv a \bmod N(\mathfrak{m})\\ a' \bmod N(\mathfrak{pm})}}\exp\left(-2\pi i\frac{a'}{N(\mathfrak{pm})}N(\mathfrak{n}_0\bar{\mathfrak{p}}^{r+1})\right)$$

$$=\lambda(\mathfrak{n}_0\bar{\mathfrak{p}}^{r})\,N(\mathfrak{n}_0\bar{\mathfrak{p}}^{r})^{-s}\exp\left(-2\pi i\frac{a}{N(\mathfrak{m})}N(\mathfrak{n}_0\bar{\mathfrak{p}}^{r})\right).$$

To do this we set $a' = a + b$, where $b \in \mathbf{Z}$ runs through the integers mod $N(\mathfrak{pm})$ such that $b \equiv 0 \bmod N(\mathfrak{m})$, say, $b = -jN(\mathfrak{m})$, $0 \le j \le N(\mathfrak{p}) - 1$. After cancelling

$$\exp\left(-2\pi i\frac{a}{N(\mathfrak{m})}N(\mathfrak{n}_0\bar{\mathfrak{p}}^{r})\right)$$

and the factor before the sum, we obtain that we must verify the equalities

$$\sum_{j=0}^{N(\mathfrak{p})-1}\exp\left(2\pi i\frac{j}{N(\mathfrak{p})}N(\mathfrak{n}_0)\right)=0,$$

$$\frac{1}{N(\mathfrak{p})}\sum_{j=0}^{N(\mathfrak{p})-1}\exp(2\pi ijN(\mathfrak{n}_0\bar{\mathfrak{p}}^{r}))=1,\quad r>0.$$

They are obvious since $\mathfrak{p}\nmid\mathfrak{n}_0$ due to the fact that $\mathfrak{m}$ is S-complete and $\bar{\mathfrak{p}}\nmid\mathfrak{n}_0$ by the definition of $\mathfrak{n}_0$, so that $N(\mathfrak{p})\nmid N(\mathfrak{n}_0)$.

1.3. **Lemma** (archimedean integral representation). *Let M be the Möbius function on integral ideals of K: $M(\mathfrak{p}_1,\ldots,\mathfrak{p}_r) = (-1)^r$ if $\mathfrak{p}_i \neq \mathfrak{p}_j$ for $i \neq j$, $\mathfrak{p}_i$ prime; $M(\vartheta) = 0$ on the remaining ideals.*

Set $\Phi_\lambda(z) = \sum_n \lambda(\mathfrak{n})\exp\{2\pi iN(\mathfrak{n})z\}$.

In the above notation, we have

$$\mu(a+\hat{\mathfrak{m}};s,\lambda)=\frac{\pi^{\nu}}{\omega^{\nu}}\frac{N(\mathfrak{m})^{s-1}}{\lambda(\bar{\mathfrak{m}})}\sum_{\vartheta|\mathfrak{m}}M(\vartheta)\lambda(\vartheta)N(\vartheta)^{-s}\int_0^{i\infty}\Phi_\lambda\left(z-\frac{a}{N(\mathfrak{m}\vartheta^{-1})}\right)\left(\frac{z}{i}\right)^s\frac{dz}{z}. \tag{7}$$

Proof. We use the known formal identity

$$\sum_{(\mathfrak{n},\mathfrak{m})=1}F(\mathfrak{n})=\sum_{n}F(\mathfrak{n})-\sum_{\mathfrak{p}|\mathfrak{m}}\sum_{n}F(\mathfrak{np})+\sum_{\substack{\mathfrak{p}_1,\mathfrak{p}_2|\mathfrak{m}\\ \mathfrak{p}_1\neq\mathfrak{p}_2}}\sum_{n}F(\mathfrak{n}\mathfrak{p}_1\mathfrak{p}_2)-\ldots$$

$$=\sum_{\mathfrak{n}}\sum_{\vartheta|\mathfrak{m}}M(\vartheta)F(\mathfrak{n}\vartheta).$$

We apply it to the right side of (4) for $\operatorname{Re} s \gg 0$ and substitute the result in (5), bearing in mind that for $\vartheta|\mathfrak{m}$

$$\sum_{\mathfrak{n}}\lambda(\mathfrak{n}\vartheta)\,N(\mathfrak{n}\vartheta)^{-s}\exp\left(-2\pi i\frac{a}{N(\mathfrak{m})}N(\mathfrak{n}\vartheta)\right)=$$

$$= \lambda(\vartheta) N(\vartheta)^{-s} \sum_{\mathfrak{n}} \lambda(\mathfrak{n}) N(\mathfrak{n})^{-s} \exp\left(-2\pi i \frac{a}{N(\mathfrak{m}\vartheta^{-1})} N(\mathfrak{n})\right)$$

$$= \lambda(\vartheta) N(\vartheta)^{-s} \frac{(2\pi)^s}{\Gamma(s)} \int_0^{i\infty} \Phi_\lambda\left(z - \frac{a}{N(\mathfrak{m}\vartheta^{-1})}\right)\left(\frac{z}{i}\right)^s \frac{dz}{z}$$

by virtue of usual formula for the Mellin transform. This proves the lemma.

Note that (for $\nu \geq 1$) $\Phi_\lambda(z)$ is a parabolic form* in the upper half-plane with respect to some congruence subgroup of $SL(2, \mathbf{Z})$ (cf. [13]). Therefore the integral representation (7) indeed makes it possible to continue our measures in s analytically.

1.4. **Lemma** (algebraicity). *For all a, $\mathfrak{m}$, $\nu \geq 1$, and k, $1 \leq k \leq \nu$, the values of $\mu(a + \hat{\mathfrak{m}}; k, \lambda)$ are algebraic.*

Proof. This result is obtained from definitions (4) and (5) if we combine three facts.

The first and main fact was proved by Damerell [1]: the values

$$\frac{\pi^{\nu-k}}{\omega^\nu} \sum_{\mathfrak{n}} \lambda(\mathfrak{n}) N(\mathfrak{n})^{-k} \chi(N(\mathfrak{n}))$$

are algebraic for all Dirichlet characters χ and all k, $\nu/2 < k \leq \nu$. Applying the standard functional equation (see §4.1 below), we obtain from this that these values are also algebraic for the left half of the critical strip $1 \leq k \leq \nu/2$.

The second fact is that any periodic function on $\mathbf{Z}$ can be expanded in Dirichlet characters. In particular, this applies to the function

$$N(\mathfrak{n}) \longmapsto \exp\left(-2\pi i \frac{a}{N(\mathfrak{m})} N(\mathfrak{n})\right),$$

and the coefficients of this expansion are algebraic. Therefore the values

$$\frac{\pi^{\nu-k}}{\omega^\nu} L(a + \hat{\mathfrak{m}}; k, \lambda)$$

are algebraic for $1 \leq k \leq \nu$.

Finally, the values of $\lambda(\overline{\mathfrak{m}})$ are integral, algebraic, and S-invertible. For let H be a natural number such that the divisor $\overline{\mathfrak{m}}^H = (\alpha)$ is principal and $\alpha \equiv 1 \bmod \mathfrak{f}$ (H exists since $(\overline{\mathfrak{m}}, \mathfrak{f}) = 1$). Then $\lambda(\overline{\mathfrak{m}}) = \sqrt[H]{\alpha^\nu}$. In particular, the divisors of $\lambda(\overline{\mathfrak{m}})$ in the field where this number lies consist only of the divisors of $\overline{\mathfrak{m}}$. This proves the lemma.

1.5. **Lemma** (integralness). *For $1 \leq k \leq \nu$ and $x \in \mathbf{Q}$ the values*

$$(\text{the denominator of } x)^{k-1} \cdot \int_0^{i\infty} \Phi_\lambda(z + x) z^{k-1} dz$$

lie in a $\mathbf{Z}$-module of finite type.

* *Editor's note. Cusp form* is more common in English.

Proof. We use the reasoning in [8]. Suppose that $x = a/b$ is in lowest terms. Suppose, say, that $x > 0$ and that

$$\frac{a}{b} = \frac{a_n}{b_n}, \frac{a_{n-1}}{b_{n-1}}, \ldots, \frac{a_0}{b_0} = \frac{0}{1}, \frac{a_{-1}}{b_{-1}} = \frac{1}{0}$$

are the successive convergents of the continued fraction for a/b. Then

$$g_j = \begin{pmatrix} a_j & (-1)^{j-1} a_{j-1} \\ b_j & (-1)^{j-1} b_{j-1} \end{pmatrix} \in SL(2, \mathbf{Z}),$$

and

$$b^{k-1} \int_0^{i\infty} \Phi_\lambda \left(z + \frac{a}{b}\right) z^{k-1} dz = b^{k-1} \int_{a/b}^{i\infty} \Phi_\lambda(z) \left(z - \frac{a}{b}\right)^{k-1} dz \in \sum_{i=0}^{k-1} \mathbf{Z} \int_{a/b}^{i\infty} \Phi_\lambda(z) z^i\, dz,$$

$$\int_{a/b}^{i\infty} \Phi_\lambda(z) z^i\, dz = -\sum_{j=0}^{n} \int_{a_{j-1}/b_{j-1}}^{a_j/b_j} \Phi_\lambda(z) z^i\, dz = -\sum_{j=0}^{n} \int_0^{i\infty} \Phi_\lambda(g_j z)(g_j z)^i\, d(g_j z).$$

The integrals in the last sum belong to the $\mathbf{Z}$-module of finite type generated by the "periods"

$$\int_0^{\infty} (\Phi_\lambda|_{(\nu+1)} g(z) z^i dz, \quad 0 \leqslant i \leqslant \nu - 1,$$

of all possible translations of the parabolic form Φ_λ via the elements $g \in SL(2, \mathbf{Z})$, and there are a finite number of these translations.

1.6. *Deduction of the theorem from the lemmas.* Lemma 1.2 establishes that the function $U \mapsto \mu(U; s, \lambda)$ is defined and additive on the base of open-closed sets in O_S and such a system of coverings, iterating which we obtain a base for the coverings Hence assertion a) of the theorem follows by means of standard and simple combinatorial arguments.

It suffices to prove assertion b) only for $U = a + \hat{\mathfrak{m}}$. Lemma 1.4 ensures the algebraicity. Formula (7) expresses $\mu(a + \hat{\mathfrak{m}}; k, \lambda)$ in the form of a linear combination of terms of the form

$$N(\mathfrak{m}\vartheta^{-1})^{k-1} \int_0^{i\infty} \Phi_\lambda \left(z - \frac{a}{N(\mathfrak{m}\vartheta^{-1})}\right) z^{k-1}\, dz.$$

For a given λ and varying k, $\mathfrak{m}$, and ϑ with $M(\vartheta) \neq 0$ each of these terms lies in a fixed $\mathbf{Z}$-module of finite type by Lemma 1.5. The denominators of the coefficients of this linear combination involve only $\lambda(\overline{\mathfrak{m}})$ for all possible S-complete ideals $\mathfrak{m}$, and $N(\vartheta)$ for the S-ideals ϑ such that $M(\vartheta) \neq 0$.

This completes the proof.

§2. The dependence of the measure on the argument k

2.1. *Differential and integral representation of the measure.* Let B be an $\hat{O}_S$-algebra that is complete in the S-adic topology and separable, and let μ be a finitely additive measure on the open-closed subsets $U \subset \hat{O}_S$ with values in B. We shall often write $\int_U d_\mu(a)$ instead of $\mu(U)$. We call a measure μ *bounded* if all the $\mu(U)$ lie in some bounded subset B.

Let $f\colon \hat{O}_S \to B$ be an S-adically continuous function. For any $U \subset \hat{O}_S$ the integral $\int_U f(a)d\mu(a)$, defined as the limit of Riemann sums, exists if μ is bounded (cf. [8], [10]). The function $U \mapsto \int_U f(a)d\mu(a)$ obviously defines a new measure μ_1, and we shall use the equivalent representation $d\mu_1(a) = f(a)d\mu(a)$.

It is easy to verify that the entire standard formalism of integration is justified for bounded measures and continuous functions.

In what follows we shall apply this formalism to measures $\mu(U;\ k,\ \lambda)$ defined by Theorem 1.1b. For the ring B_S of values of the measure we take the $\hat{O}_S$-algebra generated by the S-adic completion of the ring of all algebraic integers and the field of all algebraic numbers (in other words, B_S is the S-component of the adèle ring of the field of all algebraic numbers). The algebraic numbers are embedded in O_S in an obvious way, and we shall identify them with the image under this imbedding.

The B_S-measures $\mu(U;\ k,\ \lambda)$ for $1 \le k \le \nu$ are bounded by Theorem 1.1b. In what follows we shall deal only with their restriction to the unit group $\hat{O}_S^*$ of the ring $\hat{O}_S$. The purpose of this section is to prove the following result:

2.2. **Theorem.** *In the notation of Theorem* 1.1 *we set* $\mu_\lambda(U) = \mu(U;\ 1,\ \lambda)$ *for* $U \subset \hat{O}_S^*$. *Then for all integers* k, $1 \le k \le \nu$, *in* $\hat{O}_S^*$ *we have*

$$d\mu(a;\ k,\ \lambda) = \left(\frac{a}{i}\right)^{k-1} d\mu_\lambda(a). \tag{8}$$

Proof. Consider the sequence of classes $a_j + \hat{\mathfrak{m}}_j$ that contract to a point $a \in \hat{O}_S^*$ in the S-adic topology. It must be proved that in the topology of B_S

$$\mu(a_j + \hat{\mathfrak{m}}_j;\ k,\ \lambda) \sim \left(\frac{a_j}{i}\right)^{k-1} \mu(a_j + \hat{\mathfrak{m}}_j;\ 1,\ \lambda)$$

in the sense that the ratio of both sides tends to 1 as $j \to \infty$. To do this we apply the archimedean integral representation (7). First of all, we have

$$\begin{aligned}\int_0^{i\infty} \Phi_\lambda\left(z - \frac{a_j}{N(\mathfrak{m}_j\vartheta^{-1})}\right)\left(\frac{z}{i}\right)^k \frac{dz}{z} &= i^{-k} \int_{-a_j/N(\mathfrak{m}_j\vartheta^{-1})}^{i\infty} \Phi_\lambda(z)\left(z + \frac{a_j}{N(\mathfrak{m}_j\vartheta^{-1})}\right)^{k-1} dz \\ &= i^{-k} \sum_{l=0}^{k-1} \binom{k-1}{l} \left(\frac{a_j}{N(\mathfrak{m}_j\vartheta^{-1})}\right)^l \int_{-a_j/N(\mathfrak{m}_j\vartheta^{-1})}^{i\infty} \Phi_\lambda(z)\, z^{k-1-l}\, dz.\end{aligned} \tag{9}$$

The last integrals, as was verified in the proof of Lemma 1.5, lie in a $\mathbf{Z}$-module of finite type. On the other hand, the expressions (9) in formula (7) are multiplied by $N(\mathfrak{m}_j\vartheta^{-1})^{k-1}$ and also by factors with bounded S-denominators. This shows that to determine the asymptotic behavior as $j \to \infty$ it suffices to substitute into (7) only the term of the sum (9) for which $l = k - 1$. Thus we obtain

$$\mu(a_j + \hat{\mathfrak{m}}_j;\, k, \lambda)$$

$$\sim \frac{\pi^\nu}{\omega^\nu} \frac{1}{\lambda(\bar{\mathfrak{m}})} \sum_{\vartheta \mid \mathfrak{m}} M(\vartheta)\lambda(\vartheta) N(\vartheta)^{-1} i^{-k} a_j^{k-1} \int_{-a_j/N(\mathfrak{m}_j\vartheta^{-1})}^{i\infty} \Phi_\lambda(z)\, dz.$$

The right side is equal to $(a_j/i)^{k-1}\mu(a_j + \hat{\mathfrak{m}}_j;\, 1, \lambda)$, again by virtue of (7). This completes the proof.

§3. The S-adic integral representation of Hecke L-functions

3.1. **Theorem.** *In the former notation, for all* $1 \leq k \leq \nu$ *we have*

$$\frac{\pi^{\nu-k}}{\omega^\nu} \prod_{\mathfrak{p}\in S}\left(1 - \frac{\lambda^*(\mathfrak{p})}{N(\mathfrak{p})^k}\right)\left(1 - \frac{N(\bar{\mathfrak{p}})^{k-1}}{\lambda(\bar{\mathfrak{p}})}\right) L(k, \lambda) = \frac{2^k}{\Gamma(k)} \int_{\hat{O}_S^*} \left(\frac{a}{i}\right)^{k-1} d\mu_\lambda(a). \tag{10}$$

3.2. *Comment.* Let us compare (10) with the classical formula for the Mellin transform:

$$L(k, \lambda) = \frac{(2\pi)^k}{\Gamma(k)} \int_0^{i\infty} \Phi_\lambda(z)\left(\frac{z}{i}\right)^{k-1} d\,\frac{z}{i}. \tag{11}$$

It is clear that up to correction factors of two types (removing the transcendence and the $\mathfrak{p}$-component in the Euler product and varying the $\bar{\mathfrak{p}}$-component) $d\mu_\lambda$ should be regraded as the S-adic analog of the restriction of $\Phi_\lambda(z)dz$ to the imaginary semiaxis. Thus in this context *not functions* of an S-adic argument, *but measures* are the S-adic parallel of modular forms. The integral on the right side of (10) immediately supplies the analytic continuation in k, to which we shall return below.

3.3. **Proof.** Set $\mathfrak{m} = \prod_{\mathfrak{p}\in S}\mathfrak{p}$. Then

$$\hat{O}_S^* = \bigcup_{\substack{a \bmod N(\mathfrak{m}) \\ (a,\,\mathfrak{m})=1}} (a + \hat{\mathfrak{m}}),$$

and by Theorem 2.2

$$\frac{2^k}{\Gamma(k)} \int_{\hat{O}_S^*} \left(\frac{a}{i}\right)^{k-1} d\mu_\lambda(a) = \frac{2^k}{\Gamma(k)} \sum_{\substack{a \bmod N(\mathfrak{m}) \\ (a,\,\mathfrak{m})=1}} \mu(a + \hat{\mathfrak{m}};\, k, \lambda).$$

We compute $\sum_a \mu(a + \hat{\mathfrak{m}};\, s, \lambda)$ for $\operatorname{Re} s \gg 0$ with the help of (4) and (5), and use the fact that the analytic continuation is unique. We have

$$\frac{2^s}{\Gamma(s)} \sum_a \mu(a + \hat{\mathfrak{m}};\, s, \lambda) = \frac{\pi^{\nu-s}}{\omega^\nu} \frac{N(\mathfrak{m})^{s-1}}{\lambda(\bar{\mathfrak{m}})} \sum_{(\mathfrak{n},\mathfrak{m})=1} \lambda(\mathfrak{n}) N(\mathfrak{n})^{-s} \times$$

$$\times \sum_{\substack{a \bmod N(\mathfrak{m}) \\ (a, N(\mathfrak{m}))=1}} \exp\left(-2\pi i \frac{a}{N(\mathfrak{m})} N(\mathfrak{n})\right). \tag{12}$$

Consider an integral ideal $\mathfrak{n}$, $(\mathfrak{n}, \mathfrak{m}) = 1$, and represent it in the form $\mathfrak{n} = \mathfrak{n}_0\bar{\vartheta}$, where $\bar{\vartheta} | \bar{\mathfrak{m}}$ and $(\mathfrak{n}_0, \bar{\mathfrak{m}}\bar{\vartheta}^{-1}) = 1$. Then

$$\sum_{\substack{a \bmod N(\mathfrak{m}) \\ (a, \mathfrak{m})=1}} \exp\left(-2\pi i \frac{a}{N(\mathfrak{m})} N(\mathfrak{n})\right) = M(\mathfrak{m}\vartheta^{-1})\, \varphi(N(\vartheta)) = (-1)^{|S|} M(\vartheta)\, \varphi(N(\vartheta)).$$

Here $|S|$ is the number of elements of S, M is the Möbius function, and ϕ is the Euler function. Therefore the right side of (12) without the outer factor before the sum is equal to

$$(-1)^{|S|} \sum_{\vartheta | \mathfrak{m}} M(\vartheta)\, \varphi(N(\vartheta))\, \lambda(\bar{\vartheta})\, N(\vartheta)^{-s} \sum_{(\mathfrak{n}_0, N(\mathfrak{m})\bar{\vartheta}^{-1})=1} \lambda(\mathfrak{n}_0)\, N(\mathfrak{n}_0)^{-s}$$

$$= (-1)^{|S|} \sum_{\vartheta | \mathfrak{m}} M(\vartheta)\, \varphi(N(\vartheta))\, \lambda(\bar{\vartheta})\, N(\vartheta)^{-s} L(s, \lambda) \prod_{\mathfrak{p} \in S} \left(1 - \frac{\lambda(\mathfrak{p})}{N(\mathfrak{p})^s}\right)$$

$$\times \prod_{\mathfrak{p} | \mathfrak{m}\vartheta^{-1}} \left(1 - \frac{\lambda(\bar{\mathfrak{p}})}{N(\bar{\mathfrak{p}})^s}\right) = (-1)^{|S|} \prod_{\mathfrak{p} \in S} \left(1 - \frac{\lambda(\mathfrak{p})}{N(\mathfrak{p})^s}\right) L(s, \lambda)$$

$$\times \sum_{\vartheta | \mathfrak{m}} M(\vartheta)\, \varphi(N(\vartheta))\, \lambda(\bar{\vartheta})\, N(\vartheta)^{-s} \sum_{\vartheta_1 | \mathfrak{m}\vartheta^{-1}} M(\vartheta_1)\, \lambda(\bar{\vartheta}_1)\, N(\vartheta_1)^{-s}.$$

Let us take the inner summation sign outside, and sum first over all divisors ϑ of the product $\vartheta' = \vartheta\vartheta_1$ for a fixed ϑ' and then over all divisors $\vartheta' | \mathfrak{m}$. Bearing in mind that $\sum_{\vartheta | \vartheta'} \phi(N(\vartheta)) = N(\vartheta')$, we finally obtain

$$(-1)^{|S|} \prod_{\mathfrak{p} \in S} \left(1 - \frac{\lambda(\mathfrak{p})}{N(\mathfrak{p})^s}\right) L(s, \lambda) \prod_{\mathfrak{p} \in S} \left(1 - \frac{\lambda(\bar{\mathfrak{p}})}{N(\mathfrak{p})^{s-1}}\right). \tag{13}$$

Now we add the outer factor

$$\frac{\pi^{\nu-s}}{\omega^{\nu}} \frac{N(\mathfrak{m})^{s-1}}{\lambda(\bar{\mathfrak{m}})}$$

to (13). Set $s = k$ and distribute the divisors $N(\mathfrak{p})^{k-1}/\lambda(\bar{\mathfrak{p}})$ of the second component over the terms of the last product of (13), and then interchange the signs of all terms of this product. This removes the factor $(-1)^{|S|}$ and leads to the final formula (10).

3.4. Theorem 3.1 supplies an S-adic integral representation only for L-series with those Hecke characters whose moduli are relatively prime. However, one can treat certain characters not satisfying this condition by multiplying λ by a Dirichlet character modulo $N(\mathfrak{m}_0)$, where $\mathfrak{m}_0$ is an S-ideal.

More precisely, let $\mathfrak{m}_0$ be an S-complete ideal (here incomplete ideals could also be treated; we leave this to the reader as an exercise). Let χ be a primitive Dirichlet character mod $N(\mathfrak{m}_0)$. We denote by $\lambda\overline{\chi}$ the Hecke character $\lambda\overline{\chi}(\mathfrak{n}) = \lambda(\mathfrak{n})\overline{\chi}(N(\mathfrak{n}))$. The character χ extends to $\hat{O}_S^*$ in an obvious way.

3.5. **Theorem.** *In the above notation, for all k, $1 \le k \le \nu$, we have*

$$\frac{\pi^{\nu-k}}{\omega^{\nu}} \frac{N(\mathfrak{m}_0)^{k-1}}{\lambda(\overline{\mathfrak{m}}_0)} \tau(\chi) L(k, \lambda\overline{\chi}) = \frac{2^k}{\Gamma(k)} \int_{\hat{O}_S^*} \chi(a) \left(\frac{a}{i}\right)^{k-1} d\mu_\lambda(a),$$

where

$$\tau(\chi) = \sum_{a \bmod N(\mathfrak{m}_0)} \chi(a) \exp\left(-2\pi i \frac{a}{N(\mathfrak{m}_0)}\right).$$

Proof. As above, we represent the right side in the form

$$\frac{2^k}{\Gamma(k)} \sum_{a \bmod N(\mathfrak{m}_0)} \chi(a) \mu(a + \hat{\mathfrak{m}}_0; k, \lambda)$$

and, replacing k by s, we compute it for $\operatorname{Re} s \gg 0$ with the help of (4) and (5). We obtain

$$\frac{\pi^{\nu-s}}{\omega^{\nu}} \frac{N(\mathfrak{m}_0)^{s-1}}{\lambda(\overline{\mathfrak{m}}_0)} \sum_{(\mathfrak{n}, \mathfrak{m}_0)=1} \lambda(\mathfrak{n}) N(\mathfrak{n})^{-s} \sum_{\substack{a \bmod N(\mathfrak{m}_0) \\ (a, N(\mathfrak{m}_0))=1}} \chi(a) \exp\left(-2\pi i \frac{a}{N(\mathfrak{m}_0)} N(\mathfrak{n})\right).$$

Here the inner sum is equal to zero if $(N(\mathfrak{n}), N(\mathfrak{m}_0)) > 1$, since χ is primitive, and equal to $\chi^{-1}(N(\mathfrak{n}))\tau(\chi)$ otherwise. This completes the proof.

§4. The S-adic functional equation for Hecke L-functions

4.1. *Archimedean functional equation.* Let $M = DN(\mathfrak{f})$, where D is the discriminant of the field K. We denote by ϵ a Dirichlet character mod M such that

$$\varepsilon(a) = \left(\frac{-D}{a}\right) \frac{\lambda((a))}{a^{\nu}},$$

and by $\lambda\overline{\epsilon}$ a Hecke character of K such that

$$\lambda\overline{\varepsilon}(\mathfrak{n}) = \lambda(\mathfrak{n}) \overline{\varepsilon}(N(\mathfrak{n})).$$

There exists a constant C which is an integral power of $-i$ such that

$$\Phi_\lambda\left(-\frac{1}{Mz}\right) d\left(-\frac{1}{Mz}\right) = C(-i)^{\nu+1} M^{\frac{\nu-1}{2}} z^{\nu-1} \Phi_{\lambda\overline{\varepsilon}}(z)\, dz. \tag{14}$$

This follows from the fact that Φ_λ is a parabolic (for $\nu \ge 1$) form with respect to the group $\Gamma(M)$ belonging to the character ε and is proper with respect to Hecke operators (cf. Shimura [13] and Ogg [11]). Applying (14) in the standard way to the integral representation of the function

$$\Lambda(\lambda, s) = (2\pi)^{-s}\Gamma(s)\sum_{\mathfrak{n}}\lambda(\mathfrak{n})N(\mathfrak{n})^{-s} = \int_0^{i\infty}\Phi_\lambda(z)\left(\frac{z}{i}\right)^s z^{-1}\,dz,$$

we obtain the functional equation

$$\Lambda(\lambda, s) = CM^{\frac{\nu+1}{2}-s}\Lambda(\lambda\bar{\varepsilon}, \nu+1-s). \tag{15}$$

The purpose of this section is to prove the following S-adic analogs of formulas (14) and (15). (In the latter k can assume any S-adic values in the completion of $\mathbf{Z}$, so that it does not follow directly by combining (10) and (15).)

4.2. **Theorem.** *The following functional equations are valid*:

$$d\mu_\lambda\left(-\frac{1}{Ma}\right) = -C(-i)^{\nu+1}M^{\frac{\nu-1}{2}}a^{\nu-1}\,d\mu_{\lambda\bar{\varepsilon}}(a), \tag{16}$$

$$\int_{\hat{O}_S^*}\left(\frac{a}{i}\right)^{k-1}d\mu_\lambda(b) = CM^{\frac{\nu+1}{2}-k}\int_{\hat{O}_S^*}\left(\frac{a}{i}\right)^{\nu-k}d\mu_{\lambda\bar{\varepsilon}}(a). \tag{17}$$

Proof. Let $a, a' \in \mathbf{Z}$, $Maa' \equiv -1 \bmod N(\mathfrak{m})$, and suppose that $\mathfrak{m}$ is small in the S-adic topology. We compare $\mu(a'+\hat{\mathfrak{m}}; 1, \lambda)$ and $\mu(a+\hat{\mathfrak{m}}; 1, \lambda\bar{\varepsilon})$. By (7) we have

$$\mu(a'+\hat{\mathfrak{m}}; 1, \lambda)$$

$$= \frac{1}{i}\frac{\pi^\nu}{\omega^\nu}\frac{1}{\lambda(\bar{\mathfrak{m}})}\sum_{\vartheta|\mathfrak{m}}M(\vartheta)\lambda(\vartheta)N(\vartheta)^{-1}\int_0^{i\infty}\Phi_\lambda\left(z-\frac{a'}{N(\mathfrak{m}\vartheta^{-1})}\right)dz, \tag{18}$$

$$\mu(a+\hat{\mathfrak{m}}; 1, \lambda\bar{\varepsilon})$$

$$= \frac{1}{i}\frac{\pi^\nu}{\omega^\nu}\frac{1}{\lambda\bar{\varepsilon}(\bar{\mathfrak{m}})}\sum_{\vartheta|\mathfrak{m}}M(\vartheta)\lambda\bar{\varepsilon}(\vartheta)N(\vartheta)^{-1}\int_0^{i\infty}\Phi_{\lambda\bar{\varepsilon}}\left(z-\frac{a}{N(\mathfrak{m}\vartheta^{-1})}\right)dz. \tag{19}$$

We transform the right side of (18). Choose $U \in \mathbf{Z}$ such that $Maa' - UN(\mathfrak{m}) = -1$, and for any $\vartheta|\mathfrak{m}$ set

$$g_\vartheta = \begin{pmatrix} UN(\vartheta) & -a' \\ -Ma & N(\mathfrak{m}\vartheta^{-1}) \end{pmatrix} \in \Gamma(M).$$

Using the fact that

$$\Phi_\lambda|_{\nu+1}g_\vartheta = \varepsilon(N(\mathfrak{m}\vartheta^{-1}))\Phi_\lambda,$$

where, as usual,

$$(\Phi_\lambda|_{\nu+1}g_\vartheta)(z)=(-Maz+N(\mathfrak{m}\vartheta^{-1}))^{-(\nu+1)}\Phi_\lambda(g_\vartheta z),$$

we find that

$$\int_{-\frac{a'}{N(\mathfrak{m}\vartheta^{-1})}}^{i\infty}\Phi_\lambda(z)\,dz=\int_{g_\vartheta(0)}^{g_\vartheta(N(\mathfrak{m}\vartheta^{-1})/Ma)}\Phi_\lambda(z)\,dz=\int_0^{N(\mathfrak{m}\vartheta^{-1})/Ma}\Phi_\lambda(g_\vartheta z)\,d(g_\vartheta z)$$

$$=\int_0^{N(\mathfrak{m}\vartheta^{-1})/Ma}\varepsilon(N(\mathfrak{m}\vartheta^{-1}))(-Maz+N(\mathfrak{m}\vartheta^{-1}))^{\nu-1}\Phi_\lambda(z)\,dz$$

$$\sim\varepsilon(N(\mathfrak{m}\vartheta^{-1}))(-M)^{\nu-1}a^{\nu-1}\int_0^{N(\mathfrak{m}\vartheta^{-1})/Ma}\Phi_\lambda(z)\,z^{\nu-1}\,dz. \tag{20}$$

Here $\sim$ denotes S-adic asymptotic equivalence as $\mathfrak{m}\to 0$ and is used in the same context as in § 2.2. Expanding $(-Maz+N(\mathfrak{m}\vartheta^{-1}))^{\nu-1}$ in powers of $N(\mathfrak{m}\vartheta^{-1})$, we obtain that only the term with $(-Maz)^{\nu-1}$ can be left in the asymptotic expression, since the $N(\vartheta)$ are bounded (in (18) it is necessary to take into account only ϑ with $M(\vartheta)\neq 0$) and the integrals $\int_0^x\Phi_\lambda(z)z^i dz$, like the integrals $\int_x^{i\infty}\Phi_\lambda(z)z^i dz$, lie in a $\mathbf{Z}$-module of finite type (Lemma 1.5).

We now set $h=\begin{pmatrix}0&-1\\M&0\end{pmatrix}$ and transform the last integral in (20) using (14):

$$\int_0^{N(\mathfrak{m}\vartheta^{-1})/Ma}\Phi_\lambda(z)\,z^{\nu-1}\,dz=\int_{h(i\infty)}^{h\left(-\frac{a}{N(\mathfrak{m}\vartheta^{-1})}\right)}\Phi_\lambda(z)\,z^{\nu-1}\,dz$$

$$=-\int_{-a/N(\mathfrak{m}\vartheta^{-1})}^{i\infty}\Phi_\lambda\left(-\frac{1}{Mz}\right)\left(-\frac{1}{Mz}\right)^{\nu-1}d\left(-\frac{1}{Mz}\right)=-Ci^{\nu+1}M^{\frac{1-\nu}{2}}\int_{-a/N(\mathfrak{m}\vartheta^{-1})}^{i\infty}\Phi_{\lambda\bar{\varepsilon}}(z)\,dz. \tag{21}$$

Substituting (21) into (20), we find that

$$\int_{-a'/N(\mathfrak{m}\vartheta^{-1})}^{i\infty}\Phi_\lambda(z)\,dz\sim C(-i)^{\nu-1}\varepsilon(N(\mathfrak{m}\vartheta^{-1}))M^{\frac{\nu-1}{2}}a^{\nu-1}\int_{-a/N(\mathfrak{m}\vartheta^{-1})}^{i\infty}\Phi_{\lambda\bar{\varepsilon}}(z)\,dz.$$

We now substitute these equivalences into the right side of (18):

$$\mu(a'+\hat{\mathfrak{m}};\ 1\ \lambda)$$

$$\sim\frac{1}{i}C(-i)^{\nu-1}\varepsilon(N(\mathfrak{m}))M^{\frac{\nu-1}{2}}a^{\nu-1}\frac{\pi^\nu}{\omega^\nu}\frac{1}{\lambda(\bar{\mathfrak{m}})}\sum_{\vartheta|\mathfrak{m}}M(\vartheta)\lambda\bar{\varepsilon}(\vartheta)N(\vartheta)^{-1}$$

$$\times\int_{-a/N(\mathfrak{m}\vartheta^{-1})}^{i\infty}\Phi_{\lambda\bar{\varepsilon}}(z)\,dz.$$

comparing this formula with (19), we immediately obtain (16).

The identity (17) can be deduced from (16) by a formal computation if we bear in mind that the mapping $a \mapsto -1/Ma$ induces an automorphism of the group $\hat{O}_S^*$:

$$\int_{\hat{O}_S^*} \left(\frac{a}{i}\right)^{k-1} d\mu_\lambda(a) = \int_{\hat{O}_S^*} \left(-\frac{1}{Mai}\right)^{k-1} d\mu_\lambda\left(-\frac{1}{Ma}\right)$$

$$= -C(-i)^{\nu+1} M^{\frac{\nu-1}{2}} \int_{\hat{O}_S^*} (-iM)^{1-k} a^{\nu-k}\, d\mu_{\lambda,\bar{\varepsilon}}(a) = CM^{\frac{\nu+1}{2}-k} \int_{\hat{O}_S^*} \left(\frac{a}{i}\right)^{\nu-k} d\mu_{\lambda,\bar{\varepsilon}}(a).$$

§5. Computation of a boundary measure

In this section we compute the measure $\mu(U;\, \nu,\, \lambda)$ corresponding to the right boundary of the critical strip in terms of the Weierstrass $\wp$-function, and we undertake an original investigation of the algebraic structure of the formulas thus obtained.

5.1. Consider the Weierstrass $\wp$-function

$$\wp(z; O\omega) = \frac{1}{z^2} + \sum_{\alpha \in O \setminus \{0\}} \left(\frac{1}{(z+\alpha\omega)^2} - \frac{1}{(\alpha\omega)^2}\right)$$

and the elliptic curve E connected with it whose Weierstrass affine model is defined by the equation

$$\left(\frac{d}{dz}\wp\right)^2 = 4\wp^3 - g_2\wp - g_3. \tag{22}$$

Formulas for the invariants g_2 and g_3 can be found in the Introduction. Recall that by the choice of ω the curve E is defined over an algebraic extension of $\mathbb{Q}$.

Next we set

$$D_{S\mathfrak{f}} = \left\{\begin{array}{l}\text{fractional ideals of } K \text{ with numerator and denominator}\\ \text{relatively prime to } \left(\prod_{\mathfrak{p}\in S} \mathfrak{p}\right) \mathfrak{f}\end{array}\right\},$$

$$D^0_{S\mathfrak{f}} = \{\mathfrak{n} \in D_{S\mathfrak{f}} \mid \exists\, \alpha \in K^*, (\alpha) = \mathfrak{n},\ \alpha \equiv 1 \bmod \mathfrak{f}\}.$$

Choose integral ideals $\mathfrak{n}_k$ $(k = 1, \ldots, h)$ relatively prime to $\bar{\mathfrak{m}}$ such that

$$D_{S\mathfrak{f}} = \bigcup_{k=1}^{h} \mathfrak{n}_k^{-1} D^0_{S\mathfrak{f}}$$

disconnected union). Finally, choose h finite subsets of K, $A_k = \{\alpha_j^{(k)}\}$ $(k = 1, \ldots, h)$, such that for all k and j

$$\begin{cases} \alpha_j^{(k)} \in \bar{\mathfrak{n}}_k;\ (\alpha_j^{(k)}, \bar{\mathfrak{m}}) = 1;\ \alpha_j^{(k)} \equiv 1 \bmod \bar{\mathfrak{f}}; \\ \alpha_j^{(k)} \text{ is uniquely defined by its class mod } N(\mathfrak{m}\mathfrak{n}_k\mathfrak{f}). \quad N(\mathfrak{m}\mathfrak{n}_k\mathfrak{f}). \end{cases} \tag{23}$$

The main result of this section is the following theorem:

5.2. **Theorem.** *For $\nu \geq 3$ we have*

$$\mu(a+\hat{m};\nu,\lambda)=\frac{(-1)^{\nu}}{2^{\nu}}\frac{1}{N(\mathfrak{m})}\frac{1}{\lambda(\bar{\mathfrak{m}})}\frac{1}{N(\mathfrak{f})^{\nu}}\sum_{k=1}^{h}\frac{1}{\lambda(\mathfrak{n}_k)}$$

$$\times\sum_{\mathfrak{a}_j^{(k)}\in A_k}\exp\left(-2\pi i\frac{a}{N(\mathfrak{m}\mathfrak{n}_k)}N(\mathfrak{a}_j^{(k)})\right)\left(\frac{d^{\nu-2}}{dz^{\nu-2}}\wp\right)\left(\frac{\mathfrak{a}_j^{(k)}\omega}{N(\mathfrak{m}\mathfrak{n}_k\mathfrak{f})},O\omega\right). \tag{24}$$

Proof. For $\nu\geq 3$ we can use formulas (4) and (5) immediately:

$$\mu(a+\hat{\mathfrak{m}};\nu,\lambda)=\frac{\Gamma(\nu)}{2^{\nu}}\frac{1}{\omega^{\nu}}\frac{N(\mathfrak{m})^{\nu-1}}{\lambda(\bar{\mathfrak{m}})}\sum_{(\mathfrak{n},\mathfrak{m}\mathfrak{f})=1}\lambda(\mathfrak{n})N(\mathfrak{n})^{-\nu}\exp\left(-2\pi i\frac{a}{N(\mathfrak{m})}N(\mathfrak{n})\right). \tag{25}$$

We divide the entire region of summation over $\mathfrak{n}$ into h subsets—the intersections with $\mathfrak{n}_k^{-1}D_{S_\mathfrak{f}}^0$, $k=1,\ldots,h$. Obviously,

$$\mathfrak{n}\in\mathfrak{n}_k^{-1}D_{S\mathfrak{f}}^0\Leftrightarrow\exists\,\alpha\in\mathfrak{n}_k,\ \alpha\equiv 1\bmod\mathfrak{f},\ (\alpha,\mathfrak{m})=1,\ \mathfrak{n}=\mathfrak{n}_k^{-1}(\alpha).$$

Bearing in mind that for these $\mathfrak{n}$

$$\lambda(\mathfrak{n}_k^{-1}(\alpha))N(\mathfrak{n}_k^{-1}(\alpha))^{-\nu}=\frac{N(\mathfrak{n}_k)^{\nu}}{\lambda(\mathfrak{n}_k)}\frac{1}{\bar{\alpha}^{\nu}},$$

and redesignating $\bar{\alpha}$ by α, we find that from (25)

$$\mu(a+\hat{\mathfrak{m}};\nu,\lambda)=\frac{\Gamma(\nu)}{2^{\nu}\omega^{\nu}}\frac{N(\mathfrak{m})^{\nu-1}}{\lambda(\bar{\mathfrak{m}})}\sum_{i=1}^{h}\frac{N(\mathfrak{n}_k)^{\nu}}{\lambda(\mathfrak{n}_k)}\sum_{\substack{\alpha\in\bar{\mathfrak{n}}_k\\ \alpha\equiv 1\bmod\bar{\mathfrak{f}}\\ (\alpha,\bar{\mathfrak{m}})=1}}\frac{1}{\alpha^{\nu}}\exp\left(-2\pi i\frac{a}{N(\mathfrak{m}\mathfrak{n}_k)}N(\alpha)\right).$$

Now in the inner sum we divide the region of summation into classes mod $N(\mathfrak{m}\mathfrak{n}_k\mathfrak{f})$. In each class we write the generic term α in the form $\alpha_j^{(k)}+\beta N(\mathfrak{m}\mathfrak{n}_k\mathfrak{f})$, where $\beta\in O$ and $\alpha_j^{(k)}\in A_k$. Moreover, we bring in the factor $1/\omega^{\nu}$ from the outside. Bearing in mind that in the new notation

$$\frac{1}{(\alpha\omega)^{\nu}}=\frac{1}{N(\mathfrak{m}\mathfrak{n}_k\mathfrak{f})^{\nu}}\frac{1}{\left(\frac{\alpha_j^{(k)}}{N(\mathfrak{m}\mathfrak{n}_k\mathfrak{f})}\omega+\beta\omega\right)^{\nu}},$$

after the necessary reductions we obtain that for $\nu\geq 3$

$$\mu(a+\hat{\mathfrak{m}};\nu,\lambda)=\frac{\Gamma(\nu)}{2^{\nu}}\frac{1}{N(\mathfrak{m})}\frac{1}{\lambda(\bar{\mathfrak{m}})}\frac{1}{N(\mathfrak{f})^{\nu}}$$

$$\times\sum_{k=1}^{h}\sum_{\alpha_j^{(k)}\in A_k}\exp\left(-2\pi i\frac{a}{N(\mathfrak{m}\mathfrak{n}_k)}N(\alpha_j^{(k)})\right)\frac{(-1)^{\nu}}{(\nu-1)!}\left(\frac{d^{\nu-2}}{dz^{\nu-2}}\wp\right)\left(\frac{\alpha_j^{(k)}\omega}{N(\mathfrak{m}\mathfrak{n}_k\mathfrak{f})};O\omega\right),$$

which coincides with (24). This proves the theorem.

5.3. We now take the first steps in the algebraic investigation of formula (24). We regard the curve E (see (22)) as an elliptic curve over the ring $O[g_2, g_3, \frac{1}{2}]$ defined by the equation $Y^2 = 4X^3 - g_2X - g_3$. Differentiation d/dz passes into invariant differentiation $\partial = Yd/dX$ of the structural pencil on this curve (more precisely, on its Néron model).

As $\alpha_i^{(k)}$ runs through A_k, the point

$$(\wp, \wp')\left(\frac{\alpha_i^{(k)}\omega}{N(\mathfrak{m}\mathfrak{n}_k\mathfrak{f})}; O\omega\right)$$

runs through a finite set of points of finite order on E, which we denote by $\tilde{A}_k = \{\alpha_i^{(k)}\}$. Finally, set

$$\exp\left(-2\pi i\,\frac{N(\alpha_i^{(k)})}{N(\mathfrak{m}\mathfrak{n}_k)}\right) = \zeta(\tilde{\alpha}_i^{(k)}). \tag{26}$$

In what follows we shall consider an algebraic formula obtained from (24):

$$\mu(a + \hat{m}; \nu, \lambda) = \frac{(-1)^\nu}{2^\nu N(\mathfrak{f})^\nu \lambda(\bar{\mathfrak{m}})}\,\frac{1}{N(\mathfrak{m})}\sum_{k=1}^{h}\frac{1}{\lambda(\mathfrak{n}_k)}\sum_{\tilde{\alpha}_i^{(k)}\in\tilde{A}_k}\zeta(\tilde{\alpha}_i^{(k)})^a\partial^{\nu-2}X(\tilde{\alpha}_i^{(k)}). \tag{27}$$

The following observations will play a fundamental role in the computations in §7. The formulation takes into account that E admits complex multiplication in O. Therefore for any ideal $\mathfrak{m} \subset O$ one can speak of the period points of $\mathfrak{m}$ on E. As usual, $\mathfrak{m}$ is a complete S-ideal.

5.4. **Lemma.** a) *The set $\tilde{A}_k$ (for any $k = 1, \ldots, h$) is invariant under translations by the period points of $\mathfrak{m}$ (which generate a cyclic group of order $N(\mathfrak{m})$) but contains no point whose period would be an S-ideal.*

b) *Let $\tilde{\beta}$ be a generator of the group of period points of $\mathfrak{m}$, and let $\tilde{\alpha} \in \tilde{A}_k$. Then for $0 \le j \le N(\mathfrak{m}) - 1$ we have*

$$\zeta(\tilde{\alpha} + j\tilde{\beta}) = \zeta(\tilde{\alpha}) \cdot \zeta_a^j,$$

where ζ_a is a primitive $N(\mathfrak{m})$th root of unity (depending also on β).

Proof. a) Let $\tilde{\alpha} \in \tilde{A}_k$ with argument $\alpha\omega/N(\mathfrak{m}\mathfrak{n}_k)\mathfrak{f}$, $\alpha \in A_k$, and let $\tilde{\beta}$ be a point with argument $\beta\omega$, where $\beta \in \mathfrak{m}^{-1}$. We wish to prove that the point $\tilde{\alpha} + j\tilde{\beta}$ with argument

$$\frac{(\alpha + jN(\mathfrak{m}\mathfrak{n}_k\mathfrak{f})\beta)\omega}{N(\mathfrak{m}\mathfrak{n}_k\mathfrak{f})}$$

belongs to $\tilde{A}_k$. In other words, we are given $\alpha \in \bar{\mathfrak{n}}_k$, $\alpha \equiv 1 \bmod \bar{\mathfrak{f}}$, $(\alpha, \bar{\mathfrak{m}}) = 1$. We must prove that

$$\alpha + jN(\mathfrak{m}\mathfrak{n}_k\mathfrak{f})\beta \in \bar{\mathfrak{n}}_k,\quad \alpha + jN(\mathfrak{m}\mathfrak{n}_k\mathfrak{f})\beta \equiv 1 \bmod \bar{\mathfrak{f}},\quad (\alpha + jN(\mathfrak{m}\mathfrak{n}_k\mathfrak{f})\beta, \bar{\mathfrak{m}}) = 1$$

(see (23), the definition of A_k). Each of these three statements follows from the corresponding fact for α if we take into account that $\mathfrak{m}\beta \subset O$.

There are no S-period points in $\widetilde{A}_k$ because $(\alpha, \overline{\mathfrak{m}}) = 1$, so that it is impossible to remove the $\overline{S}$-component of the denominator of $\alpha\omega/N(\mathfrak{m}\mathfrak{n}_k \mathfrak{f})$ by multiplying this number by an S-ideal.

b) By definition (26),

$$\zeta(\widetilde{\alpha} + j\widetilde{\beta}) = \exp\left(-2\pi i \frac{N(\alpha + jN(\mathfrak{m}\mathfrak{n}_k\mathfrak{f})\beta)}{N(\mathfrak{m}\mathfrak{n}_k)}\right).$$

Bearing in mind that $\mathfrak{m}\beta \subset O$ and $\overline{\mathfrak{m}}\overline{\beta} \subset O$, we have

$$\frac{j^2 N(\mathfrak{m}\mathfrak{n}_k\mathfrak{f})^2 N(\beta)}{N(\mathfrak{m}\mathfrak{n}_k)} \in \mathbf{Z},$$

so that ζ_α in §5.4b is found by the formula

$$\zeta_\alpha = \exp(-2\pi i \operatorname{Tr}(\overline{\alpha}\beta) N(\mathfrak{f})).$$

Since $(\alpha, \overline{\mathfrak{m}}) = 1 \Rightarrow (\overline{\alpha}, \mathfrak{m}) = 1$, and since $\widetilde{\beta}$ is a generator of the group of period points of $\mathfrak{m}$, the denominator of the ideal $(\overline{\alpha}\beta)$ is precisely equal to $\mathfrak{m}$. But S contains no divisors of the discriminant of K. Therefore the denominator of $\operatorname{Tr}(\overline{\alpha}\beta)$ is precisely equal to $N(\mathfrak{m})$. *Since S contains no divisors of $N(\mathfrak{f})$, it is clear that ζ_α is a primitive root of degree $N(\mathfrak{m})$.*

§6. Theorem on analyticity: formulation and deduction from the main lemma

6.1. The remainder of the article is devoted to a study of the S-adic analytic properties of the integral in formula (10) as a function of the two arguments k and ν (of weight λ). Instead of this integral it will be more convenient for us to deal with $\int_{O_S^*} a^k d\mu(a; \nu, \lambda)$ to enable us to use the results of §5. The passage from $d\mu(a; \nu, \lambda)$ to $d\mu_2(a)$ is trivial by virtue of Theorem 2.2. This is done in Corollary 6.4.

The values of the integral lie in the ring B_S defined in §2.1. It will be convenient for us to deal with its $\mathfrak{p}$-adic components separately. More precisely, we fix $\mathfrak{p} \in S$ and set $p = N(\mathfrak{p})$. We denote by $\mathbf{C}_p$ the completion of the algebraic closure of $\mathbf{Q}_p$ and consider one of the canonical homomorphisms $B_S \to \mathbf{C}_p$ corresponding to the divisors of $\mathfrak{p}$. We fix it. We call the image of any element of B_S under this homomorphism its $\mathfrak{p}$-component. In each algebraic number field this homomorphism defines a prime ideal, which we shall sometimes call distinguished (with respect to the given $\mathfrak{p}$-component).

We shall show that the $\mathfrak{p}$-component of any integral $\int_V a^k d\mu(a; \nu(\lambda), \lambda)$, $V \subset O_S^*$, is analytic in (k, ν) when k and λ vary in appropriate sets. We begin with the determination of these sets for λ.

We fix S, $\mathfrak{f}$, and the ideals $\mathfrak{n}_k$ $(k = 1, \ldots, h)$ chosen in §5.1. We denote by q the number of elements of the residue class field of a distinguished ideal in the field $K(g_2, g_3)$.

6.2. **Lemma.** *Let $\Lambda(\mathfrak{f})$ be the set of Hecke characters* mod $\mathfrak{f}$ *of weight not less than 3. One can construct a finite partition $\Lambda(\mathfrak{f}) = \bigcup_{i\in I}\Lambda_i(\mathfrak{f})$ with the following properties:*

There exist an integer $L > 0$, a family of S-invertible algebraic integers $m_1, \ldots, m_n$, for each $i \in I$ a family of S-invertible algebraic integers $b_1^{(i)}, \ldots, b_h^{(i)}$ and a $\nu_i \in \mathbf{Z}$ such that for any $\lambda \in \Lambda_i(\mathfrak{f})$ there is a $\nu \in \mathbf{Z}$, $\nu \geq 0$, satisfying the conditions

$$\begin{cases}\nu(\lambda) = \nu_i + \nu L(p-1)(q-1), \\ \forall k,\ \lambda(\mathfrak{n}_k) = b_k^{(i)} m_k^{\nu},\ m_k \equiv 1 \bmod \mathfrak{p}.\end{cases} \tag{28}$$

Proof. Choose an $e \in \mathbf{Z}$, $e > 0$, such that for all $k = 1, \ldots, h$ there exists an $n_k' \in K$ with the property

$$\mathfrak{n}_k^e = (n_k'), \quad n_k' \equiv 1 \bmod \mathfrak{f}.$$

This is possible since $(\mathfrak{n}_k, \mathfrak{f}) = 1$. Set $n_k = \sqrt[e]{n_k'}$ (the values of the roots are arbitrary, but fixed). Then $\lambda(\mathfrak{n}_k) = \eta(\lambda, k) n_k^{\nu(\lambda)}$, where $\eta(\lambda, k)$ are eth roots of 1.

Since $\mathfrak{p} \nmid \mathfrak{n}_k$, there exists an L such that for all k we have $n_k^L \equiv a_k \bmod \mathfrak{p}$, where $a_k \in \mathbf{Z}$ (here $\mathfrak{p}$ also denotes the ideal induced by $\mathfrak{p}$ in the common field of rationality for all n_k).

We now define the partition $\Lambda(\mathfrak{f}) = \bigcup_{i\in I}\Lambda_i(\mathfrak{f})$ as the coarsest partition with the following property: for each i, as λ varies in $\Lambda_i(\mathfrak{f})$:

a) $\eta(\lambda, k)$ depends only on k, but not on λ;

b) $\nu(\lambda) \bmod L(p-1)(q-1)$ does not depend on λ.

It is clear that this partition is finite. Furthermore, when $\lambda \in \Lambda_i(\mathfrak{f})$, let $\nu(\lambda) = \nu_i + \nu L(p-1)(q-1)$, $\nu \geq 0$. Then

$$\lambda(\mathfrak{n}_k) = \eta(\lambda, k)\, n_k^{\nu_i + \nu L(p-1)(q-1)} = \eta(\lambda, k)\, n_k^{\nu_i} (n_k^{L(p-1)(q-1)})^{\nu} = b_k^{(i)} m_k^{\nu}.$$

This completes the proof, since

$$m_k = n_k^{L(p-1)(q-1)} \equiv a_k^{(p-1)(q-1)} \equiv 1 \bmod \mathfrak{p}.$$

(The reason for adding $q - 1$ will be clear later, in §7.)

We can now state the main result of this paper. The following brief notation will be convenient: for fixed $i \in I$ we denote by λ_ν any character $\lambda \in \Lambda_i(\mathfrak{f})$ with $\nu(\lambda) = \nu_i + L(p-1)(q-1)$. Although the dependence on i is not indicated and λ_ν may not be defined uniquely, this should not lead to ambiguities, since the integrals under discussion will depend only on ν (for a given i).

Further, let

$$\binom{x}{i} = \frac{x(x-1)\ldots(x-i+1)}{i!}.$$

6.3. **Theorem.** *Assume that $\mathfrak{p} \in S$ is not ramified in $K(g_2, g_3)$ and satisfies the following additional restrictions: $\mathfrak{p} \nmid 2$; the reduction of the Weierstrass model* (22) *modulo all prime divisors of $\mathfrak{p}$ is nondegenerate; and the reduction of invariant* *differentiation $\partial = d/dz = Yd/dX$ modulo all prime divisors $\mathfrak{p}$ is a nonzero invariant differentiation on the appropriate closed fiber.*

Suppose that $\{\lambda_\nu\}$, $\nu \geq 0$, runs through $\Lambda_i(\mathfrak{f})$, and let $k_0 \in \mathbf{Z}$, $k_0 \geq 0$. Then for any $V \subset \hat{O}_S^*$ the function from $\mathbf{Z} \times N$ to $\mathbf{C}_p$

$$(k, \nu) \mapsto \text{the } \mathfrak{p}\text{-component of} \int_V a^{k_0+k(p-1)} d\mu\,(a;\, \nu_i + \nu L\,(p-1)\,(q-1),\, \lambda_\nu)$$

takes integral values in $\mathbf{C}_p$ and splits into a double interpolation series

$$\sum_{l,j\geq 0} I_{lj} \binom{k}{l} \binom{\nu}{j},$$

whose coefficients satisfy the inequalities

$$|I_{lj}| \leqslant |\mathfrak{p}|^{l+j}. \tag{29}$$

Of course, all that needs to be proved is inequality (29).

It follows from this that the S-adic Hecke L-function is analytic in k and ν.

6.4. **Corollary.** *Under the hypotheses of the theorem, as $\{\lambda_\nu\}$, $\nu \geq 0$, varies in $\Lambda_i(\mathfrak{f})$ and as k varies in the progression $k_0 \bmod (p-1)$, the function*

$$(k, \nu) \mapsto \frac{(\sqrt{-1})^k \Gamma(k)}{(2\pi)^k} \frac{\pi^\nu}{\omega^\nu} \prod_{\mathfrak{p}\in S} \left(1 - \frac{\lambda(\mathfrak{p})}{N(\mathfrak{p})^k}\right)\left(1 - \frac{N(\bar{\mathfrak{p}})^{k-1}}{\lambda(\bar{\mathfrak{p}})}\right) L\,(k, \lambda)$$

can be continued to a $\mathfrak{p}$-adic analytic function of two variables with integral values in $\mathbf{C}_p$.

Proof. By Theorems 3.1 and 3.2, the right side coincides with

$$(\sqrt{-1})^k \int_{\hat{O}_S^*} \left(\frac{a}{\sqrt{-1}}\right)^{k-1} d\mu_\nu(a) = (\sqrt{-1})^k \int_{\hat{O}_S^*} \left(\frac{a}{\sqrt{-1}}\right)^{k-\nu(\lambda)} d\mu\,(a;\, \nu(\lambda),\, \lambda).$$

In each region of variation for (k, ν), obviously, $(\sqrt{-1})^{\nu(\lambda)}$ remains constant. Furthermore, the mapping $(k, \nu) \mapsto (k - \nu, \nu)$ is analytic and carries this region into another of the same type (the progression $\bmod (p-1)$ in which k takes its values changes). Therefore it suffices to verify that the integral on the right is analytic for $V = \hat{O}_S^*$. Since it can be represented by an interpolation series by Theorem 6.3, the matter reduces to verifying that this series is analytic.

Let

$$\binom{k}{l} = \frac{1}{l!} \sum_{c=0}^{l} S_l^{(c)} k^c \quad \text{and} \quad \binom{\nu}{j} = \frac{1}{j!} \sum_{d=0}^{j} S_j^{(d)} \nu^d,$$

where $S_l^{(c)}$ is Stirling's number. Formally expanding the interpolation series in powers of k and ν, we find that

$$\sum_{l,j\geqslant 0} I_{lj}\binom{k}{l}\binom{\nu}{j} = \sum_{c,d\geqslant 0} J_{c,d}k^c\nu^d,$$

$$J_{c,d} = \sum_{\substack{l\geqslant c\\ j\geqslant d}} \frac{1}{l!j!} I_{lj} S_l^{(c)} S_j^{(d)}.$$

Estimate (29) simultaneously guarantees that the series for $J_{c,d}$ converges and that $J_{c,d}$ tends to zero. More precisely, since $\text{ord}_{\mathfrak{p}}(l!j!) \leq (l+j)/(p-1)$, by virtue of (29) we have

$$|J_{c,d}| \leqslant |\mathfrak{p}|^{(c+d)\frac{p-2}{p-1}}.$$

Thus the interpolation series is analytic in a region including the unit bicylinder.

6.5. As preparation for the proof of Theorem 6.3, let us recall how the coefficients of the interpolation series are computed.

Let F be any function on the nonnegative integers with values, say, in $\mathbf{C}_p$. It can be formally represented in the form

$$F(m) = \sum_{n=0}^{\infty} \Delta_n F \cdot \binom{m}{n}, \tag{30}$$

where

$$\Delta_n F = \sum_{\nu=0}^{n} (-1)^\nu \binom{n}{\nu} F(n-\nu). \tag{31}$$

For each integer $m \geq 0$ the right side of (30) makes sense, since for $n \geq m+1$ we have $\binom{m}{n} = 0$. If the coefficients $\Delta_n F$ tend to zero in $\mathbf{C}_p$, then (30) makes sense for all $m \in \mathbf{Z}_p$ and defines a continuous function. Conversely, if F has a continuous extension to $\mathbf{Z}_p$, then $\Delta_n F \to 0$ ("Mahler's criterion"). If $\Delta_n F$ tends to zero sufficiently fast, as in our case, then the continuous extension of F is even analytic.

In the remainder of this section we shall show how estimate (29) is deduced from the following fact:

6.6. **Main Lemma.** *Suppose that $\{\lambda_\nu\}$ varies in $\Lambda_i(\mathfrak{f})$ for fixed i. Then for any open-closed subset $U \subset \hat{O}_S^*$ and any $n \geq 0$ we have*

$$\sum_{\nu=0}^{n} (-1)^\nu \binom{n}{\nu} \mu(U;\, v_i + \nu L(p-1)(q-1),\, \lambda_\nu) \equiv 0\, (\mathfrak{p}^n), \tag{32}$$

if $\mathfrak{p}$ also satisfies the additional restrictions of Theorem 6.3.

Note that the left side contains algebraic numbers from increasing fields. By divisibility by $\mathfrak{p}^n$ is meant divisibility by the corresponding induced ideal.

6.7. *Deduction of Theorem 6.3 from the main lemma.* For $U \subset \hat{O}_S^*$ and $\nu \geq 0$ set

$$\mu(U;\, v_i + \nu L(p-1)(q-1),\, \lambda_\nu) = \sum_{j=0}^{\infty} M^{(j)}(U)\binom{\nu}{j}.$$

The coefficients $M^{(j)}(U)$ are defined by (31):

$$M^{(j)}(U)=\sum_{\nu=0}^{j}(-1)^{\nu}\binom{j}{\nu}\mu(U;\nu_l+\nu L(p-1)(q-1),\lambda_\nu).$$

Obviously the $M^{(j)}$ are measures; (32) asserts that these measures decrease quickly as j grows.

Hence, dropping the indication of the $\mathfrak{p}$-component for brevity, we have

$$\int_V a^{k_0+k(p-1)}d\mu(a;\nu_l+\nu L(p-1)(q-1),\lambda_\nu)=\sum_{j=0}^{\infty}\binom{\nu}{j}\int_V a^{k_0+k(p-1)}dM^{(j)}(a).$$

Again from (31) we find that the lth interpolation coefficient for the function $k\mapsto a^{k0+k(p-1)}$ is equal to $a^{k0}(a^{p-1}-1)^l$, so that

$$a^{k_0+k(p-1)}=a^{k_0}\sum_{l=0}^{\infty}\binom{k}{l}(a^{p-1}-1)^l.$$

Therefore

$$\int_V a^{k_1+k(p-1)}dM^{(j)}(a)=\sum_{l=0}^{\infty}\binom{k}{l}\int_V a^{k_0}(a^{p-1}-1)^l\,dM^{(j)}(a).$$

Finally, the coefficients l_{lj} in (29) are found from the formulas

$$I_{lj}=\int_V a^{k_0}(a^{p-1}-1)^l dM^{(j)}(a),$$

and the estimate (29) follows from (32) and the fact that $|(a^{p-1}-1)^l|\le|\mathfrak{p}|^l$.

§7. Proof of the main lemma

7.1. *Reformulation.* We begin by substituting (27) and (28) into the left side of (32). Although (27) applies only to sets of the form $U=a+\hat{\mathfrak{m}}$, it is clear that it suffices to prove (32) for these sets. After a trivial but somewhat tedious transformation (the main step is to interchange the summations over ν and k), we obtain that we must prove the congruence

$$\frac{(-1)^{\nu_l}}{(2N(\mathfrak{f}))^{\nu_l}}\frac{1}{\lambda(\bar{\mathfrak{m}})}\sum_{k=1}^{h}\frac{1}{b_k^{(l)}}\frac{1}{N(\mathfrak{m})}\sum_{\tilde{\mathfrak{a}}_j^{(k)}\in\tilde{A}_k}\zeta(\tilde{\mathfrak{a}}_j^{(k)})^a$$

$$\times\left(\frac{\partial^{L(p-1)(q-1)}}{(2N(\mathfrak{f}))^{L(p-1)(q-1)}m_k}-1\right)^n\partial^{\nu_l-2}X(\tilde{\mathfrak{a}}_j^{(k)})\equiv 0\,(\mathfrak{p}^n).$$

The coefficient before the outer sum and the numbers $1/b_k^{(l)}$ are S-integral. In what follows we shall prove that for each fixed k the part of the sum whose representation begins with $1/N(\mathfrak{m})$ is divisible by $\mathfrak{p}^n$. We write this congruence separately, changing the notation in addition.

Let $\mathfrak{p}^m$ be the exact power of $\mathfrak{p}$ dividing $\mathfrak{m}$. We multiply the entire congruence by $\mathfrak{p}^m$ and we shall not write $1/N(\mathfrak{m}\mathfrak{p}^{-m})$. Furthermore, we drop the indication of the index k and the tilde. Instead of $\tilde{A}_k$ we shall simply write A; instead of $\tilde{\alpha}_j^{(k)}$, simply α_j. Confusion with the notation in §5.2 shall not arise, since we shall not return to it further. Instead of $\zeta(\tilde{\alpha}_j^{(k)})^a$ we shall write $\zeta(\alpha)$. All that we need from this function is axiom 7.2c below. Instead of $1/(2N(\mathfrak{f}))^{L(p-1)(q-1)}m_k$ we shall write C; instead of $\partial^{\nu_i-2}X$, simply Z. We denote by Z_α the tmaslation of Z by the point α, and by 0 the zero point. Thus we shall prove the congruence

$$\sum_{\alpha\in A} \zeta(\alpha)(C\partial^{L(p-1)(q-1)}-1)^n Z_\alpha(0) \equiv 0\,(\mathfrak{p}^{m+n}). \tag{33}$$

We shall gather together the information about the objects occurring in congruence (33) that we shall use and axiomatize the situation partially.

7.2. *Data.* a) A is a finite set of points of finite order on an elliptic curve E with complex multiplication on O. A is invariant under translations by the period points of $\mathfrak{p}^m$ but does not contain points of the order of $\mathfrak{p}^k$ for $k \geq 1$.

This follows from Lemma 5.4.

b) The Weierstrass model E has a nondegenerate nonsupersingular reduction modulo all prime divisors of $\mathfrak{p}$ in the base ring. Z is a function on a local Néron model E whose divisor of poles is concentrated on the zero section 0, and the closed fiber does not occur in a divisor of Z.

The reduction of ∂ is a nonzero invariant differentiation on the reduction of E.

This follows from the assumptions in Theorem 6.3 and the fact that $N(\mathfrak{p})$ splits completely in K (nonsupersingularity).

c) $\zeta(\alpha)$ is a function on A whose values are roots of unity. If β is a primitive point of the order of $\mathfrak{p}^m$ and $\alpha \in A$, then for all $j \in \mathbb{Z}$ we have $\zeta(\alpha + j\beta) = \zeta(\alpha)\zeta_\alpha^j$, where ζ_α is a primitive $N(\mathfrak{p})^m$th root of unity depending on α (and on β). This follows from Lemma 5.4 and the fact that $\mathfrak{p} \nmid a$. Here ζ_α corresponds to $\zeta_\alpha^{N(\mathfrak{m}\mathfrak{p}^{-m})a}$ in the notation of Lemma 5.4.

d) $C \equiv 1 \bmod \mathfrak{p}$. This follows from (28) and the fact that $2N(\mathfrak{f})$ is relatively prime to p.

We now turn to the proof of congruence (33), which will be broken up into a series of lemmas. Perhaps it is worth noting that the divisibility of the left side by $\mathfrak{p}^{m+n}$ will be established in two steps: first divisibility by $\mathfrak{p}^n$ inside some module of $\mathfrak{p}$-integral functions on E, and then divisibility by $\mathfrak{p}^m$ of the values of these functions at the zero of E.

In what follows $\mathfrak{P}$ will denote any prime divisor of $\mathfrak{p}$ in the base ring for E, and q is the order of its residue class field. It is easy to see that this is consistent with the previous definition of q if consideration is restricted to one $\mathfrak{p}$-component.

7.3. **Lemma.** *Let R_0 be the ring of rational functions on the closed fiber $E_0 = E \bmod \mathfrak{P}$ of the Néron model with a single pole at the zero point, and let R be the ring of $\mathfrak{P}$-integral rational functions on the $\mathfrak{P}$-adic Néron model E whose poles are concentrated in the kernel of the reduction, i.e. at the period points of $\mathfrak{p}^k$, $k \geq 0$.*

Then the reduction mod$\mathfrak{P}$ *induces an exact sequence*

$$0 \to \mathfrak{P}R \to R \to R_0 \to 0.$$

Proof. The arrow $R \to R_0$ is induced by the imbedding of the closed fiber E in the Néron model. Its kernel obviously contains $\mathfrak{P}R$; it coincides with $\mathfrak{P}R$, since the reduction mod$\mathfrak{P}$ is nondegenerate. It remains to prove that $R \to R_0$ is surjective.

In order to lift a function $Z_0 \in R_0$ to R, it obviously suffices to lift a divisor of it and to guarantee that it remains principal. First we somehow lift the numerator and denominator while preserving the degree by Hensel's lemma. The Jacobian sum of the points occurring in the lifted divisor will be an obstacle to its being principal. This is a point on E in the kernel of the reduction, since the reduction of the obstacle is zero. We add it to the denominator of the first lifting and 0 to its numerator. A principal divisor with the previous reduction is obtained. The ring R obviously contains a $\mathfrak{P}$-invertible function with such a divisor. Its reduction can differ from Z_0 by only a constant factor.

7.4. **Lemma.** *In the base ring for E there exists a $\mathfrak{P}$-invertible number H such that for any $Z \in R$ a $Z' \in R$ can be found with the property*

$$(\partial^{p-1} - H)Z \in \prod_{p\beta=0} Z'_\beta + \mathfrak{P}R,$$

where $p = N(\mathfrak{p})$.

Proof. Let F_0 be the field of rational functions on E_0, $H_0 \neq 0$ the Hasse invariant for (E_0, ∂_0), and ∂_0 the reduction of ∂. It is well known that $\partial_0^p = H_0\partial_0$. Hence $(\partial_0^{p-1} - H_0)F_0 \subseteq F_0^p$, and so $(\partial_0^{p-1} - H_0)R_0 \subseteq R_0 \cap F_0^p = R_0^p$. Now let $Z \in R$, $Z_0 \in R_0$ the reduction of Z, and $Z_0' \in R_0$ a function such that $(\partial_0^{p-1} - H_0)\, Z_0 = (Z_0')^p$ We denote by $Z' \in R$ any lifting of Z_0' to R according to Lemma 7.3, and let H be any lifting of H_0. Since the points β such that $\mathfrak{p}\beta = 0$ lie in the kernel of the reduction, the function $\prod_{\mathfrak{p}\beta=0} Z'_\beta \in R$ is a lifting of $(Z_0')^p$. Finally

$$(\partial^{p-1} - H)Z \in \prod_{p\beta=0} Z'_\beta + \operatorname{Ker}(R \to R_0),$$

and it remains to apply Lemma 7.3 once more.

7.5. **Lemma.** *Denote by T the following module over the $\mathfrak{P}$-adic base ring E:*

$$T = \left\{ \sum_{\alpha\in A} \zeta(\alpha) Z_\alpha \,\middle|\, Z \in R \right\}.$$

Then

$$\partial T \subseteq T, \quad (\partial^{p-1} - H)T \subseteq \mathfrak{P}T.$$

Proof. Since $\partial R \subseteq R$ and $\partial(Z_\alpha) = (\partial Z)_\alpha$, it follows that $\partial T \subseteq T$. Next, let $Z \in R$ and according to the preceding lemma let

$$(\partial^{p-1}-H)Z=\prod_{p\beta=0}Z'_\beta+Z'',\quad Z'\in R,\quad Z''\in \mathfrak{P}R.$$

Then we have

$$(\partial^{p-1}-H)\Big(\sum_{\alpha\in A}\zeta(\alpha)Z_\alpha\Big)=\sum_{\alpha\in A}\zeta(\alpha)\Big(\prod_{p\beta=0}Z'_\beta\Big)_\alpha+\sum_{\alpha\in A}\zeta(\alpha)Z''_\alpha.$$

The second sum on the right obviously lies in $\mathfrak{P}T$. Let us show that the first sum is equal to zero. Using 7.2a, we divide A into orbits under the action of the subgroup $\{j\beta\}$, where $j=0,\dots,p-1$ and β is a primitive period point of $\mathfrak{p}$. The terms of each orbit $\{\alpha+j\beta\}$ contribute nothing to the sum, because

$$\sum_{j=0}^{p-1}\zeta(\alpha+j\beta)\Big(\prod_{k=0}^{p-1}Z'_{k\beta}\Big)_{\alpha+j\beta}=\zeta(\alpha)\Big(\sum_{j=0}^{p-1}\zeta'^{j}_\alpha\Big)\prod_{k=0}^{p-1}Z'_{\alpha+k\beta}=0,$$

since ζ'_α is a primitive pth root of unity by virtue of 7.2c.

7.6. **Lemma.** *In the former notation* (C *is taken from* (33))

$$(C\partial^{L(p-1)(q-1)}-1)^nT\subseteq\mathfrak{P}^nT \tag{34}$$

Proof. First let $n=1$. By Lemma 7.5, the operator ∂^{p-1} coincides with H on $T/\mathfrak{P}T$. Therefore $\partial^{L(p-1)(q-1)}$ is the identity on $T/\mathfrak{P}T$, since $H^{q-1}\equiv 1 \bmod \mathfrak{P}$ ($q-1$ was introducted in §6.2 for this purpose). Also bearing in mind that $C\equiv 1 \bmod \mathfrak{P}$ (condition 7.2d), we obtain

$$(C\partial^{L(p-1)(q-1)}-1)T\subseteq\mathfrak{P}T.$$

An obvious induction on n completes the proof.

A glance at (33) and (34) shows that we have already proved that the left side of (33) is divisible by $\mathfrak{P}^n$, because all the elements of T take $\mathfrak{P}$-integral values at the zero point of E. This follows from 7.2a: the reduction of A does not contain the reduction of 0, and hence the reductions of $Z_\alpha(0)$ are finite.

It remains to prove that the values of the elements of T at the zero point are divisible by $\mathfrak{P}^m$.

7.7. **Lemma.** *If* $W\in T$, *then* $W(0)\equiv 0 \bmod \mathfrak{P}^m$.

Proof. We denote by $\hat{E}$ the formal completion of a local Néron $\mathfrak{P}$-model E over the closure of the maximal unramified extension of a $\mathfrak{P}$-adic base ring.

It is known (see [2] and [7]) that $\hat{E}$ is isomorphic to the formal multiplicative group of $\hat{G}_m$ over this ring. Let t be a formal parameter in the ring $\hat{G}_m$ such that $t+1$ is a character. Fix an isomorphism of $\hat{E}$ with $\hat{G}_m$, a primitive period point β of $\mathfrak{p}^m$ on E, and the value η of the function $t+1$ at β. It is clear that η is a primitive $N(\mathfrak{p})^m$th root of unity.

Let $W=\sum_{\alpha\in A}\zeta(\alpha)Z_\alpha\in T$. Since the reduction $A \bmod \mathfrak{P}$ does not contain the zero point, all the Z_α lie in the ring $\hat{E}$. Let us show that each orbit $\{\alpha+j\beta\}$, $j=0,\dots,N(\mathfrak{p})^m-1$, in A makes a contribution to W divisible by $\mathfrak{P}^m$.

Consider the expansion of Z_α in powers of t in the ring G_m:

$$Z_\alpha = \sum_{k=0}^{\infty} a_k t^k.$$

Obviously,

$$Z_{\alpha+j\beta} = \sum_{k=0}^{\infty} a_k\left(\left(t+1\right)\eta^j - 1\right)^k,$$

and, by 7.2c,

$$\sum_{j=0}^{N(\mathfrak{p})^m-1} \zeta(\alpha + j\beta)\, Z_{\alpha+j\beta}(0) = \zeta(\alpha) \sum_{j=0}^{N(\mathfrak{p})^m-1} \zeta_\alpha^j \sum_{k=0}^{\infty} a_k(\eta^j - 1)^k.$$

It now remains to remove the parentheses in $(\eta^j - 1)^k$, to sum over j internally, and to take into account that the sum $\Sigma_{j=0}^{N(\mathfrak{p})^m-1}\xi^j$, where ξ is an $N(\mathfrak{p})^m$th root of 1, is equal to zero or $N(\mathfrak{p})^m$ (for $\xi \neq 1$ or $\xi = 1$, respectively) and therefore is necessarily divisible by $\mathfrak{P}^m$.

The main lemma follows from everything that has been proved. We have established that the left side of (33) is divisible by $\mathfrak{P}^{m+n}$ for all divisors $\mathfrak{P} \mid \mathfrak{p}$ in the base ring for E, and it remains to recall that, by hypothesis, $\mathfrak{p}$ is unramified in this ring.

Received 22/APR/74

BIBLIOGRAPHY

1. R. M. Damerell, *L-functions of elliptic curves with complex multiplication.* I, Acta Arith. **17** (1970), 287–301. MR **44** #2758; erratum, **44**, p. 1633.

2. Bernard M. Dwork, *A deformation theory for the zeta function of a hypersurface,* Proc. Internat. Congr. Math. (Stockholm, 1962), Inst. Mittag-Leffler, Djursholm, 1963, pp. 247–259. MR **31** #171.

3. Kenkichi Iwasawa, *On p-adic L-functions,* Ann. of Math. (2) **89** (1969), 198–205. MR **42** #4522.

4. ———, *Lectures on p-adic L-functions,* Ann. of Math. Studies, no. 74, Princeton Univ. Press, Princeton, N. J., 1972.

5. Tomio Kubota and Heinrich Wolfgang Leopoldt, *Eine p-adische Theorie der Zetawerte.* I: *Einführung der p-adischen Dirichletschen L-Funktionen,* J. Reine Angew. Math. **214/215** (1964), 328–339. MR **29** #1199.

6. Stephen Lichtenbaum, *On the values of zeta and L-functions,* Ann. of Math. (2) **96** (1972), 336–358.

7. Jonathan Lubin and John Tate, *Formal complex multiplication in local fields,* Ann. of Math. (2) **81** (1965), 380–387. MR **30** #3094.

8. Ju. I. Manin, *Periods of parabolic forms and p-adic Hecke series,* Mat. Sb. **92** (134) (1973), 378–401 = Math. USSR Sb. **21** (1973), 371–393.

9. ———, *The values of p-adic Hecke series at integer points of the critical strip,* Mat. Sb. **93** (135) (1974), 621–626 = Math. USSR Sb. **22** (1974), 631–637.

10. Barry Mazur and H. P. F. Swinnerton-Dyer, *Arithmetic of Weil curves,* Invent. Math. **25** (1974), 1–61.

11. Andrew P. Ogg, *Modular forms and Dirichlet series,* Benjamin, New York, 1969. MR **41** #1648.

12. Jean-Pierre Serre, *Formes modulaires et fonctions zêta p-adiques*, Modular Functions of One Variable. III (Proc. Internat. Summer School, Univ. Antwerp, 1972), Lecture Notes in Math., vol. 350, Springer-Verlag, Berlin and New York, 1973, pp. 191–268.

13. Goro Shimura, *On elliptic curves with complex multiplication as factors of the Jacobians of modular function fields*, Nagoya Math. J. 43 (1971), 199–208. MR 45 #5111.

Translated by R. LENET

Мат. Сборник
Том 104(146)(1977), Вып. 4

Math. USSR Sbornik
Vol. 33(1977), No. 4

CONVOLUTIONS OF HECKE SERIES AND THEIR VALUES AT LATTICE POINTS

JU. I. MANIN AND A. A. PANČIŠKIN

ABSTRACT. An algebraicity theorem is proved for the values of convolutions of Dirichlet series connected with modular forms of various weights at lattice points of the critical strip. p-adic estimates of these values are obtained.

Bibliography: 24 titles.

§1. Introduction

1.1. Two main methods of extracting number-theoretic information from Dirichlet series are known: investigating the sums of the coefficients of a series or its values at isolated points. The asymptotic law of distribution of prime numbers and "explicit formulas" are an example of the first approach. Dirichlet formulas connecting the class number, regulator, and discriminant of a number field with the value of an L-series at unity and formulas for these values are an example of the second. After the introduction of two new extensive families of Dirichlet series–Mellin transforms of modular forms and L-series of algebraic varieties–these classical approaches also began to be applied to them. In particular, the constructions of p-adic L-functions are based on formulas for values at lattice points (see [1], [2], [11], [14]–[17] and the literature cited in these papers).

The main purpose of this article is to compute the values at certain lattice points of the convolution of two Hecke series–Mellin transforms of cusp forms for $\mathbf{SL}(2, \mathbf{Z})$. This computation clarifies the arithmetic nature of the values to such an extent that it makes it natural to attempt to carry out the p-adic construction. Unfortunately, the resulting p-adic measures grow faster than allowed in the existing theory (see [1] and the Introduction to [16]). The second result of this article is an explicit form of the functional equation for the convolutions under consideration that we require to compute half their values. We shall now state precisely some of the main assertions of this article.

1.2. Here and henceforth the following notation is adopted. Let $\Phi(s) = \sum_{n=1}^{\infty} c_n n^{-s}$ be a Dirichlet series and χ a Dirichlet character. Then, by definition, $\Phi(s, \chi) = \sum_{n=1}^{\infty} \chi(n) c_n n^{-s}$ With this notation ordinary Dirichlet series should be written as $\zeta(s, \chi)$; however, here we shall write $L(s, \chi)$, as is the custom.

Let the symbols $f(z) = \sum_1^\infty a_n e^{2\pi i n z}$ and $g(z) = \sum_1^\infty b_n e^{2\pi i n z}$ denote two cusp forms on the complex upper half-plane with respect to $\Gamma = \mathbf{SL}(2, \mathbf{Z})$ of weights k and l, respectively. We always assume that $a_1 = b_1 = 1$, that f and g are proper for all Hecke

AMS (MOS) subject classifications (1970). Primary 10H10, 12B30; Secondary 10D15.

operators, and that $k > 1$. We set $\varphi_f(s) = \sum_1^\infty a_n n^{-s}$, $\varphi_g(s) = \sum_1^\infty b_n n^{-s}$, and $L_{f,g}(s) = \sum_1^\infty a_n b_n n^{-s}$ (the convolution of φ_f and φ_g). Besides this "additive" convolution, it is also convenient to consider the "multiplicative" convolution

$$\mathfrak{L}_{f,g}(s) = \prod_p \prod_{i,j=1,2} (1 - \rho_{i,f}^{(p)} \rho_{j,g}^{(p)} p^{-s})^{-1},$$

where

$$\varphi_f(s) = \prod_p \prod_{i=1,2} (1 - \rho_{i,f}^{(p)} p^{-s})^{-1},$$

and similarly for $\varphi_g(s)$.

From the point of view of l-adic representations (see [4]), $\mathfrak{L}_{f,g}(s)$ corresponds to the tensor product of the representations connected with f and g. For any primitive Dirichlet character χ a simple formula holds which is verifiable on Euler factors:

$$\mathfrak{L}_{f,g}(s, \chi) = L(2s - k - l + 2, \chi^2) L_{f,g}(s, \chi)$$

(the first factor on the right is an ordinary Dirichlet L-function). We denote by $W(\chi)$ the standard Gauss sum of χ.

1.3. Functional Equation Theorem. *Let $p > 2$ be a prime number and χ a primitive Dirichlet character* mod p^N, $N > 1$. *Set*

$$\Lambda_{f,g}(s, \chi) = \left(\frac{2\pi}{p^N}\right)^{-2s} \Gamma\left(s - 1 + \frac{k+l}{2}\right) \Gamma\left(s + \frac{k-l}{2}\right) \mathfrak{L}_{f,g}\left(s - 1 + \frac{k+l}{2}, \chi\right),$$

$$C_\chi = \frac{2W(\chi^2)}{p^{3N-1}(p-1)} \sum_{\chi=\psi_1\psi_2}' W(\psi_1)^2 W(\psi_2)^2,$$

where Σ' is extended to distinct pairs of primitive characters. Then

$$\Lambda_{f,g}(s, \chi) = C_\chi \Lambda_{f,g}(1 - s, \bar{\chi}).$$

For the principal character this result is due to Rankin and Selberg [20], [21]. Jacquet [10] described a much more general case in a less explicit form. The case $p = 2$, and also for any composite conductor, is analyzed by our methods at the cost of lengthening the computations. Theorem 1.3 is proved in §2.

1.4. Value Theorem. *Under the previous conditions, for $m \geq 0$ set*

$$\omega_{f,g,m} = (2\pi i)^{k-l-2m} \frac{(f, f)}{(4\pi)^{1-k}(k-2)!},$$

where (f, f) is the scalar square of f in the sense of Petersson. Then for $m = 0, \ldots, k - l - 1$ the following assertions are true:

a) $\mathfrak{L}_{f,g}(k - 1 - m, \chi)\omega_{f,g,m}^{-1}$ *are algebraic numbers explicitly computable in terms of f, g and χ, and lying in the smallest extension of* $\mathbf{Q}$ *that contains the coefficients of g, the coefficients of all Hecke cusp forms of weight k for* $\mathbf{SL}(2, \mathbf{Z})$, *the values of χ, and the p^Nth roots of unity.*

b) $\operatorname{ord}_p(\mathfrak{L}_{f,g}(k - 1 - m, \chi)\omega_{f,g,m}^{-1}) \geq -N(2k - l - 2 - 2m) + O(1)$ *as* $N \to \infty$, *where* ord_p *is an absolute p-adic order.*

For the principal character χ and for $k - l > 3$ and $m = 0$ the value $\mathfrak{L}_{f,g}(k-1)$ was computed by Rankin [19].

Theorem 1.4 is proved in §§3 and 4.

§2. Proof of the functional equation theorem

2.1. The proof is based on the theory of Epstein-Siegel ζ-functions (see [22], Chapter I). The functional equation for the convolution of Hecke series in some special cases was obtained in various papers (see [18], [20] and [21]). Such a functional equation in the most general form is contained in [10] in the language of Jacquet-Langlands theory. In subsections 2.5–2.16 of this section technical methods are described that make it possible to write the functional equation explicitly for the convolution of Hecke series connected with the "newforms" of types $(\Gamma_0(N_1), \chi_1, k_1)$ and $(\Gamma_0(N_2), \chi_2, k_2)$. For our purposes only the case $(\Gamma_0(p^{2N}), \chi^2, l)$ and $(\Gamma, 1, k)$ is needed, where χ is a primitive Dirichlet character modulo p^N. We have restricted ourselves to this case, since even here the computations are quite lengthy.

Let $D_0(M)$ denotes a fundamental domain for the action of $\Gamma_0(M)$ on the upper half-plane.

2.2. LEMMA. *If χ is a primitive Dirichlet character modulo p^N, then*

$$\begin{aligned} &2(4\pi)^{-s}\Gamma(s)\,L(2s+2-k-l, \chi^2)\,L_{f,g}(s, \chi) \\ &= \int\limits_{D_0(p^{2N})} y^{s-1} f(z)\overline{g_{\bar\chi}(z)} \sum_{m\equiv 0\,(p^{2N})}{}' \frac{\chi^2(n)(mz+n)^{k-l}}{|mz+n|^{2s+2-2l}}\,dx\,dy, \end{aligned} \tag{1}$$

where Σ' denotes summation over all pairs $(m, n) \neq (0, 0)$, $z = x + iy$, and

$$g_{\bar\chi}(z) = \sum_{n=1}^{\infty} \bar\chi(n)\, b_n e^{2\pi i n z}.$$

PROOF. We have

$$(4\pi)^{-s}\Gamma(s)\sum_{n=1}^{\infty} \chi(n)\, a_n b_n n^{-s} = \int\limits_{|x|\leqslant \frac{1}{2}} y^{s-1} f(z)\overline{g_{\bar\chi}(z)}\,dx\,dy. \tag{2}$$

(We integrate term by term the product of Fourier series first with respect to x, then y, using the standard integral representation for $\Gamma(s)$.) We now use the fact that $g_{\bar\chi}(z)$ is a cusp "newform" of type $(\Gamma_0(p^{2N}), \chi^2, l)$ (see [12] and [23], Theorem 3.66). We reduce the integration in (2) to an integration over a fundamental domain:

$$\begin{aligned} &\sum_{T\in\Gamma_\infty\backslash\Gamma_0(p^{2N})}\ \int\limits_{D_0(p^{2N})} y^{s-1} f(z)\overline{g_{\bar\chi}(z)}\frac{\chi^2(d)(cz+d)^{k-l}}{|cz+d|^{2s+2-2l}}\,dx\,dy \\ &= \int\limits_{D_0(p^{2N})} y^{s-1} f(z)\overline{g_{\bar\chi}(z)} \sum{}^{*} \frac{\chi^2(n)(mz+n)^{k-l}}{|mz+n|^{2s+2-2l}}\,dx\,dy, \end{aligned} \tag{3}$$

where $T = \left(\begin{smallmatrix} a & b \\ c & d\end{smallmatrix}\right)$, Σ^* is extended to pairs (m, n) such that $m \geqslant 0$, $m \equiv 0(p^{2N})$, $(m, n) = 1$, and Γ_∞ denotes the group of matrices of the form $\pm\left(\begin{smallmatrix} 1 & * \\ 0 & 1\end{smallmatrix}\right)$. We transform the sum under the integral sign in (3) using the properties of the Möbius function:

$$\sum_{\delta|d}\mu(\delta)=\begin{cases}1, & d=1,\\ 0, & d>1,\end{cases}$$

$$2\sum{}^{*}\frac{\chi^2(n)(mz+n)^{k-l}}{|mz+n|^{2s+2-2l}}=\sum_{m\equiv 0\,(p^{2N})}{}'\sum_{\delta|(m,n)}\frac{\mu(\delta)\chi^2(n)(mz+n)^{k-l}}{|mz+n|^{2s+2-2l}}$$
$$=\sum_{\nu=1}^{\infty}\frac{\mu(\nu)\chi^2(\nu)}{\nu^{2s+2-k-l}}\sum_{m\equiv 0\,(p^{2N})}{}^{*}\frac{\chi^2(n)(mz+n)^{k-l}}{|mz+n|^{2s+2-2l}}. \tag{4}$$

where Σ' denotes summation over pairs $(m, n)\neq(0, 0)$. The lemma now follows immediately from (4).

2.3. Let us recall the main properties of Epstein-Siegel zeta-functions.

Let $u_i, v_i\in\mathbf{R}$, $i=1, 2$, and $r\in\mathbf{N}$. Then, by definition,

$$\zeta\left(s, \begin{pmatrix}u_1\\u_2\end{pmatrix}, \begin{pmatrix}v_1\\v_2\end{pmatrix}, z, r\right)$$
$$=y^s\sum_{\begin{pmatrix}m_1+u_1\\m_2+u_2\end{pmatrix}\neq\begin{pmatrix}0\\0\end{pmatrix}}e^{2\pi i(m_1u_1-m_2u_2)}\frac{(m_1+v_1+z(m_2+v_2))^r}{|m_1+v_1+z(m_2+v_2)|^{2s+r}}, \tag{5}$$

where the series on the right converges absolutely for $\mathrm{Re}(s)>1$.

The ζ-functions thus defined satisfy the following functional equation:

$$\pi^{-s}\Gamma\left(s+\frac{r}{2}\right)\zeta\left(s, \begin{pmatrix}u_1\\u_2\end{pmatrix}, \begin{pmatrix}v_1\\v_2\end{pmatrix}, z, r\right)$$
$$=e^{2\pi i(u_1v_2-v_1u_2)}\pi^{s-1}\Gamma\left(1-s+\frac{r}{2}\right)\zeta\left(1-s, \begin{pmatrix}v_1\\v_2\end{pmatrix}, \begin{pmatrix}u_1\\u_2\end{pmatrix}, z, r\right). \tag{6}$$

Now for convenience we make the change of variable $s=s_1-1+(k+l)/2$ in our integral representation in 2.2. This makes it possible to write a functional equation of the type $s_1\mapsto 1-s_1$. In what follows we denote the new variable, as before, by s. After this substitution, in terms of the Epstein-Siegel zetas equality (1) acquires the form

$$2(4\pi)^{-\frac{k+l}{2}+1}(4\pi)^{-s}\,\Gamma\left(s-1+\frac{k+l}{2}\right)\Gamma(2s, \chi^2)L_{f,g}\left(s-1+\frac{k+l}{2}, \chi\right)$$
$$=\int_{D_0(p^{2N})}y^{\frac{k+l}{2}-2}f(z)\overline{g_{\bar\chi}(z)}\left[y^s\sum_{m\equiv 0\,(p^{2N})}{}'\frac{\chi^2(n)(mz+n)^{k-l}}{|mz+n|^{2s+k-l}}\right]dx\,dy$$
$$=\sum_{a,b=0}^{p^{2N}-1}\int_{D_0(p^{2N})}y^{\frac{k+l}{2}-2}f(z)\overline{g_{\bar\chi}(z)} \tag{7}$$
$$\times\left[\frac{1}{p^{4N}}\sum_{d \bmod p^{2N}}\chi^2(d)\,e^{-2\pi i\frac{ad}{p^{2N}}}\zeta\left(s, \begin{pmatrix}-\frac{b}{p^{2N}}\\ \frac{a}{p^{2N}}\end{pmatrix}, \begin{pmatrix}0\\0\end{pmatrix}, z, k-l\right)\right]dx\,dy.$$

2.4. We multiply (7) by the Γ-factor $\pi^{-s}\Gamma(s+(k-l)/2)$ and use the functional equation (6) with $r=k-l$:

$$2(4\pi)^{1-\frac{k+l}{2}}(2\pi)^{-2s}\Gamma\left(s-1+\frac{k+l}{2}\right)\Gamma\left(s+\frac{k-l}{2}\right)L(2s,\chi^2)$$

$$\times L_{f,g}\left(s-1+\frac{k+l}{2},\chi\right)=\pi^{s-1}\Gamma\left(1-s+\frac{k+l}{2}\right)\int\limits_{D_0(p^{2N})} y^{\frac{k+l}{2}-2} f(z)\overline{g_{\bar\chi}(z)} \tag{8}$$

$$\times\left[\frac{1}{p^{4N}}\sum_{d \bmod p^{2N}}\sum_{a,b=0}^{p^{2N}-1}\chi^2(d)\,e^{-2\pi i\frac{ad}{p^{2N}}}\zeta\left(1-s,\begin{pmatrix}0\\0\end{pmatrix},\begin{pmatrix}-\frac{b}{p^{2N}}\\ \frac{a}{p^{2N}}\end{pmatrix},z,k-l\right)\right]dx\,dy.$$

2.5. The subsequent steps of the proof will consist of transforming the right side of (8). Therefore, in this equality it is convenient to replace s by $1-s$. After this substitution we write down explicitly the expressions for the zetas under the integral sign on the right side of (8). We have

$$\zeta\left(s,\begin{pmatrix}0\\0\end{pmatrix},\begin{pmatrix}-\frac{b}{p^{2N}}\\ \frac{a}{p^{2N}}\end{pmatrix},z,k-l\right)=y^s\sum_{(m_1,m_2)}{}' \frac{\left(m_1-\frac{b}{p^{2N}}+z\left(m_2+\frac{a}{p^{2N}}\right)\right)^{k-l}}{\left|m_1-\frac{b}{p^{2N}}+z\left(m_2+\frac{a}{p^{2N}}\right)\right|^{2s+k-l}}$$

$$=p^{4Ns}y^s\sum_{(m_1,m_2)}{}'\frac{(p^{2N}m_1-b+z(p^{2N}m_2+a))^{k-l}}{|p^{2N}m_1-b+z(p^{2N}m_2+a)|^{2s+k-l}} \tag{9}$$

We now substitute these expressions into (8) and sum over b. Now (8) assumes the form

$$2(4\pi)^{1-\frac{k+l}{2}}(2\pi)^{2s-2}\Gamma\left(\frac{k+l}{2}-s\right)\Gamma\left(1-s+\frac{k-l}{2}\right)L(2-2s,\chi^2)$$

$$\times L_{f,g}\left(-s+\frac{k+l}{2},\chi\right)=\pi^{-s}\Gamma\left(s+\frac{k-l}{2}\right)\int\limits_{D_0(p^{2N})} y^{\frac{k+l}{2}-2} f(z)\overline{g_{\bar\chi}(z)} \tag{10}$$

$$\times\left[p^{4Ns-4N}\sum_{d \bmod p^{2N}}\sum_{a=0}^{p^{2N}-1}\chi^2(d)\,e^{-2\pi i\frac{ad}{p^{2N}}}y^s\sum_{\substack{(m_1,m_2)\\ m_2\equiv a(p^{2N})}}{}'\frac{(m_1+m_2z)^{k-l}}{|m_1+m_2z|^{2s+k-l}}\right]dx\,dy.$$

The next step is the change of variable $z\mapsto -1/p^{2N}z$ in the integral. Let

$$H_{p^{2N}}=\begin{pmatrix}0&-1\\p^{2N}&0\end{pmatrix}$$

We note that $H_{p^{2N}}D_0(p^{2N})=\tilde D_0(p^{2N})$, where $\tilde D_0(p^{2N})$ is a fundamental domain of $\Gamma_0(p^{2N})$.

We state the other necessary information in the form of a lemma.

2.5. Lemma.

a)
$$g_{\bar\chi}\left(-\frac{1}{p^{2N}z}\right)=p^{Nl}z^l\frac{W(\bar\chi)^2}{p^N}g_\chi(z);$$

b)
$$f\left(-\frac{1}{p^{2N}z}\right)=p^{2Nk}z^kf(p^{2N}z);$$

c)
$$\operatorname{Im}\left(-\frac{1}{p^{2N}z}\right)=\frac{\operatorname{Im}(z)}{p^{2N}|z|^2}.$$

PROOF. a) follows from the fact that

$$g_{\bar\chi}|_l H_{p^{2N}}=\frac{W(\bar\chi)^2}{p^N}g_\chi$$

(see [23], Chapter 3, Proposition 3.65).

b) In the notation of [23], Chapter III, we have

$$f(p^{2N}z)=p^{-Nk}f|_k\begin{pmatrix}p^{2N}&0\\0&1\end{pmatrix},$$

$$f(p^{2N}z)|_k\begin{pmatrix}0&-1\\p^{2N}&0\end{pmatrix}=p^{-Nk}f|_k\begin{pmatrix}p^{2N}&0\\0&1\end{pmatrix}\begin{pmatrix}0&-1\\p^{2N}&0\end{pmatrix}(z)$$
$$=p^{-Nk}f|_k\begin{pmatrix}0&-p^{2N}\\p^{2N}&0\end{pmatrix}(z)=p^{-Nk}f(z),$$

whence

$$p^{-Nk}f|_k\begin{pmatrix}0&1\\p^{2N}&0\end{pmatrix}(z)=f(p^{2N}z)|_k\begin{pmatrix}0&-1\\p^{2N}&0\end{pmatrix}\begin{pmatrix}0&-1\\p^{2N}&0\end{pmatrix}=f(p^{2N}z),$$

i.e.

$$\frac{1}{z^kp^{2Nk}}f\left(-\frac{1}{p^{2N}z}\right)=f(p^{2N}z).$$

c) is obvious.

In addition, we note that the Jacobian of the transformation $z\mapsto -1/p^{2N}z$ equals $1/p^{4N}|z|^4$, and that the sum

$$y^s\sum_{m_2\equiv a(p^{2N})}{}'\frac{(m_1+m_2z)^{k-l}}{|m_1+m_2z|^{2s+k-l}}$$

after a change of variable becomes

$$z^{-(k-l)}|z|^{2s+k-l}p^{2Ns}y^s\sum_{m_1\equiv a(p^{2N})}{}'\frac{(p^{2N}m_1z-m_2)^{k-l}}{|p^{2N}m_1z-m_2|^{2s+k-l}}.$$

2.6. Taking into account all the equalities in 2.5, we rewrite the right side of (10) as follows:

$$p^{6Ns+Nk-4N}\frac{\overline{W(\bar\chi)^2}}{p^N}\pi^{-s}\Gamma\left(s+\frac{k-l}{2}\right)\int_{\tilde D_0(p^{2N})}y^{\frac{k+l}{2}-2}f(p^{2N}z)\overline{g_\chi(z)}$$
$$\times\left[\sum_{a=0}^{p^{2N}-1}\sum_{d\bmod p^{2N}}\chi^2(d)e^{-2\pi i\frac{ad}{p^{2N}}}y^s\sum_{m_2\equiv a(p^{2N})}{}'\frac{(p^{2N}m_1z-m_2)^{k-l}}{|p^{2N}m_1z-m_2|^{2s+k-l}}\right]dx\,dy. \tag{11}$$

Let us simplify the expression in brackets. To do so, we note that the sum over d in (11) is very reminiscent of a Gauss sum. We compute it with the help of the following lemma:

2.7. LEMMA a) *Let $p \neq 2$ and $N > 1$, and let χ be primitive* $\bmod p^N$. *Then χ^2 is primitive* $\bmod p^N$.

b) *Extend the character $\bar{\chi}^2(x)$ to all $x \in \mathbf{Q}$ by zero outside $\mathbf{Z}$. Then, under the conditions* a),

$$\sum_{d \bmod p^{2N}} \chi^2(d)\, e^{-2\pi i \frac{ad}{p^{2N}}} = p^N W(\chi^2)\, \bar{\chi}^2\left(\frac{a}{p^N}\right).$$

PROOF. a) We use the isomorphism $(\mathbf{Z}/p^N\mathbf{Z})^* \simeq \mathbf{Z}/(p-1)\mathbf{Z} \times \mathbf{Z}/p^{N-1}\mathbf{Z}$ for $p \neq 2$ and $N > 1$, under which a group isomorphic to $\mathbf{Z}/p^{N-1}\mathbf{Z}$ is generated by the class $1 + p \pmod{p^N}$. It follows that χ is the product of the characters χ_1, trivial on $\mathbf{Z}/p^{N-1}\mathbf{Z}$, and χ_2, trivial on $\mathbf{Z}/(p-1)\mathbf{Z}$. In this case the condition that χ be primitive is that there exists a $y \equiv 1 \pmod{p^{N-1}}$ with $\chi(y) \neq 1$. This, in turn, is equivalent to the fact that χ_2 effects an isomorphism of the subgroup generated by $1 + p \pmod{p^N}$ and the group of p^{N-1}th roots of unity. But then χ^2 also satisfies this primitivity condition.

b) follows from standard properties of Gauss sums.

We now apply this lemma to (11). By the lemma, in the sum over a we leave only the terms corresponding to those a that are divisible by p^N, and we set $a_1 = a/p^N$. We then rewrite the expression in brackets as follows:

$$\begin{aligned} p^N W(\chi^2)\, p^{-2Ns} \sum_{a_1=1}^{p^N-1} \bar{\chi}^2(a_1) \sum_{\substack{m_2 \equiv a_1 (p^N) \\ m_1 \equiv 0 (p^N)}}{}' \frac{(m_1 z - m_2)^{k-l}}{|m_1 z - m_2|^{2s+k-l}} \\ = p^N W(\chi^2)\, p^{-2Ns} \sum_{m_1 \equiv 0 (p^N)}{}' \frac{\bar{\chi}^2(m_2)(m_2 + m_1 z)^{k-l}}{|m_2 + m_1 z|^{2s+k-l}}. \end{aligned} \tag{12}$$

2.8. By analogy with Lemma 2.2, we obtain

$$\sum_{m_1 \equiv 0 (p^N)}{}' \frac{\bar{\chi}^2(m_2)(m_2 + m_1 z)^{k-l}}{|m_2 + m_1 z|^{2s+k-l}} = 2L(2s, \bar{\chi}^2) \sum_{T \in \Gamma_\infty \setminus \Gamma_0(p^N)} \frac{\bar{\chi}^2(T)\, j(T, z)^{k-l}}{|j(T, z)|^{2s+k-l}}, \tag{13}$$

where $T = \begin{pmatrix} a & b \\ c & d \end{pmatrix} \in \mathbf{SL}(2, \mathbf{Z})$, $j(T, z) = cz + d$, and $\chi^2(T) = \chi^2(d)$. We now replace the expression in brackets under the integral sign in (11) by the right side of (13). We have

$$\begin{aligned} & 2p^{4Ns+kN-3N} W(\chi^2) \frac{\overline{W(\bar{\chi})^2}}{p^N} \pi^{-s} \Gamma\left(s + \frac{k-l}{2}\right) \\ & \times \int_{\tilde{D}_0(p^{2N})} y^{s+\frac{k-l}{2}-2} f(p^{2N} z)\, \overline{g_\chi(z)} \left[\sum_{T \in \Gamma_\infty \setminus \Gamma_0(p^{2N})} \frac{\chi^2(T)\, j(T, z)^{k-l}}{|j(T, z)|^{2s+k-l}} \right] dx\, dy. \end{aligned} \tag{14}$$

2.9. We note that the integral in (14) does not depend on the choice of fundamental domain. Therefore, we shall integrate over the standard fundamental domain $D_0(p^{2N})$ and then reduce the integration over $D_0(p^{2N})$ to an integration over $D_0(p^N)$ by the relation

$$D_0(p^{2N}) = \bigcup_{R_i \in S_{p^N}^{p^{2N}}} R_i D_0(p^N),$$

where $S_{p^N}^{p^{2N}}$ denotes a system of right coset representatives of $\Gamma_0(p^{2N}) \setminus \Gamma_0(p^N)$:

$$S_{p^N}^{p^{2N}} = \left\{ \begin{pmatrix} 1 & 0 \\ up^N & 1 \end{pmatrix} \middle| \, u = 0, \ldots, p^N - 1 \right\}.$$

Moreover,

$$\int_{D_0(p^{2N})} = \sum_{R_i \in S_{p^N}^{p^{2N}}} \int_{R_i D_0(p^N)} . \tag{15}$$

2.10. We shall compute the integral in (14) using relation (15). We use the fact that

$$j(TR_i, z) = j(T, R_i z)\, j(R_i, z), \quad g_\chi(R_i z) = (g_\chi |_l R_i)(z) \cdot j(R_i, z)^l,$$
$$f(p^{2N} R_i z) = (f(p^{2N} z) |_k R_i)(z) \cdot j(R_i, z)^k, \quad dx_i\, dy_i = |(j(R_i, z)|^{-4}\, dx\, dy,$$

where $x_i = \mathrm{Re}(R_i z)$ and $y_i = \mathrm{Im}(R_i z)$.

Taking all these equalities into account, we write

$$\int_{D_0(p^N)} y^{s + \frac{k+l}{2} - 2} f(p^{2N} z)\, \overline{g_\chi(z)} \sum_{T \in \Gamma_\infty \setminus \Gamma_0(p^N)} \frac{\bar{\chi}^2(T)\, j(T, z)^{k-l}}{|j(T, z)|^{2s+k-l}}\, dx\, dy$$
$$= \int_{D_0(p^N)} y_i^{s + \frac{k+l}{2} - 2} f(p^{2N}(R_i z))\, \overline{g_\chi(R_i z)} \sum_{T \in \Gamma_\infty \setminus \Gamma_0(p^N)} \frac{\chi^2(T)\, j(T, R_i z)^{k-l}}{|j(T, R_i z)|^{2s+k-l}}\, dx_i\, dy_i \tag{16}$$
$$= \int_{D_0(p^N)} y_i^{s + \frac{k+l}{2} - 2} f(p^{2N} z)|_k R_i \cdot \overline{(g_\chi |_l R_i)(z) \sum_{T \in \Gamma_\infty \setminus \Gamma_0(p^N)} \frac{\bar{\chi}^2(T)\, j(TR_i, z)^{k-l}}{|j(TR_i, z)|^{2s+k-l}}}\, dx\, dy.$$

2.11. We now note that for all R_i

$$\sum_{T \in \Gamma_\infty \setminus \Gamma_0(p^N)} \frac{\bar{\chi}^2(T)\, j(TR_i, z)^{k-l}}{|j(TR_i, z)|^{2s+k-l}} = \sum_{T \in \Gamma_\infty \setminus \Gamma_0(p^N)} \frac{\bar{\chi}^2(T)\, j(T, z)^{k-l}}{|j(T, z)|^{2s+k-l}} . \tag{17}$$

Taking (15)–(17) into account, we rewrite (14) in the form

$$\int_{D_0(p^N)} y^{s + \frac{k+l}{2} - 2}\, \mathrm{Tr}_{(k,l)}^{\Gamma_0(p^{2N}) \setminus \Gamma_0(p^N)} (f(p^{2N} z)\, \overline{g_\chi(z)}) \sum_{T \in \Gamma_\infty \setminus \Gamma_0(p^N)} \frac{\bar{\chi}^2(T)\, j(T, z)^{k-l}}{|j(T, z)|^{2s+k-l}}\, dx\, dy, \tag{18}$$

where, by definition,

$$\mathrm{Tr}_{(k,l)}^{\Gamma_0(p^{2N}) \setminus \Gamma_0(p^N)} (f(p^{2N} z)\, \overline{g_\chi(z)}) = \sum_{R_i \in S_{p^N}^{p^{2N}}} f(p^{2N} z)|_k R_i \cdot \overline{(g_\chi |_l R_i)(z)}. \tag{19}$$

It is also possible to give an invariant definition of this trace. Let $\mathcal{H}$ be a complex space of holomorphic functions on the upper half-plane. We define the action of $\mathbf{GL}^+(2, \mathbf{Q})$ on $\mathcal{H} \otimes \mathcal{H}$ as the tensor product of an action of weight k and an action of weight l. Then the trace (19) is simply a trace in the sense of this (k, l)-action that does not depend on the choice of representatives. It is also apparent from this interpretation that (19) is invariant under the (k, l)-action of $\Gamma_0(p^N)$ convoluted with the character $\bar{\chi}^2$.

These observations now make it possible to pass in (18) to an integration over the strip $|x| < \frac{1}{2}$:

$$\int\limits_{|x|\leq 1/2} y^{s+\frac{k+l}{2}-2}\,\mathrm{Tr}_{\Gamma_0(p^{2N})}^{\Gamma_0(p^{2N})\setminus\Gamma_0(p^N)}\left(f(p^{2N}z)\,\overline{g_\chi(z)}\right)dx\,dy. \tag{20}$$

We set

$$\widetilde{L}_{f,g}(s,\bar{\chi}) = p^{kN-N}W(\chi^2)\frac{\overline{W(\chi)^2}}{p^N}\Gamma(s)^{-1}(4\pi)^s \times \int\limits_{|x|\leq \frac{1}{2}} y^{s-1}\,\mathrm{Tr}_{\Gamma_{(k,l)}}^{\Gamma_0(p^{2N})\setminus\Gamma_0(p^N)}\left(f(p^{2N}z)\,\overline{g_\chi(z)}\right)dx\,dy. \tag{21}$$

Combining (10), (14), (20), and (21), we obtain the following proposition:

2.12. Proposition. *Let*

$$\Lambda_{f,g}(s,\chi) = \left(\frac{2\pi}{p^N}\right)^{-2s}\Gamma\left(s-1+\frac{k+l}{2}\right)\Gamma\left(s+\frac{k-l}{2}\right)L(2s,\chi^2)L_{f,g}\left(s-1+\frac{k+l}{2},\chi\right),$$

$$\widetilde{\Lambda}_{f,g}(s,\chi) = \left(\frac{2\pi}{p^N}\right)^{-2s}\Gamma\left(s-1+\frac{k+l}{2}\right)\Gamma\left(s+\frac{k-l}{2}\right)L(2s,\chi^2)\widetilde{L}_{f,g}\left(s-1+\frac{k+l}{2},\chi\right).$$

Then the following functional equation holds:

$$\Lambda_{f,g}(s,\chi) = \widetilde{\Lambda}_{f,g}(1-s,\bar{\chi}).$$

The remainder of the proof is concerned with computing $\widetilde{L}_{f,g}(s,\bar{\chi})$ in terms of $L_{f,g}(s,\chi)$.

2.13. Proposition.

$$\widetilde{L}_{f,g}(s,\bar{\chi}) = \left[\frac{3W(\chi^2)}{p^{3N-1}(p-1)}\sum_{\chi=\psi_1\psi_2}{}' W(\psi_1)^2 W(\psi_2)^2\right]L_{f,g}(s,\bar{\chi})$$

(*the prime indicates that the summation runs over distinct representations* χ *in the form of a product of two primitive characters* mod p^N).

Theorem 1.3 follows immediately from Propositions 2.12 and 2.13.

Proof. By analogy with Lemma 2.6, we have

$$f(p^{2N}z) = p^{-Nk}f|_k\begin{pmatrix}0 & -1\\ p^{2N} & 0\end{pmatrix}(z),\quad g_\chi(z) = \frac{W(\chi)^2}{p^N}g_{\bar{\chi}}|_l\begin{pmatrix}0 & -1\\ p^{2N} & 0\end{pmatrix}(z).$$

It follows that

$$\mathrm{Tr}_{\Gamma_{(k,l)}}^{\Gamma_0(p^{2N})\setminus\Gamma_0(p^N)}\left(f(p^{2N}z)\,\overline{g_\chi(z)}\right)$$

$$= p^{-Nk}\frac{\overline{W(\chi)^2}}{p^N}\sum_{u=0}^{p^N-1} f|_k\begin{pmatrix}0 & -1\\ p^{2N} & 0\end{pmatrix}\begin{pmatrix}1 & 0\\ up^N & 1\end{pmatrix}g_{\bar{\chi}}|_l\begin{pmatrix}0 & -1\\ p^{2N} & 0\end{pmatrix}\begin{pmatrix}1 & 0\\ up^N & 1\end{pmatrix}.$$

In addition

$$\begin{pmatrix}0 & -1\\ p^{2N} & 0\end{pmatrix}\begin{pmatrix}1 & 0\\ up^N & 1\end{pmatrix} = \begin{pmatrix}-up^N & -1\\ p^{2N} & 0\end{pmatrix} = \begin{pmatrix}1 & -\frac{u}{p^N}\\ 0 & 1\end{pmatrix}\begin{pmatrix}0 & -1\\ p^{2N} & 0\end{pmatrix}.$$

This makes it possible to compute the trace *in two steps*: first apply

$$\sum_{u=0}^{p^N-1}\begin{pmatrix}1 & -\dfrac{u}{p^N}\\ 0 & 1\end{pmatrix},$$

and then

$$\begin{pmatrix}0 & -1\\ p^{2N} & 0\end{pmatrix}$$

(in the sense of the (k, l)-action).

2.14. Applying

$$\sum_{u=0}^{p^N-1}\begin{pmatrix}1 & -\dfrac{u}{p^N}\\ 0 & 1\end{pmatrix},$$

we get

$$\sum_{u=0}^{p^N-1} f\,|_k\begin{pmatrix}1 & -\dfrac{u}{p^N}\\ 0 & 1\end{pmatrix}\overline{g_{\bar\chi}\,|_l\begin{pmatrix}1 & -\dfrac{u}{p^N}\\ 0 & 1\end{pmatrix}} = \sum_{u=0}^{p^N-1} f\left(z-\frac{u}{p^N}\right)\overline{g_{\bar\chi}\left(z-\frac{u}{p^N}\right)}$$

$$= \sum_{u=0}^{p^N-1}\sum_{n,m=1}^{\infty}\chi(m)\,a_nb_m\exp\left(2\pi i\,(n-m)\frac{u}{p^N}\right)\exp(2\pi inz)\,\overline{\exp(2\pi imz)}$$

$$= p^N\sum_{\substack{n,m=1\\ n\equiv m(p^N)}}^{\infty}\chi(m)\,a_nb_m\exp(2\pi inz)\,\overline{\exp(2\pi imz)}.$$

2.15. Before applying

$$\begin{pmatrix}0 & -1\\ p^{2N} & 0\end{pmatrix},$$

we write the result of 2.14 in a more convenient form, namely,

$$p^N\sum_{\substack{n,m=1\\ n\equiv m(p^N)}}^{\infty}\chi(m)\,a_nb_m\exp(2\pi inz)\,\overline{\exp(2\pi imz)}$$

$$= p^N\sum_{\substack{d \bmod p^N\\ (d,p)=1}}\ \sum_{\substack{n=1\\ n\equiv d(p^N)}}^{\infty} a_n\exp(2\pi inz)\sum_{\substack{m=1\\ m\equiv d\,(p^N)}}^{\infty}\chi(m)\,b_m\,\overline{\exp(2\pi imz)}.$$

In the sum over d we have restricted ourselves to those d that are relatively prime to p, since the character $\chi(m)$ appears in the sum over $m\equiv d(p^N)$. We note that

$$\sum_{\substack{n=1\\ n\equiv d\,(pN)}}^{\infty} a_n\exp(2\pi inz) = \frac{1}{p^{N-1}(p-1)}\sum_{\psi \bmod p^N}\bar\psi(d)\sum_{n=1}^{\infty}\psi(n)\,a_n\exp(2\pi inz),$$

where the summation runs over all Dirichlet characters mod p^N This equality can be rewritten as follows:

$$\sum_{n=1}^{\infty} a_n\exp(2\pi inz) = \frac{1}{p^{N-1}(p-1)}\sum_{\psi \bmod p^N}\bar\psi(d)\,f_\psi(z),$$

where the index ψ indicates convolution with the character ψ. Similarly,

$$\sum_{\substack{m=1\\ m\equiv d(p^N)}}^{\infty} \chi(m)\, b_m\, \overline{\exp(2\pi i m z)} = \frac{1}{p^{N-1}(p-1)} \sum_{\psi_1 \bmod p^N} \psi_1(d)\, \overline{g_{\psi_1\bar{\chi}}(z)}.$$

Therefore, the result of 2.15 can be written as follows:

$$\frac{p^N}{p^{2N-2}(p-1)^2} \sum_{\substack{d \bmod p^N\\ (d,p)=1}} \sum_{\psi \bmod p^N} \sum_{\psi_1 \bmod p^N} \bar{\psi}(d)\, \psi_1(d)\, f_\psi(z)\, \overline{g_{\psi_1\bar{\chi}}(z)}.$$

After summation over d this expression assumes the form

$$\frac{p}{p-1} \sum_{\psi \bmod p^N} f_\psi(z)\, \overline{g_{\psi\bar{\chi}}(z)}.$$

2.16. We now apply

$$\begin{pmatrix} 0 & -1 \\ p^{2N} & 0 \end{pmatrix}.$$

We shall apply this matrix separately to each summand. In this connection it is necessary to examine several cases.

Case 1. $\psi \equiv \varepsilon \pmod{p^N}$ (the principal character mod p^N). We have

$$f_\varepsilon(z) = f(z) - \frac{1}{p}\sum_{u=0}^{p-1} f\,|_k \begin{pmatrix} 1 & \frac{u}{p} \\ 0 & 1 \end{pmatrix}(z).$$

Let us recall the definition of Hecke operators:

$$T_p = p^{\frac{k}{2}-1}\left[\begin{pmatrix} p & 0 \\ 0 & 1 \end{pmatrix} + \sum_{u=0}^{p-1}\begin{pmatrix} 1 & u \\ 0 & p \end{pmatrix}\right].$$

Since

$$\begin{pmatrix} 1 & \frac{u}{p} \\ 0 & 1 \end{pmatrix} = \begin{pmatrix} 1 & u \\ 0 & p \end{pmatrix}\begin{pmatrix} 1 & 0 \\ 0 & p^{-1} \end{pmatrix},$$

we obtain

$$\sum_{u=0}^{p-1} f\,|_k \begin{pmatrix} 1 & \frac{u}{p} \\ 0 & 1 \end{pmatrix} = \left[\sum_{u=0}^{p-1} f\,|_k \begin{pmatrix} 1 & u \\ 0 & p \end{pmatrix}\right]\begin{pmatrix} 1 & 0 \\ 0 & p^{-1} \end{pmatrix}$$

$$= f\,|_k\left[p^{1-\frac{k}{2}}T_p - \begin{pmatrix} p & 0 \\ 0 & 1 \end{pmatrix}\right]\begin{pmatrix} 1 & 0 \\ 0 & p^{-1} \end{pmatrix} = p^{1-\frac{k}{2}} a_p f\,|_k \begin{pmatrix} 1 & 0 \\ 0 & p^{-1} \end{pmatrix} - f\,|_k \begin{pmatrix} p & 0 \\ 0 & p^{-1} \end{pmatrix}$$

This makes it possible to write

$$f_\varepsilon\,|_k \begin{pmatrix} 0 & -1 \\ p^{2N} & 0 \end{pmatrix}$$

$$= f\,|_k \begin{pmatrix} 0 & -1 \\ p^{2N} & 0 \end{pmatrix} - p^{-\frac{k}{2}} a_p f\,|_k \begin{pmatrix} 1 & 0 \\ 0 & p^{-1} \end{pmatrix}\begin{pmatrix} 0 & -1 \\ p^{2N} & 0 \end{pmatrix} + p^{-1} f\,|_k \begin{pmatrix} p & 0 \\ 0 & p^{-1} \end{pmatrix}\begin{pmatrix} 0 & -1 \\ p^{2N} & 0 \end{pmatrix}.$$

Bearing in mind that

$$\begin{pmatrix} 0 & -1 \\ p^{2N} & 0 \end{pmatrix} = \begin{pmatrix} 0 & -1 \\ 1 & 0 \end{pmatrix} \begin{pmatrix} p^{2N} & 0 \\ 0 & 1 \end{pmatrix},$$

$$\begin{pmatrix} 1 & 0 \\ 0 & p^{-1} \end{pmatrix} \begin{pmatrix} 0 & -1 \\ p^{2N} & 0 \end{pmatrix} = \begin{pmatrix} 0 & -1 \\ p^{2N-1} & 0 \end{pmatrix} = \begin{pmatrix} 0 & -1 \\ 1 & 0 \end{pmatrix} \begin{pmatrix} p^{2N-1} & 0 \\ 0 & 1 \end{pmatrix},$$

$$\begin{pmatrix} p & 0 \\ 0 & p^{-1} \end{pmatrix} \begin{pmatrix} 0 & -1 \\ p^{2N} & 0 \end{pmatrix} = \begin{pmatrix} 0 & -p \\ p^{2N-1} & 0 \end{pmatrix} = \begin{pmatrix} 0 & -p \\ p & 0 \end{pmatrix} \begin{pmatrix} p^{2N-2} & 0 \\ 0 & 1 \end{pmatrix},$$

we obtain

$$f_\varepsilon |_k \begin{pmatrix} 0 & -1 \\ p^{2N} & 0 \end{pmatrix} (z) = p^{Nk} f(p^{2N} z) - p^{-\frac{k}{2}} a_p p^{\left(N-\frac{1}{2}\right)k} f(p^{2N-1} z) + p^{(N-1)k-1} f(p^{2N-2} z).$$

Moreover,

$$\overline{g_{\bar\chi} |_k \begin{pmatrix} 0 & -1 \\ p^{2N} & 0 \end{pmatrix}} = \frac{\overline{W(\bar\chi)^2}}{p^N} \cdot \bar{g}_\chi.$$

Case 2. $\psi = \chi$. Similarly, we obtain

$$\overline{g_\varepsilon |_l \begin{pmatrix} 0 & -1 \\ p^{2N} & 0 \end{pmatrix} (z)} = p^{Nl}\, \overline{g(p^{2N} z)} - p^{-\frac{l}{2}} b_p p^{\left(N-\frac{1}{2}\right)l}\, \overline{g(p^{2N-1} z)}$$

$$+ p^{(N-1)l-1}\, \overline{g(p^{2N-2} z)},$$

$$f_\chi |_k \begin{pmatrix} 0 & -1 \\ p^{2N} & 0 \end{pmatrix} (z) = \frac{W(\chi)^2}{p^N} f_{\bar\chi}(z).$$

Case 3. $\psi \neq \chi, \varepsilon$. In this case neither ψ nor $\psi\bar\chi$ is principal. We denote by p^{N_ψ} the conductor of ψ. The notation $W(\psi)$ will be retained for the Gauss sum modulo p^{N_ψ}. If the primitive character mod $p^{N_{\psi_0}}$ corresponding to ψ is denoted by ψ_0, then in Case 3 we have

$$f_\psi(z) = f_{\psi_0}(z), \quad g_{\psi\bar\chi}(z) = g_{(\psi\bar\chi)_0}(z).$$

We now use the fact that

$$f_\psi |_k \begin{pmatrix} 0 & -1 \\ p^{2N_\psi} & 0 \end{pmatrix} = \frac{W(\psi)^2}{p^{N_\psi}} f_{\bar\psi}, \quad \overline{g_{\psi\bar\chi} |_l \begin{pmatrix} 0 & -1 \\ p^{2N_{\psi\bar\chi}} & 0 \end{pmatrix}} = \frac{\overline{W(\psi\bar\chi)^2}}{p^{N_{\psi\bar\chi}}} \overline{g_{\bar\psi\chi}}.$$

We obtain

$$f_\psi |_k \begin{pmatrix} 0 & -1 \\ p^{2N} & 0 \end{pmatrix} (z) = f_\psi |_k \begin{pmatrix} 0 & -1 \\ p^{2N_\psi} & 0 \end{pmatrix} \begin{pmatrix} p^{2(N-N_\psi)} & 0 \\ 0 & 1 \end{pmatrix} (z)$$

$$= \frac{W(\psi)^2}{p^{N_\psi}} \cdot p^{(N-N_\psi)k} f_{\bar\psi}(p^{2(N-N_\psi)} z).$$

Similarly,

$$\overline{g_{\psi\bar\chi} |_l \begin{pmatrix} 0 & -1 \\ p^{2N} & 0 \end{pmatrix} (z)} = \frac{\overline{W(\psi\bar\chi)^2}}{p^{N_{\psi\bar\chi}}} p^{(N-N_{\psi\bar\chi})l}\, \overline{g_{\bar\psi\chi}(p^{2(N-N_{\psi\bar\chi})} z)}.$$

2.17. Computing $\tilde{L}_{f,g}(s, \bar{\chi})$ now reduces to integrating the sum

$$\frac{\overline{W(\bar{\chi})^2}}{p^N}\left[p^{Nk}f(p^{2N}z)\overline{g_\chi(z)} - a_p p^{(N-1)k}f(p^{2N-1}z)\overline{g_\chi(z)} + p^{(N-1)k-1}f(p^{2N-2}z)\overline{g_\chi(z)}\right]$$

$$+\frac{W(\chi)^2}{p^N}\Big[p^{Nl}f_{\bar{\chi}}(z)\overline{g(p^{2N}z)} - b_p p^{(N-1)l-1}f_{\bar{\chi}}(z)\overline{g(p^{2N-1}z)}$$

$$+p^{(N-1)l-1}f_{\bar{\chi}}(z)\overline{g(p^{2N-2}z)}\Big]+$$

$$+\sum_{\substack{\psi \bmod p^N\\ \psi\neq\chi,\varepsilon}}\frac{W(\psi)^2}{p^{N_\psi}}\cdot\frac{\overline{W(\psi\bar{\chi})^2}}{p^{N_{\psi\bar{\chi}}}}p^{(N-N_\psi)k+(N-N_{\psi\bar{\chi}})l}f_\psi(p^{2(N-N_\psi)}z)\overline{g_{\psi\bar{\chi}}(p^{2(N-N_{\psi\bar{\chi}})}z)}.$$

Integrating the trace in accordance with (21), we immediately see that the only nonzero contribution is made by those terms under the last summation sign and *for which $N_\psi = N$ and $N_{\psi\bar{\chi}} = N$ simultaneously*. Hence

$$\tilde{L}_{f,g}(s, \bar{\chi})$$

$$=\left[p^{kN-N}W(\chi^2)\frac{\overline{W(\bar{\chi})^2}}{p^N}p^{-kN}\frac{\overline{W(\chi)^2}}{p^N}\frac{p}{p-1}\sum_{\substack{N_\psi=N\\ N_{\psi\bar{\chi}}=N}}\frac{W(\psi)^2\,\overline{W(\psi\bar{\chi})^2}}{p^{2N}}\right]L_{f,g}(s,\chi).$$

The expression in brackets is easily simplified by recalling that $\overline{W(\chi)} = \chi(-1)W(\bar{\chi})$, and so

$$\frac{\overline{W(\bar{\chi})^2}}{p^N}\cdot\frac{\overline{W(\chi)^2}}{p^N}=1,\quad\text{and}\quad \overline{W(\psi\bar{\chi})^2}=W(\bar{\psi}\chi)^2.$$

Finally, we have

$$\tilde{L}_{f,g}(s,\bar{\chi})=\left[\frac{3W(\chi^2)}{p^{3N-1}(p-1)}\sum_{\chi=\psi_1\psi_2}{}' W(\psi_1)^2W(\psi_2)^2\right]L_{f,g}(s,\bar{\chi}).$$

§3. The operator θ

3.1. For any series $h(z) = \sum_0^\infty a_n q^n$, $q = e^{2\pi i z}$, we set

$$(\theta h)(z)=\sum_{n=1}^{\infty} n a_n q^n.$$

The Mellin transform turns this operator into a translation operator of the argument $s\mapsto s-1$. Therefore, we can use it to compute the values $\mathfrak{L}_{f,g}(s,\chi)$ at $k-l-1$ points lying to the left of the initial point $s = k-1$. The computations are based on the following modular properties of the operator θ.

Let $h(z) = \sum_1^\infty a_n q^n$ be a cusp form of type $(\Gamma_0(A), \psi, l)$, where ψ is a Dirichlet character modulo $A \geqslant 1$.

3.2. PROPOSITION. *Let $m \geqslant 0$. Then there exist positive integers $A_{m,i}^{(l)}$ such that for $T \in \Gamma_0(A)$, $T = \left(\begin{smallmatrix} a & b\\ c & d\end{smallmatrix}\right)$, the following assertions hold*:

a) $$(\theta^m h)(Tz)=\sum_{i=0}^{m}A_{m,i}^{(l)}\frac{c^{m-i}}{(2\pi i)^{m-i}}\psi(d)(cz+d)^{l+m+i}(\theta^i h)(z).$$

b) *The* $A^{(l)}_{m,i}$ *satisfy the following recurrence relations*:

(i) $A^{(l)}_{m+1,0} = (l+m)\,A^{(l)}_{m,0},\ A_{0,0} = 1,$

(ii) $A^{(l)}_{m+1,i} = (l+m+i)\,A^{(l)}_{m,i} + A^{(l)}_{m,i-1},\ i = 1, \ldots, m,$

(iii) $A^{(l)}_{m+1,m+1} = A^{(l)}_{m,m} = 1.$

PROOF. For $m = 0$ this is the condition of modularity:

$$h(Tz) = \psi(d)\,(cz+d)^l h(z).$$

Inductive step from m *to* $m+1$:

$$(\theta^{m+1}h)(Tz) = [\theta(\theta^m h)](Tz) = \left[\frac{1}{2\pi i}\frac{d}{dz}(\theta^m h)\right](Tz)$$

$$= \frac{1}{2\pi i}\frac{d}{dz}[(\theta^m h)(Tz)]\,(cz+d)^2$$

$$= \frac{1}{2\pi i}\frac{d}{dz}\left[\sum_{i=0}^{m} A^{(l)}_{m,i}\left(\frac{c}{2\pi i}\right)^{m-i}\psi(d)\,(cz+d)^{l+m+i}(\theta^i h)(z)\right](cz+d)^2$$

$$= \sum_{i=0}^{m} A^{(l)}_{m,i}\left(\frac{c}{2\pi i}\right)^{m+1-i}\psi(d)\,(cz+d)^{l+m+i+1}(l+m+i)(\theta^i h)(z)$$

$$+ \sum_{i=0}^{m} A^{(l)}_{m,i}\left(\frac{c}{2\pi i}\right)^{m-i}\psi(d)\,(cz+d)^{l+m+i+2}(\theta^{i+1}h)(z).$$

We substitute the index of summation into the second sum and regroup the terms:

$$A^{(l)}_{m,0}(l+m)\left(\frac{c}{2\pi i}\right)^{m+1}\psi(d)\,(cz+d)^{l+m+1}h(z)$$

$$+ \sum_{i=1}^{m}[(l+m+i)\,A^{(l)}_{m,i} + A^{(l)}_{m,i-1}]\left(\frac{c}{2\pi i}\right)^{m+1-i}\psi(d)\,(cz+d)^{l+m+1+i}(\theta^i h)(z)$$

$$+ A^{(l)}_{m,m}\psi(d)\,(cz+d)^{l+(m+1)+m+1}(\theta^{m+1}h)(z) = (\theta^{m+1}h)(Tz).$$

This proves the proposition.

§4. Computation of the values $\mathfrak{L}_{f,g}(s,\chi)$ at the points $s = l, l+1, \ldots, k-1$

4.1. We shall use the method of Poincaré series summation that goes back to Rankin (see [19]) to compute the values of $\mathfrak{L}_{fg}(s,\chi)$ at $(k-l)/2$ points: $s = (k+l)/2, \ldots, k-1$. At the remaining points the values will be computed by means of the functional equation 1.3.

We set

$$G_k(z,n) = \sum_{T\in\Gamma_\infty\backslash\Gamma}\frac{e^{2\pi i n Tz}}{(cz+d)^k}, \quad k \geqslant 4, \quad T = \begin{pmatrix} a & b \\ c & d\end{pmatrix}.$$

$$g_k(z,n) = n^{k-1}G_k(z,n), \quad n > 0.$$

If $f(z) = \sum_1^\infty a_n e^{2\pi i n z}$ is a cusp form of weight k with respect to $\Gamma = \mathbf{SL(2, Z)}$, then

$$\big(f(z), g_k(z,n)\big) = (4\pi)^{1-k}(k-2)!\,a_n. \tag{1}$$

From (1) and the fact that Hecke operators are Hermitian it now follows that $g_k(z,n) = g_k(z,1)|_k T_n$.

We consider the space $S_k(\Gamma)$ of cusp forms of weight k with respect to Γ. Let $\kappa = \dim S_k(\Gamma)$. There exists a basis in $S_k(\Gamma)$ consisting of normalized eigenfunctions of Hecke operators:

$$\left\{ f_k(z, j) = \sum_{n=1}^{\infty} a(n, j) q^n \mid j = 1, \ldots, \kappa \right\}.$$

In what follows we assume that $f_k(z, 1) = f(z)$ and $a(n, 1) = a_n$. As previously, χ will denote the convolution of a cusp form with the character χ, for example,

$$g_\chi(z) = \sum_{n=1}^{\infty} \chi(n) b_n q^n.$$

We consider the sum

$$\sum_{n=1}^{\infty} \chi(n) b_n n^m G_k(z, n), \tag{2}$$

where $0 \leqslant m \leqslant (k-l)/2 - 1$, $l - k \geqslant 2$. In 4.3 it will be shown that this series converges absolutely for $0 \leqslant m < (k-l)/2 - 1$.

In the case $m = (k-l)/2 - 1$ the sum of the series (2) will be defined by Hecke's method in the Appendix to §4. The essence of this method is introducing an additional complex parameter s in (2) and analytically continuing to the point $s = 0$. It turns out that if χ^2 is a nonprincipal character, then in any case the sum of the series (2) is a cusp form of weight k with respect to $\Gamma = \mathbf{SL(2, Z)}$. In our case Lemma 2.7 and the fact that χ is primitive mod p^N guarantees that χ^2 is nonprincipal. It will also be shown that all the arguments of this subsection remain valid even in the exceptional case $m = (k-l)/2 - 1$.

We first assume that $0 \leqslant m \leqslant (k-l)/2 - 2$ (for $(k-l)/2 \geqslant 2$). We expand $g_k(z, 1)$ with respect to an orthogonal Hecke basis:

$$g_k(z, 1) = \sum_{j=1}^{\kappa} \xi_j f_k(z, j).$$

Then

$$\sum_{n=1}^{\infty} \chi(n) b_n n^m G_k(z, n) = \sum_{j=1}^{\kappa} \xi_j \sum_{n=1}^{\infty} \frac{a(n, j) b_n \chi(n)}{n^{k-1-m}} f_k(z, j). \tag{3}$$

We denote the coefficient of $f_k(z, j)$ on the right by η_j:

$$\eta_j = \xi_j \sum_{n=1}^{\infty} \frac{a(n, j) b_n \chi(n)}{n^{k-1-m}} \tag{4}$$

We have

$$\begin{aligned}(4\pi)^{1-k}(k-2)! = a(1, j)(4\pi)^{1-k}(k-2)! &= (f_k(z, j), g_k(z, 1)) \\ &= \xi_j (f_k(z, j), f_k(z, j)),\end{aligned}$$

since $\{f_k(z, j) | j = 1, \ldots, \kappa\}$ is an orthogonal basis of $S_k(\Gamma)$, and $a(1, j) = 1$ for all j. Consequently,

$$\xi_j (f_k(z, j), f_k(z, j)) = (4\pi)^{1-k}(k-2)!. \tag{5}$$

We now assume that we have found an expression for η_j. It then follows from (4) and (5) that

$$\sum_{n=1}^{\infty} \frac{\chi(n) a_n b_n}{n^{k-1-m}} = \eta_1 \frac{(f, f)}{(4\pi)^{1-k}(k-2)!}. \tag{6}$$

The η_j can be found if we know the Fourier expansion

$$\sum_{n=1}^{\infty} \chi(n) b_n n^m G_k(z, n) = \sum_{n=1}^{\infty} F_{m,n} q^n. \tag{7}$$

Indeed, by virtue of (3), finding the η_j reduces to solving the system of linear equations

$$\sum_{j=1}^{\varkappa} \eta_j a(n, j) = F_{m,n_i}, \quad i = 1, \dots, \varkappa, \tag{8}$$

where the n_i are chosen so that $\{g_k(z, n_i) | i = 1, \dots, \kappa\}$ is a basis of $S_k(\Gamma)$. We set

$$A = \begin{pmatrix} a(1, n_1) & \dots & a(1, n_\varkappa) \\ \dots & & \dots \\ a(\varkappa, n_1) & \dots & a(\varkappa, n_\varkappa) \end{pmatrix}, \quad F_m = \begin{pmatrix} F_{m,n_1} \\ \cdot \\ F_{m,n_\varkappa} \end{pmatrix}, \quad \mathrm{H} = \begin{pmatrix} \eta_1 \\ \dots \\ \eta_\varkappa \end{pmatrix}.$$

Then $\det A \neq 0$, and the solution of our linear system can be written as follows:

$$\mathrm{H} = A^{-1} F_m.$$

In particular,

$$\eta_1 = \sum_{j=1}^{\varkappa} A_{1,j} F_{m,n_j},$$

where

$$A^{-1} = \begin{pmatrix} A_{1,1} & \dots & A_{1,\varkappa} \\ \dots & & \dots \\ A_{\varkappa,1} & \dots & A_{\varkappa,\varkappa} \end{pmatrix}$$

is a fixed matrix with algebraic coefficients contained in a totally real extension of $\mathbf{Q}$ generated by the coefficients of Hecke cusp forms of weight k.

Thus, our problem has been reduced to finding the Fourier expansion (7). A generalization of a lemma of Rankin (see [19], Lemma 4), which he proved for $\Gamma = \mathbf{SL(2, Z)}$, plays a key role in the computation.

Let P be a cusp point of $\Gamma_0(M)$, $M \geqslant 1$. We set

$$T_P(P) = i\infty, \ T_P \in \mathrm{SL}(2, \mathbf{Z}), \ \Gamma_P = T_P \Gamma_0(M) T_P^{-1} \cap \begin{pmatrix} 1 & \mathbf{Z} \\ 0 & 1 \end{pmatrix}, \ \omega_P = \left[\begin{pmatrix} 1 & \mathbf{Z} \\ 0 & 1 \end{pmatrix} : \Gamma_P \right],$$

and let χ be a Dirichlet character modulo M.

We define Poincaré series with respect to $\Gamma_0(M)$:

$$G_{k,M}^{(P)}(z, n) = \sum_{T \in \Gamma_P \backslash T_P \Gamma_0(M)} j(T, z)^{-k} e^{2\pi i n T z}, \quad k \geqslant 4.$$

Their main property is that if f is a cusp form of weight k with respect to $\Gamma_0(M)$,

$$(f|_k T_P^{-1})(z) = \sum_{n=1}^{\infty} a_n^P(f)\, e^{2\pi i n \frac{z}{w_P}},$$

then (see [24])

$$(f, G_{k,M}^{(P)}(z, n)) = (k-2)!\,(4\pi n)^{1-k} w_P^{k-1} a_n^P(f).$$

4.2. Lemma. *Let the sequence* $\{\beta_n\}_1^{\infty}$, $\beta_n \in \mathbf{C}$, *be such that* $\sum_{n\leqslant x}|\beta_n|^2 = O(x^{\theta})$, $\theta > 0$. *Then for* $k > \theta + 2$ *and* $h \geqslant 0$

$$\sum_{n=1}^{\infty} \beta_n G_{k,M}^{(P)}(z, n+h) = \sum_{T\in\Gamma_P\setminus T_P\Gamma_0(M)} j(T, z)^{-k} e^{2\pi i h T z} \sum_{n=1}^{\infty} \beta_n e^{2\pi i n T z}$$

(*here all series converge absolutely*).

The proof repeats the proof of Rankin's lemma.

Applying Lemma 4.2 to our case, we obtain the following proposition:

4.3. Proposition. *Let* $0 \leqslant m < (k-1)/2 - 1$. *Then*

$$\sum_{n=1}^{\infty} \chi(n) b_n n^m G_k(z, n) = \sum_{T\in\Gamma_\infty\setminus\Gamma} j(T, z)^{-k} (\theta^m g_\chi)(Tz),$$

$$\sum_{n=1}^{\infty} \chi(n) b_n n^m G_k(z, n)$$

$$= \mathrm{Tr}_k^{\Gamma_0(M^2)\setminus\Gamma}\left[\sum_{i=0}^{m} A_{m,i}^{(l)}(-1)^{m-i}\frac{(k-l-2m-1)!}{(k-l-i-m-1)!}\theta^i g_\chi(z)\cdot\theta^{m-i}E_{k-l-2m,\chi^2}^{(M^2)}(z)\right],$$

where, by definition,

1) $$E_{k-l-2m,\chi^2}^{(M^2)}(z) = \sum_{T\in\Gamma_\infty\setminus\Gamma_0(M^2)} \chi^2(T)\, j(T, z)^{-(k-l-2m)};$$

2) *the* $A_{m,i}^{(l)}$ *are the positive integers defined in* 3.2.

Proof. As is known, $\sum_{n\leqslant x}|b_n|^2 = O(x^l)$ (see [20]). Partial summation now shows that

$$\sum_{n\leqslant x} |\chi(n) b_n n^m|^2 = O(x^{l+2m}).$$

Assertion a) is now an immediate consequence of the definition of the operator θ and Lemma 4.2.

We turn to the proof of b). We have

$$\sum_{T\in\Gamma_\infty\setminus\Gamma} j(T, z)^{-k}\cdot(\theta^m g_\chi)(Tz) = \sum_{r\in\Gamma_0(M^2)\setminus\Gamma}\ \sum_{\gamma\in\Gamma_\infty\setminus\Gamma_0(M^2)} j(\gamma r, z)^{-k}(\theta^m g_\chi)(Tz),$$

where $T = \gamma z$. Using the formula $j(\gamma r, z) = j(\gamma, rz)\cdot j(r, z)$, we rewrite this as follows:

$$\begin{aligned}&\sum_{r\in\Gamma(M^2)\setminus\Gamma}\ \sum_{\gamma\in\Gamma_\infty/\Gamma_0(M^2)} j(\gamma, rz)^{-k}\cdot j(r, z)^{-k}(\theta^m g_\chi)(\gamma r z)\\ &= \sum_{r\in\Gamma_0(M^2)\setminus\Gamma}\left[\sum_{\gamma\in\Gamma_\infty\setminus\Gamma_0(M^2)} j(\gamma, z)^{-k}(\theta^m g_\chi)(\gamma z)\right]\Bigg|_k r.\end{aligned} \tag{9}$$

By Proposition 3.2, if $\gamma = \left(\begin{smallmatrix} a & b \\ c & d \end{smallmatrix}\right)$, $\gamma \in \Gamma_0(M^2)$, then

$$(\theta^m g_\chi)(\gamma z) = \sum_{i=0}^{m} A_{m,i}^{(l)} \left(\frac{c}{2\pi i}\right)^{m-i} (cz+d)^{l+m+i} \chi^2(d)\, \theta^i g_\chi(z).$$

It follows that the expression in brackets on the right side of (9) can be rewritten as follows:

$$\sum_{i=0}^{m} A_{m,i}^{(l)} (\theta^i g_\chi)(z) \sum_{\gamma \in \Gamma_\infty \setminus \Gamma_0(M^2)} \frac{\chi^2(d)\, c^{m-i}}{(2\pi i)^{m-i} (cz+d)^{k-l-i-m}}, \tag{10}$$

where $\gamma = \left(\begin{smallmatrix} a & b \\ c & d \end{smallmatrix}\right)$.

Differentiation now yields

$$\theta^{m-i}\left(\sum_{\gamma \in \Gamma_\infty \setminus \Gamma_0(M^2)} \frac{\chi^2(d)}{j(\gamma, z)^{k-l-2m}}\right)$$
$$= (-1)^{m-i} \frac{(k-l-m-i-1)!}{(k-l-2m-1)!} \cdot \frac{1}{(2\pi i)^{m-i}} \sum_{\gamma \in \Gamma_\infty \setminus \Gamma_0(M^2)} \frac{\chi^2(d)\, c^{m-i}}{(cz+d)^{k-l-i-m}}.$$

In conjunction with (9) and (10) this proves the proposition.

In what follows we shall require the following auxiliary result:

4.4. PROPOSITION. *The function*

$$\sum_{i=0}^{m} A_{m,i}^{(l)} \theta^i g_\chi(z) \sum_{\gamma \in \Gamma_\infty \setminus \Gamma_0(M^2)} \frac{\chi^2(d)\, c^{m-i}}{(2\pi i)^{m-i} (cz+d)^{k-l-i-m}}$$
$$= \sum_{\gamma \in \Gamma_\infty \setminus \Gamma_0(M^2)} j(\gamma, z)^{-k} \theta^m g_\chi(\gamma z)$$

(*where* $\gamma = \left(\begin{smallmatrix} a & b \\ c & d \end{smallmatrix}\right)$) *is a modular form of weight* k *for* $\Gamma(M^2)$.

PROOF. By Lemma 4.2, this function coincides with the sum

$$\sum_{n=1}^{\infty} \chi(n)\, b_n n^m G_{k,M^2}^{(i\infty)}(z, n),$$

where $G_{m,M^2}^{(i\infty)}(z, n)$ are the Poincaré series of level M^2 and weight k in 4.1.

We now compute the Fourier expansion of the trace in Proposition 4.3. We first show how to compute the trace in the general setting–the results of this subsection carry over to Fourier series that are not necessarily modular forms.

We restrict ourselves to the case $M = p^N$ to simplify the formulas. Let

$$F_{p^{2N}}(z) = \sum_{n=1}^{\infty} A_{p^{2N},n} q^n.$$

We require that the following Fourier expansion exist ($q = e^{2\pi i z}$):

$$G_{p^{2N}}(z) = F_{p^{2N}}\big|_k \begin{pmatrix} 0 & -1 \\ p^{2N} & 0 \end{pmatrix}(z) = \sum_{n=1}^{\infty} B_{p^{2N},n} q^n.$$

We fix a system of representatives of $\Gamma_0(p^r) \setminus \Gamma_0(p^{r-1})$, $r \geqslant 1$:

$$S_{p^r}^{p^{r-1}} = \left\{ \begin{pmatrix} 1 & 0 \\ ap^{r-1} & 1 \end{pmatrix} \middle| \; 0 \leqslant a \leqslant p-1 \right\} \cup \begin{pmatrix} 0 & -1 \\ 1 & 0 \end{pmatrix}$$

(the last term occurs only in the case $r = 1$).

We successively define

$$F_{p^r}(z) = \mathrm{Tr}_k^{\Gamma_0(p^{r+1}) \setminus \Gamma_0(p^r)} F_{p^{r+1}}(z) = \sum_{n=1}^{\infty} A_{p^r,n} q^n,$$

$$G_{p^r}(z) = F_{p^r}|_k \begin{pmatrix} 0 & -1 \\ p^r & 0 \end{pmatrix} = \sum_{n=1}^{\infty} B_{p^r,n} q^n.$$

(The equalities mean that at each stage we require that the corresponding q-expansion exist.)

4.5. Proposition. *For any r, $1 \leqslant r \leqslant 2N$,*

$$B_{p^{r-1},n} = p^{1-\frac{k}{2}} B_{p^r,pn} + (-1)^k A_{p,n}$$

(*the last term occurs only in the case $r = 1$*).

Proof. It must be verified that

$$\sum_{a=0}^{p-1} F_{p^r}|_k \begin{pmatrix} 1 & 0 \\ ap^{r-1} & 1 \end{pmatrix} \begin{pmatrix} 0 & -1 \\ p^r & 0 \end{pmatrix} + \left[F_p|_k \begin{pmatrix} 0 & -1 \\ 1 & 0 \end{pmatrix} \begin{pmatrix} 0 & -1 \\ 1 & 0 \end{pmatrix} \right]$$

$$= \sum_{a=0}^{p-1} F_{p^r}|_k \begin{pmatrix} 0 & -1 \\ p^r & 0 \end{pmatrix} \begin{pmatrix} \frac{1}{p} & -\frac{a}{p} \\ 0 & 1 \end{pmatrix} + [(-1)^k F_{p^r}]$$

(the summands in brackets occur *only in the case* $r = 1$). This is obvious:

$$\begin{pmatrix} 1 & 0 \\ ap^{r-1} & 1 \end{pmatrix} \begin{pmatrix} 0 & -1 \\ p^{r-1} & 0 \end{pmatrix} = \begin{pmatrix} 0 & -1 \\ p^r & -ap^{r-1} \end{pmatrix} = \begin{pmatrix} 0 & -1 \\ p^r & 0 \end{pmatrix} \begin{pmatrix} \frac{1}{p} & -\frac{a}{p} \\ 0 & 1 \end{pmatrix},$$

$$\begin{pmatrix} 0 & -1 \\ 1 & 0 \end{pmatrix} \begin{pmatrix} 0 & -1 \\ 1 & 0 \end{pmatrix} = \begin{pmatrix} -1 & 0 \\ 0 & -1 \end{pmatrix}.$$

Let us compute the Fourier expansion of the trace in Proposition 4.3. We first write an expression for

$$E_{k-l-2m,\chi^2}^{(M^2)}|_{k-l-2m} \begin{pmatrix} 0 & -1 \\ M^2 & 0 \end{pmatrix}$$

(the definition of $E_{k-l-2m,\chi^2}^{(M^2)}$ is found in 4.3.

4.6. Proposition.

$$E_{k-l-2m,\chi^2}^{(M^2)}|_{k-l-2m} \begin{pmatrix} 0 & -1 \\ M^2 & 0 \end{pmatrix}(z)$$

$$= \frac{1}{2M^{k-l-2m}} \sum_{\bar{d} \bmod M^2} \chi^2(\bar{d}) \sum_{\bar{c} \bmod M^2} \sum_{\bar{a} \bmod M^2} \sum_{\substack{n=1 \\ n \equiv \bar{a}^{-1}}}^{\infty} \frac{\mu(n)}{n^{k-l-2m}} G_{k-l-2m}(z; \overline{\bar{a}\bar{d}}, \overline{\bar{a}\bar{c}}; M^2), \tag{11}$$

where for $N > 1$, $a_1, a_2 \bmod N$

$$G_k(z; a_1, a_2; N) = \sum_{\substack{c\equiv a_1(N)\\ d\equiv a_2(N)}}' \frac{1}{(cz+d)^k}, \quad k > 2$$

(in Gunning's notation [3] *this is* $G_{k/2}(z; a_1, a_2, N)$*).*

PROOF. We have

$$E^{(M^2)}_{k-l-2m,\chi^2}\Big|_{k-l-2m}\begin{pmatrix}0 & -1\\ M^2 & 0\end{pmatrix}(z) = E^{(M^2)}_{k-l-2m,\chi^2}\left(-\frac{1}{M^2z}\right)(Mz)^{-(k-l-2m)}$$

$$= \frac{1}{2}\sum_{(Mc,d)=1} \frac{\chi^2(d)}{\left(M^2c\left(-\frac{1}{M^2z}\right)+d\right)^{k-l-2m}}\cdot\frac{1}{(Mz)^{k-l-2m}}$$

$$= \frac{1}{2M^{k-l-2m}}\sum_{(Mc,d)=1}\frac{\chi^2(d)}{(dz-c)^{k-l-2m}} = \frac{1}{2M^{k-l-2m}}\sum_{\bar d \bmod M^2}\chi^2(\bar d)\sum_{\substack{(c,d)=1\\ d\equiv\bar d(M^2)}}\frac{1}{(dz-c)^{k-l-2m}}$$

$$= \frac{1}{2M^{k-l-2m}}\sum_{\bar d \bmod M^2}\chi^2(\bar d)\sum_{\bar c \bmod M^2} G^{\bullet}_{k-l-2m}(z; \bar d, \bar c; M^2).$$

In the last expression the following notation is used:

$$G^{\bullet}_{k-l-2m}(z; \bar d, \bar c; M^2) = \sum_{\substack{c\equiv\bar d,\ d\equiv\bar c(M^2)\\ (d,c)=1}}\frac{1}{(cz+d)^{k-l-2m}}.$$

Using the properties of the Möbius function

$$\sum_{\delta|d}\mu(\delta) = \begin{cases}1, & \text{if } d = 1,\\ 0, & \text{if } d > 1,\end{cases}$$

we can easily express $G^{\bullet}_{k-l-2m}$ in terms of G_{k-l-2m}. Moreover, (11) can be rewritten as follows:

$$\frac{1}{2M^{k-l-2m}}\sum_{\bar d \bmod M^2}\chi^2(\bar d)\sum_{\bar c \bmod M^2}\sum_{\substack{\bar a \bmod M^2\\ (\bar a,M)=1}}\left(\sum_{\substack{n=1\\ n\equiv\bar a^{-1}(M^2)}}^{\infty}\frac{\mu(n)}{n^{k-l-2m}}\right)G_{k-l-2m}(z; \bar a\bar d, \bar a\bar c; M^2).$$

This proves the proposition.

We set

$$H^{(M^2)}_{k-l-2m,\chi^2} = E^{(M^2)}_{k-l-2m,\chi^2}\Big|_{k-l-2m}\begin{pmatrix}0 & -1\\ M^2 & 0\end{pmatrix}.$$

The purpose of the next step is to write down the Fourier expansion of $H^{(M^2)}_{k-l-2m,\chi^2}$.

4.7. PROPOSITION.

$$H^{(M^2)}_{k-l-2m,\chi^2}(z) = \frac{(2\pi i)^{k-l-2m}}{M^{k-l-2m}(k-l-2m-1)!}L(k-l-2m, \chi^2)$$

$$\times\sum_{j=1}^{\infty}\chi^2(j)\,\sigma^{\bar\chi^2}_{k-l-2m-1}(j)\,e^{2\pi i jz},$$

PROOF. The Fourier expansion of $G_{2k}(z;\, a_1, a_2;\, N)$ is known (see [3]); namely, for $k > 1$ and $N > 1$

$$G_{2k}(z;\, a_1, a_2;\, N) = A + \frac{2\,(2\pi i)^{2k}}{N^{2k}\,(2k-1)!} \sum_{n=1}^{\infty} \sigma_{2k}(n;\, a_1, a_2)\, e^{2\pi i n\cdot \frac{z}{N}}, \tag{12}$$

where

$$\sigma_r(n;\, a_1, a_2) = \sum_{\substack{d\mid n\\ nd^{-1}\equiv a_1(N)}} d^r \cos\left(\frac{2\pi a_2 d}{N}\right), \quad A = \begin{cases} \sum\limits_{n_2\equiv a_2(N)} \frac{1}{n^{2k}}, & \text{if } a_1 \equiv 0\,(N),\\ 0 & \text{otherwise.}\end{cases}$$

Therefore

$$\begin{aligned} &G_{k-l-2m}(z;\, \bar a\bar d,\, \bar a\bar c;\, M^2)\\ &= \frac{2\cdot(2\pi i)^{k-l-2m}}{M^{2(k-l-2m)}\,(k-l-2m-1)!} \sum_{j=1}^{\infty} \sigma_{k-l-2m-1}(j;\, \bar a\bar d,\, \bar a\bar c)\, e^{\frac{2\pi i\cdot z}{M^2}},\\ &\sigma_{k-l-2m-1}(j;\, \bar a\bar d,\, \bar a\bar c) = \sum_{\substack{d\mid j\\ jd^{-1}\equiv \bar a\bar c(M^2)}} d^{k-l-2m-1} \cos\frac{2\pi d\bar a\bar c}{M^2}. \end{aligned}$$

It follows that the coefficient of $e^{2\pi i j/M^2}$ in the series $H^{(M^2)}_{k-l-2m,\chi^2}$ equals

$$\begin{aligned} &\frac{(2\pi i)^{k-l-2m}}{M^{3(k-l-2m)}\,(k-l-2m-1)!} \sum_{\bar d \bmod M^2} \sum_{\bar c \bmod M^2} \sum_{\substack{\bar a \bmod M^2\\ (\bar a, M)=1}} \sum_{\substack{n=1\\ n\equiv \bar a^{-1}(M^2)}}^{\infty} \frac{\mu(n)}{n^{k-l-2m}}\\ &\times \sum_{\substack{d\mid j\\ jd^{-1}\equiv \bar a\bar d(M^2)}} \chi^2(\bar d)\, d^{k-l-2m-1} \cos\frac{2\pi \bar a\bar d}{M^2}. \end{aligned} \tag{13}$$

We note that for $j \not\equiv 0 \pmod{M^2}$ this expression equals zero, because

$$\sum_{\bar c \bmod M^2} \cos\frac{2\pi d\bar a\bar c}{M^2} = 0$$

(since $d\mid j$, $(\bar a, M) = 1$). Now let $j = j_0 M^2$. In the inner sum we leave the terms of the form $d = d_0 M^2$, $d_0 \mid j_0$. We have

$$\sum_{d_0\mid j_0} (d_0 M)^{k-l-2m-1} \sum_{\bar c \bmod M^2} \cos(2\pi \bar a d_0 \bar c) = M^{2(k-l-2m)} \sum_{\substack{d_0\mid j_0\\ j_0 d_0^{-1}\equiv \bar a\bar d(M^2)}} d_0^{k-l-2m-1}.$$

It follows from (13) and these observations that the Fourier coefficient of $e^{2\pi i j_0 z}$ in $H^{(M^2)}_{k-l-2m,\chi^2}(z)$ equals

$$\begin{aligned} &\frac{(2\pi i)^{k-l-2m}}{M^{k-l-2m}\,(k-l-2m-1)!} \sum_{\bar d \bmod M^2} \sum_{\substack{\bar a \bmod M^2\\ (\bar a, M)=1}} \sum_{\substack{n=1\\ n\equiv \bar a^{-1}(M^2)}}^{\infty} \chi^2(\bar d)\, \frac{\mu(n)}{n^{k-l-2m}}\\ &\times \sum_{\substack{d_0\mid j_0\\ j_0 d_0^{-1}\equiv \bar a\bar d(M^2)}} d_0^{k-l-2m-1}. \end{aligned} \tag{14}$$

We replace the summation over n in (14) with respect to the progression $n \equiv \bar a^{-1}(M^2)$ by the convolution with multiplicative characters ψ modulo M^2. This yields

$$\frac{(2\pi i)^{k-l-2m}}{\varphi(M^2)\,M^{k-l-2m}\,(k-l-2m-1)!}\sum_{\substack{n=1\\ n\equiv\bar a^{-1}(M^2)}}^{\infty}\frac{\mu(n)}{n^{k-l-2m}}$$

$$\times\sum_{\bar d \bmod M^2}\ \sum_{\psi \bmod M^2}\ \sum_{\substack{\bar a \bmod M^2\\ (\bar a, M)=1}}\ \sum_{\substack{d_0 | j_0\\ j_0 d_0^{-1}\equiv \bar a\bar d (M^2)}} \psi(\bar a n)\,\chi^2(\bar d)\,d_0^{k-l-2m-1}.$$

We note that in the last sum $\bar d \equiv j_0 d_0^{-1} a^{-1} \pmod{M^2}$, whence $\chi^2(\bar d) = \chi^2(j_0 d_0^{-1}\bar a^{-1})$. Summing over all indices for a fixed n, we obtain

$$\sum_{\psi \bmod M^2}\chi^2(j_0)\,\psi(n)\sum_{\substack{\bar a \bmod M^2\\ (\bar a, M)=1}}(\psi\bar\chi^2)(\bar a)\sum_{d_0 | j_0}\chi^2(d_0^{-1})\,d_0^{k-l-2m-1}.$$

But

$$\sum_{\bar a \bmod M^2}(\psi\bar\chi^2)(\bar a)=\begin{cases}0, & \text{if } \psi\neq\chi^2,\\ \varphi(M^2), & \text{if } \psi=\chi^2,\end{cases}$$

and so we finally conclude that the Fourier coefficient of $e^{2\pi i j_0 z}$ in the series $H^{(M^2)}_{k-l-2m,\chi^2}$ equals

$$\frac{(2\pi i)^{k-l-2m}}{M^{k-l-2m}\,(k-l-2m-1)!}\,\chi^2(j_0)\,\sigma^{\bar\chi^2}_{k-l-2m-1}(j_0)\,L(k-l-2m,\chi^2)^{-1}$$

This proves the proposition.

4.8. PROPOSITION. *For $r, i \geqslant 0$*

a) $$(\theta^r E^{(M^2)}_{k-l-2m,\chi^2})\left(-\frac{1}{M^2 z}\right)=\sum_{i=0}^{r}A^{(k-l-2m)}_{r,i}\,(\theta^i H^{(M^2)}_{k-l-2m,\chi^2})(z)\cdot\frac{M^{k-l-2m+2r}z^{k-l-2m+r+i}}{(2\pi i)^{r-i}};$$

b) $$(\theta^i g_\chi)\left(-\frac{1}{M^2 z}\right)=\frac{W(\chi^2)}{M}\sum_{j=0}^{i}A^{(l)}_{i,j}\,\frac{M^{l+2i}z^{l-i}}{(2\pi i)^{i-l}}\cdot(\theta^i g_{\bar\chi})(z),$$

where the $A^{(l)}_{i,j}$ are defined in 3.2.

PROOF. If we recall that

$$g_\chi|_l\begin{pmatrix}0 & -1\\ M^2 & 0\end{pmatrix}=\frac{W(\chi)^2}{M}\cdot g_{\bar\chi},\quad E^{(M^2)}_{k-l-2m,\chi^2}|_{k-l-2m}\begin{pmatrix}0 & -1\\ M^2 & 0\end{pmatrix}=H^{(M^2)}_{k-l-2m,\chi^2},$$

then the proof reduces to repeating the arguments in 3.2.

Recall that we have reduced the problem of computing the values to computing the Fourier expansion of the trace of the cusp form (15) of weight k with respect to $\Gamma_0(M^2)$:

$$\sum_{i=0}^{m}(-1)^{m-i}\frac{(k-l-2m-1)!}{(k-l-m-i-1)!}A^{(l)}_{m,i}\,(\theta^i g_\chi)(z)\cdot\theta^{m-i}E^{(M^2)}_{k-l-2m,\chi^2}(z). \tag{15}$$

In the next subsection we shall combine the results of 4.6–4.8 and write down the Fourier expansion of the function

$$\left[\sum_{i=0}^{m}(-1)^{m-i}\frac{(k-l-2m-1)!}{(k-l-m-i-1)!}A^{(l)}_{m,i}\cdot\theta^i g_\chi\cdot\theta^{m-i}E^{(M^2)}_{k-l-2m,\chi^2}\right]\Bigg|_k\begin{pmatrix}0 & -1\\ M^2 & 0\end{pmatrix}. \tag{16}$$

It will then remain only to use the results of 4.5.

We introduce the following notation:

$$F_m^{(M^2)}(z)=\sum_{n=1}^{\infty}F_{m,n}^{(M^2)}e^{2\pi inz}$$

is the modular form (15), and

$$G_m^{(M^2)}(z)=\sum_{n=1}^{\infty}G_{m,n}^{(M^2)}e^{2\pi inz}$$

is the modular form (16).

4.9. Proposition.

$$G_m^{(M^2)}=\frac{W(\chi)^2}{M}\sum_{i=0}^{m}A_{m,i}^{(l)}\frac{(-1)^{m-i}(k-l-2m-1)!}{(k-l-m-i-1)!}\theta^i g_{\bar\chi}\cdot\theta^{m-i}H_{k-l-2m,\chi^2}^{(M^2)}.$$

Proof. Using Proposition 4.8, we write down the action of the operator $|_k\left(\begin{smallmatrix}0 & -1\\ M^2 & 0\end{smallmatrix}\right)$ on (15). The resulting function is a product of two sums. In multiplying these sums we shall consider only those pairwise products which are periodic functions, in particular, which do not contain z to a nonzero power. Namely,

$$\theta^i g_\chi|_{l+2i}\begin{pmatrix}0 & -1\\ M^2 & 0\end{pmatrix}\cdot\theta^{m-i}E_{k-l-2m,\chi^2}^{(M^2)}|_{k-l-2i}\begin{pmatrix}0 & -1\\ M^2 & 0\end{pmatrix}$$
$$=\frac{W(\chi)^2}{M}\sum_{\alpha=0}^{i}\sum_{\beta=0}^{m-i}\frac{A_{m-i,\beta}^{(k-l-2m)}A_{i,\alpha}^{(l)}}{(2\pi iz)^{m-\alpha-\beta}}\theta^\alpha g_\chi\cdot\theta^\beta H_{k-l-2m,\chi^2}^{(M^2)},$$

where we consider only the terms with $\alpha=i$ and $\beta=m-i$:

$$\frac{W(\chi)^2}{M}A_{m-i,m-i}^{(k-l-2m)}\cdot A_{i,i}^{(l)}\theta^i g_{\bar\chi}\cdot\theta^{m-i}H_{k-l-2m,\chi^2}^{(M^2)}.$$

But $A_{m-i,m-i}^{(k-l-2m)}=A_{i,i}^{(l)}=1$ (see Proposition 3.2(iii)). Hence the Fourier expansion of $G_m^{(M^2)}(z)$ can be readily written down with the help of the known expansions of $\theta^i g_{\bar\chi}$ and $\theta^{m-i}H_{k-l-2m,\chi^2}^{(M^2)}$. Namely,

$$G_{m,n}^{(M^2)}=\frac{W(\chi)^2}{M}\cdot\left(\frac{2\pi i}{M}\right)^{k-l-2m}\cdot L(k-l-2m,\chi^2)$$
$$\times\sum_{i=0}^{m}A_{m,i}^{(l)}\frac{(-1)^{m-i}}{(k-l-m-i-1)!}\sum_{j_1+j_2=n}\bar\chi(j_1)\chi^2(j_2)j_1^i j_2^{m-i}\sigma_{k-l-2m-1}^{\bar\chi^2}(j_2)b_{j_1}. \tag{17}$$

4.10. From now on $M=p^N$. As in 4.5, for all r, $0\leqslant r\leqslant 2N$, we set

$$F_m^{(p^r)}(z)=\mathrm{Tr}_k^{\Gamma_0(p^N)\backslash\Gamma_0(p^r)}F_m^{(p^N)}(z)=\sum_{n=1}^{\infty}F_{m,n}^{(p^r)}e^{2\pi inz},$$

$$G_m^{(p^r)}(z)=F_m^{(p^r)}|_k\begin{pmatrix}0 & -1\\ p^r & 0\end{pmatrix}=\sum_{n=1}^{\infty}G_{m,n}^{(p^r)}e^{2\pi inz}.$$

It follows from Proposition 4.5 that

$$G_{m,n}^{(p)}=p^{(2N-1)\left(1-\frac{k}{2}\right)}G_{m,p^{2N-1}n}^{(p^{2N})}.$$

In addition, as (17) shows, $G_{m,n}^{(p)}(2\pi i)^{-(k-l-2m)}L(k-l-2m, \chi^2)^{-1}$ are algebraic numbers contained in the field generated over $\mathbf{Q}$ by the numbers b_j, $j = 1, 2, \ldots, \zeta_{p^N(p-1)}$.

4.11. In order to descend now from $\Gamma_0(p)$ to $\mathbf{SL}(2, \mathbf{Z})$, we describe the action of the operator $\left(\begin{smallmatrix} 0 & -1 \\ p & 0 \end{smallmatrix}\right)$ on the space of cusp forms of weight k with respect to $\Gamma_0(p)$.

1) The "newforms" are rearranged by this operator by multiplying by the constants of the functional equations of the corresponding Mellin transforms. These constants lie in the cyclotomic field $\mathbf{Q}(\zeta_{p(p-1)})$ (see [12]).

2) If $G(z)$ is induced from $\mathbf{SL}(2, \mathbf{Z})$, $G(z) = F(pz)$ (where $F(z)$ is a cusp form with respect to $\mathbf{SL}(2, \mathbf{Z})$), then

$$G|_k \begin{pmatrix} 0 & -1 \\ p & 0 \end{pmatrix} = p^{-\frac{k}{2}} F|_k \begin{pmatrix} p & 0 \\ 0 & 1 \end{pmatrix}\begin{pmatrix} 0 & -1 \\ p & 0 \end{pmatrix} = p^{-\frac{k}{2}} F.$$

3) If $F(z)$ is a cusp form with respect to $\mathbf{SL}(2, \mathbf{Z})$, then

$$F|_k \begin{pmatrix} 0 & -1 \\ p & 0 \end{pmatrix}(z) = p^{\frac{k}{2}} F(pz).$$

Thus, applying the operator $\left(\begin{smallmatrix} 0 & -1 \\ p & 0 \end{smallmatrix}\right)$ does not affect the nature of the algebraicity of the Fourier coefficients of the form $G_m^{(p)}(z)$. In addition, this operator can increase the denominators of the Fourier coefficients only by an integral power of p not exceeding $k/2$.

Again bearing in mind the result of Proposition 4.5, we obtain the desired assertion about the Fourier coefficients.

4.12. PROPOSITION. *For* $0 \leqslant m \leqslant (k-l)/2 - 1$

a) $F_{m,n}(2\pi i)^{-(k-l-2m)}L(k-l-2m, \chi^2)^{-1}$ *are algebraic numbers lying in the algebraic number field generated over* $\mathbf{Q}$ *by the numbers* b_n $(n = 1, 2, \ldots)$ *and* $\zeta_{p^N(p-1)}$;

b) *there exists an integer* $N_2 \geqslant 0$ *such that for any prime divisor* $\mathfrak{p}|p$ *the following products are* $\mathfrak{p}$*-integral in a field of finite degree containing the numbers in* a):

$$p^{(k-2)\left(N-\frac{1}{2}\right)+\frac{k}{2}+N(k-l-2m)+N_2} F_{m,n} L(k-l-2m, \chi^2)^{-1} \cdot (2\pi i)^{-(k-l-2m)}$$

for any $N > 1$ *and any* χ *primitive* mod p^N.

4.13. Now, using the connection between "additive" and "multiplicative" convolutions,

$$\mathfrak{L}_{f,g}(s, \chi) = L(2s-k-l+2, \chi^2) L_{f,g}(s, \chi),$$

and also the formulas (6)–(8) in 4.1, we obtain a proof of Theorem 1.4 for the points $s = (k+l)/2 + 1, \ldots, k-1$. A similar result will be proved in the Appendix to §4 for the point $s = (k+l)/2$. As we noted above, for the remaining points the theorem follows from the functional equation in 1.3.

Appendix. Computations in the exceptional case $m = (k-l)/2 - 1$

1. Using the notation of Lemma 4.2, we define the series

$$G_{k,s,N}^{(P)}(z, n) = \sum_{T \in \Gamma_P \setminus T_P \Gamma_0(N)} j(T, z)^{-k} |j(T, z)|^{-2s} e^{2\pi i \frac{nTz}{w_P}}, \quad (1)$$

which converge absolutely for $\mathrm{Re}(s) \gg 0$. In the case $N = 1$ we shall simply write $G_{k,s}(z, n)$ instead of $G_{k,s,1}^{(i\infty)}(z, n)$.

We consider the sum

$$\sum_{n=1}^{\infty} \chi(n) b_n n^{\frac{k-l}{2}-1} G_{k,s}(z, n). \tag{2}$$

The series (2) converges absolutely for $\mathrm{Re}(s) \gg 0$ (this follows from the simple estimates used in the proof of Rankin's lemma (see [19], the proof of Lemma 4) if we bear in mind that $b_n = O(n^{(l-1)/2+\varepsilon})$ (see [5]).

We introduce the notation

$$\begin{aligned}&\left(\sum_{n=1}^{\infty} \chi(n) b_n n^{\frac{k-l}{2}-1} G_{k,s}(z, n), f(z)\right)_s\\ &= \sum_{n=1}^{\infty} \chi(n) b_n n^{\frac{k-l}{2}-1} \int_D G_{k,s}(z, n) \overline{f(z)}\, y^{k-2+s}\, dx\, dy\end{aligned} \tag{3}$$

(here D is the standard fundamental domain in $\mathbf{SL}(2, \mathbf{Z})$). The following computation shows that the series on the right in (3) converges absolutely for $\mathrm{Re}(s) \gg 0$.

2. PROPOSITION. *If s is such that the series* (2) *converges absolutely, then*

$$\begin{aligned}&\left(\sum_{n=1}^{\infty} \chi(n) b_n n^{\frac{k-l}{2}-1} G_{k,s}(z, n), f(z)\right)_s\\ &= (4\pi)^{1-k+s} \Gamma(k+s-1) \sum_{n=1}^{\infty} \chi(n) a_n b_n n^{-\left(\frac{k+l}{2}+s\right)}\end{aligned}$$

(*the latter series converges absolutely for* $\mathrm{Re}(s) > 0$).

PROOF. We have

$$\begin{aligned}&\int_D G_{k,s}(z, n) \overline{f(z)}\, y^{k-2+s}\, dx\, dy\\ &= \int_D \sum_{T \in \Gamma_\infty \backslash \Gamma} \frac{e^{2\pi i n T z} y^{k-2+s}}{j(T, z)^k\, |j(T, z)|^{2s}} \overline{f(z)}\, dx\, dy\\ &= \int_{|x| \leqslant \frac{1}{2}} e^{2\pi i n z} \sum_{m=1}^{\infty} a_m \overline{e^{2\pi i m z}}\, y^{k-2+s}\, dx\, dy\end{aligned} \tag{4}$$

(the last equality holds since the series $G_{k,s}(z, n)$ converges absolutely). Furthermore,

$$\int_{|x| \leqslant \frac{1}{2}} e^{2\pi i n z} \overline{e^{2\pi i m z}}\, y^{k-2+s}\, dx\, dy = \begin{cases} 0 & \text{for } m \neq n, \\ (4\pi n)^{1-k+s}\, \Gamma(k+s-1) & \text{for } m = n. \end{cases}$$

Term-by-term integration of the series in (4) is legal. Its application completes the proof.

The plan for the subsequent operations is as follows. First, the "analytic continuation" of the series (2) to the point $s = 0$ will be found. It will be proved that the function so obtained, which we denote by

$$\sum_{n=1}^{\infty} \chi(n) b_n n^{\frac{k-l}{2}} G_k(z, n),$$

is a cusp form of weight k with respect to $\Gamma = \mathbf{SL}(2, \mathbf{Z})$. It will then be shown that the "scalar product" (3) in the continuation to the point $s = 0$ yields the Petersson scalar product

$$\left(\sum_{n=1}^{\infty} \chi(n)\, b_n n^{\frac{k-l}{2}-1} G_k(z, n), f(z)\right).$$

Next, the arguments in 4.1 can be applied to the cusp form

$$\sum_{n=1}^{\infty} \chi(n) b_n n^{\frac{k-l}{2}-1} G_k(z, n).$$

It will also be shown that the computations in 4.5–4.12 remain valid in our case when m is replaced by $(k - l)/2 - 1$.

3. First of all, for $\mathrm{Re}(s) \gg 0$

$$\sum_{n=1}^{\infty} \chi(n)\, b_n n^{\frac{k-l}{2}-1} G_{k,s}(z, n) = \sum_{T \in \Gamma_\infty \backslash \Gamma} j(T, z)^{-k} |j(T, z)|^{-2s} (\theta^{\frac{k-l}{2}-1} g_\chi)(Tz) \tag{5}$$

(a simple change in the order of summation by virtue of absolute convergence).

We now define the (k, s)-action of $\mathbf{SL}(2, \mathbf{Z})$ on complex-valued C^∞-functions on the upper half-plane: if $T = \left(\begin{smallmatrix} a & b \\ c & d \end{smallmatrix}\right)$, $T \in \mathbf{SL}(2, \mathbf{Z})$, and $h \in C^\infty(H)$, then

$$h|_{(k,s)} [T](z) = (cz + d)^{-k} |cz + d|^{-2s} h\left(\frac{az + b}{cz + d}\right).$$

4. Proposition. *Let χ be a Dirichlet character modulo M. Then*

$$\sum_{n=1}^{\infty} \chi(n)\, b_n n^{\frac{k-l}{2}-1} G_{k,s}(z, n)$$

$$= \mathrm{Tr}_{(k,s)}^{\Gamma_0(M^2)} \left[\sum_{\gamma \in \Gamma_\infty \backslash \Gamma_0(M^2)} j(\gamma, z)^{-k} |j(\gamma, z)|^{-2s} (\theta^{\frac{k-l}{2}-1} g_\chi)(\gamma z) \right]$$

(*the trace on the right is in the sense of the* (k, s)-*action*).

Proof. We separate the summation of the series on the right side of (5) into two stages:

$$\sum_{r \in \Gamma_0(M^2) \backslash \Gamma} \; \sum_{\gamma \in \Gamma_\infty \backslash \Gamma_0(M^2)} j(\gamma r, z)^{-k} |j(\gamma r, z)|^{-2s} \theta^{\frac{k-l}{2}-1} g_\chi(\gamma r z). \tag{6}$$

We now take into account that $j(\gamma r, z) = j(\gamma, rz) \cdot j(r, z)$. Then (6) can be rewritten as follows:

$$\sum_{r \in \Gamma_0(M^2) \backslash \Gamma} \left[\sum_{\gamma \in \Gamma_\infty \backslash \Gamma_0(M^2)} j(\gamma, z)^{-k} |j(\gamma, z)|^{-2s} \theta^{\frac{k-l}{2}-1} g_\chi(\gamma z) \right] \Bigg|_{(k,s)} r. \tag{7}$$

We note that the trace in the last expression does not depend on the choice of system of representatives of $\Gamma_0(M^2) \setminus \Gamma$, since the inner sum in (7) is $\Gamma_0(M^2)$-invariant in the sense of the (k, s)-action, because it coincides with the sum of the series

$$\sum_{n=1}^{\infty} \chi(n) b_n n^{\frac{k-l}{2}-1} G^{(i\infty)}_{k,s,M^2}(z, n).$$

5. We now apply the modular properties of the functions $(\theta^{(k-l)/2-1} g_\chi)(z)$. In the right side of the equality in Proposition 4 we substitute the expression for $(\theta^{(k-l)/2-1} g_\chi)(\gamma z)$ in Proposition 3.2. As a result, the series (2) assumes the form

$$\mathrm{Tr}^{\Gamma_0(M^2)\setminus\Gamma}_{(k,s)}\left[\sum_{i=0}^{\frac{k-l}{2}-1} A^{(l)}_{\frac{k-l}{2}-1,i} \frac{(-1)^{\frac{k-l}{2}-1-i}}{\left(\frac{k-l}{2}-i\right)!} \theta^i g_\chi \cdot \tilde{\theta}^{\frac{k-l}{2}-1-i} E^{(M^2)}_{2,\chi^2,s}\right], \tag{8}$$

where the following notation has been introduced:

$$E^{(M^2)}_{2,\chi^2,s}(z) = \sum_{\gamma\in\Gamma_\infty\setminus\Gamma_0(M^2)} \frac{\chi^2(d)}{(cz+d)^2|cz+d|^{2s}}, \tag{9}$$

$$\tilde{\theta}^j E^{(M^2)}_{2,\chi^2,s}(z) = \frac{(-1)^j (j+1)!}{(2\pi i)^j} \sum_{\gamma\in\Gamma_\infty\setminus\Gamma_0(M^2)} \frac{\chi^2(d) c^j}{(cz+d)^{2+j}|cz+d|^{2s}} \tag{10}$$

(here $\gamma = \binom{a\ b}{c\ d}$). We note that the series (9) and (10) converge absolutely for $\mathrm{Re}(s) > 0$, so that (8) makes it possible to determine the sum of the series (2) for any s such that $\mathrm{Re}(s) > 0$.

The following proposition justifies the notation $\tilde{\theta}$ in (10):

6. PROPOSITION.

$$\left(\frac{1}{2\pi i}\frac{\partial}{\partial z}\right)^j E^{(M^2)}_{2,\chi^2,s}(z) = \prod_{\alpha=0}^{j-1}\left(1+\frac{s}{2+\alpha}\right)(\tilde{\theta}^j E^{(M^2)}_{2,\chi^2,s})(z).$$

PROOF. We use induction on j, using the equalities

$$\frac{\partial}{\partial z}\frac{1}{[(c\bar{z}+d)(cz+d)]^s} = -s\frac{c(c\bar{z}+d)}{|cz+d|^{2s+2}} = -\frac{cs}{(cz+d)|cz+d|^{2s}},$$

whence

$$\frac{\partial}{\partial z}\frac{\chi^2(d)c^j}{(cz+d)^{2+j}|cz+d|^{2s}} = -(2+j+s)\frac{\chi^2(d)c^{j+1}}{(cz+d)^{3+j}|cz+d|^{2s}}.$$

7. PROPOSITION. *Let χ^2 be a nonprincipal Dirichlet character modulo M^2. Then*

a) *the limit $\lim_{s\to 0} E^{(M^2)}_{2,\chi^2,s}(z)$ exists and is a modular form $E^{(M^2)}_{2,\chi^2}(z)$ of weight 2 and level M^2;*

b)
$$\lim_{s\to 0}(\tilde{\theta}^j E^{(M^2)}_{2,\chi^2,s})(z) = \theta^j E^{(M^2)}_{2,\chi^2}(z).$$

PROOF. We have

$$\begin{aligned} E^{(M^2)}_{2,\chi^2,s}(z) &= \frac{1}{2}\sum_{(Mc,d)=1} \frac{\chi^2(d)}{(M^2cz+d)^2|M^2cz+d|^{2s}} \\ &= \frac{1}{2}\sum_{\bar{d} \bmod M^2} \chi^2(\bar{d}) \sum_{\substack{c\equiv 0(M^2)\\ d\equiv\bar{d}(M^2),(c,d)=1}} \frac{1}{(cz+d)^2|cz+d|^{2s}}. \end{aligned} \tag{11}$$

We transform the inner sum in (11) with the help of the Möbius function:

$$\sum_{\substack{c\equiv 0,\ d\equiv \bar d(M^2)\\ (c,d)=1}} \frac{1}{(cz+d)^2\,|cz+d|^{2s}}$$

$$=\sum_{\substack{\bar a \bmod M^2\\ (\bar a,M)=1}}\ \sum_{\substack{n=1\\ n\equiv \bar a^{-1}(M^2)}}^{\infty} \frac{\mu(n)}{n^{2+2s}} \sum_{\substack{c\equiv 0\ (M^2)\\ d\equiv \bar a\bar d(M^2)}} \frac{1}{(cz+d)^2\,|cz+d|^{2s}}.$$

Let $0\leqslant a_1<N$ and $0\leqslant a_2<N$, $N>1$. We introduce the notation

$$G_{2,s}(z;\,a_1,\,a_2;\,N)=\sum_{m_i\equiv a_i(N)}{}' \frac{1}{(m_1z+m_2)^2\,|m_1z+m_2|^{2s}} \tag{12}$$

The series (12) converges absolutely for $\mathrm{Re}(s)>0$. We can now write

$$E_{2,\chi^2,s}(z)=\frac{1}{2}\sum_{\bar d \bmod M^2}\chi^2(d)\sum_{\substack{\bar a \bmod M^2\\ (\bar a,M)=1}}\ \sum_{\substack{n=1\\ n\equiv \bar a^{-1}(M^2)}}^{\infty}\frac{\mu(n)}{n^{2+2s}}G_{2,s}(z;\,0,\,\bar a\bar d;\,M^2). \tag{13}$$

It turns out that the limit

$$G_2(z;\,a_1,\,a_2;\,N)=\lim_{s\to 0} G_{2,s}(z;\,a_1,\,a_2;\,N)$$

exists. Here $G_2(z;\,a_1,\,a_2;\,N)$ has the following Fourier expansion:

$$\begin{aligned} G_2(z;\,a_1,\,a_2;\,N) &= -\frac{2\pi i}{N^2(z-\bar z)}+\delta\left(\frac{a_1}{N}\right)\sum_{m_2\equiv a_2(N)}{}'\frac{1}{m_2^2}\\ &\quad -\frac{4\pi^2}{N^2}\sum_{j=1}^{\infty}\sigma_1(j;\,a_1,\,a_2)\,e^{2\pi i j\frac{z}{N}}, \end{aligned} \tag{14}$$

where

$$\delta(x)=\begin{cases}1, & \text{if } x \text{ is integral},\\ 0, & \text{if } x \text{ is not integral}\end{cases}\qquad \sigma_1(j;\,a_1,\,a_2)=\sum_{d|j,\ d^{-1}j\equiv a_1(N)}\cos\left(2\pi\frac{a_2 d}{N}\right)d$$

(see [7], *Werke*, p. 472). We note that this expansion coincides with the series in 4.7 (with $k=1$), differing only by a nonholomorphic term.

Let us briefly recall the proof of these facts, due to Hecke (it will be used in what follows). We consider the function

$$\varphi(z;\,s,\,a_1,\,a_2,\,N,\,u)=\sum_{m_i\equiv a_i}{}'\frac{1}{(m_1z+m_2+Nu)^2\,|m_1z+m_2+Nu|^{2s}}.$$

We expand it in a Fourier series in $e^{2\pi i u}$ and use the Poisson summation formula to find the Fourier coefficients. Then setting $u=0$, we obtain an expression for $G_{2,s}(z;\,a_1,\,a_2;\,N)$ which shows that this function is s-integral for a fixed z:

$$\begin{aligned} G_{2,s}(z;\,a_1,\,a_2;\,N) &= \delta\left(\frac{a_1}{N}\right)\sum_{m_2\equiv a_2(N)}{}'\frac{1}{|m_2|^{2+2s}}\\ &\quad+\frac{1}{N}\sum_{m_1\equiv a_1(N)}{}'\int_{-\infty}^{\infty}\frac{du}{(m_1z+u)^2\,|m_1z+u|^{2s}}\\ &\quad+\sum_{m=1}^{\infty}\sum_{m_1\equiv a_1(N)}{}'\int_{-\infty}^{\infty}\frac{e^{-2\pi i m u}\,du}{(m_1z+a_2+Nu)^2\,|m_1z+a_2+Nu|^{2s}}. \end{aligned} \tag{15}$$

The first integral can be computed easily in terms of the Γ-function. Summing yields

$$-\frac{s\sqrt{\pi}}{y^{1+s}}\frac{\Gamma\left(s+\frac{1}{2}\right)}{\Gamma(s+1)\cdot(s+2)}\sum_{m_1\equiv a_1(N)}{}' \frac{1}{|m_1|^{1+s}},$$

and the limit as $s\to 0$ equals $-\pi/yN^2$, where $z=x+iy$. The integrals in the last sum can be defined for any complex s by deforming the contour of integration, and we have a bound for the integrals, uniform in s (see [8], formula (10), and also §9), which guarantees that the series in (15) converges absolutely.

Passing to the limit as $s\to 0$, we obtain an expansion (14) that is also the Fourier expansion of the function

$$\sum_{m_1\equiv a_1(N)}{}'\ \sum_{m_2\equiv a_2(N)}{}' \frac{1}{(m_1 z+m_2)^2}-\frac{2\pi i}{N^2(z-\bar z)}-\delta\left(\frac{a_1}{N}\right)\sum_{m_2\equiv a_2(N)}{}'\frac{1}{m_2^2} \tag{16}$$

(the iterated series in (16) converges).

It is now easy to write down the Fourier expansion of $\lim_{s\to 0} E^{(M^2)}_{2,\chi^2,s}$ explicitly. After substituting (14) into (13) and passing to the limit, we see that the nonholomorphic terms are annihilated, since χ^2 is a nonprincipal character mod M^2.

Since $E^{(M^2)}_{2,\chi^2,s}(z)$ is $\Gamma_0(M^2)$-invariant under the $(2,s)$-action, it follows that

$$E^{(M^2)}_{2,\chi^2}=\lim_{s\to 0}E^{(M^2)}_{2,\chi^2,s}$$

is a modular form of level M^2 and weight 2.

The proof of the second assertion follows from the equality

$$\left(\frac{d}{dz}\right)^i E^{(M^2)}_{2,\chi^2}(z)=\lim_{s\to 0}\left(\frac{\partial}{\partial z}\right)^i E^{(M^2)}_{2,\chi^2,s}(z)$$

and Proposition 6.

8. Corollary. *The limit*

$$\lim_{s\to 0}\operatorname{Tr}^{\Gamma_0(M^2)\setminus\Gamma}_{(k,s)}\left[\sum_{i=0}^{\frac{k-l}{2}-1}A^{(l)}_{\frac{k-l}{2}-1,i}\frac{(-1)^{\frac{k-l}{2}-1-i}}{\left(\frac{k-l}{2}-i\right)!}\theta^i g_\chi\cdot\bar\theta^{\frac{k-l}{2}-1-i}E^{(M^2)}_{2,\chi^2,s}\right]$$

exists and is a cusp form of weight k with respect to $\mathbf{SL}(2,\mathbf{Z})$.

Thus, the sum of the series (2) is defined for all s such that $\mathrm{Re}(s)>0$ and is a cusp form of weight k for $s=0$.

9. Proposition.

$$\lim_{s\to 0}\left(\sum_{n=1}^{\infty}\chi(n)b_n n^{\frac{k-l}{2}-1}G_{k,s}(z,n),\ f(z)\right)_s=\left(\sum_{n=1}^{\infty}\chi(n)b_n n^{\frac{k-l}{2}-1}G_k(z,n),\ f(z)\right).$$

PROOF. We need to justify passing to the limit under the integral sign:

$$\lim_{s\to 0}\int_D \mathrm{Tr}_{(k,s)}^{\Gamma_0(M^2)\setminus\Gamma}\left[\sum_{i=0}^{\frac{k-l}{2}-1} A^{(l)}_{\frac{k-l}{2},i}\frac{(-1)^{\frac{k-l}{2}-1-i}}{\left(\frac{k-l}{2}-i\right)!}\theta^i g_\chi(z)\cdot\tilde{\theta}^{\frac{k-l}{2}-1-i}E^{(M^2)}_{2,\chi^2,s}\right]\times\overline{f(z)}\,y^{k+s-2}dxdy=\int_D \mathrm{Tr}_k^{\Gamma_0(M^2)\setminus\Gamma}$$

$$\times\left[\sum_{i=0}^{\frac{k-l}{2}-1} A^{(l)}_{\frac{k-l}{2}-1,i}\frac{(-1)^{\frac{k-l}{2}-1-i}}{\left(\frac{k-l}{2}-i\right)!}\theta^i g_\chi(z)\cdot\theta^{\frac{k-l}{2}-1-i}E^{(M^2)}_{2,\chi^2}(z)\right]\overline{f(z)}y^{k-2}dxdy. \tag{17}$$

We can now integrate over the fundamental domain $D_0(M^2)$ of the group $\Gamma_0(M^2)$; here (17) assumes the simpler form

$$\lim_{s\to 0}\int_{D_0(M^2)}\sum_{i=0}^{\frac{k-l}{2}-1} A^{(l)}_{\frac{k-l}{2}-1,i}\frac{(-1)^{\frac{k-l}{2}-1-i}}{\left(\frac{k-l}{2}-i\right)!}\theta^i g_\chi(z)\,\tilde{\theta}^{\frac{k-l}{2}-1-i}E^{(M^2)}_{2,\chi^2,s}(z)\,\overline{f(z)}\,y^{k+s-2}dxdy$$

$$=\int_{D_0(M^2)}\sum_{i=0}^{\frac{k-l}{2}-1} A^{(l)}_{\frac{k-l}{2}-1,i}\frac{(-1)^{\frac{k-l}{2}-1-i}}{\left(\frac{k-l}{2}-i\right)!}\theta^i g_\chi(z)\,\theta^{\frac{k-l}{2}-1-i}E^{(M^2)}_{2,\chi^2}(z)\,\overline{f(z)}\,y^{k+s-2}\,dxdy.$$

Since $\int_{D_0(M^2)} y^2\,dxdy < \infty$, it suffices to verify that

$$(\theta^i g_\chi)(z)\,\tilde{\theta}^{\frac{k-l}{2}-1-i}E^{(M^2)}_{2,\chi^2,s}(z)\,f(z)\,y^{k+s}$$

converges to

$$(\theta^i g_\chi)(z)\,\theta^{\frac{k-l}{2}-1-i}E^{(M^2)}_{2,\chi^2}(z)\,\overline{f(z)}\,y^k$$

uniformly in some fundamental domain (as $s\to 0$). To do so, in turn, it suffices to show that

$$\frac{d}{ds}\left[(\theta^i g_\chi)(z)\cdot\tilde{\theta}^{\frac{k-l}{2}-1-i}E^{(M^2)}_{2,\chi^2,s}(z)\,\overline{f(z)}y^{k+s}\right] \tag{18}$$

is bounded uniformly in z in some fixed fundamental domain if s varies in a neighborhood of zero.

Substituting (15) into (13), we obtain

$$E^{(M^2)}_{2,\chi^2,s}(z)=\frac{1}{2}\sum_{\bar{d}\bmod M^2}\chi^2(\bar{d})\sum_{\bar{a}\bmod M^2}\sum_{\substack{n=1\\ n\equiv\bar{a}^{-1}(M^2)}}^{\infty}\frac{\mu(n)}{n^{2+2s}}$$

$$\times\sum_{n=1}^{\infty}\sum_{m_1\equiv 0(M^2)}\int_{-\infty}^{\infty}\frac{e^{-2\pi imu}\,du}{(m_1z+\bar{a}\bar{d}+M^2u)^2\,|\,m_1z+\bar{c}\bar{d}+M^2u\,|^2}+C(s), \tag{19}$$

where

$$C(s)=\frac{1}{2}\sum_{\bar{d}\bmod M^2}\ \sum_{\substack{\bar{a}\bmod M^2\\(\bar{a},M)=1}}\chi^2(\bar{d})\sum_{\substack{n=1\\ n\equiv\bar{a}^{-1}(M^2)}}^{\infty}\frac{\mu(n)}{n^{2+2s}}\sum_{m_2\equiv\bar{a}\bar{d}}{}'\frac{1}{|m_2|^{2+2s}}.$$

Similarly, for $j > 1$

$$\widetilde{\theta}^{j}E^{(M^2)}_{2,\chi^2,s}(z)=\frac{1}{2}\frac{(-1)^j(j+1)!}{(2\pi i)^j}\sum_{\bar{d}\bmod M^2}\chi^2(\bar{d})\sum_{\bar{a}\bmod M^2}\sum_{\substack{n=1\\ n\equiv\bar{a}^{-1}(M^2)}}^{\infty}\frac{\mu(n)}{n^{2+2s}}$$
$$\times\sum_{m=1}^{\infty}\sum_{m_1\equiv 0}{}'m_1^j\int_{-\infty}^{\infty}\frac{e^{-2\pi imu}\,du}{(m_1z+\bar{a}\bar{d}+M^2u)^{2+j}\,|m_1z+\bar{a}\bar{d}+M^2u|^{2s}}.\tag{20}$$

(This can also be deduced from Proposition 6.)

We compute the derivative on the right side of (18) using (19) and (20). Estimating the derivative now reduces to estimating the integrals

$$\int_{-\infty}^{\infty}\frac{e^{-2\pi imu}\,du}{(m_1z+\bar{a}\bar{d}+M^2u)^{2+j}\,|m_1z+\bar{a}\bar{d}+M^2u|^{2s}}$$

and

$$\int_{-\infty}^{\infty}\frac{e^{-2\pi imu}\ln|m_1z+\bar{a}\bar{d}+M^2u|^2\,du}{(m_1z+\bar{a}\bar{d}+M^2u)^{2+j}\,|m_1z+\bar{a}\bar{d}+M^2u|^{2s}}$$

(the logarithm arises in differentiating with respect to s under the integral sign in (20)) and summing these estimates over m_1 and m.

An estimate of the integral

$$I(\omega,\tau;s_1,s_2)=\int_{-\infty}^{\infty}\frac{e^{i\omega u}\,du}{(\tau+u)^{s_1}(\bar{\tau}+u)^{s_2}},\qquad \omega\in\mathbf{R},$$

is given in [8] (*Werke*, p. 393). The integral

$$I_1(\omega,\tau;s_1,s_2)=\int_{-\infty}^{\infty}\frac{e^{i\omega u}\ln|\tau+u|^2\,du}{(\tau+u)^{s_1}(\bar{\tau}+u)^{s_2}}$$

is estimated similarly. Let $\tau = x + iy$. Then

$$I_1(\omega,\tau,s_1,s_2)=\int_{-\infty}^{\infty}\frac{e^{i\omega u}\ln((x+u)^2+y^2)}{(x+u+iy)^{s_1}(x+u-iy)^{s_2}}\,du$$
$$=e^{-i\omega x}\int_{-\infty}^{\infty}\frac{e^{i\omega t}\ln(t^2+y^2)}{(t+iy)^{s_1}(t-iy)^{s_2}}\,dt$$
$$=e^{-i\omega x}y^{1-s_1-s_2}\left[\int_{-\infty}^{\infty}\frac{e^{i\omega ty}\ln(t^2+1)}{(t+i)^{s_1}(t-i)^{s_2}}\,dt-2\ln y\int_{-\infty}^{\infty}\frac{e^{i\omega ty}\,dt}{(t+i)^{s_1}(t-i)^{s_2}}\right].$$

Now, if $\omega > 0$, we deform the contour of integration in the upper half-plane; but if

$\omega < 0$, in the lower half-plane:

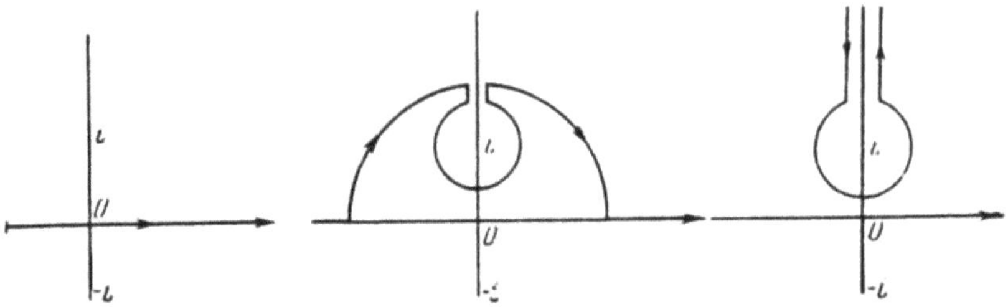

The desired estimate is as follows: if s_1 and s_2 are contained in some fixed disk, then

$$|I_1(\tau, \omega; s_1, s_2)| < ay^{-b}e^{-\frac{|\omega|y}{4}} \int_{-\infty}^{\infty} e^{-|\omega|t y}(t+c)^d\,dt,$$

where a, b, c and d are positive and do not depend on the parameters.

Substituting these estimates into the expression for the derivative, after summing we obtain the following bound for the modulus (18):

$$C_1|\theta^l g_\chi(z)\ \overline{f(z)}|\,y^k \max(y^{C_2}, y^{-C_2})$$

with some C_1, $C_2 > 0$ and $z = x + iy$.

Now, since $f(z)$ is a cusp form with respect to $\Gamma_0(M^2)$,

$$|f(z)| = \begin{cases} O(e^{-2\pi y}) & \text{as } y \to \infty, \\ O(e^{-\pi/y}) & \text{as } y \to 0 \text{ (within the bounds of the fundamental domain)}. \end{cases}$$

It follows easily that the derivative (18) is bounded. This proves the proposition.

We now note that the computations in 4.5–4.12 remain valid in our case when m is replaced by $(k-l)/2-1$. This stems from the fact that the Fourier expansion of $G_2(z; a_1, a_2; N)$ coincides with the Fourier series (12) in 4.7 (with $k = 1$) up to the nonholomorphic term $-2\pi i/N^2(z-\bar{z})$, which, however, does not affect the subsequent computations, since it disappears in the summation (in the exceptional case it is important that the character χ^2 be nonprincipal; see Lemma 2.7). This proves Theorem 1.4 completely.

Received 6/OCT/76

Bibliography

1. M. M. Višik, *Non-Archimedean measures connected with Dirichlet series*, Mat. Sb. **99 (141)** (1976), 248–260; English transl. in Math. USSR Sb. **28** (1976).

2. M. M. Višik and Ju. I. Manin, *p-adic Hecke series of imaginary quadratic fields*, Mat. Sb. **95 (137)** (1974), 357–383; English transl. in Math. USSR Sb. **24** (1974).

3. R. C. Gunning, *Lectures on modular forms*, Ann. of Math. Studies, No. 48, Princeton Univ. Press, Princeton, N. J., 1962.

4. Pierre Deligne, *Formes modulaires et représentations l-adiques*, Séminaire Bourbaki 1968/69, Exposé 355, Lecture Notes in Math., vol. 179, Springer-Verlag, Berlin and New York, 1971.

5. ———, *La conjecture de Weil*. I, Inst. Hautes Études Sci. Publ. Math. No. 43 (1974), 273–307.

6. Koji Doi and Hidehisa Naganuma, *On the functional equation of certain Dirichlet series*, Invent. Math. **9** (1969/70), 1–14.

7. Erich Hecke, *Theorie der Eisensteinschen Reihen höherer Stufe und ihre Anwendung auf Funktionentheorie und Arithmetik*, Abh. Math. Sem. Univ. Hamburg 5 (1926/27), 199–224; reprinted in his *Mathematische Werke*, Vandenhoeck & Ruprecht, Göttingen, 1959.

8. ______, *Analytische Funktionen und algebraische Zahlen*. II, Abh. Math. Sem. Univ. Hamburg 3 (1923/24), 213–236; reprinted in his *Mathematische Werke*, Vandenhoeck & Ruprecht, Göttingen, 1959.

9. ______, *Darstellung von Klassenzahlen als Perioden von Integralen* 3. *Gattung aus dem Gebiet der elliptischen Modulfunktionen*, Abh. Math. Sem. Univ. Hamburg 4 (1925/26), 211–223; reprinted in his *Mathematische Werke*, Vandenhoeck & Ruprecht, Göttingen, 1959.

10. Hervé Jacquet, *Automorphic forms on GL(2)*, Lecture Notes in Math., vol. 278, Springer-Verlag, Berlin and New York, 1972.

11. Tomio Kubota and Heinrich Wolfgang Leopoldt, *Eine p-adische Theorie der Zetawerte*. I, J. Reine Angew. Math. **214/215** (1964), 328–339.

12. Wen Ch'ing Winnie Li, *Newforms and functional equations*, Math. Ann. **212** (1975), 285–315.

13. Stephen Lichtenbaum, *On the values of zeta and L-functions*. I, Ann. of Math. (2) **96** (1972), 338–360.

14. Ju. I. Manin, *Periods of cusp forms and p-adic Hecke series*, Mat. Sb. **92** (**134**) (1973), 378–401; English transl. in Math. USSR Sb. **21** (1973).

15. ______, *The values of p-adic Hecke series at integer points of the critical strip*, Mat. Sb. **93** (**135**) (1974), 621–626; English transl. in Math. USSR Sb. **22** (1974).

16. ______, *Non-Archimedean integration and Jacquet-Langlands p-adic L-functions*, Uspehi Mat. Nauk **31** (1976), no 1 (187), 5–57; English transl. in Russian Math. Surveys **31** (1976).

17. B. Mazur and P. Swinnerton-Dyer, *Arithmetic of Weil curves*, Invent. Math. **25** (1974), 1–61.

18. A. P. Ogg, *On a convolution of L-series*, Invent. Math. **7** (1969), 297–312.

19. R. A. Rankin, *The scalar product of modular forms*, Proc. London Math. Soc. (3) **2** (1952), 198–217.

20. ______, *Contributions to the theory of Ramanujan's function $\tau(n)$ and similar arithmetical functions*. I, II, Proc. Cambridge Philos. Soc. **35** (1939), 351–356, 357–372.

21. Atle Selberg, *Bemerkung über eine Dirichletsche Reihe, die mit der Theorie der Modulformen nahe verbunden ist*, Arch. Math. Naturvid. **43** (1940), 47–50.

22. Carl Ludwig Siegel, *Lectures on advanced analytic number theory*, Tata Inst. Fund. Res., Bombay, 1965.

23. Gorô Shimura, *Introduction to the arithmetic theory of automorphic functions*, Iwanami Shoten, Tokyo; Princeton Univ. Press, Princeton, N. J., 1971.

24. Don Zagier, *Modular forms associated to real quadratic fields*, Invent. Math. **30** (1975), 1–46.

Translated by R. LENET

Invent. math. 95, 421–435 (1989)

Inventiones mathematicae

Rational points of bounded height on Fano varieties

Dedicated to A.I. Kostrikin on the occasion of his 60th birthday

Jens Franke[1], Yuri I. Manin[2], and Yuri Tschinkel[3]

[1] Universität Jena, Sekt. Math., Universitäts-Hochhaus 17.OG, DDR-6900 Jena and Dept. of Math., Moscow University

[2] Steklov Math. Institute, 42 Vavilova, 117966 GSP-1, Moscow, USSR

[3] Dept. of Math., Moscow University, 117234, Moscow, USSR

Introduction

Let F be a finite extension of $\mathbb{Q}$ endowed with a system of norms $|\cdot|_v$ such that the product formula $\prod_v |a|_v = 1$ is valid for all $a \in F$, and $\prod_{v/p} |p|_v = p^{-1}$ for a prime $p \in \mathbb{Z}$. Let V be an algebraic variety defined over F and $\mathbb{L}$ a metrized line bundle on V, i.e., a system $(L, |\cdot|_v)$ consisting of a line bundle L and a family of Banach v-adic metrics on $L \otimes F_v$ for all v's satisfying well-known conditions (cf. [1], Chap. VI). Then we can define a height function h_L: $X(F) \to \mathbb{R}$ by the formula $h_{\mathbb{L}}(x) = \prod_v |s(x)|_v^{-1}$, where s is a local section of L non-vanishing at x. Our valuations are normed in such a way that h_L remains invariant with respect to base changes $F' \supset F$.

Assume that L is ample and put

$$N(V, \mathbb{L}, H) = \mathrm{card}\{x \in V(F) \mid h_{\mathbb{L}}(x) \leqq H\}.$$

The asymptotic behavior of $N(V, \mathbb{L}, H)$ as $H \to \infty$ is a crude but essential characteristics of arithmetical properties of V. It is therefore desirable to understand its connections with algebro-geometric properties of (V, L).

Projective algebraic varieties V are divided into three large classes with respect to the behaviour of their canonical bundle ω.

(I) General type varieties, i.e., varieties with ample ω. A bold generalization of Mordell's conjecture states that the F-points on a general type variety lie on a proper Zariski-closed subset (cf. [1], p. 349, where this conjecture is deduced from a more precise Vojta's conjecture). Therefore, if $V(F)$ is infinite, there should be a proper subvariety of V which contains all F-points and whose irreducible components are not of general type. Thus the investigation of $V(F)$ may be subdivided into two steps.

a) Determine (irreducible components of) $W=\overline{V(F)}$ (Zariski closure).

b) Investigate $W(F)$.

It is now known that the moduli space M_g of stable curves of genus g is of general type for $g \geqq 24$ ([2]). A tantalizing problem arises to describe $\overline{M_g(\mathbb{Q})}$ and its part corresponding to smooth curves. A similar problem can be posed about the moduli space of principally polarized abelian varieties A_g.

We are not aware of any conjectures about values of $\dim(\overline{M_g(\mathbb{Q})})$, $\dim(\overline{A_g(\mathbb{Q})})$, and asymptotics of $N(M_g, \omega, H)$ and $N(A_g, \omega, H)$ (ω carries a natural metric investigated by Faltings).

(II) Intermediate type varieties, for which neither ω, nor ω^{-1} is ample. Arithmetics of this class is virtually unexplored, with one very important exception, that of abelian varieties. From the Mordell-Weil theorem and a refined Néron-Tate height theory one deduces that

$$N(A, \mathbb{L}, H) \sim c(\log H)^{r/2}, \tag{0.1}$$

where $r=rk(A(F))$ and c is a constant that can be expressed through $\mathrm{card}(A(F)_{\mathrm{tors}})$ and the volume of the ellipsoid $\hat{h}_L \leqq 1$ in $\mathbb{R} \otimes A(F)$. The famous Birch-Swinnerton-Dyer conjecture connects (0.1) with the behaviour of the L-function of A at the center of the critical strip.

(III) Fano varieties, i.e., varieties with ample ω^{-1} The simplest Fano variety is $\mathbb{P}^n_F$. Schanuel [3] proved that

$$N(\mathbb{P}^n_F, O(1), H) \sim c' H^{n+1}$$

Since $\omega_{\mathbb{P}^n}^{-1} = O(n+1)$, it follows that $N(\mathbb{P}^n_F, \omega^{-1}, H) \sim cH$. Prompted by this observation and some numerical data about the distribution of rational points on $x_0^3+2x_1^3+3x_2^3+4x_3^3=0$ (cf. Appendix) Yu.I. Manin conjectured that the following asymptotic behaviour

$$N(V, \omega_V^{-1}, H) \sim cH(\log H)^t \tag{0.2}$$

should be typical for Fano manifolds V with dense $V(F)$, excluding some degenerate cases (cf. below). Including a power of logarithm is necessary since it already occurs for products of projective spaces. The available information is compatible with the further conjecture that $t=rk(\mathrm{Pic}\, V)-1$.

This article is devoted to the proof of the following facts, giving some support to these conjectures.

a) Asymptotic (0.2) with error term $O(H(\log H)^{t-1})$ is stable with respect to the direct product.

b) Asymptotic (0.2) is consistent with the predictions of the Hardy-Littlewood method for complete intersections in $\mathbb{P}^n$

c) Asymptotic (0.2) is valid for generalized flag manifolds $P \backslash G$ where G is a semisimple algebraic group and P is a parabolic subgroup.

We prove a) and b) in § 1. The most technical part c) is due to J. Franke who made the crucial remark that the Dirichlet series

$$Z(s)=\sum_{x\in(P\backslash G)(F)} h_{\omega^{-1}}(x)^{-s}$$

can be identified with one of the Langlands-Eisenstein series [4]. The use of the heavy machinery of [4] is perhaps reminiscent of shooting sparrows with cannons. However, it allows one to borrow heavily from Langlands work, who established most of the analytical properties of $Z(s)$ needed to deduce (0.2) and in fact a considerably sharper result (cf. Sect. 2, Corollary to Theorem 5).

We note finally that (0.2) cannot be true without some further non-degeneracy assumptions. For example, consider the case of a smooth cubic surface $V\subset\mathbb{P}_F^3$. If one of the 27 lines on it is F-rational, we have by Schanuel's theorem

$$N(V,\omega_V^{-1},H)\geqq N(\mathbb{P}^1,\omega_V^{-1}|_{\mathbb{P}^1},H)\geqq cH^2$$

since $\omega_V^{-1}|_{\mathbb{P}^1}\simeq O_{\mathbb{P}^1}(1)$.

Generalizing this counter-example consider the following situation. Let $W\subset V$ be Fano manifolds. Suppose that for some $a>1$ we have $(\omega_V^{-1}|_W)^{\otimes a}\simeq\omega_W^{-1}$. Then (0.2) cannot be true for V and W simultaneously.

Would it suffice to exclude such "pathology" in order to ensure (0.2) or is it necessary also to assume that $\deg_{\omega^{-1}}V$ is sufficiently large?

Unfortunately, we do not know answers even for cubic or more general del Pezzo surfaces.

§ 1. Products and complete intersections

1. Products. Since $\omega_{V\times W}=\omega_V\boxtimes\omega_W$ and $h_{L_1\boxtimes L_2}(x,y)=h_{L_1}(x)\,h_{L_2}(y)$, the stability of the asymptotic behaviour (0.2) follows at once from the following elementary statement, which is probably well-known. We sketch a proof for the sake of completeness.

Let λ: $0<\lambda_1\leqq\lambda_2\leqq\dots$ and μ: $0<\mu_1\leqq\mu_2\leqq\dots$ be two real increasing sequences. Put $\lambda\mu=\{\lambda_i\mu_j$, ordered increasingly$\}$. Put $N_\lambda(H)=\operatorname{card}\{i\,|\,\lambda_i<H\}$ and similarly for $N_\lambda(H)$, $N_{\lambda\mu}(H)$.

2. Proposition. *Assume that*

$$N_\lambda(H)=c_\lambda H\log^s H+O(H\log^{s-1}H),\qquad r,s\geq 0$$
$$N_\mu(H)=c_\mu H\log^r H+O(H\log^{r-1}H).$$

Then

$$N_{\lambda\mu}(H)=B(r+1,s+1)\,c_\lambda c_\mu H\log^{r+s+1}H+O(H\log^{r+s}H),$$

where B is the beta-function.

Proof. We have

$$N_{\lambda\mu}(H)=\sum_{i=1}^{N_\lambda(H/\mu_1)} N_\mu(H/\lambda_i)=c_\mu\sum_{i=1}^{N_\lambda(H/\mu_1)}\frac{H}{\lambda_i}\log^r\left(\frac{H}{\lambda_i}\right)$$
$$+O\left(\sum_{i=1}^{N_\lambda(H/\mu_1)}\frac{H}{\lambda_i}\log^{r-1}\left(\frac{H}{\lambda_i}\right)\right). \tag{1.1}$$

Here the error term is of the same structure as the leading one with the logarithm power diminished by one. The same effect occurs repeatedly in the subsequent estimates. Therefore we shall write out explicitely only leading terms. Put

$$a(j)=\mathrm{card}\{i\,|\,\lambda_1+j\leqq\lambda_i<\lambda_1+j+1\},\qquad N=[H/\mu_1-\lambda_1]+1.$$

Then

$$\sum_{i=1}^{N_\lambda(H/\mu_1)}\frac{H}{\lambda_i}\log^r\left(\frac{H}{\lambda_i}\right)\sim\sum_{j=1}^{N}a(j)\frac{H}{j}\log^r(H/j). \tag{1.2}$$

In order to calculate (1.2), we use the Abel summation trick and estimate

$$\sum_{j=0}^{M}a(j)=N_\lambda(M+\lambda_1+1)=c_\lambda M\log^s M+O(M\log^{s-1}M).$$

In this way we obtain for the leading term of (1.1) the asymptotics

$$c_\lambda c_\mu H\sum_{j=1}^{N}j\log^r j\int_j^{j+1}x^{-2}\log^s\left(\frac{H}{x}\right)dx.$$

Finally, using a mean value theorem we can replace the sum by the integral

$$\int_1^{N+1}\log^r x(\log H-\log x)^s\,d(\log x)$$
$$\sim B(r+1,s+1)(\log H)^{r+s+1}+O((\log H)^{r+s}).$$

3. *Hardy-Littlewood method.* The adelic version of the classical Hardy-Littlewood method is explained in a very clear and conceptual way in [9] following important previous work [6–8]. We shall use here its barest rudiments, namely, a formal description of the "singular series" which in favorable cases can be proved to dominate a weighted sum of delta-functions supported by F-points of a Fano manifold V (or, rather, a cone over V).

Denote by A_F the adèle ring of F. Let W be an affine variety over F. The general Hardy-Littlewood formula is an identity

$$\sum_{x\in W(F)}\varphi(x)=\int_{W(A_F)}\varphi(x)\,\omega_{HL}(x)+R(\varphi)=P(\varphi)+R(\varphi), \tag{1.3}$$

where $\omega_{HL}(x)$ is a measure on $W(A_F)$ which will be explicitely described below in certain cases and φ belongs to a certain space of summable functions on

$W(A_F)$. The integral $P(\varphi)$ at the r.h.s. is the singular series, and getting good estimates of the error term $R(\varphi)$ is a real job for a number-theorist. We have nothing new to say about it.

Here is our small observation. Let W be a complete intersection in F^n, given by forms of degree

$$d_1, \ldots, d_m \colon f_1(x_1, \ldots, x_n) = \ldots = f_m(x_1, \ldots, x_n) = 0.$$

This means, in particular, that $W_0 = W \setminus \{0\}$ is a pointed cone over the projective complete intersection $V \subset \mathbb{P}_F^{n-1}$ defined by the same equations in homogeneous coordinates. We shall assume that V is a smooth projective manifold.

4. Proposition. a) $\omega_V^{-1} = O(n - \sum d_i)$. *In particular, V is a Fano manifold if and only if $n > \sum d_i$.*

b) *For a real $\tau > 0$ put $\tilde{\tau} = (\tilde{\tau}_v) \in A_F$, where $\tilde{\tau}_v = \tau^{[F_v:Q_v]/[F:Q]}$ if $v|\infty$, $\tilde{\tau}_v = 1$ otherwise. Let φ be a function on $W(A_F)$ and $\varphi_\tau(x_1, \ldots, x_n) = \varphi(\tilde{\tau}^{-1} x_1, \ldots, \tilde{\tau}^{-1} x_n)$. Assume that $P(\varphi_\tau)$ makes sense for all τ. Then*

$$P(\varphi_\tau) = \tau^{n - \Sigma d_i} P(\varphi_1). \tag{1.4}$$

Proof. Part a) is well known. To prove b), note that

$$\int_{W_0(A_F)} \varphi_\tau(x)\, \omega_{HL}(x) = \int_{W_0(A_F)} \varphi_1(x)\, \omega_{HL}(\tilde{\tau} x).$$

Hence it suffices to check that

$$\omega_{HL}(\tilde{\tau} x) = \tau^{n - \Sigma d_i} \omega_{HL}(x).$$

But this follows from the following description of ω_{HL} (cf. [9]). Consider $f = (f_1, \ldots, f_m)$ as a map $f\colon A^n \to A^m$ of affine spaces over F. Denote the standard volume forms by $\omega_n = dx_1 \wedge \ldots \wedge dx_n$, $\omega_m = dy_1 \wedge \ldots \wedge dy_m$. Define locally on W_0 a form Ω_{HL} by the following prescription:

$$\omega_n = \theta \wedge f^*(\omega_m), \qquad \Omega_{HL} = \theta|_{W_0}.$$

Then ω_{HL} is a measure constructed in the standard way from Ω_{HL} and the Tate-Tamagawa measure on A_F. (cf. [6, 8]). Now we have

$$\omega_n(\tilde{\tau} x) = \tau^n \omega_n(x), \qquad f^*(\omega_m)(\tilde{\tau} x) = \tau^{\Sigma d_i} f^*(\omega_m)(x)$$

since $f^*(\omega_m) = df_1 \wedge \ldots \wedge df_m$. This proves (1.4).

Finally, we must explain in which sense (1.4) conforms with conjecture (0.2). Assume for simplicity that φ_1 is the restriction to $W(A_F)$ of the characteristic function of the product of balls at infinite places and v-adic integers at finite places. Then $\sum_{x \in W_0(F)} \varphi_\tau(x)$ calculates the number of F-points of W_0 with integer coordinates whose conjugates lie in the expanded archimedean balls. For $F = \mathbb{Q}$ this is clearly equivalent to calculating the number of F-points on V with $O(1)$-height $\leq \tau$, with a correcting factor $(1 + o(1))/2\zeta(m)$ (since one really needs only points of $W_0(\mathbb{Q})$ with co-prime coordinates). Hopefully, one can use the same trick for general F. Then (1.4) will show that the number of points with $O(1)$-height $\leqq \tau$ on V grows as $c\tau^{n-\Sigma d_i}$ if the singular series $P(\varphi)$ is an asymptotic approximation to $\sum_x \varphi_r(x)$. Finally, part a) of the Proposition 4 allows one to conclude that $N(V, \omega_V^{-1}, H)$ grows linearly with H.

§ 2. Flag manifolds

1. Notation. Let G be a semisimple linear algebraic group over an algebraic number field F. We fix once and for all a minimal F-rational parabolic subgroup $P_0 \subset G$. Let P be a standard (i.e., containing P_0) parabolic subgroup. We denote by $V = P \backslash G$ the generalized flag manifold and by $\pi: G \to V$ the canonical projection.

By X^* (resp. X_*) we shall denote the group of characters (resp. cocharacters) defined over F. Any element $\chi \in X^*(P)$ defines a line bundle L_χ on V, whose sections on an open subset $U \subset V$ are

$$\Gamma(U, L_\chi) = \{f \in \Gamma(\pi^{-1}(U), O_G) \mid f(pg) = \chi(p) f(g) \text{ for all } p \in P, g \in G\}. \tag{2.1}$$

The correspondence $\chi \to L_\chi$ defines an embedding (because of $X^*(G)=0$ and Rosenlicht's theorem) of finite index (since $\operatorname{Pic}(G)$ is finite) $X^*(P) \to \operatorname{Pic} V$, hence $rk \operatorname{Pic} V = rk\, X^*(P)$.

Let A_F be the adele ring of F. We choose a maximal compact subgroup $K = \prod_v K_v \subset G(A_F)$ in such a way that $G(A_F) = P_0(A_F)K$. The bundles $L_\chi \otimes F_v$ on $V \otimes F_v$ then carry natural K_v-invariant v-adic metrics defined by the following rule. Let s be a section of L_χ on a neighbourhood of $x \in V(F_v)$ corresponding to a function f as in (1). We choose $k \in K_v$ with $\pi(k) = x$ and put

$$|s|_{x,v} = |f(k)|_v .$$

Then a height function on $V(F)$ is defined by

$$h_\chi(x) = h_{L_\chi}(x) = \prod_v |s|_{x,v}^{-1},$$

where s is a section of L_χ in a neighbourhood of x which is F-rational and non-vanishing at x. We see that a choice of K determines a metrization of L_χ.

We recall now how to describe the anticanonical bundle on V in group-theoretical terms. Let $M \supseteq M_0$ be Levi factors of P and P_0, and let $A \subseteq A_0$ be their split components. Then $X^*(M) = X^*(P)$ is a sublattice of finite index in $X^*(A)$. Let $\mathfrak{a} = X_*(A) \otimes_Z \mathbb{R}$, $\check{\mathfrak{a}} = X^*(A) \otimes_Z \mathbb{R}$. Denote by $\mathfrak{e}$ the Lie algebra of the radical of P. Let $\rho = \rho_P$ be one half the sum of the roots of A in $\mathfrak{e}$ counted with multiplicity equal to the dimension of their eigenspaces. Then $L_{-2\rho}$ is the anticanonical bundle of V.

Our main tool will be the zeta-function

$$Z_\chi(s)=\sum_{x\in V(F)} h_\chi(x)^{-s} \tag{2.2}$$

corresponding to $\chi=-2\rho$. In order to establish its analytical properties we shall express it through Eisenstein series [4].

2. Eisenstein series. For every v, we define a function $H_{P,K_v}(g)\in\mathfrak{a}$ of $g\in G(F_v)$. Consider the Iwasawa decomposition $g=pk$, $p\in P_0(F_v)$, $k\in K_v$ and put for any $\chi\in X^*(P)$:

$$\exp(\langle H_{P,K_v}(g),\chi\rangle)=|\chi(p)|_v.$$

The corresponding global function is then defined by

$$H_P(g)=\sum_v H_{P,K_v}(g_v),\qquad g=(g_v)\in G(A_F).$$

We note that the sum is finite. If confusions are impossible, the subscript K is omitted.

We now put for $\lambda\in\check{\mathfrak{a}}\otimes\mathbb{C}$, $g\in G(A)$

$$E_P^G(\lambda,g)=\sum_{\gamma\in P(F)\backslash G(F)}\exp(\langle\lambda+\rho_P,H_P(\gamma g)\rangle). \tag{2.3}$$

By the remarks following the proof of Lemma 4.1 in [4], this series absolutely converges if $\operatorname{Re}\lambda\in\rho_P+\check{\mathfrak{a}}^+$, where $\check{\mathfrak{a}}^+$ is defined in the following way.

Let $\mathfrak{a}_0=X_*(A_0)\otimes_{\mathbb{Z}}\mathbb{R}$, $\check{\mathfrak{a}}_0=X^*(A_0)\otimes\mathbb{R}$. Restriction of characters defines a natural embedding $\check{\mathfrak{a}}\to\check{\mathfrak{a}}_0$ and a natural projection $\mathfrak{a}_0\to\mathfrak{a}$. Furthermore, since $A\subset A_0$, there is a canonical embedding $\mathfrak{a}\to\mathfrak{a}_0$ which is a section of the projection defined above. Hence we get direct sum decompositions $\mathfrak{a}_0=\mathfrak{a}\oplus\mathfrak{a}_0^P$, $\check{\mathfrak{a}}_0=\check{\mathfrak{a}}\oplus\check{\mathfrak{a}}_0^P$. Let Φ_0 be the set of roots of A_0 in G, and let $\Delta_0\subset\Phi_0$ be the set of simple positive roots in the cone defined by P_0. It is known that Φ_0 is a root system. For every $\alpha\in\Phi_0$, we denote by $\check{\alpha}\in\mathfrak{a}_0$ the corresponding coroot. Put $\check{\mathfrak{a}}_0^+=\{x\in\check{\mathfrak{a}}_0\,|\,\langle x,\check{\alpha}\rangle>0$ for every $\alpha\in\Delta_0\}$. Finally, let

$$\mathfrak{a}^+=\text{interior of}\,(\check{\mathfrak{a}}\cap\text{closure of}\,\check{\mathfrak{a}}_0^+).$$

By the results of Langlands, the Eisenstein series (2.3) has a meromorphic continuation whose properties we shall recall, but first we establish how (2.2) is related to (2.3).

3. Proposition. *The zeta-function* (2.2) *absolutely converges for* $\chi\in X^*(P)\cap(-\check{\mathfrak{a}}^+)$ *(corresponding to ample bundles) and* $-(\operatorname{Re}s)\chi\in\check{\mathfrak{a}}^++2\rho$ *and is equal to* $E_P^G(-(s\chi+\rho_P),e)$, *where* $e\in G(A_F)$ *is the identity element.*

Proof. Let $g\in G(F)\subset G(A_F)$. Let us check that

$$h_\chi(\pi(g))=\exp(\langle H_P(g),\chi\rangle).$$

In fact, if $g=pk$, $p\in P_0(A_F)$, $k\in K$ and f is a function as in (2.1) defining a F-rational section of L_χ in a neighbourhood of $x=\pi(g)$ with $f(g)\neq 0$ we have

$$|s|_{v,\pi(g)}=|f(k)|_v=|\chi(p)|_v^{-1}\,|f(g)|_v.$$

Therefore, using the product formula, we get

$$\begin{aligned} h_\chi(\pi(g)) &= \prod_v |s|_{v,\pi(g)}^{-1}=\prod_v \exp(\langle H_{P,K_v}(g),\chi\rangle)\,|f(g)|_v^{-1} \\ &= \exp(\langle H_P(g),\chi\rangle) \end{aligned}$$

since $f(g)\in F^*$. Comparing this with (2.3) we get our result since $\pi\colon G\to V$ induces a surjection on F-rational points.

4. *Remarks.* a) Applying this proposition to the anticanonical class $L_{-2\rho}$, we get

$$Z_{-2\rho}(s)=\sum_{x\in V(F)} h_{-2\rho}(x)^{-s}=E_P^G((2s-1)\rho_P,e).$$

b) We can change K by putting $K'=pKp^{-1}$ for $p\in P_0(A_F)$. Indicating by a prime the corresponding zeta-function we get

$$Z'_\chi(s)=\exp(-s\langle\chi,H_P(p)\rangle)\,E_P^G(-(s\chi+\rho),p).$$

c) The constant term of E_P^G (which is comparatively easy to compute) represents a mean value of a Z-function over several choices of K.

5. Theorem. *The anticanonical zeta-function $Z_{-2\rho}(s)$ has the following properties:*

a) *It admits a meromorphic continuation to the whole complex s-plane.*

b) *It is holomorphic in the domain* $\mathrm{Re}(s)>1$, *has a pole of order* $t=rk(P)=rk(\mathrm{Pic}\,V)$ *at* $s=1$, *and no other poles with* $\mathrm{Re}(s)=1$.

c) *It has no singularities on the line* $\mathrm{Re}(s)=1/2$.

Corollary. *There is a polynomial $p(x)$ of degree t such that*

$$N(V,L_{-2\rho},H)=\mathrm{card}\{x\in V(F)\,|\,h_{L_{-2\rho}}(x)\leqq H\}=H\,p(\log H)+o(H).$$

This follows directly from the theorem by a standard Tauberian argument. It is possible to obtain for the error-term an estimate of the type $H^{1-\varepsilon}$. Comparing with the Schanuel theorem [3], *one may expect that the best value of ε is* $[F:Q]^{-1}\dim V^{-1}$.

The leading coefficient of p can be computed in terms of residua of scattering operators (cf. (2.11) below). In the case of Chevalley groups, the result can be made quite explicit (cf. (2.12)).

6. *Proof.* Part a) of the theorem follows from Proposition 3 and the general theory of Eisenstein series (cf. [4], Appendix II). In our situation, the meromorphic continuation of E_P^G is fairly easy since it is clear how to express E_P^G as a residue of $E_{P_0}^G$. The first part of b) also follows from Proposition 3. In

order to prove the second part of b) we need some results of the theory of cuspidal Eisenstein series.

Let $w \in W$, the Weyl group of Φ_0. It has a F-rational representative $w' \in N_G(A_0)$ (the normalizer of A_0 in G). We put

$$c(w, \lambda) = \int_{w' N_0(A_F) w'^{-1} \cap N_0(A_F) \backslash N_0(A_F)} \exp(\langle H_{P_0}(w'^{-1} n), \lambda + \rho_{P_0} \rangle)\, dn,$$

where N_0 is the radical of P_0. The Haar measures are normalized by

$$\int_{N_0(F) \backslash N_0(A_F)} dn = 1$$

and

$$\int_{N_0(F) \cap w' N_0(F) w'^{-1} \backslash N_0(A_F) \cap w' N_0(A_F) w'^{-1}} dn = 1.$$

Then

$$\int_{N_0(F) \backslash N_0(A_F)} E_{P_0}(\lambda, ng)\, dn = \sum_{w \in W} c(w, \lambda) \exp(\langle H_{P_0}(g), w\lambda + \rho_{P_0} \rangle).$$

By results of Langlands ([4], § 6), the functions $c(w, \cdot)$ have meromorphic continuations. They satisfy functional equations

$$c(w, \lambda)\, E^G_{P_0}(w\lambda) = E^G_{P_0}(\lambda),$$

$$c(wv, \lambda) = c(w, v\lambda)\, c(v, \lambda).$$

If we consider the partial Eisenstein series

$$E^G_{P_0}(\lambda, g) = \sum_{\gamma \in P_0(F) \backslash P(F)} \exp(\langle H_{P_0}(\gamma g), \lambda + \rho_{P_0} \rangle),$$

then

$$\int_{N_0(F) \backslash N_0(A_F)} E^P_{P_0}(\lambda, ng)\, dn = \sum_{w \in W^A} c(w, \lambda) \exp(\langle H_{P_0}(g), w\lambda + \rho_{P_0} \rangle),$$

where $W^A = \{w \in W \mid w \text{ identically acts on } \mathfrak{a}\}$. Furthermore,

$$c(w, \lambda + \vartheta) = c(w, \lambda) \tag{2.5}$$

if $w \in W^A$ and $\vartheta \in \check{\mathfrak{a}}$.

The following lemma seems to be well-known to the experts, but for the convenience of the reader we add a proof.

7. Lemma. *The singular hyperplanes of $c(w, \cdot)$ containing ρ_{P_0} are precisely the hyperplanes*

$$\langle \check{\alpha}, \lambda - \rho_{P_0} \rangle = 0, \tag{2.6}$$

where α is a simple positive root such that $w\alpha$ is a negative root. Their multiplicity is one. No other singular divisor intersects $\rho_{P_0} + i\check{\mathfrak{a}}_0$.

Proof of the lemma. First we consider the case $w = s_\alpha$ (the reflection with respect to α) with $\alpha \in \Delta_0$. By (2.5), (2.6) is the only possible singular hyperplane containing ρ_{P_0}. To prove that it is indeed singular, we may by (2.5) assume that α is the only simple root, i.e., that the F-rank of G is one. Since the constant function is square integrable on $G(F)\backslash G(A_F)$ and orthogonal to the space of cusp forms, the process of moving the contour of integration described in [4], § 7 proves that (2.6) must be singular. (In our situation, the difficult theory of [4, § 7] is superfluous and the process is quite similar to the well-known case $G=SL_2$). The fact that the pole (2.6) must be simple has been proved in [5], Lemma 98. The fact that $c(s_\alpha, \lambda)$ has no more poles on $\rho_{P_0}+i\check{\mathfrak{a}}_0$ has been proved in [4], p. 128. or [5], Theorem 7.1.

In the general case, we fix a minimal representation

$$w=s_{\alpha_1}\dots s_{\alpha_k}, \qquad \alpha_i\in\Delta_0. \tag{2.7}$$

We put $w_j=s_{\alpha_{j+1}}\dots s_{\alpha_k}$. Then

$$c(w,\lambda)=c(s_{\alpha_1}, w_1\lambda)\dots c(s_{\alpha_k},\lambda).$$

We need the following sublemma:

8. Sublemma. Let ν be the dimension of $\mathfrak{e}_\beta$, the eigenspace of the root β. If $\alpha\in\Phi_0$ is a positive root with $\alpha/2\notin\Phi_0$, then

$$\langle\check{\alpha}, \rho_{P_0}\rangle\geqq \nu_\alpha+2\nu_{2\alpha} \tag{2.8}$$

and the equality occurs if and only if $\alpha\in\Delta_0$.

Proof of the sublemma. We compare the expressions

$$s_\alpha\rho_{P_0}=\rho_{P_0}-\alpha\langle\check{\alpha},\rho_{P_0}\rangle$$

and

$$s_\alpha\rho_{P_0}=\rho_{P_0}-\sum_{\beta\in\Phi_0^+,\, s_\alpha\beta\notin\Phi_0^+}\nu_\beta\beta,$$

where Φ_0^+ is the set of positive roots. The inequality (2.8) follows at once. The equality occurs if and only if α and 2α are the only positive roots β with $s_\alpha\beta\in -\Phi_0^+$, i.e., if the length of s_α is one. In view of $\alpha/2\notin\Phi_0$, this is equivalent to $\alpha\in\Delta_0$. The proof of the sublemma is complete.

By the minimality of the expression (2.7), we have $w_j^{-1}\alpha_j\in\Phi_0^+$. The sublemma implies

$$\langle\check{\alpha}_j, w_j\rho_{P_0}\rangle\geqq\langle\check{\alpha}_j,\rho_{P_0}\rangle,$$

and equality occurs if and only if $w_j^{-1}\alpha_j\in\Delta_0$. From (2.5) and the convergence assertion in [4, Lemma 4.1] or [5, Lemma 23] it follows that the factor $c(s_{\alpha_j}, w_j\lambda)$ is regular at $i\check{\mathfrak{a}}_0+\rho_{P_0}$ unless $w_j^{-1}\alpha_j\in\Delta_0$, in which case it has a simple pole along $\langle w_j^{-1}\alpha_j, \lambda-\rho_{P_0}\rangle=0$, but no more singularities meeting $\rho_{P_0}+i\check{\mathfrak{a}}_0$. Since the roots $w_j^{-1}\alpha_j$ are precisely the positive roots β with $\beta/2\notin\Phi_0$ and $w\beta<0$, and each of them occurs only once, the Lemma 7 follows.

9. The end of the proof. Let $\Delta_0^P=\{\alpha\in\Delta_0\,|\,\alpha \text{ vanishes on } \mathfrak{a}\}$. We put

$$c_P=\lim_{\lambda\to\rho_{P_0}}\Big(\prod_{\alpha\in\Delta_0^P}\langle\check{\alpha},\lambda-\rho_{P_0}\rangle\Big)\,c(w_0^A,\lambda),$$

where w_0^A is the longest element of W^A. By a lemma of Langlands ([4, Lemma 3.6] or [5, Theorem 4]), our previous remarks imply

$$\lim_{\substack{\lambda\to 0\\ \lambda\in\check{\mathfrak{a}}_0^P}}\Big(\prod_{\alpha\in\Delta_0^P}\langle\check{\alpha},\lambda\rangle\Big)E_{P_0}^P(\lambda+\vartheta+\rho_{P_0},g)=c_P\exp(\langle H_P(g),\vartheta+2\rho_P\rangle) \tag{2.9}$$

for $\vartheta\in\check{\mathfrak{a}}$. In fact, we can use (2.4) and Lemma 7 to compare the constant terms of both sides along P_0.

Using (2.9) and the obvious relation

$$E_{P_0}^G(\lambda,g)=\sum_{\gamma\in P(F)\backslash G(F)}E_{P_0}^P(\lambda,\gamma g),\qquad \lambda\in\rho_{P_0}+\check{\mathfrak{a}}_0^+$$

we get

$$\lim_{\substack{\lambda\to 0\\ \lambda\in\check{\mathfrak{a}}_0^P}}\Big(\prod_{\alpha\in\Delta_0^P}\langle\check{\alpha},\lambda\rangle\Big)E_{P_0}^G(\lambda+\vartheta+\rho_{P_0},g)=c_P\,E_P^G(\vartheta+\rho_P,g). \tag{2.10}$$

Applying (2.10) once more with P replaced by G, we get

$$\lim_{\vartheta\to 0}\Big(\prod_{\alpha\in\Delta_0-\Delta_0^P}\langle\check{\alpha},\vartheta\rangle\Big)E_P^G(\vartheta+\rho_P,g)=c_G/c_P.$$

If we put $\vartheta=2(s-1)\rho_P$, we get the final result of our computations:

$$\lim_{s\to 1}(s-1)^{rk(P)}Z_{-2\rho}(s)=c_G/(c_P\cdot\prod_{\alpha\in\Delta_0-\Delta_0^P}(2\langle\check{\alpha},\rho_P\rangle)). \tag{2.11}$$

(It should perhaps be mentioned that in computing the right side of (2.11), one has to view ρ_P as an element of $\check{\mathfrak{a}}_0$, for in general $\check{\alpha}$ does not belong to $\mathfrak{a}$.)

By the well-known relation between the singularities of $E_{P_0}^G$ and of the functions $c(w,\cdot)$, Lemma 7 and (2.10) imply that there are at most $t=rk(P)$ singular hyperplanes of E_P^G containing ρ_F and no other singular hyperplanes meeting $\rho_{P_0}+i\check{\mathfrak{a}}_0$. It follows that $Z_{-2\rho}(s)$ has a pole of order $\leqq t$ at $s=1$, and that the t-th residue, given by (11), does not vanish. The proof of Theorem 5.b) is complete. Theorem 5.c) follows from [4, Corollary to Lemma 7.6].

10. Remarks. a) Suppose that G is of adjoint type and splits over F. We choose a Chevalley basis $\{X_\alpha\}_{\alpha\in\Phi_0}$ of the Lie algebra $\mathfrak{g}$, where Φ_0 is the root system of A_0. In the case of a non-archimedean place v, choose K_v to be the stabilizer of the corresponding Chevalley lattice in $\mathfrak{g}\otimes_F F_v$. If v is a real (resp. complex) place, we assume that the Cartan involution defining K_v is the unique automorphism (resp. anti-linear automorphism) of $\mathfrak{g}\otimes_F F_v$ which sends X_α to $X_{-\alpha}$. The functions $c(w,\lambda)$ are then given by

$$c(\omega,\lambda)=\prod_{\substack{\alpha>0\\ w\alpha<0}}\frac{\xi(\langle\check{\alpha},\lambda\rangle)}{\xi(\langle\check{\alpha},\lambda\rangle+1)}. \tag{2.12}$$

The product is over all positive roots α with $w\alpha<0$, and ξ is given by

$$\xi(s)=D^{s/2}(\pi^{-s/2}\Gamma(s/2))^{r_1}((2\pi)^{-s}\Gamma(s))^{r_2}\zeta_F(s),$$

where D is the discriminant of F and ζ_F is Dedekind's zeta function. Formula (2.12) is standard (cf. for instance [4, p. 285]). With the help of the functional equations for $c(w, \lambda)$, it is reduced to the case $G = PGL_2$, in which it is proved by a straightforward computation. Now, (2.12) implies a formula for the leading term of the asymptotics

$$\lim_{s\to 1}(s-1)^{rk(P)} Z_{-2\rho}(s)$$

$$= \prod_{\alpha\in\Phi_P - \Delta_0} \frac{\xi(\langle\check{\alpha}, \rho_{P_0}\rangle)}{\xi(\langle\check{\alpha}, \rho_{P_0}\rangle+1)} \prod_{\alpha\in\Delta_0-\Delta_0^P} \frac{\mathrm{Res}_{s=1}\,\xi}{\xi(2)\langle\check{\alpha}, 2\rho_P\rangle}.$$

where Φ_P is the set of roots of A_0 which occur in the radical of P.

Furthermore, the functional equations of the Eisenstein series imply a functional equation relating $Z_{-2\rho}(s)$ and $Z_{-2\rho}(1-s)$. In the case $V = \mathbb{P}^n(\mathbb{Q})$, this functional equation coincides with the functional equation derived by expressing $Z(s)$ as a quotient of the Epstein and the Riemann Zeta function.

b) Suppose that for every place v we are given a K_v-finite function φ_v on $(P\cap K_v)\backslash K_v$ such that $\varphi_v = 1$ for almost every v. If one wants to investigate the series

$$Z_\varphi(s) = \sum_{x\in V(F)} h_{-2\rho}(x)^{-s} \prod_v \varphi_v(x),$$

one can do so by means of Eisenstein series. For every $\lambda\in\check{\mathfrak{a}}_{\mathbb{C}}$ and every v, φ_v defines an element of $\mathrm{Ind}_{P(F_v)}^{G(F_v)}(\rho_\lambda)$, where $\rho_\lambda\colon P(F_v)\to\mathbb{C}$ is the character defined by λ. One has an Eisenstein series $E_P^G(\varphi, \lambda, g)$, $\lambda\in\check{\mathfrak{a}}_{\mathbb{C}}$, $g\in G(A_F)$ which meromorphically depends on λ, and which satisfies the equality

$$Z_\varphi(s) = E_P^G(\varphi, (2s-1)\rho, e).$$

The role of the numbers $c(w, \lambda)$ is now played by the intertwining operators $C(w, \lambda)$. The singular hyperplanes of the operator $C(w, \lambda)$ are given by (2.6) (the proof is the same as above), however the residue of $C(w, \lambda)$ along one of these hyperplanes may annihilate a part of $\mathrm{Ind}_{P(A_F)}^{G(A_F)}(\rho_\lambda)$. Parts a) and c) of 5. remain true, in b) the result is that the order of $Z(s)$ at $s=1$ is at most $rk(P)$.

Some remarks concerning the dependence of the higher residue

$$r = \lim_{s\to 1}(s-1)^{rk(P)} Z_\varphi(s) \tag{2.14}$$

on φ are perhaps in order. It is clear that r vanishes unless all hyperplanes (2.6) are indeed singular hyperplanes of $C(w_0, \lambda)$, where w_0 is the longest element of W. The operator C is a product of local intertwining operators

$$C(w_0, \lambda) = \prod_v C_v(w_0, \lambda)$$

over all places v. Let s be the finite set of places with $\varphi_v \neq 1$. For $v\notin S$, $C_v(w_0, \lambda)(\varphi_v) = c_v(w_0, \lambda)\varphi_v$, and $c(w_0, \lambda)$ is the product of $c_v(w_0, \lambda)$ over all places v. At $\lambda = \rho_{P_0}$, each $C_v(w_0, \lambda)$ is regular and different from zero. Lemma 7 implies that the singular hyperplanes of the product $\prod_{v\notin S} c_v(w_0, \lambda)$ are given by (2.6), their multiplicity is one. For each v, $C_v(w_0, \lambda)$ is regular at $\lambda = \rho_{P_0}$. It follows that r vanishes if and only if $C_v(w_0, \rho_{P_0})(\varphi_v) = 0$ for some $v\in S$.

One can prove that this happens if and only if $\int_{K_v} \varphi_v(k)\,dk=0$. If P_0 remains minimal over F_v, then our claim follows from the Langlands classification theorem. $C_v(w_0, \cdot)$ is proportional to the intertwining operator used in the proof of Langlands' theorem (cf. [10, Corollary 4.6 in Chapter IV and Proposition 2.6 in Chapter XI]), and the Langlands quotient of $\mathrm{Ind}_{P_0(F_v)}^{G(F_v)}(2\rho_{P_0})$ is the trivial one-dimensional representation. It follows that the projection of φ_v to this quotient vanishes iff $\int_{K_v} \varphi_v(k)\,dk=0$.

The same fact concerning the vanishing of $C_v(w_0, \rho_{P_0})(\varphi_v)$ can be proved by a direct computation using the integration formula which played a crucial role in the proof of Langlands' theorem. This computation goes through in the general case, i.e., without the assumption that P_0 is F_v-minimal.

Our remarks prove that (2.14) is proportional to $\int_K \varphi(k)\,dk$.

c) In a large number of cases, Lemma 7.2 in [4] can be used to prove that the poles of $Z_X(s)$ in the strip $1/2<\mathrm{Re}(s)\leqq 1$ are real. A non-real pole of $Z(s)$ in this strip gives rise to a singular hyperplane σ of $E_P^G(\cdot)$ which meets $\check{\mathfrak{a}}^+ + i\check{\mathfrak{a}}$ but for which $X(\sigma)$, the point of smallest norm in σ, is not real. For cuspidal Eisenstein series (i.e., $P=P_0$) this is impossible. In the general case one has to check that the image σ' of σ in $\check{\mathfrak{a}}+\rho_{P_0}-\rho_P \subset \check{\mathfrak{a}}_0$ belongs to the collection of hyperplanes constructed in [4, Theorem 7.7]. The main difficulty is to prove that σ' does not fall victim to a cancellation effect which is possible when residua along σ' several times occur in the process of moving the contour of integration. We don't know how to do this in general.

Appendix

Consider a primitive integer solution $x=(x_i)$ of the equation

$$x_0^3+2x_1^3+3x_2^3+4x_3^3=0$$

as a $\mathbb{Q}$-point on a minimal cubic surface V in $\mathbb{P}^3$. Put $h(x)=\sum_{i=0}^{3}|x_i|$. It is a height of x with respect to the obvious metrization of ω_V^{-1}. Denote by $x\circ y$ the third intersection point of V with the line passing through x, $y\in V(\mathbb{Q})$. Call x decomposable if for some $z\in V(Q)$ with $h(z)<h(x)$ we have $h(x\circ z)<h(x)$.

In the following table the numbers

$$N(H)=\mathrm{card}\{x\in V(\mathbb{Q})\,|\,h(x)\leqq H\},$$

$$N_i(H)=\mathrm{card}\{x\in V(\mathbb{Q})\,|\,h(x)\leqq H, x \text{ is indecomposable}\}$$

are given as a function of H. The table and the accompanying graph suggest the linear growth of both functions.

A complete table of solutions with $h(x)\leqq 200$ and their decompositions into solutions of lower height was first compiled by Yu.I. Manin. Similar lists for $h(x)\leqq 1100$ were calculated by Don Zagier to whom we are thankful for his permission to publish excepts from them.

Height ≤	Number of solutions	Number of indecomposable solutions
50	16	3
100	32	4
150	49	8
200	63	10
250	79	12
300	94	14
350	109	17
400	127	19
450	139	21
500	159	21
550	181	24
600	193	24
650	207	25
700	223	26
750	238	28
800	264	33
850	277	34
900	296	35
950	320	36
1000	340	39
1050	363	40
1100	379	42

Applying a similar search of points to a smooth cubic curve we would find that $N_i(H)$ stabilizes for H sufficiently large. This is equivalent to the Mordell-Weil theorem. Since for our cubic surface $N_i(H)$ does not show tendency to stabilize we hardly can hope that this fact generalizes. However, there still remains a possibility that all points with sufficiently large height can be generated by drawing secants and tangents starting from a finite set.

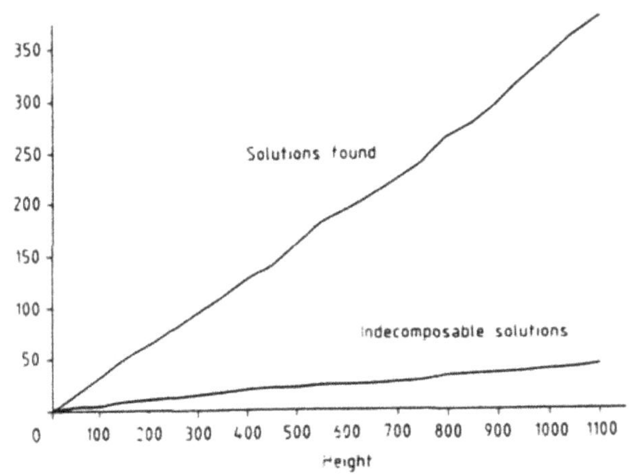

Fig. 1

References

1. Artin, M., Chai, C.-L. et al.: Arithmetic geometry. Cornell, G., Silverman, J.H. (eds.). Berlin-Heidelberg-New York: Springer 1986
2. Eisenbud, D., Harris, J.: The Kodaire dimension of the moduli space of curves of genus >23. Invent. Math. **90**, 359–388 (1987)
3. Schanuel, S.: Heights in number-fields. Bull. Soc. Math. France **107**, 433–449 (1979)
4. Langlands, R.P.: On the functional equations satisfied by Eisenstein series. (Lecture Notes in Math., Vol. 544). Berlin-Heidelberg-New York: Springer 1976
5. Harish-Chandra.: Automorphic forms on semisimple Lie groups. (Lecture notes in Math., Vol. 62). Berlin-Heidelberg-New York: Springer 1968
6. Weil, A.: Sur la formule de Siegel dans la théorie des groupes classiques. Acta Math. **113**, 1–87 (1965) [=Oe. Sci. [1965]]
7. Mars, J.G.M.: Sur l'approximation du nombre de solutions de certaines équations diophantiennes. Ann. Sci. EC. Norm. Super., IU. Ser. **6**, 357–387 (1973)
8. Igusa, J.-I.: Lectures on forms of higher degree. Tata Institute of Fund. Research. Berlin-Heidelberg-New York: Springer 1978
9. Patterson, S.J.: The Hardy-Littlewood method and Diophantine analysis in the light of Igusa's work. Math. Gött. Schriftenreihe Geom. Anal. **11**, 1–45 (1985)
10. Borel, A., Wallach, N.: Continuous cohomology, discrete subgroups, and representations of real reductive groups. (Annals of Math. Studies, Vol. 94). Princeton Univ. Press. Princeton 1980

Oblatum 14-I-1988 & 28-VI-1988

Math. Ann. 286, 27–43 (1990)

Mathematische Annalen

Sur le nombre des points rationnels de hauteur bornée des variétés algébriques

V. V. Batyrev[1] et Yu. I. Manin[2]

[1] Université de Moscow. Chair of Algebra. SU-117899 Moscow, USSR
[2] Institut Mathématique Stekloff. 42 ul. Vavilova, SU-117966 Moscow, USSR

A Hans Grauert, pour son 60e anniversaire

Introduction

Soient k un corps global, V une variété projective définie sur k, L un faisceau inversible ample sur V, h_L une hauteur exponentielle. Posons pour un sous-ensemble $U\subset V(k)$

$$\bar{\beta}_U(L)=\limsup\frac{\log\operatorname{card}\{x\in U(k)|h_L(x)\leqq H\}}{\log H}.$$

Si h_L est normalisée au moyen de mesures de Haar additives locales on a $\bar{\beta}_{\mathbf{P}^n}(\omega^{-1})=1$ (où $\omega=\mathcal{O}(-n-1)$), indépendamment de n et k.

Dans ce travail qui fait suite à [1] on cherche à généraliser cette remarque et à relier $\bar{\beta}_U(L)$ aux invariants calculables par voie géométrique.

Au n°1 nous étudions les propriétés élémentaires de $\bar{\beta}_U(L)$ ou plutôt de sa version $\beta_U(L)$. Au n°2 nous introduisons un invariant géométrique $\alpha(L)$ défini par la géométrie du cône de diviseurs effectifs du groupe de Néron-Severi. Nous décrivons quelques liens, pour la plupart conjecturaux, entre α et β au n°3. Enfin, les n°s 4 et 5 sont consacrés au cas des surfaces où la théorie de Mori permet de faire un progrès et d'évaluer $\beta_U(L)$ au moins pour des surfaces de del Pezzo de degré $\geqq 6$. En fait, on a besoin d'une variante de la théorie de Mori concernant les courbes «coextrémales» du cône dual du cône de diviseurs effectifs. Pourtant pour les surfaces la situation se simplifie et se réduit à la théorie de Mori propre.

1. Invariant arithmetique $\beta_U(L)$

1.1. *Formule du produit*. Soit k un corps global, c'est à dire une extension finie de $\mathbb{Q}$ ou de $\mathbb{F}_p(t)$. Pour un complété k_v de k, on a la valeur absolue $|\ |_v: k\to\mathbb{R}$ qui est définie comme étant le multiplicateur de la mesure de Haar additive sur k_v par rapport à l'homothétie. Soit M_k la famille complète de valeurs absolues; comme on le sait bien, on a la formule du produit

$$\prod_{v\in M_k}|x|_v=1\,,\qquad x\in k^*.$$

Soit L un espace vectoriel unidimensionnel sur k. Supposons une norme $\| \ \|_v$ donnée sur L pour toute $v \in M_k$ de telle manière que $\|al\|_v = |a|_v \|l\|_v$ si $a \in k$, $l \in L$, et que $\|l\|_v = 1$ pour chaque $l \in L \setminus \{0\}$ et presque toutes v. Alors «le volume» de L défini par

$$\|L\| = \prod_{v \in M_k} \|l\|_v, \qquad l \in L \setminus \{0\}$$

ne dépend pas du choix de l.

1.2. *Hauteur par rapport à un fibré inversible métrisé.* Soient maintenant V une variété algébrique projective sur k, L un fibré vectoriel de rang 1 sur v (que nous identifierons avec son faisceau de sections). On métrise L en choisissant une famille $\{\| \ \|_v, v \in M_k\}$ de métriques sur toutes fibres géométriques de L en points fermés de V ayant des propriétés naturelles de continuité [3].

La hauteur par rapport à $(L, \| \ \|_v)$ est la fonction $h_L: V(k) \to \mathbb{R}$ definie par

$$h_L(x) = \prod_{v \in M_k} \|s(x)\|_v^{-1}$$

où s est une section locale de L non-nul en x.

Si l'on change la famille de métriques, h_L se multiplie par $\exp(O(1))$. En général, nous ne considérons que les propriétés de h_L invariantes par rapport à un tel changement.

Si k' est une extension finie de k du degré d, $\{\| \ \|\}$ admet un prolongement canonique sur $\{V \otimes k', L \otimes k'\}$, et la restriction de $h_{L'}$, $L' = L \otimes k'$, sur $V(k)$ est égal à $h_{L'}^d$, donc dépend du choix du corps de base. En revanche, elle acquiert une propriété de stabilité importante qui est la suivante:

$$\operatorname{card}\{x \in \mathbb{P}^n(k) | h_{\omega^{-1}}(x) \leqq H\} = c(n,k) H(1 + o(1)) \text{ pour } n \geqq 1,\ H \to \infty,$$

où ω est le faisceau dualisant de $\mathbb{P}^n$. En effet, on sait [2] que si l'on calcule ce nombre par rapport à la hauteur invariante $h_L^{1/[k:\mathbb{Q}]}$ on obtient $cH^{[k:\mathbb{Q}](n+1)}(1 + o(1))$. En outre, $\omega^{-1} \simeq \mathcal{O}(n+1)$.

En général, soient L un faisceau ample métrisé sur V et $U \subsetneqq V(k)$ un sous-ensemble (par exemple, Zariski constructible ou défini par des conditions adèliques). Nous posons

$$N_U(L; H) = \operatorname{card}\{x \in U | h_L(x) \leqq H\}$$

et nous essayons de comprendre pour quels V, U, L, k on observe cet effet de croissance linéaire de $N_U(L; H)$.

Pour préciser cette notion, il est commode d'introduire la fonction zéta correspondante.

1.3. *Zéta.* On pose pour un L ample

$$Z_U(L; s) = \sum_{x \in U} h_L(x)^{-s},$$

$$\beta = \beta_U(L) = \inf\{\sigma \in \mathbb{R} | Z_U(L; s) \text{ converge pour } \operatorname{Re}(s) = \sigma\}.$$

On sait qu'alors $N_U(L;H) \sim H^\beta$ si U est infini au sens faible suivant: pour tout $\varepsilon > 0$.

$$N_U(L;H) = \begin{cases} O(H^{\beta+\varepsilon}) \\ \Omega(H^{\beta-\varepsilon}) \end{cases} \text{ lorsque } H \to \infty .$$

L'existence de $\beta_U(L)$ ne présuppose aucune propriété de continuation analytique de $Z_U(L;s)$ qui d'ailleurs n'est connue que dans des cas très particuliers. Pourtant dans ces cas on obtient des formules plus précises de type

$$N_U(L;H) = cH^\beta (\log H)^t (1 + o(1)).$$

Nous espérons que dans des conditions convenables l'invariant β admettera une description purement géométrique; de même pour t si $\beta > 0$. Au contraire, c est de la nature arithmétique, même si $U = \mathbb{P}^1(k)$.

Résumons maintenant des propriétés élémentaires de $\beta_U(L)$.

1.4. **Proposition.** a) *$\beta_U(L)$ ne dépend pas du choix de $\{\| \ \|_v\}$ sur L, ni du choix de L dans une classe $[L]$ du groupe de Néron-Severi $NS(V)$.*

b) *On a soit $0 \leqq \beta_U(L) < \infty$, soit $\beta_U(L) = -\infty$, ce dernier cas correspondant à* card$(U) < \infty$. *Si $\beta_U(L) = 0$ pour un L ample, il en est de même pour tous.*

c) *$\beta_U(L^m) = \beta_U(L)/m$, donc β_U^{-1} se prolonge à une fonction linéaire sur chaque rayon rationnel dans le cône fermé des éléments amples $N^1_+(V) \subset NS(V) \otimes \mathbb{R}$ contenant une classe $[L]$ avec $\beta_U(L) \neq 0$.*

d) *$\beta_U(L)$ se prolonge à une fonction continue définie sur $\dot{N}^1_+(V) = N^1_+(V) \backslash \partial N^1_+(V)$.*

Demonstration. On déduit toutes les assertions de propriétés standard de hauteurs (voir [3, 4]). Si l'on remplace h_L par $h'_L = h_L \exp(O(1))$, β reste le même. Si l'on remplace L par $L' \in [L]$, on a $c_1 h_L^{1-\varepsilon} \leqq h_{L'} \leqq c_2 h_L^{1+\varepsilon}$ pour tout $\varepsilon > 0$ et tous $c_1, c_2 > 0$ convenables, d'où le même résultat.

Comme $N_L(U,H) \leqq N_L(V,H) < \infty$ pour tout H, $Z_U(L;s)$ ne peut converger en un point $\sigma < 0$ que si card$(U) < \infty$, et dans ce cas, elle converge partout. D'autre part, $Z_U(L;s)$ est majorée par $Z_{\mathbb{P}^n}(O(m);s)$ pour m, n convenables, et comme on a [2]

$$\beta_{\mathbb{P}^n}(\mathcal{O}(m)) = (n+1)/m$$

on voit que $\mathcal{O} \leqq \beta_U(L) < \infty$ si card$(U) = \infty$.

L'assertion c) se déduit du fait que $h_{L^m} = h_L^m$.

Enfin, on déduit d) de l'inégalité suivante: si L_1, L_2 sont amples alors $c_1 h_{L_2}^a < h_{L_1} < c_2 h_{L_2}^b$ pour a, b, c_1, $c_2 > 0$ convenables.

Donc, en ajoutant à $[L]$ une combinaison linéaire de classes amples avec des coefficients petits on ne change $\beta_U(L)$ qu'un peu.

1.5. *La surface de croissance linéaire.* On nomme cela la (hyper)surface

$$\Sigma(U) = \{ l \in \dot{N}^1_+(V) | \beta_U(l) = 1 \}.$$

De la proposition 1.4 on déduit:

a) Si $\Sigma(U)$ est non-vide, alors chaque rayon en $\dot{N}^1_+(V)$ coupe $\Sigma(U)$ en un point. Donc $\Sigma(U)$ définit β_U sans ambiguité.

En outre, on a:

b) $\Sigma_{>}(U) := \{[L] | \beta_U(L) < 1\}$ est convexe.

En effet, si $\mathcal{O} \leqq \sigma, \tau \leqq 1, \sigma + \tau = 1$, on a l'inégalité de Hölder

$$h_L^{-\sigma}(x) h_M^{-\tau}(x) \leqq \sigma h_L(x)^{-1} + \tau h_M(x)^{-1}$$

donc $\sigma[L] + \tau[M]$ appartient à $\Sigma_{>}(U)$ en même temps que $[L]$, $[M]$.

On peut nommer U *petit* (resp. *grand) si* $\Sigma(U)$ est vide (resp. non-vide). Si V est une variété abélienne, tout $U \subset V(k)$ est petit car $h(x)$ accroit comme exponentielle d'une forme quadratique positive sur un réseau. D'autre part, tout U contenant un ouvert de Zariski de $\mathbb{P}^1(k)$ est grand. C'est le cas pour V unirationnelle sur k, $U \subset V(k)$ un ouvert de Zariski, ou bien pour des variétés de Fano contenant beaucoup de courbes rationnelles.

En appliquant ces remarques à une variété abélienne V^n avec point 0 éclaté en $\mathbb{P}^{n-1}$ on voit que $V(k)$ est grand tandis que $U = V(k) \backslash P^{n-1}(k)$ est petit, donc «presque tous» les points de V appartiennent à un fermé de Zariski de V.

Dans cet exemple, on peut computer les points au moyen d'un faisceau ample arbitraire. L'exemple suivant, dû à Serre [4] montre que l'existence d'une sous-variété qui «accumule» des points rationnels peut dépendre de L.

1.6. **Proposition.** *Soient V le plan projectif avec un point éclaté en diviseur D, $U = V \backslash D$, $\pi: V \to \mathbb{P}^2$ la contraction de D. Le groupe $NS(V)$ est engendré par $\Lambda = [\pi^*\mathcal{O}(1)]$ et $l = [\mathcal{O}(D)]$. Identifions $NS(V)$ avec $\mathbb{Z}^2$ par $a\Lambda - bl = (a, b)$. Alors $[\omega^{-1}] = (3, 1)$, $N_+^1 = \{(a, b) | a \geqq b \geqq 0\}$ et on a*

a) $\Sigma(V) = (4, 2) + \partial N_+^1$.

b) $\Sigma(U) = ((3, 1) + \partial N_{\text{eff}}^1) \cap \dot{N}_+^1$ *où N_{eff}^1 est le cône fermé $\{(a, b) | a \geqq 0,\ a \geqq b\}$ engendré par les classes de diviseurs effectifs.*

c) *Si $[L] = a\Lambda - bl \in \dot{N}_+^1$, $a > 2b$, et si l'on compte les points rationnels à l'aide de h_L, alors presque tous les points de $V(k)$ appartiennent à $D(k)$. Pour tout autre L, il n'existe aucune courbe avec cette propriété.*

Démonstration. Nous nous bornons an cas $k = \mathbb{Q}$. Choisissons des coordonnées de telle manière qu'on ait $\pi(D) = (1, 0, 0)$. Soit

$$h_{\mathcal{O}(1)}(x_1, x_2, x_3) = \max(|x_i|) \text{ pour } x_i \in \mathbb{Z},\ pgcd(x_i) = 1\,.$$

Comme $(1, 0, 0)$ est éclaté par le système linéaire de coniques passant par ce point, on a pour $x \in V \backslash D$, $\pi(x) = (x_1, x_2, x_3)$, $d = pgcd(x_2, x_3)$

$$h_{2\Lambda - l}(x) = d^{-1} \max(|x_1^\alpha x_2^\beta x_3^\gamma| \,|\, \alpha + \beta + \gamma = 2,\ \alpha \neq 2)$$

$$h_l(x) = \frac{h_{2\Lambda}(x)}{h_{2\Lambda - l}(x)} = \begin{cases} d|x_1||x_2|^{-1} & \text{si} \quad |x_1| \geqq |x_2| \geqq |x_3|; \\ d|x_1||x_3|^{-1} & \text{si} \quad |x_1| \geqq |x_3| \geqq |x_2|; \\ d & \text{si} \quad |x_1| < \max\{|x_2|, |x_3|\}\,. \end{cases}$$

Finalement

$$h_{a\Lambda - bl}(x) = \begin{cases} |x_1|^{a-b}|x_2|^b d^{-b} & \text{si} \quad |x_1| \geqq |x_2| \geqq |x_3|; \\ |x_1|^{a-b}|x_3|^b d^{-b} & \text{si} \quad |x_1| \geqq |x_3| \geqq |x_2|; \\ |x_j|^a d^{-b} & \text{si} \quad \max(|x_i|) = |x_j| \neq x_1\,. \end{cases}$$

Un calcul facile montre (voir [4]):

$$N_U(a\Lambda - bl; H) \sim \begin{cases} H^{2/(a-b)} & \text{si} \quad a<3b; \\ H^{3/a}\log(H) & \text{si} \quad a=3b; \\ H^{3/a} & \text{si} \quad a>3b. \end{cases}$$

Mais $a\Lambda - bl$ induit sur D le faisceau $\mathcal{O}(b)$. Ainsi

$$N_D(a\Lambda - bl; H) \sim H^{2/b},$$

d'où le résultat.

S'il y a une courbe irréducible accumulante $\neq D$ dans une classe $A\Lambda - Bl$, $A \geqq B > 0$, elle doit être de genre nul. Alors $a\Lambda - bl$ induit $\mathcal{O}(Aa - Bb)$ sur cette courbe, donc elle contient $\sim H^{2/(Aa-Bb)}$ points de hauteur $\leqq H$. Mais, pour $a > 2b$, V contient $\sim H^{2/b}$ de tels points, et on voit que $Aa - Bb > (2A - B)b > b$.

1.7. *Stratification arithmétique.* Considérons l'ensemble de toutes fonctions

$$S_V = \{\beta_U : N^1_+(V) \to \mathbb{R}_+ \cup \{-\infty\} \mid U \subsetneqq V \text{ ouvert de Zariski non-vide}\}.$$

Pour chaque $\beta \in S_V$ il existe un ouvert $U \subsetneqq V$ maximal tel que $\beta = \beta_U$. Si $U' \subset U''$ et $\beta_{U'} \neq \beta_{U''}$ alors presque tous les points de $U''(k)$ appartiennent à $(U'' \setminus U')(k)$ par rapport à une hauteur convenable. On espère que telles sous-variétés accumulantes admettrout une description purement géométrique (voir le par. 2).

Dans [14] Faltings considère les puissances symetriques de courbes algébriques. Il est vraisemblable que les sous-variétés accumulantes dans ce cas correspondent aux diviseurs exceptionnels.

1.8. *Question.* Si V est une variété de Fano, est il vrai que S_V est fini?

Nous croyons que S_V est infini pour des surfaces rationnelles V avec rk $\mathrm{NS}(V) \geqq 10$ et pour des surfaces $K3$ (voir n°s 3 et 5 ci-dessous).

1.9. *Question.* Que peut-on dire de S_V pour les variétés de module de courbes M_g et de variétés abéliennes A_g?

Comme M_g est de type général pour $g \geqq 24$ [5] les conjectures de Vojta [10] impliquent que $M_g(k)$ est contenu dans une sous-variété propre de M_g, c'est à dire que S_{M_g} contient la fonction constante $-\infty$.

1.10. *Question.* Quelles valeurs peut prendre β_U? Est il possible que $\beta_U(L)$ soit irrationnel pour un U ouvert et L ample?

Il serait intéressant d'appliquer à ce problème les méthodes de la théorie de fonctions récursives.

2. L'Invariant géométrique $\alpha(L)$

2.1. **Définition.** *Soit $N^1_{\mathrm{eff}}(V)$ le cône fermé engendré par $[\mathcal{O}(D)]$ pour les diviseurs effectifs D, $K = K_V = [\omega_V]$. On pose alors*

$$\alpha(L) = \inf\{\gamma \in \mathbb{R} \mid \gamma[L] + K_V \in N^1_{\mathrm{eff}}\}. \quad \square$$

Cette fonction possède des propriétés analogues à celles de β_U (voir 1.4):

a) $\alpha(L)$ est défini pour chaque ample L. Si $\alpha(L)<0$ pour un ample L, il en est de même pour tous, ce qui se passe pour $K_V \in \dot{N}^1_{\text{eff}}$ quand le modèle canonique de V est birationnel. Si $\alpha(L)=0$ pour un ample L, il en est de même pour tous, ce qui se passe pour $K_V \in \partial N^1_{\text{eff}}$. Si $\alpha(L)>0$ pour un ample L, il en est de même pour tous, ce qui se passe pour K_V en déhors de N^1_{eff}.

b) En effet, si L est ample, on a $\dim \Gamma(L^n \otimes \omega_V)>1$ pour $n \gg 0$, donc $n[L]+K_V \in N^1_{\text{eff}}$. Si $\alpha(L)<0$, on a $-[L]/n+K_V \in N^1_{\text{eff}}$ pour un $n>0$, donc $nK_v=[L]+X$, X effectif, d'où on déduit que le modèle canonique est birationnel. Le reste est clair.

c) $\alpha(L^m)=\alpha(L)/m$, donc α^{-1} est une fonction linéaire sur chaque rayon de $N^1_+(V)$ si $\alpha(L) \neq 0$.

d) α se prolonge à une fonction continue en $\dot{N}^1_+$.

Si $\alpha>0$ sur $\dot{N}^1_+$ on peut aussi décrire α par sa «surface de croissance linéaire géométrique»

$$\Sigma^\alpha=\{l \in \dot{N}^1_+(V) | \alpha(l)=1\}.$$

On a

$$\Sigma^\alpha=(-K_V+\partial N^1_{\text{eff}}) \cap \dot{N}^1_+.$$

2.2. *Exemples.* a) Soit rk $NS(V)=1$. Alors

$$\alpha(L)=-K_V/[L].$$

b) Soit $\dim(V)=1$. On a le tableau suivant;

Genre	$\alpha(L)$	$\beta(L)$
0	$-K_V/[L]>0$	$-K_V/[L]$ ou 0
1	0	0
>1	<0	$-\infty$ (Faltings)

c) La proposition 1.6 dit que $\beta_U(L)=\alpha(L)$ pour $U=V \setminus D$.

Dans le par. 5 nous généraliserons ce résultat.

2.3. *Comparaison avec la théorie de Nevanlinna.* Un faisceau inversible L est dit *presque ample* si $h^0(L^n)>cn^d$ pour $c>0$, $n \gg 0$, $d=\dim(V)$. Dans le deuxième théorème principal de la théorie de Nevanlinna [10] on utilise la fonction

$$\delta(L)=\inf\{p/q | q>0, L^p \otimes \omega_V^q \text{ est presque ample}\}.$$

On voit facilement que $\delta(L)=\alpha(L)$ pour L ample. En effet, $\delta(L) \geqq \alpha(L)$ car un faisceau presque ample est effectif. Si $\alpha(L)<\delta(L)$, choisissons un $\varepsilon>0$ rationnel tel que $\alpha(L)+2\varepsilon<\delta(L)$. On arrive à la contradiction en remarquant que

$$(\alpha(L)+2\varepsilon)[L]+K_V=\varepsilon[L]+((\alpha(L)+\varepsilon)[L]+K_V)$$

est presque ample étant la somme d'une classe ample et d'une classe effective.

2.4. *Applications de Fano.* Soit L ample, $\alpha=\alpha(L)>0$.

a) Supposons que $\alpha=p/q$ soit rationnel (pour $\dim(V)=2$ ce sera démontré en n°4 ci-dessous). Posons $L_0=L^p \otimes \omega_V^q$, $[L_0] \in \partial N^1_{\text{eff}}$.

On a $h^0(L_0^n) < cn^{d-1}$ pour $n \gg 0$, $d = \dim(V)$. En effet, soient $L = \mathcal{O}(W)$, W un diviseur effectif. On a une suite exacte $0 \to L_0^n \otimes L^{-1} \to L_0^n \to L_0^n | W \to 0$, d'où le résultat car $h^0(L_0^n \otimes L^{-1}) = 0$.

b) Supposons en outre que $R = \bigoplus_n \Gamma(V, L_0^n)$ soit de type fini (cette hypothèse est liée aux conjectures sur les modèles minimaux, voir [11]). Alors on a une application rationnelle dominante $f_L: V \to \mathrm{Proj}(R)$ qu'on peut nommer *une application de Fano*. Si $L = \omega_V^{-1}$ est ample, f_L est la projection sur un point. Si f_L admet une fibre non-singulière F, c'est une variété de Fano parce que $L_0 | F \sim \mathcal{O}_F$, sonc $\alpha[L]]|F = -K_V|F = -K_F$ est ample. En général, f_L serait un fibré en variétés de Fano généralisées.

2.5. *Sous-variétés géométriquement accumulantes.* Dans les n^{os} suivants, nous comparerons $\alpha(L)$ avec $\beta_U(L)$ pour $U \subsetneqq V$ ouvert suffisamment petit. Cela conduit à l'introduction de la notion d'*une variété géométriquement accumulante* $W \subset V$ (par rapport à un faisceau ample L) caractérisée par la propriété $\alpha(L|W) > \alpha(L)$. Pour obtenir une notion utile, il faut pourtant étendre la définition au cas de W éventuellement singulière. Soit $\bar{W}$ une normalisation de W et $\varphi: \bar{W} \to V$ le morphisme correspondant. Soit $W_0 \subseteqq \bar{W}$ la sous-variété de points lisses de $\bar{W}$. Nous posons

$$\alpha(L|W) = \inf\{p/q \mid p, q \in \mathbb{Z}, q > 0, h^0(W_0, \varphi^*(L)^p \otimes \omega_{W_0}^q) > 0\}.$$

2.6. **Théorème.** *Sur un corps k algébriquement clos de caractéristique nul, la somme Z(L) de toutes les sous-variétés géométriquement accumulantes par rapport à L est contenue dans une somme dénombrable de sous-variétés fermées propres de V.*

Démonstration. On va montrer que l'inégalité $\alpha(L|W) \leqq \alpha(L)$ est vérifiée sur une partie ouverte dense de la variété de base d'une famille de W arbitraire qui couvre V.

Plus précisément, soit S une variété non-singulière, base d'une famille $\pi: X \to S$ à fibre générique géométriquement irréductible. Soit $\dim(X) = \dim(V)$ et soit $f: X \to V$ un morphisme dominant. Nous montrerons qu'il existe une sous-variété ouverte dense $U(L) \subsetneqq S$ telle que $\alpha(L|f(X_s)) \leqq \alpha(L)$ pour tous $s \in U(L)$, $X_s = \pi^{-1}(s)$. Il suffit de construire au moyen d'une section $\sigma \in \Gamma(V, L^p \otimes \omega_V^q)$ une section non-nulle induite $\sigma_s \in \Gamma(X_s^0, L^p | X_s^0 \otimes \omega_{X_s}^q)$ pour les points s d'un ouvert dense de V. On commence par $f^*(\sigma) \in \Gamma(W, f^*(L)^p \otimes f^*(\omega_V)^q)$. On considère ensuite une suite exacte $0 \to f^*(\omega_V^q) \xrightarrow{(df)^q} \omega_U^q \to F \to 0$ sur la partie lisse U de V. On a $\mathrm{Tor}_1^{\mathcal{O}_U}(F, \mathcal{O}_{X_s}) = 0$ pour presque tous s. On a aussi un isomorphisme $\lambda_s: \omega_U^q | X_s \to \omega_{X_s}^q$ sur la partie lisse de X_s. Finalement, $f^*(\sigma)$ induit une section $((df)^q \otimes \lambda_s) f^*(s) | X_s$. □

3. Connexions entre α et β (conjectures)

En généralisant des exemples connus nous exprimerons notre intuition sur le comportement comparatif de α et β sous forme des conjectures suivantes qui sont à considérer plutôt comme des questions.

3.1. **Conjecture A.** *Pour tout* $\varepsilon > 0$ *il existe un ouvert de Zariski dense* $U = U(L, \varepsilon) \subsetneqq V$ *tel que*

$$\beta_U(L) \leqq \alpha(L) + \varepsilon.$$

3.2. **Conjecture B.** *Si V est une variété de Fano (c'est-à-dire si ω_V^{-1} est ample), alors pour toute extension finie suffisamment grande $K \supsetneqq k$ et pour tout ouvert de Zariski suffisamment petit et dense U on a*

$$\beta_U(L) = \alpha(L).$$

Ici, «suffisamment grande» signifie «contenant une extension dépendant de V, L»; «suffisamment petit» signifie «contenu dans un ouvert dependant de V, L, K».

En particulier, la conjecture B implique que pour V de Fano on a

$$\Sigma(U) = \Sigma^\alpha = (-K_V + \partial N^1_{\text{eff}}) \cap \dot{N}^1_+ .$$

Le cas crucial «$-K_V$ typiquement appartient à la surface de croissance linéaire» était un peu vaguement conjecturé en [1].

3.3. **Conjecture C.** *Soit V une variété dont la classe canonique n'appartient pas à N^1_{eff} (donc $\alpha(L) > 0$ pour tout L ample). Alors pour toute extension suffisamment grande K et pour tout ouvert de Zariski suffisamment petit et dense U on a*

$$\beta_U(L) = \alpha(L).$$

C'est une extension de la conjecture B. En effet, dans les conditions de la proposition 2.4 nous conjecturons que $\beta_U(L)$ coincide avec la même fonction calculée pour une fibre géométrique de $f: V \to \text{Proj}(R)$.

Nous allons montrer que ces conjectures constituent une interpolation entre la conjecture de Mordell généralisée pour les variétés de type général et la conjecture de croissance linéaire pour $-K_V$ sur les variétés de Fano.

3.4. **Théorème.** *La conjecture A est vraie pour une variété V avec ω_V ample si et seulement si $V(k)$ n'est pas dense de Zariski.*

Démonstration. Supposons que ω_V soit ample. Si $\alpha(L) < 0$, on a $\beta_U(L) < 0$ pour $U = U(L, \varepsilon)$ postulé, donc $U(k)$ est fini et $V(k)$ n'est pas dense. Inversement, s'il n'y a pas de k-points sur un U, on a $\beta_U(L) < -\infty < \alpha(L) + \varepsilon$. □

Parmi les cas intermédiaires entre «type général» et «Fano» la conjecture A donne une conséquence nette pour les surfaces avec $K_V = 0$.

3.5. **Théorème.** *Soit $\dim(V) = 2$, $K_V = 0$, par exemple, V est $K3$ ou une surface d'Enriques. Alors la conjecture A pour V est équivalente à la propriété «presque tous les points rationnels de V sont contenus dans l'union de courbes k-rationnelles de V" au sens suivant:*

(A'). Pour tout $\varepsilon > 0$ et L ample posons
$V(\varepsilon, L)$ = l'union de toutes courbes k-rationnelles
de V de L-degré $< 2\varepsilon^{-1}$ (ils sont en nombre fini).

Alors on a

$$N_V(L; H) = N_{V(\varepsilon, L)}(L; H) + O(H^\varepsilon).$$

Démonstration. Soit $K_V = 0$. Alors $\alpha(L) = 0$. Si la conjecture A est vraie pour V on peut trouver U de telle manière qu'on ait $\beta_U(L) \leqq \varepsilon/2$, donc $N_U(L; H) = O(H\varepsilon)$. Sans violer cette inégalité, on peut augmenter U par addition de toutes les composantes de $V\backslash U$ du genre $\geqq 1$ ou du genre 0 et de L-degré $\geqq 2\varepsilon^{-1}$. Alors $V\backslash U = V(\varepsilon, L)$.

L'implication inverse est immédiate. □

Mumford et Bogomolov ont montré qu'il y a toujours une courbe rationnelle sur une surface $K3$ (voir [8], Appendice). Bogomolov a même conjecturé que tous les points algébriques sur une telle surface appartiennent à l'union de courbes $\bar{k}$-rationnelles. Nous ne pouvons pas relier cette conjecture de Bogomolov avec notre conjecture (A').

Notons aussi que les conjectures A et A' sont vraies pour les surfaces qui sont birationnellement équivalentes aux surfaces abéliennes.

3.6. **Théorème.** *Lorsque V est une variété de Fano, la conjecture A pour V se déduit du cas spécial $L = -K_V$ de la conjecture B.*

Démonstration. La hauteur par rapport à un diviseur effectif est $>$ *const* en dehors de ce diviseur. □

3.7. *Droites extrémales.* Soit V une variété de Fano. Pour simplifier, nous supposerons que $\mathrm{NS}(V) = \mathrm{NS}(V \otimes \bar{k})$ et que toute classe de courbes extrémales au sens de Mori de V contient une courbe $B_i \subset V$ qui est définie et isomorphe à $\mathbb{P}^1$ sur k.

Soient $b_1, \ldots, b_s$ les fonctions linéaires $\mathrm{NS}(V) \otimes \mathbb{R} \to \mathbb{R}$: $b_i(l) = \langle B_i, l \rangle$ (l'indice de l'intersection). On sait que

$$N^1_+(V) = \{1 | b_i(l) \geqq 0;\ i = 1, \ldots, s\}.$$

Pour un ample $l = [L]$ et un ouvert de Zariski U on a

$$\beta_U(L) \geqq \max\{2/b_i(l) | U \cap B_i \neq \emptyset\}$$

car $\beta_U(l) \geqq \beta_{U \cap B_i}(l)$ et que L induit sur B_i le faisceau $\mathcal{O}(b_i(l))$.

Géométriquement, cela signifie que si $U \cap B_i \neq \emptyset$ pour tout i, alors $\Sigma(U)$ appartient au polyèdre fermé

$$\Delta = \{l | b_i(l) \geqq 2 \text{ pour tous } i\}.$$

D'autre part, d'après Mori, on a

$$1 \leqq b_i(\omega^{-1}) \leqq \dim(V) + 1.$$

Donc, s'il existe une courbe extrémale B_i avec $b_i(\omega^{-1}) = 1$ («une droite»), $-K_V$ certainement n'appartient pas à la surface de croissance linéaire pour V.

D'où la nécessité de prendre U suffisamment petit dans la conjecture B.

3.8. *Bornes supérieures.* Soit L un faisceau ample sur une variété projective V de dimension d.

a) Si V est non-singulière, on a $\alpha(L) \leqq d + 1$ (Mori [6]).

b) Si L est très ample et $\langle L^d \rangle \geqq 2$, on a $\beta_V(L) \leqq d + 1/2$ (voir Serre [4] et Cohen [9]).

On conjecture que lorsque L est ample et $\langle L^d \rangle \geqq 2$ on a $\alpha(L) \leqq d$ et $\beta(L) \leqq d$.

c) Si L est très ample et V est une hypersurface non-singulière, on a $\beta_V(L) \leqq d + 2/d$ (Fujiwara [12]).

d) Soit V une intersection complète non-singulière de s hypersurfaces en $\mathbb{P}^{n-1}$ (donc $d=n-s-1$), $L=\mathcal{O}(1)$, $k=\mathbb{Q}$. Alors

$$\beta_V(L) \leqq n-2s+2s^2/(n+s-1)=d-(s-1)+2s^2/(n+s-1)$$

(Shparlinski et Skorobogatov [13]).

Si V est une intersection complète *de quadriques* et que s/n est petit, cette borne s'approche à l'estimation conjecturale $\beta(-K_V) \leqq 1$ car $\omega_V^{-1}=\mathcal{O}(n-2s)$.

On démontre b), c) et d) par moyens analytiques, en utilisant le grand crible ou des estimations délicates de sommes exponentielles dûes à Deligne, Katz et Laumon.

3.9. *Espaces de drapeaux généralisés.* Soit G un groupe linéaire algébrique semi-simple admettant un sous-groupe parabolique minimal P_0 défini sur k. Soit P un sous-groupe parabolique contenant P_0, $V=P\backslash G$. C'est une variété de Fano homogène. Parmi ces variétés se trouvent les grassmanniennes; les espaces de drapeaux usuels ou isotropes, en particulier quadriques.

Des résultats de [1] on déduit:

3.10. **Théorème.** *Pour $V=P\backslash G$, on a $\beta_V=\alpha$.* □

La démonstration s'appuie sur la théorie de séries d'Eisenstein-Langlands [7]. On a:

a) $N_+^1(V)=N_{\text{eff}}^1(V)=$ une chambre de Weyl d'un espace de racine convenable.

b) Pour une normalisation de hauteurs compatible avec l'action de G sur V, $Z_V(L;s)$ admet une continuation méromorphe sur $NS(V)\otimes\mathbb{C}$, avec pôles simples en murs complexifiés de la chambre $N_+^1(V)$. Chaque réflexion de groupe de Weyl donne naissance à une equation fonctionnelle.

c) On en déduit des formules asymptotiques

$$N_V(L;H)=cH^{\alpha(L)}p(\log H)(1+O(H^{-\varepsilon}))$$

pour un $\varepsilon>0$ convenable où p est un polynôme du degré $t-1$, t la codimension maximale d'une face de $\partial N_{\text{eff}}^1$ contenant $\alpha(L)[L]+K_V$.

Nous nous bornons à indiquer la normalisation de hauteurs utilisée. Tout d'abord, on décrit $NS(P\backslash G)=\text{Pic}(P\backslash G)$ au moyen de caractères $\chi: P\to\mathbb{G}_m$, à savoir χ définit un faisceau inversible L_χ par

$$\Gamma(U,L_\chi)=\{f\in\Gamma(\pi^{-1}(U),\mathcal{O}_G) \mid f(pg)=\chi(p)f(g)\},$$
$$\pi: G\to P\backslash G, \quad p\in G, \quad g\in G.$$

On montre que les L_χ engendrent un sous-groupe d'indice fini dans $\text{Pic}(V)$.

Pour métriser tous les L_χ simultanément, on choisit un sous-groupe maximal compact adèlique $K=\prod_{v\in M_k} K_v \subset G(A_k)$ tel que $G(A_k)=P_0(A_k)K$. Ce choix donne des normes locales sur L_χ: lorsque $s\in\Gamma(U,L_\chi)$ correspond à $f\in\Gamma(\pi^{-1}(U),\mathcal{O}_G)$, on pose pour $x\in P\backslash G(k_v)$ et $x_v\in K_v$ qui se projette en x:

$$\|s(x)\|_v=|f(x_v)|_v$$

et finalement

$$h_{L_x}(x) = \prod \|s(x)\|_v^{-1}.$$

On est tenté de raffiner les conjectures B et C tenant compte du Théorème 3.10.

3.11. **Conjecture B′.** *Pour une variété de Fano V, si k est suffisamment grand et U est suffisamment petit, on a*

$$N_U(\omega_V^{-1}; H) = cH(\log H)^{t-1}(1+o(1)),$$

où $t = \mathrm{rk}(\mathrm{Pic}\, V)$.

3.12. **Conjecture C′.** *Si* K_V *n'est pas effectif, on a, avec les mêmes conditions,*

$$N_U(L; H) = cH^{\alpha(L)}(\log H)^{t(L)-1}(1+o(1)),$$

où $t(L)$ *est la codimension de la face minimale de* $\partial N^1_{\mathrm{eff}}$ *contenant* $\alpha(L)[L]+K_V$. *On conjecture ici que ce dernier point appartient à la partie polyedrale de* $\partial N^1_{\mathrm{eff}}$.

3.13. *Cas* $k = \mathbb{F}_q(C)$. Soit C une courbe non-singulière irréductible défini sur $\mathbb{F}_q$, k son corps de fonctions. On peut remplacer (V, L) défini sur k par un morphisme plat $\pi: \hat{V} \to \mathbb{C}$ et un faisceau relativement ample $\hat{L}$ sur $\hat{V}$. Un k-point x de V est représenté par un $\mathbb{F}_q$-morphisme $\hat{x}: C \to \hat{V}$ tel que $\hat{x} \circ \pi = \mathrm{id}$, et on peut prendre

$$h_L(x) = q^{\deg \hat{x}^*(\hat{L})}.$$

La fonction $Z_V(L; s)$ correspondante ne peut être toujours rationnelle en q^{-s}, à la difference de la fonction de Hasse-Weil, car déja pour une variété abélienne c'est une fonction thêta. Elle est pourtant bien rationnelle pour $V = \mathbb{P}^n$ et sans doute pour $V = P\backslash G$ générale (Harder l'a démontré pour les groupes de Chevalley [15] et Bernstein a esquissé une démonstration générale).

4. Le cas des surfaces: géométrie

4.1. *Notation.* Soit V une surface algébrique non-singulière sur un corps algébriquement clos, X un diviseur ample de V, $K = K_V$. Par abus de langage, nous noterons par X aussi la classe $[\mathcal{O}(X)]$. Rappelons que $\alpha = \alpha(X)$ est défini par

$$\alpha X + K \in \partial N^1_{\mathrm{eff}}(V).$$

Désormais nous ne considérons que le cas $\alpha \geqq 0$.

4.2. **Théorème.** *Si* $\alpha > 0$, *il existe un diviseur effectif* E *de* V *tel que* $p_a(E) = 0$, $\langle E, E \rangle = 0$ *ou* 1 *et* $\langle \alpha X + K, E \rangle = 0$. *On a donc*

$$\alpha(X) = \langle -K, E \rangle / \langle X, E \rangle.$$

En outre, on peut choisir $E = \varphi^*(E')$ *où* $\varphi: V \to W$ *est un morphisme et* E' *est un diviseur de* W *d'un des types suivants:*

i) $W = \mathbb{P}^2$, *φ est birationnel,* E' *est une droite.*

ii) *W est une courbe,* E' *est un point,* V *est un fibré en coniques de base* W.

Corollaire. *Les valeurs positives de* α *sont rationnelles de la forme* $1/n$, $2/n$ *ou* $3/n$.

Démonstration. On commence par définir $\gamma=\gamma(X)$ par

$$\gamma X+K\in\partial N^1_+(V).$$

Comme $N^1_+\subsetneqq N^1_{\text{eff}}$, on a $\gamma\geqq\alpha$. Si $\gamma>\alpha$, on peut choisir $p,q\in\mathbb{Z}_+$ de telle manière que $\alpha<p/q<\gamma$, donc que $pX+qK$ soit effectif sans être ample. On en déduit que pour $n>0$ convenable un diviseur $D\in|n(pX+qK)|$ admet une décomposition de Zariski $D=D_m+D_f$ où $\langle D_m,C\rangle\geqq 0$ pour toute courbe C de V et la partie fixe $D_f=\sum a_iC_i$, $a_i>0$, est non-nulle. On a en outre

$$\langle D_m,C_i\rangle=0 \text{ pour tous } i;$$
$$(\langle C_i,C_j\rangle) \text{ est définie négative}.$$

Donc $\langle D,C_i\rangle<0$ pour un i convenable, soit $i=1$. Comme $\langle C_1,C_1\rangle<0$ et $\langle K,C_1\rangle=(nq)^{-1}(\langle D,C_1\rangle-\langle npX,C_1\rangle)<0$, C_1 est une courbe exceptionnelle. En faisant contracter C_1 par un morphisme $\pi:V\to\bar V$ et en posant $\bar X=\pi_*(X)$ on simplifie la géométrie sans changer l'algèbre: $\bar\alpha:=\alpha(\bar X)=\alpha(X)$.

En effet, si $aX+K\in N^1_{\text{eff}}(V)$, on a

$$\pi_*(aX+K)=a\pi_*(X)+\bar K\in N^1_{\text{eff}}(\bar V),\bar K=K_{\bar V},$$

donc $\bar\alpha\leqq\alpha$. Supposons que $\bar\alpha<\alpha$ et trouvons $r,s\in\mathbb{Z}_+$ tels que $\bar\alpha<r/s<\alpha$ et qu'un $\bar D\in|r\bar X+s\bar K|$ soit effectif. Posons $d=\langle X,C_1\rangle$. On a $\pi^*(\bar K)=K-C_1$, $\pi^*(\bar X)=X+dC_1$. Ainsi

$$\pi^*(\bar D)\in|rX+sK+(dr-s)C_1|,$$
$$dr-s=\langle rX+sK,C_1\rangle=s\langle rs^{-1}X+K,C_1\rangle<s\langle pq^{-1}X+K,C_1\rangle$$
$$=(nq)^{-1}\langle D,C_1\rangle<0.$$

Ceci montre que $|rX+sK|$ contient un diviseur effectif $\pi^*(D)-(dr-s)C_1$ ce qui contredit $rs^{-1}<\alpha(X)$.

On note maintenant que si le théorème est vrai pour $(\bar V,\bar X)$, en particulier $\langle\alpha\bar X+\bar K,\bar E\rangle=0$, il est vrai pour (V,X) aussi, avec $\bar E$ remplacé par $E=\pi^*(\bar E)$.

Par descente, il reste à le démontrer dans le cas $\gamma=\alpha$. Mais dans ce cas $\alpha X+K$ appartient à la frontière du cône N^1_+ et non seulement du N^1_{eff}.

La théorie de Mori [6] ne donne de renseignements précis que sur la partie de $\partial N^1_{\text{eff}}$ appartenant au demi-espace $\langle *,K\rangle\leqq 0$. Ainsi, nous commencerons par vérifier que $\langle D,K\rangle\leqq 0$ où $D=\alpha X+K$. D'après Mori, si l'on a $\langle Y,Y\rangle>0$ pour un $Y\in N^1_{\text{eff}}$, alors $Y\in \mathring{N}^1_{\text{eff}}$. Donc $\langle D,D\rangle\leqq 0$. Mais comme $D\in\partial N^1_+$ on a aussi $\langle D,D\rangle\geqq 0$. En somme, $\langle D,D\rangle=0$ et

$$\langle\alpha X+K,K\rangle=-\alpha\langle\alpha X+K,X\rangle\leqq 0.$$

Si $\langle D,K\rangle=\langle D,X\rangle=0$, on a $D=0$ en $\mathrm{NS}(V)$, donc $-K$ est ample et $\langle\alpha X+K,E\rangle=0$ pour n'importe quelle courbe de V. Pour satisfaire $p_a(E)=0$ et $\langle E,E\rangle=0,1$ on peut prendre $E=$ une droite si $V=\mathbb{P}^2$; $E=$ une fibre d'une projection sur $\mathbb{P}^1$ si $V=\mathbb{P}^1X\mathbb{P}^1$; enfin $E=\pi^*(E')$ si $\pi:V\to W$ est un morphisme birationnel sur un modèle minimal W et E' de W est choisi comme ci-dessus.

Finalement, pour traiter le cas $\langle\alpha X+K,K\rangle<0$ on utilise le théorème de Mori. Tout d'abord, ce théorème (ou celui de Kawamata) garantit que $\alpha=pq^{-1}$ soit rationnel. On va montrer que, pour $Y=pX+qK$, on peut trouver E parmi les composantes d'un élément de $|Y|$. En effet, pour tout $0<\varepsilon<pq^{-1}$ on sait, d'après Mori [6], décomposer

$$Y=\sum_{i=1}^{n} a_i C_i + y,$$

où $a_i>0$, $y\in N^1_{\mathrm{eff}}$, C_i sont des courbes extrémales et $\langle \varepsilon X+K, y\rangle \geqq 0$. Comme $\langle Y,Y\rangle=0$ et $\langle Y,C_i\rangle \geqq 0$, $\langle Y,y\rangle \geqq 0$, on a $\langle Y,C_i\rangle=\langle Y,y\rangle=0$, d'où $0=\langle pX+qK,y\rangle=q\langle \varepsilon X+K,y\rangle+(p-q\varepsilon)\langle X,y\rangle$. Il s'ensuit que $\langle X,y\rangle=0$, donc $y=0$ parce que X est ample.

S'il existe une C_i avec $\langle C_i,C_i\rangle=0$ posons $E=C_i$.

Sinon, on a $\langle C_i,C_i\rangle=-1$ pour tous i et $\langle C_i,C_j\rangle=1$ pour i,j convenables car $\langle Y,Y\rangle=0$. Dans ce cas, $E=C_i+C_j$ convient.

Il reste à voir qu'une telle E avec $\langle E,E\rangle=0$ est une fibre d'un morphisme $\pi: V\to W$, $\dim(W)=1$. Cela se voit du fait que dans l'espace engendré par $[C_1],\dots,[C_n]$ tous les nul-vecteurs doivent être proportionnels car la restriction de $\langle\ ,\ \rangle$ sur cet espace est négative sémi-définie. Ainsi Y est proportionnel à E qui est une fibre de la projection de V sur $W=\operatorname{Proj}\left(\bigoplus_n \Gamma(O(nY))\right)$. □

On peut préciser un peu ce théorème dans des cas spéciaux, comme le suivant.

4.3. **Théorème.** *Soit V une surface de del Pezzo. Soit φ un morphisme $V\to\mathbb{P}^2$ (ou $\mathbb{P}^1$) qui contracte un système maximal de courbes exceptionnelles de V (et éventuellement projette $\mathbb{P}^1\times\mathbb{P}^1$ sur $\mathbb{P}^1$). Soit $E_\varphi=\varphi^*(E)$, E une droite (ou un point). Alors*

$$N^1_{\mathrm{eff}}(V)=\{x\in \mathrm{NS}(V)\otimes\mathbb{R} \mid \forall\varphi, \langle E_\varphi, x\rangle \geqq 0\}.$$

Corollaire. *Si X est un diviseur ample de V, on a*

$$\alpha(X)=\max_\varphi \langle -K, E_\varphi\rangle / \langle X, E_\varphi\rangle.$$

Démonstration. Posons $N=\{x \mid \forall\varphi, \langle x, E_\varphi\rangle \geqq 0\}$. Evidemment, c'est un cône convexe contenant N^1_{eff}. Il suffit donc de montrer que pour chaque point $x\in\partial N^1_{\mathrm{eff}}$ il existe une φ telle que $\langle E_\varphi, x\rangle=0$. On peut même se borner aux points rationnels d'un voisinage arbitraire de l'origine. Comme $-K\in \dot{N}^1_+$, on peut trouver un tel voisinage U que pour tout $x\in U$ rationnel il existe un diviseur ample X tel que $x=pq^{-1}X+K$. Appliquons le théorème 4.2 pour trouver une courbe E avec $\langle E,x\rangle=0$. Si $W=\mathbb{P}^2$, on a $E=E_\varphi$. Autrement, en contractant les composantes exceptionnelles de fibres de $V\to W$ on se réduit au cas où φ est un composé $V\to\mathbb{P}^1\times\mathbb{P}^1\to\mathbb{P}^1$. □

4.4. *Une description explicite.* Si $V\neq\mathbb{P}^1\times\mathbb{P}^1$, on peut obtenir V en faisant éclater $0\leqq r\leqq 8$ points de $\mathbb{P}^2$ en position suffisamment générale.

Le cas $\mathbb{P}^1\times\mathbb{P}^1$ peut être traité directement.

Pour $V=\mathbb{P}^2$ avec r points éclatés en $D_1,\dots,D_r$, on identifie $\mathrm{NS}(V)$ avec $\mathbb{Z}^{r+1}$ par

$$E=aL-b_1l_1-\dots-b_rl_r=(a,b_1,\dots,b_r);$$
$$L=[\pi^*\mathcal{O}(1)],\quad l_i=[\mathcal{O}(D_i)].$$

Les classes E_φ du théorème 4.3 vérifient

$$\langle E, E\rangle = \varepsilon, \quad \langle -K, E\rangle = 2+\varepsilon, \quad \varepsilon = 0 \text{ ou } 1;$$
$$\langle E, L\rangle \geqq 0, \quad \langle E, l_i\rangle \geqq 0.$$

Inversement, on peut montrer que chaque élément de $\mathrm{NS}(V)$ avec ces propriétés est du type E_φ. Ainsi, pour énumerer $\{E_\varphi\}$, il suffit d'énumérer les solutions entières non-négatives du système

$$a^2 - b_1^2 - \dots - b_r^2 = \varepsilon,$$
$$3a - b_1 - \dots - b_r = 2+\varepsilon.$$

Pour $r \leqq 8$, ils sont en nombre fini. Le tableau suivant contient la liste complète de ces solutions avec $r \leqq 7$ et $b_1 \geqq \dots \geqq b_r$ arrangées par la croissance de r. On a donc pour $r \leqq 7$:

$$\alpha\left(AL - \sum_{i=1}^{r} B_i l_i\right) = \max_{\substack{\sigma\in S_r \\ (a, b_i)}} \left\{\frac{2+\varepsilon}{Aa - \sum_i B_i b_{\sigma(i)}}\right\},$$

ou (a, b_i) parcourent les lignes du tableau correspondantes. Par exemple, pour $r=3$:

$$\alpha\left(AL - \sum_{i=1}^{3} B_i l_i\right) = \max\left\{\frac{3}{A}, \frac{2}{A-B_i}, \frac{3}{2A - B_1 - B_2 - B_3}\right\}.$$

Au numéro suivant, on va voir que la même formule donne β_U.

a	b_1	b_2	b_3	b_4	b_5	b_6	b_7	$2+\varepsilon$
1	0	0	0	0	0	0	0	3
1	1	0	0	0	0	0	0	2
2	1	1	1	0	0	0	0	3
2	1	1	1	1	0	0	0	2
3	2	1	1	1	1	0	0	3
3	2	1	1	1	1	1	0	2
4	2	2	2	1	1	1	0	3
5	2	2	2	2	2	2	0	3
4	3	1	1	1	1	1	1	3
5	3	2	2	2	1	1	1	3
6	3	3	2	2	2	2	1	3
7	3	3	3	3	2	2	2	3
8	3	3	3	3	3	3	3	3

5. Le cas des surfaces: arithmétique

5.1. **Théorème.** *Soit X un diviseur ample de la surface projective V tel que $\alpha = \alpha(X) > 0$. Alors pour tout ouvert de Zariski dense $U \subsetneqq V$ il existe une extension* *finie k' du corps de base tel que*

$$\beta_{U\otimes k'}(X) \geqq \alpha(X).$$

Démonstration. Nous appliquerons à (V, X) le théorème 4.2 et prendrons pour k' un corps de définition commun de morphisme φ et d'une courbe E telle que $E \cap U \neq \emptyset$, $E \simeq \mathbb{P}^1$. Nous noterons k' par k.

Comme $h_X(x) \leqq h_{\varphi_*(X)}(\varphi(x))$ en dehors d'un nombre fini de courbes il suffit de démontrer le théorème dans deux cas; a) $V = \mathbb{P}^2$; b) $V =$ un fibré en coniques lisses. Pour $V = \mathbb{P}^2$, on a $\beta_U(-K) = 1 = \alpha(-K)$ pour tout ouvert non-vide parce que c'est vrai pour $U = V$ et que le nombre de points de hauteur bornée du complément de U est négligible en comparaison avec ce nombre de U.

Pour V un fibré en coniques lisses, soit $X = -aK + bE$, E la classe d'une fibre. On a

$$\alpha(X) = \langle -K, E\rangle / \langle X, E\rangle = 1/a,$$

$$\beta_U(X) \geqq \beta_{U\cap E}(X) = \beta_E(X|E) = \beta_E(-aK_E) = 1/a. \quad \square$$

5.2. **Théorème.** *Soit V une surface de del Pezzo, $U \subsetneqq V$ un sous-ensemble du complément de l'union de courbes exceptionnelles de V. Alors*

$$\beta_U(X) \leqq \beta_U(-K)\alpha(X).$$

Démonstration. Posons $\alpha = \alpha(X)$. Comme $\alpha X \in -K + \partial N^1_{\mathrm{eff}}$, on a $h_{\alpha X} \geqq \mathrm{const}\, h_{-K}$ en dehors de composantes fixes de $|n(\alpha X + K)|$ qui sont des courbes exceptionnelles. Ainsi

$$\beta_U(\alpha X) = \beta_U(X)/\alpha \leqq \beta_U(-K). \quad \square$$

En comparant les Théorèmes 4.3, 5.1 et 5.2 on obtient

5.3. **Corollaire.** *Soit V une surface de del Pezzo dont toutes les courbes exceptionnelles sont définies sur k et soit U le complément de ces courves. Alors on a*

$$\beta_U(-K) = 1 \Leftrightarrow \forall X, \beta_U(X) = \alpha(X). \quad \square$$

Nous espérons que $\beta_U(-K) = 1$. Pour le moment, nous ne savons le démontrer que pour $k = \mathbb{Q}$, $V = \mathbb{P}^2$ éclaté en $\leqq 3$ points rationnels. En effet, on a le résultat un peu plus précis suivant qu'on peut sans doute généraliser au moins à tous k.

5.4. **Proposition.** *Pour $r \leqq 3$, on a*

$$N_U(-K_V; H) \simeq cH(\log H)^r.$$

Démonstration. Nous ne considérons que le cas $r = 3$ et nous nous bornons à obtenir une estimation à $\exp(O(1))$ près. On peut supposer que V s'obtient par éclatement de points $(1,0,0)$, $(0,1,0)$ et $(0,0,1)$. Si $x \in U$ se projette en $(x_1, x_2, x_3) \in \mathbb{P}^2$, $x_i \in \mathbb{Z}$, $pgcd(x_i) = 1$, $|x_1| \geqq |x_2| \geqq |x_3| > 0$, on a (voir la démonstration de la proposition 1.6):

$$h_{-K}(x) = |x_1|^2 |x_2| / d_1 d_2 d_3,$$

où $d_i = pgcd(x_j, x_k)$, $\{i, j, k\} = \{1, 2, 3\}$; en permutant les coordonnées on obtient la hauteur des autres points. Ainsi, il suffit de compter le nombre de solutions entières du systême

$$\begin{aligned} &|x_1|^2|x_2| \leqq Hd_1d_2d_3, \\ &0<|x_3| \leqq |x_2| \leqq |x_1|, \\ &pgcd(x_i)=1, \\ &pgcd(x_i,x_j)=d_k. \end{aligned}$$

En posant $x_1=y_1d_2d_3$, $x_2=d_1y_2d_3$, $x_3=d_1d_2y_3$ on obtient un système équivalent

$$\begin{aligned} &|y_1^2y_2d_2d_3^2| \leqq H, \\ &0<|y_3d_1d_2| \leqq |y_2d_1d_3| \leqq |y_1d_2d_3|, \\ &pgcd(y_i,y_j)=1, \quad i \neq j, \\ &pgcd(d_i,d_j)=1, \quad i \neq j, \\ &pgcd(y_i,d_i)=1. \end{aligned}$$

Le nombre de solutions ne change qu'à $\exp(O(1))$ près si l'on oublie les conditions $pgcd=1$ et ne considère que les solutions positives. Alors la seconde inégalité est équivalente à

$$0<y_3 \leqq y_2d_3/d_2, \qquad 0<d_1 \leqq y_1d_2/y_2.$$

Il y $a \simeq y_1d_3$ paires (y_3,d_1) vérifiant ces conditions. Ainsi, posant $\tau(a)=\sum_{d/a} 1$ on a

$$\begin{aligned} N_U(-K;H) &\sim \sum_{y_1^2d_3^2y_2d_2 \leqq H} y_1d_3 \sim \sum_{a^2b \leqq H} a\tau(a)\tau(b) \sim \sum_{b \leqq H} \tau(b) \sum_{a \leqq H^{1/2}b^{-1/2}} a\tau(a) \\ &\sim H \sum_{b \leqq H} \tau(b)b^{-1}\log(Hb^{-1}) \sim H(\log H)^3. \quad \square \end{aligned}$$

Bibliographie

1. Franke, J., Manin, Yu., Tschinkel, Yu.: Rational points of bounded height on Fano varieties. Invent. Math. **95**, 421–435 (1989)
2. Schanuel, S.: Heights in number fields. Bull. Soc. Math. France **107**, 433–449 (1979)
3. Cornell, G., Silverman, J.H. (eds.): Arithmetic geometry. Berlin Heidelberg New York: Springer 1986
4. Serre, J.-P.: Autour du théorème de Mordell-Weil. Cours au Collège de France, 1980–81. Braunschweig: Vieweg 1989
5. Eisenbud, D., Harris, J.: The Kodaira dimension of the moduli spaces of curves of genus >23. Invent. Math. **90**, 359–388 (1987)
6. Mori, S.: Threefolds whose canonical bundles are not numerically effective. Ann. Math. **115**, 113–176 (1982)
7. Langlands, R.P.: On the functional equations satisfied by Eisenstein series. (Lect. Notes Math., Vol. 544) Berlin Heidelberg New York: Springer 1976
8. Mori, S., Mukai, Sh.: The uniruledness of the moduli space of curves of genus 11. In: (Lecture Notes Math., Vol. 1016, pp. 334–353). Springer 1983
9. Cohen, S.D.: The distribution of Galois groups and Hilbert's irreducibility theorem. Proc. Lond. Math. Soc. **43**, 227–250 (1981)
10. Vojta, P.: Diophantine approximations and value distribution theory. (Lecture Notes in Math., Vol. 1239). Berlin Heidelberg New York: Springer 1987

11. Mori, S.: Classification of higher-dimensional varieties. In: Proc. Symp. Pure Math., Vol. 46, pp. 269–331. Providence (1987)
12. Fujiwara, M.: Upper bounds for the number of lattice points on hypersurfaces. In: Number theory and combinatorics. J. Akiyama et al. (eds.), pp. 89–96. Singapore: World Scientific 1985
13. Shparlinsky, I.E., Skorobogatov, A.N.: Exponential sums and rational points on complete intersections. Preprint 1989
14. Faltings, G.: Calculus on arithmetic surfaces. Ann. Math. **119**, 387–424 (1984)
15. Harder, G.: Chevalley groups over function fields and automorphic forms. Ann. Math. **100**, 249–306 (1974)

Reçu le 3 mai 1989

Compositio Mathematica **85**: 315–332, 1993.

Points of bounded height on del Pezzo surfaces

YU. I. MANIN[1] and YU. TSCHINKEL[2]

[1]*Steklov Mathematical Institute, 42 Vavilova, Moscow 117966, USSR*
[2]*Department of Mathematics, M.I.T., Cambridge (MA) 02139, U.S.A.*

Received 7 May 1991; accepted 14 February 1992.

0. Introduction

0.1. Heights

In this paper, we prove some results on the asymptotic behaviour of the number of algebraic points of bounded height on del Pezzo (and more general) rational surfaces. The basic (Weil) height on a coordinatized projective space over an algebraic number field k is given by the formula

$$h_{\mathcal{O}(1),k}(x_0:\cdots:x_n)=\prod_{v\in M_k}\max(|x_i|_v) \tag{0.1}$$

where the product is taken over all places v of k, and local norms $|x_i|_v$ are defined as multipliers of local additive Haar measures on k_v. More generally, we shall consider various Weil heights $h_{L,k,s',s''}$ where L is an invertible sheaf on an algebraic variety V defined over k, represented as a quotient of two ample sheaves and s' (resp. s'') are some families of sections of (L resp. L'') generating these sheaves. We shall usually denote such a height simply h_L. The chosen normalization (0.1) makes our heights non-invariant with respect to ground field extensions. In compensation, they possess the following "linear growth property". Denote by $N_{\mathbb{P}^n}(-K,H)$ the number of points in $\mathbb{P}^n(k)$ whose anticanonical height $h_{-K}(x)=h_{O(1)}(x)^{n+1}$ does not exceed H. Put $d=[k:\mathbb{Q}]$. Then

$$N_{P^n}(-K,H)=cH+\begin{cases}O(H^{1/2}\log H) & \text{if } d=n=1;\\ O(H^{1-1/d(n+1)}) & \text{otherwise.}\end{cases} \tag{0.2}$$

This is a restatement of Schanuel's theorem (cf. [Se]). The constant c depends on k, n, and the exact normalization of the height.

0.2. Del Pezzo cubic and quartic surfaces

Consider smooth surfaces V_5 (resp. V_6) over k, which are embedded into $\mathbb{P}^4$ (resp.

$\mathbb{P}^3$) as a complete intersection of two quadrics (resp. as a cubic). It is well known that over $\bar{k}$, V_5 contains 16 lines, whereas V_6 contains 27 lines. Let us call V *split* if all lines are defined already over k.

One of the results of this paper, having the most direct number-theoretical meaning, can be stated as follows. As above, put $N_V(L, H) = \operatorname{card}\{x \in V(k) | h_l(x) \leqslant H\}$. Notice that $-K_{V_a}$ is isomorphic to $\mathcal{O}(1)|V_a$.

0.3. THEOREM. (a) *For a split* $V = V_5$ *and any* $\varepsilon > 0$ *we have*

$$N_V(O(1), H) = cH^2 + \begin{cases} O(H^{5/4+\varepsilon}) & \text{if } k = \mathbb{Q}, \\ O(H^{3/2+\varepsilon}) & \text{in general.} \end{cases} \tag{0.3}$$

(b) *For a split* $V = V_6$ *and any* $\varepsilon > 0$ *we have*

$$N_V(\mathcal{O}(1), H) = cH^2 + O(H^{5/3+\varepsilon}) \quad \text{if } k = \mathbb{Q},$$

$$cH^2 \leqslant N_V(\mathcal{O}(1), H) \leqslant c'H^{2+\varepsilon} \quad \text{in general.} \tag{0.4}$$

Actually, we prove a more precise statement. Namely, the leading term cH^2 in (0.3), (0.4) counts the number of points of height $\leqslant H$ *on lines* whereas the remainder term is an estimate for $N_{U_a}(\mathcal{O}(1), H)$, where $U_a = V_a \setminus \{\text{union of lines}\}$. This structure of the leading term follows from the Schanuel theorem, and an estimate for U_a is our main concern.

0.4. Strategy of proof

More generally, let V_a be a split del Pezzo surface over k of degree $d = K_V^2 = 9 - a$ (cf. e.g. [Ma]). Denote by U_a the complement to the union of all exceptional curves (lines) on V_a. For $a \leqslant 4$ over $\mathbb{Q}$ (and for $a \leqslant 3$ in general) we prove directly in section 1 that

$$N_{U_a}(-K, H) = O(H^{1+\varepsilon}). \tag{0.5}$$

The proof uses combinatorial properties of the intersection graph of lines and arithmetics of partial (finite) heights. Then we represent V_5, V_6 as a blow up of V_3 or V_4 and apply an estimate of exceptional heights in terms of anticanonical heights, using algebro-geometric arguments.

We know of only one result of this type for cubic surfaces previously discussed in the literature. Namely, C. Hooley ([Ho]) proved by a sieve method that the number of points of $x_0^3 + x_1^3 + x_2^3 + x_3^3 = 0$ over $\mathbb{Q}$ with height $\leqslant H$ outside of lines is $O(H^{5/3+\varepsilon}$. However, this surface is not split, so that Hooley's theorem is not contained in ours.

0.5. *A perspective*

In [BaMa], we proposed the following general notions related to counting points of bounded height. Let V be a projective variety over a number field k, L an ample sheaf on V, U a quasiprojective subset of V. Put

$$\beta_U(L) = \begin{cases} \limsup \log N_U(L, H)/\log H, & \text{if } \operatorname{card}(U) = \infty; \\ -\infty & \text{otherwise.} \end{cases}$$

If $U(k)$ is infinite, there exists a unique minimal Zariski closed subset Z in U such that

$$\beta_U(L) = \beta_Z(L) > \beta_{U \setminus Z}(L).$$

Of course, Z may coincide with U. This subset is called the (minimal) *accumulating subset* (in U with respect to L). This notion does not depend on the exact choice of a height h_L used to count points.

Repeating this construction, we obtain a sequence of open subsets

$$V = V_0 \supset V_1 \supset \cdots \supset V_n \supset \cdots$$

such that $Z_i = V_i \setminus V_{i+1}$ is the minimal accumulating subset in V_i and a sequence of real numbers $\beta_i = \beta_{V_i}(L)$. The sequence $\{V_i\}$ is called an *arithmetical stratification* and β_i are called *growth orders*. If $\beta_i = 1$, we say that V_i has *a linear growth property* (with respect to L).

According to the philosophy explained in [BaMa] we expect that if the ground field is made sufficiently large, then both the arithmetical stratification and its growth order should be computable in algebro-geometric terms (although their definitions involve arithmetics). In particular, we expect that for Fano varieties the arithmetical stratification stabilizes at a finite step, and the last growth order with respect to $-K$ equals 1 (again, if the ground field is large enough).

From this viewpoint, what we prove here means in particular that if all exceptional curves on a two-dimensional Fano variety over $\mathbb{Q}$ of degree $d \geqslant 3$ are defined over $\mathbb{Q}$, they form the first accumulating subset. Moreover, the complement has the linear growth property for $d \geqslant 5$.

0.6. *Plan*

The paper is structured as follows. In Section 1, we prove directly (0.5). In Section 2, we estimate exceptional heights and deduce the Theorem 0.3. In

Section 3, we discuss the growth orders with respect to other ample sheaves, complementing the results of [BaMa]. Finally, in Section 4, we derive an approximation to the linear growth conjecture for anticanonical heights on rational surfaces obtained by blowing up rational cycles on a projective plane.

1. Finite heights and del Pezzo surfaces of degrees 5 and 6

1.1. Finite heights

Let V be a projective variety defined over k, L an invertible sheaf on it, h_L a Weil height. A choice of an isomorphism $L \simeq \mathcal{O}(D)$ allows one to decompose h_L into a product of archimedean and non-archimedean (finite) partial heights: for $x \in V \setminus D$, $h_L(x) = h_{D,\infty} h_{D,f}(x)$. In an Arakelov set up, $h_{D,f}(x)$ is just the product of the exponentiated intersection indices $\langle D, x \rangle_v$ over archimedean places $v \in M_{k,f}$. We do not want to choose a **Z**-model and hermitean metrics, so that we will rely upon the more elementary version of A. Weil's distributions. In particular, let D_i be the divisor $x_i = 0$ on $\mathbb{P}^n$ (see (0.1)). Then

$$h_{D_i,f}(x_0 : \cdots : x_n) = \prod_{v \in M_{k,f}} \max_j (|x_j/x_i|_v) \tag{1.1}$$

Our basic technique consists in studying finite heights with respect to lines on the del Pezzo surfaces (it was applied by V. Batyrev in the context of toric varieties). In order to control the resulting loss of information, we start with looking more closely at $\mathbb{P}^n$.

Let A be the ring of integers in K, A^* the group of units. Choose a family of ideals $a_1, \ldots, a_h \subset A$ representing all ideal classes, and put

$$A^{n+1}_{\text{prim}} = \{(x_0 : \cdots : x_n) \in A^{n+1} \mid \exists i, \gcd(x_0, \ldots, x_n) = a_i\}$$

A^* acts diagonally upon A^{n+1}_{prim}, and we can identify $\mathbb{P}^n(k)$ with $A^{n+1}_{\text{prim}}/A^*$ When we represent a point by its coordinates we usually take coordinates in A^{n+1}_{prim}.

From (1.1) it follows that, for $(x_0 : \cdots : x_n) \in A^{n+1}_{\text{prim}}$, $x_i \neq 0$ we have

$$h_{D_i,f} = d_i(x) N_{k/\mathbb{Q}}(x_i) \tag{1.2}$$

where $d_i : \mathbb{P}^n(k) \to \mathbb{Q}_{>0}$ is a finite-valued function. In particular, finite heights are "almost integers" for any choice of local Weil's functions (or Arakelov model).

1.2. LEMMA. *Let* $\eta : (\mathbb{P}^n \setminus \bigcup_{i=0}^n D_i)(k) \to \mathbb{Q}^{n+1}_{>0}$ *be the map*

$$\eta(x) = (h_{D_0,f}(x), \ldots, h_{D_n,f}(x)).$$

Then the number of points x with $h_{\mathcal{O}(1)}(x) \leqslant H$ having the same image $\eta(x)$ is bounded by $O(1)$, if $k = \mathbb{Q}$, and by $O(H^\varepsilon)$ for any $\varepsilon > 0$, in general.

Proof. For $k = \mathbb{Q}$, (1.2) shows that the knowledge of $h_{D_i,f}(x)$ allows us to reconstruct projective coordinates of x up to a finite bounded ambiguity.

In general, knowing the norm $N_{k/\mathbb{Q}}(x_i)$, we can reconstruct first the ideal of x_i in A in no more than $O(H^\varepsilon)$ ways. In fact, (x_i) divides $(N_{k/\mathbb{Q}}(x_i))$, and the number of ideals dividing $(N_{k/\mathbb{Q}}(x_i))$ is bounded by $C(\varepsilon)(N_{k/\mathbb{Q}}(x_i))^\varepsilon$ (look at the Dedekind zeta of k).

Fix now a family of ideals (x_i) corresponding to a given $\eta(x)$. From (1.2) one sees that a set of such points is a union of a bounded number of subsets $\{(x_0:\varepsilon_1 x_1:\varepsilon_2 x_2:\cdots:\varepsilon_n x_n)\}$ where $(x_0:\cdots:x_n)$ are fixed, and $\varepsilon_i \in A^*$ are variable (ε_0 can be killed by the overall multiplication by A^*). Now,

$$h_{\mathcal{O}(1)}(x_0:\varepsilon_1 x_1:\varepsilon_2 x_2:\cdots:\varepsilon_n x_n) = \prod_{v\in M_\infty} \max(1, |\varepsilon_i x_1/x_0|_v | i \geqslant 1) h_{D_0,f}(x_0:\cdots:x_n).$$

The left hand side is bounded by H only if $c_1 H^{-2(n-1)} \leqslant |\varepsilon_i|_v \leqslant c_2 H^2$ for all $i = 1,\ldots,n$; $v \in M_\infty$, and certain c_1, $c_2 > 0$. From the Dirichlet theorem it follows that there are no more than $O((\log H)^r)$ units with this property.

Notice that lemma 1.2 is still true if finite heights (1.2) are replaced by equivalent ones.

1.3. *Finite exceptional heights on del Pezzo surfaces*

Let now V_a be a split del Pezzo surface of degree $9 - a$ over k, E_a the set of exceptional curves (lines) on V_a, $U_a = V_a \backslash \cup_{l\in E_a} l$. Choose a family of finite exceptional heights $h_{l,f}$, $l \in E_a$. We may and will assume that they take values in $\mathbf{Z}_{>0}$. Put now

$$\tilde{\eta}(x) = (h_{l,f}(x)) \in \mathbf{Z}_{>0}^{E_a}, \qquad x \in U_a(k)$$

Generally, we compute $h_{l,f}(x)$ as follows. We represent l as an infimum (or gcd) of divisors D_i for which $h_{D_i,f}(x)$ are known (e.g. by (1.2)), and then $h_{l,f}(x) = \gcd(h_{D_i,f}(x))$.

1.4. LEMMA. *If $a \geqslant 3$ then the number of points $x \in U_a(k)$ with $h_{-K}(x) \leqslant H$ having the same image $\eta(x)$ doesn't exceed $O(1)$ for $k = \mathbb{Q}$, and $O(H^\varepsilon)$ for any $\varepsilon > 0$.*

Proof. Consider a birational morphism $\pi: V_a \to \mathbb{P}^2$ blowing down pairwise disjoint lines on V_a. Choose three lines D_1, D_2, D_3 on $\mathbb{P}^2$ joining pairwise three fundamental points of π^{-1} on $\mathbb{P}^2$. For $\{i,j,k\} = \{1,2,3\}$, let $l_i = \pi^{-1}(D_j \cap D_k)$, and $l_i' = \pi^{-1}(D_i)$ (proper inverse image).

Then, by functoriality,

$$h_{D_i,f}(\pi(x)) = h_{l'_i,f}(x)h_{l_j,f}(x)h_{l_k,f}(x)d'_i(x)$$

where d'_i are finite-valued functions. Hence, knowing $\tilde{\eta}(x)$, we can reconstruct $\eta(\pi(x))$ with a bounded indeterminacy, and then $\pi(x)$ with indeterminacy $O(1)$ for $k = \mathbb{Q}$, and $O(H^{\varepsilon})$ for any $\varepsilon > 0$ generally, in view of Lemma 1.2, where $\tilde{H}$ is a bound for $h_{\mathcal{O}(1)}(\pi(x))$. But in view of Lemma 2.2 below, one can take for $\tilde{H}$ a fixed positive power of H.

1.5. REMARK. We shall consider the family of finite heights

$$h_{l_i,f}(x) := l_i(x)$$

as a marking of the vertices of the intersection graph of E_a by positive integers. The proof of Lemma 1.4 shows that (on the set of points $x \in U_a(k)$ with $h_{-K}(x) \leqslant H$)) the total marking can be reconstructed with indeterminacy $O(1)$, resp. $O(H^{\varepsilon})$, from the marking of any complete subgraph of the form

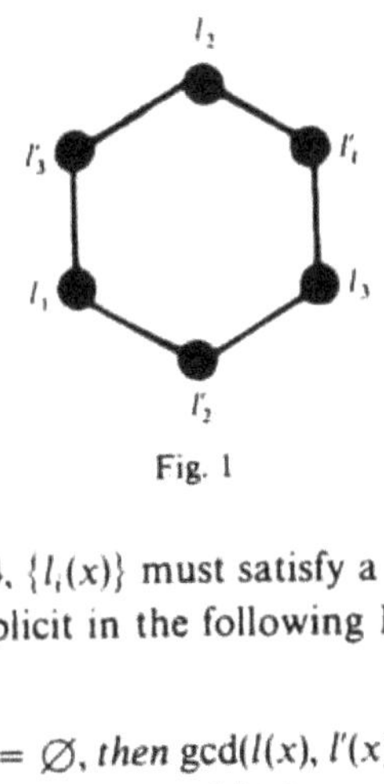

Fig. 1

This means that for $a \geqslant 4$, $\{l_i(x)\}$ must satisfy a system of strong constraints. Some of them are made explicit in the following Lemma.

1.6. LEMMA. (a). *If* $l \cap l' = \varnothing$, *then* $\gcd(l(x), l'(x))$ *is a finite valued function* (*we shall express this by saying that* $l(x)$ *and* $l'(x)$ *are almost relatively prime*).

(b). *Consider a complete subgraph* Δ *of* E_a *of the form*

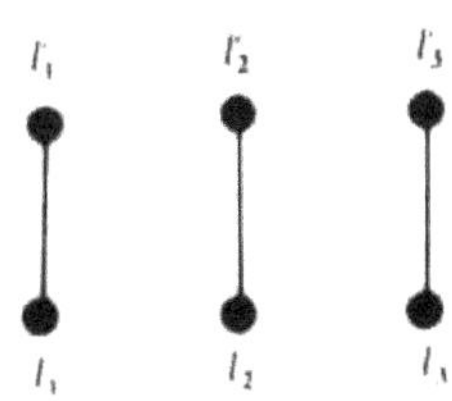

Fig. 2

Then there exist functions

$$(\sigma_1, \sigma_2, \sigma_3)\colon U_a(k) \to A^3_{\mathrm{prim}}$$

and finite-valued functions $d_i\colon U_a(k) \to \mathbb{Q}^*$, $i = 1, 2, 3$, such that

$$\sigma_1(x) + \sigma_2(x) + \sigma_3(x) = 0,$$
$$N_{k/\mathbb{Q}}(\sigma_i(x)) = d_i(x)l_i(x)l_i'(x). \tag{1.3}$$

Proof. The first statement is classical, it is due to A. Weil. In order to prove the second statement, we first notice that any complete subgraph of type Δ consists of three degenerate fibers of a morphism $\rho\colon V_a \to P^1$ representing V_a as a conic bundle. Hence we can find three sections s_1, s_2, s_3 of $\rho^*(\mathcal{O}(1))$ such that $s_1 + s_2 + s_3 = 0$ and $l_i \cup l_i' = \{s_i = 0\}$ Primitive coordinates of $\rho(x)$ with respect to these three sections define functions $\sigma_i(x)$. Formula (1.3) then follows from (1.2).

We can now prove two main results of this section.

1.7. THEOREM. *For* $V = V_3$ *over an arbitrary number field* k *we have*:

$$N_{U_3}(-K, H) = \begin{cases} O(H(\log H)^5) & \text{for } k = \mathbb{Q}, \\ O(H^{1+\varepsilon}) & \text{for any } \varepsilon > 0 \text{ in general.} \end{cases}$$

Proof. The intersection graph of E_3 is a hexagon (Fig. 1), and $-K_{V_3} = \Sigma_{i=1}^3 (l_i + l_i')$. Therefore, for $x \in U_3(k)$,

$$h_{-K}(x) = \exp(O(1)) \prod_{i=1}^{3} h_{l_i + l_i'}(x) \geqslant B \prod_{i=1}^{3} l_i(x)l_i'(x).$$

Considering $\{l_i(x), l_i'(x)\}$ as independent integer variables, we see that points x with $h_{-K}(x) \leqslant H$ define $O(H(\log H)^5)$ markings of E_3, whereas every marking corresponds to $O(1)$ (resp. $O(H^\varepsilon)$) points.

1.8. REMARK. For $k = \mathbb{Q}$, in [BaMa] it was proved that

$$N_{U_3}(-K, H) = \exp(O(1))H(\log H)^3.$$

It would be important to know the correct power of logarithm in general.

Furthermore, using the morphism $\pi\colon V_3 \to \mathbb{P}^2$, one sees that $N_{U_3}(-K, H) \geqslant \exp(O(1))H$, so that $\beta_{U_3}(-K) = 1$.

1.9. THEOREM. *For* $V = V_4$ *over* $k = \mathbb{Q}$ *and arbitrary* $\varepsilon > 0$, *we have*

$$N_{U_4}(-K, H) = O(H(\log H)^6).$$

Proof. We start with choosing finite heights $h_{l,f}: U_4(\mathbb{Q}) \to \mathbb{Z}_{>0}$, $l \in E_4$. Represent $U_4(\mathbb{Q})$ as a union of subsets $U^{(l)}$ such that

$$x \in U^{(l)} \Rightarrow l(x) = \min_{l' \in E_4} \{l'(x)\}.$$

Clearly, it suffices to prove the estimate $O(H(\log H)^6)$ for points in every subset $U^{(l)}$. Fix $l = l_0$ and consider a complete subgraph Γ of E_4 of the form

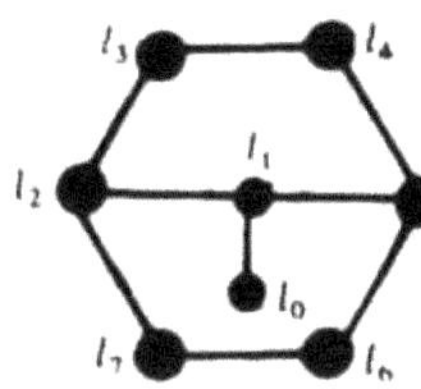

Fig. 3

Since the automorphism group of the intersection graph of E_a, $a \geqslant 3$, is transitive on vertices, it suffices to exhibit one complete subgraph isomorphic to Γ. Representing V_4 by $\mathbb{P}^2$ blown up at points $p_0, \ldots, p_3$, and denoting their preimages λ_i, and inverse image of the line joining p_i, p_j by λ_{ij}, we can describe Γ by the following identifications:

Table 1

l_0	l_1	l_2	l_3	l_4	l_5	l_6	l_7
λ_0	λ_{01}	λ_1	λ_{12}	λ_2	λ_{23}	λ_3	λ_{13}

Using the same model, one sees that $-K = \sum_{i=1}^{5} l_i$. Hence we can weaken the inequality $h_K(x) \leqslant H$ to $\prod_{i=1}^{5} l_i(x) \leqslant H$, which has $O(H(\log H)^4)$ solutions. However, we cannot prove that one can reconstruct x even with indeterminacy $O(H^\varepsilon)$ from $l_1(x), \ldots, l_5(x)$. Therefore we shall apply the following trick. Put $l_2'(x) = [l_2(x)/l_0(x)] \geqslant 1$ and weaken the inequality $h_{-K}(x) \leqslant H$ to

$$l_0(x) l_1(x) l_2'(x) l_3(x) l_4(x) l_5(x) \leqslant H \tag{1.4}$$

Consider the members in the l.h.s. of (1.4) as independent variables. The number of solutions of (1.4) is $O(H(\log(H)^5)$, but now we shall be able to reconstruct x with desired ambiguity. We shall first explain how to set $O(H^\varepsilon)$.

(i) *Reconstruction of* $l_6(x), l_7(x)$. Consider the subgraph of $\Gamma: (l_0, l_1, l_3, l_4, l_6, l_7)$. We can apply to it Lemma 1.4 and obtain the relations (here the assumption $k = \mathbb{Q}$ is used):

$$d_1(x)l_0(x)l_1(x) + d_2(x)l_3(x)l_4(x) + d_3(x)l_6(x)l_7(x) = 0$$

Knowing l_0, l_1, l_3, l_4, we can reconstruct l_6, l_7 in $O(\tau(a))$ ways where τ is the number of divisors, and $a = O(H)$. Clearly, $\max\{\tau(a)|a \leqslant H\} = O(H^\varepsilon)$ for any $\varepsilon > 0$.

(ii) *Reconstruction of* $l_2(x) \bmod l_0(x)$. To do this, consider a different complete subgraph of E_a, isomorphic to the one of Fig. 2. It is not contained in Γ. In terms of the description given at the beginning of the proof it can be identified as $(l_2, l_7; l_4, l_5; l_0, \lambda_{03})$. From the relations

$$d'_1(x)l_2(x)l_7(x) + d'_2(x)l_4(x)l_5(x) + d'_3(x)l_0(x)\lambda_{03}(x) = 0$$

it follows that knowing $l_7(x)$, $l_4(x)$, $l_5(x)$ one can reconstruct $l_2(x) \bmod l_0(x)$ up to a finite ambiguity. Here we use the fact that $l_7(x)$ and $l_0(x)$ are almost relatively prime (Lemma 1.6(a)).

(iii) *Reconstruction of* x. We can now reconstruct $l_2(x)$, because we know $[l_2(x)/l_0(x)]$ and $l_2(x) \bmod l_0(x)$. And, since we know already the marking of a hexagon, we can reconstruct x up to a finite ambiguity.

Now we will prove a sharper estimate replacing H^ε by $(\log H)^6$. We are thankful to Don Zagier for help. We start with a refinement of relations (1.3).

Split V_4 has no moduli, and we can normalise $l_i(x)$ for $x \in U_4(\mathbb{Q})$ in such a way that quadratic relations of Lemma 1.6 take a canonical form. Choose coordinates in $\mathbb{P}^2$ in such a way that $\pi: V_4 \to \mathbb{P}^2$ blows up points $P_1 = (1:0:0)$, $P_2 = (0:1:0)$, $P_3 = (0:0:1)$, $P_4 = (1:1:1)$. Let $x \in U_3(\mathbb{Q})$. Then $\pi(x)$ can be represented by $(x_1, x_2, x_3) \in \mathbf{Z}^3_{\text{prim}}$. Define the following ten integers: $d_i = \gcd(x_j, x_k)$, $y_i = x_i/d_jd_k$, $\{i, j, k\} = \{1, 2, 3\}$; $D = \gcd_{i \neq k}\{y_id_k - y_kd_i\}$; $z_j = |y_id_k - y_kd_i|/D$. One can check that they define the marking of E_4 by the respective finite heights. More precisely, with the notation of the Table 1, one has the following correspondence:

Table 2

l:	λ_0	λ_{01}	λ_1	λ_{12}	λ_2	λ_{23}	λ_3	λ_{13}	λ_{02}	λ_{03}
$l(x)$:	D	z_1	$\|y_1\|$	d_3	$\|y_2\|$	d_1	$\|y_3\|$	d_2	z_2	z_3

The symmetries of this marking are not at all obvious from the direct construction. Here is the list of all marked subgraphs of the type Δ (Lemma 1.6):

One can directly check that every quadratic relation corresponding to such a

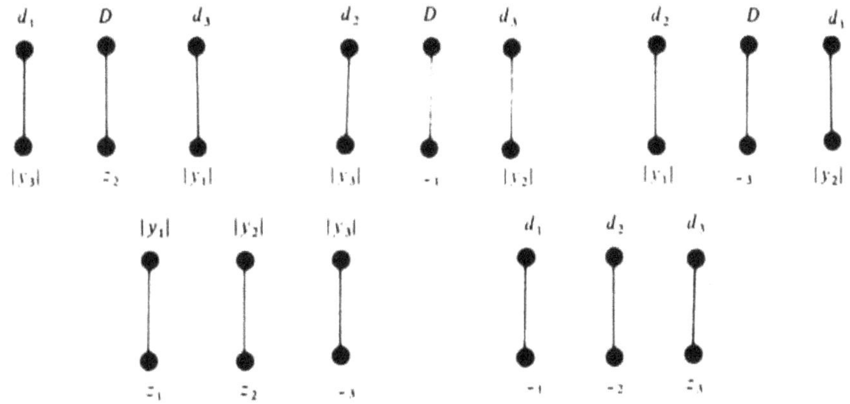

Fig. 4

subgraph can be written in the form: maximal product of two marks connected by an edge equals to the sum of two remaining products.

We can call all markings of E_4, obtained in this way, standard. In the proof of Theorem 1.9 we use a majoration of $h_{-K}(x)$ in terms of the respective marking, but actually we can reconstruct the total anticanonical height:

$$h_{-K}(x) = \exp(O(1))\max\left\{\prod_{i=1}^{5} l_i(x) | l_i = \text{vertices of a pentagon}\right\} \tag{1.5}$$

We leave (1.5) as an exercise to the reader, since we will not use it. We will make further estimates assuming that the marking $\{l_i(x)\}$ is standard.

Returning to the reconstruction procedure (i)–(iii) and writing for brevity l_i instead of former $l_i(x)$ etc. we see that

$$N_{U^{(0)}}(-K; H) \leqslant \text{const.} \sum_{l_0 l_1 l_2 l_3 l_4 l_5 \leqslant H} \tau(\delta_1 l_0 l_1 + \delta_2 l_3 l_4) \tag{1.6}$$

where $\delta_i = 1$ or -1 because our marking is standard. We first estimate the r.h.s. by

$$\begin{aligned}
&\sum_{l_0 l_1 l_3 l_4 \leqslant H} \tau(\delta_1 l_0 l_1 + \delta_2 l_3 l_4) \sum_{l_2 l_5 \leqslant (H/l_0 l_1 l_3 l_4)} 1 \\
&= \sum_{l_0 l_1 l_3 l_4 \leqslant H} \tau(\delta_1 l_0 l_1 + \delta_2 l_3 l_4) \frac{H}{l_0 l_1 l_3 l_4} (\log H + O(H)) \\
&= \left(\sum_{ab \leqslant H} \frac{\tau(a)\tau(b)\tau(\delta_1 a + \delta_2 b)}{ab}\right) (H \log H + O(1))
\end{aligned} \tag{1.7}$$

It remains to show that the r.h.s. sum in (1.7) is $O((\log H)^5)$.

Step 1. We prove that

$$g(n) := \sum_{ab \leqslant n} \tau(a)\tau(b)\tau(\delta_1 a + \delta_2 b) = O(n(\log n)^4)$$

Notice that δ_i can depend on a, b; however, it suffices to prove this estimate for constant δ_i's in two separate cases:

(i) when $\delta_1 = 1$, $\delta_2 = -1$; $a > b$;
(ii) when $\delta_1 = \delta_2 = 1$.

The second case reduces to the first one if one denotes $a + b$ by a', b by b', and observes that $a'b' \leqslant 2n$. Now

$$\sum_{\substack{ab \leqslant n \\ b<a}} \tau(a)\tau(b)\tau(a-b) = \sum_{b \leqslant n} \tau(b) \sum_{b<a\leqslant(n/b)} \tau(a)\tau(a-b); \tag{1.8}$$

For a fixed b

$$\sum_{b<a\leqslant(n/b)} \tau(a)\tau(a-b) \leqslant 4 \sum_{\substack{p\geqslant q>0 \\ r \geqslant s>0 \\ (n/b)\geqslant pq = rs+b}} 1 \leqslant 4 \sum_{0<q,\, s\leqslant\sqrt{n/b}} \sum_{\substack{0<p<(n/bq) \\ pq \equiv b \bmod s}} 1$$

$$\leqslant 4 \sum_{0<q,\, s\leqslant\sqrt{n/b}} \omega(q,s)\left(\frac{n}{bqs}+1\right) \tag{1.9}$$

where

$$\omega(q,s) = \operatorname{card}\{p(\operatorname{mod} s) | pq \equiv b(\operatorname{mod} s)\}$$

equals $d := (q, s)$ if d divides b, and 0 otherwise. We continue to estimate (1.9):

$$\leqslant 4 \sum_{d|b} d \sum_{\substack{0<q,s\leqslant\sqrt{n/b} \\ (q,s)\equiv 0(d)}} \left(\frac{n}{bqs}+1\right)$$

$$\leqslant 4 \sum_{d/b} d \sum_{0<x,y\leqslant d^{-1}\sqrt{n/b}} \left(\frac{n}{bd^2xy}+1\right)$$

$$\leqslant 4\sigma_{-1}(b)\left(\frac{n}{b}\log^2\frac{n}{b} + O\left(\frac{n}{b}\log\frac{n}{b}\right)\right) \tag{1.10}$$

where $\sigma_{-1}(b) = \Sigma_{d/b}\, 1/d$. Now (1.8) becomes

$$\leqslant 4 \sum_{b\leqslant n} \tau(b)\sigma_{-1}(b)\left(\frac{n}{b}\log^2\frac{n}{b} + O\left(\frac{n}{b}\log\frac{n}{b}\right)\right) = O(n(\log n)^4).$$

Step 2. Now apply Abel's summation:

$$\sum_{ab \leqslant H} \frac{\tau(a)\tau(b)\tau(\delta_1 a + \delta_2 b)}{ab} = O((\log H)^5).$$

2. Exceptional heights

In this section, we shall prove the following inductive estimate (cf. 0.5 for notation).

2.1. PROPOSITION. *Assume that for a given ground field k some $a \geqslant 2$ and all split del Pezzo surfaces V_a of degree $9 - a$ over k we have*

$$\beta_{U_a}(-K) \leqslant \beta_a$$

where $U_a = V_a \setminus \{\text{all lines}\}$, and β_a is a constant. Then the same is true for all split del Pezzo surfaces V_{a+1} with β_a replaced by $\beta_{a+1} = (9-a)/(8-a)\beta_a$.

For example, if we know that $\beta_4 = 1$, we deduce that $\beta_5 = 5/4$ and $\beta_6 = 5/3$; and if we know only that $\beta_3 = 1$, we get $\beta_4 = 6/5$, and $\beta_5 = 3/2$, $\beta_6 = 2$. Thus the Theorem 0.3 follows from the Proposition 2.1 and the results of the previous section.

We start with the following auxiliary result.

2.2. LEMMA. *Let $a \geqslant 2$. Denote by $\{l_1, \ldots, l_{e(a+1)}\}$ the set of all exceptional curves of the first kind (lines) on a del Pezzo surface $V = V_{a+1}$ of degree $8-a$. The class of their sum in* Pic(V) *equals* $(e(a+1)/(8-a))(-K_V)$.

Proof. Consider the formal symmetry group W_{a+1} of the configuration of lines $\{l_i\}$, that is, the group of the permutations of lines, conserving their intersection indices. From the classical identification of this group with a Weyl group, it follows that the subgroup of W_{a+1}-invariant elements is cyclic, with generator $-K$. In order to identify the coefficient, it suffices to intersect both sides with $-K$.

2.3. COROLLARY. *Choose some Weil heights h_{-K} and h_{l_i} for all i. Then there exists a constant A such that for every $x \in U_{a+1}(k)$ one can find a line $l = l(x)$ with the property*

$$h_l(x) \leqslant A(h_{-K}(x))^{1/(8-a)}$$

Proof. From the Lemma 2.2 and general properties of heights it follows that

$$\prod_{i=1}^{e(a+1)} h_{l_i}(x) = \exp(O(1))(h_{-K}(x))^{e(a+1)(8-a)} \tag{2.1}$$

Now (2.1) is obvious.

Notice that the same argument shows the existence of another exceptional height $h_{l'}$, for which

$$h_{l'}(x) \geqslant B(h_{-K}(x))^{1/(8-a)},$$

so that exceptional heights have infinitely often the same growth order as ample heights.

2.4. Proof of the proposition 2.1. Fix k, a split del Pezzo surface V_{a+1}, and some heights h_K, h_{l_i} on $V_{a+1}(k)$. Corollary 2.3 allows us to define a partition of $U(k)$ into a finite number of subsets U_l numbered by lines l such that (2.1) is valid in U_l. It suffices to prove that the number of points of $(-K_{a+1})$-height $\leqslant H$ in U_l is $O(H^{\beta_{a+1}+\varepsilon})$ where $\beta_{a+1} = ((9-a)/(8-a))\beta_a$, $K_{a+1} = K(V_{a+1})$. Embed l into a maximal system of pairwise disjoint lines on $V_{a+1}: l_1, \ldots, l_a$, $l_{a+1} = l$ (This is always possible: cf. [Ma]). Denote by $\pi: V_{a+1} \to \mathbb{P}^2$ the morphism which blows down this system. Let Λ be the class of $\pi^*(O(1))$ in $\mathrm{Pic}(V_{a+1})$. Choosing all necessary Weil heights, we have for $x \in U_l$:

$$\begin{aligned} h_{-K_{a+1}}(x) &= \exp(O(1)) h_{3\Lambda - l_1 - \cdots - l_{a+1}}(x) \\ &= \exp(O(1)) h_l(x)^{-1} h_{3\Lambda - l_1 - \cdots - l_a}(x) \geqslant B h_{-K_{a+1}}(x)^{-1/(8-a)} h_{-K_a}(\sigma(x)) \end{aligned}$$

where $\sigma: V_{a+1} \to V_a$ blows down $l_{a+1} = l$ and K_a is the canonical class of V_a. Hence for $x \in U_l$

$$h_{-K_{a+1}}(x) \geqslant C(h_{-K_a}(\sigma(x)))^{(8-a)/(9-a)}$$

Finally, by assumption, the number of points $\sigma(x)$ whose $(-K_a)$-height is bounded by H, does not exceed $O(H^{\beta_a+\varepsilon})$. This finishes the proof.

3. Growth orders with respect to other ample sheaves

Let $V = V_r$ be an arbitrary split del Pezzo surface over a numberfield k, represented as a result of blowing up $r \leqslant 8$ points in $\mathbb{P}^2(k)$, $U = U_r$ the complement to the union of lines, L an arbitrary ample invertible sheaf on V_r.

Denote by $\alpha(L)$ the (unique) rational number such that $\alpha(L)[L] + K_V$ belongs to the boundary of the cone of effective elements of $\mathrm{Pic}(V)$. In [BaMa], the following theorem was proved:

3.1. THEOREM. $\alpha(L) \leqslant \beta_U(L) \leqslant \alpha(L)\beta_U(-K)$.

It follows that for $k = \mathbb{Q}$ and $r \leqslant 4$, we have $\beta_U(L) = \alpha(L)$.

In [BaMa], we have also given an explicit formula for calculation of $\alpha(L)$, in terms of a fixed representation $\pi: V_r \to \mathbb{P}^2$ blowing down r lines. Namely, let $\Lambda = [\pi^*(\mathcal{O}(1))]$. Denote by $l_1, \ldots, l_r$ the classes of blown down lines in $\mathrm{Pic}(V)$, $L = A\Lambda - B_1 l_1 - \cdots - B_r l_r$.

Consider all classes $E = a\Lambda - b_1 l_1 - \cdots - b_r l_r$ with the following properties: E is either a preimage of a line on $\mathbb{P}^2$ under a morphism blowing down a maximal system of r pairwise non-intersecting exceptional curves on V, or a preimage of a generator of a quadric under a morphism blowing down a maximal system of $r - 1$ pairwise non-intersecting exceptional curves on V. Then

$$\alpha(L) = \max_{\varepsilon} \left\{ \frac{2 + \varepsilon}{Aa - B_1 b_1 - \cdots - B_r b_r} \right\}, \qquad \varepsilon = (E \cdot E). \tag{3.1}$$

Clearly, all classes E are contained among the solutions of the following system of diophantine equations:

$$\begin{aligned} &a^2 - b_1^2 - \cdots - b_r^2 = \varepsilon, \qquad \varepsilon = 0 \text{ or } 1; \\ &3a - b_1 - \cdots - b_r = 2 + \varepsilon, \end{aligned} \tag{3.2}$$

for which $a > 0$ and $b_i \geqslant 0$.

We shall give below the complete list of solutions of (3.2) (with $r = 8$, and b numbered in decreasing order). The table in [BaMa] contained only solutions with $r \leqslant 7$, and several solutions with $\varepsilon = 0$ were missing.

3.2. Solving 3.2. This system for $r = 8$ is equivalent to the following one:

$$\begin{aligned} &a^2 - b_1^2 - \cdots - b_8^2 - b_9^2 = -4 - 4\varepsilon; \\ &3a - b_1 - \cdots - b_8 - b_9 = 0; \\ &b_9 = 2 + \varepsilon \end{aligned}$$

From the first two equations it follows that

$$(a - 3b_1)^2 + (a - 3b_2)^2 + \cdots + (a - 3b_9)^2 = 36(1 + \varepsilon).$$

Hence we can solve (3.2) in three steps:

(i) List all $(x_1^{(i)}, \ldots, x_9^{(i)})$ for which

$$x_1^2 + \cdots + x_9^2 = 36(1 + \varepsilon)$$

and $x_j \bmod 3$ do not depend on j.

(ii) Find a 10-tuple $(a^{(i)}; b_1^{(i)}, \ldots, b_9^{(i)})$ corresponding to each entry of the list, with $x_j^{(i)} = a^{(i)} - 3b_j^{(i)}$; leave only those for which $3a^{(i)} = \sum b_j^{(i)}$

(iii). In each progression $(a^{(i)} + 3N, b_1^{(i)} + N, \ldots, b_9^{(i)} + N)$ leave only solutions with positive a and non-negative b's, one of which equals $2 + \varepsilon$. Delete $b = 2 + \varepsilon$ and renumber b's in decreasing order.

3.3. All solutions to (3.2) can be used in (3.1). To prove this one can argue as follows.

First, check directly that all classes E corresponding to the solutions of (3.2) are effective. To establish this, it suffices to present every E as a sum of exceptional classes listed in [Ma], Proposition 26.1.

Second, from Riemann-Roch it follows that $\dim|E| \geqslant 1 + \varepsilon$. Therefore, every point l of the closure of effective cone satisfies $(E \cdot l) \geqslant 0$ for every E.

Hence, even if among E's there were some hyperplanes not contained in the set of walls of this cone, they would not contribute into (3.1).

Solutions of (3.2) with $\varepsilon = 0$

a	b_1	b_2	b_3	b_4	b_5	b_6	b_7	b_8
1	1	0	0	0	0	0	0	0
2	1	1	1	1	0	0	0	0
3	2	1	1	1	1	1	0	0
4	2	2	2	1	1	1	1	0
5	2	2	2	2	2	2	1	0
4	3	1	1	1	1	1	1	1
5	3	2	2	2	1	1	1	1
6	3	3	2	2	2	2	1	1
7	3	3	3	3	2	2	2	1
8	3	3	3	3	3	3	3	1
7	4	3	2	2	2	2	2	2
8	4	3	3	3	3	2	2	2
9	4	4	3	3	3	3	3	2
10	4	4	4	4	3	3	3	3
11	4	4	4	4	4	4	4	3

Solutions of (3.2) with $\varepsilon = 1$

a	b_1	b_2	b_3	b_4	b_5	b_6	b_7	b_8
1	0	0	0	0	0	0	0	0
2	1	1	1	0	0	0	0	0
3	2	1	1	1	1	0	0	0
4	2	2	2	1	1	1	0	0
5	2	2	2	2	2	2	0	0
4	3	1	1	1	1	1	1	0
5	3	2	2	2	1	1	1	0
5	3	3	1	1	1	1	1	1
6	3	3	2	2	2	2	1	0
6	3	3	3	2	1	1	1	1
7	3	3	3	3	2	2	2	0
7	3	3	3	3	3	1	1	1
8	3	3	3	3	3	3	3	0
6	4	2	2	2	2	1	1	1
7	4	3	3	2	2	2	1	1
8	4	3	3	3	3	3	1	1
8	4	4	3	3	2	2	2	1
9	4	4	4	3	3	3	2	1
9	4	4	4	4	2	2	2	2
10	4	4	4	4	4	3	3	1
8	5	3	3	2	2	2	2	2
9	5	3	3	3	3	3	3	1
9	5	4	3	3	3	2	2	2
10	5	4	4	4	3	3	2	2
10	5	5	4	4	4	3	3	2
11	5	5	4	4	4	3	3	2
11	5	5	5	3	3	3	3	3
12	5	5	5	4	4	4	4	2
12	5	5	5	5	4	3	3	3
13	5	5	5	5	5	5	3	3
10	6	3	3	3	3	3	3	3
11	6	4	4	4	3	3	3	3
12	6	5	4	4	4	4	3	3
13	6	5	5	5	4	4	4	3
14	6	5	5	5	5	5	5	3
13	6	6	4	4	4	4	4	4
14	6	6	5	5	5	4	4	4
15	6	6	6	5	5	5	5	4
16	6	6	6	6	6	5	5	5
17	6	6	6	6	6	6	6	6
15	7	5	5	5	5	5	5	5

4. An approximation to the linear growth conjecture

In this section, we prove the following result. Consider a rational surface V which is obtained from $\mathbf{P}^2$ by blowing up a k-rational cycle z, $\pi: V \to \mathbf{P}^2$ (z may have infinitely near components).

4.1. THEOREM. *For every $\varepsilon \geqslant 0$, there exists an integer $N > 0$, an open Zariski dense subset $U \subset V$, a constant $c > 0$ and a partition*

$$U(k) = \bigcup_{\alpha=1}^{\infty} U_\alpha(k)$$

with the following properties

(a) *for all* α, card $U_\alpha(k) \leqslant N$.
(b) *for all* α,

$$\max_{x \in U_\alpha} h_{\mathcal{O}(1)}(\pi(x)) \leqslant c \min_{x \in U_\alpha} h_{\mathcal{O}(1)}(\pi(x)),$$

(c) *one can choose a point x_α in each subset in $U_\alpha(k)$ in such a way that the series*

$$\sum_\alpha h_{-K_V}(x_\alpha)^{-1-\varepsilon}$$

will converge.

Comment. For a del Pezzo surface V, we expect ([BaMa]) that $\beta_U(-K_V) = 1$. This would follow from the Theorem 4.1, if we could take $N = 1$. On the other hand, our statement holds even without assumption that V is del Pezzo. Its main interest probably lies in the method of proof, which is one more variation of the general idea that "heights with respect to the exceptional divisors should be small in average".

4.2. PROOF. Consider a large integer M and M elements $id = g_1, \ldots, g_M$ of $\mathrm{Aut}_k \mathbf{P}^2$ in a sufficiently general position. Put $z_i = g_i(z)$ where z is the cycle we have blown up to obtain V. Denote by W the result of blowing up $\cup z_i$, and by V_i the result of blowing up z_i. Obviously, g_i induces an isomorphism of $V = V_1$ with V_i. Moreover, there is a canonical birational morphism $f_i: W \to V_i$.

In order to shorten notation, we shall consider inverse image maps f_i^* on Pic-groups as embeddings, and f_i themselves as identifications outside their fundamental sets. Let K_i be the canonical class of V_i (in Pic W) and Λ the inverse image of $\mathcal{O}_{\mathbf{P}^2}(1)$. Put $Z_i = 3\Lambda + K_i$.

Denote by $P = P(M)$ the smallest integer for which the linear system

$|P\Lambda - Z_1 - \cdots - Z_M|$ is non-empty. From Riemann-Roch, it follows that $P \leqslant aM^{1/2}$ for an appropriate constant a. As in the proof of the Corollary 2.3, we obtain that for every point $x \in U(k)$ where U is the complement to the fundamental points and fixed components of $|P\Lambda - Z_1 - \cdots - Z_M|$ there exists an $i = i(x)$ such that

$$h_{Z_i}(x) \leqslant C(h_\Lambda(x))^{a/M^{1/2}} \tag{4.1}$$

Here the constant C can be taken independent on i.

Put $\eta = a/M^{1/2}$; taking large M, we can make η arbitrarily small. For a fixed i, let W_i be the set of all points of $U(k)$ for which (4.1) holds. In W_i, we have

$$h_{-K_i}(x) = \exp(O(1)) h_{Z_i}(x)^{-1} h_{3\Lambda}(x) \geqslant B h_\Lambda(x)^{3-\eta} \tag{4.2}$$

From (4.2) and Schanuel's theorem it follows that

$$\sum_{x \in W_i} h_{-K_i}(x)^{-1-\varepsilon}$$

converges for some ε tending to zero when η tends to zero.

Now, if $x \in W_1$ and if the heights are appropriately normalized, then $g_i(x) \in W_i$. It follows that we can construct the partition with the properties stated in the Theorem by collecting in one subset points $\{g_1(x), g_2(x), \ldots, g_M(x)\}$ and then deleting multiple occurrencies.

Acknowledgements

We would like to thank the Department of Mathematics of M.I.T. and the Max-Planck-Institut für Mathematik in Bonn where parts of this work were done for financial support and stimulating atmosphere. We also thank the referee who detected a mistake in the initial version of Lemma 1.2.

References

[BaMa] V. V. Batyrev, Yu. I. Manin. Sur le nombre des points rationnels de hauteur bornée des variétés algébriques. *Math. Annalen*, Bd. 286 (1990), 27–43.

[Ho] C. Hooley. On the numbers that are representable as the sum of two cubes. *Journ. für die reine und angew. Math.*, Bd 314 (1979), 147–173.

[Ma] Yu. I. Manin. *Cubic surfaces*. North-Holland, Amsterdam, 1974, 2nd edition 1986.

[Se] J.-P. Serre. *Lectures on the Mordell-Weil theorem*. Vieweg, Braunschweig-Wiesbaden, 1989.

Compositio Mathematica **85**: 37–55, 1993.

Notes on the arithmetic of Fano threefolds

YU. I. MANIN
Steklov Mathematical Institute, 42 Vavilova, Moscow 117966, Russia

Received 7 May 1991; accepted 17 October 1991

0. Introduction

0.1. Notation

This paper is a continuation of the series of works [FraMaTschi], [BaMa], [MaTschi] devoted to counting points of bounded height on algebraic varieties. We start with recalling some basic notions and notation.

Let V be an projective variety defined over a number field k, L an ample sheaf on V, h_L a height function on $V(k)$ w.r.t. L. We normalize it in such a way that it coincides up to exp(O(1))) with $H_L(x)^{[k:\mathbb{Q}]}$, where $H_L(x)$ is an extension invariant height, defined e.g. in [Se], p. 11. (In the Introduction to [FraMaTschi] we erroneously chose $H_L(x)$ although in the text actually used $h_L(x)$).

For a subset $U \subset V(k)$, put

$$Z_U(L; s) = \sum_{x \in U} h_L(x)^{-s},$$

$\beta_U(L) = \inf\{s \mid Z_U(L; s)\}$ converges$\}$.

Put also

$$N_U(L; H) = \text{card}\{x \in U \mid h_L(x) \leqslant H\}.$$

If U is finite, then $\beta_U(L) = -\infty$; otherwise

$$\beta_U(L) = \lim\sup(\log N_U(L; H)/\log H).$$

0.2. *Arithmetical stratification*

A Zariski closed subset $Z \subset V(k)$ is called *accumulating* (over k, w.r.t. L), if

$$\beta_V(L) = \beta_Z(L) > \beta_{V \setminus Z}(L).$$

For every V, there exists a sequence of Zariski open subsets

$$V = V_0 \supset V_1 \supset V_2 \supset \cdots$$

such that for every i, $V_i \backslash V_{i+1}$ is the minimal accumulating subset in V_i. We shall call this sequence the *arithmetical stratification*, and the numbers $\beta_i = \beta_{V_i}(L)$ its *growth orders*.

We can say that a variety V *has many points* if $\beta_V(L) > 0$ for some (and therefore all) ample L. The calculation of its arithmetical stratification and its growth orders is then the first natural problem. Of course, afterwards we hope to obtain some analytic continuation properties of $Z_{V_i}(L; s)$ and more precise information about $N_{V_i}(L; H)$, for example, asymptotic expressions of the type $CH^{\beta_i}(\log H)^{t_i}$ But this usually involves very subtle arithmetical problems, whereas the structure of the stratification, especially over sufficiently large ground fields, may be guessed (and sometimes proved) using algebro-geometric techniques.

In particular, we can state the following conjectures.

RATIONAL CURVE CONJECTURE (RCC). *If* $\beta_U(L) > 0$ *for a Zariski open subset* $U \subset V$, *then* U *contains an open subset of* $\mathbb{P}^1_k$.

One can construct Enriques and K3 surfaces containing infinitely many k-rational curves (it suffices to furnish one such curve and an infinite automorphism group). In [BaMa], we conjectured that arithmetical stratifications for such a surface must be infinite: unions of rational curves of maximal L-degree are consecutive accumulation subsets.

Over functional ground fields of characteristic zero, the multidimensional Mordell conjecture is proved for manifolds with ample cotangent sheaf [Mar-Des]. They do not contain rational curves.

LINEAR GROWTH CONJECTURE (LGC). *If* V *is a Fano variety* (*i.e. the anticanonical sheaf is ample*), *then the arithmetical stratification is finite, and for sufficiently large ground fields the last growth order for* $-K_V$ *equals* 1.

In [FraMaTschi], LGC was proved for homogeneous Fano varieties. It was also remarked that the circle method, when it gives asymptotic formulas, establishes LGC for certain Fano complete intersections (see e.g. [Sh1]).

In [BaMa] and [MaTschi], LGC was proved for certain two-dimensional Fano varieties (del Pezzo surfaces).

We can use LGC in order to pinpoint (parts of) accumulating subvarieties in a Fano variety V. In fact, let $W \subset V$ be a Fano subvariety. We will say that $-K_V|W$ is *too small*, if

$$K_V|W = aK_W + Z$$

in the Néron-Severi space of W, where $a < 1$, and Z is effective. For example, if V is a del Pezzo surface, and W is an exceptional curve on it, then $K_V \mid W = (1/2)K_W$, so that $-K_V \mid W$ is too small.

If LGC is valid for W, then

$$\beta_V(-K_V) \geqslant \beta_W(-K_V \mid W) \geqslant a^{-1}\beta_W(-K_W) > 1.$$

Therefore W must be contained in an accumulating subvariety (of some V_i of the arithmetical stratification w.r.t. $-K_V$), if LGC is valid also for V.

0.3. *Results*

In this paper, we discuss from this viewpoint Fano threefolds. After summing up the necessary preliminaries in Section 1, we prove in Section 2 the following lower linear bound:

0.4. THEOREM. *Let V be a Fano threefold over a number field k. Then for every Zariski open dense subset U there exists a finite extension k' of k such that if k'' contains k', then $N_{U \otimes k''}(-K; H) > c \cdot H$ for some $c > 0$ and large H, so that*

$$\beta_{U \otimes k''}(-K_V) \geqslant 1.$$

The proof is heavily based upon the classification theory of Fano threefolds, developed by Fano, Iskovskih, Shokurov, Mori and Mukai, and its main interest lies in exhibiting several different reasons for existence of many points on members of various deformation families. Actually, it establishes a stronger inequality $N_U(-K_V; H) > c \cdot H(\log H)^t$, with some $t \geqslant 0$.

In the course of the proof, we also register Fano subvarieties W with too small $-K_V \mid W$ for most of the families and make conjectures about the structure of the arithmetical stratification. In many cases, the growth order for these (conjecturally) accumulating subvarieties (with respect to $-K_V$) is known, and usually equals 2. Therefore, in order to prove that these varieties are accumulating, it suffices to establish that $\beta_U(-K_V) < 2$ for the complement U.

Section 3 is devoted to a discussion of known methods of estimating $\beta_U(-K)$ from above.

1. Preliminaries on Fano varieties

1.1. Definition and examples

In this paper, *a Fano variety* V is a smooth proper variety over a field k whose anticanonical class $-K_V$ is ample. Here are some examples.

(a) Let $V(n; d_1, \ldots, d_s)$ be a smooth complete intersection of hypersurfaces of degrees $d_1, \ldots, d_s$ in $\mathbb{P}^n$. Then the anticanonical sheaf on V is isomorphic to $\mathcal{O}(n + 1 - \Sigma d_i)$. Therefore, V is Fano iff $n + 1 - \Sigma d_i > 0$.

(b) Let G be a semisimple linear algebraic group, P a parabolic subgroup. Then the generalized flag space $P\backslash G$ is Fano. There exist also flag spaces without k-points (e.g. quadrics); they are defined by k-rational conjugacy classes of parabolic subgroups. *Every homogeneous Fano variety is a flag space*: see [Dem2].

(c) Two-dimensional Fano varieties are called *del Pezzo surfaces*. Over an algebraically closed ground field, they form 10 deformation families: (a) $\mathbb{P}^2$; (b) $\mathbb{P}^1 \times \mathbb{P}^1$; (c) S_a = the result of blowing up $1 \leqslant a \leqslant 8$ points of $\mathbb{P}^2$ in sufficiently general position: cf. [Ma].

In the next section, we shall summarily describe all 104 deformation types of Fano threefolds, discovered by prolonged efforts of Fano, Iskovskih, Shokurov, Mori, and Mukai.

1.2. *Basic invariants*

(i) *Index* $r = r(V)$ of a Fano variety is the maximal integer such that K_V is divisible by r in Pic(V). Generally, *the index is not stable with respect to ground field extensions*. For example, $r(\mathbb{P}^n) = n+1$, whereas non-trivial (having no k-points) forms of this V have index 1.

Unless stated otherwise, we shall usually give values of index over a closure of the ground field.

We have $r(V(n; d_1, \ldots, d_s)) = n + 1 - d_1 - \cdots - d_s$, while the dimension is $n - s$. The circle method works well if "the number of variables is large in comparison with degrees and number of equations", that is, if index is close to the dimension. Algebro-geometric structure of Fano varieties also tends to simplify for larger values of index.

In particular, for del Pezzo surfaces we have $r(\mathbb{P}^2) = 3$; $r(\mathbb{P}^1 \times \mathbb{P}^1) = 2$; $r(S_a) = 1$.

(ii) *Rank* $\rho = \rho(V) = \mathrm{rk}\,\mathrm{Pic}(V)$. It can also jump after a field extension, so that we shall usually give its values over a closed ground field. Over $\mathbb{C}$ it coincides with B_2. If $\dim V(n; d_1, \ldots, d_2) \geqslant 3$, $\rho(V) = 1$ according to Lefschetz. Furthermore, $\rho(S_a) = a + 1$.

(iii) *Degree* of V is $d = d(V) = (-K_V)^{\dim V}$ (self-intersection index). Usually it is more convenient to use the *reduced degree* $\delta(V) = d/r^{\dim(V)}$. For example, the projective degree $d_1 \cdots d_s$ of $V = V(n; d_1, \ldots, d_s)$ is just $\delta(V)$.

(iv) *Mori invariant* $b(V)$. Here we shall from the start extend the ground field to its closure. Put

$$b(V) = \min\{b \mid \text{through every geometric point of } V \text{ passes a rational curve } C \text{ with } (-K_V \cdot C) \leqslant b\}.$$

According to a deep theorem of Mori, $2 \leqslant b(V) \leqslant \dim(V) + 1$. Upper bound is achieved for $\mathbb{P}^n$.

Clearly, $b(V)$ is divisible by the index.

We shall now state some properties of growth orders.

1.3. PROPOSITION. *Let V be a homogeneous Fano variety having a k-point. Then V has no accumulating subvarieties w.r.t. any L. $V(k)$ is dense, and $\beta_U(-K_V) = 1$ for every dense Zariski open subset U if k is sufficiently large.*

Proof. Represent V as $P\backslash G$ where G is a semisimple linear algebraic group, P a parabolic subgroup. According to a theorem of Rosenlicht, G is k-unirational so that $G(k)$ is dense (see e.g. [BoSp]). Hence $V(k)$ is dense. Assume that $\beta_U(L) < \beta_V(L)$ for some dense U and ample L.

Cover V by a finite family of translates Ug_i, $g_i \in G(k)$. Since $g_i^*(L) \cong L$ for every i, we have $\beta_{Ug_i}(L) = \beta_U(L)$. Hence $\beta_V(L)$ cannot be larger.

The statement $\beta_U(-K_V) = 1$ is proved in [FraMaTschi] using Langlands' deep theory of Eisenstein series, in the case when G contains a Borel subgroup defined over k. □

REMARKS. (a) Actually, Langlands' theory can be used to show an asymptotic formula of the type

$$N_V(L; H) = \text{const} \cdot H^{\alpha(L)} P_L(\log H)(1 + O(H^{-\eta}))$$

where $\alpha(L)$ is a rational number defined in [BaMa], P_L a polynomial, $\eta > 0$.

(b) Let V be quasihomogeneous, i.e. having a dense orbit U with respect to an action of a linear group G on V. Then the argument above shows that $U(k)$ is dense, and U has no accumulating subvarieties with respect to any L.

1.4. PROPOSITION. *Let $f\colon W \to V$ be a birational morphism of Fano varieties, $U \subset W$ an open subset disjoint with the exceptional locus of f. Then*

$$\beta_U(-K_W) \geqslant \beta_{f(U)}(-K_V).$$

Proof. We have $-K_W = f^*(-K_V) - E$ where E is an effective divisor disjoint with U. Hence $h_E(x) > \text{const} > 0$ for $x \in U$, and

$$h_{-K_W}(x) < Ah_{f^*(-K_V)}(x) < Bh_{-K_V}(f(x))$$

for some positive constants A, B. □

1.5. PROPOSITION. *Let V be a Fano variety over a number field k, $b(V)$ its Mori invariant. Then for every Zariski dense open subset U there exists a finite extension k' of k such that if k'' contains k', then*

$$\beta_{U \otimes k''}(-K_V) \geqslant 2/b(V).$$

Proof. Choose k' in such a way that U contains a k'-point and a rational curve C with $(C \cdot -K_V) \leqslant b(V)$ passing through this point and splitting (i.e. birational to $\mathbb{P}^1$) over k'. Then

$$\beta_{U \otimes k'}(-K_V) \geqslant \beta_{C \otimes k'}(-K_V \mid C) = 2/(C \cdot -K_V). \qquad \square$$

1.6. PROPOSITION. *Let U, V be two open subsets in some Fano varieties $\bar{U}$, $\bar{V}$. Then*

$$\beta_{U \times V}(-K_{\bar{U} \times \bar{V}}) = \max(\beta_U(-K_{\bar{U}}), \beta_V(-K_{\bar{V}})).$$

Proof. This follows directly from the definition of β via the height zeta function, because $K_{\bar{U} \times \bar{V}} = p_1^*(K_{\bar{U}}) + p_2^*(K_{\bar{V}})$. $\square$

1.7. REMARK. Clearly, the Prop. 1.6 can be stated more precisely: we actually prove that $N_{U \otimes k}(-K_V, H) > \text{const} \cdot H^{2/b(V)}$ Similarly, in Prop. 1.5 we prove that

$$N_U(-K_W; H) > \text{const} \cdot N_{f(U)}(-K_V; H).$$

Therefore, when V is homogeneous, we may gain a power of logarithm.

2. Linear lower bound for Fano threefolds

In this section, we review the classification of Fano threefolds. The following is the most concise statement implying the lower bound $\beta(-K) \geqslant 1$ (stated in the Introduction) with the help of Propositions 1.3–1.6. We shall check it case by case making comments about possible accumulating subvarieties on the way.

2.1. THEOREM. *Every Fano threefold V over a closure of the ground field becomes isomorphic to a member of at least one of the following families:*

(i) *A generalized flag space $P\backslash G$.*
(ii) *A Fano variety with $b(V) = 2$.*
(iii) *A blow up of varieties of the previous two groups.*
(iv) *A direct product of $\mathbb{P}^1$ and a Fano surface.*

REMARK. A similar statement is valid in dimension 2.

Proof. Call a Fano variety *minimal*, if it is not a blow up of a Fano variety along a smooth center. Clearly, it suffices to check, that over a closed ground field of characteristic zero minimal Fano threefolds either are homogeneous, or have $b(V) = 2$. $\square$

All Fano varieties with $\rho = 1$ are minimal; their list is given in [Is1, 2] (cf. also [Mu]). Besides, there are minimal varieties with $\rho = 2, 3$: their deformation families are described in [MoMu1,2,3] together with information about admissible blowings up. Here is a summary of the relevant information, with the notation we shall use in the sequel.

A. *Varieties V with* $r \geqslant 2$, $\rho = 1$. In this group, there are two homogeneous members, $\mathbb{P}^3$ with $r = 4$, and quadric Q^3 with $r = 3$. The remaining five deformation families have $r = 2$; they are classified according to the value of the reduced degree $\delta = (-K/2)^3$, $1 \leqslant \delta \leqslant 5$. Denote by V_δ a member of the corresponding family. Rational curves C with $(C \cdot -K_V) = 2$ are called *lines* in [Is1-4]. We shall call them $(-K)$-*conics*. According to the Proposition 1.4 of Chapter III, [Is2], every V_δ is covered by $(-K)$-conics, so that $b(V) = 2$. Since K_V is divisible by two, there are no rational curves C with too small $-K_V | C$.

CONJECTURE. *If V belongs to the group A, and the ground field is sufficiently large, then* $\beta_V(-K) = 1$ *and V contains no accumulating subvarieties.*

We shall denote by $\Sigma(V)$ the scheme parametrizing $(-K)$-conics on V.

B. *Varieties with* $r = 1$, $\rho = 1$. The families $\{W_d\}$ are classified by the degree $d = (-K)^3$, taking values 2, 4, 6, 8, 10, 12, 14, 16, 18, 22. Actually, for $d = 4$ general W_4 are smooth quartics in $\mathbb{P}^4$, ((4; 4) in notation of 1.1a), but they admit flat specialization W'_4, which can be realized as a double cover of a quadric $Q^3 \subset \mathbb{P}^4$ ramified along the intersection of Q^3 with a quartic.

In this family, rational curves with $(-K \cdot C) = 2$ are called *conics*, whereas those with $(-K \cdot C) = 1$ are called *lines*.

A basic result due to Shokurov says that lines always exist (and form one-dimensional family, whose base may well be reducible and non-reduced), and conics on a W_d cover W_d, so that $b = 2$ in this group. Not all details of proofs are spelled out in the literature I was able to trace; especially the so called *hyperelliptic families* W'_4 and W_2 need some more attention; cf. below. The first basic reference is [Is2], Chapter III and Chapter II, sec. 2. Shokurov's approach was carefully worked out by Miles Reid (see [Re2], [Mu]); conics on a general W_2 are investigated in [CeVe].

CONJECTURE. *In the group B, over a sufficiently large ground field the accumulating subvariety is a union of lines and has the growth order* 2 *w.r.t.* $-K$; *the complement to it has no accumulating subvarieties, and has the growth order* 1.

C. *Varieties with* $\rho = 2, 3$. Here we refer to [MoMu1]. In this paper, all varieties with a given value of ρ are listed in the Table number ρ. We shall denote the i-th variety in this table by $\rho \cdot i$, writing ρ by Roman numerals, so that III.11 means the eleventh family with $\rho = 3$.

There are 12 minimal families in the Tables (they are easily detected by the word "none" in the last column meaning that no divisor can be blown down to a smooth curve): II: 2, 6, 8, 18, 24, 32, 34, 36; III: 1, 2, 27, 31. We exclude II.35, on which a plane can be blown down to a point.

Among these families, the following consist of homogeneous varieties: II.32 (complete flags in a projective plane), II.34 = $\mathbb{P}^1 \times \mathbb{P}^2$, III.27 = $\mathbb{P}^1 \times \mathbb{P}^1 \times \mathbb{P}^1$. Two more are quasihomogeneous: II.36 = $\mathbb{P}_{\mathbb{P}^2}(\mathcal{O} \oplus \mathcal{O}(2))$ and III.31 = $\mathbb{P}_{\mathbb{P}^1 \times \mathbb{P}^1}(\mathcal{O} \oplus \mathcal{O}(1, 1))$. Undoubtedly, for them and their forms there should exist a theory of height zeta functions as precise as that of Eisenstein series. For the time being, we see at least that $b = 2$ for them because they are line bundles.

The remaining families of this group consist of conic bundles over $\mathbb{P}^2$ or $\mathbb{P}^1 \times \mathbb{P}^1$: see [MoMu2], Theorem 1.6. Therefore, they have Mori invariant 2.

CONJECTURE. *Among the conic bundles in the group C, over sufficiently large ground fields, the accumulating subvarieties are unions of lines, and have the growth order* 2, *whereas the complement has the growth order* 1.

This completes the proof of the Theorem 2.1 (and 0.4). In particular, we could have omitted mentioning direct products, but since more precise results can be obtained for them, we notice that all Fano threefolds with $\rho \geqslant 6$ are direct products, and hence have $\rho \leqslant 10$. □

In this proof of the lower bound $\beta \geqslant 1$, we tried to minimize the algebro-geometric information quoted from the classification theory. In the rest of this section, for the sake of a future finer theory, we shall give some more details about the geometry of lines, conics, centers of admissible blowings up, and definition fields. I thank Prof. J. P. Murre for extensive list of references.

In addition, we recall the known facts about (uni)rationality which can be used to ascertain that $V(k)$ is dense whenever it is non-empty, and about birational automorphisms.

2.2. DEFINITION. (a) *A variety V over a field k is called split if* $\mathrm{Pic}(V) = \mathrm{Pic}(V \otimes \bar{k})$.

(b) *An extension $k' \supset k$ is a splitting extension, if $V \otimes k'$ is split over k'.*

It seems that most algebro-geometric arguments we use for counting points fully work only for split Fano varieties: cf. e.g. [MaTschi], where we treat split del Pezzo surfaces (for them, all exceptional curves are defined over the ground field).

2.3. *Group A*

Let V be a Fano variety with $r \geqslant 2$, $\rho = 1$. It is split over k, iff $-K_V$ is divisible by r over k. For $r = 4$, that is, V = a k-form of $\mathbb{P}^3$, we have

$$V \text{ is } k\text{-split} \Leftrightarrow V \cong \mathbb{P}^3 \Leftrightarrow V(k) \neq \emptyset.$$

Forms are classified by elements of $Br(k)$ of order 4, and their theory over number fields is well understood. Certainly, quadrics ($r = 3$) can be treated similarly.

I do not know anything about splitting fields for $r = 2$. Iskovskih describes in [Is1, 2, 4] standard models for V_δ corresponding to either $|-K_V|$, or $|H|$ where $|-K_V| = |2H|$. The latter are defined over k only for split V. Let φ_{-K} (resp. φ_H) be the corresponding map. Then we have:

(i) $\delta = 1$ *case*. $\varphi_{-K}: V_1 \to W$ is a double covering of the threefold W which is a cone over the Veronese surface F_4 in $\mathbb{P}^5$. It is ramified along a smooth divisor D, which is an intersection of W and a cubic hypersurface not passing through the vertex of W. Every such covering is a V_1.

If V_1 is split, $\varphi_H: V_1 \to \mathbb{P}^2$ is a rational map with one indeterminacy point and irreducible elliptic fibers.

(ii) $\delta = 2$ *case*. If V_2 is split, $\varphi_H: V_2 \to \mathbb{P}^3$ is a double covering ramified at a smooth quartic D. Every such covering is a V_2.

In [We2] it is proved that $\Sigma(V_2)$ is a connected surface, which is smooth iff D does not contain lines; $\mathrm{Alb}(\Sigma(V_2))$ is isomorphic to the intermediate Jacobian of V_2 of dimension 10.

(iii) $\delta = 3$ *case*. If V_3 is split, φ_H defines an isomorphism of V_3 with a smooth cubic $V(4; 3)$.

$\Sigma(V_3)$ is (generically?) smooth and irreducible.

$\mathrm{Alb}(\Sigma(V_3))$ is isomorphic to the intermediate Jacobian of W of dimension 5 ([ClGr], [AltKl]).

(iv) $\delta = 4$ *case*. If V_4 is split, φ_H defines an isomorphism of V_4 with a smooth intersection of two quadrics $V(5; 2, 2)$.

$\Sigma(V_4)$ is isomorphic to the Jacobian of a curve of genus 2, which is simultaneously the intermediate Jacobian of V_4 ([New], [Re1]).

(v) $\delta = 5$ *case*. There is only one isomorphism class of split V_5. Its $|H|$-model in $\mathbb{P}^6$ can be described as a birational image of a smooth quadric Q^3 corresponding to the linear system $|2E - C|$, where E is hyperplane section of Q^3, C a non-plane cubic curve on it.

Another description of V_5 identifies it with a section of the Grassmannian of lines in $\mathbb{P}^4$ by a general codimension 2 hyperplane.

$\Sigma(V_5)$ is (a form of) $\mathbb{P}^2$ ([Is2], Chapter III, Prop. 1.6).

Our lower bounds for point count in this group was based exclusively on the fact that every V_δ over a closed field is covered by $(-K)$-lines, i.e. rational curves of $(-K)$-degree 2.

One can in principle quantitatively control the size of the set of *k-rational* lines on V_δ, and, by implication, the subset $V_0(k) \subset V(k)$ consisting of points lying on rational lines.

PROBLEM. Count points of bounded height in $V_0(k)$.

PROBLEM. Count $V(k) \backslash V_0(k)$. One can try to obtain a lower bound by using birational automorphisms ([Is3], [Pu1], [Pu2]).

EXERCISE. Look at the proofs of (uni)rationality of these varieties in order to understand the ground fields over which this is true: V_4, V_5 (rationality); V_2, V_3 (unirationality). General V_1, V_2 are not rational: this is proved in [Beau]. No V_3 is rational ([ClGr]). I do not know whether V_1 is unirational.

2.4. *Group B*

Since $r = 1$, $\rho = 1$ in this group, all W_d are split. Another property of these varieties can be read off Mori-Mukai tables: no blowing up of a member of this group can be Fano. (Members of group A have a lot of Fano blowings up).

For a variety W of this class, we denote by $\Gamma(W)$ the base scheme parametrizing lines in W, and by $S(W)$ the subscheme of lines in W.

(i) *$d = 2$ case*. Here $\varphi_{-K}: W_2 \to \mathbb{P}^3$ is a double solid ramified along a smooth sextic surface.

Arithmetical genus of $\Gamma(W)$ is 6865; the $(-K)$-degree of $S(W)$ is 1248; every line intersects 625 lines. This is proved in [Mar], where it is also conjectured that for general W_2, $\Gamma(W)$ is smooth and irreducible.

For a generic W_2, $\Sigma(W_2)$ is a smooth irreducible surface, and $\mathrm{Alb}(\Sigma(W_2))$ is isomorphic to the intermediate Jacobian of W_2 of dimension 52. This is shown in [CeVe], where also the "non-generic" cases are described in some detail.

(ii) *$d = 4$ case*. We denote by W_4 a smooth quartic embedded in $\mathbb{P}^4$ by $|-K|$. This family admits a flat specialization on which $|-K|$ ceases to be very ample and gives instead a double covering W'_4 of a quadric Q ramified at a smooth surface of degree 8.

Collino ([Co]) shows that $\Gamma(W)$ is always one-dimensional in $\mathrm{Char}(k) \neq 2, 3$.

On W_4, $p_a(\Gamma(W)) = 1600$, $\deg(S(W)) = 320$: see [Te].

For general W_4, $\Gamma(W)$ is a smooth irreducible curve ([Co], [BarVe], [BlMu]).

Tennison also considers the special case of the Fermat quartic $x_0^4 + \cdots + x_3^4 = 0$. Here $\Gamma(W)$ is reducible and non-reduced: it consists of forty plane Fermat quartics of genus 3, each of multiplicity 2; they intersect only pairwise, and every component intersects 12 of the remaining components. In $\mathrm{Char}(k) = 3$, $\Gamma(W)$ becomes two-dimensional ([Co]).

For a general W_4, the base scheme of conics $\Sigma(W)$ is two-dimensional ([Te]) and smooth ([CoMuWe], Prop. 3.6). In $\mathrm{Char}(k) \neq 2,3$ this dimension cannot be larger ([CoMuWe]). In [Le] it is shown that $\mathrm{Alb}(\Sigma(W))$ is isomorphic to the intermediate Jacobian of W, which is of dimension 30.

[CoMuWe] is devoted to the numerical study of $\Sigma(W_4)$, in particular, 972 conics pass through a general point of W_4.

On W'_4, $\Gamma(W)$ has arithmetical genus 801; $\Sigma(W)$ has degree 320; every line intersects 81 lines ([Mar]).

There are two more cases with nice anticanonical models:

(iii) $d = 6$ *case*: $W_6 = V(5; 2, 3)$.

For a general W_6, lines are parametrized by an irreducible smooth curve (see [BlMu]). It has genus 271; the surface of lines is of degree 180; every line intersects 31 lines ([Mar]).

(iv) $d = 8$ *case*: $W_8 = V(6; 2, 2, 2)$.

For a general W_8, lines are parametrized by a smooth irreducible curve of genus 129, the surface of lines has degree 128. Every line intersects 17 lines ([BlMu] and [Mar]).

$\mathrm{Alb}(\Sigma(W_8))$ is isogenous to $J(W_8)$ (G. Welters, thesis).

(v) $W_{10} \subset \mathbb{P}^7$ which is a section of the Grassmannian of lines in $\mathbb{P}^5$, embedded into $\mathbb{P}^9$, by $\mathbb{P}^7$ (general case); $W'_{10} \subset \mathbb{P}^7$, a section of a cone over V_5 in $\mathbb{P}^7$ by a quadric (specialization) (see [Gu1] about this special subfamily).

Here $p_a(\Gamma(W)) = 71$, $\deg(S(W)) = 100$; every line intersects 11 lines ([Mar]).

(vi) $W_{12} \subset \mathbb{P}^8$

(vii) $W_{14} \subset \mathbb{P}^9$: a section of the Grassmannian of lines in $\mathbb{P}^5$, embedded into $\mathbb{P}^{14}$, by $\mathbb{P}^9$ (see [Gu2]).

For a generic W_{14}, $p_a(\Gamma(W)) = 26$, every line intersects 6 lines ([Mar]). Markushevich also investigates a more special family of W_{14} consisting of generic intersections of five Schubert varieties σ_1 in the relevant Grassmannian. For them, $\Gamma(W)$ consists of fifteen irreducible components each of which is a smooth rational curve. It has 40 double points, where the components are pairwise intersecting.

(viii)–(x) $W_{16} \subset \mathbb{P}^{10}$; $W_{18} \subset \mathbb{P}^{11}$; $W_{22} \subset \mathbb{P}^{13}$.

EXERCISE. Study the ground fields over which the following (uni)rationality statements can be proved: W_{12}, W_{16}, W_{18}, some W_{22} are rational; some W_4, W'_4, W_6, W_8, W_{10}, W_{14} are unirational but irrational (the latter statement is known for general W_{10}). We gave above information about lines on irrational W's taken from [Mar].

2.5. *Group C*

In this group, families of conics conjecturally correspond to conic bundle structures, whereas families of lines conjecturally correspond to discriminant curves on the conic bundle bases. It is worth checking this, because the geometry then is much simpler than in group B.

One can try to estimate $U(k)$ from above, where U is the complement to lines, by looking arithmetically on the families of non-split conics.

Using sieves, Serre in [Se2] proved that most fibers of such a family remain non-split, i.e. have no k-points.

What we need, is an estimate of the "volume" of split conics, that is, the coefficient $c(s)$ in the asymptotic

$$N_{C(s)}(-K_V \mid C(s), H) \approx c(s)H$$

as a function of a height of the base point s.

2.6. *Remarks*

(i) In the Group B, the conjecturally accumulating surface of lines $S(W)$ is used to calculate the intermediate Jacobian, i.e. essentially the middle motive of W. For $K3$ surfaces, deleting rational curves leads to a mixed motive, extension of W by Tate motives. Can one establish a direct connection between counting points and motives? (Over finite fields, this is of course achieved via étale cohomology).

(ii) Let k be a field of functions on a curve over an algebraically closed constant field. J. Kollàr asked me whether a Fano variety over k always has a k-point.

A natural framework for this problem is the class of C_1-fields k. For rational surfaces S, in my Moscow ICM talk it was conjectured that $S(k) \neq \emptyset$. Using the classification theory, Colliot-Thélène was able to prove this for the cases I left open. The same approach partially works for Fano threefolds V. One knows that $V(k)$ is non-empty for:

(a) Fano complete intersections of any dimension.

(b) Homogeneous Fano varieties (Springer's theorem: cf. J.-P. Serre, Cohomologie Galoisienne, Ch. III, 2.4).

(c) Fano threefolds admitting a structure of a conic bundle over k.

(d) Fano threefolds containing a rational curve or surface defined over k.

It would be interesting to investigate bad cases systematically. This will probably involve some understanding of splitting fields.

3. Non-minimal threefolds and exceptional heights

3.1. *Non-minimal Fano threefolds*

In the tables of [MoMu], there are many Fano threefolds that can be obtained by blowing up minimal homogeneous models. Since for the latter the linear growth conjecture is known in a strong form, one can try to deduce it for blowings.

In 3.2, we list some Fano blowings of $\mathbb{P}^3$ and Q^3.

In 3.3, we prove by a direct count the linear growth conjecture for $\mathbb{P}^n$ blown up along a $\mathbb{P}^m$ (over $\mathbb{Q}$). In [MaTschi], we succeeded to treat similarly $\mathbb{P}^2$ blown up along $\leqslant 4$ points. Hopefully, such methods can be applied to some more cases listed in 3.2, which was our motivation for including them.

Finally, in 3.4 we make some comments on heights with respect to the exceptional divisors, motivated by the previous discussion.

3.2. *Fano blowings up of* $\mathbb{P}^3$ *and* Q^3

In the following list, a notation of the type $X(Z) \to Y$ means that Y is obtained from X by blowing it up along Z. The list is far from complete.

$\mathbb{P}^3$ (point) $\to$ II.35 (strict transform of the twisted cubic passing through the blown up point) $\to$ III.16.

$\mathbb{P}^3$ (line) $\to$ II.33 (conic) $\to$ III.18

$\mathbb{P}^3$ (conic) $\to$ II.30.

$\mathbb{P}^3$ (plane cubic) $\to$ II.28.

$\mathbb{P}^3$ (twisted cubic) $\to$ II.27.

$\mathbb{P}^3$ (elliptic curve, intersection of two quadrics) $\to$ II.25.

$\mathbb{P}^3$ (intersection of a quadric and a cubic) $\to$ II.15.

$\mathbb{P}^3$ (intersection of two cubics) $\to$ II.4.

$\mathbb{P}^3$ (a curve of degree 6 and genus 3, intersection of cubics) $\to$ II.12.

$\mathbb{P}^3$ (a curve of degree 7 and genus 5, intersection of cubics) $\to$ II.9.

Q^3 (point) $\to$ II.30 (a point) $\to$ III.19.

Q^3 (line) $\to$ II.31.

Q^3 (conic) $\to$ II.29.

Q^3 (elliptic curve of degree 5) $\to$ II.17.

Q^3 (rational quartic spanning $\mathbb{P}^4$) $\to$ II.21.

Now we will treat the first two items of this list.

Put $k = \mathbb{Q}$. Consider a morphism $\pi: V \to \mathbb{P}^n$, blowing up a projective subspace $\mathbb{P}^m \subset \mathbb{P}^n$, $m \leqslant n-2$. Let X be the exceptional divisor on V, $U = V \backslash X$. We put $\Lambda = [\pi^*(\mathcal{O}(1))]$, $\ell = [\mathcal{O}(X)]$, the square brackets denoting classes in $\mathrm{Pic}(V)$ which we write additively. Choose an ample sheaf L on V, $[L] = a\Lambda - b\ell$, and a Weil height h_L on V. We shall consider h_L, h_Λ, h_ℓ and make the respective point counts only up to $\exp(O(1))$.

The following theorem is a generalization of Serre's result for $n = 2$, $m = 0$ ([Se1]).

3.3. THEOREM. *Up to* $\exp(O(1))$ *we have*

$$N_U(L; H) \cong \begin{cases} H^{(m+2)/(a-b)} & \text{if } (n-m-1)a < (n+1)b; \\ H^{(n+1)/a}\log(H) & \text{if } (n-m-1)a = (n+1)b; \\ H^{(n+1)/a} & \text{if } (n-m-1)a > (n+1)b. \end{cases}$$

In particular, $-K_V = (n+1)\Lambda - (n-m-1)\ell$ *is a linear growth point.*

Proof. Let $(x_0, \ldots, x_m; x_{m+1}, \ldots, x_n)$ be a coordinate system in which $\mathbb{P}^m$ is given by $x_{m+1} = \cdots = x_n = 0$. For a point $x \in V$ with $\pi(x) = (x_i) \in \mathbb{Z}^{n+1}_{\text{prim}}$, put $d = \gcd(x_{m+1}, \ldots, x_n)$. Then the height $h_{2\Lambda-\ell}(x)$ can be calculated as Weil's height of a point with projective coordinates $(x_i x_j \mid j > m)$. But the gcd of these coordinates is exactly d. It follows easily that $h_\ell(x)$ is equivalent to

$$h_\ell(x) \cong d\,\frac{|x_0| + \cdots + |x_n|}{|x_m| + \cdots + |x_n|}$$

(represent h_ℓ as $h_{2\Lambda}/h_{2\Lambda-\ell}$). It follows that if h_0 (resp. h_1) is the height of $(x_0 : \cdots : x_m)$ (resp. $(x_{m+1} : \cdots : x_n)$), then

$$h_{a\Lambda - b\ell}(x) \cong (h_0 + dh_1)^{a-b} h_1^b.$$

On the other hand, there are about $h_0^m h_1^{n-m-1}$ points with given values h_0 and h_1. It follows that

$$N_U(a\Lambda - b_\ell; H) \cong \sum_{\substack{(h_0, h_1, d)\\ (h_0 + dh_1)^{a-b} h_1^b \leqslant H}} h_0^m h_1^{m-n-1}, \tag{3.1}$$

We split (3.1) into three parts. Part $\Sigma_1 + \Sigma_2$ is taken over $dh_1 \leqslant h_0$. For a fixed $h_1 \leqslant h_0$, there are about $h_0 h_1^{-1}$ values of d. We incorporate this into the respective summand of (3.1) which then becomes $h_0^{m+1} h_1^{m-n-2}$.

Furthermore, in this domain $h_0 + dh_1$ is equivalent to h_0. Hence it suffices to sum over $h_0^{a-b} h_1^b \leqslant H$, which, together with $h_1 \leqslant h_0$, gives $h_1 \leqslant \min\{h_0, H^{1/b} h_0^{-(a-b)/b}\}$. The bound dividing two different minima is $h_0 = H^{1/a}$. Hence we can put

$$\sum\nolimits_1 = \sum_{h_0 = 1}^{H^{1/a}} h_0^{m+1} \sum_{h_1 = 1}^{h_0} h_1^{n-m-2}, \quad \sum\nolimits_2 = \sum_{h_0 = H^{1/a}}^{H^{1/(a-b)}} h_0^{m+1} \sum_{h_1 = 1}^{H^{1/b} h_0^{(b-a)/b}} h_1^{n-m-2}.$$

The sum Σ_3 is taken over $dh_1 > h_0$, where $h_0 + dh_1$ is equivalent to dh_1, so that summation can be taken over the domain $W = \{h_0 < dh_1, d^{a-b} h_1^a \leqslant H\}$. We present the result in the form

$$\sum\nolimits_3 = \sum_{h_1 = 1}^{H^{1/a}} h_1^{n-m-1} \sum_{d \leqslant H^{1/(a-b)} h_1^{a/(b-a)}} \sum_{h_0 < dh_1} h_0^m.$$

Now all three sums can be calculated directly. We obtain $\Sigma_1 \cong H^{(n+1)/a}$ However, the behaviour of $\Sigma_{2,3}$ varies in the different regions of the Pic-plane (a, b). The reason is that when we sum up one or two inner sums, we get a power of h_1 which may be < -1, -1 or > -1, depending on whether $(n-m-1)a$ is $<$, $=$, or $>(n+1)b$. We leave the rest to the reader. □

3.4. *Exceptional heights*

We will supplement the direct calculation above by certain qualitative arguments. They are centered around a vague notion that the linear growth conjecture for non-minimal Fano varieties is connected with the fact that *exceptional heights are small, at least in average.*

Concretely, let $f\colon V' \to V$ be a birational morphism of Fano varieties, isomorphic outside a divisor $E \subset V'$, $C = f(E)$. Let $U \subset V\backslash C$, $U' = f(U)$, $K_V = K$, $K_{V'} = K'$. We have $-K' = -f^*(K) - D$, $\operatorname{supp}(D) \subset E$. We put $h_D = h_{\mathcal{O}(D)}$ etc.

3.4.1. PROPOSITION. *Assume that* $\beta_U(-K) = 1$. *Put for* $\varepsilon > 0$, $N \to \infty$,

$$\sum_{\substack{x \in U' \\ h_{-K}(f(x)) \leqslant N}} h_D^{1+\varepsilon}(x) = N\psi_\varepsilon(N)$$

(notice that the number of summands is $N_U(-K; N)$ *which equals approximately* N*). Then*

$$\beta_{U'}(-K') = 1 \Leftrightarrow \sum_{N=1}^{\infty} \frac{\psi_\varepsilon(N)}{N^{1+\varepsilon}} \text{ converges for all } \varepsilon > 0.$$

Proof. This is essentially Abel summation. Clearly, $\beta_{U'}(-K') \geqslant \beta_U(-K)$. Hence equality means that the series $\sum_{x \in U'} h_{-K'}(x)^{-(1+\varepsilon)}$ converges for all $\varepsilon > 0$. This series can be rewritten as

$$\sum_{x \in U'} h_{-f^*(K)-D}(x)^{-(1+\varepsilon)} = \sum_{x \in U'} h_D(x)^{1+\varepsilon} h_{-K}(f(x))^{-1-\varepsilon} \tag{3.2}$$

We may and will assume that $h_{-K}(f(x))$ takes values $1, 2, 3, \ldots$. Put

$$a(n) = \sum_{\substack{x \in U' \\ h_{-K}(f(x)) = n}} h_D(x)^{1+\varepsilon}$$

Then (3.2) equals

$$\sum_{n=1}^{\infty} a(n) n^{-(1+\varepsilon)} = \sum_{N=1}^{\infty} \left(\sum_{n=1}^{N} a(n) \right) (N^{-1-\varepsilon} - (N+1)^{-1-\varepsilon}),$$

and the statement becomes clear. □

Notice that exceptional heights can be small only in average.

In fact, the discussion in [MaTschi], 2.3, shows that on split del Pezzo surfaces of degree $\leqslant 6$, exceptional heights infinitely often are comparable with ample heights, because the sum of all exceptional curves is proportional to the anticanonical class.

On the other hand, the argument in [MaTschi], sec. 4 shows, that exceptional heights can be infinitely often smaller than an arbitrarily small power of an ample height.

Here is an intuitive reason, explaining why for most points x, an exceptional height h_E should be small. Let us use an Arakelov definition of $h_E(x)$ representing it as a product of local heights

$$h_E(x) = \prod_{v \in M_k} h_{\mathscr{E},v}(x).$$

Here $\mathscr{E}$ denotes a divisor on an O_k-model $\mathscr{V}'$ of V', inducing E on its generic fiber V', and $h_{\mathscr{E},v}(x)$ is an exponentiated local intersection index of $\mathscr{E}$ with the section $\bar{x}$: $\mathrm{Spec}(O_k) \to \mathscr{V}'$ corresponding to x.

Now, if we were in a geometric situation (imagining $\mathrm{Spec}(O_k)$ as, say, a curve), for most x's the local intersection index would vanish, because on a blowing down $\mathscr{V}$ of $\mathscr{V}'$ the non-vanishing of an intersection index would mean that $\bar{x}$ should pass through a subset of codimension > 1.

Can one make this argument quantitative and applicable in Arakelov geometry?

Note finally that a different type of statement about smallness of exceptional heights plays a crucial role in the latest proof of Mordell's conjecture (Bombieri's version of the Vojta-Faltings argument).

It is based upon a refinement of an earlier Mumford's inequality. Let V be a smooth projective curve.

3.4.2. LEMMA. (a) *The diagonal divisor $\Delta \subset V \times V$ can be blown down iff g = genus of $V \geqslant 2$.*

(b) *There exists a map j: $V(k) \to F$ with finite fibers of bounded cardinality into an Euclidean space F such that for some constant c and all $(x, y) \in (V \times V \backslash \Delta)(k)$ we have*

$$\log h_\Delta(x, y) \leqslant g^{-1}(|j(x)|^2 + |j(y)|^2) - 2(j(x) \cdot j(y)) + c,$$

where the norms and the scalar product are taken in F.

Moreover, $|j(x)|$ as a function of x up to $O(1)$ coincides with the logarithm of an ample height function.

It would be quite interesting to try to extend Mumford's argument to multidimensional manifolds with ample tangent bundle.

References

[AltKl] A. B. Altman and S. I. Kleiman, Foundations of the theory of Fano schemes, *Comp. Math.* 34 (1977), 3–47.

[Ba] V. V. Batyrev, Toroidal Fano threefolds, *Math. USSR Izv.* 19 (1982), 13–25.

[BaMa] V. V. Batyrev and Yu. I. Manin, Sur le nombre des points rationnels de hauteur borné des variétés algébriques, *Math. Annalen* Bd. 286 (1990), 27–43.

[BarVe] W. Barth and A. van de Ven, Fano varieties of lines on hypersurfaces, *Arch. Math.* 31 (1978), 96–104.

[Beau] A. Beauville, Variétés de Prym et jacobiennes intermédiaires, *Ann. Sci. ENS* 10 (1977), 304–392.

[BlMu] S. Bloch and J. P. Murre, On the Chow group of certain types of Fano threefolds, *Comp. Math.* 39 (1979), 47–105.

[BoSp] A. Borel and T. A. Springer, Rationality properties of linear algebraic groups, in: *Algebraic groups and discontinuous subgroups*, Proc. Symp. Pure Math., vol. 9, 26–32, Providence RI, 1980.

[BoVe] L. P. Botta and A. Verra, The non-rationality of the generic Enriques threefold, *Comp. Math.* 48 (1983), 167–184.

[CeVe] G. Ceresa and A. Verra, The Abel-Jacobi isomorphism for the sextic double solid, *Pacific J. of Math.* 124 (1986), 85–105.

[Cl] C. H. Clemens, Double solids, *Adv. Math.* 47 (1983), 107–230.

[ClCoMo] H. Clemens, J. Kollár and S. Mori, Higher dimensional complex geometry, *Astérisque* 166 (1988).

[ClGr] H. Clemens and Ph. Griffiths, The intermediate Jacobian of the cubic threefold, *Ann. of Math.* 95 (1982), 281–356.

[Co] A. Collino, Lines on quartic threefolds, *J. Lond. Math. Soc.* ser. 2, 19 (1979), 257–267.

[CoMuWe] A. Collino, J. P. Murre and G. Welters, On the family of conics lying on a quartic threefold, *Rend. Sem. Mat. Univers. Polit. Torino* 38 (1980), 151–181.

[Dem1] M. Demazure, A very simple proof of Bott's theorem, *Inv. Math.* 33 (1976), 271–272.

[Dem2] M. Demazure, Automorphismes et déformations des variétés de Borel, *Inv. Math.* 39 (1977), 179–186.

[Do] R. Donagi, Group law on the intersection of two quadrics, *Ann. Scuola Norm. Sup. di Pisa* VII (1980), 217–239.

[FraMaTschi] J. Franke, Yu. I. Manin and Yu. Tschinkel, Rational points of bounded height on Fano varieties, *Inv. Math.* 95 (1989), 421–435.

[Fu] T. Fujita, On the structure of polarized manifolds with total deficiency one, *J. Math. Soc. Japan.* I, 32 (1980); II, 33 (1981), 415–434; III, 36 (1984), 75–89.

[Gu1] N. P. Gushel, On Fano varieties of genus 6, *Izv. AN SSSR, ser. mat.* 46 (1982), 1159–1174 (in Russian).

[Gu2] N. P. Gushel, On Fano varieties of genus 8, *Uspekhi* 38 (1983), 163–164 (in Russian).

[Is1] V. A. Iskovskih, Fano threefolds, I, II, *Izv. AN SSSR* 41 (1977), 516–562; 42 (1978), 504–549 (in Russian).

[Is2] V. A. Iskovskih, Anticanonical models of three-dimensional algebraic varieties, *VINITI, Sovr. Probl. Mat.* 12 (1979), 159–236 (in Russian); *J. Sov. Math.* 13, no. 6 (1980), 745–814 (in translation).

[Is3] V. A. Iskovskih, Birational automorphisms of three-dimensional algebraic varieties, ibid. 159–236 (in Russian); ibid. 815–868 (in translation).

[Is4] V. A. Iskovskih, Lectures on algebraic threefolds, Fano threefolds, Moscow University, 1988 (in Russian).

[Le] M. Letizia, The Abel-Jacobi mapping for the quartic threefold, *Inv. Math.* 75 (1984), 477–492.

[Ma] Yu. I. Manin, *Cubic Forms*, 2nd ed., North-Holland, Amsterdam, 1986.

[Mar] D. G. Markushevich, Numerical invariants of line families on certain Fano varieties, *Mat. Sbornik* 116 (1981), 265–288 (in Russian).

[Mar-Des] M. Martin-Deschamps, Propriétés de descente des variétés à fibre cotangent ample, *Ann. Inst. Fourier* 33 (1984), 39–64.

[MaTschi] Yu. I. Manin and Yu. Tschinkel, Points of bounded height on del Pezzo surfaces, 1991 (preprint).

[Me] J. Y. Merindol, Théorème de Torelli affine pour les intersections de deux quadriques, *Inv. Math.* 80 (1985), 375–416.

[Mu] J. P. Murre, Classification of Fano threefolds according to Fano and Iskovskih, in: Algebraic threefolds, Proceedings, Varenna 1981 Lect. N. Math. 947, Springer-Verlag (1982), 35–92.

[MoMu1] Sh. Mori and Sh. Mukai, Classification of Fano 3-folds with $B_2 \geqslant 2$, *Manuscripta Math.* 36, no. 2 (1981), 147–162.

[MoMu2] Sh. Mori and Sh. Mukai, On Fano threefolds with $B_2 \geqslant 2$, in: *Advanced Studies in Pure Math.* 1 (1983), Algebraic Varieties and Analytic Varieties, p. 101–129, Kinokuniya, Amsterdam-Tokyo, 1985.

[MoMu3] Sh. Mori and Sh. Mukai, Classification of Fano 3-folds with $B_2 \geqslant 2$, in: *Algebraic and Topological Theories*, Kinokuniya, Amsterdam-Tokyo, 1985, 496–548.

[Na] A. M. Nadel, The boundedness of degree of Fano varieties with Picard number one, 1990 (preprint MIT).

[New] P. E. Newstead, Stable bundles of rank 2 and odd degree over a curve of genus 2, *Topology* 7 (1968), 205–215.

[Pu1] A. V. Pukhlikov, Birational automorphisms of a three-dimensional quartic with the simplest singularity, *Mat. Sbornik* 135 (1988), 472–496 (in Russian).

[Pu2] A. V. Pukhlikov, Birational automorphisms of a double space and double quadric, *Izv. AN SSSR, ser. mat.* 52 (1988), 229–239 (in Russian).

[Re1] M. Reid, The complete intersection of two or more quadrics, Ph.D. These, Cambridge Univ., 1972.

[Re2] M. Reid, Lines on Fano threefolds according to Shokurov, *Inst. Mittag Loeffler Rep.* 11 (1980), 1–41.

[Se1] J.-P. Serre, Lectures on the Mordell-Weil theorem, Vieweg, Braunschweig-Wiesbaden, 1989.

[Se2] J.-P. Serre, *Comptes Rendus*, 1990.

[Sh1] V. V. Shokurov, Smoothness of the generic anticanonical divisor on Fano varieties, *Izv. AN SSSR, ser. mat.* 43 (1979), 430–441 (in Russian).

[Sh2] V. V. Shokurov, Existence of lines on Fano varieties. *Izv. AN SSSR, ser. mat.* 43 (1979), 922–964 (in Russian).

[Te] B. Tennison, On the quartic threefold, *Proc. Lond. Math. Soc.* 29 (1974), 714–734.

[Ti] A. S. Tikhomirov, The intermediate Jacobian of a double solid ramified at a quartic, *Izv AN SSSR, ser. mat.* 44 (1980), 1329–1377 (in Russian).

[Tyu1] A. N. Tyurin, Five lectures on threefolds, *Uspekhi* 27 (1972), 3–50 (in Russian).

[Tyu2] A. N. Tyurin, On the intersections of quadrics, *Uspekhi* 30 (1975), 51–99 (in Russian).

[Tyu3] A. N. Tyurin, Intermediate Jacobians in the theory of threefolds, in: *VINITI, Sovr. Prob. Mat.* 12 (1979), 5–57 (in Russian).

[Tyu4] A. N. Tyurin, Geometry of the Fano surface of a smooth cubic $F \subset \mathbb{P}^4$ and Torelli theorems for Fano surfaces and cubics, *Izv. AN SSSR, ser. mat.* 35 (1971), 498–529 (in Russian).

[VoKl] V. E. Voskresenskii and A. A. Klyachko, Toroidal Fano varieties and root systems, *Izv. AN SSSR, ser. mat.* 48 (1984) (in Russian); *Math. USSR Izvestiya* 24 (1985), 221–243 (in translation).

[WaWa] K. Watanabe and M. Watanabe, The classification of Fano 3-folds with torus embeddings, *Tokyo J. Math.* 5 (1982), 37–48.

[We1] G. Welters, *Abel-Jacobi isogenies for certain types of Fano threefolds*, Math. Centrum Amsterdam, 1981.

[We2] G. Welters, The Fano surface of lines on a double $\mathbb{P}^3$ with 4th order discriminant locus, Part 1, 1979 (Utrecht Univ. preprint).

[Wi] P. M. H. Wilson, Fano fourfolds of index greater than one, *Journ. reine angew. Math.* (Crelle) 379 (1987), 172–181.

PART III. DIFFERENTIAL EQUATIONS AND MATHEMATICAL PHYSICS

Volume 74A, number 3,4 PHYSICS LETTERS 12 November 1979

CONSERVATION LAWS AND LAX REPRESENTATION OF BENNEY'S LONG WAVE EQUATIONS

D.R. LEBEDEV
Institute of Theoretical and Experimental Physics, Moscow, USSR

and

Yu.I. MANIN
V.A. Steklov Mathematical Institute, Academy of Sciences, Moscow, USSR

Received 24 July 1979

A Lax-type representation is given for Benney's long wave equations. A new type of universal conserved densities is introduced.

Benney [1] has discussed the following nonlinear long wave equations describing the motion of a two-dimensional inviscid fluid with a free surface in a gravitational field:

$$u_t + uu_x - u_y \int_0^y u_x(x,\eta,t)\,d\eta + h_x = 0\,,$$

$$h_t + \Big(\int_0^h u\,dy\Big)_x = 0\,.$$

Here $-\infty \leqslant x \leqslant \infty$; $y = h(x,t)$ is the free surface over a flat rigid bottom $y = 0$; $u(x,y,t)$ is the horizontal component of the velocity. Set

$$A_n(x,t) = \int_0^h u(x,y,t)^n\,dy\,, \quad n \geqslant 0\,.$$

Benney has shown that the moments A_n satisfy the relations

$$A_{n,t} + A_{n+1,x} + nA_{n-1}A_{0,x} = 0\,, \tag{1}$$

and that this system of equations possesses an infinite number of conservation laws in the form $H_{n,t} + F_{n,x} = 0$, where H_n (F_n) is a polynomial in $A_0, ..., A_n$ ($A_0, ..., A_{n+1}$).

Benney's system was further investigated in refs. [2,3] where it was shown in particular that eqs. (1) possess a hamiltonian structure and that Benney's integrals are in involution. The first aim of this note is to supply a new and very simple formula for the conserved densities H_n and to show that eqs. (1) admit a standard Lax representation widely used in the "inverse scattering method" of solving wave equations. To state these results we introduce a formal "spectral parameter" ξ and set

$$\Phi(\xi) = \int_0^h \frac{dy}{\xi - u(x,y,t)} = \sum_{i=0}^{\infty} A_i \xi^{-(i+1)}\,. \tag{2}$$

Our formula for H_n is:

$$H_n = \mathrm{res}_\xi(\xi + \Phi(\xi))^n\,, \tag{3}$$

where res_ξ means the coefficient of ξ^{-1} in the corresponding formal series. Furthermore, define the two first-order differential operators

$$L = (1 + \Phi_\xi)\frac{\partial}{\partial x} - \Phi_x \frac{\partial}{\partial \xi}\,, \quad P = \xi \frac{\partial}{\partial x} - A_{0,x}\frac{\partial}{\partial \xi}\,. \tag{4}$$

Then the system (1) is equivalent to the following Lax-type equation:

$$L_t = [L, P]\,. \tag{5}$$

The last statement can be checked directly and eq. (3) could be derived from the results of ref. [3] (cf. I,

Theorem 0.4). But here we wish to demonstrate that Benney's equations fit very naturally into a general scheme of constructing hamiltonian systems with Lax-type representations and involutive conservation laws. This scheme has recently emerged in the work of several authors [4–7, 15] dealing with different equations. The basic discovery was that the relevant hamiltonian structure could always be identified with the canonical symplectic form on the orbits of the coadjoint representation of a convenient Lie algebra.

To be more specific we first review the essential features of the finite-dimensional situation following refs. [4] and [6]. Consider a finite-dimensional Lie algebra $\mathfrak{G}$ endowed with an invariant nondegenerate scalar product (,). Suppose that $\mathfrak{G} = \mathfrak{G}_+ \oplus \mathfrak{G}_-$, where $\mathfrak{G}_\pm$ are Lie subalgebras and $(\mathfrak{G}_+, \mathfrak{G}_+) = (\mathfrak{G}_-, \mathfrak{G}_-) = 0$. For $u \in \mathfrak{G}$ we denote by $u_\pm$ the components of u in $\mathfrak{G}_\pm$. For a differentiable function f on $\mathfrak{G}_-$ we define its gradient $df: \mathfrak{G}_- \to \mathfrak{G}_+$ by the formula

$$(v, d\,f(u)) = (d/dt)\, f(u + tv)|_{t=0} ,$$

for all $u, v \in \mathfrak{G}_-$. Identify $\mathfrak{G}_-$ with the dual space to $\mathfrak{G}_+$ via the scalar product. Then we can introduce the symplectic structure on the orbits of the coadjoint representation of $\exp(\mathfrak{G}_+)$ in $\mathfrak{G}_-$ following Kirillov [8,9] and Kostant [10]. The Poisson bracket of functions on $\mathfrak{G}_-$ is defined by the formula $\{h, f\}(u) = (u, [dh(u), df(u)])$ for all $u \in \mathfrak{G}_-$ and the flow on $\mathfrak{G}_-$ corresponding to the hamiltonian h can be defined by the vector field $u_t = [u, dh(u)]_-$. Now in certain cases we can find a very special class of hamiltonians which pairwise commute and moreover give rise to Lax-type flows in $\mathfrak{G}_-$. Namely, suppose that there exists a vector $w \in \mathfrak{G}$ such that $[w, \mathfrak{G}_+] \subset \mathfrak{G}_+$ and $[w, \mathfrak{G}_-] \subset \mathfrak{G}_-$. Call a function f on $\mathfrak{G}$ invariant if $[u, df(u)] = 0$ for all $u \in \mathfrak{G}$ (where now $df: \mathfrak{G} \to \mathfrak{G}$ is defined as earlier for functions on $\mathfrak{G}_-$). For an invariant function f set $f_w(u) = f(u + w)$ for all $u \in \mathfrak{G}_-$. Then $\{f_w, g_w\} = 0$ on $\mathfrak{G}_-$ if f, g are invariant, and the hamiltonian flow $u_t = [u, df_w(u)]_-$ can be written in the equivalent Lax form $(u + w)_t = [u + w, df_w(u)]$. To the author's knowledge this use of w was first suggested in ref. [7].

To apply this formalism to PDE we must work with the infinite-dimensional Lie algebras of the formal series $\mathcal{B}((\xi^{-1}))$ where $\mathcal{B}$ is a ring of functions of x or its formal counterpart. To treat Benney's equations we use the bracket [11]

$$[a,b]_0 = \partial_\xi a \partial_x b - \partial_\xi b \partial_x a \quad (\partial_\xi = \partial/\partial\xi,\ \partial_x = \partial/\partial x).$$

We introduce the scalar product setting $(a, b) = \int \mathrm{res}_\xi (ab)\, dx$. This means that in the functional case we must impose on elements of $\mathcal{B}$ appropriate growth conditions to ensure the convergence of the integrals. In the formal case we simply denote by $\int$ the map $\mathcal{B} \to \mathcal{B}/\partial_x\mathcal{B}$ and rephrase all the relevant definitions (like the nondegeneracy of the scalar product, the gradient etc.) in purely algebraic terms. We will not dwell upon these details here, cf. refs. [5,4,12]. Set $\mathfrak{G}_{0^+}$ = (polynomials in ξ), $\mathfrak{G}_{0^-}$ = (formal series in ξ^{-1} without constant term). Our invariant function(al)s on $\mathfrak{G}_0$ are $f_k: u \to \int \mathrm{res}\, u^k$ and for w we can take ξ. Then

$$df_{k,\xi}(u) = k(\xi + u)_+^{k-1}$$

on $\mathfrak{G}_{0^-}$. Put

$$u = \sum_{i=0}^{\infty} A_i \xi^{-(i+1)} \in \mathfrak{G}_{0^-}$$

Then the Lax equation

$$(\xi + u)_t = [\xi + u, \tfrac{1}{2}(\xi + u)_+^2]_0$$

is equivalent to Benney's flow (1) on the coefficients A_i. The conserved densities (3) are just $\mathrm{res}(\xi + u)^k$. Finally to obtain (4), (5) we simply replace in the Lax equations the elements of $\mathfrak{G}_0$ by the corresponding vector fields: $u \to (\partial_\xi u)\,\partial_x - (\partial_x u)\,\partial_\xi$.

In conclusion, we would like to mention a curious relation between Benney's equations for moments and the more conventional Lax equations $L_t = [L, P]$ where L and P are differential operators. These last equations were treated similarly in refs. [5,4] using the Lie algebra $\mathfrak{G}_1 = \mathcal{B}((\xi^{-1}))$ with the different bracket $[a, b]_1 = (a \circ b - b \circ a)$ where

$$a \circ b = (a \circ_\epsilon b)_{\epsilon=1} , \quad a \circ_\epsilon b = \sum_{k \geq 0} \frac{1}{k!} \partial_\xi^k a (\epsilon \partial_x)^k b ,$$

and considering the flows on $\mathfrak{G}_{1^+}$ instead of $\mathfrak{G}_{0^-}$. Now we can interpolate and introduce the Lie algebra $\mathfrak{G}_\epsilon$ with the bracket $[a, b]_\epsilon = \epsilon^{-1}(a \circ_\epsilon b - b \circ_\epsilon a)$. Then $[a, b]_0 = \lim_{\epsilon \to 0} [a, b]_\epsilon$ and we may suggestively call $\mathfrak{G}_0$ "the quasiclassical limit of $\mathfrak{G}_1$" (this remark is due to A. Kirillov). This observation together with the fractional powers technique [13,14,4] suggests the introduction of "the universal conserved densities"

Table 1
The first five universal densities G_n (total derivatives omitted), corresponding Benney densities H_n and Korteweg–de Vries densities K_n.

n	G_n	H_n	$A_{n-1}(u)$	K_n
1	H_1	A_0	$\frac{1}{2}u$	$\frac{1}{2}u$
2	H_2	$2A_1$	$-\frac{1}{4}u_x$	$-\frac{1}{2}u_x$
3	H_3	$3(A_2 + A_0^2)$	$-\frac{1}{8}(u^2 - u_{xx})$	$\frac{3}{8}(u^2 + u_{xx})$
4	H_4	$4(A_3 + 3A_0A_1)$	$\frac{1}{16}(6uu_x - u_{xxx})$	$-\frac{1}{4}u_{xxx}$
5	$H_5 + 5(A_0A_{0,xx} - 2A_1A_{0,x})$	$5(A_4 + 4A_0A_2 + 2A_1^2 + 2A_0^3)$	$\frac{1}{32}(2u^3 - 14uu_{xx} - 11u_x^2 + u_{xxxx})$	$\frac{5}{32}(2u^3 + 2uu_{xx} + u_x^2 + u_{xxxx})$

which can be used to generate simultaneously all Lax and Benney conservation laws. These densities are polynomials in A_i and the formal derivatives

$$A_i^{(j)} = \partial_x^j A_i, \qquad j \geqslant 1, i = 0, 1, 2, \ldots$$

defined by the formulas

$$G_n(A_i^{(j)}) = \mathrm{res}_\xi \left(\xi + \sum_{i=0}^{\infty} A_i\, \xi^{-(i+1)}\right)^{\circ n} ,$$

where $F^{\circ n}$ means $F \circ \ldots \circ F$ (n times). To get Benney's H_n we simply put $A_i^{(j)}$ for $j \geqslant 1$ in G_n. To get the Lax densities for

$$L_t = [L, P] , \quad L = \partial_x^n + u_{n-2}\, \partial_x^{n-2} + \ldots + u_0 ,$$

we consecutively calculate $A_i = A_i(u_k^{(l)})$ from the equation

$$\left(\xi + \sum_{i=0}^{\infty} A_i \xi^{-(i+1)}\right)^{\circ n} = \xi^n + u_{n-2}\xi^{n-2} + \ldots + u_0 .$$

and then insert these $A_i(u_k^{(l)})$ into G_n. In table 1 we give the first five universal densities G_n (with the total derivative terms omitted), the corresponding Benney densities H_n and the Korteweg–de Vries densities H_n with $A_n(u)$ found from the equation

$$\left(\xi + \sum_{i=0}^{\infty} A_i \xi^{-(i+1)}\right)^{\circ n} = \xi^2 + u .$$

The authors would like to thank A.A. Kirillov, M.A. Semyonov-Tyanshanskii and L.D. Faddeev for useful discussions and enlightening comments.

References

[1] D.J. Benney, Stud. Appl. Math. 52 (1973) 45.
[2] B.A. Kupershmidt and Yu.I. Manin, Funkt. Anal. 11 (1977) 31 (in Russian).
[3] B.A. Kupershmidt and Yu.I. Manin, Funkt. Anal. 12 (1978) 25 (in Russian).
[4] M. Adler, Inventions Math. 50 (1979) 219.
[5] D.R. Lebedev and Yu.I. Manin, preprint ITEP-155 (1978).
[6] A.G. Reyman, M.A. Semyonov-Tyanshanskii and I.E. Frenkel, Dokl. Akad. Nauk SSSR 247 (1979) (in Russian).
[7] S.V. Manakov, Funkt. Anal. 10 (1976) 93 (in Russian).
[8] A.A. Kirillov, Usp. Mat. Nauk 17 (1962) 57 (in Russian).
[9] A.A. Kirillov, Elements of the theory of representation (Springer, 1976).
[10] B. Kostant, Proc. US–Japan Seminar on Differential geometry (Tokyo, 1966).
[11] D.R. Lebedev, preprint ITEP-41 (1979).
[12] I.M. Gelfand and L.A. Dikii, Funkt. Anal. 12 (1979) 8 (in Russian).
[13] I.M. Gelfand and L.A. Dikii, Funkt. Anal. 10 (1976) 13 (in Russian).
[14] Yu.I. Manin, Modern problems in mathematics, Vol. 11 (Viniti, 1978) (in Russian).

Reprinted from *Funct. Anal. Appl.* **13** (1979).

GEL'FAND – DIKII HAMILTONIAN OPERATOR AND THE COADJOINT REPRESENTATION OF THE VOLTERRA GROUP

D. R. Lebedev and Yu. I. Manin

1. Introduction. We consider two linear differential operators L and P with coefficients depending on x, t, and we set $L = \partial^{n+1} + \sum_{i=0}^{n} u_i(x, t)\partial^i$, $\partial = \partial/\partial x$. By eliminating the coefficients of P, we can consider the Lax equation $\partial L/\partial t = [L, P]$ as a system of partial differential equations for the column of unknown functions $\bar{u} = (u_0, \ldots, u_n)^t$. Gel'fand and Dikii [1, 2] showed that this system can be written in the form $\bar{u}_t = B\delta H/\delta\bar{u}$, where H is a polynomial in $u_i^{(j)} = \partial^j u_i$ which depends on P, $\delta H/\delta\bar{u} = (\delta H/\delta u_0, \ldots, \delta H/\delta u_n)^t$, and finally B is the matrix differential operator

$$B = \sum_{i=1}^{\infty} B_i\partial^i - (-1)^i\partial^i \circ B_i^t, \quad (B_i)_{ab} = \binom{i+a}{a} u_{a+b+i+1}, \quad 0 \leqslant a, b \leqslant n. \tag{1}$$

Institute of Theoretical and Experimental Physics, V. A. Steklov Mathematical Institute, Academy of Sciences of the USSR. Translated from Funktsional'nyi Analiz i Ego Prilozheniya, Vol. 13, No. 4, pp. 40–46, October-December, 1979. Original article submitted November 10, 1978.

where we put u_j equal to the coefficient of ∂^j in L also for $j \geq n+1$. This Gel'fand–Dikii operator B is Hamiltonian in the sense that the expression $\{P, Q\} = \delta P/\delta \bar{u}^t B \delta Q/\delta \bar{u}$ defines a Poisson bracket on the space of polynomials in $u_i^{(j)}$ modulo exact derivatives. The proofs and motivation are contained in [1-3].

The operator B arose as the result of certain computations, and the fact that it is Hamiltonian was established in [2, 3] by direct verification of the Jacobi identity or some equivalent identity.

Due to the fact that the coefficients of B depend explicitly on unknown functions, the verification of this identity is rather complicated and does not display the real reasons for the Hamiltonian property.

The principal aim of this paper is to demonstrate that the Gel'fand–Dikii Hamiltonian structure is an analogue of a very special class of finite-dimensional symplectic structures, viz., the Kirillov structures on orbits of coadjoint representations of Lie groups. The role of the Lie group in our case is played by the group of (formal) Volterra operators.

We recall the Kirillov construction in Sec. 2 and obtain the Gel'fand–Dikii operator from a new point of view in Sec. 5 after introducing Volterra groups and their Lie algebras. In Sec. 6 we then present a new proof that B is Hamiltonian, and in Secs. 7 and 8 we explain some formal results of the works [2, 3] by means of our finite-dimensional analogy. For simplicity we restrict ourselves to the case of scalar coefficients u_i. Generalization to the matrix case is readily obtained as in [2].

We suppose that the main interest of our observation is that it opens the possibility of applying the ideology of "geometric quantization" to the Lax equations (see [7]). See also the remarks at the end of this paper.

After completing this work, we learned that analogous results have been obtained by Adler [13].

2. Kirillov Construction. We briefly recall the Kirillov construction; for more details see [6, 7].

Let G be a connected Lie group, let $\mathfrak{g}$ be its Lie algebra, and let $\mathfrak{g}^*$ be the space of linear functions on $\mathfrak{g}$. We identify $\mathfrak{g}$ with the tangent space to G at the identity. The differential of the action r_g: G → G, $r_g(x) = gxg^{-1}$ is called the adjoint representation of G on $\mathfrak{g}$. The dual representation of G on $\mathfrak{g}^*$ is called the coadjoint representation. We choose a point $u \in \mathfrak{g}^*$ and denote by Ω its orbit in the coadjoint representation. Since G acts transitively on Ω, each element $X \in \mathfrak{g}$ defines a vector field ∂_X on Ω. The mapping $X \to \partial_X$ is a homomorphism of Lie algebras. Its image generates the entire Lie algebra of vector fields. The symplectic Kirillov form ω on Ω is defined by

$$\omega(\partial_X, \partial_Y)(u) = (u, [X, Y]) \tag{2}$$

for all $X, Y \in \mathfrak{g}$, $u \in \Omega$.

The corresponding symplectic structure can be uniquely characterized by the following property. Any element $X \in \mathfrak{g}$ defines a function H_X on Ω: $H_X(u) = (u, X)$. The vector field ∂_X is Hamiltonian with Hamiltonian H_X.

We shall need the following generalization of this result. We consider a smooth mapping $\Phi: \Omega \to \mathfrak{g}$ and the vector field ∂_Φ on Ω: $\partial_{\Phi,u} = \partial_{\Phi(u),u}$ for all $u \in \Omega$, where $\partial_{\Phi,u}$ is the value of ∂_Φ at the point u. In the case where Φ is constant with image $X \in \mathfrak{g}$, we have $\partial_\Phi = \partial_X$. If Φ is the restriction to Ω of a linear mapping $\mathfrak{g}^* \to \mathfrak{g}$, then the field ∂_Φ is a generalized Euler flow.

In order to understand the commutation rule of the fields ∂_Φ, we choose a basis $\{e_1, \ldots, e_n\}$ of the algebra $\mathfrak{g}$ and set $\partial_i = \partial_{e_i}$, $\partial_\Phi = \sum_{i=1}^{n} \Phi_i \partial_i$, $\partial_\Psi = \sum_{i=1}^{n} \Psi_i \partial_i$. Then

$$[\partial_\Phi, \partial_\Psi]_u =$$

$$= \sum_{i,j=1}^{n} \Phi_i(u)\Psi_j(u)[\partial_{i,u}, \partial_{j,u}] + \sum_{i=1}^{n}(\partial_\Phi \Psi_j)(u)\partial_{j,u} - \sum_{i=1}^{n}(\partial_\Psi \Phi_i)(u)\partial_{i,u} = [\partial_{\Phi,u}, \partial_{\Psi,u}] + \sum_{i=1}^{n}(\partial_\Phi \Psi_j)(u)\partial_{j,u} - \sum_{i=1}^{n}(\partial_\Psi \Phi_i)(u)\partial_{i,u}. \tag{3}$$

We can consider the field ∂_Φ as a vector in $\mathfrak{g}$, which "rotates" in dependence on the point $u \in \Omega$. The last two sums in (3) account with the effect of this rotation.

3. Volterra Group. Below we shall consider various differential rings $\mathcal{B}$. Such a ring is an algebra over a field k of characteristic zero equipped with a differentiation $\partial: \mathcal{B} \to \mathcal{B}$. If $X \in \mathcal{B}$, we write $X^{(j)}$ in place of $\partial^j X$. The family of elements $\{u_i \mid i \in I\}$ in $\mathcal{B}$ is called differentially independent over a subring $\mathcal{B}_0 \subset \mathcal{B}$, if the family $\{u_i^{(j)} \mid i \in I, j > 0\}$ is algebraically independent over $\mathcal{B}_0$. Basic examples are: a) smooth functions on R which decay rapidly at infinity, $\partial = d/dx$; b) $\mathcal{B}_n = k[u_i^{(j)}]$, $0 < i \leq n$, $j > 0$; c) $\mathcal{B}_\infty = \varprojlim \mathcal{B}_n$, where the projective limit is taken relative to morphisms $\mathcal{B}_{n+1} \to \mathcal{B}_n$, mapping $u_{n+1}^{(j)}$ into zero. The last ring is convenient if it is necessary to consider infinite sums $\sum_{i=0}^{\infty} u_i X_i$.

If $\mathcal{B}$ is any differential ring $\mathcal{B}((\xi^{-1}))$ denotes the ring of formal Laurent series $\sum_{i\ge i_0} X_i\xi^{-i}$ over $\mathcal{B}$ with a finite number of positive powers. In this ring there are two differentiations: $\partial_\xi = \partial/\partial\xi$ and $\partial\colon \sum_i X_i\xi^{-i} \to \sum_i \partial X_i\xi^{-i}$. By means of them there is defined on $\mathcal{B}((\xi^{-1}))$ a new operation of multiplication called symbolic *multiplication:*

$$a \circ b = \sum_{k\ge 0} \frac{1}{k!}\partial_\xi^k a \partial^k b.$$

It is biadditive and associative and therefore converts $\mathcal{B}((\xi^{-1}))$ into a ring. Series with leading coefficient invertible in $\mathcal{B}$. are invertible in this ring. We can therefore consider the group

$$G(\mathcal{B}) = \{1 + \sum_{i\ge 0}(-1)^{i+1}X_i\xi^{-(i+1)} \mid X_i \in \mathcal{B}\}$$

relative to symbolic multiplication which we shall call the (formal) Volterra group. For the motivation of this terminology see [3] or [5].

4. Lie–Volterra Algebra. The bilinear part of the Volterra group law enables us to construct the Lie algebra $\mathfrak{g}(\mathcal{B}) = \{\sum_{i\ge 0}(-1)^{i+1}X_i\xi^{-(i+1)} \mid X_i \in \mathcal{B}\}$ is standard fashion with bracket $a \circ b - b \circ a$ in $\mathcal{B}((\xi^{-1}))$. The elements of $\mathfrak{g}(\mathcal{B})$ are conveniently represented by columns of coefficients $X = (X_0, X_1, \dots)^t \in \mathcal{B}^\infty$. The commutator of columns we denote by $[X \circ Y]$. Using the definition of symbolic multiplication, it is easy to obtain *the explicit formula*

$$[X \circ Y]_l = \sum_{i+j\le l-1} \binom{l-1-j}{i} X_i Y_j^{(l-i-j-1)} - \binom{l-1-i}{j} Y_j X_i^{(l-i-j-1)}. \tag{4}$$

5. Linear Functionals on $\mathfrak{g}(\mathcal{B})$. Let $X, Y \in \mathcal{B}^\infty$. We set $X^tY = \sum_{i=0}^{\infty} X_iY_i \in \mathcal{B}$. The sum is defined if one of the columns is compactly supported (i.e., almost all of its coordinates are equal to zero) or if $\mathcal{B}$ is equipped with a topology in which this series converges.

We denote by $\int$ the mapping $\mathcal{B} \to \mathcal{B}/\partial\mathcal{B}$. In this notation the correct analogue of a linear functional on the Lie algebra in the finite-dimensional case is the mapping

$$\mathfrak{g}(\mathcal{B}) \to \mathcal{B}/\partial\mathcal{B}\colon\ X = (X_i) \to \int \bar{u}^t X = \sum_{i\ge 0} u_i X_i \mod \partial\mathcal{B},$$

where $\bar{u} = (u_0, u_1, \dots)^t \in \mathcal{B}^\infty$ is some column.

We are now able to state our basic result.

THEOREM. For all $X, Y \in \mathfrak{g}(\mathcal{B})$ we have

$$\int \bar{u}^t[X \circ Y] = \int X^t BY, \tag{5}$$

where B is the Gel'fand–Dikii operator (1).*

Proof. According to formula (4),

$$\int \bar{u}^t[X\circ Y] = \int \sum_{l\ge 0} u_l [X\circ Y]_l = \int \sum_{i,j,l}\left[X_i\binom{l-1-j}{i}u_l Y_j^{(l-i-j-1)} - Y_j\binom{l-1-i}{j}u_l X_i^{(l-i-j-1)}\right].$$

The sum of the first terms is equal to $X^t\left(\sum_{a=0}^{\infty} B_a\partial^a\right)Y$. Integrating the sum of the second terms by parts, we find that

$$-\int\sum_{i,j,l}(-1)^{(l-i-j-1)}X_i\partial^{l-i-j-1}\left[\binom{l-1-i}{j}u_lY_j\right] = -\int X^t\sum_{a=0}^{\infty}(-1)^a\partial^a[B_a^tY].$$

This completes the proof.

---

* In which, however, $(B_i)_{ab} = \binom{i+a}{a}u_{a+b+i+1}$, $a, b \ge 0$.

Roughly speaking, the theorem means that the Gel'fand–Dikii operator B is the matrix of the Kirillov form on the orbit of the functional represented by $\bar{u}$ in the coadjoint representation of $G(\mathcal{B})$ (see formula (3)).

6. Hamiltonian Property of the Gel'fand–Dikii Operator. Using the theorem just proved, we are now able to give a more invariant proof of the Hamiltonian property of the operator B. The definition of this property, which is motivated by finite-dimensional analogues, is contained in Chap. I, Sec. 7 of [3] or in Sec. 1 of [4]. Its verification reduces to establishing the identity contained in Proposition 8.4, Chap. I, Sec. 8 of [3] or in Lemma 1.4 of [4] and is reproduced below in (10). In these papers and in the work of Gel'fand and Dikii [2] this identity is verified by direct computation with binomial coefficients. In the finite-dimensional case the verification of the Hamiltonian property reduces to deriving the closedness of the Kirillov form from the Jacobi identity. In order to imitate this argument, we begin with the formula

$$\int \bar{u}^t [Z \circ [X \circ Y]] = \int \bar{u}^t [[Z \circ X] \circ Y] - \int \bar{u}^t [[Z \circ Y] \circ X]. \tag{6}$$

Using (5), we rewrite the left side of (6) in the form

$$\int \bar{u}^t [Z \circ [X \circ Y]] = \int Z^t B [X \circ Y].$$

Further, in the notation of Proposition 8.4 of [3], using (4), we can rewrite $[X \circ Y]$ in the form

$$[X \circ Y] = \sum_{j \geq 0} X^t \frac{\partial B_j}{\partial \bar{u}} Y^{(j)} - X^{t(j)} \frac{\partial B_j^t}{\partial \bar{u}} Y. \tag{7}$$

Here $X^t \frac{\partial B_j}{\partial \bar{u}} Y^{(j)}$ denotes the column in $\mathcal{B}^m$ with l-th coordinate $X^t \frac{\partial B_j}{\partial u_l} Y^{(j)}$. The left side of (6) is thus equal to

$$\int Z^t B \Big(\sum_{j \geq 0} X^t \frac{\partial B_j}{\partial \bar{u}} Y^{(j)} - X^{t(j)} \frac{\partial B_j^t}{\partial \bar{u}} Y \Big). \tag{8}$$

The terms following $\int Z^t$ here coincide with the left side of the identity (23) in Proposition 8.4 of [3]. We rewrite the right side of (6) in a similar way, integrating by parts:

$$\int \bar{u}^t [[Z \circ X] \circ Y] = \int [Z \circ X]^t BY = \int \Big(\sum_{j \geq 0} Z^t \frac{\partial B_j}{\partial \bar{u}} X^{(j)} - Z^{t(j)} \frac{\partial B_j^t}{\partial \bar{u}} X \Big)^t BY =$$

$$= \int \Big(\sum_{j \geq 0} Z^t \frac{\partial B_j}{\partial \bar{u}^t} X^{(j)} - Z^{t(j)} \frac{\partial B_j^t}{\partial \bar{u}^t} X \Big) BY = \int Z^t \Big(\sum_{j \geq 0} \frac{\partial B_j}{\partial \bar{u}^t} X^{(j)} - (-1)^j \partial^j \circ \frac{\partial B_j^t}{\partial \bar{u}^t} X \Big) BY. \tag{9}$$

Here, e.g., $\partial B_j / \partial \bar{u}^t$ denotes the row of matrices with $\partial B_j / \partial u_l$ at the l-th site, while $Z^t \frac{\partial B_j}{\partial \bar{u}^t} X^{(j)}$ is the row of elements of $\mathcal{B}$ with $Z^t \frac{\partial B_j}{\partial u_l} X^{(j)}$ at the l-th site.

It is now not hard to see that the part of the last expression in (9) following $\int Z^t$ coincides with the first term of the right side of the identity (23) in Proposition 8.4 of the work [3] in somewhat different notation. The second term of the right side is obtained from the first by replacing X by Y and Y by X just as in (6). It therefore remains to verify that $\int Z^t U = 0$ implies $U = 0$. Now we may write identity (6) in any differential ring. In particular, we may choose as coordinates of $Z^t = (Z_0, Z_1, \ldots)^t$ differentially independent variables over the subring $\mathcal{B}$, containing all the components of u, X, and Y. In this case it is obvious that $Z^t U \in \operatorname{Im} \partial$ implies $U = 0$. This completes the proof.

For subsequent purposes we write out completely the identity we have proved:

$$B [X \circ Y] = \Big(\sum_{j \geq 0} \frac{\partial B_j}{\partial \bar{u}^t} X^{(j)} - (-1)^j \partial^j \circ \frac{\partial B_j^t}{\partial \bar{u}^t} X \Big) BY - (X \leftrightarrow Y). \tag{10}$$

We recall that it appeared as a formal intermediate step in the old proof of the Hamiltonian property of the operator B and was reproved above. We shall now show that this identity itself has a simple meaning, being the analogue of the fact that in the finite-dimensional case the mapping $X \to \partial_X$ is a morphism of the Lie algebra $\mathfrak{g}$ into the algebra of vector fields on an orbit of the coadjoint representation Ω.

7. Action of $\mathfrak{g}(\mathcal{B})$ on Functionals of $\bar{u}$. In this section we shall consider $u_0, u_1, \ldots$ to be differentially independent variables over a ring $\mathcal{B}$, and we set $\hat{\mathcal{B}} = \mathcal{B}[u_i^{(j)}]$ (or the appropriate completion of this ring if the number of u_i is infinite).

We denote by $\mathfrak{g}_u$ the Lie algebra of continuous differentiations of the ring $\hat{\mathcal{B}}$, which commute with ∂ and are trivial on $\mathcal{B}$. It is the formal analogue of the Lie algebra of vector fields on the orbit of the functional represented by $\bar{u}$. There is a natural bijection of $\mathfrak{g}_u$ with the part $\hat{\mathcal{B}}^\infty$: to the column $F = (F_0, F_1, \ldots)^t \in \hat{\mathcal{B}}^\infty$ there corresponds the differentiation $\partial_F: \hat{\mathcal{B}} \to \hat{\mathcal{B}}$, mapping u_i into F_i, or $\partial_F \bar{u} = F$.

Since the operator B is formally skew-symmetric, formula (5) can be rewritten in the form

$$\int \bar{u}^t [X \circ Y] = -\int (BX)^t Y.$$

Analogy with the finite-dimensional case suggests that the mapping $\mathfrak{g}(\mathcal{B}) \to \mathfrak{g}_u$: $X \mapsto -\partial_{BX}$ must be a morphism of Lie algebras, i.e., since $[X \circ Y] \mapsto -\partial_{B[X \circ Y]}$,

$$\partial_{B[X \circ Y]} = -[\partial_{BX}, \partial_{BY}]. \tag{11}$$

This is actually the case. Indeed, it suffices to verify that both sides of (11) act in the same way on $\bar{u}$. Now $\partial_{B[X \circ Y]}\bar{u} = B[X \circ Y]$, which coincides with the left side of (10). Further,

$$-[\partial_{BX}, \partial_{BY}]\bar{u} = \partial_{BY}(BX) - \partial_{BX}(BY). \tag{12}$$

In order to verify that the right sides of (12) and (10) coincide, we observe, first of all, that the matrices B_j are linear in u_j and that $[\partial_{BX}, \partial] = [\partial_{BY}, \partial] = 0$. Hence

$$\begin{aligned}\partial_{BY}(BX) &= \partial_{BY}\Big(\sum_{j\geq 0} B_j \partial^j X - (-1)^j \partial^j (B_j^t X)\Big) = \sum_{j\geq 0} \partial_{BY}(B_j \partial^j X) - (-1)^j \partial^j \partial_{BY}(B_j^t X) = \\ &= \sum_{j,s\geq 0} (BY)_s \frac{\partial B_j}{\partial u_s} \partial^j X - (-1)^j \partial^j \Big((BY)_s \frac{\partial B_j^t}{\partial u_s} X\Big) = \\ &= \sum_{j,s\geq 0} \Big[\frac{\partial B_j}{\partial u_s} X^{(j)} - (-1)^j \partial^j \circ \frac{\partial B_j^t}{\partial u_s} X\Big](BY)_s = \sum_{j\geq 0} \Big[\frac{\partial B_j}{\partial \bar{u}^t} X^{(j)} - (-1)^j \partial^j \circ \frac{\partial B_j^t}{\partial \bar{u}^t} X\Big] BY.\end{aligned}$$

Therefore, (10) is equivalent to (12).

8. Gel'fand–Dikii Lie Algebra. Our last observation concerns Theorem 1 in the paper of Gel'fand and Dikii [2]. In order to formulate it (in the scalar case), we now set $\mathcal{B} = k[u_i^{(j)}]$ (or the corresponding completion). We denote by $\mathcal{B}_0 \subset \mathcal{B}$ the subalgebra of polynomials with no linear part, and we introduce on $\mathcal{B}_0^\infty$ the structure of a Lie algebra with bracket $[,]_B$ by the formula

$$[F, G]_B = H \Leftrightarrow \partial_{BH} = -[\partial_{BF}, \partial_{BG}] \tag{13}$$

(we have changed the sign of the bracket in order that it be consistent with formula (11)). In our notation, Gel'fand and Dikii show that if $[F, G]_B = H$, then

$$H = [F \circ G] - \partial_{BF}(G) + \partial_{BG}(F). \tag{14}$$

We now observe that this case differs from that considered in the preceding section by the fact that the vector field $-\partial_{BF}$ on the orbit of $\bar{u}$ is the image of a vector $F \in \mathfrak{g}(\mathcal{B})$, which now depends on the point of the orbit. Thus, (14) should be compared with the finite-dimensional formula (3), and a glance at these two formulas shows that they express the same thing.

9. Concluding Remarks. Our results can be extended in several directions. We remark, first of all, that we essentially work with a specific formal differential group law (the Volterra group) or even with its formal Lie algebra. It is possible to consider a generalization of our construction to arbitrary differential groups. Their theory was begun by Ritt [8-11] and was recently extended by Weisfeiler [12]. In contrast to the extensively developed theory of formal Lie groups, Ritt's theory has been partially forgotten due partly to the lack of natural examples and applications. We hope that our observation will provide new impetus in this beautiful area. In particular, it has already been possible to interpret another Hamiltonian operator – the Benney operator introduced in [4] – in this fashion. This point of view may also assist in finding a way to generalize the Lax theory to the multidimensional case by means of algebraic versions of groups of pseudodifferential operators.

Returning to analysis instead of algebra, it can be observed that a unique "Volterra group" does not exist. In the space of Volterra operators, say, with generalized kernels supported above the diagonal there are various parts which are closed with respect to composition and inversion which depend also on the space of test functions. We suggest that our formal theory must be applicable to these groups with appropriate modifications. One might hope to explain in this way the known relations between the Lax equations and the systems of ordinary differential equations describing discrete chains.

Finally, we wish to mention that the role of Volterra operators has constantly been emphasized by A. B. Shabat. More attentive reading of this work might have led to an earlier understanding of the situation.

LITERATURE CITED

1. L. M. Gel'fand and L. A. Dikii, "Fractional powers of operators and Hamiltonian systems," Funks. Anal. Prilozhen., 10, No. 4, 13-20 (1976).
2. L. M. Gel'fand and L. A. Dikii, "Resolvents and Hamiltonian systems," Funkts. Anal. Prilozhen., 11, No. 2, 11-27 (1977).
3. Yu. I. Manin, "Algebraic aspects of nonlinear differential equations," in: Sov. Probl. Mat., Vol. 11, VINITI, Moscow (1978), pp. 5-152.
4. B. A. Kupershmidt and Yu. I. Manin, "The equations of long waves with a free surface. II. The Hamiltonian structure and higher equations," Funkts. Anal. Prilozhen., 12, No. 1, 25-37 (1978).
5. D. R. Lebedev, "The Zakharov–Shabat formalism and fractional powers of differential operators," Preprint ITÉF-89 (1978).
6. A. A. Kirillov, Elements of the Theory of Representations [in Russian], Nauka, Moscow (1972).
7. D. J. Simms and N. M. Woodhouse, "Lectures on geometric quantization," Springer, Lect. Notes Phys., 53 (1976).
8. J. F. Ritt, "Associative differential operations," Ann. Math. (2) 51, 756-765 (1950).
9. J. F. Ritt, "Differential groups and formal Lie theory for an infinite number of parameters," Ann. Math., (2), 52, 708-726 (1950).
10. J. F. Ritt, "Differential groups of order two," Ann. Math. (2), 53, 491-519 (1951).
11. J. F. Ritt, "Subgroups of differential groups," Ann. Math., (2), 54, 110-146 (1951).
12. B. Weisfeiler, "On Lie algebras of differential formal groups of Ritt," Bull. Am. Math. Soc., 84, No. 1, 127-130 (1978).
13. M. Adler, "On a trace functional for formal pseudodifferential operators and the symplectic structure of the Korteweg–de Vries type equations," Invent. Math., 50, No. 3, 219-248 (1979).

Commun. Math. Phys. 98, 65–77 (1985)

Communications in Mathematical Physics

A Supersymmetric Extension of the Kadomtsev-Petviashvili Hierarchy

Yu. I. Manin[1] and A. O. Radul[2]

1 Steklov Mathematical Institute, Moscow, USSR
2 Moscow State University, Moscow, USSR

Abstract. An extension of the Kadomtsev-Petviashvili hierarchy by odd variables is given. Conservation laws and formal integrability are proved.

0. Introduction

It is well known that integrable systems usually admit a natural extension by odd variables, see e.g. [1–4].

In the articles [5, 6] an infinite hierarchy of nonlinear differential equations was studied with the property that many known completely integrable systems can be obtained from this hierarchy by various reductions. It is called the Kadomtsev-Petviashvili hierarchy and it can be succinctly written in the following way. Let x be a space variable and $t_1, t_2, \ldots$ an infinite system of time variables. Denote by $u_{-1}, u_{-2}, \ldots$ an infinite set of functions depending on $x, t_1, t_2, \ldots$. Set $\partial = \partial/\partial x$ and introduce a formal pseudo-differential operator $L = \partial + \sum_{i=1}^{\infty} u_{-i}\partial^{-i}$. The Kadomtsev-Petviashvili (KP) hierarchy is

$$\partial_i L = [L^i_+, L], \quad \partial_i = \partial/\partial t_i, \tag{1}$$

where A_+ is the differential part of an operator A. If in addition $L^2 = (L^2)_+$, Eq. (1) becomes the ith equation of the Korteweg-de Vries hierarchy (KdV itself corresponding to $i=3$).

The objective of this note is to introduce a new system of equations for an infinite set of even and odd functions, depending on an even-odd pair of space variables (x, ξ) and even-odd times $(\tau_1, t_2, \tau_3, t_4, \ldots)$. We shall call this system the supersymmetric Kadomtsev-Petviashvili (SKP) hierarchy, since KP is its natural reduction. We shall show that SKP shares the standard properties of "completely integrable" systems, e.g. has infinitely many conservation laws, is formally solvable by the Zakharov-Shabat method, and can be reduced to Lax- and Gelfand-Dikii (Dickey) equations. In a subsequent publication we hope to discuss the solitons and algebraic type solutions of SKP, as well as the transformation groups for this hierarchy.

Section 1 contains definitions and statements of principal results. The proofs are given in Sect. 2.

1. Definitions and Results

1.1. Pseudodifferential Operators on the 1|1 Line

The reader may consult [7 or 8] for the background in superalgebra and supergeometry.

We fix an even variable x and odd one ξ; in general we set $\tilde{X}=0$ (respectively 1), if X is even (respectively odd). Set $\theta=\partial/\partial\xi+\xi\,\partial/\partial x$. Then $\theta^2=\frac{1}{2}[\theta,\theta]=\partial$. (We recall that the supercommutator is defined by $[X,Y]=XY-(-1)^{\tilde{X}\tilde{Y}}YX$.)

Let B be a $\mathbb{Z}_2$-graded ring, on which θ acts as an odd superderivation, $\widetilde{\theta b}=\tilde{b}+1$, $\theta(bc)=(\theta b)c+(-1)^{\tilde{b}}b\theta c$. We shall write $\theta^i b=b^{[i]}$. The ring of formal pseudodifferential operators $B((\theta^{-1}))$ consists of the formal series $L=\sum_{i\leq m} b_i\theta^i$, $b_i\in B$. It is $\mathbb{Z}_2$-graded by $\widetilde{b\theta^i}=\tilde{b}+\tilde{i}$. We set $L_+=\sum_{0\leq i\leq m} b_i\theta^i$, $L_-=\sum_{i\leq -1} b_i\theta^i$.

The multiplication in $B((\theta^{-1}))$ is best described by means of "superbinomial coefficients" $\begin{bmatrix} m\\ k\end{bmatrix}$, $k,m\in\mathbb{Z}$. First recall the ordinary coefficients: $\binom{m}{k}=0$ for $k>m$, 1 for $k=m$, $(k+1)\dots m/(m-k)!$ for $k<m$. Now we set

$$\begin{bmatrix} j\\ k\end{bmatrix}=\begin{cases}0 & \text{for}\quad k>j\quad\text{and for}\quad (j,k)\equiv(0,1)\bmod 2\\[2ex] \dbinom{[j/2]}{[k/2]} & \text{for}\quad k\leq j,\quad (j,k)\not\equiv(0,1)\bmod 2,\end{cases} \tag{2}$$

Then we define

$$\theta^j\circ a=\sum_{k\leq j}\begin{bmatrix} j\\ k\end{bmatrix}(-1)^{\tilde{a}k}a^{[j-k]}\theta^k, \tag{3}$$

$$\sum b_j\theta^j\circ\sum a_\ell\theta^\ell=\sum_{j,k,\ell}\begin{bmatrix} j\\ k\end{bmatrix}(-1)^{\tilde{a}_\ell k}b_j a_\ell^{[j-k]}\theta^{k+\ell} \tag{4}$$

[we sometimes use $\circ$ for multiplication in $B((\theta^{-1}))$ to distinguish e.g. $\theta^j\circ a$ from $\theta^j a=a^{[j]}$].

1.2. Flows with Even and Odd Times

The evolution in KP (1) is given by specifying the derivations with respect to the commuting family of the flows ∂_i. The evolution in SKP is defined with respect to a non-abelian Lie superalgebra of flows θ_i, $i\geq 1$, $\tilde{\theta}_i=\tilde{i}$ with the commutation relations

$$[\theta_{2i},\theta_{2j}]=0,\quad [\theta_{2i},\theta_{2j-1}]=0,\quad [\theta_{2i-1}\theta_{2j-1}]=2\theta_{2i+2j-2}. \tag{5}$$

The choice of (5) is motivated by the fact that $\theta_i \to \theta^i$ is a representation of (5), in the same manner as $\partial_i \to \partial^i$ is a representation of the abelian Lie algebra of KP.

Let us now introduce the even and odd time variables $(\tau_1, t_2, \tau_3, \ldots)$, and the following representation of (5) which we shall use in what follows:

$$\theta_{2i} = \partial/\partial t_{2i}, \quad \theta_{2i-1} = \partial/\partial\tau_{2i-1} + \sum_{j=1}^{\infty} \tau_{2j-1}\partial/\partial t_{2i+2j-2}. \tag{6}$$

With minor changes we could also use a more general representation

$$\theta_{2i-1} = \partial/\partial\tau_{2i-1} + \sum_{j=1}^{\infty} T^{2i}_{2j-1}(\tau)\,\partial/\partial t_{2j},$$

where T are some odd polynomials with constant coefficients such that

$$\partial/\partial\tau_{2i-1} T^{2\ell}_{2k-1} + \partial/\partial\tau_{2k-1} T^{2\ell}_{2i-1} = 2\delta_{\ell, i+k-1}.$$

1.3. The Definition of the SKP Hierarchy

Let us assume now that the Lie superalgebra (5) is represented by superderivations of the ring B, supercommuting with θ. This means that the elements of B can be informally considered as functions of x, ξ, τ_i, t_j. We shall extend the action of θ_i to $B((\theta^{-1}))$ in the obvious way.

Now we shall consider an odd pseudodifferential operator

$$\Lambda = \theta + \sum_{i=0}^{\infty} u_{-i}\theta^{-i}, \quad \tilde{u}_{-i} = \tilde{i} + 1, \quad u_i \in B,$$

and we shall call the SKP hierarchy the following system:

$$\left.\begin{aligned} \theta_{2i}\Lambda &= -[\Lambda^{2i}{}_{+}, \Lambda], \\ \theta_{2i-1}\Lambda &= -[\Lambda^{2i-1}{}_{+}, \Lambda] + 2\Lambda^{2i}. \end{aligned}\right\} \tag{7}$$

To explain the appearance of $-2\Lambda^{2i}$ in (7), recall that the ordinary KP (1) can be written also in the form $\partial_j L = -[L^j{}_{-}, L]$ since $[L^j{}_{+} + L^j{}_{-}, L] = [L^j, L] = 0$. In our case $[\Lambda^j{}_{+} + \Lambda^j{}_{-}, \Lambda] = [\Lambda^j, \Lambda] = 0$ for even j, but it is $2\Lambda^{j+1}$ for odd j. Hence the equivalent form of (7) is

$$\theta_j\Lambda = [\Lambda^j{}_{-}, \Lambda]. \tag{8}$$

Notice that (8) implies $\theta_j(u_{-1} + \frac{1}{2}u_0^{[1]}) = 0$. This is analogous to the conservation of the $(n-1)^{\text{th}}$ coefficient in the ordinary Lax equations.

To rewrite SKP once more, we introduce the formal differential

$$d = \sum_{i \geq 1} (dt_{2i}\partial/\partial t_{2i} + d\tau_{2j-1}\partial/\partial\tau_{2j-1}).$$

Using (6), we have

$$\partial/\partial\tau_{2j-1} = \theta_{2j-1} - \sum_{k=1}^{\infty} \tau_{2k-1}\theta_{2j+2k-2},$$

and therefore (7) and (8) are equivalent to the equations

$$d\Lambda = -[U_\Lambda{}^{+}, \Lambda] + 2\sum_{j=1}^{\infty} d\tau_{2j-1}\Lambda^{2j} = [U_\Lambda{}^{-}, \Lambda], \tag{9}$$

where

$$U_{\Lambda}{}^{\pm} = \sum_{i=1}^{\infty} dt_{2i}\Lambda_{\pm}^{2i} + \sum_{j=1}^{\infty} d\tau_{2j-1}\left(\Lambda_{\pm}^{2j-1} - \sum_{k=1}^{\infty} \tau_{2k-1}\Lambda_{\pm}^{2j+2k-2}\right). \tag{10}$$

1.4. The Superresidue and the Berezin Integral

From now on we assume that B is supercommutative. We set

$$\mathrm{res}_{\theta}\left(\sum_{i \leq m} b_i\theta^i\right) = b_{-1}.$$

We remind the reader that the Berezin integral $\int f(x,\xi)\,d(x,\xi)$ is well defined if $\frac{\partial}{\partial \xi} f$ as a function of x has a compact support (or quickly vanishes at infinity). In this case $\int f(x,\xi)\,d(x,\xi) = \int \left(\frac{\partial}{\partial \xi} f\right) dx$. In particular $\int f(x,\xi)\,d(x,\xi) = 0$ if $\partial f/\partial \xi = \partial g/\partial x$, where g quickly vanishes at infinity. The following lemma is an analogue of a well known fact in the theory of the KdV hierarchy (cf. [10, Chap. 2, Lemma 3.3]).

1.5. Lemma. a) *Let $P, Q \in B((\theta^{-1}))$. Then there exists a universal polynomial $F = F(P,Q)$ depending on the coefficients of P, Q and their θ-derivatives, such that*

$$\mathrm{res}_{\theta}[P,Q] = \theta F(P,Q),$$

b) $\int \mathrm{res}_{\theta}[P,Q]d(x,\xi) = 0$ *in the sense that*

$$\frac{\partial}{\partial \xi}\mathrm{res}_{\theta}[P,Q] = \frac{\partial}{\partial x}\left(1 - \xi\frac{\partial}{\partial \xi}\right)F(P,Q).$$

1.6. Theorem on the Conservation Laws. *The SKP equations for an operator Λ imply the equations*

$$\theta_i R_n + \theta S_{i,n} = 0, \quad i, n \geq 1,$$

where

$$R_n = \mathrm{res}_{\theta}\Lambda^n, \quad S_{i,n} = -F(\Lambda^i_-, \Lambda^n),$$

F being defined in Lemma 1.5.

1.7. The Zakharov-Shabat Formalism

In this section we assume that the ring B has an appropriate topology. Then it may happen that the multiplication in $B((\theta^{-1}))$ can be defined by continuity on a certain subset of the doubly infinite formal series $B\{\{\theta^{-1}\}\} = \left\{\sum_i b_i\theta^i, i \in \mathbb{Z} \mid b_i \in B\right\}$. We shall show that certain identities in such a ring make it possible to construct a solution of SKP. Consider a differential form-valued operator

$$U_{\theta} = \sum_{i=1}^{\infty} dt_{2i}\theta^{2i} + \sum_{j=1}^{\infty} d\tau_{2j-1}\left(\theta^{2j-1} - \sum_{k=1}^{\infty} \tau_{2k-1}\theta^{2j+2k-2}\right)$$

[cf. Eq. (10)]. Let $\Psi \in B\{\{\theta^{-1}\}\}$, $\tilde{\Psi}=0$. Assume that the following relations are valid:

$$d\Psi = -U_{\theta}\psi, \tag{11}$$

$$\Psi = V^{-1}Y,\ V = 1+\sum_{k=1}^{\infty} v_{-k}\theta^{-k},\ Y=\sum_{\ell=0}^{\infty} y_{\ell}\theta^{\ell}, \tag{12}$$

where $\tilde{V}=\tilde{Y}=0$. The relation (11) is a system of linear differential equations with constant coefficients, and (12) is the decomposition of a Lie superalgebra element into a product of the upper and lower triangular parts, or a solution of the formal Riemann-Hilbert problem.

1.8. Theorem. *If Y is invertible in $B\{\{\theta^{-1}\}\}$ and verifies $(Y^{-1})_{-}=0$, then $\Lambda = V\theta V^{-1}$ is a solution of SKP.*

We point out that the "dressed" operator Λ verifies the relation $u_{-1}+\frac{1}{2}u_0^{[1]}=0$.

Our next results concern the reductions of SKP to the Lax-Gelfand-Dikii hierarchies [9, 10].

1.9. Theorem on the Fractional Powers. *Let $L=\theta^N+\sum_{n\leqq N-1} u_n\theta^n$. If $N>0$, $N\equiv 1(2)$, then there exists a unique odd operator $\Lambda=\theta+\sum_{i\leqq 0} v_i\theta^i$ such that $\Lambda^N=L$.*

For $N\equiv 0 \bmod 2$ both existence and uniqueness need not be true.

1.10. Variational Formalism

In the following we shall consider the coefficients u_i of the differential operator L as differentially independent variables. To this end we set $B=A[u_i^{[j]}]$, $\widetilde{u_i^{[j]}}=\tilde{u}_i+\tilde{j}$, where $u_i^{[j]}$ are algebraically independent variables and A is a supercommutative ring with an action of θ. We shall extend the action of θ to B, setting $\bar{\theta}|A=\theta$, $\bar{\theta}u_j^{[i]}=u_j^{[i+1]}$, and in what follows we shall denote $\bar{\theta}$ again by θ. The Euler-Lagrange operators are defined by the formula

$$\frac{\delta P}{\delta u_i}=\sum_k(-1)^{\tilde{u}_i k+\frac{k(k+1)}{2}}\left(\frac{\partial P}{\partial u_i^{[k]}}\right)^{[k]}. \tag{13}$$

Furthermore, we introduce the supersymmetric Gelfand-Dikii operator (cf. [9, 10]),

$$\gamma = \Gamma - \Gamma^{*},$$

$$\Gamma_{bc}=\sum_{a\geqq 0}(-1)^{(c+1)(b+1)}\begin{bmatrix} a+b \\ b\end{bmatrix} u_{a+b+c+1}\theta^{a}. \tag{14}$$

Here Γ^{*} is defined from the adjointness property

$$\sum_{b,c}\int(\Gamma_{bc}x_c)y_b\,d(x,\xi)=\sum_{b,c}\int(\Gamma^{*}_{bc}y_c)x_b\,d(x,\xi),$$

where $\tilde{x}_i=\tilde{y}_i=\tilde{u}_i=\tilde{i}+1$. See (21) for the explicit formula.

The operator γ is a highly formal object, having an infinite order and infinite matrix coefficients, but it becomes finite if we set $u_N=1$, $u_i=0$ for $i\geqq N+1$. Denote the result by $\gamma^{(N)}$.

1.11. Theorem on the Lax Equations. *Let* $N\equiv p\equiv 1 \bmod 2$, $p\geqq N$, $L=\theta^N+\sum_{n\leqq N-1} u_n\theta^n$, $\tilde{\Lambda}=1$, $\Lambda^N=L$, $\Lambda=\theta+\sum_{i\leqq 0} v_i\theta^i$. *Moreover, let*

$$H\left(\frac{p}{N}\right)=\frac{N}{p}\operatorname{res}_\theta \Lambda^p .$$

Then the SKP equations for Λ imply the Gelfand-Dikii equations for the coefficients of L:

$$\theta_{p-N}u_b=-\sum_c \gamma_{bc}^{(N)}\frac{\delta}{\delta u_c}H\left(\frac{p}{N}\right). \tag{15}$$

2. The Proofs

2.1. Some Identities with Superbinomial Coefficients

a) The coefficients $\begin{bmatrix} j \\ k \end{bmatrix}$ for $0\leqq j,\ k\leqq 7$ constitute the matrix:

j \ k	0	1	2	3	4	5	6	7
0	1							
1	1	1						
2	1	0	1					
3	1	1	1	1				
4	1	0	2	0	1			
5	1	1	2	2	1	1		
6	1	0	3	0	3	0	1	
7	1	1	3	3	3	3	1	1

It is clear how to continue it further.

The identity $(-1)^k\cdot\binom{m}{k}\cdot k=(-1)^m\binom{-k}{-m}\cdot m$ (which follows from the definition) implies

$$\begin{bmatrix} m \\ -l-1 \end{bmatrix}=(-1)^{\left[\frac{m}{2}\right]+\left[\frac{\ell}{2}\right]+1}\begin{bmatrix} \ell \\ -m-1 \end{bmatrix} \quad \text{for } \ell,m\in\mathbb{Z}. \tag{16}$$

b) The most direct way to describe multiplication in $B((\theta^{-1}))$ is to consider (4) as the definition. Then it is necessary to prove the associativity law: $(P\circ Q)\circ R=P\circ(Q\circ R)$. Collecting together similar terms we can reduce it to a bilinear identity in $\begin{bmatrix} j \\ k \end{bmatrix}$, and then prove it directly. Another way is to use the well-known construction of multiplication in $B((\partial^{-1}))$ by means of the ordinary binomial coefficients and the fact that $B((\theta^{-1}))=B((\partial^{-1}))[\theta]$ with evident commutation relations,

$$\theta(a\,\partial^m)=a^{[1]}\theta^{2m}+(-1)^{\tilde a}a\theta^{2m+1}.$$

2.2. Proof of Lemma 1.5

The statement b) follows from a) because $\frac{\partial}{\partial\xi}\theta F(P,Q)=\frac{\partial}{\partial x}\left(1-\xi\frac{\partial}{\partial\xi}\right)F(P,Q)$.

To prove a) it is sufficient to consider the case $P=a\theta^m$, $Q=b\theta^\ell$. From (4) we find

$$a\theta^m\circ b\theta^\ell=a\sum_{k\leqq m}\begin{bmatrix}m\\k\end{bmatrix}(-1)^{\tilde b k}b^{[m-k]}\theta^{k+\ell}.$$

Hence

$$\mathrm{res}_\theta(a\theta^m\circ b\theta^\ell)=\begin{bmatrix}m\\-\ell-1\end{bmatrix}(-1)^{\tilde b(\ell+1)}ab^{[m+\ell+1]}.$$

By the symmetry

$$\mathrm{res}_\theta(b\theta^\ell\circ a\theta^m)=\begin{bmatrix}\ell\\-m-1\end{bmatrix}(-1)^{\tilde a(m+1)}ba^{[m+\ell+1]}.$$

Using these two formulas and (16) we find

$$\mathrm{res}_\theta[P,Q]=\begin{bmatrix}m\\-\ell-1\end{bmatrix}(-1)^{\tilde b(\ell+1)}(ab^{[m+\ell+1]}+(-1)^M a^{[m+\ell+1]}b), \qquad (17)$$

$$M=\tilde a(m+\ell+1)+m\ell+\left[\frac{m}{2}\right]+\left[\frac{\ell}{2}\right].$$

The right-hand side of (17) is represented as the complete θ-derivative in two different ways depending on the parity of $m+\ell+1$ $\left(\text{the case } (m,\ell)\equiv(0,0)\bmod 2\right.$ needs no attention, since then $\left.\begin{bmatrix}m\\-\ell-1\end{bmatrix}=0\right)$:

$$ab^{[m+\ell+1]}+(-1)^M a^{[m+\ell+1]}b$$

$$=\begin{cases}\theta\left(\sum_{i=0}^{m+\ell}(-1)^{\tilde a(i+1)+\left[\frac{i}{2}\right]}a^{[i]}b^{[m+\ell-i]}\right) & \text{if} \quad m+\ell+1\equiv 1\bmod 2\\ \theta^2\left(\sum_{i=0}^{\frac{m+\ell-1}{2}}(-1)^i a^{[2i]}b^{[m+\ell-2i-1]}\right) & \text{if} \quad m+\ell+1\equiv 0\bmod 2.\end{cases}$$

2.3. Proof of Lemma 1.6

From (8) we find $\theta_i\Lambda^n=[\Lambda^i{}_-,\Lambda^n]$. Therefore

$$\theta_i\,\mathrm{res}_\theta\Lambda^n=\mathrm{res}_\theta\theta_i\Lambda^n=\mathrm{res}_\theta[\Lambda^i{}_-,\Lambda^n]=\theta F(\Lambda^i{}_-,\Lambda^n).$$

2.4. Proof of Theorem 1.8

In the notation of Sect. 1.7 we construct the following operators from the data (11) and (12):

$$\Lambda = V\circ\theta\circ V^{-1}, U^{+}=(dY)\circ Y^{-1}, U^{-}=-(dV)\circ V^{-1}.$$

On the other hand, we construct from Λ "connection forms" $U_{\Lambda}^{\pm}$ by the formula (10). We shall check that if the condition of Theorem 1.8 is fulfilled, then $U^{\pm}=-U_{\Lambda}^{\mp}$. Really, using that Ψ is invertible in view of (12), we find from (11),

$$\begin{aligned}-U_{\theta}=d\Psi\circ\Psi^{-1}&=d(V^{-1}\circ Y)\circ Y^{-1}\circ V\\ &=(-V^{-1}dV\circ V^{-1}\circ Y+V^{-1}\circ dY)Y^{-1}\circ V\\ &=V^{-1}(-dV\circ V^{-1}+dY\circ Y^{-1})V=V^{-1}(U^{-}+U^{+})V.\end{aligned}$$

Hence

$$U^{+}+U^{-}=-VU_{\theta}V^{-1}=-\sum_{i=1}^{\infty}dt_{2i}\Lambda^{2i}+\sum_{j=1}^{\infty}d\tau_{2j-1}\left(\Lambda^{2j-1}-\sum_{k=1}^{\infty}\tau_{2k-1}\Lambda^{2j+2k-2}\right)$$

and, consequently, $U^{\pm}=-U_{\Lambda}^{\mp}$, because U^{+} is a differential operator, and U^{-} is an integral one. On the other hand, by the definition of Λ,

$$\begin{aligned}d\Lambda=d(V\circ\theta\circ V^{-1})&=dV\circ\theta\circ V^{-1}+V\circ\theta\circ V^{-1}\circ dV\circ V^{-1}\\ &=[dV\circ V^{-1},V\circ\theta\circ V^{-1}]=-[U^{-},\Lambda]=[U_{\Lambda}^{-},\Lambda],\end{aligned}$$

i.e. the SKP hierarchy in the form (9).

2.5. Proof of Theorem 1.9

Let us construct Λ using successive approximations. Assume that at the rth step, $r\geqq-1$, we have proved the following statement:

there exists an operator $X_r=\theta+O(1)$ such, that $X_r^N=L+O(\theta^{N-r-2})$; it is defined uniquely $\mathrm{mod}\,O(\theta^{-(r+1)})$.

For $r=-1$, evidently, $X_r=\theta$. To make the inductive step we set $X_{r+1}=X_r+x_{r+1}\theta^{-(r+1)}$ Then

$$X_{r+1}^N=X_r^N+\sum_{i=0}^{N-1}X_r^i\circ x_{r+1}\theta^{-(r+1)}\circ X_r^{N-i-1}+R_r.$$

Now we compute the right-hand side sum modulo $O(\theta^{N-r-3})$:

$$\begin{aligned}\Sigma&=\sum_{i=0}^{N-1}(-1)^i x_{r+1}\theta^{-(r+1)}\circ X_r^{N-1}\,\mathrm{mod}\,O(\theta^{N-r-3})\\ &=x_{r+1}\theta^{N-r-2}\,\mathrm{mod}\,O(\theta^{N-r-3}).\end{aligned}$$

(We use the fact that N is odd only in this place: if N is even the linear term in x_{r+1} disappears.) The remainder is the sum of the products of $j\leqq N-2$ terms X_r and of $N-j$ terms $x_{r+1}\theta^{-(r+1)}$ Therefore its order with respect to θ is not more than

$\max_{j\leqq N-2}[j-(N-j)(r+1)]=N-2r-4\leqq N-r-3$. Hence x_{r+1} is defined uniquely from the condition $X^N_{r+1}=L+O(\theta^{N-2-3})$. To finish the proof we set $\Lambda=\lim_{r\to\infty}X_r$.

We give two examples concerning even N.

a) Let $L=\theta^{2n}+\sum_{i\leqq 2n-1}u_i\theta^i$, $\tilde{L}=0$. If $\mathrm{res}_\theta L\notin\theta B$, then $L\neq\Lambda^{2n}$ for any odd Λ. In fact, $\Lambda^{2n}=\frac{1}{2}[\Lambda^{2n-1},\Lambda]$, so $\mathrm{res}_\theta\Lambda^{2n}$ should be a θ-derivative in view of Lemma 1.5.

b) In the case $L=\theta^2$ the following nonuniqueness of the square root arises: we can take Λ in the form $\theta+u_0-\frac{1}{2}u_0^{[1]}\theta^{-1}{}_{+}+\sum_{i\leqq -2}u_i\theta^i$, where the u_i satisfy the following conditions:

$$\begin{cases}u_{i,x}=0\,,\\ u_{2p-1,\xi}=0\,,\\ u_{2p,\xi}+2u_{2p-1}+\sum\limits_{\ell+m=p+1}(u_{2\ell-1}u_{2m,\xi}+u_{2\ell-1}u_{2m-1})=0\,.\end{cases}$$

It is possible, for instance, to choose arbitrary coefficients u_{2p} independent of x and then to compute u_{2p-1} by induction. In fact, the first series of equations follows from the equality

$$0=[\Lambda^2,\Lambda]=[\partial_x,\Lambda]=\sum_{i\leqq 0}u_{i,x}\theta^i\,.$$

Coefficients of $\frac{1}{2}[\Lambda,\Lambda]$ at odd powers of θ give us the following equations:

$$u_{2p-1,\xi}+\sum_{m+n=p+1}u_{2m-1}u_{2n-1,\xi}=0\,.$$

Now, inductively we get $u_{2p-1,\xi}=0$. Finally, the coefficients at $\theta^{2\ell}$, $\ell\leqq 0$, lead to the last series of equations.

2.6. More on the Variational Formalism

Using the notations from Sect. 1.10 we consider the B-module of relative differentials $\Omega=\Omega^1 B/A$. It is freely generated by the elements $\delta u_i^{[j]}$ We assume that $\tilde{\delta}=0$ (that is $\delta\tilde{P}=\tilde{P}$). The derivation θ acts on Ω by the formula $\theta(\delta P)=\delta(P^{[1]})$. Therefore we can define a bimodule $\Omega((\theta^{-1}))$ over the ring $B((\theta^{-1}))$ using the formula (4) for the exterior multiplication. The associativity axiom is verified by repeating the reasoning from Sect. 2.1b). The map $\delta: B((\theta^{-1}))\to\Omega((\theta^{-1}))$ is defined termwise. It shares the property $\delta(P\circ Q)=\delta P\circ Q+P\circ\delta Q$.

Let $\omega\in\Omega$ be a variational differential. It has a unique representation in the form $\omega=\omega_0+\theta\omega_1$, where $\omega_0\in\bigoplus B\,\delta u_i$. The existence is proved by the classical procedure of integrating by parts

$$a\,\delta u_i^{[j+1]}=(-1)^{\tilde{a}+1}(a^{[1]}\,\delta u_i^{[j]}-\theta(a\,\delta u_i^{[j]}))\,. \tag{18}$$

The uniqueness follows from the fact: if $\theta\omega_1\neq 0$, then $\theta\omega_1\notin\bigoplus B\,\delta u_i$ because the term $a\,\delta u_i^{[j]}$ in ω_1 with maximal j gives the term $a\,\delta u_i^{[j+1]}$ in $\theta\omega_1$ which cannot cancel.

It is easy to derive from (18) the following lemma, giving the invariant description of the Euler-Lagrange operators (13):

2.7. Lemma. *Let* $P \in B$ *and* $\delta P = \sum \delta u_i Q_i + \theta\omega_1$. *Then* $Q_i = \frac{\delta P}{\delta u_i}$.

To prove Theorem 1.11 we need the following facts:

2.8. Lemma. a) *Let* $\Lambda \in B((\theta^{-1}))$, $\tilde{\Lambda} = 1$, $\Lambda = \theta + \sum_{i \leqq 0} v_i\theta^i$, $p \geqq 1$. *Then*

$$\delta\, \mathrm{res}_\theta \Lambda^p \equiv \begin{Bmatrix} p\, \mathrm{res}_\theta(\delta\Lambda \circ \Lambda^{p-1}) & \text{if} \quad p \equiv 1(2) \\ 0 & \text{if} \quad p \equiv 0(2) \end{Bmatrix} \mathrm{mod}\, \theta\Omega .$$

b) *Let* $L = \Lambda^N$. *Then*

$$\delta\, \mathrm{res}_\theta \Lambda^p \equiv \frac{p}{N} \mathrm{res}_\theta(\delta L \circ \Lambda^{p-N}) \, \mathrm{mod}\, \theta\Omega \quad \text{if} \quad N \equiv p \equiv 1 \, \mathrm{mod}\, 2 .$$

Proof. a) We have

$$\delta\Lambda^p = \sum_{i=0}^{p-1} \Lambda^i \circ \delta\Lambda \circ \Lambda^{p-i-1} ,$$

$$\Lambda^i \circ \delta\Lambda \circ \Lambda^{p-i-1} = (-1)^{i(p-i)}(\delta\Lambda \circ \Lambda^{p-1} - [\delta\Lambda \circ \Lambda^{p-i-1}, \Lambda^i]) .$$

Furthermore $\sum_{i=0}^{p-1} (-1)^{i(p-i)} = p$ if $p \equiv 1(2)$ and 0 if $p \equiv 0(2)$. So the statement a) follows from Lemma 1.5, remaining undoubtedly valid when it is applied to commutators in

$$[\Omega((\theta^{-1})), B((\theta^{-1}))] .$$

b) Similarly

$$\delta\Lambda^N \circ \Lambda^{p-N} = \sum_{i=0}^{N-1} \Lambda^i \circ \delta\Lambda \circ \Lambda^{p-i-1} \equiv N\, \delta\Lambda \circ \Lambda^{p-1} \, \mathrm{mod}\, \theta\Omega \quad \text{if} \quad N \equiv 1(2) .$$

This finishes the proof.

2.9. Lemma. *Let* $N \equiv 1(2)$, $L = \theta^N + \sum_{i=0}^{N-1} u_i\theta^i$, $\tilde{L} = 1$, u_i *being independent differential variables. Let* Λ *be the odd* N^{th} *power root from* L. *constructed in Sect. 2.5. We define the differential polynomials* $v_k\left(\frac{p}{N}\right)$ *in* u_i *by the formula*

$$\Lambda^p_- = (L^{p/N})_- = \sum_{k \leqq 0} \theta^{k-1} \circ v_k\left(\frac{p}{N}\right).$$

Then if $p \equiv 1 \, \mathrm{mod}\, 2$, *we have for* $k \leqq N-1$

$$\frac{\delta}{\delta u_k} v_0\left(\frac{p}{N}\right) = \frac{p}{N}(-1)^{k+1} v_{-k}\left(\frac{p-N}{N}\right). \tag{19}$$

Proof. Since $v_0\left(\frac{p}{N}\right)$ is even, $\mathrm{res}_\theta \Lambda^p = v_0\left(\frac{p}{N}\right)$. From the previous lemma we have

$$\begin{aligned}\delta v_0\left(\frac{p}{N}\right) &= \delta\, \mathrm{res}_\theta \Lambda^p \equiv \frac{p}{N}\mathrm{res}_\theta(\delta L \circ \Lambda^{p-N})\\ &= \frac{p}{N}\mathrm{res}_\theta\left(\sum_{j=0}^{N-1}\delta u_j \theta^j \circ \sum_{k\leqq 0}\theta^{k-1}\circ v_k\left(\frac{p-N}{N}\right)\right)\\ &= \frac{p}{N}\mathrm{res}_\theta\left(\sum_{j=0}^{N-1}\delta u_j \theta^{-1}\circ v_{-j}\left(\frac{p-N}{N}\right)\right)\\ &= \frac{p}{N}\sum_{j=0}^{N-1}\delta u_j(-1)^{j+1}v_{-j}\left(\frac{p-N}{N}\right).\end{aligned}$$

To finish the proof we use Lemma 2.6.

2.10. The End of the Proof of Theorem 1.11

From the SKP equations for Λ we derive the following equations for L:

$$-\theta_{p-N}L = [L, \Lambda^{p-N}{}_-],$$

or

$$-\sum_{b=0}^{N-1}(\theta_{p-N}u_b)\theta^b = \left[\sum_{\ell=0}^{N}u_\ell\theta^\ell, \sum_{j\geqq 0}\theta^{-j-1}\circ v_{-j}\left(\frac{p-N}{N}\right)\right]. \tag{20}$$

We shall show that the coefficient at θ^b in the first member of the commutator is equal to $\sum_c \Gamma_{bc}\frac{\delta}{\delta u_c}v_0\left(\frac{p}{N}\right)$, where Γ_{bc} are defined in (14). Really

$$[L, \Lambda^{p-N}{}_-] = -[L, \Lambda^{p-N}{}_+] = [L, \Lambda^{p-N}{}_-]_+ .$$

Hence it is sufficient to compute $(L\Lambda^{p-N}{}_-)_+$ and $(\Lambda^{p-N}{}_-L)_+$. Using Lemma 2.9 we have

$$\begin{aligned}(L\Lambda^{p-N}{}_-)_+ &= \left(\sum_{\ell=0}^{N}u_\ell\theta^\ell \circ \sum_{j\geqq 0}(-1)^{j+1}\theta^{-(j+1)}\frac{\delta}{\delta u_j}H\left(\frac{p}{N}\right)\right)_+\\ &= \left(\sum_{\ell=0}^{N}\sum_{j\geqq 0}(-1)^{j+1}u_\ell\sum_b\begin{bmatrix}\ell-j-1\\ b\end{bmatrix}(-1)^{b(j+1)}\right.\\ &\quad\left.\cdot\left(\frac{\delta}{\delta u_j}H\left(\frac{p}{N}\right)\right)^{[\ell-j-1-b]}\theta^b\right)_+\\ &= \sum_{\substack{a,b,c\geqq 0\\ a+b+c+1\leqq N}}(-1)^{(c+1)(b+1)}u_{a+b+c+1}\begin{bmatrix}a+b\\ b\end{bmatrix}\left(\frac{\delta}{\delta u_c}H\left(\frac{p}{N}\right)\right)^{[a]}\theta^b.\end{aligned}$$

Here we have denoted $\ell-j-b-1=a, j=c$. Computing $(\Lambda^{p-N}{}_{-}L)_+$, we obtain at first

$$\sum_{\substack{a,b,c\geq 0\\ a+b+c+1\leq N}} \begin{bmatrix} -c-1 \\ -a-c-1 \end{bmatrix} (-1)^{ab}\left(u_{a+b+c+1}\frac{\delta}{\delta u_c}H\left(\frac{p}{N}\right)\right)^{[a]}\theta^b .$$

Then we transform the superbinomial coefficient using (16):

$$\begin{bmatrix} -c-1 \\ -a-c-1 \end{bmatrix} = (-1)^{\left[\frac{-c-1}{2}\right]+\left[\frac{a+c}{2}\right]+1}\begin{bmatrix} a+c \\ c \end{bmatrix}.$$

If a or c are even, then the sign is equal to $(-1)^{\left[\frac{a}{2}\right]}$. If a and c are both odd then the coefficients vanish. Finally

$$(\Lambda^{p-N}{}_{-}L)_+ = \sum_{\substack{a,b,c\geq 0\\ a+b+c+1\leq N}} (-1)^{\left[\frac{a}{2}\right]+ab}\begin{bmatrix} a+c \\ c \end{bmatrix}\left(u_{a+b+c+1}\frac{\delta}{\delta u_c}H\left(\frac{p}{N}\right)\right)^{[a]}\circ\theta^b . \tag{21}$$

The reader can check that the operators we have computed are adjoint.

2.11. Examples. Set $L=\Lambda^4=\Lambda^4_+=\theta^4+v_1\theta+v_0$, $\tilde{v}_1=1$, $\tilde{v}_0=0$. Then Lax's equation $L_t=[L_+^{3/2},L]$ is equivalent to

$$\left.\begin{aligned} v_{0t}&=\partial_x(v_{0xx}/4+3v_0{}^2/4+3v_1v_0{}^{[1]}/4)\,,\\ v_{1t}&=\partial_x(v_{1xx}/4+3v_1v_1{}^{[1]}/4+3v_1v_0/2)\,. \end{aligned}\right\} \tag{22}$$

Setting here $v_1=0$ we get KdV. On the other hand, setting $v_0=0$ and $v_1=w_1+\xi w_0$, $\tilde{w}_1=1$, $\tilde{w}_0=0$, we get from (22)

$$\left.\begin{aligned} w_{0t}&=\partial_x(w_{0xx}/4+3w_0{}^2/4-3w_1w_{1x}/4)\\ w_{1t}&=\partial_x(w_{1xx}/4+3w_0w_1/4)\,. \end{aligned}\right\} \tag{23}$$

Setting $w_1=0$ here we again get KdV. The first Eq. (23) after rescaling coincides with the first equation of Kupershmidt's SKdV, but the second equations are different (cf. [3]).

References

1. Saveliev, M.: Integrable supermanifolds and nonlinear systems. Preprint IHEP 84-20. Serpukhov (1984)
2. Chaichian, M., Kulish, P.: On the method of inverse scattering problem and Bäcklund transformations for supersymmetric equations. Phys. Lett. **78** B, 413 (1978)
3. Kupershmidt, B.: A super-Korteweg-de Vries equation: an integrable system. Preprint UTSI, Tullahoma (1984)
4. d'Auria, R., Sciuto, S.: Group theoretical construction of two-dimensional supersymmetric models. Nucl. Phys. B **171**, 189 (1980)
5. Date, E., Kashiwara, M., Miwa, T.: Vertex operators and τ-functions. Transformation groups for soliton equations. II. Proc. Jpn. Acad. Sci. **57** A, 387 (1981)

6. Date, E., Jimbo, M., Kashiwara, M., Miwa, T.: Operator approach to the Kadomtsev-Petviashvili equation. Transformation groups for soliton equations. III. J. Phys. Soc. Jpn. **50**, 3866 (1981)
7. Wess, J.: Supersymmetry-supergravity. In: Lecture Notes in Physics, Vol. 77. Berlin, Heidelberg, New York: Springer 1978
8. Leites, D.: Introduction to the theory of supermanifolds. Russ. Math. Surv. **35**, 3 (1980)
9. Gelfand, I., Dikii, L.: Fractional powers of operators and hamiltonian systems. Funct. Analiz **10**, 13 (1976)
10. Manin, Yu.: Algebraic aspects of differential equations. J. Sov. Math. **11**, 1 (1979)

Communicated by Ya. G. Sinai

Received June 15, 1984

Uspekhi Mat. Nauk 41:5 (1986), 85-107 Russian Math. Surveys 41:5 (1986), 33-61

The twistor transformation and algebraic-geometric constructions of solutions of the equations of field theory

M.M. Kapranov and Yu.I. Manin

CONTENTS

Introduction 33
§1. Cohomological characterization of admissible sheaves 39
§2. Admissible sheaves as direct images 43
§3. Invertible sheaves and Baker functions 45
§4. Geometry of the twistor transformation and constructions of admissible sheaves 48
§5. Twistor analysis of a generalized system with φ^4-interaction 52
References 58

Introduction

1. Statement of the problem.

The twistor transformation, introduced by Penrose's pioneering articles (see [42], [45], [6], [25]), recodes the classical fields satisfying various equations of motion and constraints as geometrical objects: complex manifolds and holomorphic bundles over them with well-defined properties. Hence direct constructions of such objects lead to constructions of various classes of solutions of equations of motion.

The first successes of the twistor transformation method were associated with the self-dual Einstein and Yang-Mills equations ([41], [48], [1]-[3], [7]). In the non-self-dual case even the very definition of the twistor transformation turned out to be non-trivial, and the development of the theory was noticeably slower here. The decisive breakthrough here was Witten's article [51], where the twistor transformation was defined simultaneously for the ordinary and the supersymmetric Yang-Mills equations in four-dimensional space-time (part of these results was simultaneously announced in [26]). The importance of this construction is confirmed by an article [52] just published, in which Witten shows that the same scheme of twistor transformation works effectively in models of supergravitation and Yang-Mills in ten-dimensional (more precisely, in 10|16-dimensional) space-time, which is now being studied intensively in connection with the theory of superstrings.

The geometrical problems themselves raised by the general (non-self-dual) twistor transformation have been less well studied.

The present article considers methods of solving the following problems to which the ordinary and supersymmetric Yang-Mills equations reduce, possibly with sources, in flat or curved space-time of dimension $4|4N$ or $10|16$.

We are given a complex-analytic manifold, or space, or superspace X. It is endowed with a family of closed subspaces $Y(u) \subset X$, which are parametrized by the points of a complex (super)space $U \ni u$. We want to construct holomorphic vector bundles on X whose restrictions to each $Y(u)$ are isomorphic to bundles of a given type (in the majority of cases they are simply trivial on $Y(u)$; an example of a completely different type is analysed by Manin and Khoang Le Min' [36]).

In all the applications we are interested in, the $Y(u)$ are compact, non-variable, and in aggregate form a locally complete variation of a closed subspace of X; X itself may be non-compact. The type of bundle over $Y(u)$ is also non-variable (instead of bundles we shall speak of locally free sheaves). The information we need about complex supergeometry is set out in the book [38]. For a generalization of the theory of deformations to the supercase see [5].

Let us give a list of the situations (X, $Y(u)$; admissible type of sheaf over $Y(u)$) to which the equations that have been studied in the literature reduce.

a) *The equations of self-duality.* Here $X = P^3$, $Y(u) = P^1$, U is a domain in the Grassmannian $G(2, T)$, $T = \mathbf{C}^4$, parametrizing the family of lines $P^1 \subset P^3 = P(T)$. The solutions of the equations of self-duality in U correspond to locally free sheaves $\mathscr{E}$ on $X(U) = \bigcup_{u\in U} Y(u)$, which are trivial when restricted to all the $Y(u)$, that is, are isomorphic to $\mathcal{O}^r_{Y(u)}$.

b) *The Yang-Mills equations.* Here $X \subset P(T) \times P(T^*)$ is the graph of the incidence relation, isomorphic to the flag space $F(1, 3, T)$ in $T = \mathbf{C}^4$; $Y(u) = P^1 \times P^1$ are quadrics lying in X (whose generators are lines in $P(T)$, $P(T^*)$); $U \subset G(2, T)$ is the domain in the Grassmannian parametrizing these quadrics. Under certain topological assumptions about U (see [38], 89) the holomorphic Yang-Mills fields on U are in one-to-one correspondence with locally free sheaves $\mathscr{E}$ on $X(U) = \bigcup_{u\in U} Y(u)$ with trivial restrictions to all $Y(u)$. If X is replaced by $X^{(3)}$ (where $X^{(i)}$ is the i-th infinitesimal neighbourhood of X in $P(T) \times P(T^*)$), then precisely the solutions of the Yang-Mills vacuum equations will be obtained on U.

c) *The supersymmetric Yang-Mills equations.* Put $T = \mathbf{C}^{4|N}$, $P^{3|N} = P(T)$, $\hat{P}^{3|N} = P(T^*)$ (dual superprojective spaces). Let $X_N \subset P^{3|N} \times \hat{P}^{3|N}$ denote the graph of the incidence relation (it is isomorphic to the superspace of flags $F(1|0, 3|N; T)$), and let $X_N^{(i)}$ be its i-th infinitesimal neighbourhood: this is an analytic superspace. The quadrics $Y(u) = P^1 \times P^1$ in X_N are parametrized by the superspace of flags $F(2|0, 2|N;\ T)$, which represents the compactified and complexified space-time for N-extended supersymmetry. Holomorphic solutions of the supersymmetric Yang-Mills equations in $U \subset F(2|0, 2|N;\ T)$ for $0 \leqslant N \leqslant 3$ correspond to locally free sheaves on $X^{(3-N)}_N(U)$, which become trivial when restricted to the quadrics. (In the supergeometry of the structure group a locally free sheaf can be a supergroup, for example $\mathrm{GL}(r|s)$: correspondingly, trivial sheaves can have mixed rank and be isomorphic to $\mathcal{O}_Y^{r|s} = \mathcal{O}_Y^r - \Pi\,\mathcal{O}_Y^s$, where Π is the change of parity functor).

d) *The connection equations for supersymmetric Yang-Mills fields.* As before, let $T = \mathbf{C}^{4|N}$, $N = 2$ or 3. We take X as the superflag manifold $F(2|1;\ T)$ for $N = 2$ or $F(2|1, 2|2, T)$ for $N = 3$. To each point $u \in F(2|0, 2|N;\ T)$ of the superspace there corresponds the incidence submanifold $Y(u) \subset X$. Here the flags, say $\mathcal{S}^{2|1} \subset \mathcal{S}^{2|2}$ and $\mathcal{S}_1^{2|0} \subset \mathcal{S}_1^{2|3}$, are called incident if $\mathcal{S}_1^{2|0} \subset \mathcal{S}^{2|1} \subset \mathcal{S}^{2|2} \subset \mathcal{S}_1^{2|3}$.

This interpretation of the connection equations was discovered by Roslyi [43].

For $N = 2$ we have $Y(u) = P^1$, and for $N = 3$ we have $Y(u) = F(1, 2;\ \mathbf{C}^3)$. In [10]–[12] these complex manifolds were called "harmonic spaces", and it was suggested that the whole quantum theory of Yang-Mills superfields could be constructed on domains of the superspaces $F(2|0, 2|1, 2|2;\ T)$ or $F(2|0, 2|1, 2|2, 2|3;\ T)$ for $N = 2, 3$ respectively. These domains are fibered over the usual Minkowski superspace and the harmonic space is the fibre over each point. Each such domain is projected into X, and the harmonic spaces go over to the family $Y(u)$.

The locally free sheaves on $X(U)$ with trivial restrictions to $Y(u)$, $u \in U$, correspond to those holomorphic Yang-Mills superfields on U that are solutions of the connection equations. For $N = 3$ the connection equations are equivalent to the equations of motion, so that we obtain an alternative algebraic-geometric formulation of the same problem as in c).

e) *The system of Yang-Mills fields and scalar fields with φ^4-self-action.* We return to the situation of b) and consider locally free sheaves $\mathcal{E}$ on $X^{(2)}(U)$ whose restrictions to the quadrics $Y(u) = P^1 \times P^1$ are isomorphic to sheaves of the form $\mathcal{O}^r(-1, 0) \oplus \mathcal{O}^s(0, -1)$. This problem is equivalent to the solution of a complicated non-linear system of equations containing the interacting Yang-Mills fields and the scalar fields in the adjoint representation of the structure group. This system is investigated in §5 of the present article. Under the stronger condition that $\mathcal{E}$ is isomorphic to $\mathcal{O}^r(-1, 0) + \mathcal{O}^s(0, -1)$ on the first neighbourhoods of the quadrics $Y(u)^{(1)} \subset X(U)$, solutions of the system with φ^4-self-action are obtained, as Khenkin has established [17].

All the problems named relate to space-time of dimension 4 or $4|4N$. The last example of Witten [52] is of this kind (see also §4).

f) *The Yang-Mills equations in ten-dimensional space.* Here U is a domain in $\mathbf{C}^{10|16}$, and X a complex supermanifold of dimension $17|8$, which parametrizes "superlight rays" of dimension $1|8$. The family $Y(u) \subset X$ consists of 8-dimensional quadrics. Locally free sheaves on $X(U)$, with trivial restrictions to $Y(u)$, $u \in U$, correspond to solutions of the supersymmetric Yang-Mills equation in a domain of ten-dimensional space-time.

g) *Yang-Mills fields interacting with supergravitation.* All the examples described in a)–f) relate to flat space-time. The inclusion of (super)gravitation in this picture alters it in the following way: X is replaced by a superspace X' of the same dimension containing a family of closed submanifolds $Y'(u) \subset X'$, which coincide with the corresponding manifolds in the flat case and are embedded in X' with the same normal sheaf. The problem is posed in the situation $(X', Y'(u))$ in the same way as in the flat case.

The problem itself of characterising supermanifolds X' and families $Y'(u)$ corresponding to solutions of the Einstein equations or supergravitation can be considered as a generalization of Penrose's construction of self-dual "gravitons". At present there is no satisfactory solution of this problem, although encouraging partial results have been obtained by Lebrun [28], [29], Manin and Penkov [33], Yasskin [46], and Witten [52]. The main difficulty is to detect what additional geometric structures on X' must be taken into account, besides $Y'(u)$. When this has been clarified, the problem of methods of construction of such "twistor spaces" will become urgent.

In this article we regard $(X', Y'(u))$ as given (not necessarily flat).

2. Methods of construction.

The following basic methods for the construction of locally free sheaves $\mathcal{E}$ on X with prescribed restrictions to $Y(u)$ are known.

A. *The method of monads.* A monad is a complex of locally free sheaves of length 3 on X: $\mathcal{F}_{-1} \xrightarrow{\alpha} \mathcal{F}_0 \xrightarrow{\beta} \mathcal{F}_1$, $\beta\alpha = 0$, $\operatorname{Ker} \alpha = 0$, $\operatorname{Im} \beta = \mathcal{F}_1$. In all applications the $\mathcal{F}_i$ are direct sums of sheaves of rank 1. The morphisms α and β can then be described by matrices. They are the basic parameters of the construction: the sheaf $\mathcal{E}$ is defined as $\operatorname{Ker} \beta/\operatorname{Im} \alpha$.

All the self-dual Yang-Mills fields have been described by the method of monads under additional conditions of reality (more precisely, reduction to a compact group) and finiteness of action (instantons) or energy (monopoles). In the last case, according to the remarkable construction of Nahm, the terms of the monad have infinite rank, and α, β are differential operators (with respect to a parameter). The use of monads over a compact base is explained by the deep results of Beilinson [4] on the structure of the derived category of sheaves over (relative) projective spaces (see also [40] and the appendix to [40]). It was first proved in [35] (see also [38], Ch. 5, §4) that solutions of the complete Yang-Mills equations, ordinary and supersymmetric, could also be constructed by the method of monads. Here the gauge group naturally turns out to be a supergroup.

B. *The method of extensions.* This consists in the construction of sheaves of rank 2 as extensions of sheaves of rank 1 and was first applied to self-dual equations in [1]. Monopole solutions admit a natural representation in this form: see Hitchin [23]. The possibilities of this approach have not been studied for higher ranks and non-self-dual problems.

C. *The method of direct images and Baker functions.* An account of constructions of this type is the main aim of the present article.

The Baker functions, widely used for the construction of solutions of non-linear wave equations in (1 + 1)-dimensional space-time (see [53], [54]), are sections of invertible sheaves with real singularities. The method of constructing them is such that in the non-compact case it allows the introduction of a dependence on arbitrary functions in elements of the Picard group. Having a supply of invertible sheaves, we can construct from them sheaves of arbitrary rank by means of the direct image operation. In addition, invertible sheaves depending on functional parameters can be used for twisting sheaves of higher rank constructed by more "rigid" means. This idea was introduced in the self-dual case by Cherednik [47].

3. The inverse Penrose transformation and the method of the inverse problem of scattering theory.

Obtaining explicit formulae for fields is not the main problem in this article, but nevertheless we shall say a few words about it and in general about the structure of the inverse Penrose transformation, emphasizing the analogy between it and the inverse scattering problem method.

Many versions of the inverse scattering problem method are known: the majority of them have the following features in common. We are interested in a non-linear system of equations $\varphi(u, x) = 0$, where the u are fields, and the x are space-time coordinates; the system is represented in the form of a necessary (and sometimes also sufficient) condition for the integrability of an auxiliary linear system $G(u, x, \psi, \lambda) = 0$. Here the ψ are unknown functions of a linear problem (for example, Baker functions), and λ denotes a family of parameters sometimes called "spectral parameters". Among the best-known examples are the Lax representations for non-linear wave equations, and the Yang-Belavin-Zakharov-Ward representation for self-dual Yang-Mills equations. After such a representation has been established, the solutions of the auxiliary linear system are either constructed explicitly or characterized by their properties, for example by their asymptotic behaviour or their singularities. In such a characterization the dependence of ψ on the parameters λ comes to the fore; moreover, the domain of values of λ is usually complexified. A typical example is scattering data.

Another method of describing the dependence of ψ on λ is more fundamental both in the construction of algebraic-geometric solutions by the inverse scattering problem method, and in the theory of the Penrose transformation. Here we use the holomorphic geometry of the vector bundle of spaces of solutions over the space of spectral parameters L. In more detail, let $E(\lambda)$ be the vector space of solutions $\psi(x, \lambda)$ of the system $G(u, x, \psi, \lambda) = 0$ for fixed u. Then $E = \bigcup_{\lambda \in L} E(\lambda)$ is just the vector bundle which it is natural to call the "generalized scattering data", and which was called admissible in subsection 1. The detailed connection between the problems introduced in subsection 1 and the ideology described above is the following. Let $X \overset{\pi_1}{\leftarrow} Z \overset{\pi_2}{\rightarrow} U$ be the dual bundle describing the family $Y(u) \subset X$ (see §4, where its structure is investigated). Then we have:

a) X plays the part of a spectral space, and the admissible sheaves $\mathscr{E}$ on X are generalized scattering data.

b) The functions ψ are the sections Ψ of the sheaf $\pi_1^*\mathscr{E}$ on Z that are lifted from X. A necessary (and locally sufficient) condition for this is the equation $\nabla_{Z/X}(\Psi) = 0$, where $\nabla_{Z/X}$ is the canonical relative connection on $\pi_1^*\mathscr{E}$. This is just the equation $G(u, x, \psi, \lambda) = 0$.

c) The fields u on U are coded on Z by the coefficients of the connection $\nabla_{Z/X}$, and also, possibly, by the global characteristics of the sheaf $\mathscr{E}$. The problem of describing u in terms of $\nabla_{Z/X}$ is easily solved for the Yang-Mills equations; projection of $\nabla_{Z/X}$ down to U turns it into the usual Yang-Mills connection. In other cases the situation may be more complicated (see §5 below and [36]).

d) Finally, the non-linear system $\varphi(u, x)$ is the projection down to U of the integrability condition $\nabla_{Z/X}^2 = 0$. The fact that such a projection gives rise to physically interesting equations is a remarkable property of Penrose's method.

4. Plan of the article.

In §1 we prove the theorem that admissible sheaves on $Y(u)$ can be characterized by the vanishing of finitely many cohomology groups. In §2 we show how to construct sheaves with such properties by means of the direct image operation. In §3 we explain that the Baker function on manifolds of arbitrary dimension is of this kind. In §4 we apply these general ideas to the problems a)–f) of subsection 1. Finally, in §5 we discuss the inverse twistor transformation and the structure of the system of equations in problem e).

5. References to the literature.

For new work on the self-dual problem (generally monopole) see [8], [23], [24], [31], [32]. The article [9] contains the first published proof of Witten's theorem [51] on the twistor transformation for $N = 3$ super-symmetric Yang-Mills equations. A more general result (see subsection 1, c)) for $N = 1$ and 2 was announced by Khenkin in [19]. In the same article [19], and also in Manin and Khenkin [34], explicit formulae can be found for the fields in space-time expressed in terms of coefficients of the $\bar{\partial}$-connection describing the admissible sheaf $\mathcal{E}$. A detailed survey of the relation between the theory of the Radon-Penrose transformation and the tangent $\bar{\partial}$-equations is given in [56]. A study from the point of view of twistor theory of bundles over $\mathbf{CP}^n$ with arbitrary type of splitting on a general line was made in [55]. An attempt at linearization of non-self-dual Yang-Mills equations was made in [57], which, however, contains an erroneous conclusion about the possibility of using the one-dimensional Riemann problem. Useful information about the subject of the present article can be found in "Twistor Newsletter", published regularly in recent years. The book [38] reviews the first stage of development of the twistor transformation method. The articles [10]–[12] first showed that this method was also suited to the quantization of supersymmetric theories in four dimensions, and [52] established its adequacy as a very interesting many-dimensional model among those now being considered.

§1. Cohomological characterization of admissible sheaves

1. Acyclic and admissible sheaves.
Let Y be a complex superspace, and $\mathcal{F}$ a sheaf of Abelian groups on it (in applications we shall generally consider coherent sheaves). The sheaf $\mathcal{F}$ is called acyclic if $H^i(Y, \mathcal{F}) = 0$ for all i.

In the situations $(X, Y(u))$ described in the Introduction, subsection 1, a)–g), classes of sheaves on X appear having restrictions of a given type to $Y(u)$. All such sheaves will be called admissible, as will their restrictions to $Y(u)$.

The main result of this section is the following criterion for admissibility.

2. ***Theorem.*** *A coherent sheaf $\mathcal{F}$ on Y is free if and only if the following sheaves are acyclic:*

a) $\mathcal{F}(-i)$ *for* $1 \leqslant i \leqslant d$, *if* $Y = P^d$.

b) $\mathcal{F}(-i)$ *for* $1 \leqslant i \leqslant d-1$ *and* $\mathcal{F} \otimes (\mathcal{S}^{\perp}/\mathcal{S})$, *if* $Y = Q^d \subset P^{d-1} = P(E)$ *(a d-dimensional quadric),* $d \geqslant 2$. *Here* $\mathcal{S} = \mathcal{O}(-1) \subset E$ *is the tautological sheaf and* $\mathcal{S}^{\perp} \supset \mathcal{S}$ *is its orthogonal complement with respect to the bilinear form on E defining the quadric.*

c) $\mathcal{F}(-1, 1)$, $\mathcal{F}(1, -1)$ *and* $\mathcal{F}(-1, -1)$, *if* $Y = Q^2 = P^1 \times P^1$.

d) $\mathcal{F} \otimes (\mathcal{S}_1)^{\otimes i_1} \otimes (\mathcal{S}_2/\mathcal{S}_1)^{\otimes i_2} \otimes \ldots \otimes (\mathcal{S}_{d-1}/\mathcal{S}_{d-2})^{\otimes i_{d-1}}$, $0 \leqslant i_k \leqslant d-k$, $(i_1, \ldots, i_{d-1}) \neq (0, \ldots, 0)$, *if* $Y = F(\mathbf{C}^d)$ *is the complete flag space. Here* $\mathcal{S}_1 \subset \ldots \subset \mathcal{S}_{d-1}$ *is the tautological flag on* Y *(similar notations are used below in* e) *and* f)).

e) $\mathcal{F} \otimes (\mathcal{S}_1)^{\otimes i_1} \otimes (\mathcal{S}_2/\mathcal{S}_1)^{\otimes i_2} \otimes \ldots \otimes (\mathcal{S}_d/\mathcal{S}_{d-1})^{\otimes i_d}$, $0 \leqslant i_k \leqslant 2d-2k+1$, $(i_1, \ldots, i_d) \neq (0, \ldots, 0)$, *if* $Y = F^{is}(E^{2d}, [\ ,\ ])$ *is the space of complete isotropic flags of the 2d-dimensional symplectic space E.*

f$_1$) *If* Y *is one of the components of the space* $F^{is}(E^{2d}, (\ ,\ ))$ *of complete isotropic flags of a 2d-dimensional orthogonal space E, then* $\mathcal{F}$ *is free if and* *only if the sheaves $\mathcal{F} \otimes \mathcal{L}_1 \otimes \ldots \otimes \mathcal{L}_d$ are acyclic, where each $\mathcal{L}_k$ runs through the set $\{(\mathcal{S}_k/\mathcal{S}_{k-1})^{\otimes i_k}: 0 \leqslant i_k \leqslant 2d-2k\} \cup \{\mathcal{S}_k^{\perp}/\mathcal{S}_k\}$, with $(\mathcal{L}_1 \otimes \ldots \otimes \mathcal{L}_d) \neq (\mathcal{O}, \ldots, \mathcal{O})$.*

f$_2$) *If $Y = F^{is}(E^{2d+1}, \langle\ ,\ \rangle)$, then $\mathcal{F}$ is free if and only if the sheaves $\mathcal{F} \otimes \mathcal{L}_1 \otimes \ldots \otimes \mathcal{L}_d$ are acyclic, where each $\mathcal{L}_k$, $1 \leqslant k \leqslant d-1$, runs through the set $\{(\mathcal{S}_k/\mathcal{S}_{k-1})^{\otimes i_k}; 0 \leqslant i_k \leqslant 2d+1-2k\} \cup \{\mathcal{S}_k^{\perp}/\mathcal{S}_k\}$, and $\mathcal{L}_d$ is either $\mathcal{O}$ or the uniquely defined "half"-sheaf* det $\mathcal{S}_d$ *in* Pic Y. *Here we only consider sets $(\mathcal{L}_1, \ldots, \mathcal{L}_d)$ different from $(\mathcal{O}, \ldots, \mathcal{O})$.*

g) *The sheaf $\mathcal{F}$ on $Y = P^1 \times P^1$ has the form $\mathcal{O}^r(-1, 0) \oplus \mathcal{O}^s(0, -1)$ if and only if $\mathcal{F}$ and $\mathcal{F}(-1, -1)$ are acyclic.*

Proof. In the case when the twisting sheaf is invertible, the necessity of the stated conditions follows immediately from standard facts about the cohomology of homogeneous spaces (see, for example, [38], Ch. I, §2). The fact that $\mathcal{S}^{\perp}/\mathcal{S}$ is acyclic in case b) follows from consideration of the exact sequences

$$0 \to \mathcal{S}^{\perp} \to E \otimes \mathcal{O} \to \mathcal{O}(1) \to 0$$

and

$$0 \to \mathcal{O}(-1) \to S^{\perp} \to S^{\perp}/S \to 0.$$

In cases f$_1$) and f$_2$) we represent $F^{is}(E, \langle\ ,\ \rangle)$ as the space of isotropic flags of the orthogonal bundle $\mathcal{S}_1^{\perp}/\mathcal{S}_1$ on the quadric $Q(E)$; after this, consideration of the Leray spectral sequence together with induction and the use of case b) will give the required necessity. Let us verify the sufficiency.

a) The first Beilinson spectral sequence ([40], Ch. II, §3) with $E_1^{pq} = H^q(P^d, \mathcal{F}(p)) \otimes \Omega^{-p}(-p)$ converges to 0 for $p+q \neq 0$ and to $\mathcal{F}$ on the line $p+q = 0$. Hence, if $\mathcal{F}(-1), \ldots, \mathcal{F}(-d)$ are acyclic we obtain an isomorphism $H^0(P^d, \mathcal{F}) \otimes \mathcal{O} \xrightarrow{\sim} \mathcal{F}$.

b) For what follows we need a more general assertion. Let E be a locally-free sheaf of rank $d+2$ over an algebraic variety X, endowed with a non-degenerate quadratic form. Let $Q(E) \subset P(E)$ be the relative quadric, $p: Q(E) \to X$ the projection, $\mathcal{O}(-1) = \mathcal{S} \subset p^*E$ the tautological sheaf, and $\mathcal{S}^{\perp} \supset \mathcal{S}$ its orthogonal complement. We shall show that every finite complex of coherent sheaves $\mathcal{F}^{\cdot}$ on $Q(E)$ such that $\mathcal{F}^{\cdot}(-i)$, $1 \leqslant i \leqslant d-1$, and $\mathcal{F}^{\cdot} \otimes (\mathcal{S}^{\perp}/\mathcal{S})$ have all hyperdirect images $\mathbf{R}^i p_*$ zero, is quasi-isomorphic to $p^*\mathcal{G}^{\cdot}$, where $\mathcal{G}^{\cdot}$ is a complex of coherent sheaves on X. (For case a) a similar assertion follows at once from the relative version of Beilinson's theorem [4].)

For the proof we may assume that there exists a linear subbundle $H \subset E$ not isotropic over any point. In fact, the assertion about $\mathcal{F}^{\cdot}$ is equivalent to the canonical morphism $p^*(Rp_*\mathcal{F}^{\cdot}) \to \mathcal{F}^{\cdot}$ being a quasi-isomorphism (Rp_* is the direct image in the derived category). This can be verified over $p^{-1}(U)$, where $U \subset X$ is a sufficiently small Zariski open set over which the desired H exists.

Consider the projection $\pi: Q(E) \to P(E_0)$, where $E_0 = E/H$, which is a finite morphism of degree 2. Also let q be the projection $P(E_0) \to X$. Then

$$\pi^*(\mathcal{O}_{P(E_0)}(-1)) = \mathcal{O}_{Q(E)}(-1). \quad \pi_*\mathcal{O}_{Q(E)} = \mathcal{O}_{P(E_0)} \oplus q^*H^* \otimes \mathcal{O}_{P(E_0)}(-1).$$

Since $\mathbf{R}^j q_* \ ((\pi_* \mathcal{F}^\cdot) \otimes \mathcal{O}_{P(E_0)}(-i)) = \mathbf{R}^j p_* \ (\mathcal{F}^\cdot \otimes \mathcal{O}_{Q(E)}(-i)) = 0$ for all j and $1 \leqslant i \leqslant d-1$, it follows from Beilinson's results [4] that $\pi_* \mathcal{F}$ is quasi-isomorphic to a complex whose k-th term has the form $q^*V_k \oplus (q^*W_k) \otimes \mathcal{O}_{P(E_0)}(-1)$, where V_k and W_k are sheaves on X. These sheaves are terms of complexes $V^\cdot$ and $W^\cdot$ quasi-isomorphic to $Rp_*\mathcal{F}^\cdot$ and $Rp_*(\mathcal{F}^\cdot \otimes \mathcal{O}_{Q(E)}(-d))[d] \otimes p^*H$ respectively. Thus we may assume that there is an exact sequence of complexes

$$(1) \qquad 0 \to q^*V^\cdot \to \pi_*\mathcal{F}^\cdot \to (q^*W^\cdot) \otimes \mathcal{O}_{P(E_0)}(-1) \to 0.$$

The natural morphism $p^*V^\cdot \sim p^*(Rp_*\mathcal{F}^\cdot) \to \mathcal{F}^\cdot$ is a quasi-isomorphism if and only if $\pi_*p^*V^\cdot \to \pi_*\mathcal{F}^\cdot$ is a quasi-isomorphism. The direct sum decomposition $\pi_*p^*V^\cdot = q^*V^\cdot \oplus q^*(V^\cdot \otimes H^*) \otimes \mathcal{O}_{P(E_0)}(-1)$ can be described as a split exact sequence from which there is a natural morphism in (1). Therefore it is sufficient to verify that the morphism $\alpha: V^\cdot \otimes H^* \to W^\cdot$ to which the composition

$$q^*(V^\cdot \otimes H^*) \otimes \mathcal{O}_{P(E_0)}(-1) \to \pi_*p^*V^\cdot \to \pi_*\mathcal{F}^\cdot \to (q^*W^\cdot) \otimes \mathcal{O}_{P(E_0)}(-1).$$

actually reduces is a quasi-isomorphism.

From the assumed vanishing of the hyperdirect images $\mathbf{R}^j p_*$ for complexes $\mathcal{F}^\cdot(-1)$ and $\mathcal{F}^\cdot \otimes (\mathcal{S}^\perp/\mathcal{S})$ a similar fact follows for the complex $\mathcal{F}^\cdot \otimes \mathcal{S}^\perp$ included in the exact sequence

$$0 \to \mathcal{F}^\cdot \otimes \mathcal{S}^\perp \to \mathcal{F}^\cdot \otimes p^*E^* \to \mathcal{F}^\cdot \otimes \mathcal{O}_{Q(E)}(1) \to 0.$$

Therefore

$$\beta: (Rq_*(\pi_*\mathcal{F}^\cdot)) \otimes E^* \sim (Rp_*\mathcal{F}^\cdot) \otimes E^* \to$$
$$\to Rp_*(\mathcal{F}^\cdot \otimes \mathcal{O}_{Q(E)}(1)) \sim Rq_*((\pi_*\mathcal{F}^\cdot) \otimes \mathcal{O}_{P(E_0)}(1))$$

is a quasi-isomorphism. On $P(E_0)$ we consider the following morphism of short exact sequences of complexes:

$$\begin{array}{ccccccccc} 0 \to & q_*(V^\cdot \otimes E_0^*) & \to & q^*(V^\cdot \otimes E^*) & \to & q^*(V^\cdot \otimes H^*) & \to 0 \\ & \downarrow & & \downarrow & & \downarrow q^*\alpha & \\ 0 \to & (q^*V^\cdot) \otimes \mathcal{O}_{P(E_0)}(1) & \to & (\pi_*\mathcal{F}^\cdot) \otimes \mathcal{O}_{P(E_0)}(1) & \to & q^*W^\cdot & \to 0. \end{array}$$

Applying the functor Rq_* to it, we obtain a morphism of exact triangles

$$\begin{array}{ccccccccc} \ldots \to & V^\cdot \otimes E_0^* & \to & (Rp_*\mathcal{F}^\cdot) \otimes E^* & \to & V^\cdot \otimes H^* & \to & V^\cdot \otimes E_0^*[1] & \to \ldots \\ & \downarrow \mathrm{id} & & \downarrow \beta & & \downarrow \alpha & & \downarrow \mathrm{id} & \\ \ldots \to & V^\cdot \otimes E_0^* & \to & Rq_*((\pi_*\mathcal{F}^\cdot) \otimes \mathcal{O}_{P(E_0)}(1)) & \to & W^\cdot & \to & V^\cdot \otimes E_0^*[1] & \to. \end{array}$$

Since id and β are quasi-isomorphisms, so is α.

c) is a special case of b), since under the identification $P^1 \times P^1 = Q^2 \subset P^3$ the restriction of the sheaf $\mathcal{O}(-1)$ is $\mathcal{O}(-1, -1)$, and

$$\mathcal{S}^{\perp}/\mathcal{S} \cong \mathcal{O}(-1, 1) + \mathcal{O}(1, -1).$$

Assertions d), e), f) are proved in a similar way. Let us analyse f_2). We shall prove the following more general assertion by induction on d. Let E be a $(2d+1)$-dimensional orthogonal bundle over an algebraic variety X, $F^{is}(E)$ the corresponding space of isotropic flags, and p: $F^{is}(E) \to X$ the projection. We assume that there is an invertible sheaf $\mathcal{H} \in \mathrm{Pic}\,(F^{is}(E))$ such that $\mathcal{H}^{-2} \otimes \det \mathcal{S}_d$ is lifted from X. If a finite complex $\mathcal{F}^{\cdot}$ of sheaves on $F^{is}(E)$ is such that $\mathbf{R}^j p_* (\mathcal{F}^{\cdot} \otimes \mathcal{L}_1 \otimes \ldots \otimes \mathcal{L}_d) = 0$ for all j, then $\mathcal{F}^{\cdot}$ is quasi-isomorphic to the complex $p^*\mathcal{G}^{\cdot}$, where $\mathcal{G}^{\cdot}$ is a complex on X. Here $(\mathcal{L}_1, \ldots, \mathcal{L}_d)$ runs through the set of appropriate choices of sheaves given in the statement of the theorem, with the difference that $\mathcal{L}_d = \mathcal{O}$ or $\mathcal{H}$.

For $d = 1$ the assertion follows from the relative version of Beilinson's theorem [4]. Assume it has been proved for some d, and let E be a $(2d+3)$-dimensional orthogonal bundle, endowed with $\mathcal{H} \in \mathrm{Pic}\,(F^{is}(E))$, as above. The space $F^{is}(E)$ is also the relative space of isotropic flags of the orthogonal bundle $\mathcal{S}_1^{\perp}/\mathcal{S}_1$ on the relative quadric $Q(E)$. $\mathcal{S}_1^{\perp}/\mathcal{S}_1$ satisfies our assumption about $\mathcal{H}$.

Let f: $F^{is}(E) \to Q(E)$ and $g: Q(E) \to X$ be the projections. Consider the complexes $Rf_* (\mathcal{F}^{\cdot} \otimes \mathcal{L}_2 \otimes \ldots \otimes \mathcal{L}_d)$ on $Q(E)$. After tensor multiplication by every possible $\mathcal{L}_1$ (that is, $\mathcal{L}_1$ is either $(\mathcal{S}_1)^{\otimes i}$, $1 \leqslant i \leqslant 2d-2$, or $\mathcal{S}_1^{\perp}/\mathcal{S}_1$) the hyperdirect images $\mathbf{R}^j g_*$ for each of these complexes are zero for all j. Hence, from the assertion proved in b), we deduce that $Rf_* (\mathcal{F}^{\cdot} \otimes \mathcal{L}_2 \otimes \ldots \otimes \mathcal{L}_d)$ are quasi-isomorphic to complexes of the form $g^*\mathcal{E}^{\cdot}$, where $\mathcal{E}^{\cdot}$ are complexes on X. Consequently, $\mathbf{R}^j p_* (\mathcal{F}^{\cdot} \otimes \mathcal{L}_2 \otimes \ldots \otimes \mathcal{L}_d)$ is isomorphic to the j-th cohomology sheaf of the corresponding complex $\mathcal{E}^{\cdot}$ on X. Since $\mathbf{R}^j p_* (\mathcal{F}^{\cdot} \otimes \mathcal{L}_2 \otimes \ldots \otimes \mathcal{L}_d) = 0$ for admissible choices of $(\mathcal{L}_2, \ldots, \mathcal{L}_d)$, we find that $\mathbf{R}^j f_* (\mathcal{F}^{\cdot} \otimes \mathcal{L}_2 \otimes \ldots \otimes \mathcal{L}_d) = 0$, that is, by the induction hypothesis, $\mathcal{F}^{\cdot} \sim f^*\mathcal{G}^{\cdot}$, where $\mathcal{G}^{\cdot}$ is a complex over $Q(E)$. Since $\mathbf{R}^j g_* (\mathcal{G}^{\cdot} \otimes \mathcal{L}_1) = \mathbf{R}^j p_* (\mathcal{F}^{\cdot} \otimes \mathcal{L}_1) = 0$ for all j, $\mathcal{G}^{\cdot}$ is quasi-isomorphic to a complex lifted from X, which is what we wanted to prove.

g) It follows from the relative Beilinson theorem that the derived category of coherent sheaves on $P^1 \times P^1$ is generated by $\mathcal{O}(i, j)$, $(i, j = 0, -1)$. Since the higher Ext's between these sheaves are zero, it follows that for any sheaf $\mathcal{F}$ there exists a complex $\mathcal{E}^{\cdot}$ quasi-isomorphic to $\mathcal{F}$ and consisting of direct sums of $\mathcal{O}(i, j)$ $(i, j = 0, 1)$. Let $\mathcal{F}$ and $\mathcal{F}(-1, -1)$ be acyclic. Then $\mathbf{H}^{\cdot}(\mathcal{E}^{\cdot}) = \mathbf{H}^{\cdot}(\mathcal{E}^{\cdot}(-1, -1)) = 0$. Let $\mathcal{E}^{\cdot}_{(i,j)}$ denote the graded subsheaf in $\mathcal{E}^{\cdot}$ whose terms are the direct sums of all summands isomorphic to $\mathcal{O}(i, j)$. It is clear that $\mathcal{E}^{\cdot}_{(0,0)}$ is a subcomplex, since the differentials $\mathcal{O} \to \mathcal{O}(i, j)$ for $(i, j) \neq (0, 0)$ can only be zero. By considering the spectral sequence $E_1^{pq} = H^q(H^p(P^1 \times P^1, \mathcal{E}^{\cdot})) \Rightarrow H^{p+q}(\mathcal{F}) = 0$ we find that $\mathcal{E}^{\cdot}_{(0,0)}$ is an exact complex. Factoring this out, we may assume that $\mathcal{E}^{\cdot}$

contains no complex of type (0, 0). The graded sheaf $\mathcal{E}^{\cdot}_{(-1,-1)}$ is naturally a factor-complex of $\mathcal{E}^{\cdot}$. By considering the analogous spectral sequence for $\mathcal{E}^{\cdot}(-1, -1)$, we find that $\mathcal{E}^{\cdot}_{(-1,-1)}$ is exact. Hence the kernel of the epimorphism $\mathcal{E}^{\cdot} \to \mathcal{E}^{\cdot}_{(-1,-1)}$ is quasi-isomorphic to $\mathcal{F}$ and consists of direct sums of only $\mathcal{O}(-1, 0)$ and $\mathcal{O}(0, -1)$. This kernel splits into a sum of complexes $\mathcal{E}^{\cdot}_{(-1,0)}$ and $\mathcal{E}^{\cdot}_{(0,-1)}$. Such a sum must be quasi-isomorphic to $\mathcal{O}^r(-1, 0) \oplus \mathcal{O}^s(0, -1)$.

3. *Remark.* It seems certain that there is a general result describing the trivial sheaves (and others too?) on manifolds of the form G/P, where G is a semisimple algebraic group and P a parabolic subgroup. Theorem 2 establishes such a result for the case when G lies in one of the four classical series (the case of incomplete flags can be analysed by means of a lift to complete flags). Considering that the flag manifold $E_8 \times E_8/B$, where B is the Borel subgroup (Kallosh [27]), must play the part of a harmonic space for a ten-dimensional heterotic string, it should be possible to find a general analogue of assertions d)–f).

§2. Admissible sheaves as direct images

1. Statement of the problem.
In this section we shall give some constructions for sets $(\tilde{Y}, Y, \varphi, \mathcal{L})$, where Y is one of the manifolds in Theorem 2 of §1, $\varphi: \tilde{Y} \to Y$ is a mapping finite at a general point, and $\mathcal{L}$ a sheaf on $\tilde{Y}$, where $\varphi_* \mathcal{L}$ is admissible in the sense of §1.

2. The case $Y = P^1$.
Let $\tilde{Y}$ be a complete algebraic curve, $\varphi: \tilde{Y} \to Y = P^1$ a finite morphism, and $\mathcal{L}$ an invertible sheaf on $\tilde{Y}$.

Put $\mathcal{L}_0 = \varphi^*(\mathcal{O}_Y(-1))$. This is an invertible sheaf on $\tilde{Y}$ of degree $-d = -\deg \varphi$. The sheaf $\varphi_* \mathcal{L}$ is trivial if and only if $H^i(\tilde{Y}, \mathcal{L} \otimes \mathcal{L}_0) = 0$ $(i = 0, 1)$. By the Riemann-Roch theorem this is equivalent to the fact that: a) $\chi(\mathcal{L} \otimes \mathcal{L}_0) = \deg \mathcal{L} - d - (g-1) = 0$, where g is the genus of $\tilde{Y}$; b) $\mathcal{L} \otimes \mathcal{L}_0$ is non-special. The first condition fixes only the degree of $\mathcal{L}$ (for given φ). The set of invertible sheaves of given degree has complex dimension q. The second condition picks out from it a dense (Zariski open) subset, at least if $\tilde{Y}$ is smooth.

Similarly, if $\operatorname{rk} \mathcal{L} = r > 1$, then $\chi(\mathcal{L} \otimes \mathcal{L}_0) = \deg \mathcal{L} - r(d + g - 1)$. In the manifold of moduli of stable locally free sheaves on $\tilde{Y}$ of rank r and degree $r(d+g-1)$ the sheaves $\mathcal{L}$ such that $\mathcal{L} \otimes \mathcal{L}_0$ is acyclic form a Zariski open subset, which is well-known to be non-empty.

3. The case $Y = P^d$.
Let $\tilde{Y}$ be an arbitrary complete d-dimensional manifold, and $\mathcal{L}_0$ an invertible sheaf on it such that the mapping $\psi: \tilde{Y} \to P^N$ defined by the sections of $\mathcal{L}_0^{-1}$ is regular, and its image has dimension d. Then $\mathcal{L}_0 = \varphi^*(\mathcal{O}(-1))$, where φ

are projections of $\psi(\widetilde{Y})$ onto the general subspace $P^d \subset P^N$. Let $\mathcal{L}$ be a sheaf on $\widetilde{Y}$; then a necessary and sufficient condition for $\varphi_* \mathcal{L}$ to be trivial is that the sheaves $\mathcal{L} \otimes \mathcal{L}_0, \ldots \mathcal{L} \otimes \mathcal{L}_0^d$ are acyclic on $\widetilde{Y}$.

Put $\widetilde{Y} = Z_1 \times \ldots \times Z_d$, where Z_i is a complete smooth curve of genus $g_i \geqslant 1$, and consider on $\widetilde{Y}$ the decomposable sheaves $\mathcal{E}_1 \otimes \ldots \otimes \mathcal{E}_d$, where $\mathcal{E}_i = \pi_i^* (\mathcal{E}'_i), \mathcal{E}'_i \in \operatorname{Pic} Z_i$, and π_i: $\widetilde{Y} \to Z_i$ the projections. Such a sheaf is acyclic if for some i the sheaf $\mathcal{E}_i$ is acyclic, that is, $\deg \mathcal{E}'_i = g_i - 1$, and $\mathcal{E}'_i$ is non-special. On the other hand, if $\deg \mathcal{E}'_i \geqslant 2g_i + 1$, then the mapping defined by means of sections of such a sheaf is regular, and its image has dimension d. Hence there exist many arithmetic progressions in $\operatorname{Pic} \widetilde{Y}$ of length d with acyclic elements and difference of type $\varphi^* (\mathcal{O}(-1))$. The degrees of their elements are constructed as follows: if $\mathcal{L}_i = \mathcal{E}_{1,i} \otimes \ldots \otimes \mathcal{E}_{d,i}$ $(i = 1, \ldots, d)$, let $\deg \mathcal{E}'_{i,i} = g_i - 1$: then $\deg \mathcal{E}'_{i,j}, j \neq i$, is found uniquely from the condition $\deg \mathcal{L}_i - \deg \mathcal{L}_{i+1} = (e_1, \ldots, e_d)$, $e_i \geqslant 2g_i + 1$ arbitrary. Let $Y_0 \subset P^d$ be a closed subvariety. Then the sheaf $\varphi^* (\mathcal{O}_{Y_0})$ is concentrated on $\varphi^{-1} (Y_0)$ and

$$\varphi_* (\mathcal{E}_1 \otimes \ldots \otimes \mathcal{E}_d \otimes \varphi^* (\mathcal{O}_{Y_0})) = \varphi_*(\mathcal{E}_1 \otimes \ldots \otimes \mathcal{E}_d) \otimes \mathcal{O}_{Y_0} = \mathcal{O}^r_{Y_0}.$$

The morphism $\widetilde{Y}_0 = \varphi^{-1}(Y_0) \to Y_0$ is finite, and in the case when it has degree r, $\mathcal{O}_{\widetilde{Y}_0} = \varphi^* \mathcal{O}_{Y_0}$. As Y_0 we may choose $P^1 \times P^1$, for example, or any other of the manifolds listed in the Introduction, subsection 1, a)–g).

4. Hypersurfaces in $Y \times P^1$.

Let $\widetilde{Y} \subset Y \times P^1$ be a hypersurface given by the equation $s = 0$, where s is a section of an invertible sheaf of the form $\pi_1^* \mathcal{E} \otimes \pi_2^* \mathcal{O}(n)$, $n \geqslant 2$: π_1: $Y \times P^1 \to Y$ and π_2: $Y \times P^1 \to P^1$ are projections. Put $\mathcal{L} = \pi_2^* (\mathcal{O}(n-1))$. Then $\varphi_* (\mathcal{L})$ is trivial, where φ: $\widetilde{Y} \to Y$ is the composition of the embedding and the projection. In the cases when Theorem 2 of §1 may be used this is easy to verify by means of the exact sequence

$$0 \to \pi_1^* \mathcal{E}^* \otimes \pi_2^* \mathcal{O}(-n) \xrightarrow{s} \mathcal{O}_{Y \times P^1} \to \mathcal{O}_{\widetilde{Y}} \to 0.$$

By taking its tensor product with $\mathcal{L} \otimes \pi_1^* \mathcal{L}_1$, where $\mathcal{L}_1$ is an acyclic sheaf on Y, we find that this sheaf is acyclic, since the first two sheaves in the resulting sequence are acyclic.

Similarly we can construct pairs $(\widetilde{Y}, \mathcal{L})$ where $\widetilde{Y} \subset Y \times Z$, Z a curve, with $\varphi_* (\mathcal{L})$ trivial.

5. Sheaves of the type $\mathcal{O}^r(-1, 0) \oplus \mathcal{O}^s(0, -1)$.

Such sheaves can be obtained as direct images in $Y = P^1 \times P^1$ by means of the following construction. Let Y' be the graph of a birational correspondence between P^2 and Y, blowing up a point of Y, and collapsing the two generators passing through this point. There is a diagram $P^2 \xleftarrow{\beta} Y' \xrightarrow{\alpha} Y$. Let φ: $\widetilde{Y} \to P^2$ be a finite morphism, and let $\mathcal{L}$ be an invertible sheaf on $\widetilde{Y}$ such that $\varphi_* (\mathcal{L})$ is trivial. Let $\widetilde{Y}' = \widetilde{Y} \times_{P^2} Y'$ and let $\mathcal{L}'$ be the lift of $\mathcal{L}$ to

$\widetilde{Y}'$ under the projection. Let l'_1, l'_2 be divisors on Y' collapsed by the morphism β. We assume that φ is not ramified over the points $\beta(l'_1)$, $\beta(l'_2)$, and that these points are sufficiently close. Then $n = \deg\varphi$ irreducible divisors $l_1^{(1)}, \ldots, l_1^{(n)}$ (respectively, $l_2^{(1)}, \ldots, l_2^{(n)}$) lie in $\widetilde{Y}'$ over l'_1 (respectively, l'_2). Assume that they are numbered so that $l_1^{(i)}$ and $l_2^{(i)}$ lie over the same sheet of the completely reducible covering of some neighbourhood $U \supset \beta(l'_1), \beta(l'_2)$ (this is possible if these points are close). Then the direct image in Y of the sheaf $\mathcal{L}' \otimes \mathcal{O}(-l_1^{(1)} - \ldots - l_1^{(r)} - l_2^{(r+1)} - \ldots - l_2^{(n)})$ is isomorphic to $\mathcal{O}(-1, 0)^r \oplus \mathcal{O}(0, -1)^{n-r}$. To prove this, let $\widetilde{\alpha}$ denote the projection $\widetilde{Y}' \to Y$. Let W_1 be the subspace in the fibre of $\varphi_*\mathcal{L}$ over $\beta(l'_1)$ spanned by the $(r+1)$-th, ..., n-th inverse images of $\varphi^{-1}(\beta(l'_1))$ with respect to the chosen numbering; similarly, let W_2 be the subspace in the fibre of $\varphi_*\mathcal{L}$ over $\beta(l'_2)$ spanned by the 1-st, ..., r-th inverse images of $\varphi^{-1}(\beta(l'_2))$. Since $\varphi_*\mathcal{L}$ is trivial, W_1 and W_2 can be regarded as subspaces in $H^0(P^2, \varphi_*\mathcal{L})$. If $\beta(l'_1)$ and $\beta(l'_2)$ are sufficiently close, then $H^0(P^2, \varphi_*\mathcal{L}) = W_1 \oplus W_2$. The direct image $\widetilde{\alpha}_*\mathcal{L}'$ is also trivial and can be naturally identified with $\mathcal{O}_{P^1\times P^1} \otimes H^0(P^2, \varphi_*\mathcal{L}) = \mathcal{O}_{P^1\times P^1} \otimes (W_1 \oplus W_2)$. With respect to this identification, the local sections of

$$\alpha_*(\mathcal{L}' \otimes \mathcal{O}(-l_1^{(1)} - \ldots - l_1^{(r)} - l_2^{(r+1)} - \ldots - l_2^{(n)}))$$

are the functions with values in $W_1 \oplus W_2$ whose values over the line $l_1 = \alpha(l'_1)$ lie in W_1, and over the line $l_2 = \alpha(l'_2)$ lie in W_2. Since the l_i are the generators of a quadric, we obtain the required result.

§3. Invertible sheaves and Baker functions

1. Motivation.

Let $(X, Y(u))$ be the situation described in the Introduction. We want to construct admissible sheaves on X (that is, sheaves with admissible restrictions to $Y(u)$) as direct images $\varphi_*\mathcal{L}$, where $\varphi: \widetilde{X} \to X$. In particular, to do this we need an effective construction for invertible sheaves on $\widetilde{X}$. If X and $\widetilde{X}$ are not compact, we should like to have a construction depending on arbitrary functions, since passage to the differential equations and the Cauchy problem for them shows that Pic $\widetilde{X}$ must be a function space.

Henceforth we shall write X instead of $\widetilde{X}$. The following linearization of the problem of describing part of Pic X is known. Let $\mathcal{O}_{X,0}$ be the sheaf of even functions on the superspace X. The invertible sheaves are classified by the elements of the group $H^1(\mathcal{O}^*_{X,0})$. They can be constructed by using the mapping $H^1(\mathcal{O}_{X,0}) \xrightarrow{\exp} H^1(\mathcal{O}^*_{X,0})$. Finally, the elements of $H^1(\mathcal{O}_{X,0})$ can be constructed with respect to any Cartier divisor $D > 0$ on X by means of the boundary mapping $\delta: H^0(\mathcal{O}_{X,0}(D)/\mathcal{O}_{X,0}) \to H^1(\mathcal{O}_{X,0})$.

Sections of the sheaf $\mathcal{O}_{X,0}(D)/\mathcal{O}_{X,0}$ are called distributions with support D. If the support D is "sufficiently non-compact", then the space of distributions is a function space, and we can obtain many of the elements of Pic X. We shall give some details.

2. Čech cocycles.
Let $X = \cup U_i$ be a sufficiently fine open covering (all the cohomology classes that interest us can be represented by cocycles on it if U_i and finite intersections of the U_i are Stein spaces).

A divisor $D > 0$ is represented by its (meromorphically invertible) local equations $d_i \in \Gamma(U_i, \mathcal{O}_{X,0})$ with the property $d_i d_j^{-1}|_{U_i \cap U_j} \in \Gamma(U_i \cap U_j, \mathcal{O}^*_{X,0})$. The sheaf $\mathcal{O}_X(D)$ consists of locally meromorphic superfunctions f with the property $d_i f|_{U_i} \in \Gamma(U_i, \mathcal{O}_X)$ for all i. In particular, $\mathcal{O}_X \subset \mathcal{O}_X(D)$. In these definitions we may allow the d_i to be meromorphic functions, and then we also obtain the non-positive divisors D. The sheaf $\mathcal{O}_X(D)$ is invertible (of rank $1|0$). Up to isomorphism its class is defined as the cohomology class of the cocycle $f_{ij} = d_i d_j^{-1} \in Z^1((U_i), \mathcal{O}^*_{X,0})$.

A distribution q with support in the divisor $D > 0$ can be represented by a set of meromorphic functions $q_i \in \Gamma(U_i, \mathcal{O}_{X,0}(D))$ with the property $q_i - q_j|_{U_i \cap U_j} \in \Gamma(U_i \cap U_j, \mathcal{O}_{X,0})$: (q_i') represents the same distribution as (q_i) if $q_i - q_i' \in \Gamma(U_i, \mathcal{O}_{X,0})$. The principal distributions are the restrictions of sections of $\Gamma(X, \mathcal{O}_{X,0}(D))$.

The distribution q corresponds to the invertible sheaf $\mathcal{L}(q)$ given by the cocycle $(\exp(q_i - q_j)|_{U_i \cap U_j})$. Its class in the group $\mathrm{Pic}\, X = H^1(X, \mathcal{O}^*_{X,0})$ is the image of q under the mapping

$$H^0(X, \mathcal{O}_{X,0}(D)/\mathcal{O}_{X,0}) \xrightarrow{\delta} H^1(X, \mathcal{O}_{X,0}) \xrightarrow{\exp} H^1(X, \mathcal{O}^*_{X,0}).$$

All these constructions have obvious functorial properties under restriction to closed subsuperspaces $Y \subset X$.

3. Baker functions.
Let q be a distribution with support in $D > 0$. A Baker function φ (with essential singularity $\exp q$) is any even function on X, meromorphic outside D and such that $\varphi/\exp q_i$ can be meromorphically extended to U_i for all i. The Baker functions corresponding to a given q obviously form a module over the ring of even meromorphic functions on X. If X is an irreducible purely even manifold, then they form a one-dimensional space.

Finite-dimensional spaces of Baker functions over the complex numbers can be obtained if the singularities of φ are restricted outside D. Let Δ be a divisor on $X - |D|$. Let $L(\Delta, q) =$ {the Baker functions with essential singularity $\exp q$ lying in $\Gamma(X - |D|, \mathcal{O}_X(\Delta))$}. Clearly $L(\Delta, q) = \Gamma(X, \mathcal{O}_{X,0}(\Delta) \otimes \mathcal{L}(q))$.

4. Baker functions and differentials.
Let φ be a Baker function with singularity $\exp q$, meromorphically invertible outside the support D. Then $\varphi^{-1}d\varphi$ is a meromorphic closed 1-form on X with $2\pi i$-integer periods (in other words, $\int_\gamma \varphi^{-1}d\varphi \in 2\pi i\mathbf{Z}$ for any closed path γ in X that does not intersect the singularities of $\varphi^{-1}d\varphi$).

Conversely, from a 1-form ω with $2\pi i$-integer logarithmic periods we can construct the distribution

$$\left(q_i(P_i) = \int_{P_i^0}^{P_i} \omega \mid P_i,\ P_i^0 \in U_i \right), \quad \text{where} \quad \omega - dq_i \in \Gamma(U_i,\ \Omega^1).$$

We shall assume further that X is a projective algebraic variety.

An everywhere holomorphic (and hence closed) 1-form on X will be called a differential of the first kind, and a closed meromorphic 1-form on X with zero residues will be called a differential of the second kind. The following result is well known.

5. ***Proposition.*** *Consider the correspondence between differentials ω of the second kind and distributions (q_i) defined by requiring that $\omega - dq_i$ is holomorphic in U_i for all i. If the divisor D is ample, this correspondence establishes a bijection between the following spaces:*

a) *Distributions with support divisible by D (that is, sections of the sheaf $\varinjlim \mathcal{O}_{X,0}(ND)/\mathcal{O}_{X,0}$) modulo those that are principal.*

b) *Differentials of the second kind regular outside $|D|$ modulo differentials of the first kind and exact differentials of meromorphic functions regular outside $|D|$.*

The dimension of both subspaces is $q = \frac{1}{2}\dim H_1(X,\ \mathbf{Q}) = \dim \operatorname{Alb} X$ (irregularity of X). ■

6. Baker functions and theta-functions.

Under the conditions of Proposition 5, let us fix a point $P_0 \in X$ and consider the Abel-Albanese mapping

$$(1) \qquad \begin{aligned} &A\colon X \to \operatorname{Alb} X = (H^0(X,\ \Omega^1))^*/H_1(X,\ \mathbf{Z}), \\ &A(P)\colon \omega \mapsto \int_{P_0}^{P} \omega \quad \text{(modulo the periods)}. \end{aligned}$$

We assume that A is an injective mapping.

There exist bases $\{\omega_1, \ldots, \omega_g\}$ of the space of differentials of the first kind and $\{a_1, \ldots, a_g;\ b_1, \ldots, b_g\}$ of the group $H_1(X,\ \mathbf{Z})$/tors, such that

$$\int_{a_k} \omega_j = \begin{cases} 0 & \text{for } j \neq k, \\ 2\pi i \delta_j \in 2\pi i \mathbf{N} & \text{for } j = k. \end{cases}$$

$B = (\int_{b_k} \omega_l)$ is a symmetric matrix. $\operatorname{Re} B > 0$. Let $(H^0(X,\ \Omega^1))^* = \mathbf{C}^g$ (with basis (e_k) dual to (ω_k)), $f_k = (B_{k1}, \ldots, B_{kg}) \in \mathbf{C}^g$, and z_k the coordinate functions on $\mathbf{C}^g$. The space V of theta-functions on $\mathbf{C}^g$ with equations

$$(2) \qquad \begin{aligned} &\theta(z + 2\pi i \delta_k e_k) = \theta(z), \\ &\theta(z + f_k) = \exp\left(-\tfrac{1}{2}B_{kk} - z_k\right)\theta(z) \end{aligned}$$

has dimension $\prod_{h=1}^{g} \delta_h$. They are lifts to $\mathbf{C}^g$ of sections of some invertible sheaf $\mathcal{E}$ on Alb X.

Suppose that the distribution q and the differential of the second kind ω_q are related as in Proposition 5, $\omega - dq_i \in \Gamma(U_i, \Omega^1)$, and in addition that ω_q is normalized: $\int_{a_i} \omega_q = 0$. Let $U = \left(\int_{b_1} \omega_q, \ldots, \int_{b_g} \omega_q\right)$, and let $\theta_1, \theta_2 \in V$ (see (2)). Then the function on X

$$\psi(P) = \left(\exp \int_{P_0}^{P} \omega_q\right) \frac{\theta_1(A(P)+U+W)}{\theta_2(A(P)+W)}, \tag{3}$$

where $W \in \mathbf{C}^g$, is one-valued if the path of integration in the exponential is the same as in $A(P)$ (see (1)).

7. ***Proposition.*** a) *For a dense open subset of points W the function $\psi(P)$ is a Baker function on X with essential singularity* $\exp q$ *and poles in Δ, where $|\Delta| = \{P\colon\ A(P) + W \in$ zeros of $\theta_2\}$.*

b) *For fixed θ_2 and W formula* (3) *defines a linear transformation $V \to L(\Delta, q)$, which is an epimorphism if $H^1(\text{Alb}\, X, \mathcal{E} \otimes J) = 0$, where J is the ideal of the subscheme $X \subset$ Alb X.*

The proof consists in examining the definitions.

Proposition 7 enables us to describe in terms of θ-functions only those Baker functions that are sections of invertible sheaves on X induced by sufficiently ample sheaves on Alb X, but clearly twisting (the introduction of "fictitious" singularities) allows any Baker function to be described in this way.

Note also that without any assumptions on the structure of the mapping A there is an isomorphism of invertible sheaves on X, $\mathcal{L}(q) \cong A^*(\mu_U^* \mathcal{E} \otimes \mathcal{E}^{-1})$, where $\mu_U\colon$ Alb $X \to$ Alb X is a shift by U. It is established by applying formula (3) to the local sections θ_i of the sheaf $\mathcal{E}$.

§4. Geometry of the twistor transformation and constructions of admissible sheaves

1. The dual bundle.

In this section we apply the constructions of admissible sheaves to the situations a)–f) described in subsection 1 of the Introduction. We begin with the following remarks. In all the cases described, space-time contains a superdomain $U = \mathbf{C}^{a|b}$, determined by natural geometrical conditions. Its twistor space X parametrizes a family of "null" superspaces $L(x) \subset U$, $x \in X$. The graph of the incidence relation between U and X will be denoted by Z. Its projections define a dual superbundle $X \xleftarrow{\pi_1} Z \xrightarrow{\pi_2} U$, where $Y(u) = \pi_1(\pi_2^{-1}(u)) \subset X$, $L(x) = \pi_2(\pi_1^{-1}(x)) \subset U$. All the $Y(u)$ are isomorphic to a single compact complex manifold Y; all the $L(x)$ are isomorphic to some $\mathbf{C}^{p|q}$ (see below). Here and below x, u, and so on, denote scheme points.

2. ***Proposition.*** *In the above situation the following assertions are valid:*

a) *The bundle π_2 is trivial: $Z \cong U \times Y$.*

b) *There exists a fibering $X \xrightarrow{\lambda} Y$ such that the family $Y(u) \subset X$ consists of its sections.*

Proof. In situations a)–e) (subsection 1 of the Introduction) we take U to be a large cell of the flag space $F(2 \mid 0, 2 \mid N; \mathbf{C}^{4|N})$ (for the details see [38], Ch. 1, §3 and Ch. 5, §1). We have $U = \mathbf{C}^{4|4N}$. Let $\mathcal{S}^{2|0} \subset \mathcal{S}^{2|N}$ be the tautological flag of sheaves. From the definition of a large cell it is clear that over U its components are free, that is, they are isomorphic to $\mathcal{O}_U^{2|0} \subset \mathcal{O}_U^{2|N}$. The supermanifold Z is in all cases some relative flag manifold parametrizing the subsheaves in these tautological sheaves, from which it follows that $Z = U \times Y$. In more detail we have:

a) $Z = P_U(\mathcal{S}^{2|0}) = U \times P^1$ $(N = 0)$.

b), c), e) $Z = P_U(\mathcal{S}^{2|0}) \times P_U(\hat{\mathcal{S}}^{2|0}) \cong U \times P^1 \times P^1$, where $\hat{\mathcal{S}}^{2|0} = (\mathcal{S}^{2|N})^{\perp} \subset \mathcal{O}_U \otimes T^*$.

d) $Z = P_U(\mathcal{S}^{2|2}/\mathcal{S}^{2|0}) = U \times P^1$ for $N = 2$, $Z = F_U(1, 2; \mathcal{S}^{2|3}/\mathcal{S}^{2|0}) = U \times F(1, 2; \mathbf{C}^3)$ for $N = 3$.

This establishes the first part of the proposition in situations a)–e). To check the second part we recall that the choice of U is fixed by the choice of a pair of subspaces $\mathcal{S}_\infty^{2|0} \subset \mathcal{S}_\infty^{2|N} \subset T$ in the twistor space. The points of U correspond to those flags of type $(2|0, 2|N)$ that are in general position with the given one. We realize the manifold Y by means of the flag complementary to that mentioned, as described earlier. The geometrical construction of the projection $\lambda : X \to Y$ consists in interpreting the point X as a flag, and, using the conditions of general position, projecting it onto the complementary flag. In more detail:

a) $X = P(T) - P(\mathcal{S}_\infty^{2|0})$; we choose a splitting $T = \mathcal{S}_\infty^{2|0} \oplus \tilde{\mathcal{S}}_\infty^{2|0}$ and a ray in T not contained in $\mathcal{S}_\infty^{2|0}$, and we project it into $\tilde{\mathcal{S}}_\infty^{2|0}$ parallel to $\mathcal{S}_\infty^{2|0}$. This gives λ: $X \to Y = P(\tilde{\mathcal{S}}_\infty^{2|0})$.

b), c), e) Similarly, we choose a splitting $T = \mathcal{S}_\infty^{2|0} \oplus \mathcal{S}_\infty^{0|N} \oplus \tilde{\mathcal{S}}_\infty^{2|0}$ such that $\mathcal{S}_\infty^{2|0} \oplus \mathcal{S}_\infty^{0|N}$ is the second component of the distinguished flag. We construct a dual splitting $T^* = \hat{\mathcal{S}}_\infty^{2|0} \oplus \hat{\mathcal{S}}_\infty^{0|N} \oplus \hat{\tilde{\mathcal{S}}}_\infty^{2|0}$ and projection λ: $X \to Y = P(\tilde{\mathcal{S}}_\infty^{2|0}) \times P(\hat{\mathcal{S}}_\infty^{2|0})$.

d) By analogy with the previous case we construct λ: $X \to Y = P(\mathcal{S}_\infty^{0|2})$ or $F(1, 2; \mathcal{S}_\infty^{0|3})$ (here it is sufficient to work with the distinguished flag).

We now turn to case f) where $U = \mathbf{C}^{10|16}$. As the flag interpretation of this case is not clear to us, we shall describe it in coordinates, following Witten [52].

The even part of U, that is, $\mathbf{C}^{10}$, is the complexification of the fundamental representation of the group $O(1, 9)$. Let $\eta_{mn}x^m x^n$ be an invariant metric. The group $O(1, 9)$ has two non-equivalent 16-dimensional spinor representations. We denote spinors from the corresponding spaces by ψ_α and η_β. We thus have two systems of Γ-matrices, $\Gamma^{m\alpha\beta} = \Gamma^{m\beta\alpha}$ and $\Gamma^m_{\alpha\beta} = \Gamma^m_{\beta\alpha}$. We can choose spinor bases so that the standard Clifford relation has the form $\Gamma^{p\alpha\beta}\Gamma^m_{\beta\gamma} + \Gamma^{m\alpha\beta}\,\Gamma^p_{\beta\gamma} = 2\eta^{mp}\delta^\alpha_\gamma$. We put $\Gamma^{mn\,\beta}_\alpha = \Gamma^m_{\alpha\gamma}\,\Gamma^{n\gamma\beta} - \Gamma^{n\gamma\beta}\Gamma^m_{\alpha\gamma}$ (repeated indices denote summation).

We denote coordinates in the complex superspace $\mathbf{C}^{10.16} = U$ by $(x^m,\ \theta^\alpha)$. The "null superspaces" $L(x) \subset U$ are given parametrically as follows. We choose a point $(x_0^m,\ \theta_0^\alpha) \in U$ and a light-like vector $(\lambda^m) \in \mathbf{C}^{10}$ (that is $\lambda^m\lambda_m = 0$). The moving parameters of $L(x) \cong \mathbf{C}^{1|8}$ are t (the even coordinate) and ε_β (the 16 odd coordinates):

$$\begin{cases} x^m = x_0^m + t\lambda^m - \Gamma^{mn}{}_\alpha{}^\beta\lambda_n\theta_0^\alpha\varepsilon_\beta, \\ \theta^\alpha = \theta_0^\alpha + \lambda_m\Gamma^{m\alpha\beta}\varepsilon_\beta. \end{cases}$$

The dimension of $L(x)$ is 1|8, not 1|16, since the kernel of the operator $\lambda_m\Gamma^m$ is eight-dimensional.

The twistor superspace X parametrizing all the $L(x)$ has dimension 17|8. The space Z is clearly isomorphic to $U \times Q$, where Q is the eight-dimensional quadric of light directions. Since the "light-like direction" of each geodesic is clearly uniquely defined (it is not altered by change of the point of reference $(x_0^m,\ \theta_0^\alpha)$), we thus obtain a projection $\lambda : X \to Q$. ■

3. Constructions of admissible sheaves.

a) Proposition 2 allows us to apply the constructions of §§2 and 3 in the following way. We fix the projection $\lambda : X \to Y$. We choose a mapping $\varphi : \tilde{Y} \to Y$ and a sheaf $\mathcal{E}$ on $\tilde{Y}$ such that $\varphi_*\mathcal{E}$ is admissible. Put $\tilde{X} = X \underset{Y}{\times} \tilde{Y}$, and let $\tilde{\mathcal{E}}$ be the lift of $\mathcal{E}$ to $\tilde{X}$, $\psi : \tilde{X} \to X$ the projection. The direct image $\psi_*(\tilde{\mathcal{E}})$ is, of course, admissible on sections $Y(u)$ of the bundle λ, but does not vary with u and hence leads only to trivial solutions of the equation. To avoid this, we replace $\tilde{\mathcal{E}}$ by $\tilde{\mathcal{E}} \otimes \mathcal{L}(q)$, where q is a distribution on $\tilde{X}$ with support in the divisor Δ lifted from $\tilde{Y}$ in such a way that the admissibility is not violated but the distribution varies with u.

Computationally it is simpler to consider instead of q its lift q' to $\tilde{Z} = U \times \tilde{Y}$. Such q' may be characterized as distributions with support in $\tilde{\pi}^{-1}(\Delta)$, satisfying the equation $d_{\tilde{Z}/\tilde{X}}(q') = 0$, where $\tilde{\pi} : \tilde{Z} \to \tilde{X}$, and the differential

$$d_{\tilde{Z}/\tilde{X}} : \mathcal{O}_{\tilde{Z}}(\tilde{\pi}^{-1}(\Delta))/\mathcal{O}_{\tilde{Z}} \to \Omega^1_{\tilde{Z}/\tilde{X}} \otimes (\mathcal{O}_{\tilde{Z}}(2\tilde{\pi}^{-1}(\Delta))/\mathcal{O}_{\tilde{Z}})$$

is induced by the relative de Rham differential.

b) If the manifold Y is regular (that is, $H^1(Y,\ \mathcal{O}_Y) = 0$), then the indicated method of construction is inappropriate, since the fields constructed in this way will take values in a commutative subalgebra of the structural Lie algebra. However, we can construct coverings $\widetilde{X} \to X$ for which the $\widetilde{Y}(u)$ covering $Y(u)$, though regular, vary with u. To do this, we take $\widetilde{X}$ to be a hypersurface in $X \times P^1$ given by a section of an invertible sheaf of the form $\pi_1^*\mathcal{E} \otimes \pi_2^*\mathcal{O}(n)$, as in §2.4. The projection down to X of the sheaf $\pi_2^*\mathcal{O}(n-1)$ will be trivial. We consider sheaves of the form $(\pi_2^*\mathcal{O}(n-1)) \otimes \mathcal{L}(q)$, where q is a distribution on $\widetilde{X}$. If q is close to zero, then the projection of this sheaf will as before be trivial on $Y(u)$, but in general non-trivial on X. As the simplest example consider the hypersurface Γ in $P^3 \times \hat{P}^3 \times P^1$ of degree $(2, 2, 2)$. Let $X = P^3(U) \times \hat{P}^3(U)$ for some $U \subset \mathbf{C}^4$ and $\widetilde{X} = \Gamma \cap (X \times P^1)$. Then the surfaces $\widetilde{Y}(u)$, which are double coverings of $Y(u) = P^1 \times P^1$, will be K3 surfaces. By applying the procedure described we obtain bundles on $P^3(U) \times \hat{P}^3(U)$, trivial on $Y(u)$, and hence solutions of the vacuum Yang-Mills equations with structure group GL_2.

4. The condition $d_{\widetilde{X}/\widetilde{Z}}(q') = 0$ in coordinates and formulae for the fields. As an example we shall give a coordinate description of the condition $d_{\widetilde{Z}/\widetilde{X}}(q') = 0$ in situation d) of the Introduction, restricting ourselves to the case $N = 3$.

Let the coordinates in the superdomain $U \subset \mathbf{C}^{4|12}$ have the form

$$(z_{\alpha\beta},\ \theta_{\alpha s},\ \dot{\theta}^t_\beta) \qquad (\alpha,\ \beta = 0,\ 1;\quad s,\ t = 1,\ 2,\ 3).$$

The spinor differentiations are determined by the formulae

$$D_{\alpha s} = \frac{\partial}{\partial\theta_{\alpha s}} + \sum_{\beta=0,\,1} \dot{\theta}^s_\beta \frac{\partial}{\partial z_{\alpha\beta}}, \qquad \dot{D}^t_\beta = \frac{\partial}{\partial\dot{\theta}^t_\beta} + \sum_{\alpha=0,\,1} \theta_{\alpha t} \frac{\partial}{\partial z_{\alpha\beta}}$$

and satisfy the relations $\{D_{\alpha s}, D_{\beta t}\} = \{\dot{D}^s_\alpha, \dot{D}^t_\beta\} = 0$, $\{D_{\alpha s}, \dot{D}^t_\beta\} = 2\delta^s_t \cdot (\partial/\partial z_{\alpha\beta})$. Here δ^s_t is the Kronecker symbol and $\{\ ,\ \}$ is the supercommutator.

The space $X(U) \subset F(2|1, 2|2;\ T)$ parametrizes the $(0|4)$-dimensional subspaces $L(x)$, the tangent spaces to which are generated by the four vector fields

$$\sum_{s=1}^{3} \lambda_s D_{\alpha s}, \qquad \sum_{t=1}^{3} \mu_t \dot{D}^t_\beta \qquad (\alpha,\ \beta = 0,\ 1)$$

for some $\{\lambda_s\}$, $\{\mu_t\}$ such that $\sum \lambda_s\mu_s = 0$. The parameters λ_s, μ_t are homogeneous coordinates of points of $Y = F(1, 2;\ \mathbf{C}^3) \subset P^2 \times \hat{P}^2$, a section of the sheaves $\mathcal{O}_Y(+1, 0)$ and $\mathcal{O}_Y(0, +1)$.

The distribution q' on $\widetilde{Z}$ with support in $\widetilde{\pi}^{-1}(\Delta)$ is given by a family of distributions $Q(z, \theta, \dot{\theta})$ on $\widetilde{Y}$, that is, by an even function $Q \in \mathcal{O}(U) \otimes H^0(\widetilde{Y}, \mathcal{O}_{\widetilde{Y}}(\Delta)/\mathcal{O}_{\widetilde{Y}})$; the condition $d_{\widetilde{Z}/\widetilde{X}}(q') = 0$ means in these terms that the section $\sum_{s=1}^{3} \lambda_s D_{\alpha s} Q(z, \theta, \dot{\theta})$ of the sheaf $(\varphi^*\mathcal{O}(1, 0))(\Delta)/\varphi^*\mathcal{O}(1, 0)$

and the section $\sum_{t=1}^{3} \mu_t \dot{D}^t_\beta Q(z, \theta, \dot{\theta})$ of the sheaf $(\varphi^*\mathcal{O}(0, 1))(\Delta)/\varphi^*\mathcal{O}(0, 1)$ are zero for all $\alpha, \beta = 1, 2$. Formulae in the spirit of [47] can be given for the potentials of the Yang-Mills superfields corresponding to the bundles constructed by the method of subsection 3 a). In fact, let $d = \deg \varphi$; for simplicity we assume that the sheaf $\mathcal{E}$ is invertible. Let $f_i(z, \theta, \dot{\theta})$ $(i = 1, ..., d)$ be a basis of the space of Baker functions $H^0(\tilde{Y}, \mathcal{E} \otimes \mathcal{L}(Q(z, \dot{\theta}, \theta)))$ depending in a (super)analytic way on $z, \theta, \dot{\theta}$. The elements of the latter space are represented as sections of $\mathcal{E}$ with essential singularities, so that they can be differentiated with respect to the parameters. From the conditions on the f_i it follows that for all $\alpha, \beta = 0, 1$ $(i = 1, ..., d)$

$$\sum_{s=1}^{3} \lambda_s D_{\alpha s} f_i \in H^0(\tilde{Y}, \mathcal{L}(Q) \otimes \mathcal{E} \otimes \varphi^*\mathcal{O}(1, 0)),$$

$$\sum_{t=1}^{3} \mu_t \dot{D}^t_\beta f_i \in H^0(\tilde{Y}, \mathcal{L}(Q) \otimes \mathcal{E} \otimes \varphi^*\mathcal{O}(0, 1)).$$

Let $\mathbf{f} = (f_1, ..., f_d)^t$. From the admissibility of $\mathcal{E} \otimes \mathcal{L}(Q)$ it follows that there exist uniquely determined odd $(d \times d)$-matrix functions $A_{\alpha s}(z, \theta, \dot{\theta})$, $\dot{A}^t_\beta(z, \theta, \dot{\theta})$ for which

$$\sum \lambda_s D_{\alpha s} \mathbf{f} = \sum \lambda_s A_{\alpha s} \mathbf{f}, \qquad \sum \mu_t \dot{D}^t_\beta \mathbf{f} = \sum \mu_t \dot{A}^t_\beta \mathbf{f}.$$

These are potentials of a field satisfying the connection equations. Since we have assumed that $\mathcal{E}$ is invertible, they can be expressed in terms of theta-functions of the manifold $\tilde{Y}$.

Similar formulae can also be obtained in the other situations. In the cases when in order to satisfy the field equations it is necessary to extend a bundle constructed on some infinitesimal neighbourhood X to the surrounding manifold X_1, this can be done as follows. We form $\tilde{X}_1 = X_1 \times_Y \tilde{Y}$ and simply look for distributions q_1 on some neighbourhood $\tilde{X}$ in $\tilde{X}_1$. These can be specified, by using a suitable dual bundle, as a family of distributions on $\tilde{Y}$ subject to linear differential equations (as above). If q is obtained by restriction of such a distribution q_1 to $\tilde{X}$, then the corresponding bundle can be extended to an (open) neighbourhood of X in X_1.

§5. Twistor analysis of a generalized system with φ^4-interaction

1. Statement of the problem.

In this section we shall carry out a twistor analysis (or rather, synthesis) of the system of fields and equations in domains $U \subset G(2, T)$, $T \cong \mathbf{C}^4$, whose solutions correspond to admissible bundles of type e) in the Introduction. These bundles are defined on $X^{(2)}(U) \subset P(T) \times P(T^*)$ and have restrictions $\mathcal{O}(-1, 0)^r \oplus \mathcal{O}(0, -1)^s$ to the quadrics $Y(u) \cong P^1 \times P^1 \subset X^{(2)}(U)$. Below we shall describe successively: a) fields on U corresponding to such bundles; b) fields and equations on $U \times U$ corresponding to the situation of "infinite extension"; c) some equations on U corresponding to extension to the second neighbourhood. Our analysis is analogous to the considerations in [38], 168–171, but is rather more complicated.

2. **Fields on U.**
Let $\mathcal{S}_+$ be the tautological sheaf on U, $\mathcal{S}_- = \mathcal{S}_+^\perp \subset T^* \otimes \mathcal{O}_U$. Let $\mathcal{F}$ denote an admissible sheaf on $X(U)$. We shall assume that U is a Stein space, so that

$$(1) \qquad \pi_1^*\mathcal{F} \cong (\pi_2^*\mathcal{E}_1)(-1,\ 0) \oplus (\pi_2^*\mathcal{E}_2)(0,\ -1),$$

where $X(U) \xleftarrow{\pi_1} Z(U) \xrightarrow{\pi_2} U$ is the bundle dual to our situation, and $\mathcal{E}_1, \mathcal{E}_2$ are locally free sheaves on U. The sheaf $\pi_1^*\mathcal{F}$ is endowed with a canonical relative connection $\nabla_{Z/X}$: $\pi_1^*\mathcal{F} \to (\pi_1^*\mathcal{F}) \otimes \Omega^1_{Z/X}$. Its diagonal elements relative to the splitting (1) when projected onto U give Yang-Mills fields $(\mathcal{E}_1, \nabla_1)$ and $(\mathcal{E}_2, \nabla_2)$: The non-diagonal elements, as can easily be seen, are (matrix) 1-forms, sections of the sheaves $\pi_2^*\mathrm{Hom}(\mathcal{E}_1, \mathcal{E}_2)(1, -1) \otimes \Omega^1_{Z/X}$ and $\pi_2^* \mathrm{Hom}(\mathcal{E}_2, \mathcal{E}_1)(-1, 1) \otimes \Omega^1_{Z/X}$. Projecting them down to U, we obtain fields

$$\varphi^+ \in \Gamma(U,\ \mathrm{sl}_2(S_+) \otimes \Lambda^2(S_-) \otimes \mathrm{Hom}(\mathcal{E}_1,\ \mathcal{E}_2)),$$
$$\varphi^- \in \Gamma(U,\ \mathrm{sl}_2(S_-) \otimes \Lambda^2(S_+) \otimes \mathrm{Hom}(\mathcal{E}_2,\ \mathcal{E}_1)).$$

Finally, using (1), we obtain a mapping

$$(2) \quad \{\text{admissible sheaves on } X(U)\} \to \to \{\text{quadruples of fields } (\mathcal{E}_i,\ \nabla_i),\ \varphi^\pm \text{ on } U\}.$$

If the intersections of the null-lines with U are connected and simply-connected, then (2) is a bijection by standard arguments: the basic reason is the lack of integrability conditions along one-dimensional fibres of π_1.

3. **Fields on $U \times U$.**
Now let $\mathcal{F}$ be an admissible sheaf on $P(U) \times \hat{P}(U)$, where $P = P(T)$, $\hat{P} = P(T^*)$. We consider the dual bundle

$$(3) \quad G(1;\ T) \times G(3;\ T) \xleftarrow{\pi_1} F(1,\ 2;\ T) \times F(2,\ 3;\ T) \to \xrightarrow{\pi_2} G(2,\ T) \times G\ (2;\ T)$$

and its restrictions to $U \times U$. The same arguments as those in the previous section establish a mapping

$$(4) \quad \{\text{admissible sheaves on } X(U)\} \to \to \{\text{quadruples of fields } (\mathcal{E}_i,\ \nabla_i),\ \Phi^\pm \text{ on } U \times U\},$$

where, as above,

$$\Phi^+ \in \Gamma(U \times U,\ \pi_{2*}(\Omega^1_{\pi_1}(1,\ -1)) \otimes \mathrm{Hom}\ (\mathcal{E}_1,\ \mathcal{E}_2)),$$
$$\Phi^- \in \Gamma(U \times U,\ \pi_{2*}(\Omega^1_{\pi_1}(-1,\ 1)) \otimes \mathrm{Hom}\ (\mathcal{E}_2,\ \mathcal{E}_1)).$$

To compute the direct images on the right, we denote the tautological flags on $F(1, 2; T)$ and $F(2, 3; T)$ by $\mathcal{S}_1 \subset \mathcal{S}_2, \mathcal{S}'_2 \subset \mathcal{S}'_3$ respectively. Then

$$\Omega^1_{\pi_1} = (\mathcal{S}_2/\mathcal{S}_1) \otimes \mathcal{S}_2^{\perp} \oplus (\mathcal{S}'_3/\mathcal{S}'_2) \otimes (\mathcal{S}'_2)^{\perp}.$$

from which we have finally, denoting the projections by $p_i : U \times U \to U$:

$$\Phi^+ \in \Gamma(U \times U,\ p_1^*\mathcal{S}_+^* \otimes p_2^*(\mathcal{S}_+ \otimes \Lambda^2\mathcal{S}_-) \otimes \mathrm{Hom}(\mathcal{E}_1, \mathcal{E}_2)),$$

$$\Phi^- \in \Gamma(U \times U,\ p_1^*(\mathcal{S}_- \otimes \Lambda^2\mathcal{S}_+) \otimes p_2^*\mathcal{S}^* \otimes \mathrm{Hom}(\mathcal{E}_2, \mathcal{E}_1)).$$

But it is clear that (4) cannot be a bijection, since the relative connection ∇_{π_1} on π_1^* satisfies a non-trivial integrability condition $\nabla^2_{\pi_1} = 0$ (the fibre of π_1 is four-dimensional).

4. Integrability conditions in coordinates.

We consider the part of the diagram (3) over $U \times U$. Let $U \subset \mathbf{C}^4$ (a large cell) and $F'(U) = \pi_2^{-1}(U)$. As in §4, we identify $F'(U)$ with $U \times U \times P^1 \times P^1$. We choose coordinates $(z_{\alpha\beta})$ in U in such a way that the metric is equal to $\det(z_{\alpha\beta})$, and fix affine coordinates (λ, μ) in $P^1 \times P^1$. The relative tangent sheaf to the fibres of π_1 is generated by the vector fields

$$v_1 = \frac{\partial}{\partial z_{00}} + \lambda \frac{\partial}{\partial z_{01}}; \quad v_2 = \frac{\partial}{\partial z_{10}} + \lambda \frac{\partial}{\partial z_{11}};$$

$$v_3 = \frac{\partial}{\partial z'_{00}} + \mu \frac{\partial}{\partial z'_{10}}; \quad v_4 = \frac{\partial}{\partial z'_{01}} + \mu \frac{\partial}{\partial z'_{11}}.$$

Since π_1 is the product of the two mappings π_{11}: $F(1, 2; T) \to G(2; T)$ and π_{12}: $F(2, 3; T) \to G(3; T)$, the sheaf $\Omega^1_{\pi_1}$ is the direct sum $\Omega^1_{\pi_{11}} \oplus \Omega^1_{\pi_{12}}$. The sections of $\Omega^1_{\pi_1}$ over the union for all z, z' of affine charts $\mathbf{A}^2 \subset P^1 \times P^1$ are given by associating a polynomial in λ and μ with coefficients depending on z, z' with each of the four vector fields written out above. Here the sections of $\Omega^1_{\pi_{11}} \subset \Omega^1_{\pi_1}$ are characterized by the fact that they are equal to zero on the second pair of fields, v_3 and v_4, while the sections of $\Omega^1_{\pi_{12}} \subset \Omega^1_{\pi_1}$ are characterized by vanishing on the first pair, v_1 and v_2. Everywhere holomorphic sections of $\Omega^1_{\pi_{11}}$ associate polynomials in λ of degree $\leqslant 1$ with the first pair: $v_1 \mapsto a_{00} + \lambda a_{01}$; $v_2 \mapsto a_{10} + \lambda a_{11}$, while everywhere holomorphic sections of $\Omega^1_{\pi_{11}}$ associate polynomials in μ of degree $\leqslant 1$ with the second pair of fields.

The sections of $\Omega^1_{\pi_1}(1, -1) = \Omega^1_{\pi_{11}}(1, -1) \oplus \Omega^1_{\pi_{12}}(1, -1)$ all come from the second term, and their values on the generating fields are as follows:

$$v_1 \mapsto 0; \quad v_2 \mapsto 0; \quad v_3 \mapsto \alpha_{00} + \lambda\alpha_{10}; \quad v_4 \mapsto \alpha_{01} + \lambda\alpha_{11}.$$

Similarly, sections of $\Omega^1_{\pi_1}(-1, 1) = \Omega^1_{\pi_{11}}(-1, 1) \oplus \Omega^1_{\pi_{12}}(-1, 1)$ come from the first term and translate as:

$$v_1 \mapsto \beta_{00} + \mu\beta_{01}; \quad v_2 \mapsto \beta_{10} + \mu\beta_{11}; \quad v_3 \mapsto 0; \quad v_4 \mapsto 0.$$

The connections in the bundles $\mathcal{E}_1$ and $\mathcal{E}_2$ are given by potentials

$$A_{ij},\ A'_{ij} \in \mathrm{gl}_r(\mathbf{C}): \qquad B_{ij},\ B'_{ij} \in \mathrm{gl}_s(\mathbf{C}) \qquad (i, j = 0, 1),$$

and the fields $\Phi^{\pm}$ are given by the components

$$\Phi^{+}_{ij} \in \mathrm{Mat}\,(r \times s); \quad \Phi^{-}_{ij} \in \mathrm{Mat}\,(s \times r) \quad (i,\ j=0,\ 1).$$

From what was said above, it follows that in coordinates (z, z', λ, μ) the connection ∇_{π_1} is given by the operators

$$\frac{\partial}{\partial z_{00}} + \lambda \frac{\partial}{\partial z_{01}} + \left(\frac{A_{00}+\lambda A_{01} \mid \Phi^{-}_{00}+\mu\Phi^{-}_{01}}{0 \mid B_{00}+\lambda B_{01}} \right),$$

$$\frac{\partial}{\partial z_{10}} + \lambda \frac{\partial}{\partial z_{11}} + \left(\frac{A_{10}+\lambda A_{11} \mid \Phi^{-}_{10}+\mu\Phi^{-}_{11}}{0 \mid B_{10}+\lambda B_{11}} \right),$$

$$\frac{\partial}{\partial z'_{00}} + \mu \frac{\partial}{\partial z'_{10}} + \left(\frac{A'_{00}+\mu A'_{10} \mid 0}{\Phi^{+}_{00}+\lambda\Phi^{+}_{10} \mid B'_{00}+\mu B'_{10}} \right),$$

$$\frac{\partial}{\partial z'_{01}} + \mu \frac{\partial}{\partial z'_{11}} + \left(\frac{A'_{01}+\mu A'_{11} \mid 0}{\Phi^{-}_{01}+\lambda\Phi^{+}_{11} \mid B'_{01}+\mu B'_{11}} \right),$$

the commutability of which for all λ, μ is equivalent to the integrability of ∇_{π_1}. We write $\nabla_{ijA} = \partial/\partial z_{ij} - A_{ij}$, $\nabla'_{ijA} = \partial/\partial z'_{ij} + A'_{ij}$, $\nabla_{ijB} = \partial/\partial z_{ij} + B_{ij}$, $\nabla'_{ijB} = \partial/\partial z'_{ij} + B'_{ij}$. $(\tilde{\nabla}_{ij}, \tilde{\nabla}'_{ij})$ will denote the operators of the natural connection in $\mathrm{Hom}(\mathcal{E}_1, \mathcal{E}_2)$ (and also in $\mathrm{Hom}(\mathcal{E}_2, \mathcal{E}_1)$: which is meant will be clear from the context). Written in detail, the integrability conditions are as follows:

1) ∇_A, ∇_B are self-dual, and ∇'_A, ∇'_B are anti-self-dual;

2) $\tilde{\nabla}_{ij}\Phi^{-}_{kl} = \tilde{\nabla}_{kj}\Phi^{-}_{il}$; $\quad \tilde{\nabla}'_{ij}\Phi^{-}_{kl} = \tilde{\nabla}'_{il}\Phi^{+}_{kj}$;

3) $[\nabla_{ijA}, \nabla'_{klA}] = -\Phi^{-}_{kl}\Phi^{-}_{jl}$; $\quad [\nabla_{ijB}, \nabla'_{klB}] = \Phi^{+}_{jl}\Phi^{-}_{ik}$;

4) $\tilde{\nabla}_{ij}\Phi^{+}_{kl} = -\tilde{\nabla}_{ik}\Phi^{+}_{jl}$; $\quad \tilde{\nabla}'_{ij}\Phi^{-}_{kl} = -\tilde{\nabla}'_{lj}\Phi^{-}_{ki}$.

5. Integrability conditions on infinitesimal neighbourhoods.

Now let $\mathcal{F}$ be an admissible sheaf on $X^{(1)}(U)$. It is not hard to see that there are no integrability conditions yet, so that it is given by two connections in bundles over the domain U (which we shall take to be diagonally embedded in $U \times U$) and fields $\Phi^{\pm}$, as in subsection 2. Let D_{ijA}, D_{ijB} be the operators of these connections. The extendability of $\mathcal{F}$ to $L^{(i)}$ means that we can extend $\Phi^{\mp}_{ij}$ from the diagonal Δ to $\Delta^{(i)}$ and find connections over $\Delta^{(i)}$ which coincide with the given ones on restriction to Δ, and are such that the integrability relations of the previous subsection are satisfied up to terms of order $(i-1)$ in the transverse direction.

If $(\nabla_{ijA}, \nabla'_{ijA})$ and $(\nabla_{ijB}, \nabla'_{ijB})$ are the connection operators over $\Delta^{(i)}$, then the fact that they coincide with the given ones over Δ means that $D_{ijA} = \nabla_{ijA} + \nabla'_{ijA}$, $D_{ijB} = \nabla_{ijB} + \nabla'_{ijB}$ on the diagonal. These conditions impose differential equations on D_{ijA}, D_{ijB}, $\Phi^{\mp}$. Let $\tilde{D}_{ij}$ be the connection operators in Hom. We shall seek equations of the form

$$\sum t_{rsijkl}\tilde{D}_{rs}\tilde{D}_{ij}\Phi^{+}_{kl} = \sum \lambda_{rsijkl}\Phi^{+}_{rs}\Phi^{-}_{ij}\Phi^{+}_{kl}$$

and similar equations for Φ^{-}, more precisely their left-hand sides, which are given by tensors $t \in (\mathbf{C}^2)^{\otimes 6}$. We shall also consider quantities of the kind

$\widetilde{D}\widetilde{D}\Phi^+$ as mappings $U \times U \to (\mathbf{C}^2)^{\otimes 6}$ and write the left-hand side in the form $t(\widetilde{D}\widetilde{D}\Phi^+)$. We shall write the symbol $\equiv$ for equivalence modulo cubic monomials $\Phi^+_{rs}\Phi^-_{ij}\Phi^+_{kl}$, so that for example $[\widetilde{\nabla}_{ij}, \widetilde{\nabla}_{kl}]\Phi^+_{rs} \equiv 0$ by virtue of the integrability relations. We have $\widetilde{D}\widetilde{D}\Phi^+ = \widetilde{\nabla}\widetilde{\nabla}\Phi^+ + \widetilde{\nabla}'\widetilde{\nabla}'\Phi^+ + (\widetilde{\nabla}'\widetilde{\nabla} + \widetilde{\nabla}\widetilde{\nabla}')\Phi^+$. Each of these three terms is subject to definite symmetries, the first two precisely, and the third modulo $\Phi^+\Phi^-\Phi^+$. To describe them concisely, we introduce notations for the following permutations of $S_6 = \mathrm{Aut}\{r, s, i, j, k, l\}$: $\tau = ijrskl$; $\gamma = rsilkj$; $\sigma = rsikjl$; $\varepsilon = rlijks$; $\alpha_{13} = isrjkl$; $\alpha_{24} = rjiskl$. Let $V \subset (\mathbf{C}^2)^{\otimes 6}$ consist of those a for which $\sigma a = -a$ and

$$(1 - \alpha_{13})(1 - \tau)a = (1 + \alpha_{24})(1 - \tau)a = 0.$$

Then $\widetilde{\nabla}\widetilde{\nabla}\Phi^+$ takes values in V. Let V' consist of those a for which $\gamma a = a$, $(1 + \alpha_{13})(1 - \tau)a = (1 - \alpha_{24})(1 - \tau)a = 0$. $\widetilde{\nabla}'\widetilde{\nabla}'\Phi^+$ takes values in V' Further we have:

$$\widetilde{\nabla}'_{rs}\widetilde{\nabla}_{ij}\Phi^+_{kl} = -\widetilde{\nabla}'_{rs}\widetilde{\nabla}_{ik}\Phi^+_{jl};\quad \widetilde{\nabla}'_{rs}\widetilde{\nabla}_{ij}\Phi^+_{kl} \equiv \widetilde{\nabla}_{ij}\widetilde{\nabla}'_{rs}\Phi^+_{kl} = \widetilde{\nabla}_{ij}\widetilde{\nabla}'_{rl}\Phi^+_{ks} \equiv \widetilde{\nabla}'_{rl}\widetilde{\nabla}_{ij}\Phi^+_{ks}.$$

This means that $\widetilde{\nabla}'\widetilde{\nabla}\Phi^+$ mod Φ^3 takes values in the space $V''_0 \subset (\mathbf{C}^2)^{\otimes 6}$, consisting of those a for which $\sigma a = -a$, $\varepsilon a = a$. Consequently $\widetilde{\nabla}'\widetilde{\nabla}\Phi^+ + \widetilde{\nabla}\widetilde{\nabla}'\Phi^+$ mod Φ^3 takes values in $V'' = (1+\tau)V''_0$. If now the tensor t is orthogonal to all three subspaces V, V', V'', then $t(\widetilde{D}\widetilde{D}\Phi^+)$ can be expressed in terms of cubic monomials, that is, we have the required equation. Since V, V', V'' are GL_2-invariant, so is $(V+V'+V'')^\perp$ (with respect to the dual action). If some representation of GL_2 appears in $(V+V'+V'')^\perp$ in a multiple form, this does not guarantee that the corresponding equations are different. We can only assert that the number of equations is not less than $\sum \dim \pi$ with respect to the irreducible representations appearing in $(V+V'+V'')^\perp$. Let us explain the structure of our representations.

6. ***Lemma.*** *As* GL_2*-modules*

$$\begin{aligned} V &\cong \Sigma^{5,1} + 2\Sigma^{4,2} + \Sigma^{3,3}, && \dim V = 12; \\ V' &\cong \Sigma^{6,0} + 3\Sigma^{5,1} + 4\Sigma^{4,2} + 2\Sigma^{3,3}, && \dim V' = 36; \\ V'' &\cong \Sigma^{5,1} + 2\Sigma^{4,2} + \Sigma^{3,3}, && \dim V'' = 12. \end{aligned}$$

Proof. In order to determine the type of V, we split $\{r, s, i, j, k, l\}$ into two parts: $\{r, i, l\}$ and $\{s, j, k\}$. The permutations appearing in the definition of V preserve these parts, that is, they lie in $S_3 \times S_3 \subset S_6$. We associate the first three indices with a two-dimensional space S, and the other three with S', so that

$$(\mathbf{C}^2)^{\otimes 6} = S^{\otimes 3} \otimes (S')^{\otimes 3} = \bigoplus_{\alpha,\beta} W_\alpha \otimes W'_\beta \otimes \Sigma^\alpha S \otimes \Sigma^\beta S',$$

where α, β run through Young diagrams with three cells and at most two rows. W_α is the representation of S_3 and Σ^α the representation of GL_2 corresponding to α.

$\Sigma^\alpha S \otimes \Sigma^\beta S'$ appears in V with multiplicity equal to the dimension of the subspace of $W_\alpha \otimes W'_\beta$ that annihilates the elements of the group ring $S_3 \times S_3$ appearing in the definition of V.

Let σ_{ij} be the standard transpositions in S_3 ($i, j = 1, 2, 3$). When we consider their action on the indices $\{r, i, l\}$ and $\{s, j, k\}$ we shall regard these as written in the above order, so that, for example, $(\sigma_{12}, 1) \in S_3 \times S_3$ interchanges r and i and does not change the second triple. Let ρ_α be a representation of S_3 in the space W_α. For each pair α, β we must find the intersection of the space $P \subset W_\alpha \otimes W'_\beta$, on which $1 \otimes \rho'(\sigma_{23}) = -1$, and the space

$$Q = W_\alpha^+(\sigma_{12}) \otimes W_\beta'^+(\sigma_{12}) + W_\alpha^-(\sigma_{12}) \otimes W_\beta'^-(\sigma_{12}) + W_\alpha^+(\sigma_{12}) \otimes W_\beta'^-(\sigma_{12}),$$

where $W_\alpha^\pm(\sigma_{ij})$ denote the eigensubspaces of $\rho(\sigma_{ij})$, and α and β can take the values $(3, 0)$ and $(2, 1)$. The space $W_{2,1}$ is identified with $\{(x_1, x_2, x_3) \in \mathbf{C}^3: \sum x_i = 0\}$ with the natural action of S_3, so that the arrangement of the $W_{2,1}^\pm(\sigma_{ij})$ becomes obvious. We examine the four cases:

a) $\alpha = (3, 0)$, $\beta = (3, 0)$. Here $P = 0 \Rightarrow P \cap Q = 0$.

b) $\alpha = (2, 1)$, $\beta = (3, 0)$. This is the same.

c) $\alpha = (3, 0)$, $\beta = (2, 1)$. Here $P = W_{3,0} \otimes W_{2,1}'^-(\sigma_{23})$, $Q = W_{3,0} \otimes W'_{2,1}$; $\dim P \cap Q = 1$.

d) $\alpha = (2, 1)$, $\beta = (2, 1)$. Here $P = W_{2,1} \otimes W_{2,1}'^-(\sigma_{23})$, $\dim P = 2$, $\dim Q = 3$ and an elementary computation shows that $\dim (P \cap Q) = 1$. Thus $V \cong \Sigma^{3,0}S \otimes \Sigma^{2,1}S' \oplus \Sigma^{2,1}S \otimes \Sigma^{2,1}S'$. Identifying S and S' with $\mathbf{C}^2$, we obtain

$$V \cong \Sigma^{5,1} + 2\Sigma^{4,2} + \Sigma^{3,3}; \quad \dim V = 12.$$

We turn to V'. We divide the indices into two parts in another way: $\{r, i, k\}$ and $\{s, j, l\}$. The permutations appearing in the definition of V' preserve these parts. Let r, i, k refer to the space S_+ and s, j, l to S_-. We keep the same notation as before, except that the transpositions σ_{ij} will refer to the new order of the indices. For each pair α, β we must find the intersection of the subspaces

$$P = W_\alpha \otimes W_\beta'^+(\sigma_{23}),$$

$$Q = W_\alpha^+(\sigma_{12}) \otimes W_\beta'^+(\sigma_{12}) + W_\alpha^-(\sigma_{12}) \otimes W_\beta'^-(\sigma_{12}) + W_\alpha^-(\sigma_{12}) \otimes W_\beta'^+(\sigma_{12}).$$

Again we examine four cases:

a) $\alpha = (3, 0)$, $\beta = (3, 0)$. Here $P = Q = W_{3,0} \otimes W'_{3,0}$, $\dim P \cap Q = 1$.

b) $\alpha = (3, 0)$, $\beta = (2, 1)$. Here $P = W_{3,0} \otimes W_{2,1}'^+(\sigma_{23})$, $Q = W_{3,0} \otimes W_{2,1}'^+(\sigma_{12})$; $P \cap Q = \{0\}$.

c) $\alpha = (2, 1)$, $\beta = (3, 0)$. Here $P = Q = W_{2,1} \otimes W'_{3,0}$; $\dim (P \cap Q) = 2$.

d) $\alpha = (2, 1)$, $\beta = (2, 1)$. Here $P = W_{2,1} \otimes W_{2,1}'^+(\sigma_{23})$; $\dim P = 2$, $\dim Q = 3$; a simple computation shows that $\dim (P \cap Q) = 1$.

Thus $V' \cong \Sigma^{3,0}S_- \otimes \Sigma^{3,0}S_- \oplus 2\Sigma^{2,1}S_+ \otimes \Sigma^{3,0}S_- \oplus \Sigma^{2,1}S_+ \otimes \Sigma^{2,1}S_-$. Identifying S_+ and S_- with $\mathbf{C}^2$, we find that $V' \cong \Sigma^{6,0} + 3\Sigma^{5,1} + 4\Sigma^{4,2} + 2\Sigma^{3,3}$; $\dim V' = 36$.

We turn to V''. It is $(1+\tau)V_0''$, where V_0'' consists of a_{rsijkl}, skew-symmetric in the pair (j, k) and symmetric in the pair (l, s), and $\tau: (r, s, i, j, k, l) \mapsto (i, j, r, s, k, l)$. We have

$$V_0'' \cong (\mathbf{C}^2)^{\otimes 2} \otimes \Lambda^2(\mathbf{C}^2) \otimes S^2(\mathbf{C}^2) \cong \Sigma^{5,1} + 2\Sigma^{4,2} + \Sigma^{3,3};$$

$\dim V_0'' = 12$. The space V_0'' intersects the kernel of $1+\tau$ only in zero, and hence as GL_2-modules $V'' \cong V_0''$. This proves the lemma.

Since $(\mathbf{C}^2)^{\otimes 6} \cong \Sigma^{6,0} + 5\Sigma^{5,1} + 9\Sigma^{4,2} + 5\Sigma^{3,3}$, it follows from the lemma that $(V + V' + V'')^{\perp}$ contains, in every case, one $\Sigma^{4,2}$ and one $\Sigma^{3,3}$. This means that there are (at least) four equations for the four functions Φ_{ij}^{+}. Apparently the sum $V + V' + V''$ is direct, and there are in fact four equations.

The left-hand sides of our equations can be reduced to a simpler form. Namely, we write the equations as $P(\widetilde{D}\widetilde{D}\Phi^+) = Q(\Phi^+\Phi^-\Phi^+)$, where $P: (\mathbf{C}^2)^{\otimes 6} \to (\mathbf{C}^2)^{\otimes 6}$ is a GL_2-invariant projection annihilating $V + V' + V''$, and

$$Q: (\mathbf{C}^2)^{\otimes 6} \to \mathrm{Im}\, P$$

is a GL_2-invariant operator. It is clear that such an equation can be replaced by $(\gamma P)(\widetilde{D}\widetilde{D}\Phi^+) = (\gamma Q)(\Phi^+\Phi^-\Phi^+)$, where $\gamma \in \mathbf{C}[S_6]^*$. By choosing a suitable form for γ, we can reduce P to a standard form, a sum of Young symmetrizers. We may thus assume that the equations for Φ^+ have the form

$$\begin{cases} \square_{4,2}\Phi^+ = Q\,(\Phi^+\Phi^-\Phi^+), \\ \square_{3,3}\Phi^+ = Q'\,(\Phi^+\Phi^-\Phi^+), \end{cases}$$

where $\square_{4,2}\Phi^+$ is the Young symmetrizer corresponding to applying $\begin{smallmatrix} rskl \\ ij \end{smallmatrix}$ to $\widetilde{D}_{rs}\widetilde{D}_{ij}\Phi_{kl}^+$; $\square_{3,3}\Phi^+$ is the Young symmetrizer corresponding to applying $\begin{smallmatrix} rsk \\ ijl \end{smallmatrix}$ to $\widetilde{D}_{rs}\widetilde{D}_{ij}\Phi_{kl}^+$. The equations for Φ^- are constructed similarly.

References

[1] M.F. Atiyah and R.S. Ward, Instantons and algebraic geometry, Comm. Math. Phys. 55 (1977), 117-124. MR 58 # 13029.

[2] ———, N.J. Hitchin, and I.M. Singer, Self-duality in four-dimensional Riemannian geometry, Proc. Roy. Soc. London Ser. A **362** (1978), 425-461. MR **80d**:53023.

[3] ———, ———, V.G. Drinfel'd, and Yu.I. Manin, Construction of instantons, Phys. Lett. A **65** (1978), 185-187. MR **82g**:81049.

[4] A.A. Beilinson, The derived category of coherent sheaves on P^N, in: *Voprosy teorii grupp i gomologicheskoi algebry* (Problems of group theory and homological algebra), no. 2, Yaroslavl State University, **1979**, 42-54.

[5] A.Yu. Vaintrob, Deformations of complex structures on supermanifolds, Funktsional. Anal. i Prilozhen. **18**:2 (1984), 59-60. MR **85k**:32041. = Functional Anal. Appl. **18** (1984), 135-136.

[6] D.E. Lerner and P.D. Sommers (eds), Complex manifold techniques in theoretical physics, Research Notes in Mathematics 32, Pitman, London 1979. MR **80m**:81004.

[7] V.G. Drinfel'd and Yu.I. Manin, Instantons and sheaves on $\mathbf{C}P^3$, Funktsional. Anal. i Prilozhen. **13**:2 (1979), 59-74. MR **81b**:14007.
= Functional Anal. Appl. **13** (1979), 124-134.

[8] S.K. Donaldson. Nahm's equation and the classification of monopoles, Preprint, Oxford 1984.

[9] J. Harnad, J. Hurtubise, M. Légaré, and S. Shnider, Constraint equations and field equations in supersymmetric $N = 3$ Yang-Mills theory, Preprint, Univ. de Montréal 1984.

[10] A. Galperin, E. Ivanov, S. Kalitzin, V. Ogievetsky, and E. Sokatchev, Unconstrained $N = 2$ matter, Yang-Mills and supergravity theories in harmonic superspace, Classical Quantum Gravity **1** (1984), 469-489.

[11] ———, ———, ———, ———, and ———, Unconstrained off-shell $N = 3$ supersymmetric Yang-Mills theory, Preprint, Joint Inst. of Nuclear Research, E 2-84-441, Dubna 1984.

[12] ———, ———, V. Ogievetsky, and E. Sokatchev, Harmonic supergraphs. Feinman rules and examples. Preprint, Joint Inst. of Nuclear Research E 2-85-128, Dubna 1985.

[13] S.G. Gindikin and G.M. Khenkin, The Penrose transformation and complex integral geometry, Current problems in mathematics, Vol. 17, 57-111, VINITI, Moscow 1981. MR **83a**:32027.

[14] ———, Integral geometry and twistors, Lecture Notes in Math. **970** (1982), 2-42. MR **84j**:32039.

[15] G.M. Khenkin [Henkin] and Yu.I. Manin. Twistor description of classical Yang-Mills-Dirac fields, Phys. Letters B **95** (1980), 405-408. MR **82f**:81045.

[16] ——— and ———, On the cohomology of twistor flag spaces, Compositio Math. **44** (1981), 103-111. MR **84m**:32042.

[17] ———, Representation of the solutions of the φ^4-equation in the form of holomorphic bundles over a twistor space, Dokl. Akad. Nauk SSSR **260** (1981). 1086-1089. MR **83h**:81050.
= Soviet Math. Dokl. **24** (1981), 415-419.

[18] ———, Yang-Mills-Higgs fields as holomorphic vector bundles, Dokl. Akad. Nauk SSSR **265** (1982), 1081-1085. MR **83m**:32032.
= Soviet Math. Dokl. **26** (1982), 224-228.

[19] ———, Yang-Mills-Higgs-Dirac fields as holomorphic vector bundles. in: *Teoretiko-gruppovye metody v fizike* (Group-theoretic methods in physics). Vol. 1. Nauka, Moscow 1983.

[20] N.J. Hitchin. Linear field equations on self-dual spaces. Proc. Roy. Soc. London Ser. A **370** (1980), 173-191. MR **81i**:81057.

[21] ———, Polygons and gravitons. Math. Proc. Cambridge Philos. Soc. **85** (1979). 465-476. MR **80h**:53047.

[22] ———, Kählerian twistor spaces. Proc. London Math. Soc. (3) **43** (1981). 133-150. MR **84b**:32014.

[23] ———, Monopoles and geodesics. Comm. Math. Phys. **83** (1982). 579-602. MR **84i**:53071.

[24] ———, On the construction of monopoles. Comm. Math. Phys. **89** (1983). 145-190. MR **84m**:53076.

[25] ———, Complex manifolds and Einstein's equations. Lecture Notes in Math. **970** (1982), 73-100. MR **84i**:32041.

[26] H. Isenberg, P.B. Yasskin. and P.S. Green. Non-self-dual gauge fields. Phys. Letters B **78** (1978). 464-468.

[27] R.E. Kallosh. Superstrings and harmonic superspace, Preprint, CERN-TH.4188/85, May 1985.

[28] C.R. Lebrun. Spaces of complex geodesics and related structures. Thesis, Oxford 1980.

[29] ———, The first formal neighbourhood of ambitwistor space for curved space-time, Lett. Math. Phys. **6** (1982), 345-354. MR **84d**:32044.

[30] D.A. Leites, Introduction to the theory of supermanifolds, Uspekhi Mat. Nauk **35**:1 (1980), 3-57. MR **81j**:58003.
= Russian Math. Surveys **35**:1 (1980), 1-64.

[31] M.K. Murray, Monopoles and spectral curves for arbitrary Lie groups, Comm. Math. Phys. **90** (1983), 263-271. MR **85e**:53094.

[32] ———, Non-abelian magnetic monopoles, Preprint MSRI-031084-7, Math. Sci. Res. Institute, Berkeley 1984.

[33] Yu.I. Manin and I.B. Penkov, Null geodesics of complex Einstein spaces. Funktsional. Anal. i Prilozhen. **16**:1 (1982), 78-79. MR **83k**:53067.
= Functional Anal. Appl. **16** (1982), 64-66.

[34] ——— and G.M. Khenkin. Yang-Mills-Dirac equations as Cauchy-Riemann equations in twistor space, Yadernaya Fiz. **35** (1984), 1610-1626. MR **85e**:32036.
= Soviet J. Nuclear Phys. **35** (1984), 941-950.

[35] ———, New exact solutions and a cohomological analysis of ordinary and supersymmetric Yang-Mills equations, Trudy Mat. Inst. Steklov **165** (1984), 98-114. MR **86j**:32059.
= Proc. Steklov Inst. Math. **165** (1984).

[36] ——— and Khoang Le Min', The Radon-Penrose transformation for the group SO(8) and instantons, Izv. Akad. Nauk SSSR Ser. Mat. **50** (1986), 195-206.

[37] ———, Supersymmetry and supergravitation in the superspace of null supergeodesics, in: *Teoretiko-gruppovye metody v fizike* (Group-theoretic methods in physics), Vol. 1, 203-208, Nauka, Moscow 1983.

[38] ———, *Kalibrovochnye polya i kompleksnaya geometriya* (Gauge fields and complex geometry), Nauka, Moscow 1984.

[39] ———, The inverse scattering transform and classical field equations of motion, in: Non-linear and turbulent processes in physics (ed. R. Sagdeev), Vol. 3, 1487-1502, Gordon and Breach, New York 1984.

[40] C. Okonek, M. Schneider, and H. Spindler, Vector bundles on complex projective spaces, Progress in Mathematics 3, Birkhäuser, Boston MA 1980. MR **81b**:14001.
Translation: *Vektornye rassloeniya na kompleksnykh proektivnykh prostranstvakh*, Mir, Moscow 1984.

[41] R. Penrose, Non-linear gravitons and curved twistor theory, Gen. Relativity Gravitation 7 (1976), 31-52. MR 55 # 11905.

[42] ———, The twistor program, Rep. Math. Phys. **12** (1977), 65-76. MR 57 # 4948.

[43] A.A. Roslyi, Connections in supersymmetric Yang-Mills theory as an integrability condition, in: *Teoretiko-gruppovye metody v fizike* (Group-theoretic methods in physics), Vol. 1, 263-268, Nauka, Moscow 1983.

[44] A.M. Semikhatov, Twistors and instantons in higher dimensions, Zh. Eksper. Teoret. Fiz. **41**:4 (1985), 165-168.
= JETP Letters **41** (1985), 201-205.

[45] V.V. Zharinov (ed.), *Tvistory i kalibrovochnye polya: sbornik statei* (Twistors and gauge fields: a collection of articles), Mir, Moscow 1983. MR **84m**:32044.

[46] P.B. Yasskin, An ambitwistor approach for gravity, Preprint, Texas University 1984.

[47] I.V. Cherednik, Finite-zone solutions to the duality equation over S^4 and two-dimensional realativistically-invariant systems, Dokl. Akad. Nauk SSSR **246** (1979), 575-578. MR **83a**:81060.
= Soviet Phys. Dokl. **24** (1979), 356-358.

[48] R.S. Ward, On self-dual gauge fields, Phys. Lett. A **61** (1977), 81-82. MR **56** # 2186.

[49] ———, Ansätze for self-dual Yang-Mills fields, Comm. Math. Phys. **80** (1981), 563-574. MR **82m**:81059.

[50] R.O. Wells, Complex geometry in mathematical physics, Presses de l'université de Montréal, Montréal 1982. MR **84f**:32037.

[51] E. Witten, An interpretation of classical Yang-Mills theory, Phys. Lett. B 78 (1978), 394-398.

[52] ———, Twistor-like transform in ten dimensions, Preprint, Princeton University, May 1985.

[53] B.A. Dubrovin, V.B. Matveev, and S.P. Novikov, Non-linear equations of Korteweg-de Vries type, finite-zone linear operators and Abelian varieties, Uspekhi Mat. Nauk **31**:1 (1976), 55-136. MR **55** # 899.
= Russian Math. Surveys **31**:1 (1976), 59-146.

[54] I.M. Krichever and S.P. Novikov, Holomorphic bundles over algebraic curves and non-linear equations, Uspekhi Mat. Nauk **35**:6 (1980), 47-68. MR **83b**:35145.
= Russian Math. Surveys **35**:6 (1976), 53-79.

[55] J. Leiterer, The Penrose transformation for bundles non-trivial on the general line, Math. Nachr. **112** (1983), 35-67. MR **85j**:32051.

[56] G.M. Khenkin [Henkin], Tangent Cauchy-Riemann equations and Yang-Mills, Higgs and Dirac fields, Proc. Internat. Congress Math. Warsaw 1983, Vol. 1, 809-827, Warsaw 1985.

[57] P. Forgacs, L. Horváth, and L. Palla, On the linearization of source-free gauge field equations, Phys. Lett. B **115** (1982), 463-467. MR **84i**:81061.

Translated by C.J. Shaddock

Steklov Mathematical Institute
of the Academy of Sciences of the USSR

Received by the Editors 10 October 1985

Reprinted from *Funct. Anal. Appl.* **20** (1986).

CRITICAL DIMENSIONS OF THE STRING THEORIES AND THE DUALIZING SHEAF ON THE MODULI SPACE OF (SUPER) CURVES

Yu. I. Manin

1. Introduction. The quantum theory of (super) strings exists at present in two entirely different mathematical fields. Under canonical quantization it appears to a mathematician as the representation theory of the algebras of Heisenberg, Virasoro, and Kac–Moody and their superextensions [1, 2]. Quantization with the help of the Polyakov path integration [3] leads to the analytic theory of algebraic (super) curves and their moduli spaces, to invariants of the type of the analytic curvature, etc. [4].

Establishment of direct mathematical connections between these two forms of a single theory is a big and important problem. The main aim of this note is to express the conjecture that the moduli space M_g of algebraic curves of genus g (for $g \to \infty$) must play the same role in the representation theory of the Virasoro algebra as the space G/B (B is the parabolic subgroup) plays in the representation theory of the semisimple Lie group G. The main result of this note is a modest computation that shows that the so-called critical dimensions for the Virasoro and the Neveu–Schwarz algebras correspond to the classes of the dualizing sheaves on the moduli spaces of curves or, respectively, of the super curves of dimension 1|1, equipped with a special additional structure.

2. Critical Dimensions. The Virasoro algebra V is generated by the generator $\{z, e_j \mid j \in \mathbf{Z}\}$, where z belongs to the center and e_j satisfy the relations

$$[e_i, e_j] = (j-i)e_{i+j} + \frac{z}{12}\delta_{i,-j}(j^3 - j).$$

The Neveu–Schwarz algebra NS is the Lie superalgebra with the even part V and the odd generators $\left\{g_k \mid k \in \mathbf{Z} + \frac{1}{2}\right\}$ with the relations

$$[e_i, g_k] = \left(k - \frac{1}{2}i\right)g_{i+k};\quad [g_k, g_l] = 2e_{k+l} + \frac{1}{3}\left(l^2 - \frac{1}{4}\right)\delta_{k,-l}z$$

(here [,] is the supercommutator). When the algebra V is realized in the state space of a string (NS for a superstring) in the D-dimensional Minkowski space–time, the eigenvalue of z turns out to be equal to D (3D/2). The critical dimension D = 26 for V (D = 10 for NS) is distinguished by several remarkable properties: by the possibility to define the action of the Lorentz algebra, by disappearance of the conformal anomalies, etc. (see also Sec. 6b below). Let $k_V = 26$, k_{NS} 15 denote the eigenvalues of z corresponding to the critical dimensions.

3. The Mumford Theorem, or 13 = 26/2. Let $p: C \to M$ be the family of smooth irreducible algebraic curves over a complex manifold M. We will denote the holomorphic tangent bundle (the form bundle) on a complex manifold X by $\mathcal{T}X$ ($\Omega^{\cdot}X$) and the sheaves of the relative objects by $\mathcal{T}X/Y$ and $\Omega^{\cdot}X/Y$. We have the exact sequence $0 \to \mathcal{T}C/M \to \mathcal{T}C \to p^*\mathcal{T}M \to 0$. The boundary homomorphism $\alpha: \mathcal{T}M \to R^1p_*(\mathcal{T}C/M)$, corresponding to it, is the Kodaira–Spencer mapping. It is convenient to consider the dual mapping, combined with the Serre duality $R^*p_*(\Omega^{\otimes 2}C/M) \to \Omega^1 M$. Following Mumford [5], let λ_n denote the class of the sheaf $\det R^*p_*(\Omega^{\otimes n}C/M)$ in Pic M. Mumford has proved that $\lambda_n = (12\binom{n}{2} + 1)\lambda_1$. In particular, λ_2 $13\lambda_1$. From this computation we get our

V. A. Steklov Mathematics Institute, Academy of Sciences of the USSR. Translated from Funktsional'nyi Analiz i Ego Prilozheniya, Vol. 20, No. 3, pp. 88-89, July-September, 1986. Original article submitted October 14, 1985.

first numeralogical coincidence: *If the Kodaira–Spencer mapping of the family C/M is an isomorphism, then the class of the dualizing sheaf* $\rho = [\det \Omega^1 M]$ *in* Pic *M is equal to* $13\lambda_1 = \frac{1}{2} k_V \lambda_1$.

Now we describe the situation in the algebraic supergeometry which corresponds to the Neveu–Schwarz superstring.

4. SUSY-Families. The data set $(p_s; C_s \to M_s; \mathcal{T}_1)$, where M_S is a smooth algebraic supervariety of dimension m|n, C_S is a smooth supervariety of dimension $m + 1|n + 1$; p_S is a smooth proper morphism, and $\mathcal{T}_1 \subset \mathcal{T} := \mathcal{T} C_s/M_s$ is a locally free locally direct subsheaf of rank 0|1 such that the Froebenius mapping $[,] \bmod \mathcal{T}_1 : \mathcal{T}_1^{\otimes 2} \to \mathcal{T}_0 := \mathcal{T}/\mathcal{T}_1$ is an isomorphism, is called *an SUSY-family of curves* (from the word "supersymmetry"). We call the composition $a_s : \mathcal{T}M \xrightarrow{a} R^1 p_{s*}(\mathcal{T}) \xrightarrow{\pi} R^1 p_{s*}(\mathcal{T}_0)$, where a is constructed as in Sec. 3 and π is the descent $\mathcal{T} \to \mathcal{T}/\mathcal{T}_1$, *the Kodaira–Spencer mapping* for the SUSY family (here there is no space to motivate this definition).

An SUSY-curve C_S is an SUSY-family over a point. Its structure is determined by the sheaf $\mathcal{N} = \mathcal{O}_{1,C_s}$ of odd functions over a smooth complete algebraic curve C $(C_S)_{red}$. The composition $\mathcal{N} \xrightarrow{d} \Omega^1 C_s \xrightarrow{\gamma} \mathcal{T}_1^*$ (γ is the dualization of the embedding $\mathcal{T}_1$) gives an isomorphism of the $\mathcal{O}_C$-modules $\mathcal{N} \simeq \mathcal{T}_1^*/\mathcal{N}\mathcal{T}_1^*$. The restriction to C gives an isomorphism $\mathcal{T}_0/\mathcal{N}\mathcal{T}_0 \simeq \mathcal{T}C$. The Froebenius form defines an isomorphism $\mathcal{N}^{\otimes 2} \simeq \Omega^1 C$. We will write $N - \Omega^{1/2}$. Thus, $C_s = (C, \mathcal{O}_C \oplus \Omega^{1/2})$: from the classical point of view, C_S consists of the curve C with fixed theta-characteristic. A general SUSY-family consists of such curves, parametrized by a supervariety.

Let M_S be a supervariety and set M $(M_S)_{red}$. Let us identify $K_0(M_S)$ with $K_0(M)$. Here $K_0(M_S)$ [$(K_0(M))$] is the Grothendieck group of Z_2-graded coherent $\mathcal{O}_{M_s}$-modules (of $\mathcal{O}_M$-modules). The identification is carried out with the help of the filtration corresponding to the powers of the ideal $(\mathcal{O}_{M_s,1})$. Setting $\mathcal{N}_M = (\mathcal{O}_{M_s,1})/(\mathcal{O}_{M_s,1})^2$ and $\mathrm{Gr}_i \mathcal{E} = \mathcal{E}(\mathcal{O}_{M_s,1})^i/\mathcal{E}(\mathcal{O}_{M_s,1})^{i+1}$, we have the following formulas for the classes of sheaves in $K_0(M_S)$: $[\mathcal{O}_{M_s}] = \sum_{i \geq 0} [S^i \mathcal{N}_M]$ and $[\mathcal{E}] = [\mathrm{gr}_0 \mathcal{E}] \cdot [\mathcal{O}_{M_s}]$.

Let us consider the SUSY-family $(p_s; C_s \to M_s; \mathcal{T}_0)$. Reducing the nilpotents, we get the family of curves p: C → M. With respect to it we can construct an element $\lambda_1 \in K_0(M)$ (in place of Pic M) as in Sec. 3. Let us set $\lambda_s = \lambda_1 \cdot [\mathcal{O}_{M_s}]$. On the other hand, let us consider the class of the dualizing sheaf $\rho_s = [\mathrm{Ber}\, \Omega^1 M]$ on M_S. Now we can demonstrate our second numerological coincidence.

5. THEOREM. If a_S is an isomorphism, then $\rho_s = \frac{1}{2} k_{NS} \lambda_s = \frac{15}{2} \lambda_s$.

Proof. At first, let us restrict C_S to $i: M \subset M_s$. It is sufficient to verify that $[\mathrm{Ber}\; i^* R^1 p_{s*}(\mathcal{T}_0)] = \lambda_1$. But this class can be computed layerwise. By Sec. 4 and the Serre duality, on a SUSY-curve we have $H^1(\mathcal{T}_0)^* = H^*(\Omega^2) \otimes H^*(\Omega^{1/2})$. Therefore, $\mathrm{Ber}\; i^* R^1 p_{s*}(\mathcal{T}_0)] = \lambda_1 - \lambda_{1/2}$. The Mumford formula is suitable for fractional n also; whence $\lambda_2 - \lambda_{3/2} = 13\lambda_1 - \frac{11}{2}\lambda_1 = \frac{15}{2}\lambda_1$.

6. Remark. a) In the geometrical representation theory of a semisimple group G, the weight lattice is identified with Pic G/B. We have associated the weight 1 on $z \in V$ (or NS) with $\lambda_1/2$ (or $\lambda_S/2$). Is it not possible to associate the second coordinate of the weight (on L_0) with the class, proportional to $\delta = [\mathcal{O}(\Delta)]$, where Δ is a divisor of the singular curves of the family? (The refined Mumford formula: $\lambda_n = \left(12\binom{n}{2} + 1\right)\lambda_1 - 2\binom{n}{2}\delta$.)

b) The appearance, namely, of the dualizing sheaves is consistent with the Feigin duality theorem [6]: $\mathrm{tor}_i^* = \mathrm{tor}_{26-i}$ for the semiinfinite homologies of the Virasoro algebra. More precisely, (h, D) is dual to (1 – h, 26 – D).

c) The realization of all M_g with the help of the Kadomtsev–Petviashvili hierarchy [7] possibly serves as an average between M_∞ and V and its superextension from [8] serves for NS.

I am deeply grateful to A. S. Shvarts, joint work with whom has given rise to this note.

LITERATURE CITED

1. J. H. Schwarz, Phys. Rep., 89, 223-305 (1982).
2. D. Gross, J. Harvey, E. Martinec, and R. Rohm, Heterotic String Theory, Preprint, Princeton (1985).
3. A. M. Polyakov, Phys. Lett., 103B, 207-210 (1981).
4. O. Alvarez, Nucl. Phys., 216, 125-184 (1983).
5. D. Mumford, Ens. Math., 24, 39-110 (1977).
6. B. L. Feigin, Usp. Mat. Nauk, 39, No. 2, 195-196 (1984).
7. T. Shiota, Characterization of Jacobian Varieties in Terms of Soliton Equations, Preprint, Harvard (1984).
8. Yu. I. Manin and A. O. Radul, Commun. Math. Phys., 98, 65-77 (1985).
9. I. B. Penkov, Inv. Math., 71, No. 3, 501-512 (1983).

Volume 172, number 2 PHYSICS LETTERS B 15 May 1986

THE PARTITION FUNCTION OF THE POLYAKOV STRING CAN BE EXPRESSED IN TERMS OF THETA-FUNCTIONS

Yu.I. MANIN

Received 4 February 1986

A formula is given for the Polyakov measure on the space of conformal classes of riemannian surfaces of genus $p \geq 2$ in terms of theta-functions.

The quantum theory of the Polyakov string [1] is defined by a path integral taken over a space of random surfaces in d-dimensional euclidean space. The partition function of the closed string is a "perturbation theory" series $Z = \Sigma_{p \geq 0} Z_p$, where the summand Z_p corresponds to p-loops. After some regularization it takes the form $Z_p = \int_{M_p} d\pi_p$, where M_p is the space of conformal classes of compact riemannian surfaces of genus p (sphere with p handles), and $d\pi_p$ a certain measure on it (see refs. [2,3]). In this note we give an expression for $d\pi_p$ in the critical dimension $d = 26$ (and for a similar fermionic string measure $d\pi_p^f$ in $d = 10$) in terms of classical theta-functions (see ref. [4]). Such an expression, which was suggested recently, resulted from a transformation of determinants of two laplacians into a sum over geodesic lengths with the help of the Selberg trace formula or its "super" analogue [3]. Our approach here is completely different. It is based upon holomorphic, rather than riemannian, geometry. This became possible after the remarkable work of Belavin and Knižnik [5], who demonstrated that $d\pi_p$ is the "modulus-squared" of a holomorphic form on M_p with known singularities at infinity, in accordance with the general philosophy of Quillen on the determinant of $\bar{\partial}$. To calculate $d\pi_p$ we utilize in an essential way the work by Mumford [6] and Faltings [7].

Our main formula for $p \geq 2$ looks as follows (the case $p = 1$ is much simpler and was known since the dual string theory was invented):

$$d\pi_p = c_p K (-\mathrm{i})^{3p-3} W_1 \bar{W}_1 \ldots W_{3p-3} \bar{W}_{3p-3} , \quad (1)$$

$$K = (\det \operatorname{Im} \pi)^{-4} |\theta(0, \tau)|^{-16}$$

$$\times \frac{|\det(\omega_j(R_k))|^{16} \Pi_{j=1}^{p} G(R_j, Q)^{16} \Pi_{j<k} G(P_j, P_k)^4}{|\det(\beta_i(P_j))|^2 |\det(\eta_{j\,a_k}^{(P_k)})|^2 \Pi_{j<k} G(R_j, R_k)^{16}} .$$

Here we denote by W_i the (1,0)-forms on M_p corresponding to a certain base w_i of holomorphic quadratic differentials on the generic riemannian surface X, τ is the matrix of periods of this surface, and θ is its theta-function. Moreover, $\log G$ is the Green kernel of an invariant laplacian on X, and P_j, R_k, Q are certain points on X.

Here are some details (for more, cf. refs. [4,7]). Let X be a riemannian surface, depending on the parameters, varying in a domain of M_p. Make a classical system of cuts a_i, b_i $(i = 1, ..., p)$ on it (turning it into a polygon). Take a base (ϕ_i) of holomorphic one-forms on it with $\int_{a_i} \phi_j = \delta_{ij}$ and set $\tau = (\tau_{ij}) = (\int_{b_j} \phi_i)$, $\omega_i = \Sigma_j (\sqrt{\operatorname{Im} \tau})^{-1}_{ij} \phi_j$. The Riemann theta-function is defined by

$$\theta(z, \tau) = \sum_{m \in Z^p} \exp[2\pi \mathrm{i}(m^t z + \tfrac{1}{2} m^t \tau m)] .$$

We put $\mu = (\mathrm{i}/2p) \Sigma_{j=1}^{p} \omega_j \wedge \bar{\omega}_j$ and introduce an invariant laplacian Δ on X by the formula $\Delta f = -(\mathrm{i}/\pi\mu) \partial\bar{\partial} f$. Now define $G(P, Q)$ by

$$f(P) = -\int_X \log G(P, +) \Delta f(Q) \mu(Q) .$$

G can be expressed in terms of theta and abelian integrals; we omit this calculation. Choose points $P_1, ..., P_{p-1}$ on X in such a way that there exists a holomorphic one-form on X with zeroes of second order at all P_i (this can be done in a finite number of ways). Denote this form by ν^2. Using ν, one can define differentials of half-integer degree on X.

We choose bases of holomorphic differentials on X as follows:

degree 3/2: $(v_1, ..., v_{2p-2})$

$$= (\nu\omega_1, ..., \nu\omega_p; \eta_1, ..., \eta_{p-2}).$$

degree 2: $(w_1, ..., w_{3p-3})$

$$= (\nu^2\omega_i; \nu\eta_j; \beta_1, ..., \beta_{p-1}).$$

For $\omega = f(z)\,dz$ put

$$|\omega(P)| = |f(P)| \lim_{Q\to P} G(P,Q)|z(P) - z(Q)|^{-1}$$

For differentials of degree m put $|\omega(P)| = |(\omega)^{1/m}(P)|^m$. We shall normalize ν by the condition $|\nu^2(P)| = \Pi_{i=1}^{p-1} G(P_i, P)$ for all P. We define local coordinates z_j at points P_j by $\nu^2 = z_j^2\,dz_j$ and put $a_j = [\nu z_j^{-1}(P_j)]^3$. Finally, we choose points $R_1, ..., R_p$ and Q on X, so that there exists a meromorphic one-form on X with zeroes of second order at R_j and a pole of second order at Q. Moreover, we take them in such a way that the divisor class $R_1 + ... + R_p - Q$ is the Riemann class Δ ([ref. [4], p. 7). This finishes the description of all notations in (1), except for c_p, which is a constant depending only on p. Presumably, one can normalize all c_p simultaneously, by looking at the asymptotic behaviour of $d\pi_p$ at a limit when X acquires a double point.

Our measure for the fermionic string has a similar structure:

$$d\pi_p^f = c_p^f L(-i)^{5p-5}$$

$$\times \prod_{j=1}^{3p-3} w_j\bar{w}_j \prod_{j=1}^{2p-2} \partial^2/\partial V_j\, \partial\bar{V}_j\,, \qquad (2)$$

$$L = (\det \operatorname{Im} \tau)^{-5/2} |\theta(0,\tau)|^{-10}$$

$$\times \frac{|\det(\omega_j(R_k))|^{10}\, \Pi_{j=1}^{p} G(R_j, Q)^{10}\, \Pi_{j<k} G(P_j, P_k)^2}{|\det(\beta_j(P_k))|^2\, \Pi_{j<k} G(R_j, R_k)^{10}}. \qquad (2\text{ cont'd})$$

This measure is defined on a complex superspace, whose underlying complex space is M_p and odd local analytic coordinates are the degree 3/2 differentials v_i on X (cf. ref. [3]). In (2), we denoted the dual base of odd vector fields on this M_p^f by $\partial/\partial V_i$. The Berezin volume element in (2) therefore is denoted by an expression of the form $dx\,dy\,\partial_\xi\,\partial_\eta$ rather than $dx\,dy\,d\xi\,d\eta$. The latter is commonly used but unfortunately due to a wrong transformation rule. The rest of the notations in (2) is the same as in (1). Strictly speaking, the formula (2) should be considered as a conjectural one, until an analogue of the Belavin–Knižnik theorem is proved for the fermionic string. Moreover, the superspace M_p^f should be completed by an additional component parametrizing the SUSY-curves with an even structural theta-characteristic (cf. ref. [8]). On this component the measure is slightly different.

We shall give proofs and a discussion in a separate publication.

I would like to thank heartily A. Belavin, A. Beilinson, V. Knižnik and A. Švarc for many helpful discussions.

References

[1] A.M. Polyakov, Phys. Lett. B 103 (1981) 207.
[2] O. Alvarez, Nucl. Phys. B216 (1983) 125.
[3] M.A. Baranov and A.S. Švarc, Pisma Zh. Eksp. Teor. Fiz. 42 (1985) 340.
[4] J.D. Fay, Lecture Notes in Mathematics, Vol. 352 (Springer, Berlin, 1973).
[5] A.A. Belavin and V.G. Knižnik, Phys. Lett. B 168 (1986) 201.
[6] D. Mumford, Ens. Math. 23 (1977) 39.
[7] G. Faltings, Ann. Math. 119 (1984) 387.
[8] Yu.I. Manin, Funkt. Anal. (1986), to be published.

Commun. Math. Phys. 107, 359–376 (1986)

Communications in Mathematical Physics

The Mumford Form and the Polyakov Measure in String Theory

A. A. Beilinson[1] and Yu. I. Manin[2]

[1] State Cardiology Center, Moscow, USSR
[2] Steklov Mathematical Institute, Moscow, USSR

Abstract. An explicit formula is derived for the Mumford form on the moduli space of algebraic curves. According to the Belavin-Knizhnik theorem, this gives a formula for the Polyakov bosonic string measure.

Introduction

1. Polyakov's String

The quantum bosonic Polyakov string theory is defined by a path integral taken over random surfaces in the (Euclidean) d-dimensional space $\mathbb{R}^d$. The partition function of the closed string has a perturbation series expansion $Z=\sum_{g\geqq 0} Z_g$:

$$\begin{aligned} Z_g &= e^{\beta(2-2g)}\int e^{-J(x,\gamma)}Dx\,D\gamma\,, \\ J(x,\gamma) &= \int_N d^2z\sqrt{|\gamma|}\,\gamma^{ab}\partial_a x^\mu \partial_b x^\mu\,. \end{aligned} \tag{1}$$

Here N is a fixed compact oriented surface of genus g (= "loop number"), z^a are local coordinates on it, $x=(x^\mu)$ is a map from N to $\mathbb{R}^d$, $\gamma_{ab}dz^a dz^b$ is a metric on N.

On the space of classical fields (x,γ) a gauge group $C\ltimes D$ acts, leaving the classical action $J(x,\gamma)$ invariant. It is a semidirect product of the diffeomorphism group D of N and of the conformal group C (= real-valued positive functions on N). Using this action, we can reduce (1) to a finite-dimensional integral in the following way. First, the integral over x's is Gaussian, hence it equals $(\det'\Delta_{0\gamma})^{-d/2}$, where $\Delta_{0\gamma}$ is the Laplace operator on the functions on N, corresponding to γ, $\det'$ its determinant without zero modes, regularized, say, by the formula $\det'\Delta_{0\gamma} = \exp(-\zeta'_\Delta(0))$. The remaining integration over the space $\operatorname{Met} N$ of γ's then reduces, via the Faddeev-Popov trick, to an integral over $\operatorname{Met} N/C\ltimes D$, which is the same as the moduli space of Riemannian surfaces (or complex algebraic curves) of genus g. As is well known, this moduli space M_g is a complex variety of complex dimension 0 for $g=0$, 1 for $g=1$, $3g-3$ for $g\geqq 2$.

Strictly speaking, this reduction is spoiled by the conformal anomaly: the regularization breaks the conformal invariance of the quantum effective action. But the anomaly cancels in the critical dimension $d=26$ (cf. [2, 3]), and one has

$$Z_g = \text{const} \int_{M_g} dv \det{}' \Delta_{2\hat{\gamma}} (\det{}' \Delta_{0\hat{\gamma}})^{-13}, \quad g \geqq 2, \tag{2}$$

where dv is the Petersson-Weil measure on the moduli space, $\Delta_{2\hat{\gamma}}$ is a Laplace operator on quadratic differentials, $\hat{\gamma}$ is a metric of constant curvature -1 in a given conformal class (for $g=1$ the formula differs slightly).

2. Results

The main result of this paper (Theorem 6, Sect. 1) is an explicit formula for the Polyakov measure $d\pi_g$ on M_g, appearing in the right-hand side of (2). Here "explicit" means "written in terms of the complex geometry of the surface itself" and not in terms of it spectral invariants. Note that certain explicit formulae for the Laplace operator determinants were given by Ray and Singer [20]. In several papers, including [5, 21] they were applied to the string measure case. In this approach, $\log \det{}' \Delta$ is transformed into a sum over lengths of closed geodesics by means of the Selberg trace formula.

The formula of our paper looks very differently and is of a different nature since it utilizes the holomorphic invariants and not the metric ones.

One may consider this formula as a generalization to the genus $g \geqq 2$ case of the well known genus 1 result:

$$\begin{aligned} Z_1 &= \int_{M_1'} \frac{i}{2} d\tau \wedge d\bar{\tau} \frac{1}{(\operatorname{Im} \tau)^{14} |\Delta(\tau)|^2}, \\ \Delta(\tau) &= e^{2\pi i \tau} \prod_{n=1}^{\infty} (1 - e^{2\pi i n \tau})^{24}, \\ M_1' &= \{\tau \mid |\tau| \geqq 1, |\operatorname{Re} \tau| \leqq \tfrac{1}{2}, \operatorname{Im} \tau > 0\}. \end{aligned} \tag{3}$$

Observe that (3) may be expressed in terms of the theta function,

$$\theta(z, \tau) = \sum_{n=-\infty}^{\infty} e^{2\pi i n z + \pi i n^2 \tau},$$

since

$$2^8 \Delta(\tau) = e^{2\pi i \tau} (\theta(0, \tau) \theta(1/2, \tau) \theta(\tau/2, \tau))^8.$$

3. How to Compute Polyakov's Measure

The computation presented here is based on the results of Mumford [8] and of Belavin and Knizhnik [7]. Here is a brief summary of their work.

The moduli space M_g is a complex algebraic variety. It has no convenient global coordinates. However, holomorphic 1-forms on M_g admit the following nice description: they correspond to the quadratic differentials on the complex curves, parametrized by the points of M_g. Let $w_1, \ldots, w_{3g-3}$ (here $g \geqq 2$) be a basis of such differentials (depending holomorphically on parameters from M_g), and let

$\omega_1, \ldots, \omega_g$ be a basis of differentials of the first kind. Mumford [8] shows, that there exists a unique, up to a constant multiple, holomorphic function F (depending on the choice of w_i, ω_j), such that

$$\mu_g = F \cdot \frac{w_1 \wedge \ldots \wedge w_{3g-3}}{(\omega_1 \wedge \ldots \wedge \omega_g)^{13}} \tag{4}$$

is a global section of the appropriate line bundle on M_g. This section is meromorphic at infinity (i.e., it is algebraic) and has a pole of order two; moreover, it has no zeroes on M_g. (Actually these properties are automatic for $g \geqq 3$, and for $g=2$ they should be imposed additionally.)

We shall call μ_g a Mumford form.

Comparing this with (2), one is lead to believe that "13" in Mumford's formula coincides with the half of the critical dimension. This was suggested by Yu. I. Manin and supported by certain evidence from the operator quantization approach.

In a remarkable paper [7] Belavin and Knizhnik pushed this much further. Namely, they proved that the Polyakov measure coincides with the modulus squared of the Mumford form. More precisely, let W_i be a 1-form on M_g, corresponding to w_i, then one has

$$d\pi_g = \text{const}\,|F|^2 \cdot (-i)^g \frac{W_1 \wedge \bar{W}_1 \ldots W_{3g-3} \wedge \bar{W}_{3g-3}}{|\det \int \omega_i \wedge \bar{\omega}_j|^{13}}. \tag{5}$$

Here bar means complex conjugate, and the integral is taken over the Riemann surface on which ω_i's are defined.

Hence to compute $d\pi_g$ it suffices to describe explicitly the Mumford form, a priori defined only by the implicit global conditions. This is the task we concentrate upon in what follows.

Note that in [14] $d\pi_g$ already was calculated in this way. This was done with the help of Faltings' paper [9], which developed in turn the ideas of Arakelov, namely the Noether theorem for arithmetic surfaces. The formula presented below is shorter than the one in [14]. We hope also that its derivation clarifies somewhat the arguments of Faltings.

The expression for the curvature form of the determinant of the Laplace operator on j-differentials, found by Belavin and Knizhnik and directly implying (5), appears to be a particular case of the similar formula for the determinants of Dirac operators on arbitrary compact manifolds, due to Bismut and Freed [17] (this was pointed out in [18]). The Bismut-Freed result (generalizing the earlier theorem of Quillen) in the context of complex Hermitian geometry may be viewed as an exact Riemann-Roch-Grothendieck formula for c_1, valid on the level of forms, and not on the cohomology level only. A similar formula for higher c_i remains to be found. We are sure it will help to understand both anomalies in physics and arithmetic geometry, in the spirit advocated in [19].

4. Superstrings

The Polyakov string has a fermion analog, described in [2, 4]. The critical dimension for the superstring is $d=10$. A computation of an analogue of $d\pi_g$ on the corresponding moduli superspaces was done in [6] by means of a supervariant of the Selberg trace formula. However, the relation with the right Polyakov measure seems not quite clear, due to two circumstances. a. The superversion of the Belavin-Knizhnik theorem is not available at the moment. b. One does not know how to sum up a Mumford superform over different spin structures before taking the modulus squared. We discuss this question briefly in Sect. 3.2.

5. The Contents of the Paper

Our basic result is, after some preliminaries, stated in Sect. 1.6. Its proof is presented in Sect. 2. The third section contains some remarks and complements.

Finally, the appendix deals with the following problem. Mumford forms μ_g are defined a priori only modulo multiplication by a constant, depending on the genus. We shall explain how one can normalize all μ_g's canonically and simultaneously. Conjecturally, one should have $e^{\beta(2-2g)}|\mu_g|^2 = d\pi_g$ with this normalization.

1. Notation and Statement of Results

1. Calculus on Riemann Surface

Let X be a compact complex Riemann surface of genus $g \geqq 1$, or, equivalently, a smooth projective algebraic curve over $\mathbb{C}$. Marking of X is a choice of a symplectic basis $(a_1, \ldots, a_g; b_1, \ldots, b_g)$ in $H_1(X, \mathbb{Z})$. This means that $(a_i, a_j) = (b_i, b_j) = 0$, $(a_i, b_j) = \delta_{ij}$. One can construct such a basis making a classical system of cuts, which turns X into a polygon with pairwise identified sides (see e.g. [12]).

The space of holomorphic 1-forms (or differentials of the first kind) on X has the Riemann basis $\omega_1, \ldots, \omega_g$ uniquely determined by the conditions $\int_{a_i} \omega_j = \delta_{ij}$, $i, j = 1, \ldots, g$. Put $\int_{b_i} \omega_j = \tau_{ij}$, $\tau = (\tau_{ij})$. One has $\tau_{ij} = \tau_{ji}$, $\operatorname{Im} \tau > 0$, hence τ lies in the Siegel upper halfspace H_g. The curve X together with its marking can be reconstructed from τ up to unique isomorphism.

Let $T \subset \mathbb{C}^g$ be a lattice, generated by columns of τ and $\mathbb{Z}^g$. The Jacobian of X is the complex torus $J = \mathbb{C}^g/T$. For each $P_0 \in X$, $z_0 \in J$ there is a standard embedding $\psi: X \to J$, mapping P_0 to z_0, defined by the formula $\psi(P) = \left(z_0 + \int_{P_0}^{P} \omega\right) \bmod T$, $\omega = (\omega_i)$. If z_i is the ith coordinate on $\mathbb{C}^g$, the T-invariant form dz_i on $\mathbb{C}^g$ may be viewed as a holomorphic (translation invariant) form on J. Any standard embedding $\psi: X \to J$ induces the same isomorphism between the space of such forms on J and the space of the differentials of the first kind on $X: \psi^*(dz_i) = \omega_i$.

The Jacobian J classifies the divisor classes of degree zero on X, or, equivalently, the isomorphism classes of invertible sheaves L on X with $c_1(L) = 0$. Namely, a sheaf $O_X(\sum a_i P_i)$ corresponds to a point $\sum a_i \psi(P_i)$, where ψ is any standard embedding.

2. Theta Function

The function $\theta: \mathbb{C}^g \times H_g \to \mathbb{C}$ is defined by the classical series

$$\theta(z, \tau) = \sum_{m \in \mathbb{Z}^g} \exp 2\pi i (m^t z + 1/2 m^t \tau m).$$

It has a kind of periodic behavior:

$$\theta(z+m,\tau)=\theta(z,\tau),$$

$$\theta(z+\tau n,\tau)=\theta(z,\tau)\exp 2\pi i(-n^t z-1/2m^t\tau n).$$

In particular, the zero divisor of $\theta(z,\tau)$ (as a function of z; τ being fixed) is T-invariant and hence comes from a divisor Θ on J.

For $\alpha=\varepsilon+\tau\delta\in\mathbb{C}^g$, put

$$\theta[\alpha](z,\tau)=\exp 2\pi i(\delta^t(z+\varepsilon)+1/2\delta^t\tau\delta)\theta(z+\varepsilon+\tau\delta,\tau).$$

If $\varepsilon,\delta\in 1/2\mathbb{Z}^g$, then the class $\alpha \bmod T\in J$ is called theta characteristics. The parity of such α is $4\varepsilon\delta \bmod 2$. The function $\theta[\alpha](z,\tau)$ is even or odd with respect to z according to the parity of α. The theta characteristics are the second order points on J. Their total number is 4^g, of which there are $2^g(2^g-1)$ even ones and $2^g(2^g+1)$ odd ones.

3. *Moduli of Curves and Moduli of Abelian Varieties*

The complex tori, corresponding to the points $\tau\in H_g$, are algebraic. They are called abelian varieties. The Siegel halfspace H_g of dimension $g(g+1)/2$ parametrizes the pairs $(A,(a_i,b_j))$, where A is a principally polarized abelian variety and (a_i,b_j) is a symplectic basis in $H_1(A,\mathbb{Z})$. Those pairs that come from (Jacobians of) curves from a closed analytic subvariety $N_g\subset H_g$ of dimension 1 for $g=1$, $3g-3$ for $g\geqq 2$. Hence for $g\geqq 4$ one has $\dim N_g<\dim H_g$.

To the same curve there corresponds many points of N_g, since we can change the marking. Such a change transforms τ by an element

$$\begin{pmatrix} A & B\\ C & D\end{pmatrix}\in \mathrm{Sp}(2g,\mathbb{Z}):\tau\mapsto(A\tau+B)(C\tau+D)^{-1}.$$

This action of $\mathrm{Sp}(2g,\mathbb{Z})$ on H_g, N_g is discrete. The space $\mathrm{Sp}(2g,\mathbb{Z})\backslash N_g=M_g$ is called the (coarse) moduli space of curves of genus g, and $\mathrm{Sp}(2g,\mathbb{Z})\backslash H_g=A_g$ is the (coarse) moduli space of (principally polarized) abelian varieties.

The functions τ_{ij}, $i\geqq j$, are global holomorphic coordinates on H_g. Locally they are coordinates at any smooth point of A_g (or a covering space of A_g). And, although the the functions τ_{ij} being restricted to N_g, become dependent (for $g\geqq 4$), at any point of N_g corresponding to a nonhyperelliptic curve X_x, one may choose $3g-3$ of τ_{ij}'s that form local coordinates at x on N_g. To be more precise, let k be the Kodaira-Spencer map identifying the quadratic differentials on X_x with the fiber of $\Omega^1 N_g$ at x. Then one has $k(\omega_a\omega_b)=(2\pi i)^{-1}d\tau_{ab}$. Below we shall describe the Mumford form in terms of (ω_a), $(\omega_a\omega_b)$. Processing it into the Polyakov measure one should replace $\omega_a\omega_b$ by $(2\pi i)^{-1}d\tau_{ab}|_{N_g}$.

4. *Theta Characteristics and Differentials of Half-Integral Weight*

For any integer d denote by J_d the variety, parametrizing the isomorphism classes of the invertible sheaves of degree d on X. This is a principal homogeneous space over J. For almost all sheaves L of degree $g-1$ one has $h^0(L)=h^1(L)=0$. More precisely, the sheaves with $h^0(L)>0$ form a divisor $\Theta_{g-1}\subset J_{g-1}$. According to the

Riemann theorem, there exists a unique isomorphism $J \to J_{g-1}$, compatible with J-action, which maps Θ to Θ_{g-1}. It identifies $0 \in J$ with the Riemann point $\Delta \in J_{g-1}$. One has $2\Delta =$ class of Ω^1 in J_{2g-2}, where Ω^1 is the sheaf of 1-forms on X (see [12, 15]). Note that Δ depends on the marking of X. In this way the set of theta characteristics transforms into the set of square roots of Ω^1. Therefore below we shall indiscriminately call "theta characteristics" a point in J, or in J_{g-1} (a half of Ω^1), or else a sheaf L_α together with an isomorphism $L_\alpha^{\otimes 2} \xrightarrow{\sim} \Omega^1$.

For any curve X there exists such an odd α (the "typical" odd one), that $h^0(L_\alpha)=1$. We choose one and fix it. We put $\Omega^{1/2}=L_\alpha$ and $\Omega^i = L_\alpha^{\otimes 2i}$ for $i \in 1/2\mathbb{Z}$.

5. Distinguished Bases

Using $\Omega^{1/2}$ we shall define now certain bases in the spaces of differentials and quadratic differentials on X. We assume that $g \geqq 2$ and that X is not hyperelliptic, if $g \geqq 3$.

a) The basis $(\varphi_0, \varphi_1, \ldots, \varphi_{g-1})$ in $\Omega^1(X)$. Put

$$\varphi_0 = \sum_{j=1}^{g} (\partial/\partial z_j \theta[\alpha])(0,\tau)\omega_j. \tag{1}$$

This differential has double zeroes on X. Assume that $\operatorname{div}\varphi_0 = \sum_{i=1}^{g-1} 2P_i$ with pairwise different P_i (this is true for an open dense subset in N_g). We have $\varphi_0 = v_\alpha^2$, where v_α is a certain 1/2-differential.

Choose local parameters t_j at P_j in such a way that $\varphi_0 = t_j^2 dt_j$. Clearly, $t_j(P) = \left(3\int_{P_j}^{P} \varphi_0\right)^{1/3}$, so that t_j are defined up to multiplication by $\sqrt[3]{1}$.

Now the conditions

$$\varphi_j = (\delta_{jk} + a_{jk}t_k)dt_k + O(t_k^3 dt_k) \text{ near } P_k;\ j,k=1,\ldots,g-1 \tag{2}$$

define the differentials φ_j uniquely. Put moreover

$$B = (B_{ij}) = \left(\int_{a_j} \varphi_{i-1}\right), \quad i,j=1,\ldots,g.$$

Clearly, $(\varphi)^t = B(\omega)^t$.

b) The basis $(w_1, \ldots, w_{3g-3})$ in $\Omega^2(X)$. For $g=2$ or for non-hyperelliptic X $(g \geqq 3)$ put

$$(w_1, \ldots, w_{3g-3}) = (\varphi_0^2, \varphi_0\varphi_1, \ldots, \varphi_0\varphi_{g-1}; \varphi_1^2, \ldots, \varphi_{g-1}^2; \\ a_{1,g-1}^{-1}\varphi_1\varphi_{g-1}, \ldots, a_{g-2,g-1}^{-1}\varphi_{g-2}\varphi_{g-1}).$$

The last group of these differentials is defined on an open dense subset of N_g, where all $a_{i,g-1}$ are invertible. Off this subset one should change this choice as explained in Sect. 2. The only important thing is that

$$w_{2g-1+j} = \delta_{jk}t_k(dt_k)^2 + O(t_k^2(dt_k)^2) \text{ near } P_k, \quad j,k=1,\ldots,g-2,$$

and also a possibility to explicitly present w_i as bilinear combinations of $\omega_a\omega_b$ for computation of the map k.

We can now write down formulae for μ_g and $d\pi_g$.

6.

Theorem. a) *For $g \geqq 2$ the Mumford form is*

$$\mu_g = \text{const}\,(\det B)^4 \frac{w_1 \wedge \ldots \wedge w_{3g-3}}{(\varphi_0 \wedge \ldots \wedge \varphi_g)^{13}}.$$

b) *In the same conditions the Polyakov measure is*

$$d\pi_g = \text{const}\,|\det B|^{-18} (\det \operatorname{Im} \tau)^{-13} W_1 \wedge \bar{W}_1 \wedge \ldots \wedge W_{3g-3} \wedge \bar{W}_{3g-3},$$

where $W_i = k(w_i)$, $k(\omega_a \omega_b) = (2\pi i)^{-1} d\tau_{ab}$.

2. Proofs

1. $\det R\pi_*$ *formalism*

Let $\pi: X \to S$ be an algebraic family of projective varieties X_s, parametrized by points $s \in S$, or, as an algebraic geometer would say, just a projective flat morphism. Consider an (algebraic coherent) sheaf L on X, flat over S; roughly speaking, this is a family of sheaves L_s on the fibers of π. If for a certain i the dimensions of the cohomology groups $H^i(X_s, L_s)$ do not depend on s, then they are fibers of the higher direct image sheaf $R^i\pi_* L$, and this sheaf is locally free. If this holds for all i, we may define a "multiplicative Euler characteristic"

$$d(L) := \bigotimes_i (\det R^i \pi_* L)^{(-1)^i} := \det R\pi_* L,$$

which is an invertible sheaf on S.

Knudsen and Mumford in [10] have shown how to define $d(L)$ with nice properties for any L flat over S, without assuming that $\dim H^i(X_s, L_s)$ does not jump. To be more precise, they proved the following result.

2.

Proposition. *For any family of projective varieties $\pi: X \to S$, any flat sheaf L on X and any isomorphism of sheaves $f: L \to L'$ one can construct an invertible sheaf $d(L)$ on S and an isomorphism $d(f): d(L) \to d(L')$ in such a way that d becomes a functor with the following properties. (Below equalities mean canonical isomorphisms.)*

a) *$d(L) = \det R\pi_* L$, if all $R^i\pi_* L$ are locally free.*

b) *$d(L)$ is compatible with base change ($=$change of parameter space S).*

c) *Let $E = (E^i, d^i)$ be a finite complex of locally free sheaves on S, whose cohomology is $R^i\pi_* L$ universally (i.e. after any base change). (See [13] for an explanation of this Grothendieck's construction). Then $d(L) = \bigotimes_i (\det E^i)^{(-1)^i}$.*

d) *For any exact triple $0 \to L' \xrightarrow{i} L \xrightarrow{j} L'' \to 0$ one has a canonical isomorphism $d(L) = d(L') \otimes d(L'')$, compatible with exact triples of exact triples and base change.*

Note that if $a: L \to L$ is multiplication by $a \in \mathbb{C}^*$, then $d(a)$ is multiplication by $a^{\chi(L)}$, where $\chi(L) = \sum (-1)^i rk R^i \pi_* L$. This explains the name "multiplicative Euler characteristics."

We refer to [10] for further details. In concrete computations below we shall not use much more than the following particular case. Assume that $R^i\pi_*L=0$ for $i\geqq 1$ and let $l=(l_1,\ldots,l_a)$ denote some free generators of $R^0\pi_*L$. Then $d(L)$ is freely generated by

$$d(l)=l_1\wedge\ldots\wedge l_a\in\det R^0\pi_*L.$$

If the same holds for the terms L, L', L'' of an exact triple, as in d) above, and if the bases are chosen in such a way that

$$l'=(l'_1,\ldots,l'_a),\qquad l=(i(l'_1,\ldots,l'_a);\, l_{a+1},\ldots,l_b),\qquad l''=j(l_{a+1},\ldots,l_b),$$

then, under the canonical isomorphism $d(L)=d(L')\otimes d(L'')$ we have $d(l)=d(l')\otimes d(l'')$.

3. The Sheaves λ_i

Now let $\pi: X\to S$ be a flat family of smooth projective curves, $g=$ genus, $\Omega=\Omega^1X/S$ the sheaf of relative 1-forms. Put $\lambda_i=d(\Omega^{\otimes i})$. We have $\lambda_i=\lambda_{1-i}$ in a canonical way [since, by Serre's duality, $d(L)=d(L^{-1}\otimes\Omega)$ for any invertible L].

The sheaves λ_i play a crucial role in the Polyakov string measure theory due to the following facts.

a) Mumford's theorem [8]: there exists a universal isomorphism $\lambda_{i+1}=\lambda_1^{6i^2+6i+1}$.

This means that we have such an isomorphism for any family of curves, and we can normalize them in a way compatible with base changes. Such a universal isomorphism is unique, up to multiplication by a constant, depending on g only, for $g\geqq 3$ (for $g=2$ the uniqueness also holds if certain assumptions on the behavior at infinity are added).

The uniqueness for $g\geqq 3$ follows from the fact that every holomorphic function on M_g is constant. To see this, consider the closure of M_g in the Satake compactification of A_g. A holomorphic function can be extended from M_g to this closure by Hartogs' theorem, since at infinity lies a subset of codimension $\geqq 2$. But the closure is compact, hence a holomorphic function on it is constant.

In the appendix we describe a canonical normalization for all g.

b) Theorem on modular families: if $\pi: X\to S$ is a locally universal family, then the Kodaira-Spencer map $R^0\pi_*\Omega^{\otimes 2}X/S\to\Omega^1S$ is an isomorphism. Passing to the determinants we get $\lambda_2=\Omega^{3g-3}S$.

The combination of a) for $i=1$ and b) gives us a universal section $\mu_g\in\Omega^{3g-3}S\otimes\lambda_1^{-13}$, called earlier the Mumford form.

c) The Belavin-Knizhnik theorem: Polyakov's measure equals $\mathrm{const}|\mu_g|^2$ in the sense explained in the Introduction.

Below we shall reprove the Mumford theorem by a method that does not use the Riemann-Roch-Grothendieck global type arguments, and allows us to trace out the behaviour of convenient generators in λ_i.

Essentially this type of argument can be found in the Faltings paper [9].

Before proceeding further we need one more algebraic-geometrical construction.

4. The Sheaves $\langle L, M\rangle$

Let X be a smooth projective curve; f, g meromorphic functions on it with disjoint divisors. Put

$$\langle f, g\rangle = f(\operatorname{div} g) = \prod_{x\in X} f(x)^{v_x(g)},$$

where $v_x(g)$ is the order of g at a point x. According to A. Weil, $\langle f, g\rangle = \langle g, f\rangle$: this generalizes the classical symmetry of the projective cross-ratio [look at the case $X = P^1$, $f = (z-a)(z-b)^{-1}$, $g = (z-b)(z-c)^{-1}$].

Assume now, that L, M are invertible sheaves on X, and s, t are their meromorphic sections with disjoint divisors. We can still define $\langle s, t\rangle = \prod_x s(x)^{v_x(t)}$ as an element of one dimensional vector space $\langle L, M\rangle_{s,t} = \bigotimes_x L_x^{\otimes v_x(t)}$, instead of the base field of complex numbers. But it appears, essentially due to the Weil symmetry, that these vector spaces admit a lot of canonical identifications. In particular, for a different couple of sections s', t' of L, M one has a canonical isomorphism $\langle L, M\rangle_{s,t} = \langle L, M\rangle_{s',t'}$. Hence we may omit the index s, t and deal with $\langle L, M\rangle$ depending on L, M only. If X, L, M depend on parameters S, then L, M form a line bundle or an invertible sheaf on S.

This formalism was developed by Deligne in [11]; here is a brief summary.

5.

Proposition. *Let $\pi: X\to S$ be a flat family of smooth projective curves. Then for each pair of invertible sheaves L, M on X and each pair of isomorphisms $\varphi: L\to L'$, $\psi: M\to M'$, one can construct an invertible sheaf $\langle L, M\rangle$ on S and an isomorphism $\langle\varphi,\psi\rangle: \langle L, M\rangle\to\langle L', M'\rangle$ with the following properties.*

a) *This construction is a bimultiplicative symmetric bifunctor, i.e. there are natural identifications*

$$\langle L, O_X\rangle = O_S, \quad \langle L_1\otimes L_2, M\rangle = \langle L_1, M\rangle\otimes\langle L_2, M\rangle,$$
$$\langle L, M\rangle = \langle M, L\rangle, \quad \langle L, M^{-1}\rangle = \langle L, M\rangle^{-1}.$$

b) *For each pair of meromorphic sections s of L, t of M with disjoint divisors flat over S, an invertible section $\langle s, t\rangle$ of $\langle L, M\rangle$ can be defined in such a way that with respect to the identifications in* a) *we have*

$$\langle s, 1\rangle = 1, \quad \langle s_1\otimes s_2, t\rangle = \langle s_1, t\rangle\otimes\langle s_2, t\rangle,$$
$$\langle s, t\rangle = (-1)^{\deg L \deg M}\langle t, s\rangle, \quad \langle s, t^{-1}\rangle = \langle s, t\rangle^{-1}.$$

Moreover, $\langle s, f\rangle = f(\operatorname{div} s)$, if f is a meromorphic function.

c) *Let $M = 0_X(D)$, where D is a relative positive divisor with structure sheaf 0_D, locally free over S. Then*

$$\langle L, 0_X(D)\rangle = \det_{0_S}(L\otimes 0_D)\otimes(\det_{0_S} 0_D)^{-1} = d(L\otimes 0_D)\otimes d(0_D)^{-1},$$

$$\langle s, t_D\rangle = \textit{determinant of the } 0_S\textit{-morphism } 0_D\to L\otimes 0_D, \textit{ sending } 1 \textit{ to } s\otimes 1.$$

Here t_D is a canonical section of $0_X(D)$, equation of D.

This data is compatible with arbitrary base change and is determined in a unique way, up to a unique isomorphism.

A relation between $d(L)$ and $\langle L, M\rangle$ is established by following Deligne's result:

6.

Lemma. *There is a canonical isomorphism*

$$\langle L, M\rangle = d(L\otimes M)\otimes d(0_X)\otimes d(L)^{-1}\otimes d(M)^{-1}.$$

The essence of this lemma is that the functor d behaves in a (non-homogeneous) quadratic way with respect to the tensor product, and $\langle\,,\rangle$ is the associated bimultiplicative functor.

For our purposes we need an explicit construction of the last isomorphism. Take $M = 0_X(-D)$. Then the exact triples

$$0\to M \xrightarrow{t_D} 0_X \to 0_D \to 0,$$

$$0\to L\otimes M \to L \to L\otimes 0_D \to 0$$

together with Proposition 2d give

$$d(0_D) = d(0_X)\otimes d(M)^{-1},$$

$$d(L\otimes 0_D) = d(L)\otimes d(L\otimes M)^{-1}.$$

Hence, by Proposition 5c we get

$$\langle L, M\rangle = \langle L, M^{-1}\rangle^{-1} = d(0_D)\otimes d(L\otimes 0_D)^{-1}$$
$$= d(L\otimes M)\otimes d(0_X)\otimes d(L)^{-1}\otimes d(M)^{-1}.$$

7. Proof of Mumford's Theorem

We shall proceed in three steps.

a) $$\lambda_{i+1} = \lambda_1\langle \Omega, \Omega\rangle^{i(i+1)/2}.$$

In fact, according to Lemma 6 and using $d(0_X) = \lambda_1$, we have

$$\langle \Omega, \Omega\rangle^i = \langle \Omega^i, \Omega\rangle = d(\Omega^{i+1})d(0_X)d(\Omega^i)^{-1}d(\Omega)^{-1} = \lambda_{i+1}\lambda_i^{-1}.$$

Now induction by i shows what we want. (From now on we sometimes omit $\otimes$ in notation.)

b) Assume that on the family $\pi: X\to S$ a relative theta characteristic is given, utilizing which we define $\lambda_i = d(\Omega^{1/2\otimes 2i})$ for half-integral i's. Then

$$\langle \Omega, \Omega\rangle = \lambda_1^8\lambda_{1/2}^{-8}.$$

In fact,

$$\langle \Omega, \Omega\rangle = \langle \Omega^{1/2}, \Omega^{1/2}\rangle^4$$

and, by Lemma 6,

$$\langle \Omega^{1/2}, \Omega^{1/2}\rangle = d(\Omega)d(0_X)d(\Omega^{1/2})^{-2} = \lambda_1^2\lambda_{1/2}^{-2}.$$

c) Under the same assumptions, $\lambda_1^4\lambda_{1/2}^8 = 0_S$. Unfortunately, we could not find a proof local by S and are bound to reproduce a global and analytic argument due to Faltings.[1] It consists in a reinterpretation of λ_1 and $\lambda_{1/2}$ in terms of Jacobian and in the subsequent proof of the corresponding fact for a universal family of all abelian varieties.

To do the first reduction, denote by $\Theta_{g-1} \subset J_{g-1}$ the relative theta divisor of the family and let $\alpha: S \to J_{g-1}$ be the relative theta characteristics, corresponding to $\Omega^{1/2}$. Finally, let $e: S \to J$ be the identity section. Then

$$\lambda_{1/2} = \alpha^* O(-\Theta_{g-1}), \qquad \lambda_1 = e^*(\Omega^g J/S).$$

For the second step recall a classical result on the behaviour of $\theta(z,\tau)$ under the simultaneous action of $\mathrm{Sp}(2g, \mathbb{Z})$ upon z and τ. The group $\mathrm{Sp}(2g, \mathbb{Z})$ consists of integral matrices $\begin{pmatrix} A & B \\ C & D \end{pmatrix}$ conserving the form $a(x_1, x_2; y_1, y_2) = x_1^t y_2 - x_2^t y_1$; $x_i, y_j \in \mathbb{Z}^g$. Let $\Gamma_{1,2}$ be the subgroup, conserving also $x_1^t x_2 \bmod 2$ (this means that the diagonal elements of A^tC, B^tD are even).

Then for all $T = \begin{pmatrix} A & B \\ C & D \end{pmatrix} \in \Gamma_{1,2}$ one has

$$\begin{aligned} &\theta(((C\tau + D)^t)^{-1} z, (A\tau + B)(C\tau + D)^{-1}) \\ &= \zeta \det(C\tau + D)^{1/2} \exp(iz^t(C\tau + D)^{-1} Cz)\theta(z, \tau), \end{aligned}$$

where ζ is a 8th root of unity, depending on T.

The proof, given in [12], consists of two steps. First, one checks this relation for a simple generator system of $\Gamma_{1,2}$; to do this, one needs only periodicity and Poisson formula. Second, one establishes a group property of the transformation formula.

Namely, the transformation formula is equivalent to the $\Gamma_{1,2}$-invariance of certain expressions involving theta constants, in particular

$$\exp(2\pi i n^t \tau n)\theta^8(z - 1/2(m + \tau n), \tau)(dz_1 \wedge \ldots \wedge dz_g)^4|_{z=0}. \tag{3}$$

This invariance can be restated in geometric terms.

Let τ vary in the Siegel half-space H_g. Instead of $A_g = \mathrm{Sp}(2g, \mathbb{Z})\backslash H_g$ consider a space A_g', parametrizing pairs (abelian variety, degree one symmetric ample line bundle). Every such pair essentially is $(A, O(-\Theta_\alpha))$, where α is a point of second order on A. Let $\mathscr{A} \to A_g'$ be the universal family of such pairs, $e: A_g' \to \mathscr{A}$ its identity section. Then the expression (3) defines a canonical invertible section of the sheaf

$$e^*(O(-\Theta_\alpha)^8 \otimes (\Omega^g \mathscr{A}/A_g')^4),$$

i.e. an isomorphism of this sheaf with $0_{A_g'}$.

Restricting this isomorphism to M_g', parametrizing pairs $(J, O(-\Theta_\alpha))$, we get finally $\lambda_{1/2}^8 \lambda_1^4 = 0_{M_g'}$.

d) End of proof.

Putting together a), b), c), we obtain

$$\lambda_{i+1} = \lambda_1 \langle \Omega, \Omega \rangle^{i(i+1)/2} = \lambda_1(\lambda_1^{12})^{i(i+1)/2} = \lambda_1^{6i^2+6i+1}.$$

[1] A similar argument may also be found in [7]

The independence of this isomorphism of the choice of $\Omega^{1/2}$ follows for $g \geqq 3$ from global topological considerations. For a precise statement see the appendix.

8. Distinguished Bases

The only advantage of this proof with respect to Mumford's original one (Riemann-Roch-Grothendieck plus global topological arguments) is that it gives rise to explicit formulae. To derive them, we shall now construct distinguished bases in the sheaves $R^k\pi_*\Omega^{i/2}$, hence in $\lambda_{i/2}$, compatible with various canonical identifications. In this subsection we deal with the family $\pi: X \to M_g^-$, where M_g^- is a subspace of M_g', parametrizing pairs (curve, odd theta characteristics L_α with $h^0(L_\alpha)=1$. The base M_g^- projects onto M_g; $\Omega^{1/2}$ is, by definition, L_α.

a) A basis in $\lambda_{1/2}$. The section $v_\alpha = \varphi_0^{1/2}$, defined by (1), Sect. 1, generates $\pi_*\Omega^{1/2}$. Serre's duality identifies $(R^1\pi_*\Omega^{1/2})^{-1}$ with $\pi_*\Omega^{1/2}$. Hence $\lambda_{1/2} = (\pi_*\Omega^{1/2})^2$ acquires a generator $d_{1/2} = v_\alpha^2$.

b) A basis in λ_1. Choose a basis $(\varphi_0, \ldots, \varphi_{g-1})$ in $\pi_*\Omega^1$ according to (2), Sect. 1. Identify $R^1\pi_*\Omega^1$ with $0_{M_g^-}$ by the Serre duality. The distinguished generator of λ_1 is $d_1 = \varphi_0 \wedge \ldots \wedge \varphi_{g-1}$.

Recall that B is a matrix, transforming ω into φ (Sect. 1, N°. 5).

We assert now that the isomorphism $0_{M_g^-} = \lambda_{1/2}^8\lambda_1^4$ maps the unit section of $0_{M_g^-}$ to

$$(v_\alpha^2)^8(\omega_1 \wedge \ldots \wedge \omega_g)^4 = d_{1/2}^8 d_1^4 (\det B)^{-4}. \tag{4}$$

To see this, denote by $p: X \times J_{g-1} \to J_{g-1}$ the projection, by L the universal sheaf of degree $g-1$ on $X \times J_{g-1}$. Then $d(L)$, constructed with respect to p, is canonically isomorphic to $O(-\Theta_{g-1})$. In fact, on the complement $J_{g-1} \backslash \Theta_{g-1}$ we have $d(L) = 0_J$, and the unit section t of this sheaf has a pole of the first order at Θ_{g-1}. At a simple point $\alpha \in \Theta_{g-1}$ this isomorphism is given by the formula $t_\alpha \mapsto (ss')_\alpha$, where $s \in H^0(X, L_\alpha)$, $s' \in H^1(X, L_\alpha)^* = H^0(X, \Omega^1 \otimes L^{-1})$ are such that $(ss')_\alpha = (dt)_\alpha$. In particular, if α is a theta characteristic, t_α goes into $(\sqrt{dt_\alpha})^2$. But this is precisely v_α^2 from (1), Sect. 1.

c) A basis in $\lambda_{3/2}$. The sheaf $\pi_*\Omega^{3/2}$ has the distinguished basis

$$(v_\alpha\varphi_0, v_\alpha\varphi_1, \ldots, v_\alpha\varphi_{g-1}; \psi_1, \ldots, \psi_{g-2}),$$

where ψ_j fulfill the conditions

$$\psi_j = \delta_{jk}(dt_k)^{3/2} + O(t_k(dt_k)^{3/2}) \text{ near } P_k, \tag{5}$$

$j, k = 1, \ldots, g-2$. Observe that the principal part of ψ_j near P_{g-1} is determined from the equation $\sum_1^{g-1} \operatorname{res}_{P_k}(v_\alpha^{-1}\psi_j) = 0$. Clearly (5) determines ψ_j up to differentials vanishing at all P_k, but one normalizes ψ_j in a unique way, as in (2), Sect. 1, if one demands the vanishing of the term $v_\alpha^{3/2} = t_k^3(dt_k)^{3/2}$ in the expansion of ψ_j near P_k. The other choices below can be normalized similarly.

The distinguished generator of $\lambda_{3/2}$ is

$$d_{3/2} = v_\alpha\varphi_0 \wedge v_\alpha\varphi_1 \wedge \ldots \wedge \psi_{g-2}.$$

d) A basis in $\lambda_{(i+1)/2}$, $i \geqq 3$. Since one has $R^1\pi_*\Omega^{i/2} = 0$ for $i \geqq 3$, the above game may be continued further inductively. One declares that the distinguished basis in $\pi_*\Omega^{(i+1)/2}$ is one in $\pi_*\Omega^{i/2}$, multiplied by v_α, plus the differentials with principal parts $\delta_{jk}(dt_k)^{(i+1)/2}$ at P_k, $j, k = 1, \ldots, g-1$.

The distinguished generator of $\lambda_{(i+1)/2}$ is

$$d_{(i+1)/2} = \text{the wedge product of the elements of the distinguished basis of } \pi_*\Omega^{(i+1)/2}.$$

e) A basis in $\langle \Omega^{1/2}, \Omega^{1/2} \rangle$. Put $D = \sum_1^{g-1} P_k$. According to Proposition 5c, we have $\langle \Omega^{1/2}, O(D) \rangle = \bigotimes_1^{g-1} \Omega_{P_i}^{1/2}$, and $\langle v_\alpha, 1 \rangle$ under this isomorphism maps into $\bigotimes_1^{g-1} v_{P_i}$, where v_{P_i} is $v_\alpha \bmod t_i v$, i.e. $t_i dt_i^{1/2} \bmod t_i^2$. The isomorphism $O(D) \xrightarrow{v_\alpha} \Omega^{1/2}$, $1 \mapsto v_\alpha$, sends this class into the distinguished generator of $\langle \Omega^{1/2}, \Omega^{1/2} \rangle$, which we denote

$$\langle \mathrm{id}, v_\alpha \rangle \left(\bigotimes_1^{g-1} v_{P_i} \right) = \prod_1^{g-1} \langle v_{P_i}, v \rangle = v.$$

Now the formal properties of the functors d and $\langle\, , \rangle$ imply, that

$$d_{i+1} = d_i v^{4i} \text{ via identification } \lambda_{i+1} = \lambda_i \langle \Omega, \Omega \rangle^i \quad (\text{N}^\circ.\ 7\text{a}) \tag{6}$$

$$d_1^2 = d_{1/2}^2 v \text{ via identification } \lambda_1^2 = \lambda_{1/2}^2 \langle \Omega^{1/2}, \Omega^{1/2} \rangle \quad (\text{N}^\circ.\ 7\text{b}) \tag{7}$$

9. Proof of Theorem 6, Sect. 1

Putting together (6), (7), and (4), we get formally

$$d_2 = d_1 v^4 = d_1^9 d_{1/2}^{-8} = d_1^{13} (\det B)^{-4},$$

which implies the statement a) of the theorem. To deduce b), note that

$$\det \left(\frac{i}{2} \int_X \omega_a \wedge \bar{\omega}_b \right) = \det \operatorname{Im} \tau,$$

hence

$$\det \left(\frac{i}{2} \int_X \varphi_a \wedge \bar{\varphi}_b \right) = \det |B|^2 \det \operatorname{Im} \tau.$$

3. Remarks and Complements

1. Admissible Metrics

The formula for the Polyakov measure, given in [14], was derived by almost the same method as the one above. The difference between two formulae owes partly to a different choice of distinguished bases, but mainly to the fact in [14] the machine of special hermitian metrics on invertible sheaves was systematically used. Since it may have a wider use in various contexts, we add here some hints on the derivation of the formula in [14].

Let $(L, |\,|)$ be an invertible sheaf with hermitian metrics on the complex variety of M. Its curvature form is the C^∞ (1, 1)-form $\partial\bar{\partial} \log |s|^2$, where s is an arbitrary local invertible section of L. This form is closed, and its cohomology class equals $2\pi i c_1(L)$. If M is compact, one gets in this way a bijection

$$\begin{bmatrix}(1,1)\text{-forms} \\ \text{in the class } c_1(L)\end{bmatrix} \leftrightarrow \begin{bmatrix}\text{metrics on } L \\ \text{up to proportionality}\end{bmatrix}.$$

On a Riemann surface X of genus $g \geqq 1$ there is a canonical $(1,1)$-form

$$\varrho_X = i/2g\omega^t(\operatorname{Im}\tau)^{-1} \wedge \omega,$$

where ω is the Riemann basis. Therefore any invertible sheaf L admits metrics with the curvature form proportional to ϱ_X. Such metrics are called admissible. For example, the sheaf $\mathcal{O}(P)$ has a canonical admissible metric, defined by $|t_P|(Q) = G(P,Q)$, where G is the exponentiated Green function of the Laplace operator $\Delta = -\frac{i}{\pi\varrho_X}\partial\bar{\partial}$. To be more precise,

$$f(P) = -\int_X \log G(P,Q)\Delta f(Q)\varrho_X(Q) \quad \text{for} \quad \int_X f\varrho_X = 0,$$

$$\int_X \log G(P,Q)\varrho_X(Q) = 0.$$

Faltings [9] has shown, that if an admissible metric on a sheaf L is given, then $d(L)$ can be endowed with a canonical metric that transforms in an explicit way under the canonical isomorphisms of Sect. 2. One can then use it instead of distinguished bases to compute $|\mu_X|^2 = d\pi_g$ (but not μ_X itself).

For example, the isomorphism

$$d(L) = d(L(-\sum P_i)) \otimes \left(\bigotimes_i L_{P_i}\right),$$

which we used constantly, becomes an isometry, if $d(L)$ and $d(L(-\sum P_i)$ are endowed with Faltings' metrics, while $\otimes L_{P_i}$ gets the metrics

$$\otimes(\text{induced metrics on } L_{P_i}) \otimes \prod_{i<j} G(P_i, P_j)^{-1}.$$

Such arguments, together with another result of Faltings on the metric change via $\lambda_1^{13} \to \lambda_2$ give the formula of [14].

2. *The Problem of Measure for Superstrings*

One hopes that Polyakov's measure for the superstring in the critical dimension 10 should also be computable in the same vein, through an intermediary Mumford-Berezin form on a superalgebraic moduli space M_g^s of dimension $3g-3|2g-2$ (for $g \geqq 2$). At the moment, however, only some parts of the whole picture are on a mathematically sound base.

a) It is unknown (at least, to the authors), whether an analogue of the Belavin-Knizhnik theorem is true.

b) The supervariety M_g^s, or rather an infinite covering of it, was constructed by one of us (Yu. I. M.) by means of Schottky's superuniformization, but not all techniques needed for transposition of calculations of Sect. 2 are worked out yet.

At our present level of knowledge, we can calculate at least the analog of the Mumford form on the "component" version of the superspace M_g^s, which parametrizes certain (1|1)-dimensional superalgebraic curves with superconformal structure in the sense of [6], having odd typical structural theta characteristics. By definition, the underlying space of this superspace is M_g^- (see Sect. 2, N°. 8), and the sheaf of holomorphic superfunctions is $\Lambda^{\cdot}(\pi_*\Omega^{3/2})$, where $\pi: X \to M_g^-$ is the universal family.

Denote this supervariety by $\tilde{M}_g^s$ and consider the family of supercurves $\tilde{\pi}: \tilde{X} \to M_g^s$, where odd superfunctions on $\tilde{X}$ are generated by the functions, lifted from M_g^s, and the sections of $\Omega^{1/2}$, lifted from X.

On $\tilde{M}_g^s$ there are analogs of Mumford's sheaves

$$\Lambda_i = \operatorname{Ber} R\tilde{\pi}_*((\Omega^s)^{i/2}),$$

where $(\Omega^s)^{1/2}$ comes from $\Omega^{1/2}$. The results of Sect. 2 give us a canonical isomorphism on M_g^-, $(\lambda_1\lambda_{1/2}^{-1})^5 = \lambda_2\lambda_{1/2}^{-1}$, which after lifting to M_g^s becomes an isomorphism $\Lambda_1^5 = \Lambda_3$.

The section of $\Lambda_1^{-5}\Lambda_3$, corresponding to 1, is a component analogue of a "true" Mumford's form. The second formula of [14] gives its modulus squared. The computations of Sect. 2 give an explicit formula for this section itself in terms of distinguished bases.

3. Distinguished Bases and Canonical Models of Algebraic Curves

If a curve X of genus $g \geqq 3$ is not hyperelliptic, then the distinguished basis $(\varphi_0, \ldots, \varphi_{g-1})$ embeds it into a coordinatized projective space P^{g-1}.

An advantage of this choice over Petri's one (see [16, pp. 123–135]) is that this embedding is defined only up to finite ambiguity. In fact, to define it, one fixes a typical odd theta characteristic and $g-1$ cubic roots of unity. The formula for φ_0 in terms of the theta function seems to introduce a transcendental element into this construction, but one may well avoid it. E.g. we can leave φ_0 defined up to multiplication by c, and $\varphi_j, j \geqq 1$, up to multiplication by $c^{-1/3}$ [the normalization (2), Sect. 1] is crucial here).

The ideal of equations for X in this embedding is generated for $g \geqq 4$ by quadratic relations between φ_j's. (excepting plane quintics and trigonal curves).

The basic quadratic relations can be written in the form

$$\varphi_k\varphi_l = \sum_{i=1}^{3g-3} a_{kl}^i w_i, \qquad k, l = 0, \ldots, g-1.$$

Here (w_i) is the distinguished basis of quadratic differentials, and coefficients a_{kl}^i can be calculated through the expansion coefficients of φ_i by v_a near P_k.

This refinement of the Petri method may be of independent algebro-geometric interest.

Appendix. Simultaneous Normalization of Mumford's Forms

In the main body of the paper Mumford's forms μ_g were defined only up to multiplication by a constant, depending on g. In this appendix we shall show how to normalize μ_g in a canonical coherent way by purely algebraic means. Hopefully, in this normalization $d\pi_g = e^{\beta(2-2g)}|\mu_g|^2$ for a certain (zero?) value of β. This conjecture is equivalent to a factorization property of $d\pi_g$.

To achieve this goal, we need to fix the behaviour of μ_g when a curve acquires a singularity. Let us first introduce some notation. Below "curve" means "projective algebraic curve, possibly reducible or even disconnected, with quadratic singularities only." For a curve X, let $\Omega^1 X$ be the sheaf of 1-forms on it and ω_X the dualizing sheaf (which is always invertible). They are related by a canonical morphism $i: \Omega^1 X \to \omega)_X$, which is an isomorphism off singular points. The image of i coincides with $\pi_* \Omega^1 \tilde{X}$, where $\pi: \tilde{X} \to X$ is the normalization map, pulling apart the double points. Now let α be such a double point; x_1, x_2 parameters of two branches, crossing at α. The one-dimensional fibers of $\operatorname{Ker} i$, $\operatorname{Coker} i$ at α are generated by $x_1 dx_2$, $x_1^{-1}dx_1 - x_2^{-1}dx_2$ respectively. Put

$$\lambda_j(X) = \det R\Gamma(X, \omega_X^{\otimes j}),\ j \in \mathbb{Z}; \quad \tilde{\lambda}(X) = \det R\Gamma(X, \Omega^1 X),$$
$$\delta(X) = \lambda_1(X)\tilde{\lambda}(X)^{-1}.$$

These are one-dimensional vector spaces with the following properties.

i) They vary in a holomorphic way, when X varies. The symmetries of X act upon them.

ii) There is a canonical element $\det i \in \delta(X)$, vanishing iff X is singular.

iii) (Serre's duality). $\lambda_j(X) = \lambda_{1-j}(X)$.

iv) (Disjoint sums). If $X = X_1 \sqcup X_2$, then there are canonical identifications $\lambda_j(X) = \lambda_j(X_1) \otimes \lambda_j(X_2)$, and similarly for $\tilde{\lambda}, \delta$.

For λ_j this comes from the exact sequence $0 \to \omega_1^{\otimes j} \to \omega_X^{\otimes 1} \to \omega_{X_2}^{\otimes j} \to 0$; similarly for $\tilde{\lambda}$; and δ expresses through $\lambda_1, \tilde{\lambda}$. Observe that this isomorphism for $\lambda_j, \tilde{\lambda}$ multiplies by $(-1)^{\chi(0_{X_1})\chi(0_{X_2})}$, if one changes the order of X_1, X_2; while one for δ does not depend on this order.

v) (Glueing points). Let X be a curve, α_1, α_2 smooth points on it. Denote by $\bar{X}$ a curve, obtained from X by glueing these points together into a point α. Put $l_i = (\Omega^1 X)_{\alpha_i}$. Then there are canonical identifications

$$\lambda_j(\bar{X}) = \lambda_j(X)(l_1 l_2)^{-j(j-1)/2}, \quad \tilde{\lambda}(\bar{X}) = \tilde{\lambda}(X) l_1 l_2,$$
$$\delta(\bar{X}) = \delta(X)(l_1 l_2)^{-1}.$$

To see this, consider the glueing map $\pi: X \to \bar{X}$, and let x_1, x_2 be parameters at α_1, α_2. This data defines the following exact sequences:

a) $$0 \to \omega_{\bar{X}}^{\otimes j} \to \pi_*(\omega_X(\alpha_1 + \alpha_2))^{\otimes j} \xrightarrow{\varphi} \mathbb{C}_\alpha \to 0.$$

b) $$0 \to \omega_X^{\otimes j} \to (\omega_X(\alpha_1 + \alpha_2))^{\otimes j} \to 0_X/m_{\alpha_1}^j \oplus 0_X/m_{\alpha_2}^j \to 0.$$

c) $$0 \to (l_1 l_2)_\alpha \xrightarrow{\psi} \Omega^1 \bar{X} \to \pi_* \Omega^1 X \to 0.$$

Here the maps φ and ψ are defined by

$$\varphi(f_1(x_1^{-1}dx_1)^j + f_2(x_2^{-1}dx_2)^j) = f_1(\alpha_1) + (-1)^{j-1} f_2(\alpha_2),$$
$$\psi(c(dx_1)_{\alpha_1}(dx_2)_{\alpha_2}) = c x_1 dx_2.$$

Applying to these exact sequences $\det R\Gamma$, we get:

$$\lambda_j(\bar{X}) \overset{(a)}{=} \det R\Gamma(X, \omega_X(\alpha_1+\alpha_2)^{\otimes j}) \overset{(b)}{=} \lambda_j(X)\det(0_X/m^j_{\alpha_1}\oplus 0_X/m^j_{\alpha_2})$$

$$= \lambda_j(X)\bigotimes_{a=j}^{0}\det(m^{a-1}_{\alpha_1}/m^a_{\alpha_1}\oplus m^{a-1}_{\alpha_2}/m^a_{\alpha_2}) = \lambda_j(X)(l_1l_2)^{j(j-1)/2},$$

$$\tilde{\lambda}(\bar{X}) \overset{(c)}{=} \tilde{\lambda}(X)l_1l_2 .$$

The isomorphism for δ comes from ones for $\lambda_1, \tilde{\lambda}$. Again, the isomorphisms for $\lambda_j, \tilde{\lambda}$ change sign, when the order of α_1, α_2 is reversed.

vi) (Projective line). If X is a smooth curve of genus zero, then $\lambda_j(X)=\tilde{\lambda}(X) =\delta(X)=\mathbb{C}$ canonically.

Indeed, by ii), iii) it suffices to look at the case $\lambda_j, j\geqq 0$. Choose a point $x\in X$. Consider an exact sequence

$$0\to\omega_X^{\otimes j}\to\omega_X^{\otimes j}((2j-1)x)\to F\to 0,$$

where $F=\omega_X^{\otimes j}((2j-1)x)/\omega_X^{\otimes j}$. The middle sheaf is acyclic, being isomorphic to $0_X(-1)$, hence $\lambda_j(X)=(\det F)^{-1}$. But $\det F=\mathbb{C}$, under the isomorphism, mapping 1 to $t^{-1}(dt)^j\wedge\ldots\wedge t^{-2j+1}(dt)^j$, where t is an arbitrary parameter at x. This isomorphism does not depend on choices made to construct it.

Now put

$$\bar{\lambda}_j=\lambda_j\delta^{-j(j-1)/2}, \qquad \nu_j=\bar{\lambda}_j\bar{\lambda}_1^{-6j^2+6j-1}.$$

Clearly, the above implies

vii) The properties i), iii), iv), vi) hold for $\bar{\lambda}_j, \nu_j$. In the situation v) we have $\bar{\lambda}_j(\bar{X}) =\bar{\lambda}_j(X)$, $\nu_j(\bar{X})=\nu_j(X)$ ("factorization property").

We can state now the main result of this appendix.

Proposition. *There is a unique identification* $\nu_j(X)=\mathbb{C}$ *compatible with* i), iv), *and* v). *It is then automatically compatible with* iii) *and for* $X=P^1$ *coincides with* vi). *The element* $\mu^{(j)}(X)\in\nu_j(X)$ *is the normalized Mumford form (or, rather, its value at* X*).*

Sketch of Proof. Assume first that such μ exists. According to iv), v) it is defined by its values on smooth connected X. But any such X is a member of a family of curves with compact connected parameter space, in which P^1 with a number of double points also occurs. (Take, e.g. the Deligne-Mumford modular family.) This shows that $\mu(P^1)$, together with i), v), determines μ uniquely. It remains to see that $\mu(P^1)$ coincides with 1 in normalization vi). Consider a constant family $P^1\times A^1$ and blow up a point in a fiber over $0\in A^1$. In this new family the fiber over 0 is $P^1\sqcup P^1$ with two points glued together, i.e. $P^1\vee P^1$. We have $\mu(P^1\vee P^1)=\mu(P^1)\cdot\mu(P^1)$, and $\mu(P^1\vee P^1)=\lim\limits_{t\to 0}\mu(P^1_t)$. This forces $\mu(P^1)$ to be 1.

As for existence of μ, one can prove it by a simple induction on genus, using the structure of Deligne-Mumford moduli space at infinity, together with iv), v) to normalize the Mumford forms.

Acknowledgements. Our deep thanks are due to A. A. Belavin and A. S. Švarc for very illuminating discussions of the subject of this work.

References

1. Polyakov, A.M.: Quantum geometry of bosonic strings. Phys. Lett. **103B**, 207–210 (1981)
2. Fradkin, E.S., Tseytlin, A.A.: Quantization of two-dimensional supergravity and critical dimensions in string models. Phys. Lett. **106B**, 63–68 (1981)
3. Alvarez, O.: Theory of strings with boundaries: fluctuations, topology and quantum geometry. Nucl. Phys. B **216**, 125–184 (1983)
4. Di Vecchia, P., Durhuus, B., Olesen, P., Petersen, J.L.: Fermionic strings with boundary terms. Nucl. Phys. B **207**, 77–95 (1982)
5. Baranov, M.A., Svarc, A.S.: On the multiloop contribution in the string theory. Pisma ZETP **42**:8, 340–342 (1985)
6. Baranov, M.A., Manin, Yu.I., Frolov, I.V., Svarc, A.S.: Multiloop contribution in fermionic string and Selberg supertrace formula. Yad. Fiz. **39**:2, 211–215 (1986)
7. Belavin, A.A., Knizhnik, V.A.: Complex geometry and quantum string theory. Preprint Landau Institute, 32, 1986
8. Mumford, D.: Stability of projective varieties. Enseign. Math. **23**, 39–100 (1977)
9. Faltings, G.: Calculus on arithmetic surfaces. Ann. Math. **119**:2, 387–424 (1984)
10. Knudsen, F., Mumford, D.: The projectivity of the moduli space of stable curves. I. Preliminaries on "det" and "Div". Math. Scand. **39**:1, 19–35 (1976)
11. Deligne, P.: La formule de dualité globale. In: Lecture Notes in Mathematics, Vol. **305**, pp. 481–587. Berlin, Heidelberg, New York: Springer 1970
12. Mumford, D.: Tata lectures on theta. I, II. Boston: Birkhäuser 1984
13. Mumford, D.: Abelian varieties. Bombay: Oxford UP 1970
14. Manin, Yu.I.: The partition function of the Polyakov string can be expressed through theta function. Pisma ZETP **43**:4, 161–163 (1986)
15. Fay, J.D.: Theta function on Riemann surfaces. In: Springer Lecture Notes in Mathematics, Vol. **352**. Berlin, Heidelberg, New York: Springer 1975
16. Arbarello, E., Cornalba, M., Griffiths, P.A., Harris, J.: Geometry of algebraic curves, Vol. **1**. Berlin, Heidelberg, New York: Springer 1985
17. Bismut, J.-M., Freed, D.S.: The analysis of elliptic families. I, II, Commun. Math. Phys. **106**, 159–176 (1986) and to appear
18. Bost, J.B., Jolicoeur, T.: A holomorphy property and critical dimension in string theory from an index theorem. Preprint Saclay PhT/86-28, 1986
19. Manin, Yu.I.: New dimensions in geometry. In: Lecture Notes in Mathematics, Vol. **1111**, pp. 59–101. Berlin, Heidelberg, New York: 1985
20. Ray, D., Singer, I.M.: Analytic torsion for complex manifolds. Ann. Math. **98**, 154–180 (1973)
21. D'Hoker, E., Phong, D.H.: On determinants of Laplacians on Riemann surfaces. Commun. Math. Phys. **104**, 537–545 (1986)

Communicated by A. Jaffe

Received July 28, 1986

Note added in proof. Since this paper was written we became aware of a very interesting letter by P. Deligne to D. Quillen, dated back to 20 June 1985, where a refinement of the Grothendieck-Riemann-Roch for c_1 of direct images of rank 0 virtual vector bundles over Riemann surfaces was given and the normalized Mumford forms were defined.

Ann. Inst. Fourier, Grenoble
37, 4 (1987), 191-205.

SOME REMARKS ON KOSZUL ALGEBRAS AND QUANTUM GROUPS

by Yu. I. MANIN

Introduction.

This note is devoted to algebras defined by quadratic relations. Such algebras arise in various contexts and have interesting homological properties whose investigation is based upon Koszul complexes: cf. [2], [3], [6].

My interest in them is connected with recent work of L. D. Faddeev and collaborators, V. G. Drinfeld and M. Jimbo who have introduced some remarkable Hopf algebras, non commutative and non cocommutative, or « quantum groups »: cf. [1], Drinfeld's Berkeley talk with ample bibliography. The coordinate rings of these groups are defined mainly by quadratic relations of very specific type.

The main observation of this note (Section 2, Th. 4) consists in realization, that the category of quadratic algebras is endowed with a natural « pseudo tensor product » $\bullet$ and a corresponding internal *Hom* functor. After a dualization this leads to large supply of « quantum semi-groups » of coendomorphisms of arbitrary quadratic algebras together with a supply of their comodules (Section 2, n. 9). At the same time a Koszul complex is associated with each morphism of quadratic algebras (Section 2, n. 5).

Section 1 gives a detailed description of quantum GL(2) as an automorphism « group » of a quantum plane. This description discovered by Yu. Kobozev was the starting point for this work. Section 3 gives two constructions of quantum determinants. In Section 4 we explain how to carry over the previous constructions into certain tensor categories of linear spaces (« the Yang-Baxter categories »).

Key-words: Koszul algebra - Tensor category - Quadratic algebra - Quantum groups - Yang-Baxter equation.

1. Quantum group GL(2): a motivating example.

1. *Notation.* – We fix once and for all a ground field k over which all tensor products are taken. For a linear k-space V, T(V) means its tensor algebra. For a subset $R \subset T(V)$ we denote by T(V)/(R) the quotient algebra with respect to the ideal generated by R. If $V = \bigoplus_i kx_i$, we write

$$T(V)/(R) = k[x_i] \quad \text{with relations} \quad r = 0 \quad \text{for} \quad r \in R$$

etc. As in [1], it is sometimes suggestive to imagine a ring A = T(V)/(R) as a coordinate ring, i.e. a ring of functions on an imaginary space of « noncommutative geometry », or « quantum space » Spec A.

2. *Two quantum planes.* – Let $q \in k$, $q \neq 0$. The quantum plane Spec $A_q(2|0)$ is defined by the ring

$$A_q(2|0) = k[x, y] \quad \text{with relation} \quad xy = q^{-1}yx. \tag{1}$$

We shall need also the quantum plane

$$\begin{aligned} &A_q(0|2) = k[\xi, \eta] \quad \text{with relations} \\ &\xi\eta = -q\eta\xi, \qquad \xi^2 = \eta^2 = 0. \end{aligned} \tag{2}$$

Since relations (1), (2) are homogeneous, both rings are graded, with generators of degree 1. The dimensions of their homogeneous components are the same ones, as for commutative polynomials of two variables in case (1) and for a grassmannian algebra of two generators in case (2).

3. *Quantum matrices.* – The coordinate ring of the manifold of quantum matrices $\begin{pmatrix} a & b \\ c & d \end{pmatrix}$ is defined by the relations

$$\begin{aligned} &M_q(2) = k[a,b,c,d], \\ &ab = q^{-1}ba, \quad ac = q^{-1}ca, \quad cd = q^{-1}dc, \quad bd = q^{-1}db, \\ &bc = cb, \quad ad - da = (q^{-1} - q)bc. \end{aligned} \tag{3}$$

These relations first emerged in a fairly indirect way: cf. [1], formulas (16)-(19). The next Proposition shows that they can be naturally interpreted as defining the quantum space of linear endomorphisms of relations (1) and (2).

4. PROPOSITION. – *Let* (x, y) (resp. (ξ, η)) *be the generic solutions of* (1), (resp. (2)). *Let* a, b, c, d *commute with* x, y, ξ, η. *Put*

$$\begin{pmatrix} x' \\ y' \end{pmatrix} = \begin{pmatrix} a & b \\ c & d \end{pmatrix}\begin{pmatrix} x \\ y \end{pmatrix}, \quad \begin{pmatrix} x'' \\ y'' \end{pmatrix} = \begin{pmatrix} a & c \\ b & d \end{pmatrix}\begin{pmatrix} x \\ y \end{pmatrix}, \quad \begin{pmatrix} \xi' \\ \eta' \end{pmatrix} = \begin{pmatrix} a & b \\ c & d \end{pmatrix}\begin{pmatrix} \xi \\ \eta \end{pmatrix}.$$

If $q^2 \neq -1$, *the following conditions are equivalent*

(i) (x', y') *and* (x'', y'') *verify* (1).

(ii) (x', y') *verify* (1) *and* (ξ', η') *verify* (2).

(iii) (a,b,c,d) *verify* (3).

Proof. – The relation $x'y' = q^{-1}y'x'$ means

$$(ax+by)(cx+dy) = q^{-1}(cx+dy)(ax+by).$$

Taking into account that a, b, c, d commute with x, y and comparing coefficients, we get

$$(3)' \qquad \begin{aligned} x^2 &: ac = q^{-1}ca, \\ y^2 &: bd = q^{-1}db, \\ xy &: ad - da = q^{-1}cb - qbc. \end{aligned}$$

Exchanging here b and c we get the relations, equivalent to $x''y'' = q^{-1}y''x''$:

$$(3)'' \quad ab = q^{-1}ba, \quad cd = q^{-1}dc, \quad ad - da = q^{-1}bc - qcb.$$

Comparing the last relations in (3)′ and (3)″ we obtain

$$(q+q^{-1})(bc-cb) = 0 \Rightarrow bc = cb, \quad \text{if} \quad q^2 \neq -1.$$

Hence (3)′ and (3)″ together are equivalent to (3).

Finally, a similar direct calculation shows that the relations (2) for ξ', η' are equivalent to (3)″.

This proposition, due to Yu. Kobozev, shows without further calculations the two main properties of relations (3):

5. *Multiplicativity.* – Let (a, b, c, d) and (a', b', c', d') separately verify (3) and pairwise commute among themselves. Then the matrix elements of $\begin{pmatrix} a & b \\ c & d \end{pmatrix}\begin{pmatrix} a' & b' \\ c' & d' \end{pmatrix}$ also verify (3).

6. *Quantum determinant.* – In the same conditions put

$$\mathrm{DET}_q\begin{pmatrix} a & b \\ c & d \end{pmatrix} = ad - q^{-1}bc = da - qcb. \tag{4}$$

Then

$$\mathrm{DET}_q\begin{pmatrix} a & b \\ c & d \end{pmatrix}\begin{pmatrix} a' & b' \\ c' & d' \end{pmatrix} = \mathrm{DET}_q\begin{pmatrix} a & b \\ c & d \end{pmatrix}\mathrm{DET}_q\begin{pmatrix} a' & b' \\ c' & b' \end{pmatrix}.$$

In fact, if (3) (or even only (3)") hold, we have in notation of Proposition 4:

$$\xi'\eta' = \mathrm{DET}_q\begin{pmatrix} a & b \\ c & d \end{pmatrix}\xi\eta.$$

7. *Quantum group* $G'L_q(2)$. – Its quantum coordinate ring can be obtained from that of $M_q(2)$ by inverting $\mathrm{DET}_q\begin{pmatrix} a & b \\ c & d \end{pmatrix}$. Similarly, adding the relation $\mathrm{DET}_q = 1$, we get quantum $SL(2)$.

8. *Language of Hopf algebras.* – Proposition 4 in more intrinsic form says that the map

$$\Delta : M_q(2) \to M_q(2) \otimes M_q(2), \qquad \Delta\begin{pmatrix} a & b \\ c & d \end{pmatrix} = \begin{pmatrix} a & b \\ c & d \end{pmatrix} \otimes \begin{pmatrix} a & b \\ c & d \end{pmatrix}$$

makes of $M_q(2)$ a Hopf algebra (without antipode), and the maps

$$\delta : A_q(2|0) \to M_q(2) \otimes A_q(2|0), \qquad \delta\begin{pmatrix} x \\ y \end{pmatrix} = \begin{pmatrix} a & b \\ c & d \end{pmatrix} \otimes \begin{pmatrix} x \\ y \end{pmatrix}$$

$$\delta' : A_q(2|0) \to M_q(2) \otimes A_q(2|0), \qquad \delta'\begin{pmatrix} x \\ y \end{pmatrix} = \begin{pmatrix} a & c \\ b & d \end{pmatrix} \otimes \begin{pmatrix} x \\ y \end{pmatrix}$$

$$\delta^{\pi} : A_q(0|2) \to M_q(2) \otimes A_q(0|2), \qquad \delta^{\pi}\begin{pmatrix} \xi \\ \eta \end{pmatrix} = \begin{pmatrix} a & b \\ c & d \end{pmatrix} \otimes \begin{pmatrix} \xi \\ \eta \end{pmatrix}$$

define comodules over this Hopf algebra.

2. Quadratic algebras.

1. *Quadratic algebras.* – Quadratic algebra is an associative **Z**-graded k-algebra $A = \sum_{i=0}^{\infty} A_i$ with the following properties:

a) $A_0 = k$, $\dim A_1 < \infty$.

b) A is generated by A_1 over k, and the ideal of relations between elements of A_1 (i.e. the kernel of the homomorphism $T(A_1) \to A$) is generated by a subspace $R(A) \subset A_1 \otimes A_1$

(Priddy [2] calls such a ring « homogeneous prekoszul algebra » and endows A_1 with an additional Z-grading).

Morphism of quadratic algebras $f: A \to B$ is a k-homomorphism preserving gradings. There is a bijection between such morphisms and k-linear maps $f_1 : A_1 \to B_1$ for which $(f_1 \otimes f_1)(R(A)) \subset R(B)$. We denote by QA the category of quadratic algebras.

It is often convenient to write A as

$$A \leftrightarrow \{A_1, \; R(A) \subset A_1 \otimes A_1\}.$$

For example,

$$A_q(2|0) \leftrightarrow \{kx \otimes ky, \; k(x \otimes y - q^{-1} y \otimes x)\}.$$

Algebras $A_q(0|2)$ and $M_q(2)$ are also quadratic.

2. *Operations on quadratic algebras.* – Let A, B be two quadratic algebras. Put

$$A \circ B \leftrightarrow \{A_1 \otimes B_1, \; S_{(23)}(R(A) \otimes B_1^{\otimes 2} + A_1^{\otimes 2} \otimes R(B))\} \tag{5}$$

$$A \bullet B \leftrightarrow \{A_1 \otimes B_1, \; S_{(23)}(R(A) \otimes R(B))\} \tag{6}$$

$$A^! \leftrightarrow \{A_1^*, \; R(A)^\perp\}. \tag{7}$$

In (6) we have $R(A) \otimes R(B) \subset A_1 \otimes A_1 \otimes B_1 \otimes B_1$ while $R(A \bullet B) \subset A_1 \otimes B_1 \otimes A_1 \otimes B_1$, and $S_{(23)}$ in the relevant rearrangement operator (for $\sigma \in \mathfrak{S}_p$ we write here

$$S_\sigma(a_1 \otimes \ldots \otimes a_n) = a_{\sigma^{-1}(1)} \otimes \cdots \otimes a_{\sigma^{-1}(n)}).$$

For the similar reason we put $S_{(23)}$ in (5). In (7) V^* means a dual space of V, and the corresponding pairing $\langle v^*, v\rangle$ is also called a contraction. We identify $(V \otimes W)^*$ with $V^* \otimes W^*$, etc. By definition, $R(A)^\perp$ is then the annihilator of $R(A)$.

3. *Properties of operations.* – A usual tensor product of quadratic algebras (over k) is again a quadratic algebra :

$$A \otimes B \leftrightarrow \{A_1 \oplus B_1, \; R(A) \oplus R(B) \oplus [A_1, B_1]\},$$

where $[A_1, B_1]$ is generated by

$$a \otimes b - b \otimes a \in A_1 \otimes B_1 \oplus B_1 \otimes A_1 .$$

However, $\circ$ and $\bullet$ are in some respects more relevant in QA, e.g. the comultiplication for natural quadratic Hopf algebras is a morphism $A \to A \circ A$. (Since there is a canonical ring homomorphism $A \circ B \to A \otimes B$ doubling the degree, this defines also a comultiplication in the usual sense).

Both products are endowed with functorial associativity and commutativity constraints :

$$(A \circ B) \circ C = A \circ (B \circ C), \qquad (A \bullet B) \bullet C = A \bullet (B \bullet C),$$
$$A \circ B \simeq B \circ A, \qquad A \bullet B \simeq B \bullet A,$$

which are induced by the standard maps on generators of degree 1. Morphisms $f : A \to A'$ and $g : B \to B'$ define morphisms $f \circ g : A \circ B \to A' \circ B'$ and $f \bullet g : A \bullet B \to A' \bullet B'$, mapping $a \otimes b$ to $f(a) \otimes g(b)$ for $a \in A_1$, $b \in B_1$.

A morphism $f : A \to B$ defines a morphism $f^! : B^! \to A^!$, for which $\langle f^!(b), a \rangle = \langle f(a), b \rangle$ if $a \in A_1$, $b \in B_1^*$ This defines an equivalence of categories $! : QA \to QA^{0p}$. Besides, there are natural isomorphisms

$$(A \bullet B)^! = A^! \circ B^!, \qquad (A \circ B)^! = A^! \bullet B^!$$

Finally, since $R(A) \otimes R(B) = R(A) \otimes B_1^{\otimes 2} \cap A_1^{\otimes 2} \otimes R(B)$, there is a canonical morphism $A \bullet B \to A \circ B$.

The following simple fact is our main result. It shows that $(QA, \bullet)$ is a (non additive) tensor category with internal *Hom* in the terminology of [5] (cf. also Section 4 below).

4. THEOREM. – *a) There is a functorial isomorphism*

$$\mathrm{Hom}\,(A \bullet B, C) = \mathrm{Hom}\,(A, B^! \circ C)$$

identifying a map $f : A_1 \otimes B_1 \to C_1$ *with a map* $g : A_1 \to B_1^* \otimes C_1$ *if* $\langle g(a), b \rangle = f(a \otimes b)$ *for all* $a \in A_1$, $b \in B_1$ (l.h.s. *denotes contraction with respect to* B_1).

b) Let $K = k[\varepsilon]$, $\varepsilon^2 = 0$. *Then* K *is a unit object of* $(QA, \bullet)$. *In particular, for all A there is a canonical isomorphism* $K \bullet A \simeq A$.

Proof. – *a*) We must check that if f, g are related as in the statement the following conditions are equivalent :

$$(f\otimes f)\, S_{(23)}(R(A)\otimes(B)) \subset R(C)),$$
$$(g\otimes g)R(A) \subset S_{(23)}(R(B)^{\perp} \otimes C_1^{\otimes 2} + B_1^{*\otimes 2} \otimes R(C).$$

But they are respectively equivalent to

$\langle R(C)^{\perp}, (f\otimes f)S_{(23)}(R(A)\otimes R(B))\rangle = 0$ (contraction w.r.t. $C_1 \otimes C_1$)
$\langle R(B)\otimes R(C)^{\perp}, (g\otimes g)R(A)\rangle = 0$ (contraction w.r.t. $B_1\otimes B_1\otimes C_1^{*}\otimes C_1^{*}$).

In their turn, each of these last orthogonality relations mean that if we start with an element of $R(A)$, apply $g \otimes g$ and then contract consecutively with arbitrary elements of $R(B)$ and $R(C)^{\perp}$, we shall get zero.

b) By definition

$$K\bullet A \leftrightarrow \{\varepsilon\otimes A_1, S_{(23)}(\varepsilon^{\otimes 2}\otimes R(A))\},$$

whence an isomorphism $a \mapsto \varepsilon \otimes a$ for $a \in A_1$. □

The rest of this Section is devoted to formal consequences of this theorem.

5. *Generalized Koszul complexes*. – Put $A = K$ in Th. 4*a*. We get

$$\mathrm{Hom}\,(B,C) = \mathrm{Hom}\,(K, B^{!}\circ C).$$

Denote by d_f the image of $\varepsilon \in K$ under the morphism $K^{!} \to B^{!}\circ C$ corresponding to $f: B \to C$. We have $d_f^2 = 0$. Therefore with each morphism f in QA is associated a complex

$$K^{\bullet}(f) = \{B^{!}\otimes C, \text{ right multiplication by } d_f\}.$$

For $B = C = A$, $f = \mathrm{id}$ we get one of Koszul complexes associated with a quadratic algebra A :

$$K^{\bullet}(A) = \{A^{!}\otimes A, \delta\}.$$

In order to define the second complex put $A' = \bigoplus_i (A_i^{!})*$ where $A_i^{!}$ is the i-th component of $A^{!}$ This space is a right $A^{!}$-module. Put

$$K_{\bullet}(A) = A \otimes A' \subset \mathrm{Hom}_{\text{mod-}A}(A^{!}\otimes A, A)$$

and define the differential d by $(df)(a) = f(a\delta)$, $a \in A^{!} \otimes A$.

The following result is proved in [2] (cf. also [3]):

6. THEOREM-DEFINITION. – *A quadratic algebra* A *is called a Koszul algebra if the following equivalent properties are verified :*

a) $K_\bullet(A)$ *is acyclic.*

b) Algebra $\mathrm{Ext}^*_A(k,k)$ *is generated by* $\mathrm{Ext}^1_A(k,k) \simeq A^*_1$.

c) Algebra $\mathrm{Ext}^*_A(k,k)$ *is naturally isomorphic to* $A^!$ □

Using a sufficient condition for this property found by Priddy, the existence of a PBW-basis ([2], Th. 5.3), one can prove that coordinate rings of quantum spaces and their (co)endomorphism spaces considered below are Koszul algebras.

7. *Internal Hom.* – As in general formalism of tensor categories (see [5]) we put $\mathit{Hom}(B,C) = B^! \circ C$. In particular,

$$B^! = \mathit{Hom}(B, K^!).$$

This can be checked directly using $K^! = k[t]$, or by dualization of Th. 4*b*.

8. *Internal multiplication.* – By the general properties of Hom, the following internal product maps are defined:

$$\begin{aligned} &\mathit{Hom}(B,C) \bullet B \to C, \\ &\mathit{Hom}(B,C) \bullet \mathit{Hom}(C,D) \to \mathit{Hom}(B,D) \end{aligned} \tag{8}$$

with evident associativity properties (cf. [5], (1.6.2)).

9. *Dualization and internal comultiplication.* – We define the algebra of internal cohomomorphisms by

$$\mathit{hom}(B,C) = \mathit{Hom}(B^!,C^!)^! = B^! \bullet C.$$

Applying ! to (8), we get internal comultiplication maps

$$\begin{aligned} &\Delta_{BC} : C \to \mathit{hom}(B,C) \circ B, \\ &\Delta_{BCD} : \mathit{hom}(B,D) \to \mathit{hom}(B,C) \circ \mathit{hom}(C,D). \end{aligned} \tag{9}$$

In particular, $\mathit{hom}(A,A) = \mathit{end}(A)$ is endowed with a Hopf algebra structure

$$\Delta_A = \Delta_{AAA} : \mathit{end}(A) \to \mathit{end}(A) \circ \mathit{end}(A),$$

while algebras A, *hom* (A, B) and their homogeneous components furnish plenty natural comodules over *end* (A).

This is our generalization of the main construction of Section 1. More precisely, *end* $A_q(2|0)$ is defined by a half of relations, defining $M_q(2)$. The complementary relations stem from *end* $A_q(2|0)'$ since $A_q(2|0)' \simeq A_q(0|2)$.

3. Determinant.

In this section we shall consider two natural constructions of determinants for quantum semigroups *end* (A). The first one is applicable to algebras A, which are « similar » to usual Grassmann algebras and imitates the conventional definition of det. The second one is universal and imitates the homological description of Berezinian in commutative superalgebra, but its properties are poorly known. (In the context of Yang-Baxter categories it was considered by Lyubashenko [4].)

1. *Quantum grassmannian algebras (q.g.a).* – We shall call a quadratic algebra A a q.g.a. of dimension n, if $\dim A_n = 1$ and $A_m = 0$ for $m > n$. From the construction of the morphism (9), $\Delta : A \to end\,(A) \circ A$ one sees that $\Delta(A_i) \subset (end\,(A))_i \otimes A_i$ for any i. In particular, we can define an element $DET_A \in (end\,(A))_n$ for a q.g.a. of dimension n by the formula $\Delta(a) = DET_A \otimes a$, where $a \in A_n$ is a generator. It enjoys the comultiplicativity property $\Delta(DET_A) = DET_A \otimes DET_A$.

2. *Example.* – Let $A = k[X_1, \ldots, X_n]$ with relations $X_i^2 = 0$, $X_iX_j = -qX_jX_i$ for $i < j$. One easily sees that A is an n-dimensional q.g.a. The determinant DET_A lies in the ring *end* (A) generated by Y_i^j corresponding to $X_i \otimes X^j$, where $\{X^j\}$ is the dual basis to $\{X_i\}$. The formula for calculating DET_A is

$$\prod_{i=1}^{n}\left(\sum_{j=1}^{n} Y_i^j X_j\right) = DET_A . \prod_{j=1}^{n} X_j \tag{10}$$

where $[Y_i^j, X_k] = 0$. Therefore

$$DET_A = \sum_{s \in \mathfrak{S}_n} (-q)^{l(s)} Y_1^{s(1)} \ldots Y_1^{s(n)}$$

(cf. [4] and the formula in [1] following (19). However *end* (A) has fewer relations than (16)-(19), as our discussion in Sec. 1 shows).

3. *Example* (V. G. Drinfeld [7]). – Let k be a quotient field of a local ring O. Consider anticommutation relations

$$X^k X^l + X^l X^k = \sum_{i,j} c^{kl}_{ij} X^i X^j \tag{11}$$

where c^{kl}_{ij} are some elements of the maximal ideal in O enjoying the symmetry conditions $c^{kl}_{ij} = c^{lk}_{ij} = -c^{kl}_{ji}$. Clearly, for $c^{kl}_{ij} = 0$ we get a usual Grassmann algebra. In [7] Drinfeld calculated conditions on c^{kl}_{ij} which should be verified in order that (11) define a q.g.a. over O (and *a fortiori* over k).

Put

$$b^{j_1 j_2 j_3}_{i_1 i_2 i_3} = \sum_j c^{j_1 j_2}_{i_1 j}\, c^{j j_3}_{i_2 i_3}\,,$$

$$b = (b^{j_1 j_2 j_3}_{i_1 i_2 i_3}), \quad a = b(1-b/3)^{-1}.$$

Then

$$\text{(11) define a q.g.a. over } O \Leftrightarrow \operatorname{alt}(i)\ \operatorname{sym}(j)\ a^{j_1 j_2 j_3}_{i_1 i_2 i_3} = 0\,.$$

The dimension of A in this case equals n and we obtain a quantum determinant in *end* (A).

In [7] it is shown also that q.g.a. (11) are Koszul.

4. *Question.* – Let A be a q.g.a. Can one make of *end* $(A)/(\mathrm{DET}_A - 1)$ a Hopf algebra with antipode, i.e. a quantum group SL(A) ?

One can similarly define a DET with respect to any component A_n if $\dim A_n = 1$.

5. *Homological determinant.* – Let A be a quadratic algebra. Consider $A^! \otimes A$ as a quadratic algebra, put $E(A) = \mathit{end}\,(A^! \otimes A)$ and construct the structure comultiplication

$$\Delta : A^! \otimes A \to E(A) \circ (A^! \otimes A) \to E(A) \otimes A^! \otimes A\,.$$

Put $\Delta(\delta) = \delta'$, where $\delta \in A_1^* \otimes A_1$ corresponds to id. In $E(A) \otimes A^! \otimes A$ there is also an element $1 \otimes \delta$ with square zero. Although δ' and $1 \otimes \delta$ do not coincide, there exists a minimal ideal $I \subset E(A)$ with the property

$$\delta' \equiv 1 \otimes \delta \mod I \otimes A^! \otimes A\,.$$

Put $\tilde{E}(A) = E(A)/I$ and further

$$H(A) = H^*((A^! \otimes A), \delta),$$
$$\tilde{E}(A) \otimes H(A) = H^*(\tilde{E}(A) \otimes A^! \otimes A, 1 \otimes \delta).$$

Finally, define the cohomological determinant as a map

$$\mathrm{DETH}_A = H^*(\Delta \bmod I) : H(A) \to \tilde{E}(A) \otimes H(A).$$

If $\dim H(A) = 1$, we can consider this as an element

$$\mathrm{DETH}_A \in \tilde{E}(A).$$

Therefore a natural problem arises: to study the class of quadratic algebras A for which the Koszul complex $K^\bullet(A)$ has one-dimensional cohomology (the same in Yang-Baxter categories: cf. below).

4. Tensor categories and quadratic algebras.

1. *Tensor categories.* – Recall that a tensor category [5] is a couple $(C, \otimes)$ consisting of a category C, a functor $\otimes : C \times C \to C$, $(X, Y) \mapsto X \otimes Y$ and two additional functor isomorphisms

$$\psi_{X,Y,Z} : X \otimes (Y \otimes Z) \to (X \otimes Y) \otimes Z$$

(associativity constraint) and

$$\varphi_{X,Y} : X \otimes Y \to Y \otimes X, \qquad \varphi_{X,Y}\varphi_{Y,X} = \mathrm{id}$$

(commutativity constraint). Several axioms are imposed on this data, the most important for us being a compatibility diagram. A unit object of a tensor category consists of an object $U \in C$ and an isomorphism u : $U \to U \otimes U$ such that the functor $X \to U \otimes X$ is an equivalence. Internal *Hom* (X, Y) represents the functor $T \mapsto \mathrm{Hom}(T \otimes Y, Y)$. For further details see [5].

2. *YB-categories.* – A tensor category $\mathcal{S} = (C, \otimes)$ is called a YB-category (or Yang-Baxter category, or vectorsymmetry, cf. Lyubashenko [4]) if the following conditions are fulfilled:

a) C is a subcategory of finite-dimensional vector spaces over a field k.

b) $\otimes$ in C coincides with tensor product over k, and $\psi_{X,Y,Z}(x \otimes (y \otimes z)) = (x \otimes y) \otimes z$ is the usual associativity constraint.

c) $(k, 1 \mapsto 1 \otimes 1)$ is a unit object, for all $X, Y \in C$ there exists *Hom* (X,Y) and the structure diagram *Hom* $(X,Y) \otimes X \to Y$ is isomorphic to the standard diagram $Y \otimes X^* \otimes X \to Y$.

It follows that essentially new data distinguishing the YB-category $(C, \otimes)$ from C consists of a family of isomorphisms $S_{(1,2)}$: $X \otimes Y \to Y \otimes X$ (former $\varphi_{X,Y}$) verifying the equations

$$S_{(1,2)}S_{(1,2)} = \mathrm{id},$$
$$S_{(1,2)(3)}S_{(1)(2,3)}S_{(1,2)(3)} = S_{(1)(2,3)}S_{(1,2)(3)}S_{(1)(2,3)}.$$

Physicists call the last equation the Yang-Baxter, or triangle, relation. In the context of tensor categories it is essentially the compatibility diagram for two constraints.

Each object X of a YB-category $\mathcal{S}$ over a field of characteristics zero defines a quadratic algebra « $\mathcal{S}$-symmetric algebra of X ».

3. Proposition. — *a*) *For each* n *one can define a representation* $\rho_n^{\mathcal{S}} : \mathfrak{S}_n \to GL(X^{\otimes n})$ *by the following prescription : to calculate* $\rho_n(s)$, *decompose* s *into a product of transpositions and construct the product of the corresponding commutativity operators* S.

b) *Put*

$$A \leftrightarrow \{X, \mathrm{Im}(1 - S_{(1,2)}) \subset X \otimes X\}.$$

Then the natural map $T(X) \to A$ *induces isomorphisms* $(X^{\otimes n})^{\mathfrak{S}_n} \simeq A_n$. □

These symmetric algebras in YB-categories are the closest analogs of polynomial rings in noncommutative algebra. Their internal coendomorphism algebras should be considered as (coordinate rings of) quantum matrices.

4. *Quadratic* $\mathcal{S}$*-algebras*. — There is however a more interesting way to connect the constructions of Sec. 2 with YB-categories. One can first change definitions and then the resultats in order to generalize, say, th. 2.4 to arbitrary YB-category. It was in this way that the main notions of superalgebra were formed.

Fix $\mathcal{S}$ and call a quadratic $\mathcal{S}$-algebra such a quadratic algebra A that $A_1 \in \mathcal{S}$ and $R(A) \subset A_1 \otimes A_1$ is a subobject in $\mathcal{S}$. By definition, an $\mathcal{S}$-morphism of quadratic $\mathcal{S}$-algebras should be induced by a morphism $A_1 \to B_1$ in $\mathcal{S}$

We now define $\mathcal{S}$-operations on the quadratic $\mathcal{S}$-algebras $A \underset{\mathcal{S}}{\bigcirc} B$, $A \underset{\mathcal{S}}{\bullet} B$ by the same formulas (5), (6) where however the operator $\underset{\mathcal{S}}{S_{(23)}}$ should now be understood as the commutativity constraint in $\mathcal{S}$ The definition of A' changes in an indirect way via a new identification of $(V \otimes W)^*$ with $V^* \otimes W^*$ All the main statements and constructions of Sec. 2 remain valid in the new context.

In particular, we get the Koszul $\mathcal{S}$-complexes which are also complexes in the usual sense. However, the new Hopf-$\mathcal{S}$-algebras will not in general be usual Hopf algebras since $\underset{\mathcal{S}}{\bigcirc}$ and $\underset{\mathcal{S}}{\otimes}$ in general will not coincide with $\bigcirc$ and $\otimes$ (the multiplication rule in the product algebra changes in a nontrivial way due to the S-operator).

5. *Example.* – Let $V = \bigoplus_{i=1}^{n} kX_i$ and A is generated by X_i subject to the relations $X_iX_j = a_{ij}X_jX_i$ for all i, j where $a_{ij} \in k^*$. It will be an $\mathcal{S}_3$-symmetric algebra in the relevant YB-category $\mathcal{S}_3$ generated by V with the symmetry operator $X_i \otimes X_j \mapsto a_{ij}X_j \otimes X_i$, if $a_{ij}a_{ji} = 1$ for $i < j$ and $a_{ii} = \pm 1$. But its Hopf algebra of internal coendomorphisms $end_{\mathcal{S}} A = A' \underset{\mathcal{S}}{\bullet} A$ can be naturally considered in at least three different YB-categories $\mathcal{S}_1$, $\mathcal{S}_2$, $\mathcal{S}_3$.

Version 1. $\mathcal{S}_1$ = the category of finite dimensional vector spaces with the standard symmetry.

(12) $A \leftrightarrow \{V; r_{ij} = X_i \otimes X_j - a_{ij}X_j \otimes X_i, i < j;$
$r_{ii} = X_i \otimes X_i$ for $a_{ii} = -1\}$

(13) $A' \leftrightarrow \{V = \otimes kX^i; r^{kl} = X^k \otimes X^l + a_{kl}^{-1}X^l \otimes X^k;$
$r^{kk} = X^k \otimes X^k$ for $a_{kk} = 1\}$.

Denote by Y_i^k the image of $X^k \otimes X_i$ in $A' \bullet A$. By definition, we get the following relations for Y_i^k:

$S_{(23)}(r^{kk} \otimes r_{ii})$: $(Y_i^k)^2 = 0$ for $a_{ii} = -1$, $a_{kk} = 1$.

$S_{(23)}(r^{kl} \otimes r_{ii})$: $Y_i^kY_i^l + a_{kl}^{-1}Y_i^lY_i^k = 0$ for $k < l$, $a_{ii} = -1$.

$S_{(23)}(r^{kk} \otimes r_{ij})$: $Y_i^kY_j^k - a_{ij}Y_j^kY_i^k = 0$ for $i < j$, $a_{kk} = 1$.

$S_{(23)}(r^{kl} \otimes r_{ij})$: $Y_i^kY_j^l + a_{kl}^{-1}Y_i^lY_j^k - a_{ij}Y_j^kY_i^l - a_{ij}a_{kl}^{-1}Y_j^lY_i^k = 0$

for $i < j$, $k < l$.

Version 2. $\mathscr{S}_2$ = category of Z_2-graded vector spaces, morphisms = linear maps conserving grading, $s_{(12)}(v \otimes w) = (-1)^{\tilde{v}\tilde{w}} w \otimes v$, where $v \in V_{\tilde{v}}$, $w \in W_{\tilde{w}}$.

In the previous notation put $\tilde{X}_i = 0$ for $a_{ii} = 1$, 1 for $a_{ii} = -1$. As earlier, put $\langle X^k, X_i \rangle = \delta_i^k$, $\tilde{X}_k = \tilde{X}^k$, $\tilde{Y}_i^k = \tilde{X}_i + \tilde{X}_k$. For simplicity we shall write $\tilde{i} = \tilde{X}_i$, $\tilde{ik} = \tilde{X}_i \tilde{X}_k$. Then $A^! \underset{\mathscr{S}_2}{\bullet} A$ is defined by the relations

$$S_{(23)}(r^{kk} \otimes r_{ii}):\ (Y_i^k)^2 = 0 \text{ for } \tilde{X}_i = 1,\ \tilde{X}_k = 0.$$

$$S_{(23)}(r^{kl} \otimes r_{ii}):\ (-1)^{\tilde{l}} Y_i^k Y_i^l + (-1)^{\tilde{k}} a_{kl}^{-1} Y_i^l Y_i^k = 0 \text{ for } k < l,\quad \tilde{X}_i = 1.$$

$$S_{(23)}(r^{kk} \otimes r_{ij}):\ Y_i^k Y_j^k - a_{ij} Y_{jl}^k Y_i^k = 0 \text{ for } i < j,\ \tilde{X}_k = 0.$$

$$S_{(23)}(r^{kl} \otimes r_{ij}):\ (-1)^{\tilde{l}\tilde{i}} Y_i^k Y_j^l + (-1)^{\tilde{k}\tilde{i}} a_{kl}^{-1} Y_i^l Y_j^k - (-1)^{\tilde{l}\tilde{j}} a_{ij} Y_j^k Y_i^l - (-1)^{\tilde{k}\tilde{j}} a_{ij} a_{kl}^{-1} Y_j^l Y_i^k = 0$$

for $i < j$, $k < l$.

As in Sec. 1, we can add to these relations those ones which correspond to the transposed matrix Y_i^k (in version 1) or to the supertransposed one (in version 2): $(Y^{st})_i^k = (-1)^{\tilde{ik} + \tilde{i}} Y_k^i$. In this way we shall get various versions of the quantum (super) matrix semigroup.

We leave to the reader the last version. We recall only that first one should calculate in $\mathscr{S}_3$ a commutation rule $X^j \otimes X_i \mapsto A_{ij} X_i \otimes X^j$, whose coefficients are defined by the condition that the contraction $V^* \otimes V \to k$ be a morphism in $\mathscr{S}_3$.

BIBLIOGRAPHY

[1] V. G. Drinfeld, Quantum groups. *Zap. Naučn. sem. LOMI*, vol. 155 (1986), 18-49 (in russian).

[2] S. B. Priddy, Koszul resolutions, *Trans. AMS*, 152-1 (1970), 39-60.

[3] C. Löfwall, On the subalgebra generated by one-dimensional elements in the Yoneda Ext-algebra, *Springer Lecture Notes in Math.*, vol. 1183 (1986), 291-338.

[4] V. V. Lyubashenko, Hopf algebras and vector-symmetries, *Uspekhi Mat. Nauk*, 41-5 (1986), 185-186 (in russian).

[5] P. Deligne, J. Milne, Taonakian categories, *Springer lecture Notes in Math.*, vol. 900 (1982), 101-228.
[6] A. A. Beilinson, V. Ginsburg, Mixed categories, Ext-duality and representations, 1986, preprint.
[7] V. G. Drinfeld, On quadratic commutation relations in the quasiclassic limit, in: *Mat. Fizika i Funkc. Analiz, Kiev, Naukova Dumka* (1986), 25-33 (in russian).

Yu. I. Manin,
Steklov Institute of Mathematics
Vavilova 42
Moscow 117966 GSP 966 (SSSR).

REFLECTIONS ON ARITHMETICAL PHYSICS*

Yu.I. Manin

Il en est qui, face à cela, se contentent de hausser les épaules d'un air désabusé et de parier qu'il n'y a rien à tirer de tout cela, sauf des rêves. Ils oublient, ou ignorent, que notre science, et toute science, serait bien peu de choses, si depuis ses origines elle n'avait été nourrie des rêves et des visions de ceux qui s'y adonnent avec passion.
A.*Grothendieck* [0], p.18

The development of theoretical physics in the last quarter of the twentieth century is guided by a very romantic system of values. Aspiring to describe fundamental processes at the Planck scale physicists are bound to lose any direct connection with observable world. In this social context the sophisticated mathematics emerging in the string theory ceases to be only a technical tool needed to calculate some measurable effects and becomes a matter of principle.

Today at least some of us are again nurturing an ancient Platonic feeling that mathematical ideas are somehow predestined to describe physical world, however remote from reality their origins may seem to be.

From this viewpoint one should perversely expect number theory to become the most applicable branch of mathematics.

*) *Dedicated to Alexander Grothendieck on his 60th birthday.*

ISBN 0-12-218100-X

Indeed, one clearly discerns the trend to at least include the number theory into universe of discourse of modern theoretical physics.

The author of these notes was amazed and pleased by the discovery that in order to calculate the Polyakov measure in string theory one can use Faltings' calculation of a specific number theoretical function called "height" (cf. [1], [2],[3]).

Sasha Polyakov told me that after Faltings' talk at the Berkeley International Congress of Mathematicians Ed Witten bought all books in number theory available at a bookshop across the road. (I did not check this with Ed - se non è vero, è ben trovato).

Therefore it is high time to present some musings of a professional number-theorist and an amateur-physicist on the controversial subject of arithmetical physics.

First of all, can one calculate anything physical by unambigously number-theoretical means ? I believe that the answer should be affirmative. Look at one of the beautiful formulas discovered by Euler:

$$\pi^2/6 \qquad \prod_{p\ \mathrm{prime}} (1 - p^{-2})^{-1} \tag{1}$$

The r.h.s. unmistakably belongs to number theory: primes $p = 2,3,5,7,11, \ldots$ are one of its main concerns. I dare say that the l.h.s. involving π is a physical constant although it may need some arguing to convince the reader. In fact, π can be (and had been) measured, just like water boiling temperature or the Earth Equator length. One may say that Euclidean geometry where π appears as a mathematical constant is actually kinematics of ideal solids valid in the macroscopic approximation of flat gravitational vacuum.

In order to understand (1) better it is useful to recall some facts on primes. Classically, a prime number p is defined as a positive integer having no divisors except of 1 and itself. Each integer can be decomposed into a product of primes in a unique way; there is an infinity of primes; they are distributed quite irregularly; the simplest asymptotic formula for the number of primes not exceeding N is N/log N.

This is not however the viewpoint we need here.

A modern explanation of the role of primes is given by Ostrowski's theorem: primes describe all reasonable ways to define a notion of continuity on the set of rational numbers Q (in addition to the traditional one).

To be more precise, define a function $|a|_p$ of $a \in Q$ by $|a|_p = p^{-n}$, if $a = p^n c d^{-1}$ where c,d are integers nondivisible by p. It verifies the usual properties of a norm: $|ab|_p = |a|_p |b|_p$, $|a + b|_p \leq |a|_p + |b|_p$ (in fact, even $\leq \max(|a|_p, |b|_p)$ the so called non-Archimedean traingle inequality). Therefore it defines on Q a topology in which $a_i \to 0$ if $|a_i|_p \to 0$. This topology is called the p-adic one. Addition and multiplication being continuous p-adically, one can define in a standard way the notion of a Cauchy sequence and the set of limits of such sequences which are called p-adic numbers.

The set of p-adic numbers Q_p is a new analog of the set of reals R which can also be constructed in this way with the help of the usual absolute value $|a|$ which is conveniently denoted as $|a|_\infty$. Ostrowski's theorem says that any norm, or valuation, on Q defines the same topology as $|\cdot|_\infty$ or $|\cdot|_p$ for some prime p.

Of course, the properties of Q_p are in many respects very different from those of R. The main reason is that the topology of Q_p is quite unlike that of R: p-adic numbers form a Cantor set, or a "fractal" [4]. Nevertheless, many chapters of classical calculus and geometry were developed over p-adic numbers. For a beautiful introduction see [5].

The physical world as we know it can be described very efficiently by mathematics based upon R (and its further extension C, complex numbers). Recently it was suggested that at the Planck scale a p-adic topology may become more relevant [6]. This however raises a question: why should one particular p be physically marked ? Isn't it more reasonable to believe in a democracy of all available topologies ? (Or at least of all p-adic ones since R is certainly "the first among equals" being defined by the only Archimedean norm).

It appears that Euler's identity (1) and a whole series of similar facts can be explained in a way furnishing a very suggestive picture of such democracy.

Let us start with an almost evident formula $|a|_\infty \quad \prod_p |a|_p^{-1}$. It means that the knowledge of the ordinary absolute value of a rational number is equivalent to that of all p-adic absolute values simultaneously. Or, totally democratically, $\prod_v |a|_v \quad 1$, where $v - \infty$ or $p - 2,3,5, \ldots$ Formulas of this kind abound in the number theory and are called product formulas, reciprocity laws etc.

In the last product formula we considered a rational number a in turn as a real, 2-adic, 3-adic ... number. Let us introduce more generally the set of infinite vectors $(a_\infty, a_2, a_3, \ldots)$ where $a_\infty \in R$, $a_p \in Q_p$. Such a vector, with a restriction $|a_p|_p - 1$ for all sufficiently large p, is called an adèle. This term was coined by Claude Chevalley around 1940, together with the word idèle meaning an invertible adèle (i.e. $a_v \neq 0$ for all v, $|a_p|_p = 1$ for large p). Etymology of these words is doubtful; idèle presumably stems from ideal while adèle means additive idèle. Anyway, among number theorists these are now common words.

Let us now imagine the first steps of mathematics in which the notion of a real number is replaced by that of adèle. An adèle $a = (a_v)$ has components, real one a_∞ and p-adic ones a_p for all p's. The set of all adèles is a topological ring A_Q with componentwise addition and multiplication. Its topology combines Archimedean and fractal properties. Rational numbers Q are imbedded into A_Q diagonally: $a \equiv (a, a, a, \ldots)$. A simple but important observation is that $Q \subset A_Q$ is a discrete subgroup, like $Z \subset R$. In other words, a sequence of rational numbers cannot converge in all v-adic topologies simultaneously (e.g. if it converges to zero p-adically for all p's it must consist of infinitely growing integers).

Recalling that $R/Z \sim U(1)$ is a circle we arrive to the notion of adèlic circle:

$$A_Q/Q = (R \times \prod_p Z_p)/Z \quad , \tag{2}$$

where Z_p is the set of p-adic integers ($a \in Z_p \Leftrightarrow |a|_p \leq 1$). We see from (2) that the adèlic circle is a mixture of U(1) and a compact topological totally discontinuous group which can be also described as a "lattice approximation limit" for U(1), i.e. $\varprojlim_n Z/nZ$. Thus we encounter again the Archimedean and fractal properties combined in one object. Fourier analysis based upon A_Q/Q rather than U(1) ties together in a very nice way the usual and the finite Fourier transforms: cf. Tate's thesis [7].

Going one step further we may define the simplest non-commutative adèlic group $SL_2(A_Q)$ which is essentially the set of infinite matrix vectors

$$\{ \begin{pmatrix} a_v & b_v \\ c_v & d_v \end{pmatrix} \in SL_2(Q_v) \mid v = \infty, 2, 3, 5, \ldots \}$$

such that (a_v), (b_v), (c_v), (d_v) are adèles. Again, $SL_2(Q)$ is discrete in $SL_2(A_Q)$. With the help of a left invariant differential form on SL_2 and norms $|a|_v$ we can define a left invariant measure $dm = \prod_v dm_v$ on $SL_2(A_Q)$ in the same way as on its classical component $SL_2(R)$. If we normalize dm by the condition $\int_{SL_2(A_Q)/SL_2(Q)} dm \quad 1$ and then calculate this integral componentwise we get finally a beautiful explanation of (1):

$$1 = \int_{SL_2(A_Q)/SL_2(Q)} dm = \int_{SL_2(R)/SL_2(Z)} dm_\infty \times \prod_p \int_{SL_2(Z_p)} dm_p \, , \tag{3}$$

$$\int_{SL_2(R)/SL_2(Z)} dm_\infty \qquad \pi^2/6 \quad , \quad \int_{SL_2(Z_p)} dm_p \qquad 1 - p^{-2} \quad . \tag{4}$$

Here (3) is established in the same way as (2), the Archimedean part of (4) is proved by a Poisson summation trick, and the p-part of (4) follows from the fact that SL_2 over a finite field of p elements consists of p^3-p^2 points, so that the relative amount of points in $SL_2(Z_p)$ with respect to Z_p^3 is just $1 - p^{-2}$.

Now we can see the following pattern:

- (at least some) essential notions of real and complex calculus and geometry have their adèlic counterparts;
- adèlic objects have a strong tendency to be simpler than their Archimedean components, e.g. the adèlic fundamental domains of arithmetical discrete subgroups of semisimple groups usually have volume 1 (the Siegel-Tamagawa-Weil philosophy, cf. [15]);
- due to this fact and to the "product formulas" like (2) or (3) embodying the idea of democracy for all topologies, an information on the real component of an adèlic object can be read off *either* from this real component *or* from the product of p-adic components for all p's.

With some strain, one can generalize and state the following principle which is the main conjecture of this talk.

On the fundamental level our world is neither real, nor p-adic, it is adèlic. For some reasons, reflecting the physical nature of our kind of living matter (e.g. the fact that we are built of massive particles), we tend to project the adèlic picture onto its real side. We can equally well spiritually project it upon its non-Archimedean side and calculate most important things arithmetically.

The relation between "real" and "arithmetical" pictures of the world is that of complementarity, like the relation between conjugate observables in quantum mechanics.

Of course, one is not obliged to take this metaphysics seriously. A skeptical reader can still use it as a guiding principle in a mathematical study of the structure of string theory.

I shall describe below some recent work which seems promising in this respect.

To start with, a reinterpretation of the calculation of Polyakov's measure in [1] shows [3] that if we take a point x of the moduli space M_g which has algebraic coordinates, then the density of this measure with respect to a canonical one equals the inverse Archimedean part of a function called height of x.

A remarkable property of the height compatible with our philosophy is that it is defined by a product of factors corresponding to all valuations of the field in which the coordinates of x lie.

I conjecture that one can *define on the space of adèlic points of a universal moduli space an adèlic Polyakov measure whose Archimedean component is the ordinary Polyakov measure. Hopefully, the corresponding total adèlic volume would be computable as in (3), (4), thus giving an arithmetical expression of the string partition function.*

If this hope comes true we shall be able to speak on adèlic strings with some justification. Of course, the main reason for believing in this is the remarkable appearance in the string theory of algebraic varieties (moduli spaces) and measures on them (Mumford forms) which are intrinsically defined over integers and not just over R or C.

In order to explain some details we need an extension of the picture we worked with up to now. Number theory studies not only rational numbers Q but all algebraic numbers $\overline{Q}$ i.e., roots of polynomials with rational coefficients. It is convenient to work with smaller number fields K i.e., with finite-dimensional Q-subspaces of $\overline{Q}$ containing 1 and closed under multiplication. For such a field K one can prove a generalization of the Ostrowski theorem, describing all valuations of K. Since $Q \subset K$, each such valuation w induces on Q a valuation v which is equivalent either to $|\cdot|_\infty$ or to $|\cdot|_p$. We say that w extends, or divides v. Now the following facts are true: a) each valuation of Q extends to a finite number of valuations of K; b) valuations extending $|\cdot|_\infty$ correspond to various embeddings of K into complex numbers. After this theorem has been proved, one can define w-adic numbers K_w, adèles A_K, idèles J_K and all the other objects we considered previously "over Q".

Suppose now that we have a linear space L over K endowed with norms $||\cdot||_w$, one for each valuation w of K, such that $||al||_w = |a|_w||l||_w$ for $a \in K$, $l \in L$ and $||l||_w = 1$ for almost all w if $l \neq 0$. We can then define the height of $l \in L$ by the formula

$$h(l) = \prod_w ||l||_w \quad .$$

From the product formula $\prod_w |a|_w = 1$ for $a \in K$ one sees that h(l) in fact depends only on the ray Kl in L i.e., the height is a function on the projective space associated with L.

Since the moduli space M_g is in fact defined by algebraic equations with integer coefficients we can embed M_g into such a projective space. If this embedding is constructed with the help of Mumford's determinant vector bundles, we obtain in this way the height connected with the Polyakov measure.

The total height, unlike its Archimedean part, is defined only for points with algebraic coordinates, which, although dense in M_g, do not look very appealing physically. However, in a recent work [8] it was established that just these points appear in a natural way as lattice points in a string lattice approximation scheme.

The situation is as follows. In string lattice approximation one replaces a smooth Riemann surface, i.e., a string world sheet with a metric ds^2, by a triangulated metric surface which is essentially given by combinatorial data listing, say, vertices and lengths of arcs connecting some pairs of vertices. This is a two-dimensional version of the Regge calculus in general relativity.

Consider now only equilaterally triangulated compact oriented surfaces with all arcs of the length 1 (changing this value leaves the surface in the same conformal class). One easily proves that such a surface is endowed with a complex structure compatible with the metric (first delete vertices then extend the obtained complex structure which is possible because the sum of angles at each vertex equals $n\pi/6$ for an integer n). Therefore such a surface defines a point in M_g. The main theorem of [8] foreseen by A.Grothendieck [0] states that in this way we obtain precisely all algebraic points. Thus the number theoretic picture reflects very well a combinatorial-metric picture. An outstanding problem now is establishing further ties between these two descriptions, in particular, calculating height in metric terms.

We now come to the last point of our discussion.

The most extensive generalization of Euler's formula (1) is connected with calculation of adèlic volume of homogeneous spaces of the type $H(A_K)/H(K)$ where H is a semisimple algebraic group and K is an algebraic number field. (We treated briefly the case $H - SL_2$). Similar calculations for other kinds of algebraic varieties, such as M_g, present great difficulties. Why can we then hope to deal arithmetically with M_g ?

A possible way out is remarkably connected with a new approach to another baffling property of the Polyakov partition function, namely, that it is inherently a perturbative expansion. It was suggested by several authors that one should work instead with a sort of Universal moduli space $\hat{M}$ embodying all M_g's (and something else).

In [9], motivated by this idea and operator approach, I conjectured that this $\hat{M}$ should be a homogeneous space with respect to the Virasoro algebra.

This conjecture was recently proved by four groups [10] - [13]. They all use the same basic construction due to Sato and Segal-Wilson. In this construction $\hat{M}$ is an infinite Grassmannian and the "moduli part" of $\hat{M}$ parametrizes triples (X,p,z) where X is a complex Riemann surface, p a point on it and z a local coordinate at this point.

If one succeeds to define the nonperturbative path integral as an integral over $\hat{M}$ it may well become accessible to the arithmetical treatment. To this end we shall need an extension of the Tamagawa-Weil theory to infinite-dimensional groups, like Kac-Moody ones and $GL(\infty)$. (Remark in passing that $vol(SL(n, R)/SL(n,Z)) = \zeta(2) \ldots \zeta(n)$ has a well defined limit for $n \to \infty$. Is it possible to obtain it as a volume for $n \quad \infty$?).

In conclusion I wuuld like to describe very briefly some preoccupations of number theorists which may be relevant to the program of arithmetization of physics.

Number theory has its own Grand Unification Group: it is the so called Galois group $G \quad Gal(\overline{Q}/Q)$ which consists of all permutations of algebraic numbers among themselves conserving algebraic relations between them with rational coefficients. It is an infinite topological group of "fractal" type whose structure is very complex and in a sense embodies all arithmetics, if one takes into account also some canonical central extensions of its subgroups, the so called Weil groups. To illustrate this assertion, we only note that the maximal abelian quotient G^{ab} is just $\prod_p Z_p^*$ so that primes appear again in a completely unexpected way, essentially as generators of G^{ab}. In studying the representations of G and its closed subgroups a number theorist meets with automorphic and modular functions in much the same way (or, rather, dual one) as a physicist studying the representations of Kac-Moody and Virasoro algebras. A series of deep conjectures due to Langlands [14] connect the representation theory of G with that of groups $H(A_K)$ via modular forms and their Mellin transforms.

I sincerely hope that such remarkable coincidences are not fortuitous.

Finally, I am pleased and proud to dedicate this note to Alexander Grothendieck whose insights immensely influenced mathematics and start to influence physics as well.

References

[0]. Grothendieck A. Esquisse d'un programme. Preprint 1984

[1]. Manin Yu.I. The partition function of the Polyakov string can be expressed in terms of theta functions. Phys.Lett. B172 (1986), 184-186.

[2]. Faltings G. Calculus on arithmetic surfaces. Ann. of Math., 119 (1984), 387-424.

[3]. Smit D.-J. String theory and algebraic geometry of moduli spaces. Preprint ITPh, Utrecht (1987).

[4]. Mandelbrot B. *Fractals*. Freeman, San-Francisco, 1977.

[5]. Koblitz N. *p-adic numbers, p-adic analysis and zeta-functions*. Springer, Heidelberg, 1977.

[6]. Volovich I.V. a) Number theory as the ultimate physical theory. Preprint CERN-TH 4781/87.
b) p-adic space-time and string theory. TMPh, 71:3 (1987), 337-340 (in Russian).

[7]. Tate J. Fourier analysis in number fields and Hecke's zeta-functions. In: *Algebraic Number Theory*, ed. by J.W.S. Cassels, A.Fröhlich. Academic Press, 1967, 305-347.

[8]. Voevodsky V.A., Shabat G.B. Equilateral triangulations of Riemann surfaces and curves over algebraic number fields. Preprint, 1987 (in Russian).

[9]. Manin Yu.I. Critical dimensions of string theories and the dualizing sheaf on moduli spaces of (super)curves. Funkc.Analiz i ego Pril., 20:3 (1986), 88-89 (in Russian).

[10]. Kontsevich M.L. Virasoro algebra and Teichmüller spaces. Funkc.Analiz i ego Pril., 21:2 (1987), 78-79 (in Russian).

[11]. Beilinson A.A., Schechtman V.V. Determinant bundles and Virasoro algebras. Preprint (1987).

[12]. Kawamoto N., Namikawa Yu., Tsuchiya A., Yamada Y. Geometric realization of conformal field theory on Riemann surfaces. Preprint, Nagoya University (1987).

[13]. Arbarello E., De Concini C., Kac V., Procesi C. Moduli spaces of curves and representation theory. Manuscript (1987).

[14]. Langlands R.P. L-functions and automorphic representations. In: *Proc. ICM 1978*, vol.1, Helsinki 1980, 165-176.

[15]. Kneser M. Semi-simple algebraic groups. In: *Algebraic Number Theory*, ed. by J.W.S. Cassels, A. Fröhlich, Academic Press, 1977, 250-265.

Steklov Mathematical Institute
Ul. Vavilova 42, Moscow 117 333

ТЕОРЕТИЧЕСКАЯ
И МАТЕМАТИЧЕСКАЯ
ФИЗИКА
Том 92, N º3
сентябрь, 1992

 Yu. I. Manin

NOTES ON QUANTUM GROUPS AND QUANTUM DE RHAM COMPLEXES

De Rham complexes of quantum spaces and quantum groups are studied by means of extension of "universal co-action" technique to the differential algebras.

To the memory of M. C. Polivanov

§0. INTRODUCTION. BASIC EXAMPLES

0.1. Wess-Zumino de Rham complex of the quantum plane. This paper is devoted to an extension of the approach to quantum groups suggested in [1] and [2]. Namely, we replace the category of associative algebras considered there by a category of differential graded algebras treated as "de Rham complexes of quantum spaces".

This extension was explained in my Kyoto lectures in May 1990, and afterwards elaborated in several courses in Moscow, Utrecht, Cambridge, and at 1991 Bonn's Arbeitstagung. When a preliminary version of these notes was written, I became aware of G. Maltsiniotis' articles [3, 4] who has suggested the same generalization. Some of the examples were calculated independently by G. Maltsiniotis and participants of my Moscow seminar E. Demidov, E. Mukhin, D. Zhdanovich: see §3. The general structure results of §§1, 2 may have an independent interest.

A non-commutative differential calculus of the type considered here was first suggested by S. L. Woronowicz and developed by him and his collaborators: see [5–8], and the references therein. However, our treatment was principally motivated by the preprint by J. Wess and B. Zumino [9]. To explain its results, I will start with the basic example $GL_q(2)$.

Let k be a ground field, $q \in k^*$. By definition, the ring of polynomial functions $F = F[GL_q(2)]$ is a Hopf algebra which can be described in the following way. As a k-algebra, it is generated by a, b, c, d and a formal inverse of a central element

$$D = DET_q \begin{pmatrix} a & b \\ c & d \end{pmatrix} = ad - q^{-1}bc \ ,$$

where a, b, c, d satisfy the following commutation relations:

$$ab = q^{-1}ba, \quad ac = q^{-1}ca, \quad cd = q^{-1}dc, \quad bd = q^{-1}db \ , \\ bc = cb, \quad ad - da = (q^{-1} - q)bc \ . \tag{0.1}$$

The comultiplication $\Delta : F \to F \otimes F$ is defined by

$$\Delta \begin{pmatrix} a & b \\ c & d \end{pmatrix} = \begin{pmatrix} a & b \\ c & d \end{pmatrix} \otimes \begin{pmatrix} a & b \\ c & d \end{pmatrix} \tag{0.2}$$

where the tensor product in the r. h. s. denotes the usual product of matrices in which products like ab are replaced by $a \otimes b$. The counit is given by

$$\varepsilon \begin{pmatrix} a & b \\ c & d \end{pmatrix} = \begin{pmatrix} 1 & 0 \\ 0 & 1 \end{pmatrix} . \tag{0.3}$$

Finally, the antipode map $i: F \to F$ is

$$i \begin{pmatrix} a & b \\ c & d \end{pmatrix} = D^{-1} \begin{pmatrix} d & -qb \\ -c/q & a \end{pmatrix} . \tag{0.4}$$

Although it can be checked directly that all these structures are well defined and satisfy the Hopf algebra axioms, the computations are tedious and not very enlightening.

A more conceptual approach consists in introducing two quantum planes $A_q^{2|0}$ and $A_q^{0|2}$, with function rings

$$F[A_q^{2|0}] = k\langle x, y\rangle/(xy - q^{-1}yx), \tag{0.5}$$

$$F[A_q^{0|2}] = k\langle \xi, \eta\rangle/(\xi^2, \eta^2, \xi\eta + q\eta\xi) , \tag{0.6}$$

and obtaining (0.1) as solution of the following universal problem (for $q^2 \neq -1$): the coordinate change

$$\begin{pmatrix} x' \\ y' \end{pmatrix} = \begin{pmatrix} a & b \\ c & d \end{pmatrix} \otimes \begin{pmatrix} x \\ y \end{pmatrix} , \quad \begin{pmatrix} \xi' \\ \eta' \end{pmatrix} = \begin{pmatrix} a & b \\ c & d \end{pmatrix} \otimes \begin{pmatrix} \xi \\ \eta \end{pmatrix} \tag{0.7}$$

should be compatible with (0.5), (0.6).

In this way, $GL_q(2)$ emerges as a "quantum automorphism group" of a pair of "non-commutative linear spaces", and all its properties can be directly derived from this definition. In particular, (0.2) expresses the composition of two "automorphisms", and D can be calculated from the formula $\xi'\eta' = D\xi\eta$, exactly as in the classical definition of the determinant via a volume form.

In [2] (cf. also [10], Ch. IV) this construction was generalized: it was shown that $F[A_q^{2|0}]$ can be replaced by an arbitrary quadratic algebra (or even arbitrary algebra with a fixed system of generators) considered as a function ring on an abstract quantum linear space (or rather cone).

Until recently, however, it was unclear, why do we need in this construction *two* quantum spaces, and not just one.

A beautiful answer was given by J. Wess and B. Zumino ([9]). Namely, they suggested to consider $F[A_q^{2|0}]$ and $F[A_q^{0|2}]$ as parts of a differential graded algebra $\Omega[A_q^{2|0}]$, *the quantum de Rham complex* of the quantum plane $A_q^{2|0}$.

More precisely, $\Omega[A_q^{2|0}]$ is generated by (x, y, ξ, η) over k and graded by (ξ, η)-degree. To the commutation rules (0.5), (0.6) Wess and Zumino add new cross-commutation relations between (x, y) and (ξ, η) respectively:

$$\begin{aligned} x\xi &= q^{-2}\xi x, & x\eta &= q^{-1}\eta x + (q^{-2} - 1)\xi y \\ y\xi &= q^{-1}\xi y, & y\eta &= q^{-2}\eta y . \end{aligned} \tag{0.8}$$

The differential d is uniquely defined by the conditions

$$dx = \xi, \quad dy = \eta, \quad d^2 = 0 \tag{0.9}$$

and the usual (not quantized !) Leibniz formula $d(fg) = df.g + (-1)^{\deg f} f dg$.

Actually, there are *two* sets of good cross-commutation relations between coordinates and differentials: the other one can be obtained from (0.8) by interchanging $x \leftrightarrow y, \xi \leftrightarrow \eta$, and replacing q by q^{-1}.

These sets are the only ones compatible with the action (0.7) of $GL_q(2)$ and satisfying the Wess-Zumino condition

(WZ): ***The algebra of the differential forms as a k-space must be the tensor product of the function algebra and algebra of differential forms with constant coefficients.***

Compatibility with the action of $GL_q(2)$ formally means that (0.7) defines a graded differential algebra homomorphism

$$\Omega[A_q^{2|0}] \to F[GL_q(2)] \otimes \Omega[A_q^{2|0}] \tag{0.10}$$

(on $F[GL_q(2)]$, the differential vanishes).

Now the next step is almost obvious. In (0.10), the quantum plane is represented by its de Rham complex, whereas the quantum group is represented only by its function ring. We would like to have also $\Omega(GL_q(2))$ instead of $F[GL_q(2)]$ in the r. h. s. of (0.10). In order to construct it, we must redo the theory of [2] from scratch, starting with the differential graded algebras (DGA) as abstract quantum de Rham complexes and constructing their automorphism objects which will then be de Rham complexes of quantum groups, so that e. g. (0.10) becomes replaced by a universal DGA-morphism

$$\Omega[A_q^{2|0}] \to \Omega[GL_q(2)] \otimes \Omega[A_q^{2|0}] \ . \tag{0.11}$$

Actually, this is a largely formal undertaking, since it involves only work with tensor algebra in more general rigid tensor categories than that of linear spaces. But it is not the end of the story.

The point is that the universal DGA W obtained by this construction (before localization making it Hopf) is not a deformation of the classical de Rham complex of the ring of matrix coefficients. In fact, it has exponential growth order. We are thus facing the same problem of "missing relations" in a new guise. However, it can now be solved without introducing new objects: a calculation shows that W has a unique quotient which is a bialgebra satisfying (WZ). We explain this calculation in some detail in §3, in a slightly generalized setting, for the two-parametric deformation of $GL(2)$. There are two more quotients satisfying (WZ) which are not bialgebras; all in all, we obtain six de Rham complexes of quantum matrices $M_{p,q}(2)$, because the same construction applies to the second de Rham complex of Wess-Zumino.

This approach also enriches our understanding of quantum groups viewed as deformations of the universal enveloping algebras (cf. [11]). In fact, classicaly Lie algebras consist of vector fields, which can be defined algebraically as derivations of function algebras: linear maps ∂ satisfying $\partial(fg) = \partial f.g + f\partial g$. This form of Leibniz formula cannot, however, be taken as a definition of a vector field in non-commutative geometry: in the correct definition the second summand is badly twisted (see (2.5) below) because it is a by-product of the usual Leibniz formula for the total differential *and* the cross-commutation relations between functions and differentials. The role of the (WZ)-condition in this context is that it provides the necessary relations. We devote §§3 and 4 to the general study of their structure. A somewhat unexpected outcome is the appearance of quantum matrix semigroups classifying Wess-Zumino type "skew products": see Theorem 1.3.

This chapter of non-commutative differential geometry can be profitably compared with Connes' approach based upon his universal de Rham complex and "cycles" ([12]). Calculating the cohomology of our de Rham complexes in the simplest examples, one sees that Connes' definition of a cycle can and should be quantized, by replacing in his condition $\int[\omega,\omega'] = 0$ the usual commutators by twisted ones. However, we have no universal definition for the latters: practically speaking, they are calculated each time anew using the Wess-Zumino type arguments. It is possible that that the correct framework for this type of complexes is the tensor algebra over braided (instead of tensor) categories where a new version of the cyclic cohomology might be defined: see [13, 14].

Before turning to the general formalism, we will give the basic formulas of one-dimensional quantum differential geometry which are both amusing and instructive.

0.2. One-dimensional de Rham complex. This is the complex induced by the cross-commutation rules (0.8) on the "axes" of our quantum plane:

$$\Omega = k[x, dx]_v; \quad (dx)^2 = 0; \quad v dx.x = v^{-1} x.dx \tag{0.12}$$

(we replaced q^{-1} by v^2 in order to conform with notation in Lusztig's papers on representations of quantum groups). For $n \geq 1$, we have

$$\begin{aligned} d(x^n) &= dx.x^{n-1} + x.dx.x^{n-2} + \dots \\ &= dx(1 + v^2 + \dots + v^{2n-2})x^{n-1} = v^{n-1}[n]_v dx.x^{n-1} \quad , \end{aligned} \tag{0.13}$$

where $[n]_v = \frac{v^n - v^{-n}}{v - v^{-1}}$ are Gaussian, or quantized, integers. If we put $[0]_v = 0$ and $[-n]_v = -[n]_v$ for positive n, the same formula (0.13) will describe the natural extension of d to $k[x, x^{-1}, dx]_v$. (If $v = \pm 1$, we of course put $v^{n-1}[n]_v = n$).

0.3. Cohomology. If v is not a root of unity, $[n]_v \neq 0$ for $n \neq 0$. Assume in addition that all $[n]_v$, $n \neq 0$, are invertible in the ground ring k. Then (0.13) shows that, as in the classical case,

$$H^\bullet(k[x, dx]_v) \cong k; \quad H^\bullet(k[x, x^{-1}, dx]_v) \cong k \oplus k dx.x^{-1} \quad . \tag{0.14}$$

Notice that this may happen also in finite characteristics, namely, when v is transcendental over $\mathbf{F}_p$, and when k contains $\mathbf{F}_p[[n]_v^{-1} \mid n \in \mathbf{Z} \setminus \{0\}]$.

For $\omega = dx.f$, $f \in k[x, x^{-1}]$, consider the residue functional

$$\int dx \left(\sum a_i x^i\right) = a_{-1} \quad .$$

It satisfies two relations:

$$\int dg = 0; \qquad \int [dx.x^n, x^m]_{v^m} = 0 \quad ,$$

where, by definition, $[x, y]_w = wxy - w^{-1}yx$. It is the presence of such twisted commutators in the second formula that distinguishes $(k[x, x^{-1}, dx]_v, \int)$ from a usual Connes' cycle.

We shall often meet such commutators later on. Notice that the last relation in (0.12) is $[dx, x]_v = 0$.

Now, let v^2 be a primitive root of unity of degree $l > 1$. Then

$$(0.15) \qquad [1]_v \neq 0, \dots, [l-1]_v \neq 0; \qquad [l]_v = 0; \qquad [n+lm]_v = v^{lm}[n]_v \ .$$

Therefore, assuming again that non-vanishing $[n]_v$ are invertible, we have

$$(0.16) \qquad H^*(k[x, x^{-1}, dx]_v) \cong \oplus_{m\in \mathbf{Z}}(kx^{ml} \oplus kdx.x^{ml-1}) \ .$$

In the classical case, such is the structure of the de Rham cohomology in characteristic l. In the quantum context, k may have any characteristic, including zero.

This is probably the simplest example of a general phenomenon: quantizarion at special parameter values tends to reproduce some effects that we are accustomed to see as characteristic-dependent, in all characteristics. Here is one more example: characteristic-independent Frobenius.

0.4. Gauss binomial coefficients. Assume that

$$(0.17) \qquad [x, y]_v = 0$$

for certain elements x, y of an associative k-algebra. Then the following binomial formula is valid:

$$(0.18) \qquad (x+y)^n = \sum_{j=0}^{n} v^{j(n-j} \begin{bmatrix} n \\ j \end{bmatrix}_v x^{n-j} y^j \ ,$$

where for $0 \leq j \leq n$ we put

$$(0.19) \qquad \begin{bmatrix} n \\ j \end{bmatrix}_v = \frac{[n]!_v}{[j]!_v[n-j]_v} \ [n]!_v = [1]_v \dots [n]_v, \ [0]!_v = 1 \ .$$

0.4.1. Proof of (0.18). Using (0.19), we can directly check that

$$\begin{bmatrix} n+1 \\ j \end{bmatrix}_v = v^j \begin{bmatrix} n \\ j \end{bmatrix}_v + v^{-n-1} \begin{bmatrix} n \\ j-1 \end{bmatrix}_v \ .$$

Multiplying this by t^j and summing over $j = 0, \dots, n+1$, we get

$$F_{n=1}(t) = F_n(vt)(1 + v^{-n}t) \ ,$$

where

$$F_n(t) := \sum_{j=0}^{n} \begin{bmatrix} n \\ j \end{bmatrix}_v t^j \ .$$

Hence by induction

$$F_n(t) = \prod_{k=0}^{n-1}(1 + v^{2k-n+1}t) \ .$$

Replacing t by $v^{n-1}t$, we get

$$\sum_{j=0}^{n} \begin{bmatrix} n \\ j \end{bmatrix}_v v^{j(n-1)}t^j = \prod_{k=0}^{n-1}(1 + v^{2k}t) \ .$$

Now, if $[x,y]_v = 0$, by formally inverting x we obtain

$$(x+y)^n = x^n \prod_{i=0}^{n-1}(1+v^{2i}x^{-1}y) = x^n \sum_{j=0}^{n-1} \begin{bmatrix} n \\ j \end{bmatrix}_v v^{j(n-1)}(x^{-1}y)^j ,$$

which leads to (0.18) because $(x^{-1}y)^j = v^{-j(j-1)}x^{-j}y^j$.

Assume now that v^2 is a primitive root of unity of degree $l > 1$. Then from (0.15) and (0.18) we obtain a simple Frobenius type formula valid for $[x,y]_v = 0$:

$$(x+y)^l = x^l + y^l \ . \tag{0.20}$$

Notice also that if l is odd, we have

$$(xy)^l = v^{l(l-1)}x^l y^l = x^l y^l = y^l x^l \ . \tag{0.21}$$

On a deeper level, these formulas lead to:

a). Existence of "unramified coverings" of classical simple groups in the category of quantum groups, icluding lifts of Frobenius morphisms to characteristic zero. This can be vaguely interpreted as existence of hidden fundamental groups of the classical algebraic groups, which become visible only in non-commutative geometry.

b). A parallelism between the representation theory of quantum groups in characteristic zero at roots of unity , and that of classical groups in finite characteristic, investigated by Lusztig.

0.5. Frobenius map for $GL_q(2)$. In this subsection, we return to the notation of 0.1. Let q be a primitive root of unity of odd degree l, a,b,c,d the matrix elements of $GL_q(2)$. Then we have ([15]):

i). a^l, b^l, c^l, d^l are central in $F[GL_q(2)]$.

ii). $\Delta\begin{pmatrix} a^l & b^l \\ c^l & d^l \end{pmatrix} = \begin{pmatrix} a^l & b^l \\ c^l & d^l \end{pmatrix} \otimes \begin{pmatrix} a^l & b^l \\ c^l & d^l \end{pmatrix}$.

iii). $\det\begin{pmatrix} a^l & b^l \\ c^l & d^l \end{pmatrix} = \left[DET_q\begin{pmatrix} a & b \\ c & d \end{pmatrix}\right]^l$.

In the language of non-commutative geometry, this can be rewritten as an exact sequence of quantum groups

$$1 \to H_q \to GL_q(2) \to GL_1(2) = GL(2) \to 1 \ , \tag{0.22}$$

where the third arrow is the Frobenius morphism Φ_l. The kernel H_q is a finite-dimensional Hopf algebra defined by relations (0.1) to which are added

$$a^l = d^l = 1, \ b^l = c^l = 0 \ .$$

It is flat over $\mathbf{Z}[q,q^{-1}]$ which is a natural definition ring for the whole setting. If l is an odd prime, (0.22) can be reduced modulo its prime divisor in this ring, and Φ_l reduces to a usual Frobenius morphism of $GL(2)$.

We sketch here a proof of i) and ii). Obviously, b^l and c^l commute with a,b,c,d, and a^l commutes with b,c. It takes slightly more efforts to check that $a^l d = da^l$. Since the determinant D is central (for all q), we have

$$a^l d = a^{l-1}(D + q^{-1}bc) = Da^{l-1} + q^{1-2l}bca^{l-1} =$$
$$(D+qbc)a^{l-1} = da^l \ .$$

One can treat d^l similarly.

In order to check ii), that is, to prove that, say, $\Delta(a^l) = \Delta(a)^l$, one applies (0.20) to the matrix elements of $\Delta\begin{pmatrix} a & b \\ c & d \end{pmatrix}$. This is possible, because $\Delta(a) = a\otimes a + b\otimes c$, and

$$(a\otimes a)(b\otimes c) = q^{-2}(b\otimes c)(a\otimes a) \ ,$$

where q^{-2} is still a primitive root of degree l.

After this digression, we return to one-dimensional differential geometry.

0.6. Differential operators. Define $\partial_v : A = k[x, x^{-1}] \to A$ by $df = dx.\partial_v f$. When $[n]_v$ for all $n \neq 0$ are invertible, the ring of differential operators can be defined as the ring generated by ∂_v and multiplications. Otherwise "quantized divided powers" $\partial_v^i/[i]!_v$ should be introduced as primary objects.

Let σ be the automorphism of A over k with $\sigma(x) = v^2x$. Then

$$fdx = dx.\sigma(f)$$

so that

$$\partial_v(fg) = \partial_v f.g + \sigma(f)\partial_v g \ . \tag{0.23}$$

In particular, $\partial_v \circ x - v^2 x \circ \partial_v = 1$, or

$$[\partial_v, x]_{v^{-1}} = v^{-1}\mathrm{id} \tag{0.24}$$

as operators on A. This is a quantized version of the Heisenberg commutation relation.

From (0.23) one easily deduces that if $[i]_v$ are invertible for $i \neq 0$, then the ring of differential operators is a free A-module freely generated by $1, \partial_v, \partial_v^2, \dots$.

0.7. Vector fields. The space of vector fields $A\partial_v$ is spanned by $L_n = vx^{n+1}\partial_v$. The usual commutation relations now deform to

$$[L_n, L_m]_{v^{n-m}} = [m-n]_v L_{n+m} \tag{0.25}$$

(apply both parts to x^r).

Notice that the parameter v^{n-m} of quantized commutator in (0.25) varies with n, m. A quantized version of the Virasoro algebra with zero central charge may be defined as an associative algebra generated by abstract symbols L_n subject to (0.25). However, if we want to define on it a comultiplication, we must take into account one more complication. Namely, the usual coalgebra structure upon $U(g)$, where g is a Lie algebra of vector fields, $\Delta(X) = X\otimes 1 + 1\otimes X$, is just a translation of the Leibniz rule $X(fg) = Xf.g + fXg$. But in our case it is replaced by (0.23):

$$L_n(fg) = L_n f.g + \sigma(f).L_n g, \ \sigma(fg) = \sigma(f)\sigma(g) \ .$$

This means that to obtain a closed formula for Δ, we must add σ to $\{L_i\}$. But then one can delete $\{L_0\}$ which can be expressed via $K_0 = \sqrt{\sigma}$:

$$K_0(x^r) = v^r x^r, \ \Delta(K_0) = K_0\otimes K_0 \ ,$$

$$L_0(x^r) = v^r[r]_v x^r = \frac{K_0^2 - 1}{v - v^{-1}} x^r \ .$$

Summarizing, we have (assuming v and $v-v^{-1}$ invertible) the following commutation relations and comultiplication rules:

$$\Delta(L_i)=L_i\otimes 1+K_0^2\otimes L_i;\ \Delta(K_0)=K_0\otimes K_0; \tag{0.26}$$

$$[L_n,L_m]_{v^{n-m}}=\begin{cases}[m-n]_vL_{n+m}, & \text{for } n+m\neq 0;\\ -[2n]_v\frac{K_0^2-1}{v-v^{-1}}, & \text{for } n+m=0;\end{cases} \tag{0.27}$$

$$K_0L_nK_0^{-1}=v^nL_n. \tag{0.28}$$

However, the problem of abstract compatibility of the comultiplication with the commutation relations is somewhat tricky. In 2.6 below, we recall a general principle connecting comultiplication with Leibniz formulas. It depends, however, on a non-degeneracy assumption which generally fails for our differential operator rings. Therefore, we do not claim that an abstract algebra defined by (0.27), (0.28) admits the comultiplication (0.26). The same remark applies to the next subsection.

0.8. Twisted $U(sl(2))$. For $v=1$, $\{L_n,L_0,L_{-n}\}$ generate a Lie subalgebra isomorphic to $sl(2)$. In our case, we take $\{L_n,K_0,L_{-n}\}$. In order to get the commutation relations in a more symmetric form, put (for a fixed $n>0$):

$$e=-v^{-n/2}[2n]_v^{-1}L_n,\qquad f=K_0^{-1}L_{-n}, k=K_0$$

(in this subsection, k has this meaning, and not that of a ground ring). Then we have

$$[e,f]_{v^{3n/2}}=\frac{k-k^{-1}}{v-v^{-1}}; \tag{0.29}$$

$$kek^{-1}=v^ne;\qquad kfk^{-1}=v^{-n}f; \tag{0.30}$$

$$\Delta(e)=e\otimes 1+k^2\otimes e;\qquad \Delta(f)=f\otimes k^{-1}+k\otimes f \tag{0.31}$$

This should be compared with the usual $U_{v^2}(sl(2))$:

$$[E,F]=\frac{K-K^{-1}}{v-v^{-1}}\ ; \tag{0.32}$$

$$KEK^{-1}=v^2E;\ KFK^{-1}=v^{-2}F\ ; \tag{0.33}$$

$$\Delta(E)=E\otimes 1+K\otimes E;\ \Delta(F)=F\otimes K^{-1}+1\otimes F;\ \Delta(K)=K\otimes K. \tag{0.34}$$

The principal difference is between (0.29) and (0.32): the commutator in (0.29) is twisted. We can also obtain a usual commutator in our setting, but then the r. h. s. of (0.32) becomes spoiled: put

$$\dot e=-[2n]_v^{-1}K_0^{-2}L_n,\qquad \dot f=K_0^{-2}L_{-n},\ \dot k=K_0^2\ ;$$

then

$$[\dot e,\dot f]=\frac{\dot k^{-2}-1}{v-v^{-1}}\ ; \tag{0.35}$$

$$\dot k\dot e\dot k^{-1}=v^{2n}\dot e;\qquad \dot k\dot f\dot k^{-1}=v^{-2n}\dot f; \tag{0.36}$$

$$\Delta(\dot e)=\dot e\otimes\dot k^{-1}+1\otimes\dot e;\qquad \Delta(\dot f)=\dot f\otimes\dot k^{-1}+1\otimes\dot f\ . \tag{0.37}$$

0.9. The standard $U_{v^2}(g_A)$. For reader's convenience, we reproduce here the Drinfeld-Jimbo presentation of $U_{v^2}(g_A)$, where g_A is the Lie algebra corresponding to a symmetrizable generalized Cartan matrix A.

Recall that $A = (a_{ij})$, $1 \le i, j \le n$, $a_{ii} = 2$, $a_{ij} \le 0$ for $i \ne j$; $d_i a_{ij} = d_j a_{ji}$, $d_i \in \mathbf{Z}$; $\gcd(d_i) = 1$.

We need n triples $(E_i, F_i, K_i^{\pm 1}), i = 1, \dots, n$. They satisfy analogues of (0.32)-(0.34), and more generally

$$K_i K_j = K_j K_i \ ; \tag{0.38}$$

$$K_i E_j K_i^{-1} = v^{d_i a_{ij}} E_j;\ K_i F_j K_i^{-1} = v^{-d_i a_{ij}} \ ; \tag{0.39}$$

$$E_i F_j - F_j E_i = \delta_{ij} \frac{K_i - K_i^{-1}}{v^{d_i} - v^{-d_i}} \ . \tag{0.40}$$

But the most essential novelty is the deformed Serre's relations:

$$\sum_{r+s=1-a_{ij}} (-1)^s \begin{bmatrix} 1 - a_{ij} \\ s \end{bmatrix}_{v^{d_i}} E_i^r E_j E_i^s = 0,\ i \ne j \ ; \tag{0.41}$$

and similarly for $\{F_i\}$.

We cannot, of course, "explain" these relations using only one-dimensional de Rham complex. A very intriguing new approach to them was recently developed by A. A. Beilinson, G. Lusztig, and R. MacPherson et al. in Duke MJ.

It is based upon the observation that, for $q = v^2$ a prime power, $v^{n-1}[n]_v$ coincides with the number of $\mathbf{F}_q$-points of $\mathbf{P}^{n-1}$. A sophisticated version of this simple remark, combined with an earlier work by Ringel, gives a geometric interpretation of (0.41) in terms of the geometry of flag manifolds over finite fields. This suggests a new kind of relations between the finite characteristic geometry and non-commutative geometry.

This paper is a slightly revised version of the Max-Planck-Institut preprint MPI/91-60. The list of references was completed by several papers that treat closely related matters: see especially [16, 17, 14, 18–20].

Acknowledgement. Parts of this work were done when the author had financial support of the Netherlands Mathematical Society, Utrecht University, and the Max-Planck Institut für Mathematik in Bonn. I am grateful to these institutions for support and the excellent working conditions.

§1. SKEW PRODUCTS

1.0. Notation. We denote by k a $\mathbf{Z}_2$-graded supercommutative ring. All k-modules are $\mathbf{Z}_2$-graded; $\tilde{a}$ denotes the $\mathbf{Z}_2$-degree, or parity, of a. All morphisms are homogeneous; we consider also odd morphisms of k-modules. All k-modules are assumed to be free; we consider mostly direct free submodules. All k-algebras are (super-)central. The tensor product of two k-algebras is endowed with multiplication $(a \otimes b)(a' \otimes b') = (-1)^{\tilde{a}'\tilde{b}} aa' \otimes bb'$.

1.1. Skew products. Let $A, B \subset C$ br two subalgebras of an associative k-algebra C. We shall say that they are skew-commuting if the multiplication maps induce injections of k-modules $m: A \otimes B \to C$, $m: B \otimes A \to C$ with the same image.

This image is then a subalgebra of C denoted $AB = BA$. If it coincides with C, we say that C is a skew product of A and B. Obviously, $A \otimes B$ is a skew product.

Given A, B, we want to classify their skew products. Intuitively, this amounts to classifying "good" cross-commutation relations between elements of the type $a \otimes 1$ and $1 \otimes b$ in $A \otimes B$ defining new multiplication on $A \otimes B$.

1.2. Basic construction. Assume that a skew product $C = AB = BA$ is given. It defines (and is defined by) the k-linear isomorphism $r = r_C : B \otimes A \to A \otimes B$ induced by multiplication in C. Consider for every $b \in B$ two operators $\tau_b : A \to A \otimes B$, $\sigma_b : A \to B \otimes A$ defined by

$$\tau_b(a) = r(b \otimes a); \qquad \sigma_b(a) = (-1)^{\dot{a}\dot{b}} r^{-1}(a \otimes b) .$$

Clearly, τ_b and σ_b are left k-linear in a, of parity $\dot{b}$. Their dependence of b is even left k-linear.

We want to consider τ_b (resp. σ_b) as elements of k-algebras $\mathbf{End}_{k-mod}(A) \otimes B$ (resp. $B \otimes \mathbf{End}_{k-mod}(A)^{op}$) where **End** is the algebra of both even and odd endomorphisms.

We will work everything out in coordinates. Choose a $\mathbf{Z}_2$-graded free k-basis $\{b_i\}$, $i \in I$, of B. For variable $a \in A$, denote by $\tau_{b,j}(a)$ coefficients in the following equivalent expressions:

$$\tau_b(a) = r(b \otimes a) = \sum_j (-1)^{\dot{b}_j \dot{a}} \tau_{b,j}(a) \otimes b_j \qquad (\text{in } A \otimes B),$$

$$(1.1) \qquad ba = \sum_j (-1)^{\dot{b}_j \dot{a}} \tau_{b,j}(a) b_j \qquad (\text{in } C)$$

and write (1.1) symbolically as

$$(1.2) \qquad \tau_b = \sum_j \tau_{b,j} \otimes b_j .$$

Similarly, put

$$(1.3) \qquad \sigma_b = \sum_i b_i \otimes \sigma_{b,i}$$

meaning that for all $a \in A$,

$$(-1)^{\dot{a}\dot{b}} \sigma_b(a) = r^{-1}(a \otimes b) = \sum_i (-1)^{\dot{a}\dot{b}} b_i \otimes \sigma_{b,i}(a),$$

$$(1.4) \qquad ab = \sum_i (-1)^{\dot{a}\dot{b}} b_i \sigma_{b,i}(a) .$$

The sums (1.2), (1.3) are generally infinite. However, they are finite, if B is finite dimensional, or if B is graded by finite dimensional submodules B^n, $AB^n = B^n A$, and $\{b_i\}$ is homogeneous with respect to this grading. In the sequel, we shall assume one of these conditions.

1.2.1. LEMMA. *$\tau_{b,j}$ and $\sigma_{b,j}$ are k-mod morphisms $A \to A$ of parity $\dot{b} + \dot{b}_j$.*

PROOF. For $c \in k$, we have from (1.1):

$$\tau_b(ca) = r(b \otimes ca) = (-1)^{\dot{b}\dot{c}} c\, r(b \otimes a) = (-1)^{\dot{b}\dot{c}} c \sum_j (-1)^{\dot{b}_j \dot{a}} \tau_{b,j}(a) \otimes b_j .$$

On the other hand,

$$\tau_b(ca) = \sum_j (-1)^{\dot{b}_j(\dot{c}+\dot{a})} \tau_{b,j}(ca) \otimes b_j \ ,$$

so that $\tau_{b,j}(ca) = (-1)^{(\dot{b}+\dot{b}_j)\dot{c}} \tau_{b,j}(a)$.

Similarly, from (1.4),

$$\begin{aligned}\sigma_b(ca) &= (-1)^{\dot{c}+\dot{a}\dot{b}} r^{-1}(ca \otimes b) = c(-1)^{(\dot{a}+\dot{c})\dot{b}} r^{-1}(a \otimes b)\\ &= c\sum_i (-1)^{\dot{c}\dot{b}} b_i \otimes \sigma_{b,i}(a) = c(-1)^{\dot{c}\dot{b}} \sigma_b(a) \ ,\end{aligned}$$

and from (1.3)

$$\sigma_b(ca) = \sum_i b_i \otimes \sigma_{b,i}(ca) \ ,$$

so that

$$\sigma_{b,i}(ca) = (-1)^{\dot{c}(\dot{b}+\dot{b}_i)} c\sigma_{b,i}(a) \ .$$

1.2.2. Lemma. *The maps*

$$\tau : B \to \mathbf{End}_{k\text{-}mod}(A) \otimes B : \ b \mapsto \tau_b \ , \tag{1.5}$$

$$\sigma : B \to B \otimes \mathbf{End}_{k\text{-}mod}(A)^{op} : \ b \mapsto \sigma_b \tag{1.6}$$

are $\mathbf{Z}_2$-graded ring homomorphisms, such that for $c \in k$, $\tau_c = 1 \otimes c = c \otimes 1 = \sigma_c$.

Proof. We first remark that these maps are invariantly defined. In the definition of C^{op} for a $\mathbf{Z}_2$-graded ring C, we of course introduce signs: C^{op} has the same additive group as C, and multiplication $c_1 * c_2 = (-1)^{\dot{c}_1\dot{c}_2} c_2 c_1$.

Now, identifying via multiplication $B \otimes A$ and $A \otimes B$ with their images in C, we have in view of (1.1):

$$\begin{aligned}\tau_{b'b}(a) &= r(b'b \otimes a) = b'ba = \sum_l (-1)^{\dot{b}_l\dot{a}} b'\tau_{b,l}(a) b_l\\ &= \sum_l (-1)^{\dot{b}_l\dot{a}} \sum_j (-1)^{\dot{b}_j \dot{\tau}_{b,l}(a)} \tau_{b',j}(\tau_{b,l}(a)) \otimes b_j b_l \ .\end{aligned}$$

On the other hand, by (1.2):

$$\begin{aligned}(\tau_{b'}\tau_b)(a) &= \left(\sum_j \tau_{b',j} \otimes b_j\right)\left(\sum_l \tau_{b,l} \otimes b_l\right)(a)\\ &= \sum_{j,l} (-1)^{\dot{b}_j \dot{\tau}_{b,l}} (\tau_{b',j}\tau_{b,l} \otimes b_j b_l)(a)\\ &= \sum_{j,l} (-1)^{\dot{b}_j(\dot{b}+\dot{b}_l)+\dot{a}(\dot{b}_j+\dot{b}_l)} \tau_{b',j}\tau_{b,l}(a) \otimes b_j b_l \ .\end{aligned}$$

It remains to compare signs:

$$\tilde{b}_l\tilde{a} + \tilde{b}_j(\tilde{a} + \tilde{b} + \tilde{b}_l) = \tilde{b}_j(\tilde{b} + \tilde{b}_l) + \tilde{a}(\tilde{b}_j + \tilde{b}_l) \ .$$

Similarly, by (1.4),

$$\begin{aligned}\sigma_{b'b}(a) &= (-1)^{\tilde{a}(\tilde{b}+\tilde{b}')} r^{-1}(a\otimes b'b) = (-1)^{\tilde{a}(\tilde{b}+\tilde{b}')}(ab')b \\ &= (-1)^{\tilde{a}(\tilde{b}+\tilde{b}')}(-1)^{\tilde{a}\tilde{b}'}\sum_i b_i(\sigma_{b',i}(a)b) \\ &= (-1)^{\tilde{a}\tilde{b}}\sum_i\sum_j(-1)^{(\tilde{b}'+\tilde{b}_i+\tilde{a})\tilde{b}} b_i b_j\otimes\sigma_{b,j}(\sigma_{b',i}(a)) \\ &= \sum_{i,j}(-1)^{(\tilde{b}'+\tilde{b}_i)\tilde{b}} b_i b_j\otimes\sigma_{b,j}\sigma_{b',i}(a)\ .\end{aligned}$$

Finally, by (1.3), taking into account the reverse multiplication in $\mathbf{End}(A)^{op}$:

$$\begin{aligned}(\sigma_{b'}\sigma_b)(a) &= \Big(\sum_i b_i\otimes\sigma_{b',i}\Big)\Big(\sum_j b_j\otimes\sigma_{b,j}\Big)(a) \\ &= \sum_{i,j}(-1)^{\tilde{b}_j(\tilde{b}'+\tilde{b}_i)} b_i b_j\otimes(-1)^{(\tilde{b}'+\tilde{b}_i)(\tilde{b}+\tilde{b}_j)}\sigma_{b,j}\sigma_{b',i}(a)\ .\end{aligned}$$

1.2.3. Structural action. Consider now the two universal algebra morphisms, which are automatically universal bialgebra coactions, linear (and homogeneous) upon $\{b_i\}$:

$$\delta_l : B\to E_l(B)\otimes B:\qquad \delta_l(b_i)=\sum_j z_{ij}\otimes b_j\,,$$

$$\delta_r : B\to B\otimes E_r(B):\qquad \delta_r(b_i)=\sum_j b_j\otimes \dot{z}_{ji}\ .$$

Here $z_{ij}\in E_l(B)$ (resp. $\dot{z}_{ji}\in E_r(B)$) are algebra generators arranged in a format of a multiplicative matrix. In particular, the parity of z_{ij}, $\dot{z}_{ji}$ is $\tilde{b}_i+\tilde{b}_j$.

There exists a natural map

$$E_r(B)\to E_l(B):\qquad \dot{z}_{ji}\mapsto(-1)^{(\tilde{b}_j+\tilde{b}_i)\tilde{b}_j} z_{ij}$$

("supertransposition") which is an isomorphism of bialgebras

$$st:\qquad (E_r(B),\Delta_r)\to(E_l(B),\Delta_l^{op})\ ,$$

where Δ_r, Δ_l are the respective comultiplications.

Using (1.5), (1.6), and universality, one sees that instead of r, σ one can give the two algebra homomorphisms

$$\begin{aligned}\lambda:&\qquad E_l(B)\to\mathbf{End}_{k\text{-}mod}(A);\\ \rho:&\qquad E_r(B)^{op}\to\mathbf{End}_{k\text{-}mod}(A)\end{aligned}$$

This means that, starting with a skew product of A and B, we have constructed actions of $E_l(B)$ and $E_r(B)^{op}$ upon the left k-module A. We shall call them ***structural actions.***

The comultiplication map $\Delta_r:\quad E_r(B)\to E_r(B)\otimes E_r(B)$ is also a comultiplicatuon map for $E_r(B)^{op}$. Using it, we can define the action ϕ of $E_r(B)^{op}$ upon $A\otimes A$: for $e\in E_r(B)^{op}$, $c\in A\otimes A$, put

$$\phi(e)(c)=(\rho\otimes\rho)(\Delta(e))(c)\ .$$

1.2.4. Lemma. *The multiplication morphism $m : A \otimes A \to A$ is a morphism of $E_r(B)^{op}$-modules.*

Proof. For a basis $\{b_j\}$ of B, put as above

$$\delta_r(b_j) = \sum_i b_i \otimes \hat{z}_{ij}, \qquad \hat{z}_{ij} \in E_r(B)^{op}, \tag{1.7}$$

and

$$\sigma_{ij} = \sigma_{b_j,i} .$$

Comparing (1.7) and (1.3) one sees that $\hat{z}_{ij}$ acts upon A by the operator $\rho(\hat{z}_{ij}) = \sigma_{ij}$. To prove the Lemma, we must check that for every $z \in E(B)^{op}$, $c \in A \otimes A$, we have

$$m[\Delta_r(z)(c)] = z(m(c)) \tag{1.8}$$

(we should have written here $\rho(z)$ instead of z, and $(\rho \otimes \rho)(\Delta(z)$ instead of $\Delta(z)$).

First, it suffices to verify (1.8) for generators $z = \hat{z}_{ij}$. In fact, both sides are linear in z, and if (1.8) holds for z_1, z_2 and all c, it holds also for $z_1 z_2$ and all c:

$$m[\Delta(z_1 z_2)(c)] = m[\Delta(z_1)(\Delta(z_2)(c))] = z_1(m[\Delta(z_2)(c)]) = z_1 z_2 m(c) .$$

Second, it suffices to check (1.8) for decomposable $c = f \otimes g$, because both sides are biadditive in c. Now,

$$\Delta(\hat{z}_{ij}) = \sum_k \hat{z}_{ik} \otimes \hat{z}_{kj} .$$

Therefore, (1.8) reduces to the equality of the following two expressions:

$$\begin{aligned} m[\Delta(\hat{z}_{ij})(f \otimes g)] &= m\left[\left(\sum_k \hat{z}_{ik} \otimes \hat{z}_{kj}\right)(f \otimes g)\right] \\ &= \sum_k (-1)^{(\tilde{b}_k + \tilde{b}_j)\tilde{f}} \sigma_{ik}(f)\sigma_{kj}(g), \\ \hat{z}_{ij}(m(f \otimes g)) &= \sigma_{ij}(fg) . \end{aligned}$$

To this end we apply (1.4) in turn to the cases $b = b_j$, and $a = g, f, fg$:

$$\begin{aligned} (fg)(b_j) = f(gb_j) &= \sum_k (-1)^{\tilde{g}\tilde{b}_j} (fb_k)\sigma_{kj}(g) \\ &= \sum_k (-1)^{\tilde{g}\tilde{b}_j} \sum_i (-1)^{\tilde{f}\tilde{b}_k} b_i \sigma_{ik}(f)\sigma_{kj}(g); \\ (fg)b_j &= \sum_i (-1)^{(\tilde{f}+\tilde{g})\tilde{b}_j} b_i \sigma_{ij}(fg) . \end{aligned}$$

Remark. For operators τ, one can prove a similar identity (but notice different orders of f, g in two sides):

$$\tau_{ik}[m(f \otimes g)] = m[\Delta(\tau_{ik})(g \otimes f)] .$$

1.2.5. Lemma. *Put*

$$\sigma_{ij} = \sigma_{b_j,i}, \quad \tau_{kl} = \tau_{b_k,l}, \quad \hat{\tau}_{kl} = (-1)^{\dot{b}_k(\dot{b}_k+\dot{b}_l)}\tau_{lk} \ .$$

Then we have

$$\sum_j \sigma_{kj}\hat{\tau}_{ji} = \delta_{ki} id_A \ , \tag{1.9}$$

$$\sum_j \hat{\tau}_{kj}\sigma_{ji} = \delta_{ki} id_A . \tag{1.10}$$

Proof. Rewrite $b_i a$ by first applying (1.1) and then (1.4):

$$b_i a = \sum_j (-1)^{\dot{b}_j \dot{a}} \tau_{ij}(a) b_j = \sum_j (-1)^{\dot{b}_j \dot{a}} \sum_k (-1)^{\dot{b}_j(\dot{a}+\dot{b}_i+\dot{b}_j)} b_k \sigma_{kj}(\tau_{ij}(a)) \ .$$

This gives (1.9).

Similarly, rewrite ab_i by first applying (1.4) and afterwards (1.1):

$$ab_i = \sum_j (-1)^{\dot{a}\dot{b}_i} b_j \sigma_{ji}(a) = \sum_j (-1)^{\dot{a}\dot{b}_i} \sum_k (-1)^{\dot{b}_k(\dot{a}+\dot{b}_i+\dot{b}_j)} \tau_{jk}\sigma_{ji}(a) b_k \ .$$

This gives (1.10).

We can now state the main result of this section.

1.3. Theorem. *The construction described above establishes a bijection between the following two sets:*

(SP) Skew products of A and B (compatible with the gradation of B).

(EA) Actions of $E_r(B)^{op}$ upon A, defined by an invertible operator-valued matrix (σ_{kj}), such that $m : A \otimes A \to A$ is a morphism of $E_r(B)^{op}$-modules. Here $(E_r(B), \Delta_r)$ is the universal coalgebra coacting upon B (compatibly with gradation).

Proof. We have already constructed the map (SP)$\to$(EA). It is injective because the knowledge of an $E_r(B)^{op}$-action allows one to reconstruct σ and then the crosscommutation relations in $B \otimes A$ corresponding to the given skew product structure.

It remains to show that this map is surjective. This means that, starting with an action, we must define an associative multiplication, say, on $B \otimes A$, and to check that it defines a skew product. Obviously, the multiplication must be defined by the formula

$$(b \otimes a)(b' \otimes a') = (m_B \otimes m_A)(b \otimes \sigma_{b'}(a) \otimes a') \ ,$$

where σ is reconstructed from the action. The rest of the proof consists of checks that have essentially be made already in the course of proving our Lemmas 1.2.1 - 1.2.4. We will only show associativity in some detail.

It suffices to establish that the product of three elements of the type $b_j \otimes a$ does not depend on the bracket configuration. We will omit the tensor product sign. Take the three elements $b_j a'$, $b_l a''$, $b_n a'''$ We can assume that $b_j = 1$, $a''' = 1$, because they will cancel in the identity. So finally we want

$$(a' b_l a'') b_n = a'(b_l a'' b_n) \ .$$

First calculate the l.h.s. expression:

$$(a'b_l a'')b_n = \sum_k (-1)^{\dot{a}'\dot{b}_l} b_k \sigma_{kl}(a')a''b_n$$
$$= \sum_k (-1)^{\dot{a}'\dot{b}_l} b_k \sum_m (-1)^{\dot{b}_n(\dot{a}''+\dot{b}_k+\dot{b}_l+\dot{a}')} b_m \sigma_{mn}[\sigma_{kl}(a')a'']$$

and take into account that

$$\sigma_{mn}[\sigma_{kl}(a')a''] = \sum_i (-1)^{(\dot{b}_i+\dot{b}_n)(\dot{a}'+\dot{b}_k+\dot{b}_l)} (\sigma_{mi}\sigma_{kl})(a')\sigma_{in}(a'') .$$

Second, calculate the r.h.s. expression:

$$a'(b_l a'' b_n) = a'b_l \sum_i (-1)^{\dot{a}''\dot{b}_n} b_i \sigma_{in}(a'')$$
$$= \sum_k (-1)^{\dot{a}'\dot{b}_l} b_k \sigma_{kl}(a') \sum_i (-1)^{\dot{a}''\dot{b}_n} b_i \sigma_{in}(a'') .$$

It remains to pass b_i to the left beyond $\sigma_{kl}(a')$. We will obtain a sum of members of the type $b_m(\sigma_{mi}\sigma_{kl})(a')\sigma_{in}(a'')$ with some signs. A direct verification shows that signs will be the same as for the l.h.s.

1.4. Remarks. The following questions deserve a further study.

a). In the context of de Rham complexes of the type considered here, A and B are usually both graded so that the same skew product can be obtained dually from the action of $E_r(A)^{op}$ on B. A particularly symmetric situation arises if the same coalgebra acts and coacts upon A and B simultaneously. (Notice however that the differential breaks the symmetry).

b). If the action of $E_r(B)^{op}$ factorizes through the Hopf envelope, the matrix (σ_{ij}) is automatically invertible (apply the antipode map). The case when this Hopf algebra is commutative, i.e. consists of functions on a group scheme acting on B is especially interesting.

c). The quantization parameters in our examples can be viewed as describing the structural action. Hence intuitively introduction of $E_r(B)^{op}$ can be imagined as "a quantization of Planck's constant".

§2. QUANTUM DE RHAM COMPLEXES AND DIFFERENTIAL OPERATORS

2.0. Notation. In this section, we study differential $(\mathbf{Z}, \mathbf{Z}_2)$-graded k-algebras Ω. We call the $\mathbf{Z}_2$-degree parity, and the $\mathbf{Z}$-degree dimension, denote by Ω^i the k-module of elements of dimension i, and assume that $\Omega^0 = k$, $\Omega^{<0} = 0$. We denote sometimes Ω^0 by A ("functions").

Differential d is of degree (1,1), and is left k-linear in the sense of superalgebra. The Leibniz formula $d(ab) = (da)b + (-1)^{\dot{a}} adb$ involves parity rather than dimension.

We assume also given a bigraded k-submodule $B = \oplus_{i=0}^{\infty} B^i$ ("differential forms with constant coeficients") where all B^i are free of finite rank. The strongest condition we impose on B is that it is a quadratic subalgebra of Ω consisting of closed elements, such that Ω is a skew product of A and B. Weaker conditions sufficient for establishing some of the algebraic properties of Ω are stated at appropriate places below, and marked C1 etc.

2.1. Differentials and vector fields. Here we assume the following.

C1. *Multiplication in Ω induces linear isomorphisms* $B^1 \otimes A \to \Omega^1 \leftarrow A \otimes B^1$.

Let $\{b_i\}$ be a homogeneous basis of B^1. Then, as in §1, we can define operators $\sigma_{ij}, \tau_{ij} : A \to A$ by

$$fb_j = \sum_i (-1)^{\tilde{f}\tilde{b}_j} b_i \sigma_{ij}(f); \tag{2.1}$$

$$b_i f = \sum_j (-1)^{\tilde{f}\tilde{b}_j} \tau_{ij}(f) b_j. \tag{2.2}$$

where $f \in A$.

Moreover, we can define a family of "quantum vector fields" $\partial_i : A \to A$ dual to $\{b_i\}$ by

$$df = \sum_{i=1}^{n} b_i \partial_i f, \ f \in A \ . \tag{2.3}$$

Clearly, ∂_i is left k-linear of parity $\tilde{b}_i + 1$. From the Leibniz formula for d,

$$d(\omega\nu) = d\omega.\nu + (-1)^{\tilde{\omega}} \omega d\nu \tag{2.4}$$

and (2.1), (2.3) one obtains the twisted Leibniz formulas for ∂_i: if $f, g \in A$,

$$\partial_i(fg) = \partial_i f.g + \sum_j (-1)^{\tilde{f}\tilde{b}_j} \sigma_{ij}(f) \partial_j(g) \ . \tag{2.5}$$

We supplement (2.5) by the respective "Leibniz formulas" for σ_{ij} and τ_{ij}:

$$\sigma_{ij}(fg) = \sum_k (-1)^{(\tilde{b}_j + \tilde{b}_k)\tilde{f}} \sigma_{ik}(f) \sigma_{kj}(g); \tag{2.6}$$

$$\tau_{ij}(fg) = \sum_k (-1)^{(\tilde{b}_j + \tilde{b}_k)\tilde{g}} \tau_{ij}(f) \tau_{jk}(g) \ . \tag{2.7}$$

They follow from the identities

$$(fg)b_j = f(gb_j), \ b_j(fg) = (b_j f)g) \ .$$

2.2. Basic problem. In the setting of 2.0, 2.1, we can construct an algebra U of operators on A generated by ∂_i, σ_{ij} and possibly also τ_{ij} and some left or right multiplications by elements of A. It is natural to consider such algebras as a replacement in non-commutative geometry of Lie algebras generated by vector fields, and generally of rings of differential operators. We want to understand the following problems.

a). What are the relations between $\sigma_{ij}, \tau_{ij}, \partial_i$? (The relations between these operators and multiplications are given by Leibniz formulas).

b). Is it possible to define the structure of a coalgebra upon U in such a way that its comultiplication Δ on the generators would be given by the formulas compatible with (2.5), (2.6):

$$\Delta(\partial_i) = \partial_i \otimes 1 + \sum_j \sigma_{ij} \otimes \partial_j. \tag{2.8}$$

$$\Delta(\sigma_{ij}) = \sum_k \sigma_{ik} \otimes \sigma_{kj} \ . \tag{2.9}$$

In other words, we want $m: A\otimes A \to A$ to be a morphism of U-modules, the structure of an U-module upon $A\otimes A$ being defined via Δ.

The results of §1 give a partial answer to these questions. Namely, if Ω is a skew product of A and B, we have described the "universal part" of the relations between σ_{ij}. More precisely, we know the relations between generators of $E_r(B)^{op}$, and they are universal in the sense that they do not depend neither on A nor on the particular choice of the skew product structure AB. And (2.9) actually defines a comultiplication compatible with the Leibniz formulas (Lemma 1.2.4).

We proceed now to describe relations between ∂_i themselves, and cross-commutation relations between ∂_i, and σ_{ij}, τ_{ij}, again trying to understand only the universal part, and mostly assuming quadraticity. In 2.6 we will address U directly.

2.3. Quadratic relations between vector fields. In the setting of 2.1, denote by D a free k-module given together with an *odd* non-degenerate pairing $\langle|\rangle: B^1\otimes_k D\to k$. The sign rules are:

$$\langle \alpha b|\partial\rangle = \alpha\langle b|\partial\rangle = (-1)^{\tilde\alpha(\tilde b+1)}\langle b|\alpha\partial\rangle;\quad \alpha\in k,\ b\in B^1,\ \partial\in D \tag{2.10}$$

(mnemonically, $|$ is odd). Parity of $\langle b|\partial\rangle$ is $\tilde b+\tilde\partial+1$.

Next, introduce an even pairing

$$(\ ,\ ): (B^1)^{\otimes 2}\otimes_k D^{\otimes 2}\to k$$

by the formula

$$(b\otimes b', \partial\otimes\partial') = (-1)^{(\tilde\partial+\tilde b')(\tilde\partial'+1)}\langle b|\partial'\rangle\langle b'|\partial\rangle\ . \tag{2.11}$$

It is well defined. In fact, using (2.10) we check first that the r.h.s actually depends only on $b\otimes b'$ and $\partial\otimes\partial'$ that is, gives the same result when evaluated upon $b\alpha\otimes b'$ and $b\otimes\alpha b'$ etc. Moving α from b to b' at the r.h.s. we get an extra sign:

$$[(-1)^{\tilde\alpha(\tilde\partial'+1)}]^2 = 1\ .$$

Similarly, replacing $\partial\alpha\otimes\partial'$ by $\partial\otimes\alpha\partial'$:

$$(-1)^{\tilde\alpha(\tilde\partial'+1)+\tilde\alpha(\tilde\partial+\tilde b')}(-1)^{\tilde\alpha(\tilde\partial+1+\tilde b'+\tilde\partial')} = 1\ .$$

Notice that (2.11) is k-linear in $b\otimes b'$ but only semilinear in $\partial\otimes\partial'$:

$$(b\otimes b'\alpha, \partial\otimes\partial') = (-1)^{\tilde\alpha}(b\otimes b', \alpha\partial\otimes\partial')\ . \tag{2.12}$$

We can make it bilinear by considering instead ${}^tD^{\otimes 2}$ which is $D^{\otimes 2}$ as an additive group, but has a twisted left k-module structure:

$$\alpha\partial\otimes\partial'\ (\text{in } {}^tD^{\otimes 2}) = (-1)^{\alpha}\alpha\partial\otimes\partial'\ (\text{in } D^{\otimes 2})\ .$$

Clearly, ${}^tD^{\otimes 2}$ and $D^{\otimes 2}$ have the same lattices of submodules and the same sets of bases of submodules.

Let now $\{b_i\}$, $\{D_j\}$ be dual bases of B^1, $D: \langle b_i|D_j\rangle = \delta_{ij}$. Then, from (2.11),

$$(b_i\otimes b_j, D_k\otimes D_l) = (-1)^{\tilde b_i}\delta_{il}\delta_{jk}\ . \tag{2.13}$$

Denote by $R\subset (B^1)^{\otimes 2}$ the k-submodule of quadratic relations, that is, the kernel of the multiplication map $(B^1)^{\otimes 2}\to\Omega$. Assume that R is a free direct submodule, and define $R_D\subset D^{\otimes 2}$ as orthogonal complement to R w.r.t. (2.11). Denote by B^2 the image of $(B^1)^{\otimes 2}$ in Ω and impose the next condition:

C2. *Multiplication in Ω induces a linear injection $B^2\otimes A\to\Omega$.*

Then we have:

2.3.1. PROPOSITION. *Choose dual bases $(b_i|D_j) = \delta_{ij}$ as above. Assume that $db_i = 0$. Define $\partial_j : A \to A$ by (2.3). Then every element of R_D vanishes (as an operator on A) when evaluated on $\{\partial_i\}$ instead of $\{D_i\}$.*

Comment. Assume that B is a quadratic graded superalgebra generated by B^1. Denote by B^τ the "odd dual opposite" quadratic superalgebra which is defined as $T(D)/(R_D)$ where $T(D)$ is the tensor algebra of D. Our Prop. 2.3.1 says that the operator algebra generated by $\{\partial_j\}$ is a quotient of B^τ.

On the other hand, $E_r(B)^{op}$ in this case can be described as $(B^! \bullet B)^{op}$ where $B^!$ is the even dual quadratic superalgebra, and the "black product" $\bullet$ is defined via tensor multiplication of relation modules (see [1, 2, 21]). For the sake of completeness, and in order to fix all signs in the presence of odd variables and constants, we shall reproduce below in 2.4 a coordinate description.

The simplest classical case is $A = k[x_1, ..., x_r]$, $B = \Lambda[dx_1, ..., dx_r]$. Proposition 2.3.1 generalizes the fact that $\partial/\partial x_i$ pairwise commute whereas dx_i anticommute.

PROOF. From (2.3) we see that for every $f \in A$

$$0 = d^2 f = \sum_{ij} (-1)^{\tilde{i}} b_i b_j \partial_j \partial_i f \ . \tag{2.14}$$

Choose a free $\mathbf{Z}_2$-graded k-graded right k-basis $r_\alpha \in (B^1)^{\otimes 2}$ of R, and complement it by elements $s_\beta \in (B^1)^{\otimes 2}$ such that $\{r_\alpha, s_\beta\}$ form a right k-basis of $(B^1)^{\otimes 2}$; $\alpha = 1, ..., m; \beta = m+1, ..., n^2$.

Denote by $\{S_\alpha, R_\beta\}$ the dual left basis of ${}^t D^{\otimes 2}$ such that

$$(r_\alpha, R_\beta) = (S_\alpha, s_\beta) = 0; (r_{\alpha_1}, S_{\alpha_2}) = \delta_{\alpha_1, \alpha_2}; \quad (s_{\beta_1}, R_{\beta_2}) = \delta_{\beta_1, \beta_2} \ . \tag{2.15}$$

In particular, R_β form a basis of $R_D \subset D^{\otimes 2}$. Then we have in $(B^1)^{\otimes 2} \otimes {}^t D^{\otimes 2}$:

$$\sum_{ij} (-1)^{\tilde{b}_i} b_i \otimes b_j \otimes D_j \otimes D_i = \sum_\alpha r_\alpha \otimes S_\alpha + \sum_\beta s_\beta \otimes R_\beta \tag{2.16}$$

because of (2.13) and (2.15). Denote by $S_\alpha(\partial)$ (resp. $R_\beta(\partial)$) the operators obtained by replacing D_i's by ∂_i's. Similarly, denote by $r_\alpha(b)$ (resp. $s_\beta(b)$) the image of r_α (resp. s_β) in B^2 Then $r_\alpha(b) = 0$, and $s_\beta(b)$ are right linearly independent over A. From (2.14) and (2.16) we see that

$$\sum_\beta s_\beta(b) R_\beta(\partial) f = 0$$

for all f, so that $R_\beta(\partial) = 0$.

2.4. Quadratic relations between τ_{ij}, and σ_{ij}. Keeping notation of 2.3, put

$$r_\alpha = \sum_{ij} c_\alpha^{ij} b_i \otimes b_j, \ c_\alpha^{ij} \in k \ . \tag{2.17}$$

Denote by B^{1*} the right even dual B^1, given together with an even non-degenerate scalar product with the obvious sign conventions. Extend it to an even scalar product $(B^1)^{\otimes 2} \otimes (B^{1*})^{\otimes 2} \to k$ by the rule

$$(b_1 \otimes b_2, b^1 \otimes b^2) = (-1)^{\tilde{b}^1 \tilde{b}_2} (b_1, b^1)(b_2, b^2) \ .$$

Denote by $B^!$ the quadratic algebra $T(B^{1*})/(R^c)$ where R^c is the orthogonal complement to R.

Let $\{b_i\}$, $\{b^j\}$ be dual bases in B^1, B^{1*}. As above, choose a free basis $\{r_\alpha, s_\beta\}$ of $B^1\otimes B^1$, and let $\{s^\alpha, r^\beta\}$ be the dual basis of $B^{1*}\otimes B^{1*}$. Then $\{r^\beta\}$ is a basis of R^c. Put

$$r^\beta = \sum_{kl} b^k\otimes b^l c^\beta_{kl}, \quad c^\beta_{kl}\in k \tag{2.18}$$

We have then for every k, l

$$b_k\otimes b_l = \sum_\beta (-1)^{\tilde{b}_k\tilde{b}_l} c^\beta_{kl} s_\beta \quad \bmod R\ . \tag{2.19}$$

To see it, take the scalar product of both sides with all r^β. Put now

$$E = T(B^1\otimes B^{1*})/(S_{(23)}(R\otimes R^c))\ .$$

This algebra has generators $b_i\otimes b^j = z_{ij}$ satisfying relations $S_{(23)}(r_\alpha\otimes r^\beta)$, or explicitely

$$\sum_{ijkl} c^{ij}_\alpha(-1)^{\tilde{b}_j\tilde{b}_k} z_{ik}z_{jl}c^\beta_{kl} = 0\ . \tag{2.20}$$

Put $B = T(B^1)/(R)$. Then the map $\delta:\ B\to E\otimes B$ transforming $\{b_i\}$ by (z_{ij}) is the universal left coaction.

2.4.1. PROPOSITION. *$\{\tau_{ik}\}$ satisfy (2.20), whereas $\{\sigma_{ik}\}$ satisfy the opposite supertransposed relations.*

PROOF. We have, using (2.19):

$$\begin{aligned} 0 = r_\alpha(b)f &= \sum_{ij} c^{ij}_\alpha b_i b_j f = \sum_{ij} c^{ij}_\alpha b_i \sum_l (-1)^{\tilde{f}\tilde{b}_l}\tau_{jl}(f)b_l \\ &= \sum_\alpha^{ij}\sum_l (-1)^{\tilde{f}\tilde{b}_l}\sum_k (-1)^{(\tilde{f}+\tilde{\tau}_{jl})\tilde{b}_k}\tau_{ik}\tau_{jl}(f)b_k b_l \\ &= \sum_{ijkl\beta} c^{ij}_\alpha(-1)^{\tilde{f}(\tilde{b}_k+\tilde{b}_l)+\tilde{b}_j\tilde{b}_k}\tau_{ik}\tau_{jl}(f)c^\beta_{kl}s_\beta(b)\ . \end{aligned}$$

It remains to simplify the sign using

$$\tilde{c}^\beta_{kl} = \tilde{r}^\beta + \tilde{b}_k + \tilde{b}_l$$

and to take into account that s_β are right A-independent. One can treat σ similarly.

Finally, I state relations between τ_{ij} and ∂_k. Unfortunately, I was unable to interpret them conceptually.

2.5.1. PROPOSITION. *For i, β fixed, $f \in A$, we have*

$$\sum_{jkl}(-1)^{\tilde{f}\tilde{b}_j+(\tilde{\partial}_k+\tilde{b}_i+\tilde{f})\tilde{b}_l}\tau_{kl}(\partial_k(\tau_{ij}(f)))c^{\beta}_{lj}$$

$$=\sum_{jkl}(-1)^{\tilde{b}_i+\tilde{b}_k(\tilde{f}+\tilde{\partial}_j)+(\tilde{b}_j+\tilde{\partial}_j+\tilde{f})\tilde{b}_l}\tau_{il}(\tau_{jk}(\partial_j(f)))c^{\beta}_{lk} .$$

There are similar relations for σ. To prove them, one should differentiate (2.1) and (2.2).

2.6. Bialgebra structures on operator rings. Returning now to the setting of 2.2, let us discuss the comultiplication problem in relarion to Leibniz formula. This is well-known, and we include it for the sake of completeness.

Generally, let A be a k-algebra, and $U \subset \mathbf{End}_k(A)$ a k-module of left linear operators upon A. Assume that A admits an augmentation map $\epsilon: A \to k$, and for $u \in U$, $f \in A$ put

$$\langle u, f\rangle = \epsilon(u(f)) . \tag{2.21}$$

2.6.1. PROPOSITION. *Assume that for every $u \in U$, there exist $u_i, v_i \in U$, $i = 1, \ldots, n$ such that for all $f, g \in A$*

$$u(fg) = \sum_i(-1)^{\tilde{f}\tilde{v}_i}u_i(f)v_i(g) . \tag{2.22}$$

Assume also that (2.21) has trivial left kernel. Then U admits a unique comultiplication map $\Delta: U \to U \otimes U$ with the property

$$(\Delta(u), f\otimes g) = \langle u, fg\rangle \tag{2.23}$$

for all $f, g \in A$. It is coassociative, and admits a counit $\epsilon_U: u \mapsto \epsilon(u1)$. If, in addition, U is an algebra. then (U, Δ, ϵ_U) is a bialgebra.

SKETCH OF PROOF. Existence follows from (2.22): put $\Delta(u) = \sum u_i \otimes v_i$. Uniqueness follows from non-degeneracy. We leave the rest to the reader.

2.6.2. Example. For $A = k[x, x^{-1}]_v$ (see sec. 0.2), $U = \oplus_{i\neq 0}kL_i \oplus K_0^2$ (see sec. 0.6), $\langle u, f\rangle$ has trivial left kernel precisely when v is not a root of unity.

§3. EXAMPLES CONTINUED: DE RHAM COMPLEXES OF QUANTUM SPACES AND $GL_{Q,\lambda}(n)$

3.1. Two-parametric family of (2,2)-quantum matrices. Here we assume k pure even. Choose two invertible even elements $p, q \in k$. The ring of polynomial functions E on the non-commutative quantum space $M_{p,q}(2)$ is generated by even variables a, b, c, d with the following commutation relations:

$$\begin{aligned} &ab = p^{-1}ba; \quad ac = q^{-1}ca; \quad ad = da + (q^{-1} - p)cb; \\ &bc = pq^{-1}cb; \quad bd = q^{-1}db; \quad cd = p^{-1}dc . \end{aligned} \tag{3.1}$$

This ring is a bialgebra with the standard comultiplication and counit. For $pq \neq -1$, it can be defined as universally coacting algebra on the function algebras of two quantum planes:

$$A_q : k\langle x, y\rangle/(xy - q^{-1}yx); \; x, y \quad \text{even}; \tag{3.2}$$

$$B_p : \; k\langle \xi, \eta\rangle/(\xi^2, \eta^2, \xi\eta + p\eta\xi); \xi, \eta \quad \text{odd} . \tag{3.3}$$

Adding the relation $ad - p^{-1}bc = 1$, we get the coordinate ring of $SL_{p,q}(2)$.

Order the generators by $d < b < c < a$. The Diamond Lemma is applicable, and shows that monomials $d^k c^l b^m a^n$ form a k-basis of E. (Of course, a similar statement is true and can be easily checked also for A_q, B_p.)

A slight generalization of the Wess-Zumino construction gives the following result.

3.2. PROPOSITION. *There are precisely two skew products Ω of A_q and B_p, admitting a differential with $dx = \xi$, $dy = \eta$, such that all these structures are compatible with the coaction of E (with zero differential).*

SKETCH OF PROOF. We actually prove slightly more: the skew product property should be postulated only on forms of degree two and three in (x, y, ξ, η); the full strength then follows from the Diamond Lemma.

Comparing dimensions, we see that we must find a system of four cross-commutation relations expressing $x\xi, x\eta, y\xi, y\eta$ linearly via $\xi x, \xi y, \eta x, \eta y$. (Actually, one relation is known: $d(xy - q^{-1}yx) = 0$). We have sixteen indeterminate coefficients which are successively constrained by the rest of conditions. It is convenient to use them in the following order.

i). Apply the differential to the cross-commutation relations. The result should be a consequence of (3.3). This eliminates about half of coefficients.

ii). Use compatibility with the coaction of E. This leaves one parameter free. Of course, this can also be explained conceptually, by analyzing the decomposition of the space

$$(kx \oplus ky) \otimes (k\xi \oplus k\eta) \oplus (k\xi \oplus k\eta) \otimes (kx \oplus ky)$$

with respect to the coaction of E. The result is:

$$x\xi = [(p + q^{-1})r - 1]\xi x; \qquad y\eta = [(p + q^{-1})r - 1]\eta y;$$
$$x\eta = (rq^{-1} - 1)\xi y + pq^{-1}r\eta x; \quad y\xi = r\xi y + (pr - 1)\eta x \tag{3.4}$$

where r is the last undetermined coefficient.

iii). Constraint r by checking linear relations between cubic monomials. In practice, it suffices to resolve the "overlap ambiguity" $x\eta\xi$, that is, to reduce this monomial first as $x(\eta\xi)$ and second as $(x\eta)\xi$. In both cases, the result is proportional to $\xi\eta x$, but the coefficients coincide iff $r^2 - (p^{-1} + q)r + qp^{-1} = 0$, that is, $r = p^{-1}$ or $r = q$. The net result is:

Variant 1: $\Omega^{(1)}_{p,q}$

$$x\xi = (pq)^{-1}\xi x; \qquad y\eta = \eta y;$$
$$x\eta = (p^{-1}q^{-1} - 1)\xi y + q^{-1}\eta x; \quad y\xi = p^{-1}\xi y \, . \tag{3.5}$$

Variant 2: $\Omega^{(2)}_{p,q}$

$$x\xi = pq\xi x; \quad y\eta = pq\eta y;$$
$$x\eta = p\eta x; \quad y\xi = q\xi y + (pq - 1)\eta x \, . \tag{3.6}$$

Notice that (3.5) and (3.6) are connected by the isomorphism interchanging x and y, ξ and η, p and p^{-1}, which is compatible with the isomorphism interchanging a and d, b and c.

iv). Finally, one checks that the combined relations (3.2), (3.3) and either (3.5), or (3.6), satisfy all the conditions of the Diamond Lemma, that is, all ambiguities of cubic monomials are resolvable (with respect to any of the orders $\xi < \eta < x < y$ or $x < y < \xi < \eta$.

We now turn to the calculation of the de Rham complex of $M_{p,q}(2)$. We start with determining the universal differential quadratic algebra $W^{(i)}_{p,q}$ left coacting upon $\Omega^{(i)}_{p,q}$. Denote $\alpha, \beta, \gamma, \delta$ the differentials of a, b, c, d respectively. We assume $pq \neq -1$.

3.3. Proposition. *$W^{(1)}_{p,q}$ is defined by the relations (3.1) and the following ones:*
a). Relations between differentials of matrix coefficients:

$$\begin{gathered}\alpha^2 = \beta^2 = \gamma^2 = \delta^2 = 0;\\ \beta\alpha = -q^{-1}\alpha\beta; \quad \gamma\alpha = -p^{-1}\alpha\gamma;\\ \delta\gamma = -q^{-1}\gamma\delta; \quad \delta\beta = -p^{-1}\beta\delta \ ;\end{gathered} \tag{3.7}$$

$$pq\delta\alpha + p\gamma\beta = -\alpha\delta - q\beta\gamma \ . \tag{3.8}$$

b). Cross-commutation relations:

$$\begin{gathered}a\alpha = (pq)^{-1}\alpha a;\\ a\beta = (p^{-1}q^{-1} - 1)\alpha b + p^{-1}\beta a;\\ a\gamma = q^{-1}\gamma a + (p^{-1}q^{-1} - 1)\alpha c \ ;\end{gathered}$$

$$\begin{gathered}b\alpha = q^{-1}\alpha b;\\ b\beta = (pq)^{-1}\beta b;\\ b\delta = q^{-1}\delta b + (p^{-1}q{-1} - 1)\beta d \ ;\end{gathered}$$

$$\begin{gathered}c\alpha = p^{-1}\alpha c;\\ c\gamma = (pq)^{-1}\gamma c;\\ c\delta = p^{-1}\delta c + (p^{-1}q^{-1} - 1)\gamma d \ ;\end{gathered}$$

$$\begin{gathered}d\beta = p^{-1}\beta d;\\ d\gamma = q^{-1}\gamma d;\\ d\delta = (pq)^{-1}\delta d;\end{gathered} \tag{3.9}$$

$$\begin{gathered}a\delta + pc\beta = (p^{-1}q^{-1} - 1)\alpha d + (q^{-1} - p)\gamma b + p^{-1}\beta c + \delta a;\\ pqd\alpha + qb\gamma = \alpha d + p\gamma b;\\ pqd\alpha + pc\beta = \alpha d + q\beta c;\end{gathered} \tag{3.10}$$

$W^{(2)}_{p,q}$ can be obtained from here by the involution described above.

Comment. The proof consists of direct calculations which we will omit. The answer is not quite satisfactory because we want the de Rham complex of $M_{p,q}(2)$ to be of the same size as that of the commutative polynomial ring of four variables. For this to hold in degree two, we miss one quadratic relation between differentials and one cross-comutation relation. They are supplied by the following result, again stressing the role of skew products.

3.4. THEOREM. *For $pq \neq 1$, $W^{(i)}_{p,q}$ has precisely three quadratic differential quotients for which quadratic and cubic components are freely spanned by the lexicographically ordered monomials in $\alpha < \beta < \gamma < \delta < d < c < b < a$. Every such quotient is a skew product of $k[a,b,c,d]$ and $k[\alpha,\beta,\gamma,\delta]$ (modulo relations), and their Hilbert series are the same as in the commutative case. For $pq = 1$, these quotients coincide.*

For $W^{(1)}_{p,q}$, the missing relations are as follows:

Variant 1:

$$d\alpha = \alpha d;\ \delta\alpha = -\alpha\delta\ . \tag{3.11}$$

Variant 2:

$$c\beta = p^2\beta c;\ \gamma\beta = -p^2\beta\gamma\ . \tag{3.12}$$

Variant 3:

$$b\gamma = q^2\gamma b; \beta\gamma = -q^2\gamma\beta\ . \tag{3.13}$$

SKETCH OF PROOF. we start with writing a missing cross-commutation relation, say, for $b\gamma$ as a linear combination of the lexicographically ordered monomials, and then check the resolution of overlaps in cubic monomials. Differentiating the answer, we get a missing relation between the differentials. Finally, we check the conditions of the Diamond Lemma for cubic overlaps.

3.5. Remarks. The long calculations needed to check the Theorem 3.4 were made by D. Zhdanovich and the author. The Proposition 3.3 was independently proved by G. Maltsiniotis [3], who has also discovered the missing relations (3.11), but not (3.12), (3.13). The reason was that Maltsiniotis was only looking for the DGA-quotients of $W^{(1)}_{p,q}$ compatible with the coalgebra structure, and thus leading to the de Rham complex of $GL_{p,q}(2)$. Only (3.11) satisfy this condition. (The existence of $W^{(2)}_{p,q}$ also is not mentioned by Maltsiniotis).

D. Zhdanovich remarked that (3.7), (3.8), and $(3.11)_{\alpha,\delta}$ determine the algebra of the differential forms with constant coefficients which is naturally isomorphic to $M_{q,p}(2)^!$ (even quadratic dual). This does not hold for the other two quotients.

The following remark is also due to D. Zhdanovich. Relations (3.12) and (3.13) can be obtained by imitating the Wess-Zumino approach. More precisely, let us *start* with an algebra of functions determined by (3.1), and its differentials defined by (3.7), (3.8), and either $(3.12)_{\gamma,\beta}$ or $(3.13)_{\gamma,\beta}$. Let us then construct the universal bialgebra Fcoacting upon these algebras (by the same (4,4)-matrix on coordinates and their differentials). Then the cross-commutation relations (3.9), (3.10), and $(3.12)_{c,\beta}$ (resp. $(3.14)_{b,\gamma}$) are uniquely defined by the Wess-Zumino type conditions.

Question. Is F a flat deformation of the polynomial algebra of four variables?

3.6. n-dimensional Wess-Zumino de Rham complex. The following generalization of the DGA (3.5) was considered by several people, among them G. Maltsiniotis, E. Demidov, E. Mukhin, after the discovery by M. Artin, J. Tate, A. Sudbery et. al. of the correct pair of n-dimensional quantum spaces leading to the construction of many-parametric quantum group $GL_{P,Q}(n)$ by the universal coaction method. This algebra is generated by x_i, $i = 1,\ldots,n$; $\xi_i = dx_i$, subject to the following commutation relations:

$$x_i x_j = q_{ji}^{-1} x_j x_i, \qquad i<j; \tag{3.14}$$

$$\xi_i^2 = \xi_i\xi_j + \lambda_j q_{ji}^{-1}\xi_j\xi_i = 0, \qquad i<j; \tag{3.15}$$

$$\xi_i x_i = \lambda_i x_i \xi_i; \tag{3.16}$$

$$\xi_i x_j = \lambda_j q_{ji}^{-1} x_j \xi_i; \qquad i<j; \tag{3.17}$$

$$x_i\xi_j = q_{ji}^{-1}\xi_j x_i - (1-\lambda_j^{-1})\xi_i x_j; \quad i<j\ . \tag{3.18}$$

Here $q_{ij}(i<j), \lambda_j$ are invertible quantization parameters. One can check without serious difficulties that this DGA $\Omega_{Q,\lambda}$ is well defined and is a flat deformation of the de Rham complex of the polynomial ring in n variables. In fact, the lexicographically ordered monomials in x, ξ form its free k-basis.

G.Maltsionitis has applied the universal coaction construction in the DGA-category to $\Omega_{Q,\lambda}$ and established the following result [4]:

3.7. THEOREM. *The universal coacting DGA-algebra for $\Omega_{Q,\lambda}$ has a quotient which is a flat deformation of the standard de Rham complex of the polynomials in matrix entries iff $\lambda_j = \lambda$ does not depend on j. This quotient is then unique, and is a skew product of two subalgebras, generated by matrix entries, and their differentials, respectively.*

For a list of relations, see [4].

3.8. Question. Do there exist other quotients, with the standard Hilbert function, generalizing (3.12) and (3.13) to general n? Up to know, problems of this type have been treated by a direct application of the Diamond Lemma. G. Maltsionitis says it took him eighty pages of calculations to prove Theorem 3.7. About forty pages were spent to the discovery of (3.12) and (3.13). Clearly, a more intelligent approach is highly desirable.

3.9. Exercise. Calculate operator algebras discussed in §2 in the examplea above. In particular, calculate the algebras generated by σ_{ij} and $x_i\partial_j$ as possible analogs of $U(gl(n))$.

§4. COHOMOLOGY OF ELEMENTARY EXTENSIONS

4.1. Elementary extensions. Let $U \subset V$ be an embedding of differential algebras, $t \in V$ an even element, dt its differential. We shall say that V is *an elementary extension of* V by (t, dt), if the following conditions are satisfied:

i). $\{t^i; dt.t^j\}$ for $i, j \geq 0$ form a free basis of V as a right U-module.

ii). $(dt)^2 = 0$; $tdt = v^2 dt.t$ for an invertible element $v \in k$.

We can then slightly generalize the calculations of 0.2, 0.3 and prove the following result.

4.2. THEOREM. *Assume that all non-vanishing $[i]_v$ are invertible in k.*

a). If v is not a root of unity, then the embeding $U \to V$ defines the isomorphism of the cohomology spaces $H^\bullet(U) \to H^\bullet(V)$.

b). If v is a primitive root of unity of degree l, then $H^\bullet(V)$ as a right $H^\bullet(V)$-module is freely generated by the cohomology classes of cocycles t^{jl}, $j \geq 0$, and $dt.t^{jl-1}$, $j \geq 1$.

PROOF. Write a generic element of V as

$$s = f + \sum_{i\geq 1} t^i g_i + \sum_{j\geq 0} dt.t^j h_j; \quad f, g_i, h_j \in U\ . \tag{4.1}$$

Then, by (0.13)

$$ds = df + \sum_{i\geq 1} t^i dg_i + \sum_{j\geq 0} dt.t^j(v^j[j+1]_v g_{j+1} - dh_j) \ . \tag{4.2}$$

Hence s is a cycle iff f, g_i $(i \geq 1)$ are cycles, and

$$v^j[j+1]_v g_{j+1} = dh_j \ . \tag{4.3}$$

Now, if all $[j+1]_v$ for $j \geq 0$ are non-vanishing and hence invertible, then the last summands of (4.1) add up to a boundary, because

$$\begin{aligned} t^{j+1}g_{j+1} + dt.t^j h_j &= v^{-j}[j+1]_v^{-1} t^{j+1} dh_j + dt.t^j h_j \\ &= v^{-j}[j+1]_v^{-1} d(t^{j+1}h_j) \ . \end{aligned}$$

Hence the cohomology class of s coincides with that of f, a cycle in U, and from (4.2) one easily sees that $H^\bullet(U) \to H^\bullet(V)$ is also injective.

If $[j+1]_v = 0$ precisely for $j+1 \equiv 0 \mod l$, the same argument shows that s is equivalent to

$$s' = f + \sum_{j\geq 1} t^{lj} g_{lj} + \sum_{j\geq 1} dt.t^{lj-1} h_{lj-1} \ ,$$

where f, g_{lj}, h_{lj-1} are cycles in U.

Then (4.2) shows that these cycles are defined uniquely modulo boundaries in U.

4.3. Application to the n-dimensional Wess-Zumino-de Rham complex. Let Ω_n be the DGA described in 3.6. One easily checks that the natural embedding $\Omega_{n-1} \subset \Omega_n$ (with compatible sets of quantization constants) makes Ω_n a left elementary extension of Ω_{n-1}, for which $v_n^2 = \lambda_n^{-1}$, which allows us to calculate $H^\bullet(\Omega_n)$. From the Theorem 3.7 one infers that Ω_n is "quantum homogeneous" exactly when $\lambda_1 = \ldots \lambda_n = \lambda$, so that the cohomology of Ω_n is then determined by this unique parameter.

Actually, Ω_n is a skew product of Ω_{n-1} and $k[x_n, dx_n]_{v_n}$. To check this, it suffices to prove that the standard ordered monomials in x_i, ξ_i can be expressed as linear combinations of monomials ordered by decreasing indices. To put ξ_n to the leftmost place, one should use (3.18) successively.

4.4. Question. Can one apply a similar reasoning to the de Rham complexes of matrix bialgebras described in §3?

References

[1] *Manin Yu.* Some remarks on Koszul algebras and quantum groups. Ann. Inst. Fourier. 1987. V. 37. № 4. P. 191-205.

[2] *Manin Yu.* Quantum groups and non-commutative geometry. CRM, Université de Montréal, 1988.

[3] *Maltsiniotis G.* Groupes quantiques et structures différentielles. C. R. Acad. Sci. 1990. V. 311. P. 831-834.

[4] *Maltsiniotis G.* Calcul différentiel sur le groupe linéaire quantique. Preprint, 1990.

[5] *Woronowicz S.* Differential calculus on compact matrix pseudogroups (quantum groups). Commun. Math. Phys. 1989. V. 122. P. 125-170.

[6] *Podleś P.* Differential calculus on matrix spheres. Lett. Math. Phys. 1991. V. 22.

[7] *Pusz W.* Twisted canonical anticommutation relations. Rep. Math. Phys. 1989. V. 27. P. 349-360.

[8] *Pusz W.* On the implementation of $S_\mu U(2)$ action in the irreducible representations of twisted canonical commutation relations. Lett. in Math. Phys. 1991. V. 21. P. 59-67.
[9] *Wess J., Zumino B.* Covariant differential calculus on the quantum hyperplane. Nucl. Phys. 1990 (vol. in honor of R. Stora). V. 18B. P. 301.
[10] *Manin Yu.* Topics in non-commutative geometry. Princeton University Press, 1991.
[11] *Drinfeld V.* Quantum groups. Proc. Int. Math. Berkeley. V. 1. 1986. P. 798-820.
[12] *Connes A.* Non-commutative differential geometry. Publ. Math. de l'IHES. 1985. V. 62. P. 41-144.
[13] *Baez J.* R-commutative geometry and quantization of Poisson algebras. Preprint, Wellesley College, 1991.
[14] *Baez J.* Differential calculi for quantum vector spaces with Hecke-type relations. Preprint, Wellesley College, 1991.
[15] *Parshall B., Wang J.-P.* Quantum linear groups I, II. Preprint, 1989.
[16] *Tsygan B.* Notes on differential forms on quantum groups. Preprint, Penn State, 1991.
[17] *Demidov E.* Multiparametric quantum deformations of the general linear supergroup (in Russian). PhD thesis. Moscow University, 1992.
[18] *Ogievetsky O.* Differential operators on quantum spaces for $GL_q(n)$ and $SO_q(n)$. Preprint MPI-Ph/91-103,
[19] *Jurco B.* Differential calculus on quantized simple Lie groups. Lett. Math. Phys. 1991. V. 22.
[20] *Schmüdgen K.* Covariant differential calculi on quantum spaces. Preprint. Leipzig, 1991.
[21] *Manin Yu.* Multiparametric quantum deformation of the general linear supergroup. Commun. Math. Phys. 1989. V. 123. P. 163-175.

Математический институт им.В.А.Стеклова
Российской академии наук

Поступила в редакцию
17.VI.1992г.

Ю. И. Манин

ЗАМЕЧАНИЯ О КВАНТОВЫХ ГРУППАХ И КВАНТОВЫХ КОМПЛЕКСАХ ДЕ РАМА

Изучаются комплексы де Рама квантовых пространств и квантовых групп с помощью обобщения техники "универсального кодействия" на дифференциальные алгебры.

PART IV. MISCELLANY

Mathematics as Metaphor

*Yuri Ivanovich Manin**

Steklov Mathematical Institute, 42 Vavilova, 117966 GSP-1, Moscow, USSR

あかあかやあかあかあかやあかあかや
あかやあかあかあかあかや月

Myōo (1173–1232)

Ordre. [...] Je sais un peu ce que c'est et combien peu de gens l'entendent. Nulle science humaine ne le peut garder. Saint Thomas ne l'a pas gardé. La mathématique le garde, mais elle est inutile en sa profondeur.

Pascal, Pensées

Introduction

When H. Poincaré first published in 1902 his book *La Science et l'hypothèse*, it became a bestseller. The first chapter of this book was devoted to the nature of mathematical reasoning. Poincaré discussed an old philosophical controversy: can mathematical knowledge be reduced to long chains of tautological transformations of some basic ("synthetic") truths or does it contain other things as well? He argued that the creative power of mathematics was due to a free choice of the initial hypotheses-definitions which were later constrained by comparing deductions with the observable world.

We seem to be much less interested in philosophical subtleties than were Poincaré's contemporaries. It is not that science itself has become less popular. Books like S. Weinberg's *The First Three Minutes* and S.W. Hawking's *A Brief History of Time* sell by the thousands and are favorably reviewed in widely distributed newspapers. What has changed is the general mood. The paradoxality of the new physical theories is perceived less dramatically and more pragmatically. (The perception of the visual arts evolved in much the same way: the first exhibitions of the Impressionists were a kind of spiritual revolution, but each new wave of the post-war avant-garde immediately acquired family traits of academism.)

* Delivered by Barry Mazur.

Proceedings of the International Congress of Mathematicians, Kyoto, Japan, 1990

In this atmosphere, the heated discussions of bygone days about the foundational crisis of mathematics and the nature of infinity seem almost irrelevant and certainly inappropriate. The audience responds in a much livelier way to opinions about school education or to a new generation of computers.

That is why I have decided to present in this section an unpretentious essay in which our science is considered as a specialized dialect of the natural language, and its functioning as a special case of speech. This implies certain suggestions about high school and university training.

Metaphorism

The word "metaphor" is used here in a non-technical sense, which is best rendered by the following quotations from James P. Carse's book *Finite and Infinite Games*:

> "Metaphor is the joining of like to unlike such that one can never become the other."
>
> "At its root all language has the character of metaphor, because no matter what it intends to do, it remains language, and remains absolutely unlike whatever it is about."
>
> "The unspeakability of nature is the very possibility of language."

Considering mathematics as a metaphor, I want to stress that the interpretation of mathematical knowledge is a highly creative act. In a way, mathematics is a novel about Nature and Humankind. One cannot tell precisely what mathematics teaches us just as one cannot tell what exactly we are taught by *War and Peace*. The teaching itself is submerged in the act of re-thinking this teaching.

This opinion seemingly disagrees with the time-honored tradition of applied mathematics in scientific and technological calculations.

In fact, I want only to restore a certain balance between the technological and the human sides of mathematics.

Two Examples

Let me try to illustrate the metaphoric potential of mathematics by discussing two disjoint subjects: Kolmogorov complexity and the "Dictator Theorem" of K. Arrow.

(i) The Kolmogorov complexity of a natural number N is the length of the shortest program P that can generate N, or the length of the shortest code of N. The reader should imagine a way of coding integers which is a partial recursive function $f(P)$ whose values are whole numbers. Kolmogorov's theorem states that among all such functions there exist ones that are most economical in the following sense: if $C_f(N)$ is the minimal value of P such that $f(P) = N$, then $C_f(N) \leq$ const.$C_g(N)$, where const. depends only on f and g but not on N.

Since P can be reconstructed from its binary notation, the length $K_f(N)$ of the shortest program generating N is bounded by $\log_2 C_f(N)$. This function, or rather the class of all such functions defined up to a bounded summand, is the Kolmogorov complexity.

First, $K(N) \leq \log N + \text{const}$. This conforms nicely with the historical successes of positional notation systems which have provided us with number generating programs of logarithmic length. However, there are arbitrarily large integers whose Kolmogorov complexity is much smaller than the length of their notation. e.g., $K(10^N) \leq K(N) + \text{const}$. In general, when we use large numbers, we seem to use only those numbers whose Kolmogorov complexity is relatively small. Even decimals of π which are, probably, the longest well-defined numbers ever produced by mathematicians, are Kolmogorov simple, because $K([10^N \pi]) \leq \log N + \text{const}$. In general, small Kolmogorov complexity = high degree of organization.

On the other hand, almost all integers N have complexity close to $\log N$. For example, if $f(P) = N$ for an optimal f, then $K(P)$ is equivalent to $\log P$. Such integers have many remarkable properties which we usually connect with "randomness."

Second, Kolmogorov complexity can be easily defined for discrete objects which are not numbers, such as Russian or English texts. Therefore, *War and Peace* has a rather well defined measure of complexity; its indeterminacy is connected with the choice of an optimal method of coding and seems to be quite small if one chooses one of the small number of reasonable codings.

Question: Is *War and Peace* a highly organized or an almost random combinatorial object?

Third, Kolmogorov complexity is a non-computable function. More precisely, if f is optimal, then there is no recursive function $G(N)$ which differs from $C(N)$ by $\exp(O(1))$. One can only bound complexity by computable functions.

I feel that Kolmogorov complexity is a notion that must be kept in mind in any discussion of the nature of human knowledge.

As long as the content of our knowledge is expressed symbolically (verbally, digitally, ...) there are physical restrictions on the volume of information that can be kept and handled. We always rely on methods for compressing information. Kolmogorov complexity puts absolute restrictions on the efficiency of such compressions. When we speak, say, of physical laws, expressed by the equations of motion, we mean that a precise description of the behavior of a physical system can be obtained by translating these laws into a computer program. But the complexity of laws that we can discover and use is clearly bounded. Can we be sure that there are no laws of arbitrarily high complexity governing even "elementary" systems?

At this point, our discussion becomes totally non-mathematical. Since this is a mathematically-minded audience, I must stop here. But such is the fate of every metaphor.

(ii) Arrow's Dictator Theorem was discovered around 1950. Mathematically, it is a combinatorial statement describing certain functions with values in binary relations. Intuitively, it is a formalized discussion of the problem of Social Choice. Suppose that a lawmaker has to establish a law which governs the processing of individual wills of voters into a collective decision. If the problem is to choose one of two alternatives, the standard solution is to rely on the majority of votes. However, there are usually more than two alternatives (think of funds-allocation problems), and voters may be asked to order them according to their preferences. What should be the algorithm that extracts the collective preference from any set of individual preferences? Arrow considered algorithms satisfying some natural and democratic axioms (e.g., when everybody prefers A to B, the society prefers A to B). Nevertheless, he discovered that when there are more than two alternatives, the only way to achieve a solution is to nominate a member of the society ("the Dictator") and to make his personal preference the social one. (Actually this is one of the versions of Arrow's theorem discovered later. Also, it refers to the case of a finite society; in the infinite case, the social decisions can be made by ultrafilters, appropriately called "the ruling hierarchies.")

In a way, this theorem illustrates the content of Jean-Jacques Rousseau's idea of a Contrat Social.

The fundamental inconsistency of the image of the ideal democratic choice can be illustrated by the following story referring to three voters and three alternatives. It is the story of three knights-errant at a crossroads with a stone before them. The inscription on the stone prophesies only losses: he who goes to the left will lose his sword; he who goes to the right will lose his horse; he who goes straight will lose his head. The knights dismount and take counsel. In a Russian version of this story the knights have names and personalities: the youngest and most ardent is Alyosha

Popovich, the oldest and wisest is Dobrynya Nikitich, and the middle brother, a slow peasant, is Ilya Muromets. Alyosha values his sword more than his horse, and his horse more than his head; Dobrynya values most his head, then his sword, then his horse; and Ilya prefers his horse to his head, and his head to his sword.

The reader will note that the three individual preference orders constitute one and the same cyclic order on the set of alternatives. As a result, one can decide by majority the choice between any two alternatives, but the union of these decisions will be inconsistent: the democratic procedure cannot provide us with a well-ordered list. The knights sigh and delegate the decision-making power to Dobrynya.

Does Arrow's theorem tell us something that we did not know beforehand? Yes, I think it does if we are ready to discuss it seriously, that is, to look closely at the combinatorial proof, to imagine the possible real-life content of various assumptions and the elementary logical steps made on the way - briefly, to enhance our imprecise imagination by the rigid logic of mathematical reasoning. We can understand better, for example, some tricks of policy making and some pitfalls into which society can easily, even wholeheartedly, fall (like accepting without questioning a list of alternatives imposed by a ruling hierarchy, although precisely the compilation of this list can be the central issue of the social decision-making).

At this stage, we come to the main topic of our discussion: what distinguishes a mathematical discourse from a natural language discourse, why the Pascalian "ordre" came to rule over our specialized symbolic activities, and whether it is truly so "useless in its profundity"?

Language and Mathematics

A very interesting chapter of the interaction between mathematics and the humanities started about thirty years ago when automatic translation was first seriously attempted. These first attempts were a painful failure, at least for many optimists who believed that in this area there are no fundamental obstacles and it remains only to overcome technical difficulties connected with the sheer amount of information to be processed. In other words, they took for granted that translation is, in principle, performed by a not very complex algorithm which needs only to be made explicit and then implemented as a computer program.

This assumption is a nice example of a mathematical metaphor (actually, an instance of the general "computer metaphor" used in the brain sciences).

This metaphor proved to be extremely fruitful for theoretical linguistics because it forced linguists to start describing vocabulary, semantics, accidence, and syntax of human languages with an unprecendented degree of explicitness and completeness. Through this program, some totally new notions and tools were discovered.

However, the successes of automatic translation were (and still are) scanty. It became clear that written human speech is extremely bad input data for any algorithmic processing planned as translation or even as a deduction. (I add this proviso because there is nothing special in human speech considered as material for, say, statistical studies).

This fact can be considered as a universal property of human languages, and it deserves some attention. One must first of all reject as too naive the usual explanation that the universe of meanings of a human language is too vast and too poorly structured to admit a well organized metalanguage that describes it. The point is that even if we severely restrict this universe to the subset of arithmetic of small integer quantities, we still face the same difficulty. In fact, this difficulty was a decisive reason for the crystallization of the whole system of arithmetical notation and of the basic algorithms of calculation, and, later on, of symbolic algebra. Even the vocabulary of elementary arithmetic in human language is basically archaic: the finite natural series of primitive societies "one, two, three, indefinitely many" is reproduced on an exponential scale in our "thousand, million, billion, zillion." The expressions for relatively small numbers like "1989" are actually names of the decimal notation and not of the numbers themselves.

Viète's algebra was superior to the semi-verbal algebra of Diophantus not because it could express new meanings but because of its incomparably greater susceptibility to algorithmic processing ("identical transformations" of our high school algebra).

The rupture of the intuitive and emotional ties between a text and its producer/user so characteristic of the language of science was compensated for by the new computational automatisms. In their (albeit restricted) domain they proved to be infinitely more efficient than the traditional Platonian and Aristotelian culture of everyday-language discourse. Why then are our scientific papers still written as a disorganized mixture of words and formulas? Partly because we still need those emotional ties; partly because some meanings (like human values) are best rendered in human language. But even as a medium of scientific speech, human language has some inherent advantages: by appealing to the spatial and qualitative imagination, it helps us to understand "structurally stable" properties, such as the number of free parameters (dimension), existence of extrema, symmetries. To put it bluntly, it makes possible the metaphorical use of science.

6

Metaphor and Proof

The views professed here can be considered in relation to the high school and graduate curriculum.

The general mathematical education of the first half of this century was application-oriented. It provided the basic minimum for handling problems of everyday life and for a smooth transition to the study of engineering and scientific calculations at the college level. The break of this curriculum with the activity of professional mathematicians became more and more pronounced. As is well known, this brought a reaction in the form of New Math in the USA and of similar programs in other countries. These programs introduced into high school mathematics the notions and principles borrowed from the professionals: set theory, axiomatic presentation of proofs, and the culture of strict definitions.

New Math became widely accepted, but its expansion was accompanied by protesting voices which in the 70s and 80s merged into a loud chorus. The critics disagreed with the basic arguments of the New Math proponents. Leaving aside the objections based on data from the cognitive sciences and from learning psychology, I shall only recall those concerning the general evaluation of the role of proof in mathematics.

The one pole is represented by the well known statement of Nicolas Bourbaki: 'Dépuis les Grécs, qui dit Mathématiques, dit démonstrations." Following this perception, rigorous proof was made a matter of principle in the New Math programs. It was argued that a) a proof helps us to understand a mathematical fact; b) a rigorous proof is the most essential component of modern professional mathematics; c) mathematics has universally recognized criteria of rigor.

These views were extensively criticised, e.g. by Gila Hanna in the book "Rigorous Proof in Mathematics Education," OISE Press, Ontario, 1983. In particular, Gila Hanna pointed out that mathematicians are far from unanimous in accepting the criteria of rigor (referring to quarrels between logicists, formalists and intuitionists), and that working mathematicians constantly break all the rules in the book.

In my opinion, this is irrelevant.

What is relevant is the imbalance among various basic values which is produced by the emphasis on proof. Proof itself is a derivative of the notion of "truth." There are values besides truth, among them "activities," "beauty" and "understanding," which are essential in high school teaching as well as later. A teacher (or a university professor) who neglects these values fails tragically. Unfortunately, this also is not universally recognized. A sociological analysis of the controversies around the Catastrophe Theory of René Thom shows that what provoked such criticism was the shift of orientation from formal truth to understanding. Of course, Catastrophe Theory is one of the developed mathematical metaphors and should only be judged as such.

Pedagogically, a proof is just one of the genres of a mathematical text. There are many different genres: a calculation, an annotated sketch, a computer program, a description of an algorithmic language, or the neglected genre of a discussion of the connections between a formal definition and intuitive notions. Every genre has its own laws, and, in particular, its laws of rigor, and the only reason why they are not codified is because they have not been paid special attention.

The key task of a teacher is to demonstrate in the restricted area of his or her course the variety of mathematical activities and of the underlying value orientations. Of course, this variety is hierarchically organized. The goals may vary from achieving an elementary arithmetical and logical literacy to programming skills, and from the simplest everyday problems to the principles of modern scientific thinking. In the spectrum of these goals, the emphasis on the norms of "rigorous proof" can safely occupy a peripheral position.

Having said all this, I must stress that my argument by no means undermines the ideal of rigorous mathematical reasoning. This ideal is a fundamental principle of mathematics, and in this sense Bourbaki is certainly right. Since it had no external object of study, and was based on the consensus of a restricted circle of devotees, mathematics could not develop without permanent control by rigid rules. The applicability of mathematics in the strict sense of the word (like its indispensability in the Apollo project) is due to our ability to control series of symbolic manipulations of fantastic length.

The existence of this ideal is far more essential than its unattainability. The freedom of mathematics (G. Cantor) can only develop within the limits of iron necessity. The hardware of modern computers is an incarnation of this necessity.

Metaphor helps a human being to breathe in this rarefied atmosphere of the gods.

8

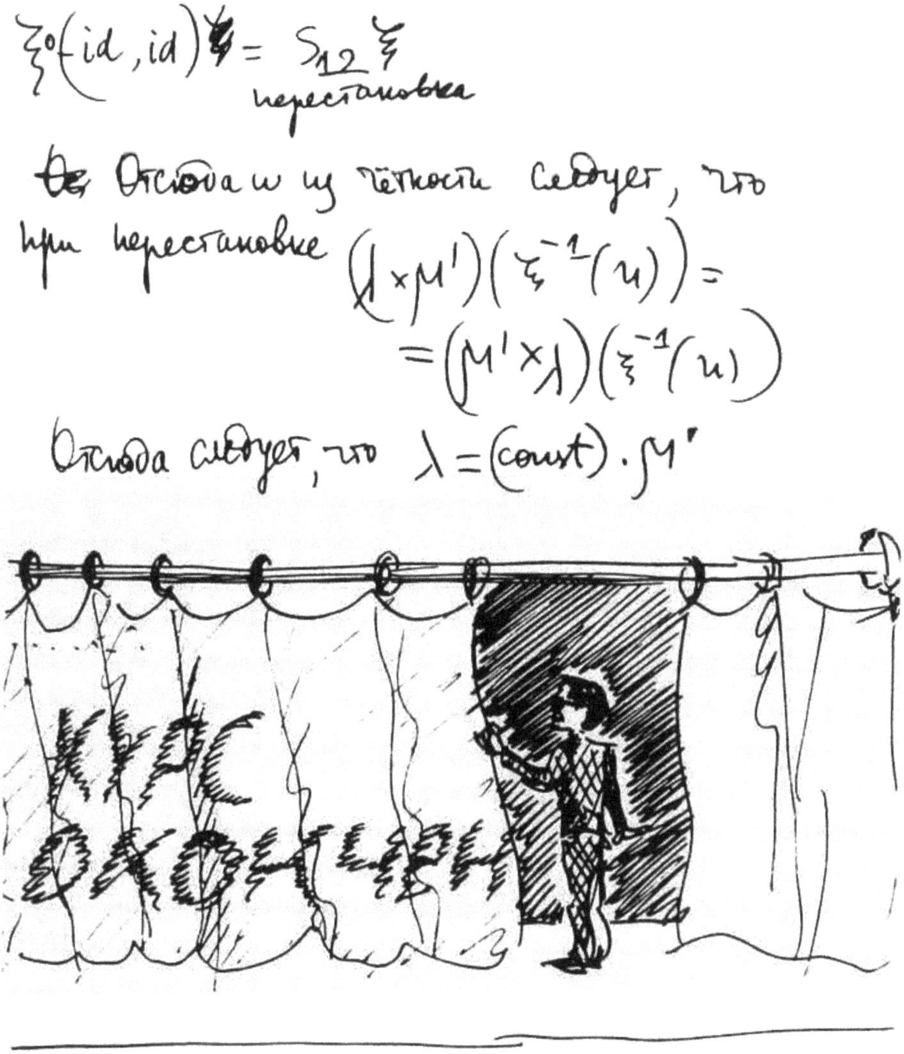

Drawing by S. Arakelov on the last page of his lecture notes: Yu. I. Manin finishing a course on Abelian Varieties.

RELATED REFERENCES AND COMMENTS

PAPER 1

The Hasse-Witt matrix of an algebraic curve.

The mod p results proved in this paper constitute now a very small part of the p-adic cohomological picture established or conjectured for varieties of arbitrary dimension. For the early stages of this theory see [M] and [K]. The Witt vectors cohomology (due to Serre) used in my paper must be replaced by the crystalline cohomology sketched by A. Grothendieck [Gr] and developed by P. Berthelot, W. Messing, J.-M. Fontaine et al.

The more elementary aspects of the relationship between counting points modulo p and differential equations for periods have re-emerged on various occasions in the theory of elliptic genus, algorithmic problems over finite fields, linear codes etc.

[B] P. Berthelot. *Cohomologie cristalline des schémas de caractéristique* $p > 0$. Springer LN in Math., 407 (1974).

[BBM] P. Berthelot, L. Breen, W. Messing. *Théorie de Dieudonné cristalline II.* Springer LN in Math., 930 (1982).

[Gr] A. Grothendieck. *Crystals and the De Rham cohomology of schemes (notes by I. Coates and O. Jussila).* In: Dix Exposés sur la Cohomologie de Schémas, North Holland, 1968.

[H] T. Honda. *Algebraic differential equations.* Symp. Math., v. 24 (1981), Ac. Press, 169–204.

[K] N. Katz. *Algebraic solutions of differential equations (p-curvature and the Hodge filtration).* Inv. Math., 18 (1972), 1–118.

[M] B. Mazur. *Frobenius and the Hodge filtration.* Bull. AMS, 78:5 (1972), 653–667.

[Y] N. Yui. *Formal groups and some arithmetical properties of elliptic curves.* In: Springer LN in Math., 732 (1979), 630–658.

PAPER 2

Rational points of algebraic curves over functional fields.

Let V be an algebraic variety defined over complex field, ω a closed meromorphic or holomorphic differential form on V. Integrating the form along cycles lying in the complement of the singularities of ω we obtain its periods, which are the most important transcendental invariants of V. When V is replaced by an algebraic system of varieties V_t, its periods become functions of algebraic parameters t which satisfy an overdetermined system of linear differential equations, classically called the Picard–Fuchs equations. In more modern language, the flat connection on the bundle of the fiber cohomology H making locally constant cocycles horizontal, defines the structure of a lisse $\mathcal{D}$-module on the sheaf $\mathcal{H}$ of sections of H. A. Grothendieck in [Gro] called it the Gauss–Manin connection. Its general construction was elaborated in [KO]. The Picard–Fuchs equations themselves arise at the first step of a resolution of $\mathcal{H}$ by free $\mathcal{D}$-modules. In my paper an extension of this construction played an essential role. Namely, the Picard–Fuchs relations were applied to the integrals not over cycles but rather over chains with algebraic boundary. This is a development of the classical theory of normal functions.

It seems that the paper appeared at just the right time to stimulate a resurrection of the classical techniques of differential equations in the algebro–geometric context.

Shortly afterwards, P. Griffiths embarked upon a thorough investigation of analysis and geometry of periods. For a very readable account of the first period of this theory, see [Gri]. In order to follow the later developments, it is important to realize that a fiber space whose fibers are the cohomology spaces of a family of smooth projective algebraic varieties has a rich structure, now called a variation of Hodge structure. The canonical connection is only a part of it governing the first order deformations via the Griffiths transversality condition. The integral structure gives rise to the intermediate Jacobians, the problem of singularities includes the regular singularity theorem, the monodromy theorem, and a lot more, especially after Deligne's discovery of the mixed Hodge structure supported by the cohomology of any (possibly singular and non–compact) algebraic cohomology. For this see the report [BZ].

I would like to mention also the construction of the canonical connection on the homotopy spaces ([N–A]) of fibers, and a version in non–commutative geometry ([Ge]).

The functional Mordell-Weil theorem was subsequently reproved by different methods in the works of H. Grauert, A. Parshin, S. Arakelov, P. Vojta. A. Parshin has shown that it follows from the Shafarevich's finiteness conjecture which was established first in the geometric case, and then by G. Faltings in the arithmetical situation.

The method of horizontal derivation, if it is to be applied at all in the arithmetic case, must be considerably modified. Recently such a p–adic modification was suggested in the work of A. Buium ([B2], [B3]) who has even formulated and proved an analog of the "Kernel Theorem" for the abelian varieties over p–adic numbers.

An error in my proof of this theorem was discovered by Coleman ([Co]) and corrected by Chai ([Ch]). Shortly before the publication of this correction, Coleman

([Co]) and I showed that a weaker version of the Kernel Theorem which can be salvaged without much difficulty, still suffices for the proof of the functional Mordell conjecture.

[BZ] J.-L. Brylinski, S. Zucker. *An overview of recent advances in Hodge theory.* In: Several Complex Variables VI, Enc. Math. Sci. 69, Springer Verlag, 1990, 39–142.

[B1] A. Buium. *Effective bound for the geometric Lang conjecture.* Duke Math. J., 71:2 (1993), 475–499.

[B2] A. Buium. *Geometry of p-jets I, II.* Preprints MPI 94-127, 94-128.

[B3] A. Buium. *Differential characters of abelian varieties over p-adic fields.* Preprint MPI, 1995.

[BV1] A. Buium, F. Voloch. *Reduction of the Manin map modulo p.* (to appear in Crelle's J.)

[BV2] A. Buium, F. Voloch. *The Mordell conjecture in characteristic p: an explicit bound.* Preprint.

[Ch] Ching-Li Chai. *A note on Manin's Theorem of the Kernel.* American J. Math., 113 (1991), 387–389.

[Co] R. Coleman. *Manin's proof of the Mordell conjecture over function fields.* Enseign. Math., 36 (1990), 393–427.

[Ge] E. Getzler. *Cartan homotopy formulas and the Gauss-Manin connection in cyclic homology..* Israel Math. Conf. Proc., 7 (1993), 65–78.

[Gri] Ph. Griffiths. *Periods of integrals on algebraic manifolds: summary of main results and discussion of open problems.* Bull. AMS, 76 (1970), 228–296.

[Gro] A. Grothendieck. *Crystals and the de Rham cohomology of schemes.* Notes by J. Coates and O. Jussila. In: Dix exposés sur la cohomologie des schémas, North Holland, Amsterdam, and Masson. Paris, 1968, 306–358.

[KO] N. Katz, T. Oda. *On the differentiation of the De Rham cohomology classes with respect to parameters.* J. Math. Kyoto Univ., 8 (1968), 199–213.

[N-A] V. Navarro-Aznar. *Sur la connexion de Gauss-Manin en homotopie rationnelle.* Ann. Sci. ENS, (4) 26:1 (1993), 99–148.

[P] F. Pham. *Singularités des systèmes algébriques de Gauss-Manin.* Progr. in Math. 2, Birkhäuser, Boston, 1979.

[V] F. Voloch. *Explicit p-descent for elliptic curves in characteristic p.* Comp. math., 74 (1990), 247–258.

PAPER 4

Correspondences, motives, and monoidal transforms.

Motives are among the most remarkable discoveries made by A. Grothendieck (see [Gr], [S]), and have stimulated a wealth of research (see [DMO]). Grothendieck gave some talks on motives, but did not write any formal account of this theory.

My paper was the first publication on this subject. Grothendieck generously lectured to me personally, when I visited the IHES late in the Spring of 1967 when all seminars had ended.

For a recent survey of much more sophisticated results on motives, see [Mot]. Unfortunately, Grothendieck's Standard Conjectures (see [Kl]) remain unproved.

Not much has been done by direct methods. However, a beautiful new theorem using a version of the Identity Principle was recently proved in [GS] and [GN–A], answering a question of Serre ([S]).

The important problem of defining the category of mixed motives was recently addressed by Voevodsky ([V]) in the setting of triangulated categories, and also by M. Hanamura [H].

[DMO] P. Deligne, J. S. Milne, A. Ogus. *Hodge cycles, motives, and Shimura varieties.* Springer LN in Math., vol. 900 (1982).

[GS] H. Gillet, C. Soulé. *Descent, Motives and K-theory.* Preprint, 1995.

[Gr] A. Grothendieck. *Récoltes et semailles.* Preprint Montpellier, 1989.

[GN–A] F. Guillén, V. Navarro–Aznar. *Un critère d'extension d'un foncteur défini sur les schémas lisses.* Preprint, 1995.

[H] M. Hanamura. *Mixed motives and algebraic cycles I, II.* MPI preprint, 1996.

[J] U. Jannsen. *Motives, numerical equivalence, and semisimplicity.* Inv. Math., 197 (1992), 447–452.

[Kl] S. Kleiman. *Motives.* In: Algebraic Geometry, Oslo, 1970, ed. F. Oort, Walters-Noordhoff, Groningen, 1972, 53–82.

[K] K. Künnemann. *Chow-Motive von abelschen schemata und die Fouriertransformation.* Pub. Math. Institut, Universität Münster, 1992.

[Mot] *Motives.* Ed. by U. Jannsen, S. Kleiman, J.-P. Serre, Proc. Symp. Pure Math., vol. 55, pp. 1–2, Providence, 1994.

[S] J.-P. Serre. *Motifs.* Astérisque, vol. 198–200, 1991, 333–349.

[V] V. Voevodsky. *Triangulated category of motives over a field.* Preprint Harvard, 1994.

PAPER 5

Three-dimensional quartics and counterexamples to the Lüroth conjecture (with V.A.Iskovskih).

During the 1967 Colloquium on Algebraic Geometry I gave a talk in the Tata Institute suggesting three approaches to the then open Lüroth problem. Namely, to establish the non-rationality of, say, a threefold V one can try to calculate its group of birational automorphisms and to show that it is too small (Fano's method), or to look at its 3-motive (represented, for example, by its intermediate Jacobian), or finally to try to show that its Brauer group (torsion in the integral 3-cohomology) is non-trivial. In the next two years all three approaches were successfully applied to the various classes of threefolds. Our paper was chronologically the first; shortly afterwards appeared [CG] and then [AM]. The intermediate Jacobian technique developed by C. H. Clemens and P. Griffiths led in 1984 to the spectacular series of the stably rational (hence unirational) but irrational threefolds, produced by A. Beauville, J.-L. Colliot-Thélène, and H. P. F. Swinnerton-Dyer ([BC-TS]).

The modernized Fano method was applied to a thorough investigation of the birational automorphism groups of three-dimensional Fano varieties: see [I], [IP] and references therein.

An extension to higher dimensions proved to be quite difficult. The first successful generalization has been achieved by A. Pukhlikov ([P1], [P2]).

[AM] M. Artin, D. Mumford. *Some elementary examples of unirational varieties which are not rational.* Proc. Lond. Math. Soc. (3), 25 (1972), 75–95.

[BC–TS] A. Beauville, J.-L. Colliot-Thélène, H. P. F. Swinnerton-Dyer. *Variétés stablement rationnelles non rationnelles.* Ann. of Math., 121 (1985), 283-318.

[CG] C. H. Clemens, P. Griffiths. *The intermediate Jacobian of the cubic threefold.* Ann. of Math., (2), 95 (1972), 281-356.

[I] V. Iskovskikh. *Birational automorphisms of algebraic threefolds.* In: Current Problems of Mathematics, 12 VINITI, Moscow 1979, 159–236.

[IP] V. Iskovskikh, A. Pukhlikov. *Birational automorphisms of Fano varieties.* In: Geometry of complex projective varieties, Mediterranean, Renk, 1993, 191–201.

[P1] A. Pukhlikov. *Birational isomorphisms of four-dimensional quintics.* Inv. Math., 87 (1987), 303–329.

[P2] A. Pukhlikov. *Remark on the theorem by V. A. Iskovskikh and Yu. I. Manin on three-dimensional quartics.* (Russian.) In: Trudy of Steklov Instutute, 208 (1995), 278–289.

PAPER 6

New directions in geometry.

This address written for one of the famous Bonn's Arbeitstagungs apparently stimulated the development of higher dimensional Arakelov geometry which has by now reached a very high level of sophistication: see [GS1–3] and [F].

M. Atiyah who presented the paper at the Arbeitstagung in my absence stressed in his published comments [A] the importance of the quantum extension of the classical geometric picture. It did not take long to vindicate his expectations. Among subsequent developments, I especially enjoyed the role of Arakelov geometry in string physics, in particular, in the calculation of Polyakov's measure on the moduli spaces of curves (see below, Comments to the Papers 26–27). Of course, the supergeometry invented by physicists from the start was meant to be the classical counterpart of a quantum field theoretical picture. In [M], [P], [V] the interested reader can find some further mathematics.

[A] M. Atiyah. *Commentary on the article of Yu. I. Manin: "New dimensions in Geometry"*. In: Springer Lecture Notes in Math., 1111 (1984), 103–109.

[F] G. Faltings. *Lectures on the arithmetic Riemann-Roch theorem.* Ann. Math. Studies, 127 (1992).

[GS1] H. Gillet, C. Soulé. *Un théorème de Riemann-Roch arithmétique.* C. R. Ac. Sci. Paris I, 309 (1989), 929–932.

[GS2] H. Gillet, C. Soulé. *Arithmetic intersection theory.* Publ. Math. IHES, 72 (1990), 94-174.

[GS3] H. Gillet, C. Soulé. *An arithmetic Riemann-Roch theorem.* Inv. Math., 110 (1992), 473–543.

[M] Yu. Manin. *Topics in non-commutative geometry.* Princeton University Press, 1991.

[P] I. Penkov. *$\mathcal{D}$-modules on supermanifolds.* Inv. Math. 71 (1983), 501–512.

[V] A. Vaintrob. *Deformations of complex superspaces and coherent sheaves on them.* J. of Soviet Math., 51 (1990), 2140–2188.

PAPER 7

Arrangements of hyperplanes, higher braid groups, and higher Bruhat orders (with V. V. Schechtman).

This paper originated from our desire to clarify the structure of the so-called Zamolodchikov tetrahedron equations which generalize the famous triangle equations upon which the theory of quantum groups is based. The connection between the discriminantal arrangements and Zamolodchikov equations is explained in [KV4] following [MS], where new constant solutions of these equations are constructed. The structure involved turned out to be rather rich. Kapranov and Voevodsky immersed it into the theory of polycategories with additional structures generalizing tensor product (monoidal and braided polycategories). The last conjecture of our paper was proved in [KV1].

[KV1] M. Kapranov, V. Voevodsky. *Free n-category generated by a cube, oriented matroids, and higher Bruhat orders.* Funk. Analisys and Appl., 25:1 (1990), 50–52.

[KV2] M. Kapranov, V. Voevodsky. *Combinatorial-geometric aspects of polycategory theory: pasting schemes and higher Bruhat orders.* Cahiers de Topologie et Géom. Diff. Catégoriques, 32:1 (1991), 11–27.

[KV3] M. Kapranov, V. Voevodsky. *Braided monoidal 2-categories and Manin–Schechtman higher braid groups.* J. Pure and Appl. Algebra, 92 (1994), 241-267.

[KV4] M. Kapranov, V. Voevodsky. *2-Categories and Zamolodchikov tetrahedra equations.* In: Proc. Symp. in Pure Math., vol. 56 (1994), Part 2, 177–259. AMS, Rhode–Island.

[L] R. Lawrence. *A presentation for Manin-Schechtman higher braid groups.* Preprint 1992.

[MS] Yu. Manin, V. Schechtman. *Arrangements of real hyperplanes and Zamolodchikov equations.* In: Group Theoretical Methods in Physics, vol. 1 (Yurmala, 1985), 151–165, VNU Sci. Press, Utrecht, 1986.

[Z] G. Ziegler. *Higher Bruhat orders and cyclic hyperplane arrangements.* Topology, 32:2 (1993), 259–279.

PAPER 8

The p-torsion of elliptic curves is uniformly bounded.

The argument of this paper basically shows that the modular curve $X_0(p^n)$ over a given algebraic number field k has only a finite number of rational points if n is big enough. Of course, it now follows from Faltings' solution of Mordell's conjecture that this is true whenever the genus of $X_0(p^n)$ is larger than 1.

More subtle boundedness question over $\mathbf{Q}$ was treated by B. Mazur in [M]. Using his theory of Eisenstein quotients of modular Jacobians he obtained a complete classification of torsion groups over $\mathbf{Q}$.

Recently, a breakthrough for general number fields has been achieved by L. Merel [Me]. Combining Mazur's technique, Kolyvagin's method, and the theory of modular symbols, he proved a strong uniform boundedness theorem over extensions of given degree of an arbitrary number field.

[K] S. Kamienny. *Torsion points on elliptic curves and q-coefficients of modular forms.* Inv. Math., 109:2 (1992), 221–229.

[M] B. Mazur. *Modular curves and the Eisenstein ideal.* Publ. Math. IHES, 47 (1978), 33–186.

[Me] L. Merel. *Bornes pour la torsion des courbes elliptiques sur les corps de nombres.* Preprint, 1994.

PAPER 9

Le groupe de Brauer-Grothendieck en géométrie diophantienne.

The construction described in this talk remains the only known general obstruction for the passage from local to global solvability in arithmetic. Some effort has been devoted to proving that this is the only obstruction in some classes of varieties, and to possible applications of such statements. For Abelian varieties, this is equivalent to the finiteness of the Tate–Shafarevich group. D. Kanevski ([K]) has derived some beautiful consequences of this conjecture for cubic hypersurfaces. The article [CTK–S] investigates diagonal cubic surfaces and checks the conjecture numerically for small enough coefficients. For a general report, see [CT].

Hardly anybody believes that there are no more obstructions, say, in the class of all smooth manifolds (or even smooth projective hypersurfaces), Somewhat surprisingly, nobody has produced an example where this obstruction vanishes but there are still no global points. Only recently a conditional result of this type was obtained by P. Sarnak and L. Wang. They construct in [SarW] hypersurfaces which yield, "either a violation of the Hasse principle which is not accounted for by the Brauer–Manin obstruction, or a violation of Lang's conjecture (that a hyperbolic variety defined over **Q** has finitely many rational points)."

For a marginally related but important and difficult result (using Margulis' machinery), see [G]: for a reductive group G over a number field k, $G(k)/(R-$ *equivalence*) is finite. (The point is that Brauer equivalence majorizes the R-equivalence).

[CT] J.-L. Colliot-Thélène. *L'arithmétique des variétés rationnlles.* Ann. Fac. Sci. de Toulouse (6), **I**, (1992), 295–336.

[C–TKS] J.-L. Colliot-Thélène, D. Kanevsky, J.-J. Sansuc. *Arithmétique des surfaces cubiques diagonales.* In: Springer LNM, 1290 (1987), 1–108.

[C–TS1] J.-L. Colliot-Thélène, J.-J. Sansuc. *Sur le principe de Hasse et l'approximation faible, et sur une hypothèse de Schinzel.* Acta Arithm., 41 (1982), 33–53.

[C–TS2] J.-L. Colliot-Thélène, J.-J. Sansuc. *La descente sur les variétés rationnelles II.* Duke Mj 54 (1987), 375–492.

[C–TSk] J.-L. Colliot-Thélène, A. N. Skorobogatov. *Approximation faible pour les intersections de deux quadriques en dimension 3.* C. R. Ac. Sci. Paris, (I) 314 (1992), 127–132.

[C–TSw–D] J.-L. Colliot-Thélène, Sir Peter Swinnerton-Dyer. *Hasse principle and weak approximation for pencils of Severi-Brauer and similar varieties.* J. reine u. angew. Math., 453 (1994), 49–112.

[G] Ph. Gille. *Torseurs sur la droite affine et R-equivalence.* Thése, Univ. de Paris-Sud, 1994.

[H1] D. Harari. *Groupe de Brauer de certaines hypersurfaces.* Astérisque 209 (1992), 203–214.

[H2] D. Harari. *Méthode de fibrations et obstruction de Manin.* Duke Math. J., 75:1 (1994), 221–260.

[K] D. Kanevski. *Application of the conjecture on the Manin obstruction to various Diophantine problems.* Astérisque 147–148 (1987), 307–314.

[Sa1] Sh. Saito. *Some observations on motivic cohomologies of arithmetic schemes.* Inv. Math., 98 (1989), 371–414.

[Sa2] Sh. Saito. *On Brauer–Manin equivalence for zero cycles on varieties over local fields.* Proc. Kinosaki Conference on Algebraic Geometry, 1994, 10–16.

[Sal1] P. Salberger. *On the arithmetic of some bundle surfaces.* In: Séminaire de Théorie de Nombres, Paris 1985–86, ed. C. Goldstein, Birkhäuser, 1987, 182–197.

[Sal2] P. Salberger. *Some new Hasse principles for conic bundle surfaces.* In: Séminaire de Théorie de Nombres, Paris 1987–88, ed. C. Goldstein, Birkhäuser, 1990, 283–305.

[S] J.-J. Sansuc. *Groupe de Brauer et arithmétique des groupes algébriques linéaires sur un corps de nombres.* J. r. u. angew. Math., 327 (1981), 12–80.

[SarW] P. Sarnak, L. Wang. *Some hypersurfaces in* P^4 *and the Hasse principle.* Preprint, 1995.

[Se] J.-P. Serre. *Spécialisation des éléments de groupe de* $Br_2(\mathbf{Q}(T_1,\ldots,T_n))$. C. R. Ac. Sci. Paris, (I) 311, (1990), 397–402.

[Sk] A. N. Skorobogatov. *Arithmetic on certain quadric bundles of relative dimension two I.* J. reine u. angew. Math., 407 (1990), 57–74.

PAPER 11

Height on families of abelian varieties (with Ju. G. Zarhin).

Results of this type are useful in various effectivity problems. In particular, they are necessary in the implementation of my conditional algorithm ([M], [Z]) for calculating a basis of rational points on an elliptic curve (assuming Taniyama–Weil and Birch–Swinnerton–Dyer conjectures). For this, see [GZ].

[GZ] J. Gebel, H. Zimmer. *Computing the Mordell–Weil group of an elliptic curve over Q*. CRM Proc. and Lect. Notes, vol. 4 (1994), 61–83.

[M] Yu. I. Manin. *Cyclotomic fields and modular curves.* Russian Math. Surveys, 26:6 (1971), 7–78.

[S] J. H. Silverman. *The difference between the Weil height and the canonical height on elliptic curves.* Math. Comp. 55 (1990), 723–743.

[Z] H. G. Zimmer. *On Manin's conditional algorithm.* Bull. Math. Soc. France Mém. 49–50 (1977), 211–224.

PAPER 12

Periods of p-adic Schottky groups (with V.G.Drinfeld).

For some related work see [G] and [GP]. I returned to the Schottky uniformization in the classical case in [M] trying to reverse Mumford's analogy and to consider the Schottky theory as Arakelov geometry at the infinite prime. The analogy goes remarkably far and enlivens the somewhat formal landscape at the infinite prime. However, a multidimensional version seems hard to come by, whereas the classical Arakelov picture of arithmetical surfaces admits a very rich multidimensional extension.

The two papers of Mumford which this note connects were very influential. They stimulated Drinfeld's theory of p-adic symmetric domains and the theory of p-adic uniformization of Shimura varieties.

[G] L. Gerritzen. *Periods and Gauss-Manin connection for families of p-adic Schottky groups.* Math. Ann. 275:3 (1986), 425–453.

[GP] L. Gerritzen, M. van der Put. *Schottky groups and Mumford curves.* Springer LN in Math., 817 (1980).

[M] Yu. Manin. *Three-dimensional hyperbolic geometry as ∞-adic Arakelov geometry.* Inv. Math., 104 (1991), 223-244.

PAPERS 10, 13, 14, 15

Parabolic points and zeta-functions of modular curves.

Periods of parabolic forms and p-adic Hecke series.

The values of p-adic Hecke series at integer points.

p-adic Hecke series of imaginary quadratic fields (with M.M. Vishik.)

These papers are dedicated to the study of modular symbols and their applications.

The role of modular symbols is determined by the fact that they constitute an additional structure on the Betti realization of the cohomology of modular varieties. According to the Taniyama–Shimura–Weil–Langlands philosophy, these modular motives tend to have very strong universality properties with respect to arithmetical L-functions.

In [Dr] it is shown that not only on modular elliptic curves but on all modular curves, divisors of degree zero supported by cusps have finite order in the Jacobian. The paper [E] reinterprets this statement and the proof in terms of mixed Hodge theory. In his famous paper [Be], A. Beilinson, among other results, generalizes this to the K-theory of Kuga–Sato varieties and proves this generalization by explicitly constructing elements of K-groups of open subsets of these varieties similar to modular units.

A thorough geometric and algebraic investigation of the modular symbols of arbitrary weight was made by V. Shokurov ([Sh1]-[Sh4].) In [M] the technique of modular symbols was extended to Hilbert modular schemes; the formalism in this context deserves further study. See also [Ku1] and [Ku2].

Modular symbols constitute a powerful tool for all sorts of numerical calculations dealing with coefficients of modular forms, values of L-functions which are their Mellin transforms, arithmetic of elliptic curves (via modular uniformization and the Birch–Swinnerton–Dyer conjecture.) See Cremona's book [C] for some background and extensive tables. See also [Sk], [Z], [ZK].

In [BP-R] modular symbols are used to numerically check the supersingular p-adic Birch–Swinnerton–Dyer conjecture.

In addition to modular symbols, there are different techniques for studying L-values and p-adic L-functions: see [D], [HS], [P], [K], and references therein.

[Be] A. Beilinson. *Higher regulators and values of L-functions.* J. Soviet Math., 30 (1985), 2036–2070.

[BP-R] D. Bernardi, B. Perrin-Riou. *Variante p-adique de la conjecture de Birch et Swinnerton-Dyer (le cas supersingulier).* C. R. Ac. Sci., (I) 317 (1993), 227–232.

[C] J. E. Cremona. *Algorithms for modular elliptic curves.* Cambridge UP, 1992.

[D] P. Deligne. *Valeurs de fonctions L et périodes d'intégrales.* Proc. Symp. Pure Math., 33 (1979), 313–346.

[Dr] V. G. Drinfeld. *Two theorems on modular curves.* Func. An. 'and its Applic., 7:2 (1973), 155–156.

[E] R. Elkik. *Le théorème de Manin-Drinfeld.* Astérisque, 183 (1990), 59–67.

[GS] P. Colmez, L. Schneps. *p-adic interpolation of special values of Hecke L-functions.* Comp. Math., 82:2 (1992), 143–187.

[HS] G. Harder, N. Schappacher. *Special values of Hecke L-functions and Abelian integrals.* Springer LN in Math., 1111 (1985), 17–49.

[K] N. Katz. *p-adic interpolation of real analytic Eisenstein series.* Ann. Math. 104 (1976), 459–571.

[Ku1] P. Kurchanov. *Cohomology of discrete groups and Dirichlet series associated with cusp forms of Jacquet-Langlands.* Soviet Math. Izvestiya, 42 (1978), 588–601.

[Ku2] P. Kurchanov. *Local measures connected with cusp forms of Jacquet-Langlands.* Soviet Math., Sbornik, 108 (1979), 483–503.

[M] Yu. Manin. *Non-Archimedean integration and p-adic Jacquet-Langlands series.* Russian Math. Surveys, 31:1 (1976), 5–57.

[Me1] L. Merel. *Opérateurs de Hecke pour $\Gamma_0(N)$ et fractions continues.* Ann. Inst. Fourier, 41(3), (1991)

[Me2] L. Merel. *Intersections sur les courbes modulaires.* Manuscripta Math., 80 (1993), 283–289.

[P] A. Panchishkin. *Non-Archimedean L-functions of Siegel and Hilbert modular forms.* Springer LN in Math., 1471 (1991).

[Sh1] V. Shokurov. *Holomorphic differential forms of higher degree on Kuga's modular varieties.* Math. USSR Sbornik, 30:1 (1976), 119–142.

[Sh2] V. Shokurov. *Modular symbols of arbitrary weight.* Func. Anal. and its appl., 10:1 (1976), 85–86.

[Sh3] V. Shokurov. *Shimura integrals of cusp forms.* Math. USSR Izvestiya, 16:3 (1981), 603–646.

[Sh4] V. Shokurov. *The study of the homology of Kuga varieties.* Math. USSR Izvestiya, 16:2 (1981), 399–418.

[Sk] N.-P. Skoruppa. *Explicit formulas for the Fourier coefficients of Jacobi and elliptic modular forms.* Inv. Math., 102 (1990), 501–520.

[Z] D. Zagier. *Hecke operators and periods of modular forms.* In: Israel Math. Conf. Proc., vol. 3, (1990), I. I. Piatetski-Shapiro Festschrift, Part II, 321–336.

[ZK] D. Zagier, W. Kohnen. *Modular forms with rational periods.* In: Modular forms, ed. R. A. Rankin, Ellis Horwood, Chichester, 1984, 197–249.

PAPERS 17, 18, 19, 20

Rational points of bounded height on Fano varieties (with J. Franke, Yu. Tschinkel).

Sur le nombre des points rationnels de hauteur bornée des variétés algébriques (with V. V. Batyrev).

Points of bounded height on del Pezzo surfaces (with Yu. Tschinkel).

Notes on the arithmetic of Fano threefolds.

The Hardy–Littlewood circle method can be applied to certain problems of the following kind. Consider a system of homogeneous polynomial equations

$$V: \quad F_i(x_0,\dots,x_n) = 0,\ i = 1,\dots r;\ F_i \in \mathbf{Z}[x_0,\dots,x_n];\ \deg F_i = d_i.$$

We wish to find an asymptotic formula for the number of primitive integral solutions of this system of bounded but growing size, for example, fitting in a dilated box $B_H = HB_1$ as $H \to \infty$.

The basic requirements for applicability of the circle method are:

a) $n+1-\sum_i^r d_i$ is positive;

b) V has sufficiently many points locally over $\mathbf{R}$ and all $\mathbf{Q}_p$.

Formally, it is easy to predict the leading term of the asymptotics. On the contrary, many technical problems arise in estimating the remainder term, which are overcome by adopting more restrictions. Some of them must be rather arbitrary, other might be essential for the truth of the result, but nobody understands much about which are which.

Whenever the technique works, it shows that the number of the primitive solutions of the system in question with max $(|x_i|) \le H$ is $CH^{n+1-\sum_i^r d_i}(1+o(1))$ where C is a product of local densities.

Our papers constitute the first steps of a program which is an extension of this theorem in the wider context of Diophantine geometry.

Roughly speaking, the generalization proceeded along the following lines.

a). The system of equations V was replaced by an arbitrary projective Fano manifold over a number field k. This means that the anticanonical class on it is ample. Notice that a smooth complete intersection such as V above is Fano precisely when $n+1-\sum_i^r d_i > 0$.

b). The "size" of a solution was interpreted as the height h_L of a k-point $x \in V(k)$ with respect to a metrized line bundle L, and the dependence of the asymptotic (for the number of solutions of bounded size) on L and h_L was scrutinized.

c). The need to delete some accumulating subvarieties of V before counting was recognized, and the geometry of accumulation was discussed. Without this, no meaningful general picture would emerge.

Below I list some recent papers discussing various aspects of this program. E. Peyre in [P] suggests the value of the constant in the asymptotic formula for an anticanonical height, and checks it in a number of cases. V. Batyrev and Yu. Tschinkel in [BT1] prove the asymptotic formula for anisotropic toric varieties, and extend their result to the general case in [BT2]. M. Strauch in [S] treats the case of certain twisted products of generalized flag spaces. One interesting feature of their results is that the growth formula makes sense and remains true even if the anticanonical class is only effective, but not ample. P. Swinnerton–Dyer and D. R. Heath–Brown discuss cubic surfaces, in particular, the influence of the weak approximation failure. [HB] suggests a mean–value theorem for the constants appearing in the respective asymptotic formulas.

[BT1] V. Batyrev, Yu. Tschinkel. *Rational points of bounded height on compactifications of anisotropic tori.* Int. Math. Res. Notes, 12 (1995), 591–635.

[BT2] V. Batyrev, Yu. Tschinkel. *Manin's conjecture for toric varieties.* Preprint, 1995.

[H–B] D. R. Heath–Brown. *The density of zeros of forms for which weak approximation fails.* Math. Comp. 59 (1992), 613–623.

[P] E. Peyre. *Hauteurs et mesures de Tamagawa sur les variétés de Fano.* Duke MJ, 79:1 (1995), 101–218.

[S] M. Strauch. *Rationale Punkte von beschränkter Höhe auf getwisteten Produkten von veralgemeinerten Fahnenvarietäten.* Diplomarbeit, Bonn, 1995.

[Sw–D] P. Swinnerton-Dyer. *Counting rational points on cubic surfaces.* Contemp. Math., 162 (1994), 371–379.

PAPERS 21, 22, 23

Conservation laws and Lax representation of Benney's long wave equations (with D. R. Lebedev).

Gelfand-Dikii Hamiltonian operator and the co-adjoint representation of the Volterra group (with D.R.Lebedev).

A supersymmetric extension of the Kadomtsev-Petviashvili hierarchy (with A.O.Radul).

The first of these papers studies what is now called the dispersionless limit of hydrodynamic equations.

The group–theoretic explanation of the first Hamiltonian structure given in the second paper was simultaneously discovered by M. Adler. Discussions of this topic with Drinfeld influenced his theory of Poisson structures on Lie groups. The literature dedicated to this subject is too vast to be cited here. I restrict myself to the recent report [KZ] which works in a setting quite close to our original one, explains the second Gelfand–Dickey Hamiltonian structure in the same vein, and contains an extensive list of references.

M. Mulase ([Mu]) investigated the Super KP introduced in our joint paper with A. Radul. One of his main results is the proof of the formal solvability of the associated linear system (equations (11), for which our SKP is a compatibility condition.) More precisely, he generalized to a formal infinite–dimensional context the classical fact that through any point of a bundle with a flat connection passes a unique horizontal section, and supplied an appropriate analog of the Birkhoff decomposition (12). In addition, Mulase has shown that a modified KP can be obtained from SKP by eliminating odd variables (as against reducing them, which gives the usual KP.) For symmetries of SKP, see [P], [T1], [T2].

[KZ] B. Khesin, I. Zakharevich. *Lie-Poisson group of pseudo-differential symbols.* Comm. Math. Phys., 171:3 (1995), 475–530.

[Mu] M. Mulase. *Solvability of the Super KP equation and a generalization of the Birkhoff decomposition.* Inv. Math., 92:1 (1988), 1–46.

[P] S. Panda. *The bi-Hamiltonian structures of the Manin-Radul super KP hierarchy.* Phys. Lett. B, 291:1–2 (1992), 77–82.

[R] J. M. Rabin. *The geometry of the super KP flows.* Comm. Math. Phys., 137 (1991), 533–552.

[T1] K. Takasaki. *Symmetries of the Super KP Hierarchy.* Letters in Math. Phys., 17 (1989), 351–357.

[T2] K. Takasaki. *Integrable systems in gauge theory, Kähler geometry and super KP hierarchies — symmetries and algebraic point of view.* In: Proc. Int. Congr. Math., Kyoto, 1990, vol 2, 1205–1214.

[UY] K. Ueno, M. Yamada. *Supersymmetric extension of the Kadomtsev–Petviashvili hierarchy and the universal super Grassmann manifold.* In: Conformal Field Theory and Solvable Lattice Models, ed. M. Jimbo, T. Miwa, A. Tsuchiya, Tokyo, 1988, 373–391.

PAPER 24

The twistor transformation and algebraic-geometric constructions of solutions of the equations of field theory (with M.M.Kapranov).

This paper is both a summary of our previous work on twistor technique in the theory of non-linear differential equations and extension of it to new situations. See [N] for a further study of multidimensional Baker–Akhiezer functions, and [ELB], [M] for some related developments.

[ELB] M. G. Eastwood, C. R. LeBrun. *Flattening complex manifolds: curvature and Kodaira-Spencer maps.*

[M] S. Merkulov. *Relative deformation theory and differential geometry.* Preprint Odense, 1993.

[N] A. Nakayashiki. *Structure of Baker–Akhieser module of principally polarized abelian varieties and associated integrable systems.* Preprint RIMS–650, Kyoto, 1989.

PAPER 25

Critical dimensions of string theories and a dualizing sheaf on the moduli space of (super)curves.

This naive note gave rise to a small theory. M. Kontsevich [K] and A. Beilinson were the first to make the vague conjecture stated in the first paragraph precise. The basic picture suggested in [K] looks as follows. Instead of the moduli space M_g of smooth curves of genus g one should consider the enhanced moduli space $M := M_{g,1,par}$ where the subscript 1 refers to a marked point on a curve, and the subscript *par* refers to a choice of a formal parameter at this point. M is a smooth infinite dimensional variety endowed with a canonical action of the Virasoro algebra (with zero central charge) which is transitive in an appropriate sense. The positive part of Virasoro changes the location of the point on the curve, the negative part changes the moduli of the curve. The moduli space itself then looks like the space of double coset classes of a "Virasoro group" with respect to the analogs of a Borel and a discrete subgroup. A relationship with the representation theory of the Virasoro algebra is then established by extending its action to the Mumford determinant bundles λ_j in such a way that the central charge becomes $c_j = 6j^2 - 6j + 1$. This explains the numerology of our note.

For an extension of this picture in various directions, in particular to the supersymmetric curves, see [BMS], [BS]. The representation theoretical aspects of the picture are stressed in [ACKP].

[ACKP] E. Arbarello, C. De Concini, V. Kac, G. Procesi. *Moduli spaces of curves and representation theory.* Comm. Math. Phys. 117 (1988), 1-36.

[BMS] A. Beilinson, Yu. Manin, V. Schechtman. *Sheaves of the Virasoro and Neveu-Schwarz algebras.* In: K-theory, Arithmetic and Geometry, Ed. by Yu. Manin, Springer LN in Math., 1289 (1987), 52-66.

[BS] A. Beilinson, V. Schechtman. *Determinant bundles and Virasoro algebras.* Comm. Math. Phys., 118 (1988), 651-701.

[K] M. Kontsevich. *Virasoro algebras and Teichmüller spaces.* Func. Analysis and its applications, 21:2 (1987), 156-157.

[M] Yu. Manin. *Neveu-Schwarz sheaves and differential equations for Mumford superforms.* Journ. of Geometry and Physics, 5:2 (1988), 161-181.

PAPERS 26, 27

The partition function of the Polyakov string can be expressed in terms of theta functions.

The Mumford form and the Polyakov measure in string theory (with A.A.Beilinson).

Polyakov's perturbative quantum string theory involves a summation over compact metrized Riemann surfaces of arbitrary topology, replacing the Feynmann graphs of the perturbative quantum field theory. The term corresponding to a given genus is a path integral over an infinite-dimensional space involving in particular the action of the diffeomorphism group. As Polyakov discovered, a formal manipulation with this integral allows one to reduce it to a finite dimensional integral over a moduli space of complex Riemann surfaces. In an influential paper, A. Belavin and V. Knizhnik identified in principle the integration measure by applying Mumford's study of determinantal sheaves. The papers reprinted here furnished two rather explicit formulas for this measure. The first note expresses it by means of Arakelov–Faltings metrics in the Arakelov geometry of Jacobians, the second produces a slightly more precise version (the Mumford form itself, whose modulus squared is the Polyakov measure), by more geometric tools.

J. Fay's book [F] gives a thorough mathematical treatment of many questions related to the study of Polyakov measure, in particular, of the main results in the physical paper [ABNMV].

For a superversion of this theory, the reader is referred to [M] and [V].

[ABMNV] L. Alvarez-Gaume, J.-B. Bost, G. Moore, Ph. Nelson, C. Vafa. *Bosonization on higher genus Riemann surfaces.* Comm. Math. Phys., 112 (1987), 503–552.

[F] J. Fay. *Kernel Functions, Analytic Torsion, and Moduli Spaces.* AMS Publications, vol. 92, Nr. 464 (March 1992).

[M] Yu. Manin. *Neveu-Schwarz sheaves and differential equations for Mumford superforms.* J. Geom. and Phys., 5:2 (1988), 161–181.

[V] A. Voronov. *A formula for the Mumford measure in the superstring theory.* Funkc. Anal. and its Appl., 22:2 (1987) 67–68.

PAPERS 28, 30

Some remarks on Koszul algebras and quantum groups.

Notes on quantum groups and quantum de Rham complexes.

The approach to quantum groups introduced here stresses their natural definition as generalized symmetries of non-commutative, or quantum, spaces. Both types of objects are described by coordinates, or rather coordinate rings, which become non-commutative. This viewpoint is dual to the more traditional one due to V. G. Drinfeld and M. Jimbo which stresses deformations of universal enveloping algebras (cf. also [W].) It has some intuitive advantages which have been exploited by those physicists which have been studying quantized Minkowski spaces and groups, quantum gauge groups, Schrödinger equations on quantum spaces, etc. The second paper shows that the differential calculus can be naturally quantized using the same formalism over the category of complexes.

In another direction, the two products and the !-duality in the category of quadratic algebras studied in the first paper were recently extended to *quadratic operads* in the beautiful paper [GK]. It explained in a precise way and unified several observations made in algebra on various occasions, e. g. the duality between the categories of commutative and Lie algebras, and an old theory due to D. Lazard of formal groups over operadic algebras (called by him *analyseurs*.)

[AST] M. Artin, W. Schelters, J. Tate. *Quantum deformations of* GL_n. Comm. in Pure and Appl. Math., 44 (1990), 879–895.

[B] J. Baez. *R-commutative geometry and quantization of Poisson algebras.* Adv. Math., 95:1 (1992), 61–91.

[GK] V. Ginzburg, M. Kapranov. *Koszul duality for operads.* Duke Math. Journ., 76:1 (1994), 203–272. (see also Erratum).

[M] G. Maltsiniotis. *Le langage des espaces et groupes quantiques.* Comm. Math. Phys., 151 (1993), 275–302.

[PW] B. Parshall, J. P. Wang. *Quantum linear groups.* Mem. Amer. Math. Soc., vol. 439 (1991).

[S] A. Schirrmacher. *Remarks on the use of R-matrices.* In: Quantum groups and related topics, ed. R. Gielerak, J. Lukierski, Z. Popowicz, Kluwer 1992, 55–65.

[T] M. Takeuchi. *Quantum orthogonal and symplectic groups and their embeddings into quantum GL.* Proc. Lond. Jap. Ac. Sci., ser. A, Math. Sci., 65 (1989), 55–58.

[Ts] B. Tsygan. *Notes on differential forms on quantum groups.* Selecta Math. f. Sovietica, 12:1 (1993), 75–103

[W] S. L. Woronowicz. *Compact matrix pseudogroups.* Comm. Math. Phys., 111 (1987), 613–665.

List of Publications by Yu. I. Manin

1956

1. On cubic congruences to a prime modulus. *Russian: Izv. AN SSSR, ser. mat., 20:6 (1956), pp. 673-678. English: AMS Translations, ser. 2, 13 (1960), 1-7.*

1958

2. Algebraic curves over fields with differentiation. *Russian: Izv. AN SSSR, ser. mat., 22:6 (1958), 737-756. English: AMS Translations, ser. 2, 37 (1964), 59-78.*

1959

3. On the moduli of a field of algebraic functions. *Russian: Doklady AN SSSR, 125:3 (1959), 488-491.*

1961

4. The Hasse–Witt matrix of an algebraic curve. *Russian: Izv. AN SSSR, ser. mat., 25:1 (1961), 153-172. English: AMS Translations, ser.2, vol. 45 (1965), 245-264.*
5. On the Diophantine equations over functional fields. *Russian: Doklady AN SSSR, 139:4 (1961), 806-809.*
6. On the ramified coverings of algebraic curves. *Russian: Izv. AN SSSR, ser. mat., 25:6 (1961), 789-796.*
7. On the theory of abelian varieties. *Russian: PhD Thesis, Steklov Math. Institute, Moscow, 1961, 66 pp.*

1962

8. On the theory of abelian varieties over a field of finite characteristic. *Russian: Izv. AN SSSR, ser. mat., 26:2 (1962), 281-292.*
9. A remark on Lie p–algebras. *Russian: Sibirskii Mat. Journal, 3:3 (1962), 479-480.*
10. An elementary proof of Hasse's theorem. *Russian: Chapter 10 in "Elementary Methods in Analytic Number Theory", by A. O. Gelfond and Yu. V. Linnik, Moscow, Fizmatgiz, 1962. 13 pp.*
11. Two–dimensional formal abelian groups. *Russian: Doklady AN SSSR, 143:1 (1962), 35-37.*
12. On the classification of the formal abelian groups. *Russian: Doklady AN SSSR, 144:3 (1962), 490-492.*
13. On the geometric constructions with compasses and ruler. *Russian: Chapter 6 in "Encyclopaedia of Elementary Mathematics" vol. 6, Fizmatgiz, 1962, 14 pp.*

1963

14. Rational points of algebraic curves over functional fields. *Russian: Izv. AN SSSR, ser. mat. 27:6 (1963), 1395-1440. English: AMS Translations ser. 2, vol. 50 (1966), 189-234.*

15. The theory of commutative formal groups over fields of finite characteristic. *Russian: Uspekhi Mat. Nauk, 18:6 (1963), 3-90. English: Russian Math. Surveys, 18:6 (1963), 1-83.*

16. A proof of the analogue of the Mordell conjecture for curves over functional fields. *Russian: Doklady AN SSSR, 152:5 (1963), 1061-1063.*

17. On the arithmetic of rational surfaces. *Russian: Doklady AN SSSR, 152:1 (1963), 47-49. English: Soviet Math. Dokl., 4 (1963), 1243-1247.*

1964

18. The Tate height of points on an abelian variety, its variants and applications. *Russian: Izv. AN SSSR, ser. mat., 28:6 (1964), 1363-1390. English: AMS Translations, ser. 2, 59 (1966), 82-110.*

19. Diophantine equations and algebraic geometry. *Russian: In: Proc. of the 4th All-Union Math. Congress, vol.2, Fizmatgiz, 1964, 15-21.*

20. Rational points on algebraic curves. *Russian: Uspekhi Mat. Nauk, 19:6 (1964), 83-87.*

1965

21. Moduli fuchsiani. *Annali Scuola Norm. Sup. di Pisa, 19 (1965), 113-126.*

22. Minimal models of ruled and rational surfaces (with Yu. R. Vainberg). *Russian: In: Algebraic surfaces, ed. by I.R. Shafarevich, Trudy MIAN, 75 (1965), 75-91.*

23. Algebraic topology of algebraic varieties. *Russian: Uspekhi Mat. Nauk, 20:6 (1965), 3-12.*

1966

24. Rational surfaces over perfect fields. *Russian: Publ. Math. IHES, vol. 30 (1966), 415-457. English: AMS Translations (ser. 2), 84 (1969), 137-186.*

25. Differential forms and sections of elliptic pencils. *Russian: In: Contemporary Problems of the Theory of Analytic Functions, Nauka, Moscow, 1966, 224-229.*

26. Two theorems on rational surfaces. In: Simp. Int. di Geom. Algebrica, Roma, 1966, 198-207.

1967

27. Rational G-surfaces. *Russian: Doklady AN SSSR, 1175:1 (1967), 28-30.*

28. Rational surfaces over perfect fields. II. *Russian: Mat. Sbornik, 72:2 (1967), 161-192. English: Math. USSR Sbornik, 1 (1967), 141-168.*

1968

29. Correspondences, motives, and monoidal transforms. *Russian: Mat. Sbornik, 77:4 (1968), 475-507. English: Mathematics of the USSR Sbornik, vol. 6 (1968), 439-470.*

30. Rational surfaces and Galois cohomology. *In: Proc. Moscow ICM, MIR, Moscow 1968, 495-509.*

31. On some groups related to cubic surfaces. *In: Algebraic geometry. Tata Press, Bombay, 1968, 255-263.*

32. Cubic hypersurfaces. I. Quasigroups of point classes. *Russian: Izv. AN SSSR, ser. mat., 32:6 (1968), 1223-1244. English: Math USSR - Izvestia, 2 (1968), 1171-1191.*

33. Lectures on Algebraic Geometry. *Comp. Center, Moscow University, 1968, 185 pp.*

1969

34. The p-torsion of elliptic curves is uniformly bounded. *Russian: Izv. AN SSSR, ser. mat., 33:3 (1969), 459-465. English: Math. USSR Izvestia, vol. 3 (1969), 433-438.*

35. Hypersurfaces cubiques. II. Automorphismes birationnelles en dimension deux. *Inv. Math., 6 (1969), 334-352.*

36. Cubic hypersurfaces. III. Moufang loops and Brauer equivalence. *Russian: Mat. Sbornik, 79:2 (1969), 155-170. English: Math. USSR-Sbornik, 24:5 (1969), 1-89.*

37. Lectures on the K-functor in algebraic geometry. *Russian: Uspekhi Mat. Nauk, 24:5 (1969), 3-86. English: Russian Math. Surveys, 24:5 (1969), 1-89.*

38. Comments on the five Hilbert's problems. *Russian: In "Hilbert Problems", Nauka, Moscow, 1969, 154-162, 171-181, 196-199.*

39. Regular elements in the Cremona group. *Russian: Mat. Zametki, 5:2 (1969), 145-148.*

1970

40. The refined structure of the Néron-Tate height. *Russian: Mat. Sbornik, 83:3 (1970). 331-348. English: Math. USSR-Sbornik, 12 (1970), 325-342.*

41. Lectyres on Algebraic Geometry I: Affine Schemes. *Russian: Moscow University, 1970, 133 pp.*

1971

42. Le groupe de Brauer-Grothendieck en géométrie diophantienne. *In: Actes Congr. Int. Math. Nice, Gauthier-Villars 1971, vol. 1, p. 401-411.*

43. Three-dimensional quartics and counterexamples to the Lüroth conjecture (with V.A.Iskovskih). *Russian: Mat. Sbornik, 86:1 (1971) 140-166. English: Math. USSR-Sbornik, vol. 15 (1972), 141-166.*

44. Cyclotomic fields and modular curves. *Russian: Uspekhi Mat. Nauk, 26:6 (1971), 7-71. English: Russian Math. Surveys, 26: (1971), 7-78.*

45. Mordell–Weil theorem *Russian: Appendix to "Abelian Varieties" by D. Mumford, Mir, Moscow , 1971, 17 pp.*

1972

46. Parabolic points and zeta-functions of modular curves. *Russian: Izv. AN SSSR, ser. mat. 36:1 (1972), 19–66. English: Math. USSR Izvestija, publ. by AMS, vol. 6, No. 1 (1972), 19–64.*

47. Cubic Forms: Algebra, Geometry, Arithmetic. *Russian: Nauka, Moscow, 1972, 304 pp.*

1973

48. Height on families of abelian varieties (with Ju. G. Zarhin). *Russian: Mat. Sbornik, 89:2 (1973), 171–181 . English: Math. USSR Sbornik, 18 (1972), 169–179.*

49. Periods of p-adic Schottky groups (with V.G.Drinfeld). *Journ. f. d. reine u. angew. Math., vol. 262/263 (1973), 239–247.*

50. Periods of parabolic forms and p-adic Hecke series. *Russian: Mat. Sbornik, 92:3 (1973), 378–401. English: Math. USSR Sbornik, 21:3 (1973), 371–393.*

51. Explicit formulas for the eigenvalues of Hecke operators. *Acta Arithmetica, 24 (1973), 239–249.*

52. Hilbert's tenth problem. *Russian: In: Sovr. Probl. Mat., 1 (1973), 5–37. English: Journ of Soviet Math.*

1974

53. The values of p-adic Hecke series at integer points of the critical strip. *Russian: Mat. Sbornik, 93:4 (1974), 621–626. English: Math. USSR Sbornik, 22:4 (1974), 631–637.*

54. p-adic Hecke series of imaginary quadratic fields (with M.M.Vishik). *Russian: Mat. Sbornik, 95:3 (1974), 357–383. English: Math. USSR Sbornik, 24:3 (1974), 345–371.*

55. p-adic automorphic functions. *Russian: In: Sovrem. Probl. Mat., vol. 3 (1974), 5–92. English: Journ. of Soviet Math., 5 (1976), 279–333.*

56. Lectures on Mathematical Logic. *Russian: Moscow Institute of Electronic Engineering, 1974, Part 1, 133 pp., Part 2, 69 pp.*

57. Cubic Forms: Algebra, Geometry, Arithmetic. *North Holland, Amsterdam, 1974, 292 pp.*

1975

58. Continuum problem. *Russian: In: Sovrem. Probl. Mat., vol. 5 (1975), 5–72.*

59. Gödel's Theorem. *Russian: Priroda, 12 (1975), 80–87.*

1976

60. Non–Archimedean integration and p-adic Jacquet–Langlands L-series. *Russian: Uspekhi Mat. Nauk, 31:1 (1976), 5–54. English: Russian Math. Surveys, 31:1 (1976), 5–57.*

1977

61. Poisson brackets and the kernel of the variational derivation in the formal calculus of variations (with I. M. Gelfand, M. A. Shubin). *Russian: Funkc. Anal. i ego Prilozen., 10:4 (1977), 31–42. English: Func. Anal. Appl., 10:4 (1977).*

62. Convolutions of Hecke series and their values at lattice points (with A.A.Panchishkin). *Russian: Mat. Sbornik, 104:4 (1977), 617–651. English: Math. USSR Sbornik, 33:4 (1977), 539–571.*

63. Long wave equations with a free surface . I. Conservation laws and solutions (with B. Kupershmidt). *Russian: Funkc. Anal. i ego Prilozen., 11:3 (1977), 31–42. English: Func. Anal. Appl., 11:3 (1977), 188–197.*

64. A Course in Mathematical Logic. *Springer Verlag, 1977. XIII+286 pp.*

65. Assioma/Postulato. *In: Enciclopaedia Einaudi, vol. 1 (1977), 992–1010, Torino, Einaudi.*

66. Applicazioni. *In: Enciclopaedia Einaudi, vol. 1 (1977), 701–743, Torino, Einaudi.*

67. Men and Signs. *Russian: In: Priroda, 5 (1977), 150–152.*

1978

68. Long wave equations with a free surface II. The Hamiltonian structure and the higher equations (with B. Kupershmidt). *Russian: Funkcional. Anal. i ego Prilozen., 12:1 (1978), 25–37. English: Func. Anal. Appl., 12:1 (1978), 20–29.*

69. Matrix solitons and vector bundles over singular curves. *Russian: Func. Anal. i ego Prilozen., 12:4 (1978), 53–63. English: Func. Anal. Appl., 12:4 (1978).*

70. Algebraic aspects of non-linear differential equations. *Russian: In: Sovrem. Probl. Mat., vol. 11 (1978), 5–152. English: Journ. of Soviet Math., 11 (1979), 1–122.*

71. Self-dual Yang-Mills fields over a sphere (with V. G. Drinfeld). *Russian: Func. Anal. i ego Prilozen., 12:2 (1978), 78–79. English: Func. Anal. Appl., 12:2 (1978).*

72. On the locally free sheaves on CP^3 connected with Yang-Mills fields (with V. G. Drinfeld). *Russian: Uspekhi Mat. Nauk, 33:3 (1978), 241–242.*

73. Construction of instantons (with M.F.Atiyah, V.G.Drinfeld, N.J.Hitchin). *Phys. Lett. A, 65:3 (1978), 185–187.*

74. Instantons and sheaves on CP^3 (with V.G.Drinfeld). *In: Springer Lecture Notes in Math., vol. 732 (1978), 60–81.*

75. A description of instantons (with V.G.Drinfeld). *Comm. Math. Phys., v. 63 (1978), 177–192.*

76. A description of instantons II (with V.G.Drinfeld). *Russian: In: Proc. of Int. Seminar on the Physics of High Energy and Quantum Field Theory, Serpoukhov, 1978, 71–92.*

77. Modular forms and number theory. *In: Proc. Int. Cogr. of Math., Helsinki 1978, vol. 1, 177–186.*

78. Continuo/Discreto. *In: Enciclopaedia Einaudi, vol. 3 (1978), 935–986, Torino, Einaudi.*

79. Dualitá (with I.M.Gelfand). *In: Enciclopaedia Einaudi, vol. 5 (1978), 126-178, Torino, Einaudi.*

80. Divisibilitá. *In: Enciclopaedia Einaudi, vol. 5 (1978), 14-37, Torino, Einaudi.*

1979

81. Conservation laws and Lax representation of Benney's long wave equations (with D. R. Lebedev). *Phys. Lett. A, 74 (1979), 154-156.*

82. Gelfand-Dikii Hamiltonian operator and the co-adjoint representation of the Volterra group (with D.R.Lebedev). *Russian: Func. Anal. i ego Prilozhen., 13:4 (1979), 40-46. English: Func. Anal. Appl., 13:4 (1979), 268-273.*

83. Yang-Mills fields, instantons, tensor product of instantons (with V.G.Drinfeld). *Russian). Yad. Fiz., 29:6 (1979), 1646-1653 English: Soviet J. Nucl. Phys., 29:6 (1979), 845-849.*

84. Modular forms and number theory. *In: Proc. Int. Math. Congr. 1978, Helsinki 1979, vol. 1, 177-186.*

85. Mathematics and Physics. *Russian: Znaniye, Moscow, 1979, 63 pp.*

86. Provable and Unprovable. *Russian: Sov. Radio, 1979, 166 pp.*

87. Insieme. *In: Enciclopaedia Einaudi, vol. 7 (1979), 744-776, Torino, Einaudi.*

88. Razionale/Algebrico/Transcedente. *In: Enciclopaedia Einaudi, vol. 11 (1979), 603-628, Torino, Einaudi.*

89. A new encounter with Alice. *Russian: In: Priroda, 7 (1979), 118-120.*

1980

90. An instanton is determined by its complex singularities (with A.A.Beilinson, S.I.Gelfand). *Russian: Funkc. Analiz i ego Priloz., 14:2 (1980), 48-49. English: Func. Anal. Appl. 14:2 (1980).*

91. Twistor description of classical Yang-Mills fields. *Phys. Lett. B, 95:3-4 (1980), 405-408.*

92. Benney's long wave equations II. The Lax representation and conservation laws (with D. R. Lebedev). *Russian: Zap. Nauchn. Sem. LOMI, 96 (1980), 169-178.*

93. Penrose transform and classical Yang-Mills fields. *Russian: In: Group Theoretical Methods in Physics, Nauka, 1980, vol. 2, 133-144.*

94. Methods of algebraic geometry in modern mathematical physics (with V. G. Drinfeld, I. Krichever, S. P. Novikov). *Math. Phys. Review, 1 (1980), Harwood Academic, Chur, 3-57.*

95. Computable and Uncomputable. *Russian: Sov. Radio, 1980, 128 pp.*

96. Linear Algebra and Geometry (with A. I. Kostrikin). *Russian: Moscow University, 1980, 319 pp.*

1981

97. Gauge fields and holomorphic geometry. *Russian: In: Sovr. Probl. Mat., 17 (1981), 3-55. English: Journal of Soviet Math.*

98. Hidden symmetries of long waves. *Physica D, 3:1-2 (1981), 400-409.*

99. On the cohomology of twistor flag spaces (with G.M.Henkin). *Compositio Math., 44 (1981), 103-111.*

100. Expanding constructive universes. *In: Springer Lect. Notes in Comp. Sci., 122 (1981), 255-260.*

101. Simmetria (with I. M. Gelfand). *In: Enciclopaedia Einaudi, vol. 12 (1981), 916-943, Torino, Einaudi.*

102. Strutture matematiche. *In: Enciclopaedia Einaudi, vol. 13 (1981), 765-798, Torino, Einaudi.*

103. Natural language in scientific texts. *Russian: In: Structure of Text-81, Moscow, Nauka, 1981, 25-27.*

104. Mathematics and Physics. *Birkhäuser, Boston, 1981, 100 pp.*

105. Lo Dimostrable e Indemostrable. *Mir, Moscow, 1981, 262 pp.*

1982

106. Null-geodesics of complex Einstein spaces (with I.B.Penkov). *Russian: Funk. Analiz i ego Priloz., 16:1 (1982), 78-79. English: Func. Anal. Appl., 16:1 (1982).*

107. Yang-Mills-Dirac equations as Cauchy-Riemann equations on the twistor space (with G. M. Henkin). *Russian: Yad. Fiz., 35:6 (1982), 1610-1626.*

108. Gauge fields and cohomology of spaces of null-geodesics. *In: Gauge Theories, Fundamental Interactions and Related Results, Birkhäuser, 1982, 231-234.*

109. What is the maximum number of points on a curve over F_2 ? *Journ. Fac. Sci. Univ. Tokyo, ser. 1A, 28:3 (1982), 715-720.*

110. Remarks on algebraic supermanifolds. *Russian: In: Algebra (for the 90th anniversary of O. Yu. Schmidt), Moscow University, 1982, 95-101.*

111. Mathematics and Physics. *Bulgarian: Nauka i Izkusstvo, Sofia, 1982, 56 pp.*

112. A Trilogy about Mathematics. *Russian: In: Znaniye-Sila, 3 (1982), 32.*

1983

113. Flag superspaces and supersymmetric Yang-Mills equations. *In: Arithmetic and Geometry, papers to honor I.R.Shafarevich, vol. 2, 175-198, Birkhauser, 1983.*

114. Supersymmetry and supergravity in the superspace of null supergeodesics. *Russian: In: Group Theoretical Methods in Physics, Nauka, Moscow, 1983, vol. 1, 203-208.*

115. Geometric ideas in the field theory. *Russian: In: Geometric ideas in physics, Mir, 1983, 5-16.*

116. Tynyanov and Griboyedov. *Russian: Revue des Etudes Slaves, t. LV, f. 3 (1983), 507-521. Paris.*

1984

117. Geometry of supergravity and Schubert supercells. *Russian: Zapiski Sem. LOMI, 133 (1984), 160-176.*

118. Grassmannians and flags in supergeometry. *Russian: In: Problems of Modern Analysis, Moscow University, 1984, 83-101.*

119. Schubert supercells. *Russian: Func. Analiz i ego Priloz., 18:4 (1984), 75–76. English: Func. Anal. Appl., 18:4 (1984), 329–330.*

120. New exact solutions and a cohomological analysis of the classical and supersymmetric Yang–Mills equations. *Russian: Trudy MIAN, 165 (1984), 98–114.*

121. Holomorphic supergeometry and Yang–Mills superfields. *Russian: In: Sovr. Probl. Mat., 24 (1984), 3–80. English: Journal of Soviet Math.*

122. New directions in geometry. *Russian: Uspekhi Mat. Nauk, 39:6 (1984), 47–73. English: Russian Math. Surveys, 39:6 (1984), 51–83, and Springer Lecture Notes in Math., 1111, 1985, 59-101.*

123. The inverse scattering transform and classical equations of motion. *In: Nonlinear and Turbulent Processes in Physics, Harwood Academic, Chur, 1984, vol. 3, 1487–1502.*

124. Linear codes and modular curves (with S. G. Vladut). *Russian: In: Sovr. Probl. Mat., vol. 25 (1984), 209–257. English: J. Soviet Math., 30 (1985), 2611–2643.*

125. Some applications of algebraic geometry. *Russian: Trudy Mian, 168 (1984), 110–131.*

126. Gauge Fields and Complex Geometry. *Russian: Nauka, Moscow, 1984. 335 pp.*

1985

127. A supersymmetric extension of the Kadomtsev-Petviashvili hierarchy (with A.O.Radul). *Comm. Math. Phys., 98 (1985), 65–77.*

128. Sun, a Poor Totem. *Russian: In: Priroda, 6 (1985), 123–127.*

129. Geometry unbound. *Science, 9 (1985), 89–91.*

1986

130. The twistor transformation and algebraic–geometric constructions of solutions of the equations of field theory (with M.M.Kapranov). *Russian: Uspekhi Mat. Nauk, 41:5 (1986), 85–107. English: Russian Math. Surveys, 41:5 (1986), 33–61.*

131. On the higher Bruhat orders related to the symmetric group. *Russian: Func. Anal. i ego Prilozhen., 20:2 (1986), 74–75. English: Func. Anal. Appl., 20:2 (1986), 148–150.*

132. Critical dimensions of string theories and a dualizing sheaf on the moduli space of (super)curves. *Russian: Funkc. Anal. i ego Prilozhen., 20:3 (1986), 88–89. English: Func. Anal. Appl., 20:3 (1986), 244–246.*

133. The partition function of the Polyakov string can be expressed in terms of theta functions. *Phys. Lett. B, 172(2) (1986), 184–185.*

134. The Mumford form and the Polyakov measure in string theory (with A.A.Beilinson). *Comm. Math. Phys., 107:3 (1986), 359–376.*

135. The Radon-Penrose transformation for the group $SO(8)$ and instantons (with Hoang Le Minh). *Russian: Izv. AN SSSR, ser. mat., 50:1 (1986),195–206. English: Math. USSR Izvestia, 50 (1), 1986, 189–200.*

136. The multiloop contribution in the fermionic string (with M. A. Baranov, I. V. Frolov, A. S. Schwartz). *Russian: Yad. Fiz., 43:4 (1986), 1053–1055. English: Sov. J. Nucl. Phys., 43 (1986), 670–671.*

137. Arrangements of real hyperplanes and the Zamolodchikov equations (with V.V.Shekhtman). *Russian: In: Group Theoretical Methods in Physics, vol. 1 (Jurmala, 1985), 316–325. Nauka, Moscow, 1986. English: Group Theoretical Methods in Physics, vol. 1 (Yurmala, 1985), 151–165, VNU Sci. Press, Utrecht, 1986.*

138. Rational varieties: algebra, geometry, and arithmetic (with M. Tsfasman). *Russian: Uspekhi mat. Nauk, 41:2 (1986), 43–94. English: Russian Math. Surveys, 41:2 (1986), 51–116.*

139. Cubic Forms: Algebra, Geometry, Arithmetic. 2nd edition, revised and completed. *North Holland, Amsterdam, 1986, 326 pp.*

140. Bevezetés a kiszámithatósag matematikai elméletébe. *(Hungarian translation of Computable and Uncomputable.) Müszaki Könyvkiadó, Budapest, 1986, 144 pp.*

1987

141. Quantum strings and algebraic curves. *In: Proc. Int. Congr. of Math., Berkeley, 1986, AMS, Providence RI, 1987, vol. 2, 1286-1295.*

142. Some remarks on Koszul algebras and quantum groups. *Ann. Inst. Fourier, 37:4 (1987), 191–205.*

143. Values of Selberg's zeta function at integer points (with A.A.Beilinson). *Russian: Func. Analiz i ego Prilozen., 21:1 (1987), 68-69. English: Func. Anal. Appl., 21:1 (1987), 58–60.*

144. Sheaves of the Virasoro and Neveu-Schwarz algebras (with A.A. Beilinson, V.V. Shechtman). *In: Springer Lecture Notes in Math., 1289 (1987), 52–66.*

145. A superanalog of the Selberg trace formula and multiloop contribution for fermionic string (with M. A. Baranov, I. V. Frolov, A. S. Schwartz). *Comm. Math. Phys., 111 (1987), 373–392.*

146. On the problem of the early stages of speech and consciousness (phylogenesis). *Russian: In: Intellectual Processes and their Modeling, Nauka, Moscow, 1987, 154-178.*

147. "It is love." *Russian: In: Priroda, 4 (1987), 118–120.*

148. The mythological trickster in psychology and history of culture. *Russian: In: Priroda, 7 (1987), 42–52.*

1988

149. Quantum groups and non-commutative geometry. *Montreal University Press, 1988. 88 p.*

150. Neveu–Schwarz sheaves and differential equations for Mumford superforms. *J. of Geometry and Physics, 5:2 (1988), 161–181.*

151. Brouwer Memorial Lecture 1987. *Nieuw Arch. Wisk. (4) 6 (1988), no. 1–2, 1–6.*

152. Elements of supergeometry (with A. A. Voronov, I. B. Penkov). *Russian: In: Sovr. Probl. Mat., vol. 32 (1988), 3–25.*

153. Supercell partitions of flag superspaces (with A. A. Voronov). *Russian: In: Sovr. Probl. Mat., vol. 32 (1988), 27–70.*

154. Methods of the Homological Algebra I. Introduction to the Cohomology Theory and Derived Categories (with S. I. Gelfand). *Russian: Nauka, Moscow, 1988, 416 pp.*

155. Gauge Field Theory and Complex Geometry. *Springer Verlag, 1988. 295 pp.*

156. Homological Algebra (with S. I. Gelfand). *Russian: Sovr. Probl. Mat., vol. 38 (1988), 238 pp.*

1989

157. Arrangements of hyperplanes, higher braid groups, and higher Bruhat orders (with V.V.Schechtman). *In: Advanced Studies in Pure Math., vol. 17, Algebraic Number Theory, 289-308. Academic Press, 1989.*

158. Rational points of bounded height on Fano varieties (with J.Franke, Yu. Tschinkel). *Inv. Math., 95 (1989), 421-435.*

159. Multiparametric quantum deformation of the general linear supergroup. *Comm. Math. Phys., 123 (1989), 123-135.*

160. Determinants of Laplacians on Riemann surfaces. *In: Conformal Invariance and string theory (Poiana Brasov, 1987), Academic Press, Boston, MA, 1989, 285-291.*

161. Reflections on arithmetical physics. *In: Conformal Invariance and string theory (Poiana Brasov, 1987), Academic Press, Boston, MA, 1989, 293-303.*

162. The formalism of left and right connections on supermanifolds (with I.B.Penkov). *In: Lectures on Supermanifolds, Geometrical Methods and Conformal Groups, World Scientific, 1989, 3-13.*

163. Strings. *Math. Intelligencer, 11:2 (1989), 59-65.*

164. Multiparametric quantum deformation of the general linear supergroup. *Comm. Math. Phys., 123 (1989), 123-135.*

165. Letter to the editor. *Russian: Izv. AN SSSR, ser. mat. 53:2 (1989). English: Math. USSR Izvestija, 34:2 (1990), 465-466.*

166. Linear Algebra and Geometry (with A. I. Kostrikin). *Gordon and Breach, 1989, 309 pp.*

167. Elementary Particles: Mathematics, Physics and Philosophy (with I. Yu. Kobzarev). *Reidel, Dordrecht, 1989, 227 pp.*

1990

168. Sur le nombre des points rationnels de hauteur bornée des variétés algébriques (with V.V.Batyrev). *Math. Annalen, 286 (1990), 27-43.*

169. Non-standard quantum deformations of $GL(n)$ and constant solutions of the Yang-Baxter equation (with E.E.Demidov, E.E.Mukhin, D.V.Zhdanovich). *Progress of Theor. Phys. Supplement, 102 (1990), 203-218.*

170. Quantized theta-functions. *Progress of Theor. Phys. Supplement, 102 (1990).*

171. Number Theory (with A. A. Panchishkin). *Russian: Sovr. Probl. Mat., vol. 49 (1990), 348 pp.*

1991

172. Three-dimensional hyperbolic geometry as ∞-adic Arakelov geometry. *Inv. Math., 104 (1991), 223-244.*

173. Quantum Groups. *Nederl. Wetensk. Verslag, Afd. Naturkunde, 100 (1991), 55-68.*

174. Topics in Non-commutative Geometry. *Princeton University Press, 1991, 163 pp.*

1992

175. Notes on quantum groups and quantum de Rham complexes. *Teoreticheskaya i Matematicheskaya Fizika, 92:3 (1992), 425-450.*

176. Archetype of Empty City. *Russian: In: Arbor Mundi, ed. by E. Meletinsky, 1 (1992), 28-34.*

1993

177. Notes on the arithmetic of Fano threefolds. *Comp. Math. 85 (1993), 37-55.*

178. Points of bounded height on del Pezzo surfaces (with Yu. Tschinkel). *Comp. Math. 85 (1993), 315-332.*

1994

179. Homological Algebra (with S. I. Gelfand). *Enc. of Math. Sci., vol. 38, Springer Verlag, 1994, 222 pp.*

180. Gromov-Witten classes, quantum cohomology, and enumerative geometry (with M. Kontsevich). *Comm. Math. Phys., 164:3 (1994), 525-562.*

181. Problems on rational points and rational curves on algebraic varieties. *Preprint MPI/94-86, 34 pp.*

182. Mordell-Weil problem for cubic surfaces. *Preprint, 6 pp.*

183. Algebraic groups in non-commutative geometry. *Preprint, 11 pp.*

1995

184. Lectures on zeta functions and motives (according to Deninger and Kurokawa). *In: Columbia University Number Theory Seminar, Astérisque 228 (1995), 121-164.*

185. Number Theory I. Fundamental Problems, Ideas and Theories. (with A. Panchishkin). *Enc. of Math. Sci., vol. 49, Springer Verlag, 1995, 303 pp.*

186. Generating functions in algebraic geometry and sums over trees. *In: The Moduli Space of Curves, ed. by R. Dijkgraaf, C. Faber, G. van der Geer. Progress in Math., vol. 129. Birkhäuser, 1995, 401-417.*

187. Quantum cohomology of a product (with M. Kontsevich, Appendix by R. Kaufmann). *Preprint MPI/95-19, 29 pp.*

188. Stacks of stable maps and Gromov-Witten invariants (with K. Behrend). *Preprint MPI/95-56, 50 pp.*

189. Truth, rigor and common sense. *Preprint MPI/95-107, 10 pp.*

190. Automorphic pseudodifferential operators (with P. B. Cohen and D. Zagier). *Preprint MPI/95-132, 24 pp.*

191. Distribution of rational points on Fano varieties. *Preprint, 7 pp.*

1996

192. Higher Weil–Petersson volumes of moduli spaces of stable n-pointed curves. *Preprint, 21 pp.*

193. Interrelations between Mathematics and Physics. *Preprint, 10 pp.*

www.ingramcontent.com/pod-product-compliance
Lightning Source LLC
LaVergne TN
LVHW070904300926
849162LV00012B/1853